BETA

β

Mathematics Handbook

Concepts Theorems Methods
Algorithms Formulas Graphs Tables

Lennart Råde Bertil Westergren

SECOND EDITION

Studentlitteratur

CRC

CRC Press
Boca Raton Ann Arbor London Tokyo

Second edition 1990

First published in Sweden in 1989 by
Studentlitteratur, Åkergränden 1, S-221 00 Lund

Distributed worldwide outside Scandinavia by
CRC Press, Inc, 2000 Corporate Blvd., NW.,
Boca Raton, Florida 33431
Direct all inquiries to the distributor

Library of Congress Cataloging-in-Publication Data
A CIP catalog record for this book is available from the Library of Congress

ISBN 0-8493-7758-7

Printed in Sweden
Studentlitteratur, Lund, 1992

Contents

Preface

The BETA handbook covers basic areas of mathematics, numerical analysis, probability and statistics and various applications. The handbook is intended for students and teachers of mathematics, science and engineering and for professionals working in these areas. The aim of the handbook is to provide useful information in a lucid and accessible form in a moderately large volume. The handbook concentrates on definitions, results, formulas, graphs, figures and tables and emphasizes concepts and methods with applications in technology and science. As a preparation for BETA one of the authors has previously published ALPHA, which is a mathematics handbook primarly intended for students at the pre-university level.

The BETA handbook is organised in 19 chapters starting with basic concepts in discrete mathematics and ending with chapters on probability and statistics and a miscellaneous chapter. Crossreferences and an extensive index help the user to find required information. We have not included numerical tables of functions which are available on most scientific calculators and pocket computers. We have treated one variable and multivariable calculus in different chapters, because students, usually, meet these areas in different courses. In formulating theorems and results sometimes all assumptions are not explicitly stated. With regard to programming languages we have only included instructions and commands from the BASIC language. This language is well suited for modest mathematical problems and is available on most pocket and micro computers.

We are happy to have been able to draw on the expertise of several of our colleagues. Our thanks are especially due to Johan Karlsson, Jan Petersson, Rolf Pettersson and Thomas Weibull. We also want to thank Christer Borell, Kenneth Eriksson, Carl-Henrik Fant, Kjell Holmåker, Lars Hörnström, Jacques de Maré and Bo Nilsson for their helpful assistance.

Some tables and graphs have been copied with permission from publishers, whose courtesy is here acknowledged. We are thus indebted to the American Statistical Association for permission to use the table of Gurland-Tripathis corrections factors in section 18.2 and the table of the Kolmogorov-Smirnov test in section 18.7, to the American Society for Quality Control for permission to use the table for construction of single acceptance sampling control plans in section 18.8 (copyright 1952 American Society for Quality Control), to McGraw-Hill Book Company for permission to use the table on tolerance limits for the normal distribution in section 18.4 (originally published in Eisenhart, et. al: Techniques of Statistical Analysis, 1947) and to Pergamon Press for permission to use the graph of the Erlang Loss Formula in section 17.6 (originally published in L. Kosten, Stochastic Theory of Service Systems, 1973).

We shall be grateful for any suggestions about changes, additions, or deletions, as well for corrections. It is finally our hope that many users will find the BETA handbook a useful guide to the world of mathematics.

Preface to the second edition

In the second edition of the BETA handbook corrections and a large number of additions have been made throughout the book. The major differences between the first and second editions are the following.

Chapter 1 now gives a more complete treatment of basic concepts and methods in discrete mathematics (logic, algebraic structures and graph theory). There is a new section on codes. To chapter 2 has been added a treatment of real numbers in different bases, to chapter 4 a new section on complex matrices and to chapter 9 a new section on linear difference equations. Tables of integrals and Fourier transforms in chapter 7 and 13 have been enlarged. Chapter 12 now gives more information about Chebyshev polynomials, Bessel functions, and exponential, sine, cosine, and Fresnel integrals. Chapter 13 has a new section on Hankel och Hilbert transforms and chapter 16 a new section on numerical summation. In chapter 18 tables for Bartlett's test and the Studentized range have been included. These tables are published with permission from the American Statistical Association and the Biometrika Trustees. Furthermore, this chapter now also has a section on factorial experimental design and a statistical glossary.

We want to thank Juliusz Brzezinski and Eskil Johnson for helpful suggestions.

Lennart Råde, Bertil Westergren

1 Fundamentals. Discrete Mathematics

1.1 Logic

Statement calculus

Connectives

Disjunction	$P \vee Q$	P or Q
Biconditional	$P \leftrightarrow Q$	P if and only if Q
Conditional	$P \rightarrow Q$	If P then Q
Conjunction	$P \wedge Q$	P and Q
Negation	$\sim P$ or $\neg P$	Not P

Truth tables (F=false, T=true)

P	Q	P∨Q	P∧Q	P→Q	P↔Q
T	T	T	T	T	T
T	F	T	F	F	F
F	T	T	F	T	F
F	F	F	F	T	T

P and $\neg P$ have opposite truth values.

Tautologies

A *tautology* is true for all possible assignments of truth values to its components.

A tautology is also called a *universally valid formula* and a *logical truth.* A statement formula which is false for all possible assignments of truth values to its components is called a *contradiction.*

Tautological equivalences $\Leftrightarrow$

$\neg\neg P \Leftrightarrow P$	(double negation)
$P \wedge Q \Leftrightarrow Q \wedge P$	
$P \vee Q \Leftrightarrow Q \vee P$	
$(P \wedge Q) \wedge R \Leftrightarrow P \wedge (Q \wedge R)$	
$(P \vee Q) \vee R \Leftrightarrow P \vee (Q \vee R)$	
$P \wedge (Q \vee R) \Leftrightarrow (P \wedge Q) \vee (P \wedge R)$	(Distributive laws)
$P \vee (Q \wedge R) \Leftrightarrow (P \vee Q) \wedge (P \vee R)$	
$\neg(P \wedge Q) \Leftrightarrow \neg P \vee \neg Q$	(De Morgan laws)
$\neg(P \vee Q) \Leftrightarrow \neg P \wedge \neg Q$	
$P \vee P \Leftrightarrow P$	
$P \wedge P \Leftrightarrow P$	
$R \vee (P \wedge \neg P) \Leftrightarrow R$	
$R \wedge (P \vee \neg P) \Leftrightarrow R$	
$P \rightarrow Q \Leftrightarrow \neg P \vee Q$	
$\neg(P \rightarrow Q) \Leftrightarrow P \wedge \neg Q$	
$P \rightarrow Q \Leftrightarrow (\neg Q \rightarrow \neg P)$	
$P \rightarrow (Q \rightarrow R) \Leftrightarrow ((P \wedge Q) \rightarrow R)$	
$\neg(P \leftrightarrow Q) \Leftrightarrow (P \leftrightarrow \neg Q)$	
$(P \leftrightarrow Q) \Leftrightarrow (P \rightarrow Q) \wedge (Q \rightarrow P)$	
$(P \leftrightarrow Q) \Leftrightarrow (P \wedge Q) \vee (\neg P \wedge \neg Q)$	

Tautological implications $\Rightarrow$

$P \wedge Q \Rightarrow P$	(simplification)
$P \wedge Q \Rightarrow Q$	
$P \Rightarrow P \vee Q$	(addition)
$Q \Rightarrow P \vee Q$	
$\neg P \Rightarrow (P \rightarrow Q)$	
$Q \Rightarrow (P \rightarrow Q)$	
$\neg(P \rightarrow Q) \Rightarrow P$	
$\neg(P \rightarrow Q) \Rightarrow \neg Q$	
$\neg P \wedge (P \vee Q) \Rightarrow Q$	(disjunctive syllogism)
$P \wedge (P \rightarrow Q) \Rightarrow Q$	(modus ponens)
$\neg Q \wedge (P \rightarrow Q) \Rightarrow \neg P$	(modus tollens)
$(P \rightarrow Q) \wedge (Q \rightarrow R) \Rightarrow (P \rightarrow R)$	(hypothetical syllogism)
$(P \vee Q) \wedge (P \rightarrow R) \wedge (Q \rightarrow R) \Rightarrow R$	(dilemma)

T $\Leftrightarrow$ any tautology **F** $\Leftrightarrow$ any condradiction

Exclusive OR, NAND and NOR

The connective *exclusive or* is denoted "$\veebar$" and is defined so that $P \veebar Q$ is true whenever either P or Q, but not both are true.

The connective *NAND* is denoted by "$\uparrow$" and is defined so that

$$P \uparrow Q \Leftrightarrow \neg(P \wedge Q)$$

The connective *NOR* is denoted by "$\downarrow$" and is defined so that

$$P \downarrow Q \Leftrightarrow \neg(P \vee Q)$$

Tautological equivalences

$P \bar{\vee} Q \Leftrightarrow Q \bar{\vee} P$
$(P \bar{\vee} Q) \bar{\vee} R \Leftrightarrow P \bar{\vee} (Q \bar{\vee} R)$
$P \wedge (Q \bar{\vee} R) \Leftrightarrow (P \wedge Q) \bar{\vee} (P \wedge R)$
$(P \bar{\vee} Q) \Leftrightarrow ((P \wedge \neg Q) \vee (\neg P \wedge Q))$
$P \bar{\vee} Q \Leftrightarrow \neg (P \leftrightarrow Q)$

$P \uparrow Q \Leftrightarrow Q \uparrow P$
$P \downarrow Q \Leftrightarrow Q \downarrow P$

$P \uparrow (Q \uparrow R) \Leftrightarrow \neg P \vee (Q \wedge R)$
$(P \uparrow Q) \uparrow R \Leftrightarrow (P \wedge Q) \vee \neg R$
$P \downarrow (Q \downarrow R) \Leftrightarrow \neg P \wedge (Q \vee R)$
$(P \downarrow Q) \downarrow R \Leftrightarrow (P \vee Q) \wedge \neg R$

Truth table

P	Q	$P \bar{\vee} Q$	$P \uparrow Q$	$P \downarrow Q$
T	T	F	F	F
T	F	T	T	F
F	T	T	T	F
F	F	F	T	T

The connectives $(\neg, \wedge)$ and $(\neg, \vee)$ can be expressed in terms of $\uparrow$ alone or in terms of $\downarrow$ alone.

$$\neg P \Leftrightarrow P \uparrow P \qquad P \wedge Q \Leftrightarrow \neg (P \uparrow Q) \qquad P \vee Q \Leftrightarrow \neg P \uparrow \neg Q$$

$$\neg P \Leftrightarrow P \downarrow P \qquad P \wedge Q \Leftrightarrow \neg P \downarrow \neg Q \qquad P \vee Q \Leftrightarrow \neg (P \downarrow Q)$$

Duality

Consider formulas containing $\vee$, $\wedge$ and $\neg$. Two formulas A and A^* are *duals* of each other if either one is obtained from the other by replacing $\wedge$ by $\vee$ and vice versa.

A generalisation of De Morgan's laws:

$$\neg A(P_1, P_2, \ldots, P_n) \Leftrightarrow A^*(\neg P_1, \neg P_2, \ldots, \neg P_n).$$

Here P_i are the *atomic* variables in the duals A and A^*.

Normal forms

If (for example) P, Q and R are statement variables, then the eight (in general 2^n) formulas $P \wedge Q \wedge R$, $P \wedge Q \wedge \neg R$, $P \wedge \neg Q \wedge R$, $P \wedge \neg Q \wedge \neg R$, $\neg P \wedge Q \wedge R$, $\neg P \wedge Q \wedge \neg R$, $\neg P \wedge \neg Q \wedge R$ and $\neg P \wedge \neg Q \wedge \neg R$ are the *minterms* of P, Q and R. Every statement formula A is equivalent to a disjunction of minterms, called its *principal disjunctive normal form* or *sum-of-product form.* Similarly A is equivalent to a conjunction of *maxterms* called its *principal conjunctive normal form* or *product-of-sum form.* (Cf. Boolean algebra, sec. 1.4).

Example (Cf. example of Boolean algebra sec. 1.4)

If P, Q, R are the atomic variables, write equivalent sum-of-products and product-of-sums of A and $\neg A$ if $A = (P \wedge Q) \vee (Q \wedge \neg R)$.

Solution. (Using $S \vee \neg S \Leftrightarrow \mathbf{T}$ and distributive laws).

1. $A \Leftrightarrow (P \wedge Q \wedge (R \vee \neg R)) \vee ((P \vee \neg P) \wedge Q \wedge \neg R) \Leftrightarrow (P \wedge Q \wedge R) \vee (P \wedge Q \wedge \neg R) \vee \vee (P \wedge Q \wedge \neg R) \vee (\neg P \wedge Q \wedge \neg R) \Leftrightarrow (P \wedge Q \wedge R) \vee (P \wedge Q \wedge \neg R) \vee (\neg P \wedge Q \wedge \neg R)$
2. $\neg A \Leftrightarrow$ [The remaining minterms] $\Leftrightarrow$

   $(P \wedge \neg Q \wedge R) \vee (P \wedge \neg Q \wedge \neg R) \vee (\neg P \wedge Q \wedge R) \vee (\neg P \wedge \neg Q \wedge R) \vee (\neg P \wedge \neg Q \wedge \neg R)$
3. $A \Leftrightarrow \neg(\neg A) \Leftrightarrow$ [Duality, see above] $\Leftrightarrow (\neg P \vee Q \vee \neg R) \wedge (\neg P \vee Q \vee R) \wedge \wedge (P \vee \neg Q \vee \neg R) \wedge (P \vee Q \vee \neg R) \wedge (P \vee Q \vee R)$
4. $\neg A \Leftrightarrow (\neg P \vee \neg Q \vee \neg R) \wedge (\neg P \vee \neg Q \vee R) \wedge (P \vee \neg Q \vee R)$

1.1

Predicate calculus

Quantifiers

Universal quantifier	$\forall x$	For all x, ...
Existential quantifier	$\exists x$	There exists an x such that

Valid formulas for quantifiers

$$(\exists x)(P(x) \vee Q(x)) \Leftrightarrow (\exists x)P(x) \vee (\exists x)Q(x)$$
$$(\forall x)(P(x) \wedge Q(x)) \Leftrightarrow (\forall x)P(x) \wedge (\forall x)Q(x)$$
$$\neg(\exists x)P(x) \Leftrightarrow (\forall x)\neg P(x)$$
$$\neg(\forall x)P(x) \Leftrightarrow (\exists x)\neg P(x)$$
$$(\forall x)P(x) \vee (\forall x)Q(x) \Rightarrow (\forall x)(P(x) \vee Q(x))$$
$$(\exists x)(P(x) \wedge Q(x)) \Rightarrow (\exists x)P(x) \wedge (\exists x)Q(x)$$
$$(\forall x)(P \vee Q(x)) \Leftrightarrow P \vee (\forall x)Q(x)$$
$$(\exists x)(P \wedge Q(x)) \Leftrightarrow P \wedge (\exists x)Q(x)$$
$$(\forall x)P(x) \to Q \Leftrightarrow (\exists x)(P(x) \to Q)$$
$$(\exists x)P(x) \to Q \Leftrightarrow (\forall x)(P(x) \to Q)$$
$$P \to (\forall x)Q(x) \Leftrightarrow (\forall x)(P \to Q(x))$$
$$P \to (\exists x)Q(x) \Leftrightarrow (\exists x)(P \to Q(x))$$

Formulas for two quantifiers

$$(\forall x)(\forall y)P(x, y) \Leftrightarrow (\forall y)(\forall x)P(x, y)$$
$$(\forall x)(\forall y)P(x, y) \Rightarrow (\exists y)(\forall x)P(x, y)$$
$$(\forall y)(\forall x)P(x, y) \Rightarrow (\exists x)(\forall y)P(x, y)$$
$$(\exists y)(\forall x)P(x, y) \Rightarrow (\forall x)(\exists y)P(x, y)$$
$$(\exists x)(\forall y)P(x, y) \Rightarrow (\forall y)(\exists x)P(x, y)$$
$$(\forall x)(\exists y)P(x, y) \Rightarrow (\exists y)(\exists x)P(x, y)$$
$$(\forall y)(\exists x)P(x, y) \Rightarrow (\exists x)(\exists y)P(x, y)$$
$$(\exists x)(\exists y)P(x, y) \Leftrightarrow (\exists y)(\exists x)P(x, y)$$

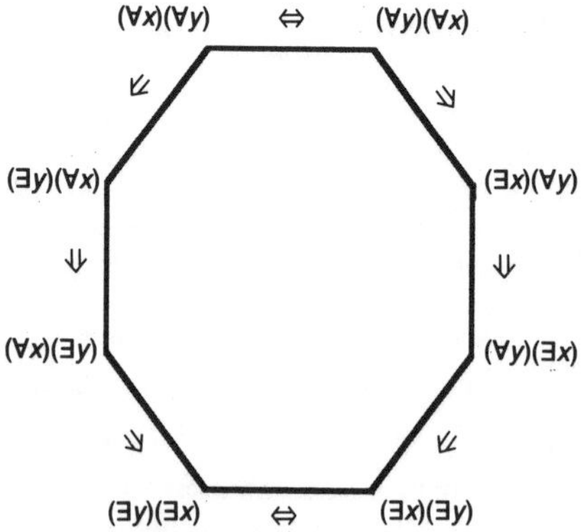

Methods of proof

Some proof methods

Statement to be proved	Proof method	Procedure
Q	Modus ponens	P $\underline{P \rightarrow Q}$ $\therefore Q$
$\neg P$	Modus tollens	$\neg Q$ $\underline{P \rightarrow Q}$ $\therefore \neg P$
Q	Disjunctive syllogism	$P \vee Q$ $\underline{\neg P}$ $\therefore Q$
$P(a)$	Universal instantiation	$\underline{(\forall x)P(x)}$ $\therefore P(a)$
$P \Rightarrow Q$	Direct proofs	Show that Q is true if P is true
$P \Rightarrow Q$	Indirect proofs	Show that $\neg Q \Rightarrow \neg P$
$P \Leftrightarrow Q$	Implication proof	Show that $P \Rightarrow Q$ and $Q \Rightarrow P$
$P \Leftrightarrow Q$	Equivalence proofs	Show that $R \Leftrightarrow S$ where $(R \Leftrightarrow S) \Leftrightarrow (P \Leftrightarrow Q)$
P	Contradiction	Assume P is false and derive a contradiction
$\neg(\exists x)P(x)$	Contradiction	Assume $(\exists x)P(x)$ and derive a contradiction
$(\exists x)P(x)$	Constructive proofs	Exhibit a such that $P(a)$ is true
$(\exists x)P(x)$	Nonconstructive proofs	Show that $\neg(\exists x)P(x)$ implies a contradiction
$\neg(\forall x)P(x)$	Counterexample	Show that $(\exists x)\neg P(x)$
$(\forall x)P(x)$	Universal generalization	Show that $P(a)$ is true for an arbitrary a

Proof by induction

A proof by induction that $P(n)$ is true for alla positive integers n proceeds in two steps.

1) Prove that $P(1)$ is true.
2) Prove that $(\forall n)(P(n) \Rightarrow P(n+1))$

Example.

We want to prove that $\sum_{i=1}^{n} i^2=n(n+1)(2n+1)/6$.

1) The formula obviously holds for $n=1$.
2) We make the induction hypothesis that $\sum_{i=1}^{n} i^2=n(n+1)(2n+1)/6$ for some positive integer n.

   This implies

   $\sum_{i=1}^{n+1} i^2= \sum_{i=1}^{n} i^2+(n+1)^2=n(n+1)(2n+1)/6+(n+1)^2=(n+1)(n+2)(2n+3)/6$

   This is the formula to be proved for $(n+1)$. Thus the formula holds for all positive integers n.

1.2 Set Theory

Relations between sets

Notation: $x\in A$, the element x belongs to the set A
$x\notin A$, the element x does not belong to the set A

Let A and B be sets and Ω the universal set. Then A is a *subset* of B,

$$A\subset B,$$

if

$$(\forall x)(x\in A \Rightarrow x\in B).$$

(Sometimes the notation "$A\subseteq B$" is used and then "$A\subset B$" means that $A\subseteq B$ and $A\neq B$.)

The set B is a *superset* to A, $B\supset A$, if $A\subset B$.

The sets A and B are *equal*, $A=B$, if $A\subset B \wedge B\supset A$.

The *empty set* is denoted by $\emptyset$.

$$\emptyset\subset A\subset\Omega$$

$$A\subset A$$

$$(A\subset B)\wedge(B\subset C) \Rightarrow A\subset C$$

The *power set* $\mathcal{P}(\Omega)$ is the set of all subsets of Ω. If Ω has n elements, then $\mathcal{P}(\Omega)$ has 2^n elements.

1.2

Operations with sets. Set algebra

Operation	Notation	Definition	A B
Union	$A \cup B$	$\{x \in \Omega;\ x \in A \vee x \in B\}$	
Intersection	$A \cap B$	$\{x \in \Omega;\ x \in A \wedge x \in B\}$	
Difference	$A \setminus B$	$\{x \in \Omega;\ x \in A \wedge x \notin B\}$	
Symmetric difference	$A \triangle B$	$\{x \in \Omega;\ x \in A \triangledown x \in B\}$	
Complementation	A^c, A' or $\complement A$	$\{x \in \Omega;\ x \notin A\}$	

Commutative laws

$$A \cup B = B \cup A \qquad A \cap B = B \cap A$$

Associative laws

$$(A \cup B) \cup C = A \cup (B \cup C) \qquad (A \cap B) \cap C = A \cap (B \cap C)$$

Distributive laws

$$A \cup (B \cap C) = (A \cup B) \cap (A \cup C) \qquad A \cap (B \cup C) = (A \cap B) \cup (A \cap C)$$

Complementation

$$\emptyset^c = \Omega \qquad \Omega^c = \emptyset \qquad (A^c)^c = A$$

$$A \cup A^c = \Omega \qquad A \cap A^c = \emptyset$$

De Morgan laws

$$(A \cup B)^c = A^c \cap B^c \qquad (A \cap B)^c = A^c \cup B^c$$

Symmetric difference

$$A \triangle B = B \triangle A$$

$$(A \triangle B) \triangle C = A \triangle (B \triangle C)$$

$$A \triangle \emptyset = A$$

$$A \triangle A = \emptyset$$

$$A \triangle B = (A \cap B^c) \cup (B \cap A^c)$$

Cartesian product

The *Cartesian product* $A \times B$ of A and B is

$$A \times B = \{(a, b);\ a \in A \wedge b \in B\}.$$

Here (a, b) is the *ordered pair* with first component a and second component b.

$$A \times (B \cup C) = (A \times B) \cup (A \times C)$$

$$A \times (B \cap C) = (A \times B) \cap (A \times C)$$

$$(A \cup B) \times C = (A \times C) \cup (B \times C)$$

$$(A \cap B) \times C = (A \times C) \cap (B \times C)$$

The set of all functions from A to B is denoted B^A.

Cardinal numbers

Let $c(A)$ denote the *cardinal number* of a set A.
Writing $A \sim B$ if there is a bijection between A and B, then

$c(A) = c(B) \Leftrightarrow A \sim B$
$c(A) < c(B) \Leftrightarrow A \nsim B$ and there exists $B_1 \subset B$ such that $A \sim B_1$
$c(A) = n$ if A is finite with n elements.
$\aleph_0 = c(Q)$ = cardinality of a countable set.
$\aleph = c(R)$ = cardinality of a continuum e.g. the set of all continuous functions $R \to R$.
$2^{\aleph}$ = cardinality of the set of all functions $R \to R$.
$2^{c(A)}$ = cardinality of the set of all subsets of A.

$c(A) + c(B) = c(A \cup B)$ if $A \cap B = \varnothing$	$a^b a^c = a^{b+c}$
$c(A)c(B) = c(A \times B)$	$(a^b)^c = a^{bc}$
$c(A)^{c(B)} = c(A^B)$	(a, b, c cardinal numbers)
$c(A) < 2^{c(A)};\ n < \aleph_0 < 2^{\aleph_0} = \aleph < 2^{\aleph}$	
$a_1 + a_2 + \ldots = \aleph_0\ (a_i \in Z^+)$	$a_1 a_2 \ldots = \aleph\ (a_i \in Z^+, a_i > 1)$
$\aleph_0 + \aleph_0 + \ldots + \aleph_0 = n\aleph_0 = \aleph_0\ (n = 1,2,3\ldots)$	$\aleph + \aleph + \ldots + \aleph = n\aleph = \aleph\ (n = 1,2,3\ldots)$
$\aleph_0 + \aleph_0 + \ldots = \aleph_0 \aleph_0 = \aleph_0$	$\aleph + \aleph + \ldots = \aleph_0 \aleph = \aleph$
$\aleph_0 \aleph_0 \ldots \aleph_0 = (\aleph_0)^n = \aleph_0\ (n = 1,2,3\ldots)$	$\aleph \aleph \ldots \aleph = \aleph^n = \aleph\ (n = 1,2,3\ldots)$
$\aleph_0 \aleph_0 \ldots = (\aleph_0)^{\aleph_0} = \aleph$	$\aleph \aleph \ldots = (\aleph)^{\aleph_0} = \aleph$
$2^{\aleph_0} = n^{\aleph_0} = (\aleph_0)^{\aleph_0} = \aleph\ (n = 2,3\ldots)$	$2^{\aleph} = n^{\aleph} = (\aleph_0)^{\aleph} = \aleph^{\aleph} = 2^{\aleph}\ (n = 2,3\ldots)$
$(\aleph_0)^{\aleph} = 2^{\aleph}$	

Alphabets and languages

An *alphabet* L is a finite nonempty set of symbols.

Let L^* be the set of all strings (words) of elements in L including the empty string λ. A *language* over L is a subset of L^*.

Let A and B be languages over L. Then the *set product* AB is the language

$$AB=\{xy;\ x\in A,\ y\in B\}.$$

$$(AB)C=A(BC)$$

$$A(B\cup C)=AB\cup AC \qquad (B\cup C)A=BA\cup CA$$

$$A(B\cap C)\subset AB\cap AC \qquad (B\cap C)A\subset BA\cap CA$$

1.3 Binary Relations and Functions

Basic definitions

A *binary relation* R on $A\times B$ or from A to B is a subset of $A\times B$. A binary relation on A is a subset of $A\times A$.

Notation: $(x, y)\in R \Leftrightarrow xRy \quad (x, y)\notin R \Leftrightarrow x\not R y$

The *domain* D and the *range* (codomain) C are defined as

$$D=\{x;\ (\exists y)(x, y)\in R\},$$

$$C=\{y;\ (\exists x)(x, y)\in R\}.$$

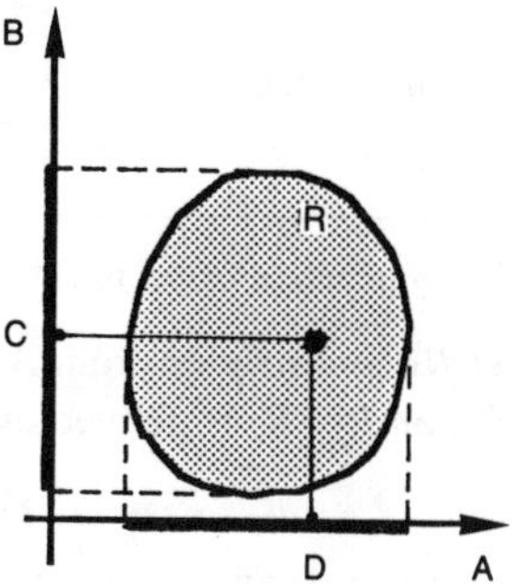

A relation R can be portrayed with a digraph.

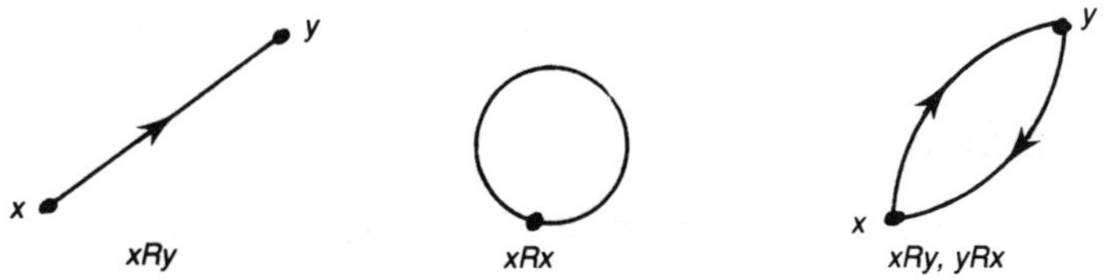

The *converse* R^{-1} of a relation R is the relation

$$R^{-1}=\{(y, x);\ (x, y)\in R\}.$$

The digraph of R^{-1} is obtained from that of R by reversing the direction of all arrows.

$$(R^{-1})^{-1}=R \qquad (A\times B)^{-1}=B\times A$$

$$(R_1\cup R_2)^{-1}=R_1^{-1}\cup R_2^{-1} \qquad (R_1\cap R_2)^{-1}=R_1^{-1}\cap R_2^{-1}$$

Properties of relations on A		
Property	Definition	Digraph
Reflexive	xRx for every $x \in A$	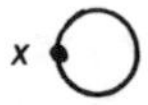
Symmetric	$xRy \Rightarrow yRx$ for all $x, y \in A$	
Transitive	$xRy,\ yRz \Rightarrow xRz$ for all $x, y, z \in A$	
Irreflexive	$x \not R x$ for every $x \in A$	
Antisymmetric	$x \neq y,\ xRy \Rightarrow y \not R x$ for all $x, y \in A$	

The *transitive closure* is $R^+ = R \cup R^2 \cup R^3 \cup \ldots$ ($R^2 = R \circ R$ etc.)

Let R_1 be a relation from A to B and R_2 a relation from B to C. Then the *composite* relation $R_1 \circ R_2$ is defined as follows.

$$R_1 \circ R_2 = \{(x, z);\ x \in A,\ z \in C,\ (\exists y)(y \in B,\ (x, y) \in R_1,\ (y, z) \in R_2)\}$$

$$(R_1 \circ R_2) \circ R_3 = R_1 \circ (R_2 \circ R_3) \qquad (R_1 \circ R_2)^{-1} = R_2^{-1} \circ R_1^{-1}$$

Relation (incidence) matrices

The *relation* (*incidence*) matrix $M = M_R = (r_{ij})$ of a relation R on a finite set A is defined by

$$r_{ij} = \begin{cases} 1 & \text{if } x_i R x_j \\ 0 & \text{if } x_i \not R x_j \end{cases}$$

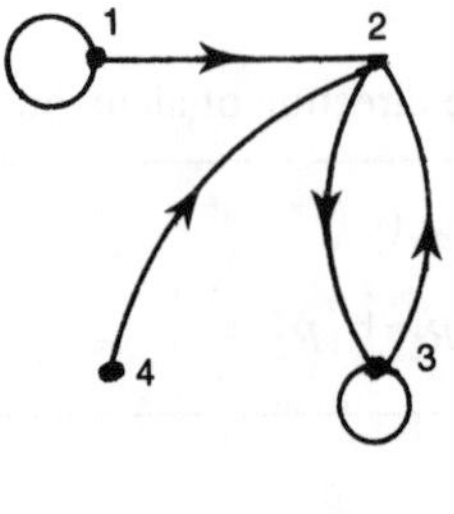

$$\begin{bmatrix} 1 & 1 & 0 & 0 \\ 0 & 0 & 1 & 0 \\ 0 & 1 & 1 & 0 \\ 0 & 1 & 0 & 0 \end{bmatrix}$$

Properties of relation matrices

1. Converse relation: $M_{R^{-1}}=(M_R)^t$ (transpose)
2. Composite relation: $M_{R\circ S}=M_RM_S$ with Boolean arithmetic (i e usual matrix multiplication but with the special rule $1+1=1$).
3. Reflexive relation: $r_{ii}=1$, all i.
4. Symmetric relation: $M^t=M$, i e $r_{ij}=r_{ji}$, all i, j.
5. Transitive relation: $M^2\leqslant M$, i e $[M^2]_{ij}\leqslant[M]_{ij}$, all i, j.
   (R is reflexive)$\Rightarrow$(R is transitive$\Leftrightarrow M^2=M$).
6. Irreflexive relation: $r_{ii}=0$, all i.
7. Antisymmetric relation: $r_{ij}=1\Rightarrow r_{ji}=0$, $i\neq j$.
8. Union: $M_{R\cup S}=M_R\vee M_S$, i e $[M_{R\cup S}]_{ij}=r_{ij}+s_{ij}$ (Boolean addition, i e $1+1=1$)
9. Intersection: $M_{R\cap S}=M_R\wedge M_S$ i e $[M_{R\cap S}]_{ij}=r_{ij}s_{ij}$
10. Transitive closure: $M^+=M_R\vee M_{R^2}\vee M_{R^3}\vee\ldots$

Particular relations

Type of relation on A	Definition
Equivalence relation	Reflexive, symmetric and transitive
Partial order	Reflexive, antisymmetric and transitive
Compatibility relation	Reflexive and symmetric
Quasi order	Transitive and irreflexive
Linear order or total order	Partial order and xRy or yRx for all x, $y\in A$
Well ordering	Linear order and every nonempty subset of A has a least element

Equivalence relations

If R is an equivalence relation the *R-equivalence class* generated by $x\in A$ is $[x]_R=\{y;\ xRy\}$. The equivalence classes constitute a partition of A. They are mutually disjoint and their union is A.

Partially ordered sets

Let $(P, \leqslant)$ be a partially ordered set ($x\leqslant y \Leftrightarrow xRy$).

1. *Hasse diagram.* The partial ordering $\leqslant$ on a set P can be represented by a *Hasse diagram.* In such a diagram an element is represented by a small cirle or a dot. If $x<y$, then x is at a lower level than y, and there exists a line from x *upwards* to y (either directly or via other elements).
2. Let $A\subset P$. An element $x\in P$ is an *upper bound* [*lower bound*] for A if $a\leqslant x$ [$x\leqslant a$] for all $a\in A$.

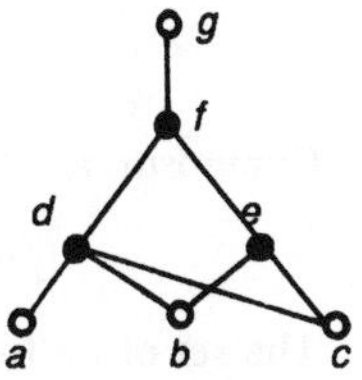

3. Let $A \subset P$. An element $x \in P$ is a *least upper bound* (LUB) or *supremum* (sup) for A if x is an upper bound for A and $x \leq y$ for any upper bound y for A. Analogously, an element $x \in P$ is a *greatest lower bound* (GLB) or *infimum* (inf) for A if x is a lower bound for A and $y \leq x$ for any lower bound y for A.

Example

In the above Hasse diagram let $P=\{a, b, c, d, e, f, g\}$ and $A=\{d, e, f\}$. Then

(*i*) $a<d$, $c<f$ but neither $a<e$ nor $d<a$.
(*ii*) b and c are the lower bounds for A.
(*iii*) Neither A nor P has an inf.
(*iv*) f and g are the upper bounds for A.
(*v*) $\sup A=f$ and $\sup P=g$.

Functions

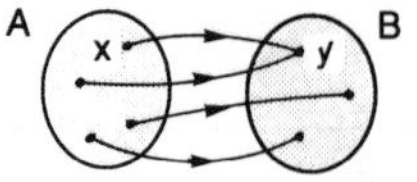

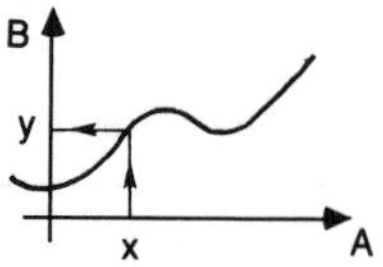

A *function* from A to B, $f\colon A \to B$, is a relation with the property: To each $x \in A$ is uniquely assigned a $y \in B$.

Notation: $y=f(x)$ or $x \overset{f}{\to} y$.

$D_f=A=$*domain* of f; $R_f=\{f(x)\colon x \in A\}=$*range* of f.

If $f\colon A \to B$ and $C \subset A$, then the *image* of C under f is

$$f(C)=\{f(x);\ x \in C\}$$

$$f(A \cup B)=f(A) \cup f(B) \qquad f(A \cap B) \subset f(A) \cap f(B)$$

If D is a subset of B then

$$f^{-1}(D)=\{x;\ f(x) \in D\}$$

is the *inverse image* of D under f.

$$f^{-1}(A \cup B)=f^{-1}(A) \cup f^{-1}(B) \qquad f^{-1}(A \cap B)=f^{-1}(A) \cap f^{-1}(B)$$

If $f\colon A \to B$ and $g\colon B \to C$ are functions then the *composite* function $g \circ f$ is the function from A to C such that

$$g \circ f(x)=g(f(x)),\ x \in A$$

Composition of functions is associative.

$$(f \circ g) \circ h=f \circ (g \circ h)$$

The set of all functions from A to B is denoted B^A.

Properties of functions $f\colon A \to B$		
Property	Definition	Graph
Surjective or onto	$f(A)=B$	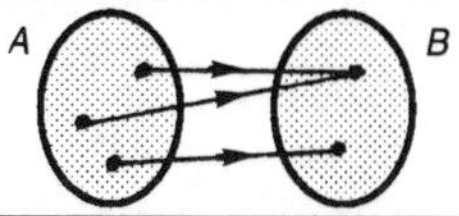
Injective or one-to-one	$x \neq x' \Rightarrow f(x) \neq f(x')$ for all $x, x' \in A$	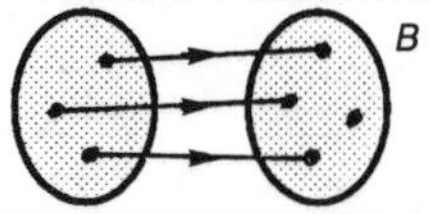
Bijective or one-to-one and onto	Surjective and injective	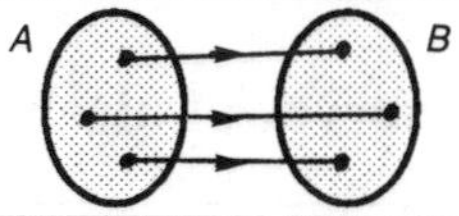

If $f\colon A \to B$ is bijective then the *inverse* function f^{-1} is the converse relation f^{-1}. The inverse function f^{-1} is a bijective function from B to A.

$$y=f(x) \Leftrightarrow x=f^{-1}(y)$$

$$f^{-1}(f(x))=x \text{ for } x \in A \qquad f(f^{-1}(x))=x \text{ for } x \in B$$

$$(g \circ f)^{-1}=f^{-1} \circ g^{-1}$$

1.4 Algebraic Structures

Basic algebraic structures

A *binary operation* $*$ on a set S is a function $*\colon S \times S \to S$. The element in S assigned to (x,y) is denoted $x*y$. The operation $*$ is

1. *commutative* if $x*y = y*x$, all $x, y \in S$.
2. *associative if* $x*(y*z) = (x*y)*z$, all $x, y, z \in S$.
3. *distributive* over the operation $\circ$ if $x*(y \circ z) = (x*y)\circ(x*z)$, all $x, y, z \in S$.

The operation $*$ has

4. an *identity* element e if $x*e = e*x = x$, all $x \in S$.
5. a *zero* element 0 if $x*0 = 0*x = 0$, all $x \in S$.

The element $x \in S$

6. has an *inverse* x^{-1} if $x*x^{-1} = x^{-1}*x = e$
7. is *idempotent* if $x*x = x$.

Direct product

If $(A,*)$ and $(B,\circ)$ are algebraic systems of the same type, then the *direct product* of these systems is the algebraic system $(A \times B, \lozenge)$ with $(a_1,b_1)\lozenge(a_2,b_2) = (a_1 * a_2, b_1 \circ b_2)$, $a_i \in A$, $b_i \in B$.

A survey of algebraic structures

Algebraic structure	Definition
Semigroup $(S, *)$	$*$ is associative
Monoid $(S, *)$ or $(S, *, e)$	Semigroup with identity element e such that $e * x = x * e = x$, all x (e is unique)
Group $(S, *)$ or $(S, *, e)$	Monoid such that every element x has a unique inverse x^{-1} such that $x^{-1} * x = x * x^{-1} = e$
Abelian group or commutative group $(S, *)$	Group such that $*$ is commutative
Ring $(S, +, \cdot)$	$(S, +)$ is an abelian group and $(S, \cdot)$ is a semigroup and $a \cdot (b+c) = a \cdot b + a \cdot c$, $(b+c) \cdot a = b \cdot a + c \cdot a$
Field $(S, +, \cdot)$	Ring such that the nonzero elements form an abelian group under multiplication $\cdot$
Lattice $(S, \leqslant)$	$(S, \leqslant)$ is a partially ordered set (poset) such that every pair x, y of elements in S have a greatest lower bound GLB and a least upper bound LUB
Boolean algebra $(S, +, \cdot, ', 0, 1)$	The binary operations $+$ and $\cdot$ are commutative and associative and distribute over each other. The elements 0 and 1 are identity elements for $+$ and $\cdot$ and $x + x' = 1$ and $x \cdot x' = 0$ (x' is the complement of x.)

Algebraic structure	Concrete examples
Semigroup	The integers under addition (multiplication) The languages over an alphabet under set product The set of binary relations on a set under composition
Monoid	The real numbers under addition with 0 as identity The real numbers under multiplication with 1 as identity The power set $\wp(\Omega)$ under union with $\varnothing$ as identity
Group	The set of permutations of a set under composition The set of symmetries of a regular polygon $(Z_n, +_n)$ where $+_n$ is addition modulo n
Abelian group	The integers under addition The set of rational numbers $\neq 0$ under multiplication
Ring	The set of integers (rational numbers, even numbers, real numbers, complex numbers) under addition and multiplication $(Z_n, +_n, \times_n)$ where $+_n$ and $\times_n$ are addition and multiplication modulo n
Field	The set of rational, real or complex numbers under addition and multiplication $(Z_n, +_n, \times_n)$ where n is prime
Lattice	The power set $\wp(\Omega)$ under $\subset$, the subset relation The set of positive integers under D, where D is the "divides" relation, xDy if x divides y. The GLB and LUB of x and y is defined as the GCD and LCM of x and y
Boolean algebra	$(\wp(\Omega), \cup, \cap$, complementation, $\varnothing, \Omega)$ $(S, \wedge, \vee, \neg, F, T)$ where S is the set of equivalence classes of statement formulas in n statements and F and T denotes contradiction and tautology respectively $S=\{0, 1\}$ with usual definition of addition and multiplication with the exception of $1+1=1$ (Boolean addition and multiplication) and with $0'=1$, $1'=0$

Homomorphisms and isomorphisms

Let S_1 and S_2 be algebraic structures of the same type. A mapping $g: S_1 \to S_2$ is called a *homomorphism* if g preserves the algebraic structure.

Semigroups

Let $(S_1, *_1)$ and $(S_2, *_2)$ be semigroups. A mapping $g: S_1 \to S_2$ such that for all $x, y \in S_1$

$$g(x *_1 y) = g(x) *_2 g(y)$$

is called a *semigroup homomorphism.*

Monoids

Let $(S_1, *_1)$ and $(S_2, *_2)$ be monoids with identity elements e_1 and e_2. A semigroup homomorphism $g: S_1 \to S_2$ such that

$$g(e_1) = e_2$$

is called a *monoid homomorphism.*

Groups

Let $(S_1, *_1)$ and $(S_2, *_2)$ be groups. A mapping $g: S_1 \to S_2$ such that for all $x, y \in S_1$

$$g(x *_1 y) = g(x) *_2 g(y)$$

is called a *group homomorphism.* If follows from group properties that

$$g(e_1) = e_2 \text{ and } g(x^{-1}) = g(x)^{-1} \text{ for every } x \in S_1.$$

Rings

Let $(S_1, +_1, \circ_1)$ and $(S_2, +_2, \circ_2)$ be rings. A mapping $g: S_1 \to S_2$ such that for all x, $y \in S_1$

$$g(x +_1 y) = g(x) +_2 g(y) \text{ and } g(x \circ_1 y) = g(x) \circ_2 g(y)$$

is called a *ring homomorphism.*

Lattices

Let $(S_1, \leqslant_1)$ and $(S_2, \leqslant_2)$ be lattices. A mapping $g: S_1 \to S_2$ such that for all $x, y \in S_1$

$$g(\mathrm{GLB}(x, y)) = \mathrm{GLB}(g(x), g(y))$$

$$g(\mathrm{LUB}(x, y)) = \mathrm{LUB}(g(x), g(y))$$

is called a *lattice homomorphism.*

It follows from lattice properties that

$$x \leqslant_1 y \Rightarrow g(x) \leqslant_2 g(y)$$

Boolean algebras

Let $(S_1, +_1, \circ_1, ', 0_1, 1_1)$ and $(S_2, +_2, \circ_2, ', 0_2, 1_2)$ be Boolean algebras. A mapping $g: S_1 \to S_2$ such that for all $x, y \in S_1$

$$g(x+_1y)=g(x)+_2g(y) \qquad g(x\circ_1 y)=g(x)\circ_2 g(y)$$
$$g(0_1)=0_2 \qquad g(1_1)=1_2 \qquad g(a')=g(a)'$$

is called a *Boolean homomorphism.*

To prove that $g: S_1 \to S_2$ is a Boolean homomorphism it is sufficient to prove that

$$g(x+_1y)=g(x)+_2g(y) \text{ and } g(a')=g(a)'$$

Further "morphisms"

A homomorphism $g: S_1 \to S_2$ is

(*i*) an *epimorphism* if g is onto (surjective).
(*ii*) a *monomorphism* if g is one-to-one (injective)
(*iii*) an *isomorphism* if g is one-to-one and onto (bijective). The inverse of a bijective homomorphism is a homomorphism. If there exists an isomorphism between two algebraic structures they are called *isomorphic.*

An isomorphism is

(*iiia*) an *endomorphism* if $S_2 \subset S_1$
(*iiib*) an *automorphism* if $S_2 = S_1$

Further properties of algebraic structures

Groups

1. *Definition.* A *group* $(G, *)$ or G is an algebraic system such that
   (*i*) $x*(y*z)=(x*y)*z$, all $x, y, z \in G$.
   (*ii*) There is an identity element $e \in G$ such that $x*e=e*x=x$, all $x \in G$.
   (*iii*) For every $x \in G$, there exists an inverse $x^{-1} \in G$ such that $x^{-1}*x=x*x^{-1}=e$.
   The group is *abelian* if $*$ is commutative.
2. *Subgroup.* $(S, *)$ is a subgroup of $(G, *)$ if $S \subset G$ and
   (*i*) $a, b \in S \Rightarrow a*b \in S$ (*ii*) $e \in S$ (*iii*) $a \in S \Rightarrow a^{-1} \in S$.
   Note: S is a subgroup $\Leftrightarrow a, b \in S \Rightarrow a*b^{-1} \in S$ for all $a, b \in S$.
3. If g is a group homomorphism (see above) from $(G, *)$ to $(H, \circ)$, then the *kernel* $\ker(g)=\{x \in G: g(x)=e_H\}$ of g is a subgroup of G.
4. *Cayley's theorem.* Any finite group of n elements is isomorphic to a permutation group of degree n.
5. Let $(H, *)$ be a subgroup of $(G, *)$.
   For any $a \in G$, the set $aH=\{a*h: h \in H\}$ is called a *left coset* of H in G.
   The element a is a *representative element* of aH.
   Similarly. $Ha=\{h*a: h \in H\}$ is a *right coset.*
   If G is abelian then $aH=Ha$, all $a \in G$.

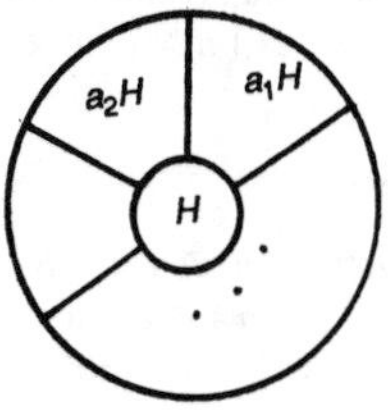

The set of left [right] cosets of H in G forms a partition of G, i e every element of G belongs to exactly one left [right] coset (equivalence class).
Note: $b \in aH \Leftrightarrow a^{-1}*b \in H$; $c \in Ha \Leftrightarrow c*a^{-1} \in H$.

6. The *order* $|G|$ of a finite group is the number of elements of G.
   Lagrange's theorem: The order of a subgroup of a finite group divides the order of the group.
   The *index* of a subgroup H of G is $|G|/|H|$=Number of cosets.
   Note: If the order of G is n then $a^n=e$, all $a \in G$. The *cyclic* subgroup $\{a^k\colon a \in G, k=1, 2, \ldots\}$ generated by a is of order m, where m is the smallest number such that $a^m=e$. Here m divides n.
7. If G is finite and abelian and if the prime number p divides the order of G, then there exists an element $a \in G$ of order p, i e $a^p=e$.
8. *Normal subgroups.* A subgroup $(N,*)$ of $(G,*)$ is called *normal if* $aN=Na$ for all $a \in G$.
   Quotient groups. If N is a normal subgroup of G, let G/N denote the set of left (or right) cosets of N in G. With the operation • defined by $(aN)\bullet(bN)=(a*b)N$, G/N is a group, called the *quotient group* (or *factor group*) of G by N, and $|G/N|=|G|/|N|$.

   Theorems

   a. If $(G,*)$ and $(H,\circ)$ are groups and $g\colon G \to H$ is a homomorphism, then the kernel of g is a normal subgroup of G.
   b. (*Fundamental theorem of group homomorphisms*). If g is a homomorphism from $(G,*)$ to $(H,\circ)$ then $G/\ker(g)$ is isomorphic to $g(G)$.
9. *Group multiplication tables.* The binary operation $*$ in a group can be represented by a table.

Example. There are two non-isomorphic groups of order 4:

$*$	e	a	b	c
e	e	a	b	c
a	a	b	c	e
b	b	c	e	a
c	c	e	a	b

$\{e, b\}$ is a subgroup
Cyclic: $a^2=b$, $a^3=c$, $a^4=e$

$*$	e	a	b	c
e	e	a	b	c
a	a	e	c	b
b	b	c	e	a
c	c	b	a	e

$\{e, a\}$, $\{e, b\}$, $\{e, c\}$ are subgroups
Non-cyclic

Group order	1	2	3	4	5	6	7	8	9	10	11	12	13	14	15	16	17	18	19	20
Number of non-isomorphic groups	1	1	1	2	1	2	1	5	2	2	1	5	1	2	1	14	1	5	1	5

10. *Sylow groups.* If the order of G is $p^k m$, where p is prime and does not divide m, then G has a subgroup (called *Sylow p-subgroup*) of order p^k. The number n_p of Sylow p-groups (*i*) divides m and (*ii*) $n_p \equiv 1 \pmod{p}$.

Rings

1. *Definition.* A *ring* $(R, +, \cdot)$ or R is an algebraic system such that
   (*i*) $(R, +)$ is an abelian group with identity element 0 and inverse $-x$ of x, i e $x+y=y+x$, $(x+y)+z=x+(y+z)$, $x+0=0+x=x$, $x+(-x)=(-x)+x=0$.

(*ii*) The operation $\cdot$ is associative and distributive over $+$, i e $(x \cdot y) \cdot z = x \cdot (y \cdot z)$, $x \cdot (y+z) = x \cdot y + x \cdot z$, $(x+y) \cdot z = x \cdot z + y \cdot z$.

(Note, below $\cdot$ is often omitted writing $xy = x \cdot y$.) The ring is *commutative* if $\cdot$ is commutative.

2. The ring has a "1" if $\cdot$ has an identity element, i e $x \cdot 1 = 1 \cdot x = x$.

3. A commutative ring with a "1" is an *integral domain* if for $x \neq 0$, $xy = xz \Rightarrow y = z$. (If $xy = 0$ with x and $y \neq 0$, then x and y are called *zero divisors*.)

4. $(A, +, \cdot)$ is a *subring* of $(R, +, \cdot)$ if $A \subset R$ and if

   (*i*) $x, y \in A \Rightarrow x+y \in A$ and $xy \in A$, (*ii*) $0 \in A$, (*iii*) $x \in A \Rightarrow -x \in A$

5. *A subring* $(I, +, \cdot)$ of $(R, +, \cdot)$ is an *ideal* of R if $ax \in I$ and $xa \in I$ for all $a \in I$ and all $x \in R$. If R is commutative, then the smallest ideal of R containing an element a is $(a) = \{xa: x \in R\}$. Ideals of this form are called *principal ideals*.

   An ideal $M \neq R$ is a *maximal* ideal if whenever I is an ideal of R such that $M \subset I \subset R$, then either $I = M$ or $I = R$.

   An ideal P of a commutative ring R is a *prime* ideal if $ab \in P$, $a, b \in R \Rightarrow a \in P$ or $b \in P$.

6. *Quotient rings*. $R/I = \{$cosets of I relative $+\} = \{x+I: x \in R\}$ is a ring itself with addition $\oplus$ and multiplication $\otimes$ defined by $(x+I) \oplus (y+I) = (x+y)+I$ and $(x+I) \otimes (y+I) = xy+I$.

   The ideal I is the zero element of the quotient ring.

7. If R is a commutative ring and P is an ideal of R then P is prime $\Leftrightarrow R/P$ is an integral domain.

8. If R is a commutative ring with "1" and M is an ideal of R then M is maximal $\Leftrightarrow R/M$ is a field.

9. *Fundamental homomorphism theorem for rings*. If R and S are rings and $g: R \to S$ is a homomorphism (see above) with $I = \ker(g)$, then $\phi: R/I \to S$ defined by $\phi(x+I) = g(x)$ is an isomorphism of R/I onto $g(R)$.

10. The *characteristic* of a ring R is the least positive integer n such that $na = a+a+\ldots+a = 0$ for all $a \in R$. If such a number does not exist then the characteristic is 0.

> If D is an integral domain
> (*i*) then the characterstic is either 0 or a prime p.
> (*ii*) of characteristic 0, then D contains a subring isomorphic to $\mathbf{Z}$.
> (*iii*) of characteristic p, then D contains a subring isomorphic to $\mathbf{Z}_p$.

Fields

1. *Definition*. A *field* $(F, +, \cdot)$ or F is an algebraic system such that

   (*i*) $(F, +, \cdot)$ is a commutative ring with a "1", i e

   $x+y = y+x$, $(x+y)+z = x+(y+z)$, $x+0 = x$, $x+(-x) = 0$,
   $xy = yx$, $(xy)z = x(yz)$, $x(y+z) = xy+xz$, $1x = x$

   (*ii*) $(F - \{0\}, \cdot)$ is a group, i e for any $x \neq 0$ in F there exists a multiplicative inverse x^{-1} such that $xx^{-1} = x^{-1}x = 1$.

2. The following inclusions hold: Fields$\subset$Integral domains$\subset$Commutative rings$\subset$Rings.

3. *Subfield*. $(A, +, \cdot)$ is a *subfield* of $(F, +, \cdot)$ if $A \subset F$ and

   (*i*) $a, b \in A \Rightarrow a+b \in A$ and $ab \in A$
   (*ii*) $0 \in A$ and $1 \in A$
   (*iii*) $a \in A \Rightarrow -a \in A$ and $a \in A$, $a \neq 0 \Rightarrow a^{-1} \in A$

4. A field has only the ideals F and $\{0\}$.
5. (*i*) If F is finite and has characteristic p (p prime) then F has p^n elements for some positive integer n.
   (*ii*) For any prime number p and any positive integer n there exists a field with p^n elements.
6. (*i*) The field E is an *extension* of the field F if E contains a subfield isomorphic to F.
   (*ii*) If S is a subset of E then $F(S)$ denotes the smallest subfield of E that contains both S and F. For example,
   (*a*) $\boldsymbol{R}(i)=\boldsymbol{C}$.
   (*b*) $\boldsymbol{Q}(\sqrt{2})=\{a+b\sqrt{2}\colon a,\ b\in\boldsymbol{Q}\}$
   (*c*) $\boldsymbol{Q}(\pi)=\{(a_0+a_1\pi+a_2\pi^2+\ldots+a_n\pi^n)/(b_0+b\pi+b_2\pi^2+\ldots+b_m\pi^m)\colon a_i,\ b_i\in Q$, not all $b_i=0\}$

Integers modulo *n*

The set $\boldsymbol{Z}_n$ consists of the congruence classes modulo n which are

$$[k]=\{\ldots,\ k-2n,\ k-n,\ k,\ k+n,\ k+2n,\ \ldots\},\ k=0,\ 1,\ \ldots,\ n-1.$$

The integers $0, 1, \ldots, n-1$ are representative elements of each congruence class, respectively.

1. *Operations:* $[k_1]+_n[k_2]=[k_1+k_2]$ $[k_1]\times_n[k_2]=[k_1k_2]$
2. $[k_1]=[m_1],\ [k_2]=[m_2]\Rightarrow$ (*i*) $[k_1+k_2]=[m_1+m_2]$ (*ii*) $[k_1k_2]=[m_1m_2]$
3. (*i*) $\boldsymbol{Z}_n$ is a cyclic group under $+_n$
   (*ii*) $\boldsymbol{Z}_n$ is a ring under $+_n$ and $\times_n$ [$\boldsymbol{Z}_n$ is a field $\Leftrightarrow$ n is prime.]

Polynomial rings

If F is a field then $F[x]$ denotes the set of polynomials $\sum_{k=0}^{n} a_kx^k$, $a_k\in F$, n arbitrary.

1. Every ideal I of the ring $(F[x], +, \cdot)$ is a principal ideal of the form $I=(g(x))=\{f(x)g(x)\colon f(x)\in F[x]\}$ for some fixed $g(x)\in I$.
2. In $F[x]$ the division algorithm holds, i e if $f(x),\ g(x)\in F[x]$ there exist $q(x)$ (the quotient) and $r(x)$ (the remainder) in $F[x]$ such that $f(x)=q(x)g(x)+r(x)$ with degree $r(x)<$ degree $g(x)$.
3. A polynomial $p(x)\in F[x]$ of degree ≥ 1 is *irreducible* if it is not a product of two polynomials of lower degree.
4. *Factor theorem.* If $p(x)\in F[x]$ and $a\in F$, then $(x-a)$ is a factor of $p(x)\Leftrightarrow p(a)=0$.

Example
The polynomial x^2+1 is

(*a*) reducible in $\boldsymbol{C}[x]$ because $x^2+1=(x-i)(x+i)$
(b) irreducible in $\boldsymbol{R}[x]$ because x^2+1 has no zero in $\boldsymbol{R}$
(*c*) reducible in $\boldsymbol{Z}_2[x]$ because $x^2+1=(x+1)^2$.

5. The quotient ring $F[x]/(p(x))$ is a field $\Leftrightarrow$ $p(x)$ is irreducible over F.
6. Let $p(x)$ in $F[x]$ be irreducible over F. Then $F=F[x]/(p(x))$ is a field extension of F and $p(x)$ has a root in F.
7. If $p(x)$ is any polynomial of positive degree over F, then $p(x)$ has a root in some extension $\overline{F}$ of F.

8. If $p(x)=a_0+a_1x+\ldots+a_nx^n$ is a polynomial over F and $I=(p(x))$, then each element of $F[x]/I$ can be written uniquely as $I+(b_0+b_1x+\ldots+b_{n-1}x^{n-1})$, $b_i \in F$.

Lattices

Definition. A *lattice* $(L, \leqslant)$ or L

(*i*) is a partially ordered set, i e reflexive, antisymmetric and transitive,
(*ii*) has for every pair $a, b \in L$ a greatest lower bound (GLB) or *meet* or *product* and a least upper bound (LUB) or *join* or *sum.*

Notation: GLB $\{a, b\}=a*b=a \wedge b=a \cdot b=ab$
LUB $\{a, b\}=a \oplus b=a \vee b=a+b$

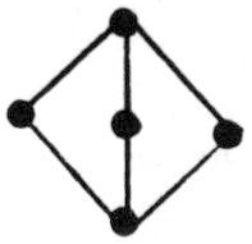

A lattice

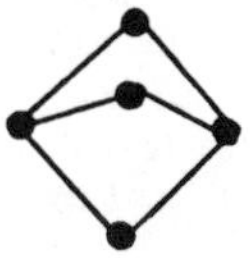

Not a lattice

Properties
For any $a, b, c \in L$:

1. $aa=a$	$a+a=a$	(Idempotent)
2. $ab=ba$	$a+b=b+a$	(Commutative)
3. $(ab)c=a(bc)$	$(a+b)+c=a+(b+c)$	(Associative)
4. $a(a+b)=a$	$a+(ab)=a$	(Absorption)
5. $a+(bc)\leqslant(a+b)(a+c)$	$a(b+c)\geqslant(ab)+(ac)$	(Distributive inequalities)

If equality holds, then the lattice is called *distributive.*

6. $a\leqslant b \Leftrightarrow ab=a \Leftrightarrow a+b=b$
7. $a\leqslant b \Rightarrow ac\leqslant bc$ and $a+c\leqslant b+c$
8. $a\leqslant b$ or $a\leqslant c \Rightarrow a\leqslant b+c$ $\quad$ $a\leqslant b$ and $a\leqslant c \Rightarrow a\leqslant bc$
$a\geqslant b$ or $a\geqslant c \Rightarrow a\geqslant b+c$ $\quad$ $a\geqslant b$ and $a\geqslant c \Rightarrow a\geqslant bc$
9. A lattice is *complete* if every non-empty subset has a LUB and a GLB.
Notation: $0=$The least element of L, $1=$The greatest element of L.
10. An element $b \in L$ is a *complement* of $a \in L$ if $ab=0$ and $a+b=1$. If every element of L has a complement the lattice is called *complemented.* (Complemented, distributive lattices, see Boolean algebras below.)

Boolean algebras

A *Boolean algebra* $(B, +, \cdot, ', 0, 1)$ or B is a complemented, distributed lattice. The (unique) complement of a is denoted a'.

Notation: $a \cdot b=ab=\text{GLB}\{a, b\}$, $a+b=\text{LUB}\{a, b\}$.
(Advice: To understand the following laws, think of the Boolean algebra of a power set $\mathcal{P}(S)$ with $A+B=A \cup B$, $A \cdot B=A \cap B$, $A'=S-A$, $0=\emptyset$, $1=S$ and $A\leqslant B \Leftrightarrow A\subset B$).

Properties

For any $a, b, c \in B$:

1.	$aa=a$	$a+a=a$	(Idempotent)
2.	$ab=ba$	$a+b=b+a$	(Commutative)
3.	$(ab)c=a(bc)$	$(a+b)+c=a+(b+c)$	(Associative)
4.	$a(a+b)=a$	$a+(ab)=a$	(Absorption)
5.	$a+(bc)=(a+b)(a+c)$	$a(b+c)=(ab)+(ac)$	(Distributive)
6.	$(ab)+(bc)+(ca)=(a+b)(b+c)(c+a)$		
7.	$ab=ac$ and $a+b=a+c \Rightarrow b=c$		
8.	$0 \leq a \leq 1,\ a \cdot 0=0,\ a+1=1,\ a \cdot 1=a,\ a+0=a$		
9.	$a \cdot a'=0,\ a+a'=1,\ 0'=1,\ 1'=0$		(Complement)
10.	$(a \cdot b)'=a'+b',\ (a+b)'=a' \cdot b'$		(De Morgan)
11.	$a \leq b \Leftrightarrow ab=a \Leftrightarrow a+b=b$ $a \leq b \Leftrightarrow ab'=0 \Leftrightarrow b' \leq a' \Leftrightarrow a'+b=1$		

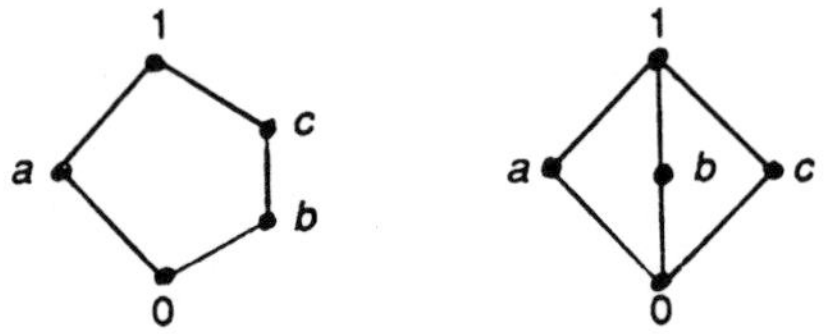

Not Boolean algebras.
Distributive laws not valid.
Complement of a not unique.

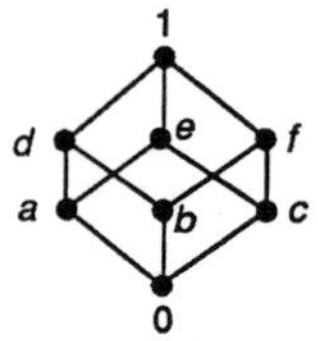

Boolean algebra.
Distributive laws valid.
$a'=f,\ e'=b$ unique.

Smallest Boolean algebra $B=\{0,1\}$:

+	0	1
0	0	1
1	1	1

·	0	1
0	0	0
1	0	1

Minterms

The Boolean algebra B constituted by *minterms* (*atoms*) $a_1, a_2, \ldots, a_n$ is defined by (taking the example $n=5$)

(*i*) Each element a of B is a sum of some minterms, e.g. $a=a_1+a_3+a_4$, $1=a_1+a_2+a_3+a_4+a_5$ (sum of *all* minterms).
(*ii*) The complement $a'=a_2+a_5$ (sum of the remaining minterms).
(*iii*) Sum +: Remember $a_i+a_i=a_i$.
(*iv*) Product ·: Remember $a_i \cdot a_i=a_i$ and $a_i \cdot a_j=0,\ i \neq j$.

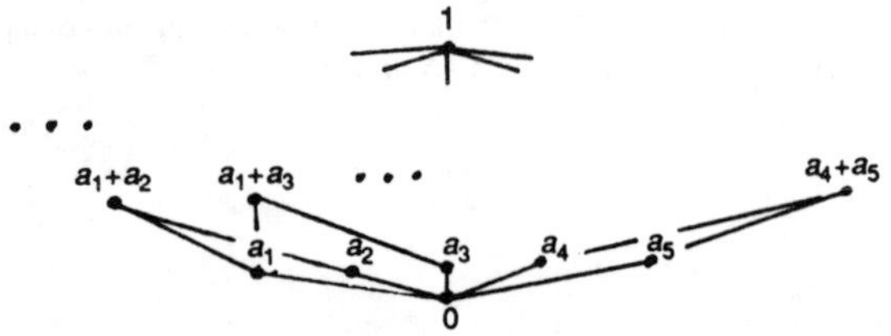

Remark. The above Boolean algebra is isomorphic to a free Boolean algebra if $n=2^k$, k integer.

Free Boolean algebras

In a *free* Boolean algebra generated by n variables $x_1, x_2, ..., x_n$, the elements are (finite) combinations of the x_i, x_i' and + and $\cdot$. The elements can be written uniquely as a sum of *minterms* $x_1^{\alpha_1}x_2^{\alpha_2} ... x_n^{\alpha_n}$, its *disjunctive normal form* (or similarly as a product of *maxterms*), where α_i=0 or 1 and $x_i^0=x_i'$, $x_i^1=x_i$. Sum of all minterms=1.

Remark. The number of minterms is 2^n and the number of free Boolean expressions is 2^{2^n}.

Duals

The *dual* $\beta(x_1, x_2, ..., x_n)$ of a Boolean expression $\alpha(x_1, x_2, ..., x_n)$ is obtained by interchanging the operations $\cdot$ and +.

Note: $[\alpha(x_1, x_2, ..., x_n)]'=\beta(x_1', x_2', ..., x_n')$ (Cf. 10. above).

Example (Cf. Example of sec. 1.1)

Consider the free Boolean algebra generated by x, y and z. Find the sum of minterms and product of maxterms of b and b' if $b=xy+yz'$.

Solution

$b\ =xy(z+z')+yz'(x+x')=xyz+xyz'+xyz'+x'yz'=xyz+xyz'+x'yz'$

$(=\text{min}_7+\text{min}_6+\text{min}_2=\cdot\ 2, 6, 7)$

$b'\ =(\cdot\ 0, 1, 3, 4, 5)=x'y'z'+x'y'z+x'yz+xy'z'+xy'z$

$b\ =(x')'=+\ 0, 1,\ 3, 4, 5=(x+y+z)(x+y+z')(x+y'+z')(x'+y+z)(x'+y+z')$

$b'\ =+2, 6, 7=(x'+y'+z')(x'+y'+z)(x+y'+z)$

If $B=\{0, 1\}$, the *values* of the above *Boolean function* are given in the table

x	y	z	xy	yz'	$b=xy+yz'$	b'
1	1	1	1	0	1	0
1	1	0	1	1	1	0
1	0	1	0	0	0	1
1	0	0	0	0	0	1
0	1	1	0	0	0	1
0	1	0	0	1	1	0
0	0	1	0	0	0	1
0	0	0	0	0	0	1

Minimization of Boolean polynomials

Given a Boolean polynomial (expression, function), how to reduce it to its simplest form? This problem may be solved using succesively the reduction idea $xyz+xyz'=$ $=xy(z+z')=xy$.

A systematic way of simplifying a Boolean polynomial (*McCluskey's* method) is illustrated by the example (written in disjunctive normal form):

$$b=xyzw+xyz'w+x'yzw+x'yz'w'+xy'zw'+xyzw'+x'yzw'$$

Enumerate the atoms (minterms) in a first column (a). Beginning from the top, compare the terms.

Those which differ by only one variable and its complement are reduced, and the reduced new terms are listed in the next column (*b*). This procedure is repeated until no more reductions can be made. Finally go backwards in the columns and pick up terms until all atoms are covered.

(*a*)	(*b*)	(*c*)
1 $xyzw$	1,3 yzw	1,3,5,7 yz
2 $xy'z'w$	1,5 xyz	1,3,5,7 yz
3 $x'yzw$	3,7 $x'yz$	4,5,6,7 yw'
4 $x'yz'w'$	4,6 $yz'w'$	(No 2 is missing)
5 $xy'zw'$	4,7 $x'yw'$	
6 $xyz'w'$	5,7 yzw'	
7 $x'yzw'$		

Thus, $b=yz+yw'+xy'z'w$

Logic design

Input–output table

Gate	Input		Output	**Logic gate symbols (IEC 612-12)**
	a	b		
Inverter ¬, −	0 1		1 0	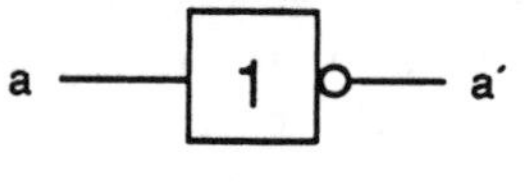
AND gate ∧, ·	0 0 1 1	0 1 0 1	0 0 0 1	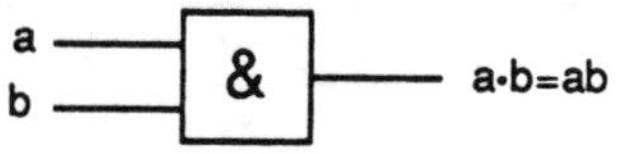
OR gate ∨, +	0 0 1 1	0 1 0 1	0 1 1 1	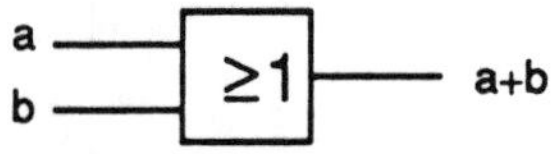
Exclusive Or gate $\overline{\vee}$, ⊕	0 0 1 1	0 1 0 1	0 1 1 0	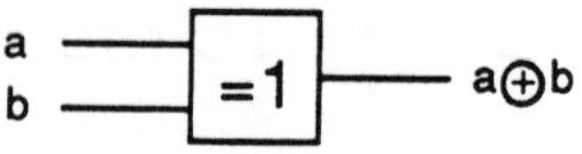
NAND gate ↑	0 0 1 1	0 1 0 1	1 1 1 0	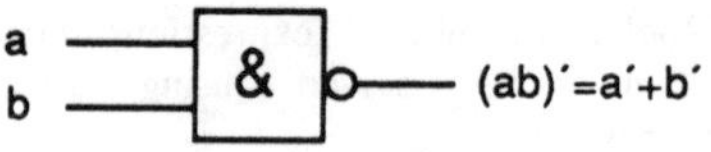
NOR gate ↓	0 0 1 1	0 1 0 1	1 0 0 0	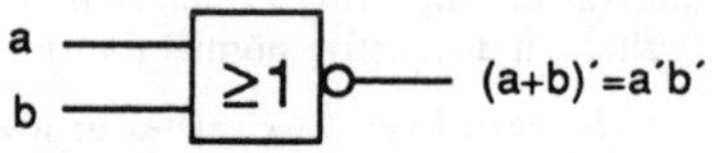

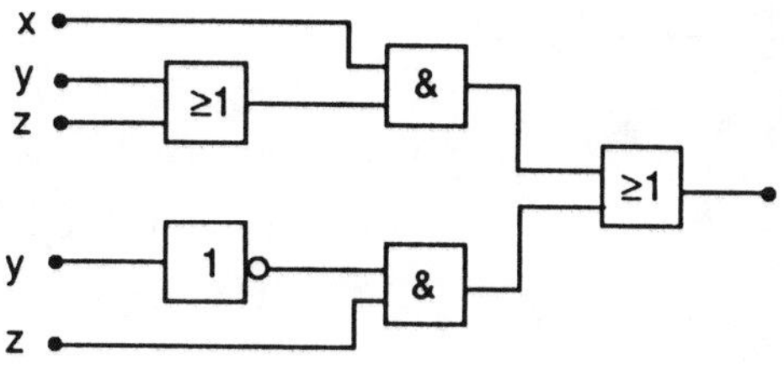

Realization of the Boolean expression $x(y+z) + y'z$

1.5 Graph Theory

A *graph G* is an ordered tripple (V, E, φ), where

V is the set of *nodes* or vertices,
E is the set of *edges* and
φ is a mapping from E to ordered or unordered pairs of V.

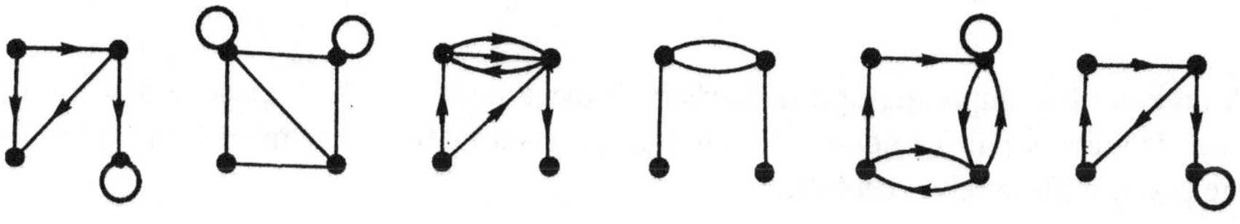

Directed graph or digraph | Undirected graph | Multigraphs | Simple graphs

An edge of a graph which is associated with an ordered pair of nodes is *directed.* An edge which is associated with an unordered pair of nodes is *undirected.* A graph with every edge directed (undirected) is called a *directed (undirected) graph.* A graph is *simple* if there is at most one edge between all pairs of nodes. If the graph has parallel edges it is called a *multigraph.*

The *converse* $G^{-1}=(v, E^{-1})$ of a digraph $G=(V, E)$ is a digraph in which E^{-1} is the converse of the relation E (i e the arrows have opposite direction).

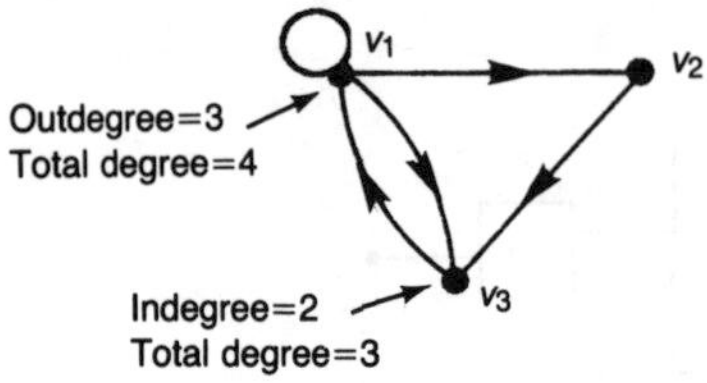

$v_1\ v_2\ v_3$	node simple path of length 2
$v_1\ v_3\ v_1\ v_2$	edge simple path of length 3
$v_1\ v_2\ v_3\ v_1$	simple cycle of length 3

For a directed graph the *outdegree* of a node v is the number of edges which have v as their *initial* node and the *indegree* of v is the number of edges which have v as their *terminal* node. The *total degree* of v is the sum of the outdegree and the indegree of v. For an undirected graph the degree of a node v is the numbers of edges, which are incident with v.

For a digraph a *path* is a sequence of edges such that the terminal node of every edge is the initial node of the next. The *length* of a path is the number of edges in the sequence. A path is *node (edge) simple* if all nodes (edges) in the path are different. A path with origin and end in the same node is called a *cycle.*

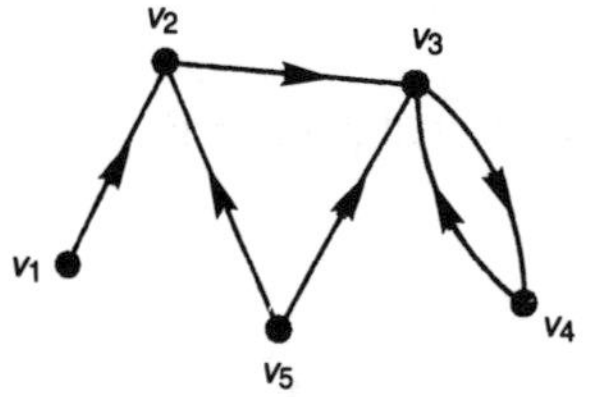

Node v_4 is reachable from v_1.

Node v_5 is not reachable from any other node.

A node u of a simple digraph is *reachable* from a node v if there exists a path from v to u. If for any pair of nodes of such a graph each node is reachable from the other, the graph is (strongly) *connected.*

For a simple digraph a maximal strongly connected subgraph is called a *strong component,* and every node lies in exactly one strong component. Thus the strong componenets constitute a partition of the digraph.

Matrix representations

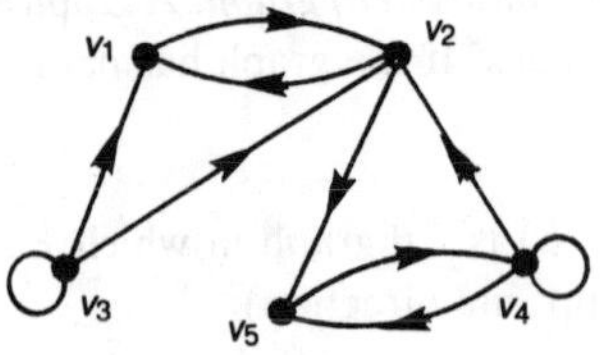

$$A: \begin{bmatrix} 0 & 1 & 0 & 0 & 0 \\ 1 & 0 & 0 & 0 & 1 \\ 1 & 1 & 1 & 0 & 0 \\ 0 & 1 & 0 & 1 & 1 \\ 0 & 0 & 0 & 1 & 0 \end{bmatrix}$$

For a simple digraph (V, E, φ) with $V=\{v_1, v_2, \ldots, v_n\}$ the *adjacency matrix* $A=[a_{ij}]$ is the $n\times n$ matrix such that

$$a_{ij}=\begin{cases}1 \text{ if } (v_i, v_j)\in E\\ 0 \text{ otherwise}\end{cases}$$

Let A^k be the *kth* power of A. The element in row i and column j of A^k is the number of paths of length k from node v_i to node v_j.

The *path (connection, reachability) matrix* $P=[p_{ij}]$ of a simple digraph with nodes $\{v_1, v_2, \ldots, v_n\}$ is defined such that

$$p_{ij}=\begin{cases}1 \text{ if there is a path from } v_i \text{ to } v_j \text{ or } i=j\\ 0 \text{ otherwise}\end{cases}$$

Then

$$P=(I+A)^{(n-1)}=I+A+A^{(2)}+\ldots+A^{(n-1)}$$

where $A^{(k)}$ denotes the Boolean power matrix (with entries 0 or 1) calculated with Boolean matrix addition and multiplication.

Properties

Converse digraph. If G is represented by A then G^{-1} is represented by A^t (the transpose of A). For a symmetric (or for an undirected) graph, $A=A^t$.

1. Two nodes v_i and v_j are in the same strong component $\Leftrightarrow$ $[P\wedge P^t]_{ij}=p_{ij}p_{ji}=1$.
2. $[AA^t]_{ij}$ = number of nodes which are *terminal* nodes of edges from both v_i and v_j
3. $[A^tA]_{ij}$ = number of nodes which are *initial* nodes, whose edges terminate in both v_i and v_j
4. $[A^k]_{ij}$ = number of paths of *length* k from v_i to v_j.
5. $[A^{(k)}]_{ij}=1 \Rightarrow$ There is a path of length k from v_i to v_j.
   $[A^{(k)}]_{ij}=0 \Rightarrow$ There is no path of length k from v_i to v_j.

Trees

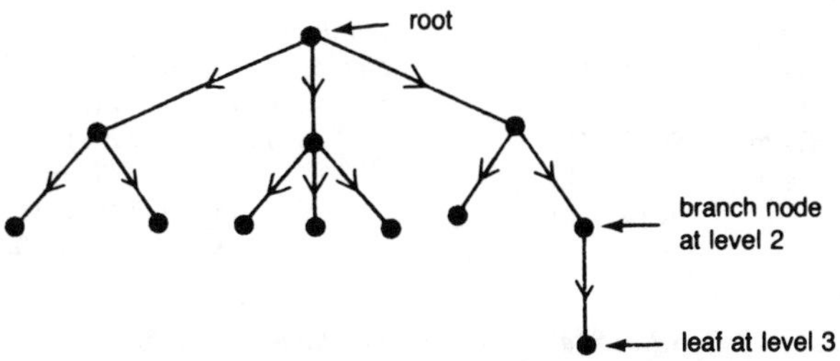

A *directed tree* is a digraph without cycles such that exactly one node called the *root* has indegree 0 while all other nodes have indegree 1. Nodes with outdegree 0 are called *leaves* or *terminal nodes.* All other nodes are called *branch nodes.* The *level* of a node is the length of its path from the root.

Weighted digraphs

A *weighted digraph* is a digraph in which each directed edge (v_i, v_j) is assigned a positive number (the weight) $w_{ij}=w(v_i, v_j)$. If there is no edge from v_i to v_j then $w_{ij}=\infty$. The graph can be represented by a *weighted adjacency matrix* $W=(w_{ij})$. The weight of a path is the sum of the weights of the edges occuring in the path.

$$W=\begin{bmatrix} \infty & 2 & 6 & 5 & \infty \\ \infty & \infty & 4 & \infty & 7 \\ \infty & \infty & \infty & 1 & 2 \\ \infty & \infty & \infty & \infty & 1 \\ \infty & \infty & \infty & \infty & \infty \end{bmatrix} \qquad \begin{array}{l} a=v_1 \\ b=v_2 \\ c=v_3 \\ d=v_4 \\ z=v_5 \end{array}$$

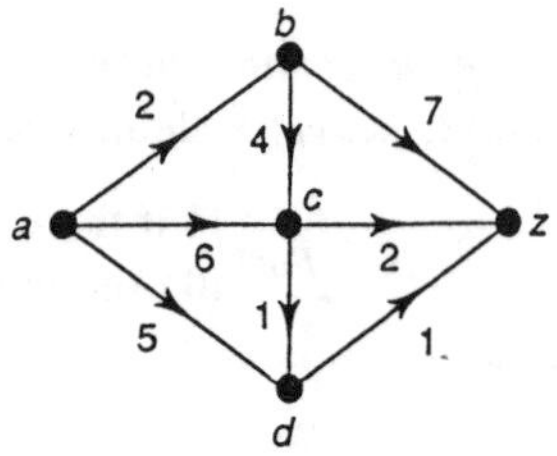

Dijkstra's algorithm for finding the shortest path

To find the shortest path from a to z.

The instructions in the brackets [...] are not needed if only the *minimum weight* of a path from a to z is sought.

Notation: TL=Temporary Label, PL=Permanent Label, SP=Shortest Path.

Step 0. Set $PL(a)=0$ and $V=a$, $TL(x)=\infty$, $x\neq a$. Here and below the node which most recently is assigned a PL is denoted V.
[$SP(a)=\{a\}$, $SP(x)=\varnothing$, all $x\neq a$]

Step 1. (Assignments of new TL.) Set for all x without PL new TLs by
$TL(x)=\min(\text{old } TL(x), PL(V)+w(V, x))$
Let y be that node with the smallest TL. Set $V=y$ and change $TL(y)$ to $PL(y)$.
[(*i*) If $TL(x)$ is *not* changed, do not change $SP(x)$. (*ii*) If $TL(x)$ is changed, set $SP(x)=\{SP(V), x\}$]

Step 2. (*i*) If $TL(V)=\infty$, then there is no path from a to z. Stop.
(*ii*) If $V=z$ then $PL(z)$ is the weight of the shortest path from a to z. Stop.
[The shortest path is $SP(z)$.]
(*iii*) Return to *step* 1.

Example Consider the weighted graph above. Let $a=v_1$ and $z=v_5$.

Step 0. $PL(a)=0$ $V=a$ $SP(a)=\{a\}$
$TL(b)=\infty$ $SP(b)=\varnothing$
$TL(c)=\infty$ $SP(c)=\varnothing$
$TL(d)=\infty$ $SP(d)=\varnothing$
$TL(z)=\infty$ $SP(z)=\varnothing$
$\therefore V=a$ $PL(V)=\{a\}$

Interation 1.
Step 1. $PL(a)=0$ $SP(a)=\{a\}$
$TL(b)=2$ $SP(b)=\{a, b\}$
$TL(c)=6$ $SP(c)=\{a, c\}$
$TL(d)=5$ $SP(d)=\{a, d\}$
$TL(z)=\infty$ $SP(z)=\varnothing$
$\therefore V=b$, $PL(V)=2$ $SP(V)=\{a, b\}$

Iteration 2.
Step 1. $PL(a)=0$ $SP(a)=\{a\}$
$PL(b)=2$ $SP(b)=\{a, b\}$
$TL(c)=6$ $SP(c)=\{a, c\}$
$TL(d)=5$ $SP(d)=\{a, d\}$
$TL(z)=9$ $SP(z)=\varnothing$
$\therefore V=d$, $PL(V)=5$ $SP(V)=\{a, d\}$

Iteration 3.
Step 1. $PL(a)=0$ $SP(a)=\{a\}$
$PL(b)=2$ $SP(b)=\{a, b\}$
$TL(c)=6$ $SP(c)=\{a, c\}$
$PL(d)=5$ $SP(d)=\{a, d\}$
$TL(z)=6$ $SP(z)=\{a, d, z\}$
$\therefore V=z$, $PL(V)=6$. Stop.
Shortest path$=\{a, d, z\}$ with weight 6.

1.6 Codes

Matrix group codes

Below, arithmetic *modulo* 2 is used, i.e. $0+0=0$, $0+1=1+0=1$, $1+1=0$ and $0\cdot 0=0\cdot 1=1\cdot 0=0$, $1\cdot 1=1$.

Notation: $\mathbf{Z}_2^n=\{\text{binary } n\text{-tuples}\}=\{\mathbf{a}=(a_1, a_2, \ldots, a_n)\colon a_i=0 \text{ or } 1, \text{ all } i\}$

1. (*i*) An (m, n)-code K is a one-to-one function $K\colon X\subset\mathbf{Z}_2^m\to\mathbf{Z}_2^n$, $n\geq m$
2. The *Hamming distance* $H(\mathbf{a}, \mathbf{b})$ between $\mathbf{a}, \mathbf{b}$ in $\mathbf{Z}_2^n$ is defined by

$$H(\mathbf{a}, \mathbf{b})=\sum_{i=1}^{n}(a_i+b_i)=\text{number of coordinates for which } a_i \text{ and } b_i \text{ are different.}$$

The *weight* of a is $\sum_{i=1}^{n} a_i$.

3. A code $K\colon \mathbf{Z}_2^m\to\mathbf{Z}_2^n$ is a *matrix code* if $\mathbf{x}=\mathbf{x}'A$, where the matrix A is of type (m, n).
4. The range of K is the set of *code words.*
5. A *code* $K\colon \mathbf{Z}_2^m\to\mathbf{Z}_2^n$ is a *group code* if the code words in $\mathbf{Z}_2^n$ form an additive group.
6. A code *detects* k errors if the minimum distance between the code words is at least $k+1$. A code *corrects* k errors if the minimum distance between the code words is at least $2k+1$.
7. The *weight* of a code is the minimum distance between the code words=the minimal number of rows in the control (decoding) matrix C whose sum is the zero vector.

Encoding and decoding matrices

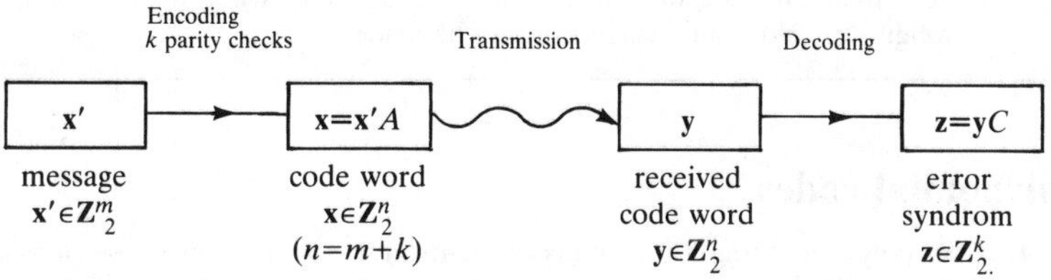

The *encoding* matrix: $A=\underbrace{[\underset{(m,m)}{I_m}\;\; \underset{(m,k)}{Q}]}_{(m,n)}$ (normalized form)

The *decoding* matrix: $C=\begin{bmatrix} Q \\ I_k \end{bmatrix}\begin{matrix}(m,k)\\(k,k)\end{matrix}\Bigg\}(n,k)$

(Sometimes C^t is called the *parity check matrix*).

Decoding procedure:

1. If $\mathbf{z}=\mathbf{0}$: y is a code word.
2. If $\mathbf{z}\neq\mathbf{0}$: Find $\mathbf{y}_1$ of *minimal* weight (the coset leader) with the same error syndrom $\mathbf{z}$. Then $\mathbf{y}+\mathbf{y}_1$ is (probably) the transmitted code word and the first m coordinates the meassage. (Alternatively, a decoding table may be used.)

8. *Hamming codes.* In a *Hamming code* the rows of $C=C_n$ are formed by all binary r-ruples$\neq\mathbf{0}$, i.e. $n=2^r-1$. For example,

$$(C_3)^t=\begin{bmatrix}1&1&0\\1&0&1\end{bmatrix}\qquad (C_7)^t=\begin{bmatrix}1&1&1&0&1&0&0\\1&1&0&1&0&1&0\\1&0&1&1&0&0&1\end{bmatrix}$$

Example

Given $A=\begin{bmatrix}1&0&0&1&1&0\\0&1&0&1&0&1\\0&0&1&1&1&1\end{bmatrix}$ i.e. $m=3$, $k=3$, $n=6$.

Decode (*a*) $\mathbf{y}=(110011)$, (*b*) $\mathbf{y}=(011111)$ and (*c*) $\mathbf{y}=(111111)$ and find the message.

Solution:

Control matrix $C=\begin{bmatrix}1&1&0\\1&0&1\\1&1&1\\1&0&0\\0&1&0\\0&0&1\end{bmatrix}$

Weight of $C=3$, row(1)+row(4)+row(5)=(000). The code detects 2 errors and corrects 1 error.

(*a*) $\mathbf{z}=\mathbf{y}C=(000)$. Thus, y is a code word. Message $\mathbf{x}'=(110)$.
(*b*) $\mathbf{z}=\mathbf{y}C=(101)$. $\mathbf{y}_1=(010000)\Rightarrow\mathbf{y}_1C=(101)$. Thus, $\mathbf{y}+\mathbf{y}_1=(001111)$ and $\mathbf{x}'=(001)$
(*c*) $\mathbf{z}=\mathbf{y}C=(011)$. No code word of weight 1 has this error syndrom. The error has weight $\geqslant 2$. No "safe" correction can be made.

Polynomial codes

Consider the polynomial ring $\mathbf{Z}_2[x]$ with coefficients in the field $\mathbf{Z}_2$. In taking sums and products of polynomials in $\mathbf{Z}_2[x]$ the laws of $\mathbf{Z}_2$ have to be respected. [E.g. $(1+x+x^2)+(1+x+x^3)=x^2+x^3$ and $(1+x+x^2)(1+x+x^3)=1+x^4+x^5$.]

Identifying $f(x)\in\mathbf{Z}_2[x]$ with the binary vector (sequence) of the coefficients of f [e.g. $1+x+x^3 \leftrightarrow(1101)$], the following definition can be made:

Definition. Given the *generator polynomial* $g(x)=a_0+a_1x+\ldots+a_kx^k\leftrightarrow(a_0a_1\ \ldots\ a_k)$, the *polynomial code* K: $\mathbf{Z}_2^m\to\mathbf{Z}_2^n$ $(n=m+k)$ maps $\boldsymbol{x}=p(x)\in\mathbf{Z}_2^n$ to $\mathbf{y}=q(x)\in\mathbf{Z}_2^n$ by $q(x)=g(x)p(x)$.

The last equality can also be written $(q_0q_1\ \ldots\ q_{n-1})=(p_0p_1\ \ldots\ p_{m-1})G$, *where*

$$G=\begin{bmatrix}a_0a_1\ \ldots\ldots\ldots\ldots\ a_k\ 0\ \ldots\ldots\ 0\\ 0\ a_0a_1\ \ldots\ldots\ldots\ldots\ a_k\ 0\ \ldots\ 0\\ \ldots\ldots\ldots\ldots\ldots\ldots\ldots\ldots\\ 0\ \ldots\ldots\ 0\ a_0a_1\ \ldots\ldots\ a_k\end{bmatrix}\quad\text{of type } (m,\ n)$$

Thus, a polynomial code is a matrix code (not necessarily mormalized).

Example
K: $\mathbf{Z}_2^2 \to \mathbf{Z}_5^2$, $g(x)=1+x^2+x^3$.

$\mathbf{x}=(x_0x_1) \leftrightarrow x_0+x_1x \frown (x_0+x_1x)(1+x^2+x^3)=x_0+x_1x+x_0x^2+(x_0+x_1)x^3+x_1x^4$
$\leftrightarrow (x_0,x_1,x_0,x_0+x_1,x_1)=\mathbf{y}$

Thus, (00)⌢(00000), (01)⌢(01011), (10)⌢(10110), (11)⌢(11101)

BCH-codes

(BCH=Bose, Ray-Chaudhuri, Hocquenghem)
Algebraic concepts and notation, see sec. 1.4.

Let $\bar{\mathbf{Z}}_2$ denote a field which is an extension of $\mathbf{Z}_2$ such that any polynomial with coefficients in $\mathbf{Z}_2$ has all its zeros in $\bar{\mathbf{Z}}_2$.

9. $(x_1+x_2+\ldots+x_n)^2=x_1^2+x_2^2+\ldots+x_n^2$ since $1+1=0$.
10. Two irreducible polynomials in $\mathbf{Z}_2[x]$ with a common zero in $\bar{\mathbf{Z}}_2$ are equal.
11. Let $\alpha \in \bar{\mathbf{Z}}_2$ be a zero of an irreducible polynomial $g(x) \in \mathbf{Z}_2[x]$ of degree r. Then for any positive j
    (*i*) α^j is a zero of some irreducible polynomial $g_j(x) \in \mathbf{Z}_2[x]$ of degree $\leq r$.
    (*ii*) $g_j(x)$ divides $1+x^{2^r-1}$.
    (*iii*) degree $g_j(x)$ divides r.
12. The *exponent* of an irreducible polynomial $g(x) \in \mathbf{Z}_2[x]$ is the least positive integer e such that $g(x)$ divides $1+x^e$.
    Note: $e \leq 2^r-1$, where r=degree $g(x)$ and e divides 2^r-1.
13. For any positive integer r there exist an irreducible polynomial of degree r and exponent 2^r-1. Such a polynomial is called *primitive.*
14. If $g(x)$ is a primitive polynomial of degree $\leq r$ and $\alpha \in \bar{\mathbf{Z}}_2$ is a zero of $g(x)$, then $\alpha^0=1$, α, α^2, ..., α^{2^r-2} are different (and $\alpha^{2^r-1}=1$).
15. $g(\alpha)=0 \Rightarrow g(\alpha^2)=g(\alpha^4)=g(\alpha^8)=\ldots=0$ (Cf. 9. above).

Definition and construction of BCH-codes

1° Decide a minimal distance $2t+1$ between the code words and an integer r such that $2^r>2t+1$.

2° Choose a primitive polynomial $g_1(x)$ of degree r (see table) and denote by $\alpha \in \bar{\mathbf{Z}}_2$ a zero of $g_1(x)$.

3° Construct (cf example below) irreducible polynomials $g_2(x)$, ..., $g_{2t}(x)$ of degree $\leq r$ with zeros α^2, ..., α^{2t}, respectively.

4° Let $g(x)$ of degree k (k is always $\leq tr$) be the least common multiple of the polynomials $g_1(x)$, ..., $g_{2t}(x)$ (i.e. product of *all different* of the polynomials $g_1(x)$, ..., $g_{2t}(x)$).

5° The *BCH-code* is that polynomial code K: $\mathbf{Z}_2^m \to \mathbf{Z}_2^n$ generated by $g(x)$.
Here, $n=2^r-1$, $m=n-k$ and the weight of the code is at least 2^t+1.

Example

1° According to the above notation, take $t=2$, $r=4$ (and thus $n=15$).

2° Choose $g_1(x)=1+x^3+x^4$. (Cf the table of irreducible polynomials below.)

3° Construction of the polynomials $g_2(x)$, $g_3(x)$, $g_4(x)$: $g_1(x)=g_2(x)=g_4(x)$ since $g_1(\alpha)=$ $=g_1(\alpha^2)=g_1(\alpha^4)=0$ (see 10 and 15 above). It remains to construct $g_3(x)$ with $g_3(\alpha^3)=0$. One way to do that is the following (another way is to use the table below):

Set $g_3(x)=1+Ax+Bx^2+Cx^3+Dx^4$. Now, $g_1(\alpha)=0 \Rightarrow \alpha^4+\alpha^3+1=0 \Rightarrow$ $\alpha^4=1+\alpha^3$. Therefore, recursively,
$\alpha^5=\alpha\alpha^4=\alpha(1+\alpha^3)=\alpha+\alpha^4=1+\alpha+\alpha^3$
$\alpha^6=\alpha\alpha^5=\alpha(1+\alpha+\alpha^3)=\alpha+\alpha^2+\alpha^4=1+\alpha+\alpha^2+\alpha^3$
$\alpha^9=\alpha^3\alpha^6=\alpha^3(1+\alpha+\alpha^2+\alpha^3)=\alpha^3+\alpha^4+\alpha^5+\alpha^6=1+\alpha^2$
$\alpha^{12}=\alpha^3\alpha^9=\alpha^3(1+\alpha^2)=\alpha^3+\alpha^5=1+\alpha$
$g_3(\alpha^3)=0 \Rightarrow 1+A\alpha^3+B(1+\alpha+\alpha^2+\alpha^3)+C(1+\alpha^2)+D(1+\alpha)=0.$

Identifying coefficients yields
$1+B+C+D=0,\ B+D=0,\ B+C=0,\ A+B=0 \Rightarrow A=B=C=D=1.$

Thus, $g_3(x)=1+x+x^2+x^3+x^4$.

4° $g(x)=g_1(x)g_3(x)=(1+x^3+x^4)(1+x+x^2+x^3+x^4)=1+x+x^2+x^4+x^8$

5° Degree $g(x)=8=k$, $m=7$.
E.g. $(1100100) \leftrightarrow 1+x+x^4 \rightharpoondown (1+x+x^4)(1+x+x^2+x^3+x^4+x^8)=1+x^3+x^6+x^9+x^{12}$ $\leftrightarrow (100100100100100)$.

Examples of known BCH-codes

m	n	t	m	n	t	m	n	t
4	7	1	16	31	3	36	63	5
5	15	3	18	63	10	64	127	10
6	31	7	21	31	2	92	127	5
7	15	2	24	63	7	139	255	15
11	15	1	26	31	1	215	255	5
11	31	5	30	63	6	231	255	3

Table of irreducible polynomials in $Z_2[x]$

Explanations

1. Given a polynomial $p(x)=\sum_{k=0}^{n} a_k x^k \in \mathbf{Z}_2[x]$ the *reciprocal* polynomial $p^*(x)$ is defined by $p^*(x)=\sum_{k=0}^{n} a_k x^{n-k}$ [e.g. $p(x)=a+bx+cx^2 \leftrightarrow (a,b,c) \Rightarrow p^*(x) \leftrightarrow (a,b,c)^*=(c,b,a) \leftrightarrow$ $\leftrightarrow c+bx+ax^2$.]
2. If $p(x)$ is irreducible, [primitive] and has a zero α, then $p^*(x)$ is irreducible, [primitive] and has zero α^{-1}. Therefore, in the table below only *one* in the pair $p(x)$ and $p^*(x)$ is listed, i.e. *the sequence of coefficients may be read either from the left or from the right to obtain an irreducible polynomial.*

3. For each degree, let α be a zero of the first listed polynomial. The entry following the bold **j** is the minimum polynomial of α^j. How to find the minimum polynomial not listed, use 10, and 15, above and $\alpha^{2^r-1}=1$.

   [Example: Let α denote a zero of the polynomial (100101) $\leftrightarrow 1+x^3+x^5$ of degree 5 (see table). By the table it is clear that α^3 is a zero of (111101) and α^5 a zero of (110111). Then the minimum polynomial of (remember $\alpha^{31}=1$)

   $\alpha, \alpha^2, \alpha^4, \alpha^8, \alpha^{16}, \alpha^{32}=\alpha$ is $p_1(x)=(100101)$

   $\alpha^3, \alpha^6, \alpha^{12}, \alpha^{24}, (\alpha^{48}=\alpha^{48-31}=)\alpha^{17}$ is $p_3(x)=(111101)$

   $\alpha^5, \alpha^{10}, \alpha^{20}, (\alpha^{40}=)\alpha^9, \alpha^{18}$ is $p_5(x)=(110111)$

   Using the zeros of the reciprocal polynomials, the minimum polynomial of

   $(\alpha^{-1}=)\alpha^{30}, (\alpha^{-2}=)\alpha^{29}, (\alpha^{-4}=)\alpha^{27}, (\alpha^{-8}=)\alpha^{23}, (\alpha^{-16}=)\alpha^{15}$ is $p_1^*(x)=(101001)$

   $(\alpha^{-3}=)\alpha^{28}, (\alpha^{-6}=)\alpha^{25}, (\alpha^{-12}=)\alpha^{19}, (\alpha^{38}=)\alpha^7, \alpha^{14}$ is $p_3^*(x)=(101111)$

   $(\alpha^{-5}=)\alpha^{26}, (\alpha^{-10}=)\alpha^{21}, (\alpha^{-20}=)\alpha^{11}, \alpha^{22}, (\alpha^{44}=)\alpha^{13}$ is $p_5^*(x)=(111011)$]

4. The exponent e of the irreducible $p(x)$ of degree m can be calculated by $e=(2^m-1)/(GCD(2^m-1, j))$. (In factorization of 2^m-1 the table of factorization of Mersenne numbers in section 2.2 may be used.)

   [E.g. The exponent of (1001001) is $e=(2^6-1)/(GCD(2^6-1,7)=63/GCD(63,7)=9$].

TABLE OF ALL IRREDUCIBLE POLYNOMIALS IN $Z_2[x]$ OF DEGREE ≤ 10

(P = Primitive, NP = Non-primitive)

Degree 2 ($\alpha^3 = 1$)			
1 111 (*P*)			
Degree 3 ($\alpha^7 = 1$)			
1 1011 (*P*)	[Note: (1011)* = 1101 (*P*) is also irreducible etc. below. See 2. above.]		
Degree 4 ($\alpha^{15} = 1$)			
1 10011 (*P*)	**3** 11111 (*NP*)	**5** 111	
Degree 5 ($\alpha^{31} = 1$)			
1 100 101 (*P*)	**3** 111 101 (*P*)	**5** 110 111 (*P*)	
Degree 6 ($\alpha^{63} = 1$)			
1 100 0011 (*P*)	**3** 101 0111 (*NP*)	**5** 110 0111 (*P*)	**7** 100 1001 (*NP*)
9 1101	**11** 110 1101 (*P*)	**21** 111	

Degree 7 ($\alpha^{127} = 1$)

1 1000 1001 (*P*)	**3** 1000 1111 (*P*)	**5** 1001 1101 (*P*)	**7** 1111 0111 (*P*)
9 1011 1111 (*P*)	**11** 1101 0101 (*P*)	**13** 1000 0011 (*P*)	**19** 1100 1011 (*P*)
21 1110 0101 (*P*)			

Degree 8 ($\alpha^{255} = 1$)

1 1000 11101 (*P*)	**3** 1011 10111 (*NP*)	**5** 1111 10011 (*NP*)
7 1011 01001 (*P*)	**9** 1101 11101 (*NP*)	**11** 1111 00111 (*P*)
13 1101 01011 (*P*)	**15** 1110 10111 (*NP*)	**17** 10011
19 1011 00101 (*P*)	**21** 1100 01011 (*NP*)	**23** 1011 00011 (*P*)
25 1000 11011 (*NP*)	**27** 1001 11111 (*NP*)	**37** 1010 11111 (*P*)
43 1110 00011 (*P*)	**45** 1001 11001 (*NP*)	**51** 11111
85 111		

Degree 9 ($\alpha^{511} = 1$)

1 10000 10001 (*P*)	**3** 10010 11001 (*P*)	**5** 11001 10001 (*P*)
7 10100 11001 (*NP*)	**9** 11000 10011 (*P*)	**11** 10001 01101 (*P*)
13 10011 10111 (*P*)	**15** 11011 00001 (*P*)	**17** 10110 11011 (*P*)
19 11100 00101 (*P*)	**21** 10000 10111 (*NP*)	**23** 11111 01001 (*P*)
25 11111 00011 (*P*)	**27** 11100 01111 (*P*)	**29** 11011 01011 (*P*)
35 11000 00001 (*NP*)	**37** 10011 01111 (*P*)	**39** 11110 01101 (*P*)
41 11011 10011 (*P*)	**43** 11110 01011 (*P*)	**45** 10011 11101 (*P*)
51 11110 10101 (*P*)	**53** 10100 10101 (*P*)	**55** 10101 11101 (*P*)
73 1011	**75** 11111 11011 (*P*)	**77** 11010 01001 (*NP*)
83 11000 10101 (*P*)	**85** 10101 10111 (*P*)	

Degree 10 ($\alpha^{1023} = 1$)

1 10000 001001 (*P*)	**3** 10000 001111 (*NP*)	**5** 10100 001101 (*P*)
7 11111 111001 (*P*)	**9** 10010 101111 (*NP*)	**11** 10000 110101 (*NP*)
13 10001 101111 (*P*)	**15** 10110 101011 (*NP*)	**17** 11101 001101 (*P*)
19 10111 111011 (*P*)	**21** 11111 101011 (*P*)	**23** 10000 011011 (*P*)
25 10100 100011 (*P*)	**27** 11101 111011 (*NP*)	**29** 10100 110001 (*P*)
31 11000 100011 (*NP*)	**33** 111101	**35** 11000 010011 (*P*)
37 11101 100011 (*P*)	**39** 10001 000111 (*NP*)	**41** 10111 100101 (*P*)
43 10100 011001 (*P*)	**45** 11000 110001 (*NP*)	**47** 11001 111111 (*P*)
49 11101 010101 (*P*)	**51** 10101 100111 (*NP*)	**53** 10110 001111 (*P*)
55 11100 101011 (*NP*)	**57** 11001 010001 (*NP*)	**59** 11100 111001 (*P*)
69 10111 000001 (*NP*)	**71** 11011 010011 (*P*)	**73** 11101 000111 (*P*)
75 10100 011111 (*NP*)	**77** 10100 001011 (*NP*)	**83** 11110 010011 (*P*)
85 10111 000111 (*P*)	**87** 10011 001001 (*NP*)	**89** 10011 010111 (*P*)
91 11010 110101 (*P*)	**93** 11111 111111 (*NP*)	**99** 110111
101 10000 101101 (*P*)	**103** 11101 111101 (*P*)	**105** 11110 000111 (*NP*)
107 11001 111001 (*P*)	**109** 10000 100111 (*P*)	**147** 10011 101101 (*NP*)
149 11000 010101 (*P*)	**155** 10010 101001 (*NP*)	**165** 101001
171 11011 001101 (*NP*)	**173** 11011 011111 (*P*)	**179** 11010 001001 (*P*)
341 111		

2 Algebra

2.1 Basic Algebra of Real Numbers

Sum and product symbols

$$\sum_{i=1}^{n} x_i = \sum_{k=1}^{n} x_k = x_1 + x_2 + \ldots + x_n \qquad \prod_{k=1}^{n} x_k = x_1 x_2 \ldots x_n$$

Algebraic laws

$a+b=b+a \quad ab=ba$ (*commutative laws*)

$(a+b)+c=a+(b+c) \quad (ab)c=a(bc)$ (*associative laws*)

$a(b+c)=ab+ac \quad (a+b)(c+d)=ac+ad+bc+bd$ (*distributive laws*)

$$\frac{a}{b} \pm \frac{c}{d} = \frac{ad \pm bc}{bd} \qquad \frac{a}{b} \cdot \frac{c}{d} = \frac{ac}{bd} \qquad \frac{\frac{a}{b}}{\frac{c}{d}} = \frac{a}{b} \cdot \frac{d}{c} = \frac{ad}{bc}$$

$a+(+b)=a-(-b)=a+b \qquad a+(-b)=a-(+b)=a-b$

$(+a)(+b)=(-a)(-b)=ab \qquad (+a)(-b)=(-a)(+b)=-ab$

$$\frac{+a}{+b} = \frac{-a}{-b} = \frac{a}{b} \qquad \frac{+a}{-b} = \frac{-a}{+b} = -\frac{a}{b}$$ (*laws of sign*)

$a<b,\ b<c \Rightarrow a<c \qquad a<b \Leftrightarrow a+c<b+c$

$a<b,\ c>0 \Rightarrow ac<bc \qquad a<b,\ c<0 \Rightarrow ac>bc \qquad a<b \Leftrightarrow -a>-b$ (*laws of order*)

Powers and roots

Powers

m, n integers:

$a^n = a \cdot a \ldots a$ (n times) $\qquad a^{-n} = \dfrac{1}{a^n} \qquad a^0 = 1$

$a^m \cdot a^n = a^{m+n} \qquad \dfrac{a^m}{a^n} = a^{m-n} \qquad (a^m)^n = a^{mn}$

$(ab)^n = a^n b^n \qquad \left(\dfrac{a}{b}\right)^n = \dfrac{a^n}{b^n} \qquad (-a)^n = \begin{cases} a^n, & n \text{ even} \\ -a^n, & n \text{ odd} \end{cases}$

Powers n^k

n	n^3	n^4	n^5	n^6	n^7	n^8	n^9	n^{10}
1	1	1	1	1	1	1	1	1
2	8	16	32	64	128	256	512	1 024
3	27	81	243	729	2 187	6 561	19 683	59 049
4	64	256	1 024	4 096	16 384	65 536	262 144	1 048 576
5	125	625	3 125	15 625	78 125	390 625	1 953 125	9 765 625
6	216	1 296	7 776	46 656	279 936	1 679 616	10 077 696	60 466 176
7	343	2 401	16 807	117 649	823 543	5 764 801	40 353 607	282 475 249
8	512	4 096	32 768	262 144	2 097 152	16 777 216	134 217 728	1 073 741 824
9	729	6 561	59 049	531 441	4 782 969	43 046 721	387 420 489	3 486 784 401
10	1 000	10 000	100 000	1 000 000	10 000 000	100 000 000	1 000 000 000	10 000 000 000
11	1 331	14 641	161 051	1 771 561	19 487 171	214 358 881	2 357 947 691	25 937 424 601
12	1 728	20 736	248 832	2 985 984	35 831 808	429 981 696	5 159 780 352	61 917 364 224

Sum of powers, see sec. 8.6.

Roots

$$x=\sqrt[n]{a}=a^{1/n} \Leftrightarrow x^n=a \qquad (a,\ x\geqslant 0)$$

$$\sqrt[n]{-a}=-\sqrt[n]{a},\ n \text{ odd} \qquad (a\geqslant 0)$$

$$a^{\frac{m}{n}}=\sqrt[n]{a^m}=(\sqrt[n]{a}\,)^m \qquad \sqrt[n]{ab}=\sqrt[n]{a}\,\sqrt[n]{b} \qquad \sqrt[n]{\frac{a}{b}}=\frac{\sqrt[n]{a}}{\sqrt[n]{b}}$$

$$\sqrt[n]{a^n b}=a\sqrt[n]{b} \qquad \sqrt[n]{a}\,\sqrt[m]{a}=\sqrt[mn]{a^{m+n}}$$

The binomial and multinomial theorems

The binomial theorem

$$(a+b)^n=\sum_{k=0}^{n}\binom{n}{k}a^{n-k}b^k=\binom{n}{0}a^n+\binom{n}{1}a^{n-1}b+\ldots+\binom{n}{n-1}ab^{n-1}+\binom{n}{n}b^n$$

The multinomial theorem

$$(a_1+\ldots+a_m)^n=\sum_{k_1+\ldots+k_m=n}\frac{n!}{k_1!\ \ldots\ k_m!}{a_1}^{k_1}\ldots\ {a_m}^{k_m}$$

$(a+b)^2=a^2+2ab+b^2$ $\qquad$ $(a-b)^2=a^2-2ab+b^2$

$(a+b)^3=a^3+3a^2b+3ab^2+b^3$ $\qquad$ $(a-b)^3=a^3-3a^2b+3ab^2-b^3$

$(a+b+c)^2=a^2+b^2+c^2+2ab+2ac+2bc$

$(a-b-c)^2=a^2+b^2+c^2-2ab-2ac+2bc$

Factorizations

$ab+ac=a(b+c)$ $\qquad$ $a^2-b^2=(a-b)(a+b)$

$a^3-b^3=(a-b)(a^2+ab+b^2)$ $\qquad$ $a^3+b^3=(a+b)(a^2-ab+b^2)$

$a^4-b^4=(a-b)(a+b)(a^2+b^2)$ $\qquad$ $a^4+b^4=(a^2+\sqrt{2}ab+b^2)(a^2-\sqrt{2}ab+b^2)$

$a^n-b^n=(a-b)(a^{n-1}+a^{n-2}b+\ldots+ab^{n-2}+b^{n-1})$

Factorials and binomial coefficients

$$n!=1\cdot 2\cdot 3\ldots n \qquad 0!=1 \qquad (\textit{factorials})$$

$$(2n-1)!!=1\cdot 3\cdot 5\ldots(2n-1) \qquad (2n)!!=2\cdot 4\cdot 6\ldots 2n \quad (\textit{semifactorials})$$

$$\binom{n}{k}=\frac{n(n-1)\ldots(n-k+1)}{k!}=\frac{n!}{k!(n-k)!} \qquad (\textit{binomial coefficients})$$

Factorials

n	***n!***	***n***	***n!***
1	1	11	39 916 800
2	2	12	479 001 600
3	6	13	6 227 020 800
4	24	14	87 178 291 200
5	120	15	1 307 674 368 000
6	720	16	20 922 789 888 000
7	5 040	17	355 687 428 096 000
8	40 320	18	6 402 373 705 728 000
9	362 880	19	121 645 100 408 832 000
10	3 628 800	20	2 432 902 008 176 640 000

$$n!\sim\sqrt{2\pi}\, n^{n+1/2}\, e^{-n} \text{ as } n\to\infty \qquad (\textit{Stirling's formula})$$

$$\binom{n}{n-k}=\binom{n}{k} \qquad \binom{n}{k}+\binom{n}{k+1}=\binom{n+1}{k+1}=\binom{n}{k}+\binom{n-1}{k}+\ldots+\binom{k}{k}$$

$$\binom{n}{0}+\binom{n+1}{1}+\binom{n+2}{2}+\ldots+\binom{n+k}{k}=\binom{n+k+1}{k}$$

$$\binom{m}{0}\binom{n}{k}+\binom{m}{1}\binom{n}{k-1}+\ldots+\binom{m}{k}\binom{n}{0}=\binom{m+n}{k}$$

$$\binom{n}{0}+\binom{n}{1}+\ldots+\binom{n}{n}=2^n \qquad \binom{n}{0}^2+\binom{n}{1}^2+\ldots+\binom{n}{n}^2=\binom{2n}{n}$$

$$\binom{n}{0}-\binom{n}{1}+\ldots+(-1)^n\binom{n}{n}=0$$

Pascal's triangle of binomial coefficients

n	1	$(a+b)^n$
1	1 1	$a+b$
2	1 2 1	$a^2+2ab+b^2$
3	1 3 3 1	$a^3+3a^2b+3ab^2+b^3$
4	1 4 6 4 1	$a^4+4a^3b+6a^2b^2+4ab^3+b^4$
5	1 5 10 10 5 1	$a^5+5a^4b+10a^3b^2+10a^2b^3+5ab^4+b^5$
6	1 6 15 20 15 6 1	…
7	1 7 21 35 35 21 7 1	
8	1 8 28 56 70 56 28 8 1	
9	1 9 36 84 126 126 84 36 9 1	
10	1 10 45 120 210 252 210 120 45 10 1	

Each number is the sum of the two numbers above to the right and to the left.

Absolute value

For each real number x the *absolute value* $|x|$ is defined by

$$|x|=\begin{cases} x \text{ if } x\geqslant 0 \\ -x \text{ if } x\leqslant 0 \end{cases}$$

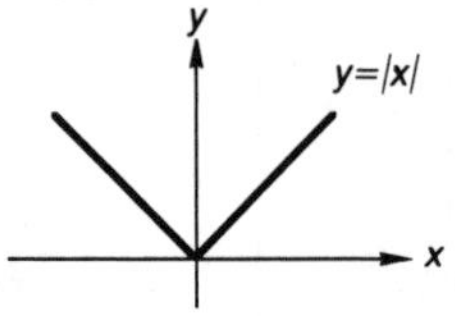

Means

Given n real numbers $x_1, x_2, \ldots, x_n$.

Arithmetic mean $A=\dfrac{x_1+\ldots+x_n}{n}$

Geometric mean $G=\sqrt[n]{x_1 \ldots x_n} \quad (x_i>0)$

Harmonic mean H: $\dfrac{1}{H}=\dfrac{1}{n}\left(\dfrac{1}{x_1}+\ldots+\dfrac{1}{x_n}\right) \quad (x_i>0)$

Weighted means $(\lambda_1+\ldots+\lambda_n=1,\ \lambda_i>0,\ x_i>0)$:

$$A_\lambda=\lambda_1x_1+\ldots+\lambda_nx_n,\quad G_\lambda=x_1^{\lambda_1}\ldots x_n^{\lambda_n}$$

$$H\leqslant G\leqslant A \quad (\text{equality} \Leftrightarrow x_1=x_2=\ldots=x_n) \qquad G_\lambda\leqslant A_\lambda$$

Some inequalities

1. $|xy| \leq \frac{1}{2}(x^2+y^2)$ $\qquad$ $|xy| \leq \frac{1}{2}\left(\varepsilon x^2+\frac{1}{\varepsilon}y^2\right)$, any $\varepsilon>0$

2. *The triangle inequality.*

$$\big||x|-|y|\big| \leq |x \pm y| \leq |x|+|y| \qquad \left|\sum_{k=1}^{n} x_k\right| \leq \sum_{k=1}^{n} |x_k|$$

3. *Hölder's inequality.* If $\frac{1}{p}+\frac{1}{q}=1$, p, $q>1$ then

$$\sum_{k=1}^{n} |x_k y_k| \leq \left\{\sum_{k=1}^{n} |x_k|^p\right\}^{\frac{1}{p}} \cdot \left\{\sum_{k=1}^{n} |y_k|^q\right\}^{\frac{1}{q}}$$

4. *Cauchy's inequality.*

$$\left(\sum_{k=1}^{n} x_k y_k\right)^2 \leq \sum_{k=1}^{n} x_k^2 \sum_{k=1}^{n} y_k^2 \quad (\text{equality} \Leftrightarrow y_k=cx_k)$$

5. *Minkowski's inequality.* If $p>1$, x_k, $y_k>0$ then

$$\left(\sum_{k=1}^{n} (x_k+y_k)^p\right)^{\frac{1}{p}} \leq \left(\sum_{k=1}^{n} x_k^p\right)^{\frac{1}{p}} + \left(\sum_{k=1}^{n} y_k^p\right)^{\frac{1}{p}}$$

(equality $\Leftrightarrow y_k=cx_k$)

Percent (%)

Growth factor

The *growth factor* associated with a change of p % is $(1+p/100)$. The value q_n of a quantity after n succesive changes of p_k % ($k=1, \ldots, n$) is

$$q_n=q_o\left(1+\frac{p_1}{100}\right)\left(1+\frac{p_2}{100}\right)\ldots\left(1+\frac{p_n}{100}\right)$$

Calculation of interests

p=rate of interest

1. (*Compounded interest*) Value of capital c after t years:

$$c\left(1+\frac{p}{100}\right)^t$$

2. (*Present value*). The capital c falls due in t years.

$$\textit{Present value: } \frac{c}{\left(1+\frac{p}{100}\right)^t}$$

3. (*Annuity*). Yearly instalment on a loan c, payed during t years by equal amounts:

$$c \cdot \frac{\left(1+\frac{p}{100}\right)^t \cdot \frac{p}{100}}{\left(1+\frac{p}{100}\right)^t - 1}$$

4. (*Inflation*). An inflation of p % implies a money value decrease of

$$\frac{100p}{100+p} \%$$

2.2 Number Theory

Number systems

$\boldsymbol{N}$	natural numbers
$\boldsymbol{Z}$	integers, ($\boldsymbol{Z}^+$ positive integers)
$\boldsymbol{Q}$	rational numbers
$\boldsymbol{R}$	real numbers
$\boldsymbol{C}$	complex numbers

Natural numbers.

$\boldsymbol{N}=\{0, 1, 2, 3, ...\}$. Sometimes 0 is omitted.

Integers.

$\boldsymbol{Z}=\{0, \pm 1, \pm 2, ...\}$

Rational numbers.

$\boldsymbol{Q}=\{p/q: p, q \in \boldsymbol{Z}, q \neq 0\}$.

The numbers in $\boldsymbol{Q}$ can be represented by a finite or a periodic decimal expansion. $\boldsymbol{Q}$ is *countable,* i.e. there exists a one-to-one correspondence between $\boldsymbol{Q}$ and $\boldsymbol{N}$.

Real numbers.

$\boldsymbol{R}=\{\text{real numbers}\}$.

Real numbers which are not rational are called *irrational.* Every irrational number can be represented by an infinite non-periodic decimal expansion. *Algebraic numbers* are solutions of an equation of the form $a_nx^n+...+a_0=0$, $a_k \in \boldsymbol{Z}$. *Transcendental* are those numbers in $\boldsymbol{R}$ which are not algebraic. $\boldsymbol{R}$ is not countable. (Example: 4/7 is rational, $\sqrt{5}$ is algebraic and irrational, e and π are transcendental.)

Complex numbers.

$\boldsymbol{C}=\{x+iy : x, y \in \boldsymbol{R}\}$, where i is the imaginary unit, i.e. $i^2=-1$.

The supremum axiom

For any nonempty bounded subset S of $\boldsymbol{R}$ there exist unique numbers $G=\sup S$ and $g=\inf S$ such that:

(*i*) $g \leq x \leq G$, all $x \in S$
(*ii*) For any $\varepsilon > 0$ there exist $x_1 \in S$ and $x_2 \in S$ such that

$$x_1 > G - \varepsilon \text{ and } x_2 < g + \varepsilon$$

Primes and prime number factorizations

Theorems on prime numbers

1. For every positive integer n exists a prime factor of $n!+1$ exceeding n.
2. Every prime factor of $p_1 p_2 \ldots p_n + 1$ where $p_1, p_2, \ldots, p_n$ are prime differs from each of $p_1, p_2, \ldots, p_n$.
3. There are infinitely many primes. (*Euclid*)
4. For every positive integer $n \geq 2$ there exists a string of n consecutive composite integers.
5. If a and b are relatively prime then the arithmetic sequence $an+b$, $n=1, 2, \ldots$ contains an infinite number of primes. (*Lejeune-Dirichlet*)

The following conjectures have not been proved.

6. Every even number ≥ 6 is the sum of two odd primes (*the Goldbach conjecture*).
7. There exist infinitely many prime twins. Prime twins are pairs like (3, 5), (5, 7), and (2087, 2089).

Unique factorization theorem

Every integer >1 is either a prime or a product of uniquely determined primes.

The function $\pi(x)$

The function value $\pi(x)$ is the number of primes which are less than or equal to x.

x	100	200	300	400	500	600	700	800	900
$\pi(x)$	25	46	62	78	95	109	125	139	154
x	1000	2000	3000	4000	5000	6000	7000	8000	9000
$\pi(x)$	168	303	430	550	669	783	900	1007	1117
x	10000	20000	30000	40000	50000	60000	70000	80000	90000
$\pi(x)$	1229	2262	3245	4203	5133	6057	6935	7837	8713

x	10^5	10^6	10^7	10^8	10^9	10^{10}
$\pi(x)$	9592	78498	664579	5761455	50847534	455052512

Asymptotic behavior: $\pi(x) \sim \dfrac{x}{\ln x}$ as $x \to \infty$

2.2

The first 400 prime numbers

2	179	419	661	947	1229	1523	1823	2131	2437
3	181	421	673	953	1231	1531	1831	2137	2441
5	191	431	677	967	1237	1543	1847	2141	2447
7	193	433	683	971	1249	1549	1861	2143	2459
11	197	439	691	977	1259	1553	1867	2153	2467
13	199	443	701	983	1277	1559	1871	2161	2473
17	211	449	709	991	1279	1567	1873	2179	2477
19	223	457	719	997	1283	1571	1877	2203	2503
23	227	461	727	1009	1289	1579	1879	2207	2521
29	229	463	733	1013	1291	1583	1889	2213	2531
31	233	467	739	1019	1297	1597	1901	2221	2539
37	239	479	743	1021	1301	1601	1907	2237	2543
41	241	487	751	1031	1303	1607	1913	2239	2549
43	251	491	757	1033	1307	1609	1931	2243	2551
47	257	499	761	1039	1319	1613	1933	2251	2557
53	263	503	769	1049	1321	1619	1949	2267	2579
59	269	509	773	1051	1327	1621	1951	2269	2591
61	271	521	787	1061	1361	1627	1973	2273	2593
67	277	523	797	1063	1367	1637	1979	2281	2609
71	281	541	809	1069	1373	1657	1987	2287	2617
73	283	547	811	1087	1381	1663	1993	2293	2621
79	293	557	821	1091	1399	1667	1997	2297	2633
83	307	563	823	1093	1409	1669	1999	2309	2647
89	311	569	827	1097	1423	1693	2003	2311	2657
97	313	571	829	1103	1427	1697	2011	2333	2659
101	317	577	839	1109	1429	1699	2017	2339	2663
103	331	587	853	1117	1433	1709	2027	2341	2671
107	337	593	857	1123	1439	1721	2029	2347	2677
109	347	599	859	1129	1447	1723	2039	2351	2683
113	349	601	863	1151	1451	1733	2053	2357	2687
127	353	607	877	1153	1453	1741	2063	2371	2689
131	359	613	881	1163	1459	1747	2069	2377	2693
137	367	617	883	1171	1471	1753	2081	2381	2699
139	373	619	887	1181	1481	1759	2083	2383	2707
149	379	631	907	1187	1483	1777	2087	2389	2711
151	383	641	911	1193	1487	1783	2089	2393	2713
157	389	643	919	1201	1489	1787	2099	2399	2719
163	397	647	929	1213	1493	1789	2111	2411	2729
167	401	653	937	1217	1499	1801	2113	2417	2731
173	409	659	941	1223	1511	1811	2129	2423	2741

2.2

Prime Number Factorizations from 1 to 999

n	0	1	2	3	4	5	6	7	8	9
0	...	–	–	–	2^2	–	$2\cdot3$	–	2^3	3^2
1	$2\cdot5$	–	$2^2\cdot3$	–	$2\cdot7$	$3\cdot5$	2^4	–	$2\cdot3^2$	–
2	$2^2\cdot5$	$3\cdot7$	$2\cdot11$	–	$2^3\cdot3$	5^2	$2\cdot13$	3^3	$2^2\cdot7$	–
3	$2\cdot3\cdot5$	–	2^5	$3\cdot11$	$2\cdot17$	$5\cdot7$	$2^2\cdot3^2$	–	$2\cdot19$	$3\cdot13$
4	$2^3\cdot5$	–	$2\cdot3\cdot7$	–	$2^2\cdot11$	$3^2\cdot5$	$2\cdot23$	–	$2^4\cdot3$	7^2
5	$2\cdot5^2$	$3\cdot17$	$2^2\cdot13$	–	$2\cdot3^3$	$5\cdot11$	$2^3\cdot7$	$3\cdot19$	$2\cdot29$	–
6	$2^2\cdot3\cdot5$	–	$2\cdot31$	$3^2\cdot7$	2^6	$5\cdot13$	$2\cdot3\cdot11$	–	$2^2\cdot17$	$3\cdot23$
7	$2\cdot5\cdot7$	–	$2^3\cdot3^2$	–	$2\cdot37$	$3\cdot5^2$	$2^2\cdot19$	$7\cdot11$	$2\cdot3\cdot13$	–
8	$2^4\cdot5$	3^4	$2\cdot41$	–	$2^2\cdot3\cdot7$	$5\cdot17$	$2\cdot43$	$3\cdot29$	$2^3\cdot11$	–
9	$2\cdot3^2\cdot5$	$7\cdot13$	$2^2\cdot23$	$3\cdot31$	$2\cdot47$	$5\cdot19$	$2^5\cdot3$	–	$2\cdot7^2$	$3^2\cdot11$
10	$2^2\cdot5^2$	–	$2\cdot3\cdot17$	–	$2^3\cdot13$	$3\cdot5\cdot7$	$2\cdot53$	–	$2^2\cdot3^3$	–
11	$2\cdot5\cdot11$	$3\cdot37$	$2^4\cdot7$	–	$2\cdot3\cdot19$	$5\cdot23$	$2^2\cdot29$	$3^2\cdot13$	$2\cdot59$	$7\cdot17$
12	$2^3\cdot3\cdot5$	11^2	$2\cdot61$	$3\cdot41$	$2^2\cdot31$	5^3	$2\cdot3^2\cdot7$	–	2^7	$3\cdot43$
13	$2\cdot5\cdot13$	–	$2^2\cdot3\cdot11$	$7\cdot19$	$2\cdot67$	$3^3\cdot5$	$2^3\cdot17$	–	$2\cdot3\cdot23$	–
14	$2^2\cdot5\cdot7$	$3\cdot47$	$2\cdot71$	$11\cdot13$	$2^4\cdot3^2$	$5\cdot29$	$2\cdot73$	$3\cdot7^2$	$2^2\cdot37$	–
15	$2\cdot3\cdot5^2$	–	$2^3\cdot19$	$3^2\cdot17$	$2\cdot7\cdot11$	$5\cdot31$	$2^2\cdot3\cdot13$	–	$2\cdot79$	$3\cdot53$
16	$2^5\cdot5$	$7\cdot23$	$2\cdot3^4$	–	$2^2\cdot41$	$3\cdot5\cdot11$	$2\cdot83$	–	$2^3\cdot3\cdot7$	13^2
17	$2\cdot5\cdot17$	$3^2\cdot19$	$2^2\cdot43$	–	$2\cdot3\cdot29$	$5^2\cdot7$	$2^4\cdot11$	$3\cdot59$	$2\cdot89$	–
18	$2^2\cdot3^2\cdot5$	–	$2\cdot7\cdot13$	$3\cdot61$	$2^3\cdot23$	$5\cdot37$	$2\cdot3\cdot31$	$11\cdot17$	$2^2\cdot47$	$3^3\cdot7$
19	$2\cdot5\cdot19$	–	$2^6\cdot3$	–	$2\cdot97$	$3\cdot5\cdot13$	$2^2\cdot7^2$	–	$2\cdot3^2\cdot11$	–
20	$2^3\cdot5^2$	$3\cdot67$	$2\cdot101$	$7\cdot29$	$2^2\cdot3\cdot17$	$5\cdot41$	$2\cdot103$	$3^2\cdot23$	$2^4\cdot13$	$11\cdot19$
21	$2\cdot3\cdot5\cdot7$	–	$2^2\cdot53$	$3\cdot71$	$2\cdot107$	$5\cdot43$	$2^3\cdot3^3$	$7\cdot31$	$2\cdot109$	$3\cdot73$
22	$2^2\cdot5\cdot11$	$13\cdot17$	$2\cdot3\cdot37$	–	$2^5\cdot7$	$3^2\cdot5^2$	$2\cdot113$	–	$2^2\cdot3\cdot19$	–
23	$2\cdot5\cdot23$	$3\cdot7\cdot11$	$2^3\cdot29$	–	$2\cdot3^2\cdot13$	$5\cdot47$	$2^2\cdot59$	$3\cdot79$	$2\cdot7\cdot17$	–
24	$2^4\cdot3\cdot5$	–	$2\cdot11^2$	3^5	$2^2\cdot61$	$5\cdot7^2$	$2\cdot3\cdot41$	$13\cdot19$	$2^3\cdot31$	$3\cdot83$
25	$2\cdot5^3$	–	$2^2\cdot3^2\cdot7$	$11\cdot23$	$2\cdot127$	$3\cdot5\cdot17$	2^8	–	$2\cdot3\cdot43$	$7\cdot37$
26	$2^2\cdot5\cdot13$	$3^2\cdot29$	$2\cdot131$	–	$2^3\cdot3\cdot11$	$5\cdot53$	$2\cdot7\cdot19$	$3\cdot89$	$2^2\cdot67$	–
27	$2\cdot3^3\cdot5$	–	$2^4\cdot17$	$3\cdot7\cdot13$	$2\cdot137$	$5^2\cdot11$	$2^2\cdot3\cdot23$	–	$2\cdot139$	$3^2\cdot31$
28	$2^3\cdot5\cdot7$	–	$2\cdot3\cdot47$	–	$2^2\cdot71$	$3\cdot5\cdot19$	$2\cdot11\cdot13$	$7\cdot41$	$2^5\cdot3^2$	17^2
29	$2\cdot5\cdot29$	$3\cdot97$	$2^2\cdot73$	–	$2\cdot3\cdot7^2$	$5\cdot59$	$2^3\cdot37$	$3^3\cdot11$	$2\cdot149$	$13\cdot23$
30	$2^2\cdot3\cdot5^2$	$7\cdot43$	$2\cdot151$	$3\cdot101$	$2^4\cdot19$	$5\cdot61$	$2\cdot3^2\cdot17$	–	$2^2\cdot7\cdot11$	$3\cdot103$
31	$2\cdot5\cdot31$	–	$2^3\cdot3\cdot13$	–	$2\cdot157$	$3^2\cdot5\cdot7$	$2^2\cdot79$	–	$2\cdot3\cdot53$	$11\cdot29$
32	$2^6\cdot5$	$3\cdot107$	$2\cdot7\cdot23$	$17\cdot19$	$2^2\cdot3^4$	$5^2\cdot13$	$2\cdot163$	$3\cdot109$	$2^3\cdot41$	$7\cdot47$
33	$2\cdot3\cdot5\cdot11$	–	$2^2\cdot83$	$3^2\cdot37$	$2\cdot167$	$5\cdot67$	$2^4\cdot3\cdot7$	–	$2\cdot13^2$	$3\cdot113$
34	$2^2\cdot5\cdot17$	$11\cdot31$	$2\cdot3^2\cdot19$	7^3	$2^3\cdot43$	$3\cdot5\cdot23$	$2\cdot173$	–	$2^2\cdot3\cdot29$	–
35	$2\cdot5^2\cdot7$	$3^3\cdot13$	$2^5\cdot11$	–	$2\cdot3\cdot59$	$5\cdot71$	$2^2\cdot89$	$3\cdot7\cdot17$	$2\cdot179$	–
36	$2^3\cdot3^2\cdot5$	19^2	$2\cdot181$	$3\cdot11^2$	$2^2\cdot7\cdot13$	$5\cdot73$	$2\cdot3\cdot61$	–	$2^4\cdot23$	$3^2\cdot41$
37	$2\cdot5\cdot37$	$7\cdot53$	$2^2\cdot3\cdot31$	–	$2\cdot11\cdot17$	$3\cdot5^3$	$2^3\cdot47$	$13\cdot29$	$2\cdot3^3\cdot7$	–
38	$2^2\cdot5\cdot19$	$3\cdot127$	$2\cdot191$	–	$2^7\cdot3$	$5\cdot7\cdot11$	$2\cdot193$	$3^2\cdot43$	$2^2\cdot97$	–
39	$2\cdot3\cdot5\cdot13$	$17\cdot23$	$2^3\cdot7^2$	$3\cdot131$	$2\cdot197$	$5\cdot79$	$2^2\cdot3^2\cdot11$	–	$2\cdot199$	$3\cdot7\cdot19$
40	$2^4\cdot5^2$	–	$2\cdot3\cdot67$	$13\cdot31$	$2^2\cdot101$	$3^4\cdot5$	$2\cdot7\cdot29$	$11\cdot37$	$2^3\cdot3\cdot17$	–
41	$2\cdot5\cdot41$	$3\cdot137$	$2^2\cdot103$	$7\cdot59$	$2\cdot3^2\cdot23$	$5\cdot83$	$2^5\cdot13$	$3\cdot139$	$2\cdot11\cdot19$	–
42	$2^2\cdot3\cdot5\cdot7$	–	$2\cdot211$	$3^2\cdot47$	$2^3\cdot53$	$5^2\cdot17$	$2\cdot3\cdot71$	$7\cdot61$	$2^2\cdot107$	$3\cdot11\cdot13$
43	$2\cdot5\cdot43$	–	$2^4\cdot3^3$	–	$2\cdot7\cdot31$	$3\cdot5\cdot29$	$2^2\cdot109$	$19\cdot23$	$2\cdot3\cdot73$	–
44	$2^3\cdot5\cdot11$	$3^2\cdot7^2$	$2\cdot13\cdot17$	–	$2^2\cdot3\cdot37$	$5\cdot89$	$2\cdot223$	$3\cdot149$	$2^6\cdot7$	–
45	$2\cdot3^2\cdot5^2$	$11\cdot41$	$2^2\cdot113$	$3\cdot151$	$2\cdot227$	$5\cdot7\cdot13$	$2^3\cdot3\cdot19$	–	$2\cdot229$	$3^3\cdot17$
46	$2^2\cdot5\cdot23$	–	$2\cdot3\cdot7\cdot11$	–	$2^4\cdot29$	$3\cdot5\cdot31$	$2\cdot233$	–	$2^2\cdot3^4\cdot13$	$7\cdot67$
47	$2\cdot5\cdot47$	$3\cdot157$	$2^3\cdot59$	$11\cdot43$	$2\cdot3\cdot79$	$5^2\cdot19$	$2^2\cdot7\cdot17$	$3^2\cdot53$	$2\cdot239$	–
48	$2^5\cdot3\cdot5$	$13\cdot37$	$2\cdot241$	$3\cdot7\cdot23$	$2^2\cdot11^2$	$5\cdot97$	$2\cdot3^5$	–	$2^3\cdot61$	$3\cdot163$
49	$2\cdot5\cdot7^2$	–	$2^2\cdot3\cdot41$	$17\cdot29$	$2\cdot13\cdot19$	$3^2\cdot5\cdot11$	$2^4\cdot31$	$7\cdot71$	$2\cdot3\cdot83$	–

E.g. $432=2^4\cdot3^3$

Factorizations (continued)

n	0	1	2	3	4	5	6	7	8	9
50	$2^2 \cdot 5^3$	$3 \cdot 167$	$2 \cdot 251$	–	$2^3 \cdot 3^2 \cdot 7$	$5 \cdot 101$	$2 \cdot 11 \cdot 23$	$3 \cdot 13^2$	$2^2 \cdot 127$	–
51	$2 \cdot 3 \cdot 5 \cdot 17$	$7 \cdot 73$	2^9	$3^3 \cdot 19$	$2 \cdot 257$	$5 \cdot 103$	$2^2 \cdot 3 \cdot 43$	$11 \cdot 47$	$2 \cdot 7 \cdot 37$	$3 \cdot 173$
52	$2^3 \cdot 5 \cdot 13$	–	$2 \cdot 3^2 \cdot 29$	–	$2^2 \cdot 131$	$3 \cdot 5^2 \cdot 7$	$2 \cdot 263$	$17 \cdot 31$	$2^4 \cdot 3 \cdot 11$	23^2
53	$2 \cdot 5 \cdot 53$	$3^2 \cdot 59$	$2^2 \cdot 7 \cdot 19$	$13 \cdot 41$	$2 \cdot 3 \cdot 89$	$5 \cdot 107$	$2^3 \cdot 67$	$3 \cdot 179$	$2 \cdot 269$	$7^2 \cdot 11$
54	$2^2 \cdot 3^3 \cdot 5$	–	$2 \cdot 271$	$3 \cdot 181$	$2^5 \cdot 17$	$5 \cdot 109$	$2 \cdot 3 \cdot 7 \cdot 13$	–	$2^2 \cdot 137$	$3^2 \cdot 61$
55	$2 \cdot 5^2 \cdot 11$	$19 \cdot 29$	$2^3 \cdot 3 \cdot 23$	$7 \cdot 79$	$2 \cdot 277$	$3 \cdot 5 \cdot 37$	$2^2 \cdot 139$	–	$2 \cdot 3^2 \cdot 31$	$13 \cdot 43$
56	$2^4 \cdot 5 \cdot 7$	$3 \cdot 11 \cdot 17$	$2 \cdot 281$	–	$2^2 \cdot 3 \cdot 47$	$5 \cdot 113$	$2 \cdot 283$	$3^4 \cdot 7$	$2^3 \cdot 71$	–
57	$2 \cdot 3 \cdot 5 \cdot 19$	–	$2^2 \cdot 11 \cdot 13$	$3 \cdot 191$	$2 \cdot 7 \cdot 41$	$5^2 \cdot 23$	$2^6 \cdot 3^2$	–	$2 \cdot 17^2$	$3 \cdot 193$
58	$2^2 \cdot 5 \cdot 29$	$7 \cdot 83$	$2 \cdot 3 \cdot 97$	$11 \cdot 53$	$2^3 \cdot 73$	$3^2 \cdot 5 \cdot 13$	$2 \cdot 293$	–	$2^2 \cdot 3 \cdot 7^2$	$19 \cdot 31$
59	$2 \cdot 5 \cdot 59$	$3 \cdot 197$	$2^4 \cdot 37$	–	$2 \cdot 3^3 \cdot 11$	$5 \cdot 7 \cdot 17$	$2^2 \cdot 149$	$3 \cdot 199$	$2 \cdot 13 \cdot 23$	–
60	$2^3 \cdot 3 \cdot 5^2$	–	$2 \cdot 7 \cdot 43$	$3^2 \cdot 67$	$2^2 \cdot 151$	$5 \cdot 11^2$	$2 \cdot 3 \cdot 101$	–	$2^5 \cdot 19$	$3 \cdot 7 \cdot 29$
61	$2 \cdot 5 \cdot 61$	$13 \cdot 47$	$2^2 \cdot 3^2 \cdot 17$	–	$2 \cdot 307$	$3 \cdot 5 \cdot 41$	$2^3 \cdot 7 \cdot 11$	–	$2 \cdot 3 \cdot 103$	–
62	$2^2 \cdot 5 \cdot 31$	$3^3 \cdot 23$	$2 \cdot 311$	$7 \cdot 89$	$2^4 \cdot 3 \cdot 13$	5^4	$2 \cdot 313$	$3 \cdot 11 \cdot 19$	$2^2 \cdot 157$	$17 \cdot 37$
63	$2 \cdot 3^2 \cdot 5 \cdot 7$	–	$2^3 \cdot 79$	$3 \cdot 211$	$2 \cdot 317$	$5 \cdot 127$	$2^2 \cdot 3 \cdot 53$	$7^2 \cdot 13$	$2 \cdot 11 \cdot 29$	$3^2 \cdot 71$
64	$2^7 \cdot 5$	–	$2 \cdot 3 \cdot 107$	–	$2^2 \cdot 7 \cdot 23$	$3 \cdot 5 \cdot 43$	$2 \cdot 17 \cdot 19$	–	$2^3 \cdot 3^4$	$11 \cdot 59$
65	$2 \cdot 5^2 \cdot 13$	$3 \cdot 7 \cdot 31$	$2^2 \cdot 163$	–	$2 \cdot 3 \cdot 109$	$5 \cdot 131$	$2^4 \cdot 41$	$3^2 \cdot 73$	$2 \cdot 7 \cdot 47$	–
66	$2^2 \cdot 3 \cdot 5 \cdot 11$	–	$2 \cdot 331$	$3 \cdot 13 \cdot 17$	$2^3 \cdot 83$	$5 \cdot 7 \cdot 19$	$2 \cdot 3^2 \cdot 37$	$23 \cdot 29$	$2^2 \cdot 167$	$3 \cdot 223$
67	$2 \cdot 5 \cdot 67$	$11 \cdot 61$	$2^5 \cdot 3 \cdot 7$	–	$2 \cdot 337$	$3^3 \cdot 5^2$	$2^2 \cdot 13^2$	–	$2 \cdot 3 \cdot 113$	$7 \cdot 97$
68	$2^3 \cdot 5 \cdot 17$	$3 \cdot 227$	$2 \cdot 11 \cdot 31$	–	$2^2 \cdot 3^2 \cdot 19$	$5 \cdot 137$	$2 \cdot 7^3$	$3 \cdot 229$	$2^4 \cdot 43$	$13 \cdot 53$
69	$2 \cdot 3 \cdot 5 \cdot 23$	–	$2^2 \cdot 173$	$3^2 \cdot 7 \cdot 11$	$2 \cdot 347$	$5 \cdot 139$	$2^3 \cdot 3 \cdot 29$	$17 \cdot 41$	$2 \cdot 349$	$3 \cdot 233$
70	$2^2 \cdot 5^2 \cdot 7$	–	$2 \cdot 3^3 \cdot 13$	$19 \cdot 37$	$2^6 \cdot 11$	$3 \cdot 5 \cdot 47$	$2 \cdot 353$	$7 \cdot 101$	$2^2 \cdot 3 \cdot 59$	–
71	$2 \cdot 5 \cdot 71$	$3^2 \cdot 79$	$2^3 \cdot 89$	$23 \cdot 31$	$2 \cdot 3 \cdot 7 \cdot 17$	$5 \cdot 11 \cdot 13$	$2^2 \cdot 179$	$3 \cdot 239$	$2 \cdot 359$	–
72	$2^4 \cdot 3^2 \cdot 5$	$7 \cdot 103$	$2 \cdot 19^2$	$3 \cdot 241$	$2^2 \cdot 181$	$5^2 \cdot 29$	$2 \cdot 3 \cdot 11^2$	–	$2^3 \cdot 7 \cdot 13$	3^6
73	$2 \cdot 5 \cdot 73$	$17 \cdot 43$	$2^2 \cdot 3 \cdot 61$	–	$2 \cdot 367$	$3 \cdot 5 \cdot 7^2$	$2^5 \cdot 23$	$11 \cdot 67$	$2 \cdot 3^2 \cdot 41$	–
74	$2^2 \cdot 5 \cdot 37$	$3 \cdot 13 \cdot 19$	$2 \cdot 7 \cdot 53$	–	$2^3 \cdot 3 \cdot 31$	$5 \cdot 149$	$2 \cdot 373$	$3^2 \cdot 83$	$2^2 \cdot 11 \cdot 17$	$7 \cdot 107$
75	$2 \cdot 3 \cdot 5^3$	–	$2^4 \cdot 47$	$3 \cdot 251$	$2 \cdot 13 \cdot 29$	$5 \cdot 151$	$2^2 \cdot 3^3 \cdot 7$	–	$2 \cdot 379$	$3 \cdot 11 \cdot 23$
76	$2^3 \cdot 5 \cdot 19$	–	$2 \cdot 3 \cdot 127$	$7 \cdot 109$	$2^2 \cdot 191$	$3^2 \cdot 5 \cdot 17$	$2 \cdot 383$	$13 \cdot 59$	$2^8 \cdot 3$	–
77	$2 \cdot 5 \cdot 7 \cdot 11$	$3 \cdot 257$	$2^2 \cdot 193$	–	$2 \cdot 3^2 \cdot 43$	$5^2 \cdot 31$	$2^3 \cdot 97$	$3 \cdot 7 \cdot 37$	$2 \cdot 389$	$19 \cdot 41$
78	$2^2 \cdot 3 \cdot 5 \cdot 13$	$11 \cdot 71$	$2 \cdot 17 \cdot 23$	$3^3 \cdot 29$	$2^4 \cdot 7^2$	$5 \cdot 157$	$2 \cdot 3 \cdot 131$	–	$2^2 \cdot 197$	$3 \cdot 263$
79	$2 \cdot 5 \cdot 79$	$7 \cdot 113$	$2^3 \cdot 3^2 \cdot 11$	$13 \cdot 61$	$2 \cdot 397$	$3 \cdot 5 \cdot 53$	$2^2 \cdot 199$	–	$2 \cdot 3 \cdot 7 \cdot 19$	$17 \cdot 47$
80	$2^5 \cdot 5^2$	$3^2 \cdot 89$	$2 \cdot 401$	$11 \cdot 73$	$2^2 \cdot 3 \cdot 67$	$5 \cdot 7 \cdot 23$	$2 \cdot 13 \cdot 31$	$3 \cdot 269$	$2^3 \cdot 101$	–
81	$2 \cdot 3^4 \cdot 5$	–	$2^2 \cdot 7 \cdot 29$	$3 \cdot 271$	$2 \cdot 11 \cdot 37$	$5 \cdot 163$	$2^4 \cdot 3 \cdot 17$	$19 \cdot 43$	$2 \cdot 409$	$3^2 \cdot 7 \cdot 13$
82	$2^2 \cdot 5 \cdot 41$	–	$2 \cdot 3 \cdot 137$	–	$2^3 \cdot 103$	$3 \cdot 5^2 \cdot 11$	$2 \cdot 7 \cdot 59$	–	$2^2 \cdot 3^2 \cdot 23$	–
83	$2 \cdot 5 \cdot 83$	$3 \cdot 277$	$2^6 \cdot 13$	$7^2 \cdot 17$	$2 \cdot 3 \cdot 139$	$5 \cdot 167$	$2^2 \cdot 11 \cdot 19$	$3^3 \cdot 31$	$2 \cdot 419$	–
84	$2^3 \cdot 3 \cdot 5 \cdot 7$	29^2	$2 \cdot 421$	$3 \cdot 281$	$2^2 \cdot 211$	$5 \cdot 13^2$	$2 \cdot 3^2 \cdot 47$	$7 \cdot 11^2$	$2^4 \cdot 53$	$3 \cdot 283$
85	$2 \cdot 5^2 \cdot 17$	$23 \cdot 37$	$2^2 \cdot 3 \cdot 71$	–	$2 \cdot 7 \cdot 61$	$3^2 \cdot 5 \cdot 19$	$2^3 \cdot 107$	–	$2 \cdot 3 \cdot 11 \cdot 13$	–
86	$2^2 \cdot 5 \cdot 43$	$3 \cdot 7 \cdot 41$	$2 \cdot 431$	–	$2^5 \cdot 3^3$	$5 \cdot 173$	$2 \cdot 433$	$3 \cdot 17^2$	$2^2 \cdot 7 \cdot 31$	$11 \cdot 79$
87	$2 \cdot 3 \cdot 5 \cdot 29$	$13 \cdot 67$	$2^3 \cdot 109$	$3^2 \cdot 97$	$2 \cdot 19 \cdot 23$	$5^3 \cdot 7$	$2^2 \cdot 3 \cdot 73$	–	$2 \cdot 439$	$3 \cdot 293$
88	$2^4 \cdot 5 \cdot 11$	–	$2 \cdot 3^2 \cdot 7^2$	–	$2^2 \cdot 13 \cdot 17$	$3 \cdot 5 \cdot 59$	$2 \cdot 443$	–	$2^3 \cdot 3 \cdot 37$	$7 \cdot 127$
89	$2 \cdot 5 \cdot 89$	$3^4 \cdot 11$	$2^2 \cdot 223$	$19 \cdot 47$	$2 \cdot 3 \cdot 149$	$5 \cdot 179$	$2^7 \cdot 7$	$3 \cdot 13 \cdot 23$	$2 \cdot 449$	$29 \cdot 31$
90	$2^2 \cdot 3^2 \cdot 5^2$	$17 \cdot 53$	$2 \cdot 11 \cdot 41$	$3 \cdot 7 \cdot 43$	$2^3 \cdot 113$	$5 \cdot 181$	$2 \cdot 3 \cdot 151$	–	$2^2 \cdot 227$	$3^2 \cdot 101$
91	$2 \cdot 5 \cdot 7 \cdot 13$	–	$2^4 \cdot 3 \cdot 19$	$11 \cdot 83$	$2 \cdot 457$	$3 \cdot 5 \cdot 61$	$2^2 \cdot 229$	$7 \cdot 131$	$2 \cdot 3^3 \cdot 17$	–
92	$2^3 \cdot 5 \cdot 23$	$3 \cdot 307$	$2 \cdot 461$	$13 \cdot 71$	$2^2 \cdot 3 \cdot 7 \cdot 11$	$5^2 \cdot 37$	$2 \cdot 463$	$3^2 \cdot 103$	$2^2 \cdot 29$	–
93	$2 \cdot 3 \cdot 5 \cdot 31$	$7^2 \cdot 19$	$2^2 \cdot 233$	$3 \cdot 311$	$2 \cdot 467$	$5 \cdot 11 \cdot 17$	$2^3 \cdot 3^2 \cdot 13$	–	$2 \cdot 7 \cdot 67$	$3 \cdot 313$
94	$2^2 \cdot 5 \cdot 47$	–	$2 \cdot 3 \cdot 157$	$23 \cdot 41$	$2^4 \cdot 59$	$3^3 \cdot 5 \cdot 7$	$2 \cdot 11 \cdot 43$	–	$2^2 \cdot 3 \cdot 79$	$13 \cdot 73$
95	$2 \cdot 5^2 \cdot 19$	$3 \cdot 317$	$2^3 \cdot 7 \cdot 17$	–	$2 \cdot 3^2 \cdot 53$	$5 \cdot 191$	$2^2 \cdot 239$	$3 \cdot 11 \cdot 29$	$2 \cdot 479$	$7 \cdot 137$
96	$2^6 \cdot 3 \cdot 5$	31^2	$2 \cdot 13 \cdot 37$	$3^2 \cdot 107$	$2^2 \cdot 241$	$5 \cdot 193$	$2 \cdot 3 \cdot 7 \cdot 23$	–	$2^3 \cdot 11^2$	$3 \cdot 17 \cdot 19$
97	$2 \cdot 5 \cdot 97$	–	$2^2 \cdot 3^5$	$7 \cdot 139$	$2 \cdot 487$	$3 \cdot 5^2 \cdot 13$	$2^4 \cdot 61$	–	$2 \cdot 3 \cdot 163$	$11 \cdot 89$
98	$2^2 \cdot 5 \cdot 7^2$	$3^2 \cdot 109$	$2 \cdot 491$	–	$2^3 \cdot 3 \cdot 41$	$5 \cdot 197$	$2 \cdot 17 \cdot 29$	$3 \cdot 7 \cdot 47$	$2^2 \cdot 13 \cdot 19$	$23 \cdot 43$
99	$2 \cdot 3^2 \cdot 5 \cdot 11$	–	$2^5 \cdot 31$	$3 \cdot 331$	$2 \cdot 7 \cdot 71$	$5 \cdot 199$	$2^2 \cdot 3 \cdot 83$	–	$2 \cdot 499$	$3^3 \cdot 37$

Least Common Multiple (LCM)

Let $[a_1, \ldots, a_n]$ denote the *least common multiple* of the integers $a_1, \ldots, a_n$. One method of finding that number is: Prime number factorize $a_1, \ldots, a_n$. Then form the product of these primes raised to the greatest power in which they appear.

Example.

Determine $A=[18, 24, 30]$. $18=2\cdot 3^2$, $24=2^3\cdot 3$, $30=2\cdot 3\cdot 5$. Thus, $A=2^3\cdot 3^2\cdot 5=360$.

Greatest Common Divisor (GCD)

Let (a, b) denote the *greatest common divisor* of a and b. If $(a, b)=1$ the numbers are *relatively prime*. One method (*Euclid's algorithm*) of finding (a, b) is:

Assuming $a>b$ and dividing a by b yields $a=q_1b+r_1$, $0\leq r_1<b$. Dividing b by r_1 gives $b=q_2r_1+$ $+r_2$, $0\leq r_2<r_1$. Continuing like this, let r_k be the first remainder which equals 0. Then $(a, b)=r_{k-1}$.

Example.

Determine (112, 42). By the above algorithm:

$112=2\cdot 42+28$, $42=1\cdot 28+14$, $28=2\cdot 14+0$. Thus, $(112, 42)=14$.

Note. $(a, b)\cdot [a, b]=ab$

Modulo

If m, n and p are integers, then m and n are congruent modulo p, $m=n \bmod(p)$, if $m-n$ is a multiple of p, i.e. m/p and n/p have equal remainders.

$m_1=n_1 \bmod(p)$, $m_2=n_2 \bmod(p) \Rightarrow$

(*i*) $m_1\pm m_2=(n_1\pm n_2) \bmod(p)$ (*ii*) $m_1m_2=(n_1n_2) \bmod(p)$

Diophantine equations

A *Diophantine* equation has integer coefficients and integer solutions.

As an example the equation

$$(*) \qquad ax+by=c, \quad a, b, c\in \mathbf{Z}$$

has integer solutions x and y if an only if (a, b) divides c. In particular, $ax+by=1$ is solvable $\Leftrightarrow (a, b)=1$.

If x_0, y_0 is a particular solution of $(*)$ then the general solution is

$$x=x_0+nb/(a, b), \quad y=y_0-na/(a, b), \quad n\in \mathbf{Z}$$

Mersenne numbers $M_n = 2^n - 1$

Numbers of the form $2^n - 1$ are called *Mersenne numbers*. The following table gives the prime factorizations of the first 40 Mersenne numbers.

n	Prime factors of M_n
2	3
3	7
4	$3 \cdot 5$
5	31
6	$3^2 \cdot 7$
7	127
8	$3 \cdot 5 \cdot 17$
9	$7 \cdot 73$
10	$3 \cdot 11 \cdot 31$
11	$23 \cdot 89$
12	$3^2 \cdot 5 \cdot 7 \cdot 13$
13	8191
14	$3 \cdot 43 \cdot 127$
15	$7 \cdot 31 \cdot 151$
16	$3 \cdot 5 \cdot 17 \cdot 257$
17	131 071
18	$3^3 \cdot 7 \cdot 19 \cdot 73$
19	524287
20	$3 \cdot 5^2 \cdot 11 \cdot 31 \cdot 41$
21	$7^2 \cdot 127 \cdot 337$

n	Prime factors of M_n
22	$3 \cdot 23 \cdot 89 \cdot 683$
23	$47 \cdot 178481$
24	$3^2 \cdot 5 \cdot 7 \cdot 13 \cdot 17 \cdot 241$
25	$31 \cdot 601 \cdot 1801$
26	$3 \cdot 2731 \cdot 8191$
27	$7 \cdot 73 \cdot 262657$
28	$3 \cdot 5 \cdot 29 \cdot 43 \cdot 113 \cdot 127$
29	$233 \cdot 1103 \cdot 2089$
30	$3^2 \cdot 7 \cdot 11 \cdot 31 \cdot 151 \cdot 331$
31	2147483647
32	$3 \cdot 5 \cdot 17 \cdot 257 \cdot 65537$
33	$7 \cdot 23 \cdot 89 \cdot 599479$
34	$3 \cdot 43691 \cdot 131071$
35	$31 \cdot 71 \cdot 127 \cdot 122921$
36	$3^3 \cdot 5 \cdot 7 \cdot 13 \cdot 19 \cdot 37 \cdot 73 \cdot 109$
37	$223 \cdot 616318177$
38	$3 \cdot 174763 \cdot 524287$
39	$7 \cdot 79 \cdot 8191 \cdot 121369$
40	$3 \cdot 5^2 \cdot 11 \cdot 17 \cdot 31 \cdot 41 \cdot 61681$
41	$13367 \cdot 164511353$

Mersenne primes: If $2^p - 1$ is prime then p is prime.

Fermat primes: If $2^p + 1$ is prime then p is a power of 2.

The following are some prime Mersenne numbers.

M_{61} = 2305 843 009 213 693 951
M_{89} = 618 970 019 642 690 137 449 562 111
M_{107} = 162 259 276 829 213 363 391 578 010 288 127
M_{127} = 170 141 183 460 469 231 731 687 303 715 884 105 727
M_{216091} is the (1986) largest known prime number.

Fibonacci numbers

The nth Fibonacci number is denoted by F_n. These numbers are defined by the formulas.

$$F_1=1 \qquad F_2=1 \qquad F_{n+2}=F_n+F_{n+1},\ n\geq 1$$

(Cf. sec. 9.5 for an explicit formula.)

The following table gives prime factorizations of the 50 first Fibonacci numbers.

n	F_n	F_n	n	F_n	F_n
1	1	1	26	121 393	$233 \cdot 521$
2	1	1	27	196 418	$2 \cdot 17 \cdot 53 \cdot 109$
3	2	2	28	317 811	$3 \cdot 13 \cdot 29 \cdot 281$
4	3	3	29	514 229	514229
5	5	5	30	832 040	$2^3 \cdot 5 \cdot 11 \cdot 31 \cdot 61$
6	8	2^3	31	1 346 269	$557 \cdot 2417$
7	13	13	32	2 178 309	$3 \cdot 7 \cdot 47 \cdot 2207$
8	21	$3 \cdot 7$	33	3 524 578	$2 \cdot 89 \cdot 19801$
9	34	$2 \cdot 17$	34	5 702 887	$1597 \cdot 3571$
10	55	$5 \cdot 11$	35	9 227 465	$5 \cdot 13 \cdot 141961$
11	89	89	36	14 930 352	$2^4 \cdot 3^3 \cdot 17 \cdot 19 \cdot 107$
12	144	$2^4 \cdot 3^2$	37	24 157 817	$73. 149 \cdot 2221$
13	233	233	38	39 088 169	$37 \cdot 113 \cdot 9349$
14	377	$13 \cdot 29$	39	63 245 986	$2 \cdot 233 \cdot 135721$
15	610	$2 \cdot 5 \cdot 61$	40	102 334 155	$3 \cdot 5 \cdot 7 \cdot 11 \cdot 41 \cdot 2161$
16	987	$3 \cdot 7 \cdot 47$	41	165 580 141	$2789 \cdot 59369$
17	1597	1597	42	267 914 296	$2^3 \cdot 13 \cdot 29 \cdot 211 \cdot 421$
18	2584	$2^3 \cdot 17 \cdot 19$	43	433 494 437	433494437
19	4181	$37 \cdot 113$	44	701 408 733	$3 \cdot 43 \cdot 89 \cdot 199 \cdot 307$
20	6765	$3 \cdot 5 \cdot 11 \cdot 41$	45	1 134 903 170	$2 \cdot 5 \cdot 17 \cdot 61 \cdot 109 \cdot 441$
21	10 946	$2 \cdot 13 \cdot 421$	46	1 836 311 903	$139 \cdot 461 \cdot 28657$
22	17 711	$89 \cdot 199$	47	2 971 215 073	2971215073
23	28 657	28 657	48	4 807 526 976	$2^6 \cdot 3^2 \cdot 7 \cdot 23 \cdot 47 \cdot 1103$
24	46 368	$2^5 \cdot 3^2 \cdot 7 \cdot 23$	49	7 778 742 049	$13 \cdot 97 \cdot 6168709$
25	75 025	$5^2 \cdot 3001$	50	12 586 269 025	$5^2 \cdot 11 \cdot 101 \cdot 151 \cdot 3001$

Real numbers in different number bases

The positional system

Every real number x may be written

$$x=x_mB^m+x_{m-1}B^{m-1}+\ldots+x_0B^0+x_{-1}B^{-1}+\ldots=(x_mx_{m-1}\ \ldots\ x_0.\ x_{-1}\ \ldots)_B$$

where the natural number $B>1$ is the *base,* and each *digit* x_i is one of the numbers 0, 1, ..., $B-1$. As an example the number x="thirtysix and three eighths", written in the *decimal* and *binary* systems respectively, becomes

$$x=3\cdot 10^1+6\cdot 10^0+3\cdot 10^{-1}+7\cdot 10^{-2}+5\cdot 10^{-3}=(36.375)_{10}$$

$$x=1\cdot 2^5+0\cdot 2^4+0\cdot 2^3+1\cdot 2^2+0\cdot 2^1+0\cdot 2^0+0\cdot 2^{-1}+1\cdot 2^{-2}+1\cdot 2^{-3}= =(100100.011)_2.$$

Conversion Algorithms

a ($B\to 10$). When converting a number $X=(X_mX_{m-1}\ \ldots\ X_0\ .\ X_{-1}\ \ldots)_B$ given in a system with base B to the decimal system, compute

$$X=X_mB^m+X_{m-1}B^{m-1}+\ldots+X_0+X_{-1}B^{-1}+\ldots$$

b ($10\to B$). When converting a positive number X given in the decimal system to a system with base B, the *integer part* Y of X and the *fractional part* Z of X are treated separately. (Below, the example $X=(12345.6789)_{10}$ and $B=8$ is treated together with the method description.)

The integer part Y	**Example**
	$Y=12345$, $B=8$
(*i*) Divide Y by B. If the quotient is Q_1 and the remainder R_1 (R_1 is one of the integers 0, 1, ..., $B-1$), then R_1 is the *first* digit *from the right* of Y in the new base B.	$Y/8=1543+1/8$, i.e. $Q_1=1543$ $R_1=1$
(*ii*) Divide Q_1 by B. If the quotient is Q_2 and the remainder R_2, then R_2 is the *second* digit *from the right.*	$Q_1/8=192+7/8$, i.e. $Q_2=192$ $R_2=7$ $Q_3=24$ $R_3=0$
(*iii*) Proceed accordingly until the quotient becomes zero.	$Q_4=3$ $R_4=0$ $Q_5=0$ $R_5=3$ Thus, $Y=(30071)_8$
The fractional part Z	$Z=0.6789$, $B=8$
(*i*) Multiply Z by B. If the integer part of the product is I_1 (I_1 is one of the integers 0, 1, ..., $B-1$) and the new fractional part F_1, then I_1 is the *first* digit of the fractional part Z in the base B.	$Z\cdot 8=5.4312$ i.e. $I_1=5$ $F_1=0.4312$
(*ii*) Multiply F_1 by B. If the integer part is I_2 and the fractional part F_2, then I_2 is the *second* fractional digit in the base B.	$F_1\cdot 8=3.4496$ i.e. $I_2=3$ $F_2=0.4496$
(*iii*) Proceed accordingly until the producet becomes an integer, or until the desired number of fractional digits have been computed.	$I_3=3$ $F_3=0.5968$ $I_4=4$ $F_4=0.7744$ $I_5=6$ $F_5=0.1952$ etc. Thus, $Z\approx(0.5335)_8$ and $X\approx(30071.5335)_8$

Binary System (Digits 0 and 1)

Addition: 0+0=0 0+1=1+0=1 1+1=10

Multiplication: 0·0=0·1=1·0=0 1·1=1

Powers of 2 in decimal scale

$n=$	$2^n=$	2^{-n}
0	1	1
1	2	0.5
2	4	0.25
3	8	0.125
4	16	0.0625
5	32	0.03125
6	64	0.015625
7	128	0.007812 5
8	256	0.003906 25
9	512	0.001953 125
10	1024	0.000976 5625
11	2048	0.000488 28125
12	4096	0.000244 140625
13	8192	0.000122 070312 5
14	16384	0.000061 035156 25
15	32768	0.000030 517578 125
16	65536	0.000015 258789 0625
17	131072	0.000007 629394 53125
18	262144	0.000003 814697 265625
19	524288	0.000001 907348 632812 5
20	1 048576	0.000000 953674 316406 25
21	2 097152	0.000000 476837 158203 125
22	4 194304	0.000000 238418 579101 5625
23	8 388608	0.000000 119209 289550 78125
24	16 777216	0.000000 059604 644775 390625
25	33 554432	0.000000 029802 322387 695312 5
26	67 108864	0.000000 014901 161193 847656 25
27	134 217728	0.000000 007450 580596 923828 125
28	268 435456	0.000000 003725 290298 461914 0625
29	536 870912	0.000000 001862 645149 230957 03125
30	1073 741824	0.000000 000931 322574 615478 515625
31	2147 483648	0.000000 000465 661287 307739 257812 5
32	4294 967296	0.000000 000232 830643 653869 628906 25
33	8589 934592	0.000000 000116 415321 826934 814453 125
34	17179 869184	0.000000 000058 207660 913467 407226 5625
35	34359 738368	0.000000 000029 103830 456733 703613 28125
36	68719 476736	0.000000 000014 551915 228366 851806 640625
37	137438 953472	0.000000 000007 275957 614183 425903 320312 5
38	274877 906944	0.000000 000003 637978 807091 712951 660156 25
39	549755 813888	0.000000 000001 818989 403545 856475 830078 125
40	1 099511 627776	0.000000 000000 909494 701772 928237 915039 0625
41	2 199023 255552	0.000000 000000 454747 350886 464118 957519 53125
42	4 398046 511104	0.000000 000000 227373 675443 232059 478759 765625
43	8 796093 022208	0.000000 000000 113686 837721 616029 739379 882812 5
44	17 592186 044416	0.000000 000000 056843 418860 808014 869689 941406 25
45	35 184372 088832	0.000000 000000 028421 709430 404007 434844 970703 125
46	70 368744 177664	0.000000 000000 014210 854715 202003 717422 485351 5625
47	140 737488 355328	0.000000 000000 007105 427357 601001 858711 242675 78125
48	281 474976 710656	0.000000 000000 003552 713678 800500 929355 621337 890625
49	562 949953 421312	0.000000 000000 001776 356839 400250 464677 810668 945312 5
50	1125 899906 842624	0.000000 000000 000888 178419 700125 232338 905334 472656 25

Hexadecimal system

(Digits: 0, 1, 2, 3, 4, 5, 6, 7, 8, 9, A=10, B=11, C=12, D=13 , E=14, and F=15)

Addition table

	1	**2**	**3**	**4**	**5**	**6**	**7**	**8**	**9**	**A**	**B**	**C**	**D**	**E**	**F**	
1	2	3	4	5	6	7	8	9	A	B	C	D	E	F	10	**1**
2	3	4	5	6	7	8	9	A	B	C	D	E	F	10	11	**2**
3	4	5	6	7	8	9	A	B	C	D	E	F	10	11	12	**3**
4	5	6	7	8	9	A	B	C	D	E	F	10	11	12	13	**4**
5	6	7	8	9	A	B	C	D	E	F	10	11	12	13	14	**5**
6	7	8	9	A	B	C	D	E	F	10	11	12	13	14	15	**6**
7	8	9	A	B	C	D	E	F	10	11	12	13	14	15	16	**7**
8	9	A	B	C	D	E	F	10	11	12	13	14	15	16	17	**8**
9	A	B	C	D	E	F	10	11	12	13	14	15	16	17	18	**9**
A	B	C	D	E	F	10	11	12	13	14	15	16	17	18	19	**A**
B	C	D	E	F	10	11	12	13	14	15	16	17	18	19	1A	**B**
C	D	E	F	10	11	12	13	14	15	16	17	18	19	1A	1B	**C**
D	E	F	10	11	12	13	14	15	16	17	18	19	1A	1B	1C	**D**
E	F	10	11	12	13	14	15	16	17	18	19	1A	1B	1C	1D	**E**
F	10	11	12	13	14	15	16	17	18	19	1A	1B	1C	1D	1E	**F**
	1	**2**	**3**	**4**	**5**	**6**	**7**	**8**	**9**	**A**	**B**	**C**	**D**	**E**	**F**	

E.g. B + 6 = 11

Multiplication table

	1	**2**	**3**	**4**	**5**	**6**	**7**	**8**	**9**	**A**	**B**	**C**	**D**	**E**	**F**	
1	1	2	3	4	5	6	7	8	9	A	B	C	D	E	F	**1**
2	2	4	6	8	A	C	E	10	12	14	16	18	1A	1C	1E	**2**
3	3	6	9	C	F	12	15	18	1B	1E	21	24	27	2A	2D	**3**
4	4	8	C	10	14	18	1C	20	24	28	2C	30	34	38	3C	**4**
5	5	A	F	14	19	1E	23	28	2D	32	37	3C	41	46	4B	**5**
6	6	C	12	18	1E	24	2A	30	36	3C	42	48	4E	54	5A	**6**
7	7	E	15	1C	23	2A	31	38	3F	46	4D	54	5B	62	69	**7**
8	8	10	18	20	28	30	38	40	48	50	58	60	68	70	78	**8**
9	9	12	1B	24	2D	36	3F	48	51	5A	63	6C	75	7E	87	**9**
A	A	14	1E	28	32	3C	46	50	5A	64	6E	78	82	8C	96	**A**
B	B	16	21	2C	37	42	4D	58	63	6E	79	84	8F	9A	A5	**B**
C	C	18	24	30	3C	48	54	60	6C	78	84	90	9C	A8	B4	**C**
D	D	1A	27	34	41	4E	5B	68	75	82	8F	9C	A9	B6	C3	**D**
E	E	1C	2A	38	46	54	62	70	7E	8C	9A	A8	B6	C4	D2	**E**
F	F	1E	2D	3C	4B	5A	69	78	87	96	A5	B4	C3	D2	E1	**F**
	1	**2**	**3**	**4**	**5**	**6**	**7**	**8**	**9**	**A**	**B**	**C**	**D**	**E**	**F**	

E.g. B · 6 = 42

Special numbers in different number bases

B=2:
π = 11.001001 000011 111101 101010 100010 001000 010110 100011...
e = 10.101101 111110 000101 010001 011000 101000 101011 101101...
γ = 0.100100 111100 010001 100111 111000 110111 110110 110110...
$\sqrt{2}$ = 1.011010 100000 100111 100110 011001 111111 001110 111100...
ln2 = 0.101100 010111 001000 010111 111101 111101 000111 001111...

B=3:
π = 10.010211 012222...
e = 2.201101 121221..
γ = 0.120120 210100...
$\sqrt{2}$ = 1.102011 221222...
ln2 = 0.200201 022012...

B=12:
π = 3.184809 493B91...
e = 2.875236 069821...
γ = 0.6B1518 8A6760...
$\sqrt{2}$ = 1.4B7917 0A07B8...
ln2 = 0.839912 483369...

B=8:
π = 3.110375 524210 264302...
e = 2.557605 213050 535512...
γ = 0.447421 477067 666061...
$\sqrt{2}$ = 1.324047 463177 167462...
ln2 = 0.542710 277574 071736 ...

B=16:
π = 3.243F6A 8885A3...
e = 2.B7E151 628AED...
γ = 0.93C467 E37DB0...
$\sqrt{2}$ = 1.6A09E6 67F3BC...
ln2 = 0.B17217 F7D1CF...

Powers of 16 in decimal scale

(Digits: 0,1,2,3,4,5,6,7,8,9,A=10,B=11,C=12,D=13,E=14,F=15)

n=	16^n =	16^{-n} =
0	1	1
1	16	0.0625
2	256	0.0039 0625
3	4096	0.0002 4414 0625
4	65536	0.0000 1525 8789 0625
5	1 048576	0.0000 0095 3674 3164 0625
6	16 777216	0.0000 0005 9604 6447 7539 0625
7	268 435456	0.0000 0000 3725 2902 9846 1914 0625
8	4294 967296	0.0000 0000 0232 8306 4365 3869 6289 0625
9	68719 476736	0.0000 0000 0014 5519 1522 8366 8518 0664 0625
10	1 099511 627776	0.0000 0000 0000 9094 9470 1772 9282 3791 5039 0625
11	17 592186 044416	0.0000 0000 0000 0568 4341 8860 8080 1486 9689 9414 0625
12	281 474976 710656	0.0000 0000 0000 0035 5271 3678 8005 0092 9355 6213 3789 0625
13	4503 599627 370496	
14	72057 594037 927936	
15	1 152921 504606 846976	
16	18 446744 073709 551616	
17	295 147905 179352 825856	
18	4722 366482 869645 213696	
19	75557 863725 914323 419136	
20	1 208925 819614 629174 706176	

Powers of 10 in hexadecimal scale

n=	10^n =	10^{-n} = (approximately with 16 fractional places)
Dec	Hex	Hex
0	1	1
1	A	0.1999 9999 9999 999A
2	64	0.028F 5C28 F5C2 8F5C
3	3E8	0.0041 8937 4BC6 A7F0
4	2710	0.0006 8DB8 BAC7 10CB
5	1 86A0	0.0000 A7C5 AC47 1B48
6	F 4240	0.0000 10C6 F7A0 B5EE
7	98 9680	0.0000 01AD 7F29 ABCB
8	5F5 E100	0.0000 002A F31D C461
9	3B9A CA00	0.0000 0004 4B82 FA0A
10	2 540B E400	0.0000 0000 6DF3 7F67
11	17 4876 E800	0.0000 0000 0AFE BFF1
12	E8 D4A5 1000	0.0000 0000 0119 7998
13	918 4E72 A000	0.0000 0000 001C 25C2
14	5AF3 107A 4000	0.0000 0000 0002 D093
15	3 8D7E A4C6 8000	0.0000 0000 0000 480F
16	23 86F2 6FC1 0000	0.0000 0000 0000 0735

2.3 Complex Numbers

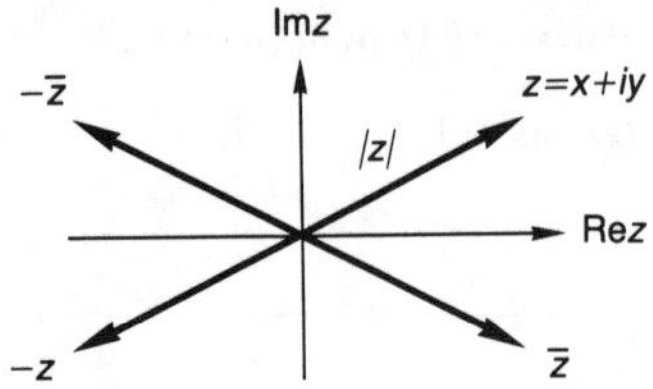

Imaginary unit i

$$i^2=-1$$

The imaginary unit is sometimes denoted j.

Rectangular form

Complex numbers z have the form $z=x+iy$, x and y real.

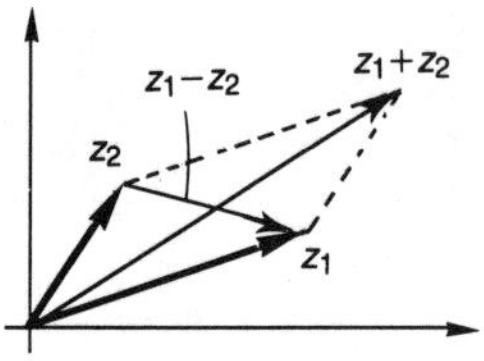

$x=\text{Re}z$ (*real part*); $y=\text{Im}z$ (*imaginary part*)

$\bar{z}=x-iy$ (*conjugate* of z)

$|z|=\sqrt{x^2+y^2}$ (*modulus* of z)

$|z_1-z_2|$=distance between the points z_1 and z_2.

$$z_1+z_2=(x_1+iy_1)+(x_2+iy_2)=(x_1+x_2)+i(y_1+y_2)$$

$$z_1-z_2=(x_1+iy_1)-(x_2+iy_2)=(x_1-x_2)+i(y_1-y_2)$$

$$z_1\cdot z_2=(x_1+iy_1)(x_2+iy_2)=(x_1x_2-y_1y_2)+i(x_1y_2+x_2y_1)$$

$$\frac{z_1}{z_2}=\frac{x_1+iy_1}{x_2+iy_2}=\frac{(x_1+iy_1)(x_2-iy_2)}{(x_2+iy_2)(x_2-iy_2)}=\frac{(x_1x_2+y_1y_2)+i(x_2y_1-x_1y_2)}{x_2^2+y_2^2}$$

$$\bar{\bar{z}}=z \qquad z\bar{z}=|z|^2 \qquad \overline{z_1\pm z_2}=\bar{z}_1\pm\bar{z}_2 \qquad \overline{z_1z_2}=\bar{z}_1\cdot\bar{z}_2 \qquad \overline{\left(\frac{z_1}{z_2}\right)}=\frac{\bar{z}_1}{\bar{z}_2}$$

$$\big||z_1|-|z_2|\big|\leqslant|z_1\pm z_2|\leqslant|z_1|+|z_2| \qquad |z_1z_2|=|z_1|\cdot|z_2| \qquad \left|\frac{z_1}{z_2}\right|=\frac{|z_1|}{|z_2|}$$

Polar form

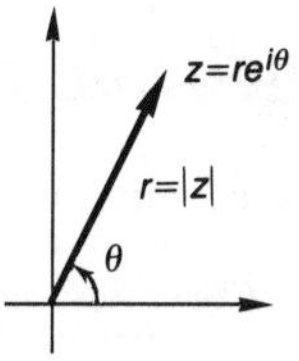

r=modulus of z, θ=argument of z

$z=x+iy=r(\cos\theta+i\sin\theta)=re^{i\theta}$

$$\begin{cases} x=r\cos\theta \\ y=r\sin\theta \end{cases} \qquad \begin{cases} r=\sqrt{x^2+y^2} \\ \tan\theta=\dfrac{y}{x} \Leftrightarrow \theta=\arctan\dfrac{y}{x}+n\pi \ (n=0 \text{ if } x>0,\ n=1 \text{ if } x<0) \end{cases}$$

$z_1\cdot z_2=r_1e^{i\theta_1}\cdot r_2e^{i\theta_2}=r_1r_2e^{i(\theta_1+\theta_2)}$	$\arg(z_1z_2)=\arg z_1+\arg z_2$
(moduli multiplied, arguments added)	
$\dfrac{z_1}{z_2}=\dfrac{r_1e^{i\theta_1}}{r_2e^{i\theta_2}}=\dfrac{r_1}{r_2}e^{i(\theta_1-\theta_2)}$	$\arg\dfrac{z_1}{z_2}=\arg z_1-\arg z_2$
(moduli divided, arguments subtracted)	
$z^n=(re^{i\theta})^n=r^ne^{in\theta}$	$\arg z^n=n\arg z$

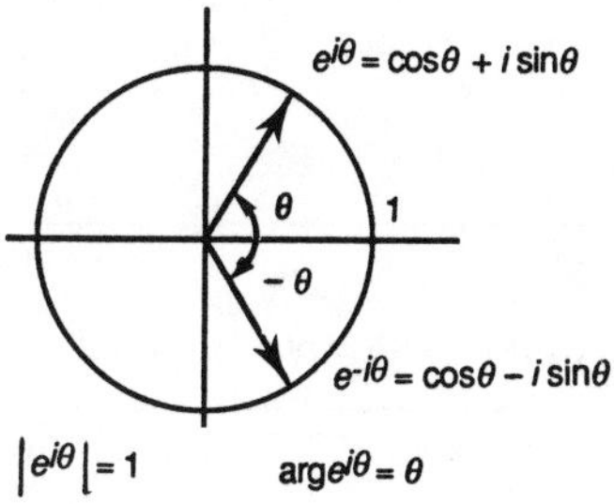

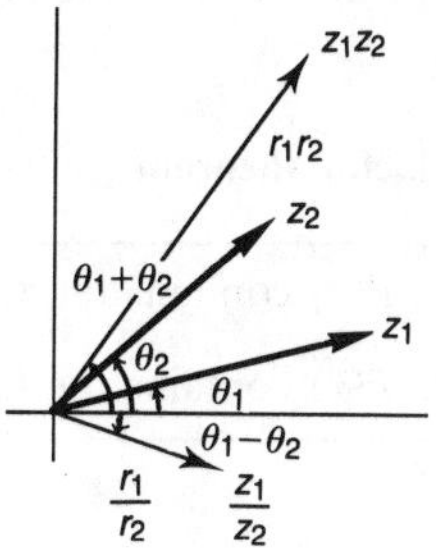

de Moivre's formula

$$(\cos\theta+i\sin\theta)^n=\cos n\theta+i\sin n\theta$$

The Euler formulas

$$\cos\theta=\frac{e^{i\theta}+e^{-i\theta}}{2} \qquad \sin\theta=\frac{e^{i\theta}-e^{-i\theta}}{2i}$$

Complex analysis, see chapt. 14.

2.4 Algebraic Equations

An *algebraic equation* is of the form

(2.1) $$P(z)=a_nz^n+a_{n-1}z^{n-1}+\ldots+a_1z+a_0=0 \quad (a_i \text{ complex numbers})$$

The *degree* of the equation is n (if $a_n \neq 0$).

Zeros and roots

A number r is a *zero* of *multiplicity* m of a polynomial $P(z)$ if there is a polynomial $Q(z)$ with $Q(r)\neq 0$ such that

$$P(z)=(z-r)^m \, Q(z)$$

Also, r is a *root* of multiplicity m of the equation $P(z)=0$.

(An algorithm for finding $Q(z)$, see sec. 5.2).

If r is a root of multiplicity m ($m \geq 1$) of the equation $P(z)=0$, then r is a root of multiplicity $m-1$ of the equation $P'(z)=0$.

The factor theorem

1. $P(z)$ contains the factor $z-r \Leftrightarrow P(r)=0$
2. $P(z)$ contains the factor $(z-r)^m \Leftrightarrow P(r)=P'(r)=\ldots=P^{(m-1)}(r)=0$

The fundamental theorem of algebra

An algebraic equation $P(z)=0$ of degree n has exactly n roots (including multiplicity). If the roots are $r_1, \ldots, r_n$, then

$$P(z)=a_n(z-r_1) \ldots (z-r_n)$$

Relationship between roots and coefficients

If $r_1, \ldots, r_n$ are the roots of (2.1) then

$$\begin{cases} r_1+r_2+\ldots+r_n=-\dfrac{a_{n-1}}{a_n} \\ r_1r_2+r_1r_3+\ldots+r_{n-1}r_n=\sum\limits_{i<j} r_ir_j=\dfrac{a_{n-2}}{a_n} \\ \ldots \\ r_1r_2 \ldots r_n=(-1)^n \dfrac{a_0}{a_n} \end{cases}$$

Equations with real coefficients

Assume that all a_i of (2.1) are real.

1. If r is a non-real root of (2.1) then so is $\bar{r}$, i.e. $P(r)=0 \Rightarrow P(\bar{r})=0$.
2. $P(z)$ can be factorized into real polynomials of degree at most two.
3. If all a_i are integers and if $r=p/q$ (p and q having no common divisor) is a rational root of (2.1), then p divides a_o and q divides a_n.
4. The number of positive real roots (including multiplicity) of (2.1) either equals the number of sign changes of the sequence $a_0, a_1, \ldots, a_n$ or equals this number minus an even number. If all roots of the equation are real the first case always applies. (*Descartes' rule of signs.*)

Quadratic equations

$ax^2+bx+c=0$	$x^2+px+q=0$
$x=\dfrac{-b\pm\sqrt{b^2-4ac}}{2a}$	$x=-\dfrac{p}{2}\pm\sqrt{\left(\dfrac{p}{2}\right)^2-q}$

$b^2-4ac>0 \Rightarrow$ two unequal real roots
$b^2-4ac<0 \Rightarrow$ two unequal complex roots $\quad (\pm\sqrt{-\mathrm{d}}=\pm \mathrm{i}\sqrt{\mathrm{d}})$
$b^2-4ac=0 \Rightarrow$ the roots are real and equal

The expression b^2-4ac is called the *discriminant.*
Let x_1 and x_2 be roots of the equation $x^2+px+q=0$. Then

$$\begin{cases} x_1+x_2=-p \\ x_1x_2=q \end{cases}$$

Cubic equations

The equation $\qquad az^3+bz^2+cz+d=0$

is by the substitution $z=x-b/3a$ reduced to

(2.2) $\qquad x^3+px+q=0$

Set $\qquad D=\left(\dfrac{p}{3}\right)^3+\left(\dfrac{q}{2}\right)^2.$

Then (2.2) has (*i*) one real root if $D>0$, (*ii*) three real roots of which at least two are equal if $D=0$, (*iii*) three distinct real roots if $D<0$. Put

$$u=\sqrt[3]{-\frac{q}{2}+\sqrt{D}}, \qquad v=\sqrt[3]{-\frac{q}{2}-\sqrt{D}}.$$

The roots of (2.2) are

$$x_1=u+v \quad x_{2,3}=-\frac{u+v}{2}\pm\frac{u-v}{2}i\sqrt{3} \quad (\textit{Cardano's formula})$$

If x_1, x_2, x_3 are roots of the equation $x^3+rx^2+sx+t=0$ then

$$\begin{cases} x_1+x_2+x_3=-r \\ x_1x_2+x_1x_3+x_2x_3=s \\ x_1x_2x_3=-t \end{cases}$$

Binomic equations

A *binomic equation* is of the form

$$z^n=c, \; c \text{ complex number}$$

1. Special case $n=2$: $z^2=a+ib$.

   Roots:
$$z=\pm\sqrt{a+ib}=\begin{cases} \pm\left[\sqrt{\frac{r+a}{2}}+i\sqrt{\frac{r-a}{2}}\right], & b\geq 0 \\ \pm\left[\sqrt{\frac{r+a}{2}}-i\sqrt{\frac{r-a}{2}}\right], & b\leq 0 \end{cases}, \qquad r=\sqrt{a^2+b^2}$$

2. General case: Solution in *polar* form: Set $c=re^{i\theta}$

$$z^n=c=re^{i(\theta+2k\pi)}$$

Roots: $$z=\sqrt[n]{r}e^{i(\theta+2k\pi)/n}=\sqrt[n]{r}\left(\cos\frac{\theta+2k\pi}{n}+i\sin\frac{\theta+2k\pi}{n}\right),$$

$k=0, 1, \ldots, n-1$

Note. The roots form a *regular n-gon*.

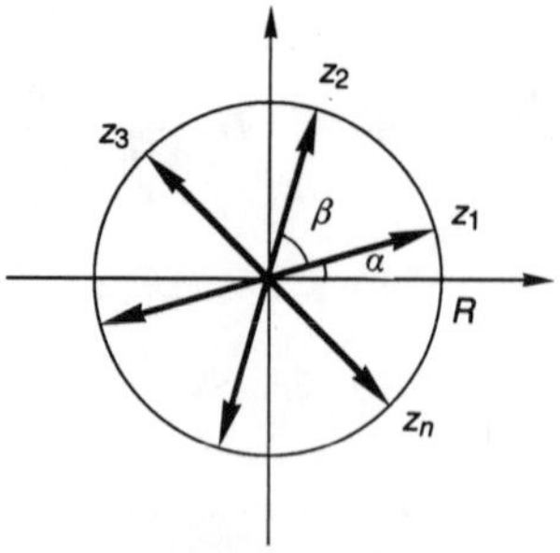

$\alpha=\frac{\theta}{n}$, $\beta=\frac{2\pi}{n}$, $R=\sqrt[n]{r}$

3 Geometry and Trigonometry

3.1 Plane Figures

Triangles

a, b, c=sides, *α, β, γ*=angles, *h*=altitude, *m*=median, *s*=bisector, $2p=a+b+c=$ =perimeter, *R*=circumradius, *r*=inradius, *A*=area.

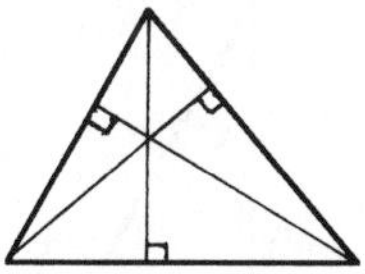

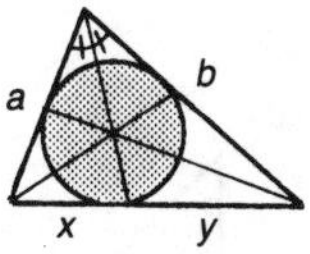

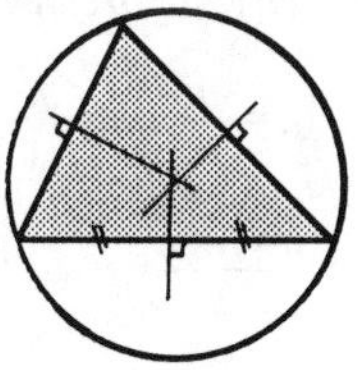

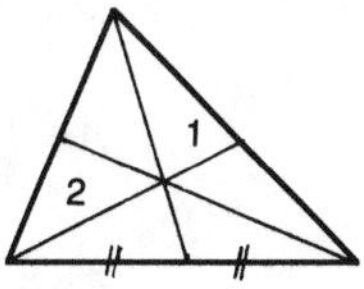

$$\frac{x}{y} = \frac{a}{b}$$

Theorems

1. (a) $\alpha+\beta+\gamma=180°$
   (b) $\alpha<\beta<\gamma \Leftrightarrow a<b<c$

2. Each of the following triples are concurrent in one point:
   (a) the altitudes
   (b) the angle bisectors (center of the inscribed circle)
   (c) the perpendicular bisectors (center of the circumscribed circle)
   (d) the medians (dividing each other in the proportions 2:1).

3. Two triangles are congruent if the following corresponding elements are equal.
   (a) three sides
   (b) one angle and the sides including the angle
   (c) one side and the two angles adjacent to it.

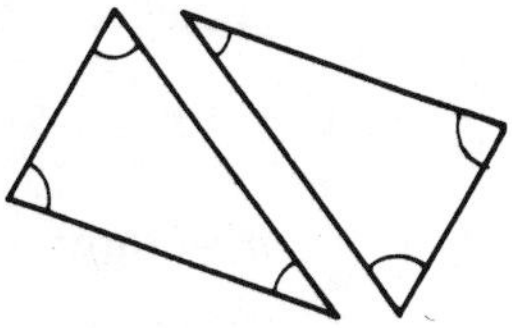

4. Two triangles are similar

   $\left(\text{i.e. } \frac{a}{a'}=\frac{b}{b'}=\frac{c}{c'}, \text{ and } \alpha=\alpha', \beta=\beta', \gamma=\gamma'\right)$

   if one of the following conditions is satisfied:

(*i*) the sides are proportional ($a:b:c=a':b':c'$)
(*ii*) an angle of one equals an angle of the other and the sides including these angles are proportional ($\alpha=\alpha'$, $b/c=b'/c'$)
(*iii*) two angles in one are equal to two angles in the other triangle ($\alpha=\alpha'$, $\beta=\beta'$).

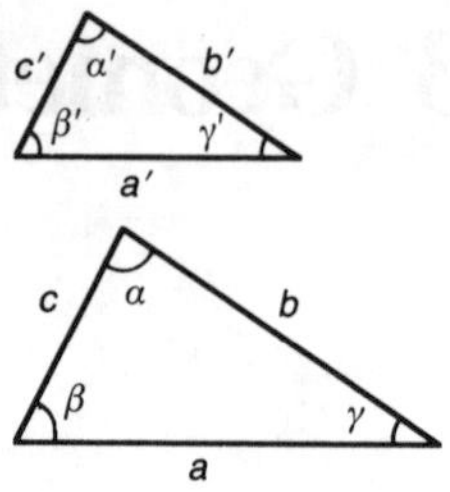

5. If two triangles are similar then

$$\frac{A}{A'}=\left(\frac{a}{a'}\right)^2=\left(\frac{h}{h'}\right)^2=\ldots$$

Mensuration formulas

Right triangle

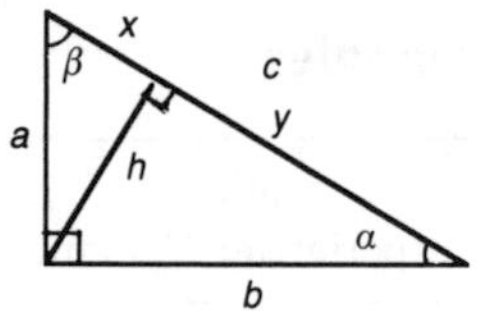

$c^2=a^2+b^2$ (*Pythagorean relation*)

$$A=\frac{ab}{2}\qquad h=\frac{ab}{c},\ x=\frac{a^2}{c},\ y=\frac{b^2}{c}$$

$$R=\frac{c}{2}\qquad r=\frac{a+b-c}{2}$$

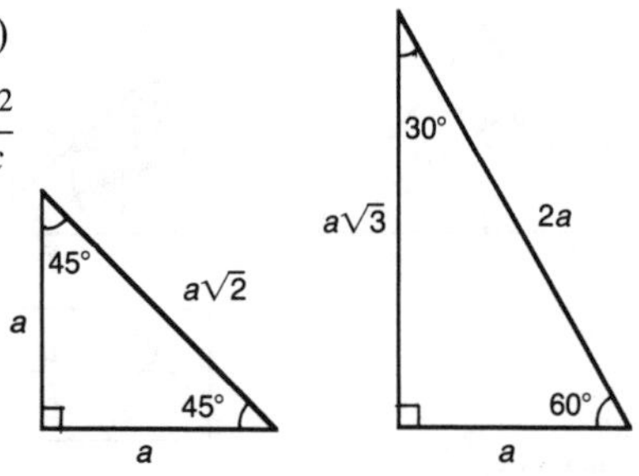

Equilateral triangle

$$A=\frac{a^2\sqrt{3}}{4}=\frac{h^2}{\sqrt{3}}$$

$$h=\frac{a\sqrt{3}}{2}\quad R=\frac{a}{\sqrt{3}}\quad r=\frac{a}{2\sqrt{3}}$$

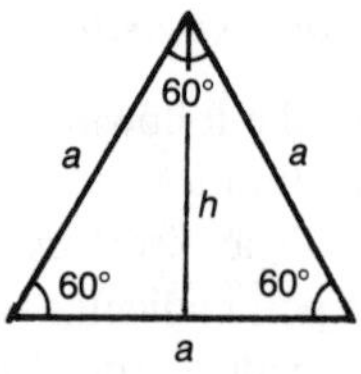

General triangle

$$A=\frac{ah}{2}=\frac{bc\sin\alpha}{2}=\sqrt{p(p-a)(p-b)(p-c)}$$
(*Heron's formula*)

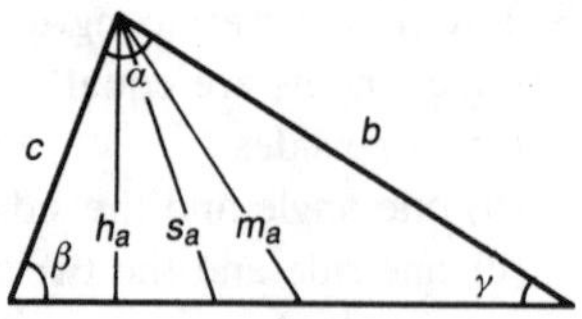

$$h_a=c\sin\beta=\frac{2\sqrt{p(p-a)(p-b)(p-c)}}{a}$$

$$m_a=\frac{1}{2}\sqrt{2b^2+2c^2-a^2}\qquad s_a=\sqrt{bc\left[1-\left(\frac{a}{b+c}\right)^2\right]}$$

$$R=\frac{abc}{4A}\qquad r=\frac{2A}{a+b+c}=\frac{A}{p}$$

Triangle trigonometry

Trigonometric formulas, see sec. 5.4.

1. Right triangle.

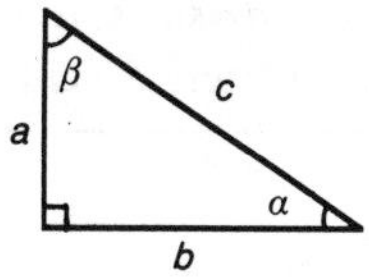

(1) $\sin\alpha=\frac{a}{c}$, $\cos\alpha=\frac{b}{c}$, $\tan\alpha=\frac{a}{b}$, $\cot\alpha=\frac{b}{a}$

2. General triangle.

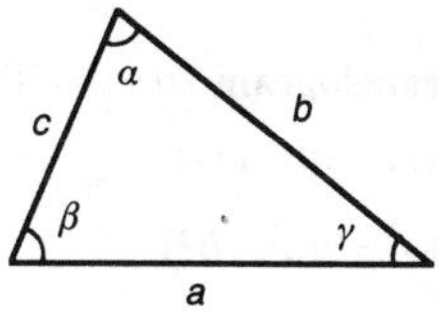

(2) $\frac{\sin\alpha}{a}=\frac{\sin\beta}{b}=\frac{\sin\gamma}{c}=\frac{1}{2R}$ (*law of sines*)

(3) $a^2=b^2+c^2-2bc\cos\alpha$ (*law of cosines*)

$$\frac{a+b}{a-b}=\frac{\tan\frac{\alpha+\beta}{2}}{\tan\frac{\alpha-\beta}{2}}$$ (*law of tangents*)

$$A=\frac{bc\sin\alpha}{2}$$ (*area formula*)

(4) $\alpha+\beta+\gamma=180°$ (*angle relation*)

Solution of plane triangles

1. Right triangle: Use (1).
2. General triangle:

	Given		Method: Find
1.	Three sides	a, b, c	α, β, γ from (3) and (4)
2.	Two sides and the included angle	b, c, α	a from (3); β (if $b<c$) from (2); γ from (4)
3.	Two sides and an opposite angle	b, c, β	γ from (2); α from (4); a from (2). (Possibly two solutions)
4.	One side and two angles	a, β, γ	α from (4); b, c from (2)

Quadrilaterals

> a, b, c, d=sides, $\alpha, \beta, \gamma, \delta$=angles, h=altitude, e, f=diagonals, $2p=a+b+c+d=$ =perimeter, R=circumradius, r=inradius, A=area

Square

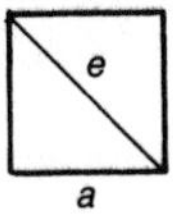

$A=a^2=\frac{e^2}{2}$ $\quad R=\frac{a}{\sqrt{2}}$

$e=a\sqrt{2}$ $\quad r=\frac{a}{2}$

Rectangle

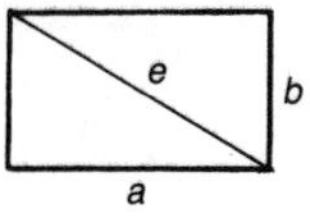

$A=ab$

$e=\sqrt{a^2+b^2}$ $\quad R=\frac{e}{2}$

Parallelogram ($\alpha+\beta=180°$)

$A=ah=ab\sin\alpha$ $\quad h=b\sin\alpha$

$e^2+f^2=2(a^2+b^2)$ $\quad$ (*parallelogram law*)

$e=\sqrt{a^2+b^2+2ab\cos\alpha}$ $\quad f=\sqrt{a^2+b^2-2ab\cos\alpha}$

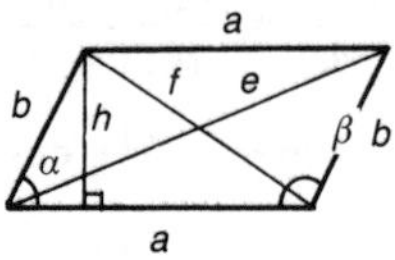

Rhombus

$A=ah=a^2\sin\alpha=\frac{1}{2}ef$ $\quad e^2+f^2=4a^2$

$e=2a\cos\frac{\alpha}{2}$ $\quad f=2a\sin\frac{\alpha}{2}$

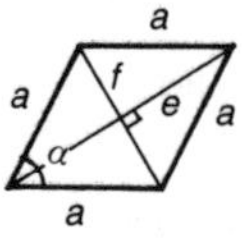

Trapezoid

$A=\frac{(a+c)h}{2}$ $\quad h=d\sin\alpha=b\sin\beta$

$e=\sqrt{a^2+b^2-2ab\cos\beta}$ $\quad f=\sqrt{a^2+d^2-2ad\cos\alpha}$

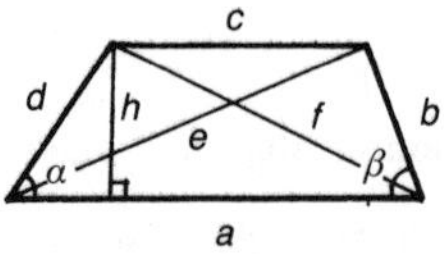

General quadrilateral

$\alpha+\beta+\gamma+\delta=360°$

$\theta=90° \Leftrightarrow a^2+c^2=b^2+d^2$

$A=\frac{1}{2}ef\sin\theta=\frac{1}{4}(b^2+d^2-a^2-c^2)\tan\theta=$

$=\frac{1}{4}\sqrt{4e^2f^2-(b^2+d^2-a^2-c^2)^2}$

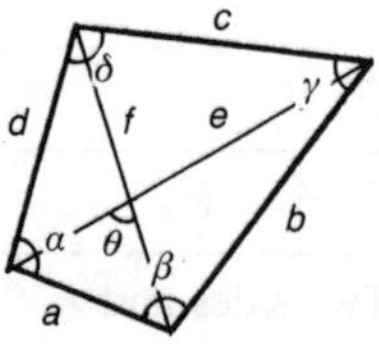

Tangent – quadrilateral

$a+c=b+d$ $\quad A=pr$

If $\alpha+\gamma=\beta+\delta$ then

$A=\sqrt{abcd}$

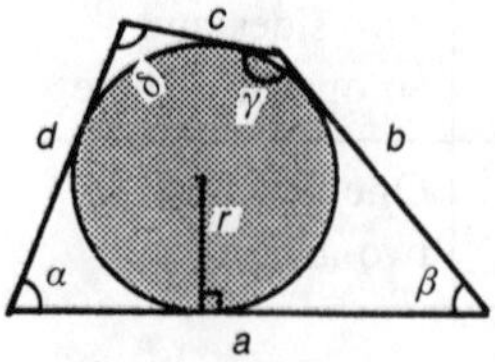

Secant – quadrilateral

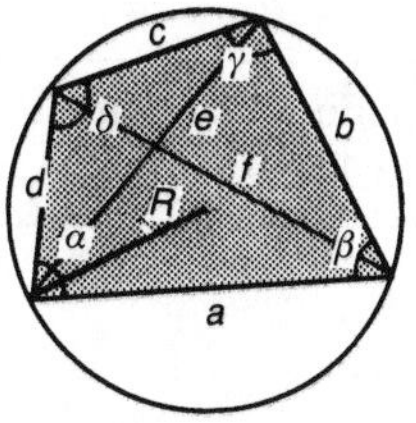

$$\alpha+\gamma=\beta+\delta=180°$$

$$A=\sqrt{(p-a)(p-b)(p-c)(p-d)}$$

$$R=\frac{1}{4}\sqrt{\frac{(ac+bd)(ad+bc)(ab+cd)}{(p-a)(p-b)(p-c)(p-d)}}$$

$$e=\sqrt{\frac{(ad+bc)(ac+bd)}{ab+cd}} \qquad ef=ac+bd$$

Polygons

General polygon (*n* corners):

Sum of angles $=(n-2)\cdot 180°$ Number of diagonals $=\frac{n(n-3)}{2}$

Regular polygon (*n* corners):

a=side, α=angle, R=circumradius, r=inradius, A=area

$$\alpha=\frac{n-2}{n}\cdot 180° \qquad A=\frac{1}{4}na^2\cot\frac{180°}{n}$$

$$r=\frac{a}{2}\cot\frac{180°}{n} \qquad R=\frac{a}{2\sin\frac{180°}{n}}$$

Circles

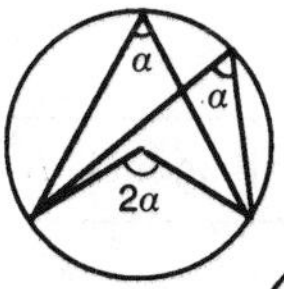

Theorems

1. (a) Inscribed angles on the same arc are equal.
   (b) An inscribed angle is equal to half the central angle having the same arc.

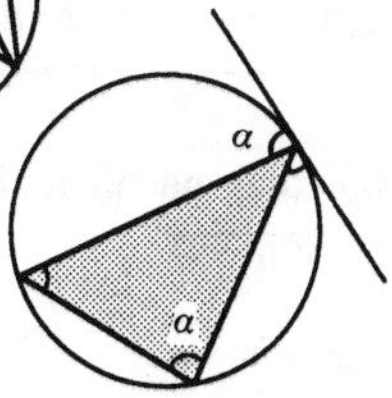

2. The angle between a tangent and a chord is equal to the angle in the opposite segment.
3. When two chords intersect the product of the two line segments making up each chord are equal.

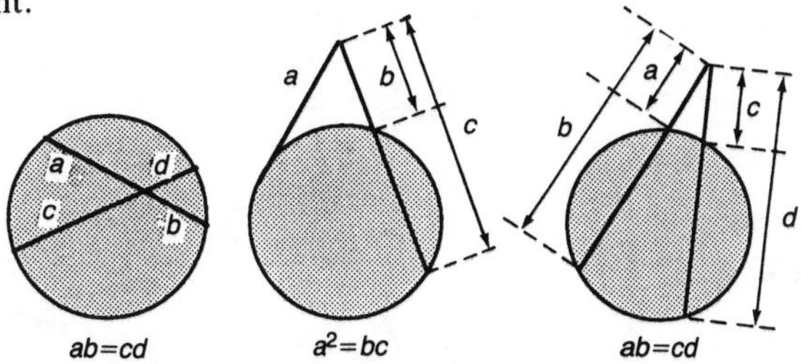

Mensuration formulas

r=radius, d=diameter, c=circumference, s=length of arc, α in radians, π=3,14159 ...

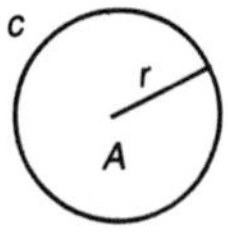

Circle

$c=2\pi r=\pi d$

$A=\pi r^2=\frac{\pi d^2}{4}$

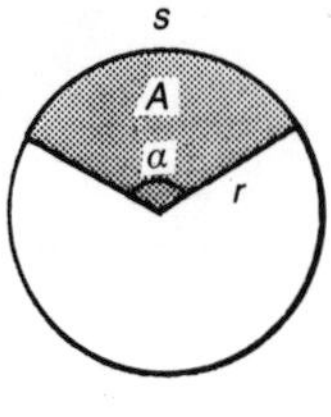

Sector

$s=\alpha r$

$A=\frac{sr}{2}=\frac{\alpha r^2}{2}$

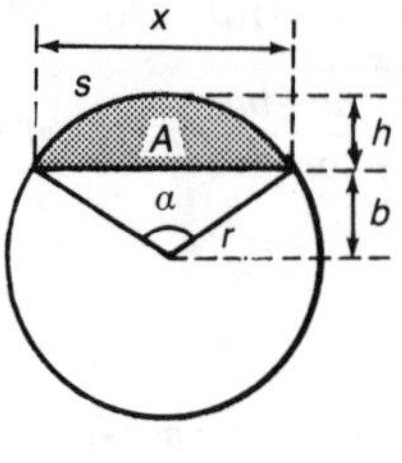

Segment

$x=2r\sin\frac{\alpha}{2}$

$h=r\left(1-\cos\frac{\alpha}{2}\right)$

$h(2r-h)=\left(\frac{x}{2}\right)^2$

$h\approx\frac{x^2}{8r}\ (h<<r)$

$A=\frac{r^2}{2}(\alpha-\sin\alpha)=\frac{1}{2}(rs-bx)$

3.2 Solids

Polyhedra

a, b, c=edges, d=diagonal, B=area of base, S=total surface area, V=volume

Rectangular parallelepiped (box)

$d=\sqrt{a^2+b^2+c^2}$

$S=2(ab+bc+ac)$ $\qquad V=abc$

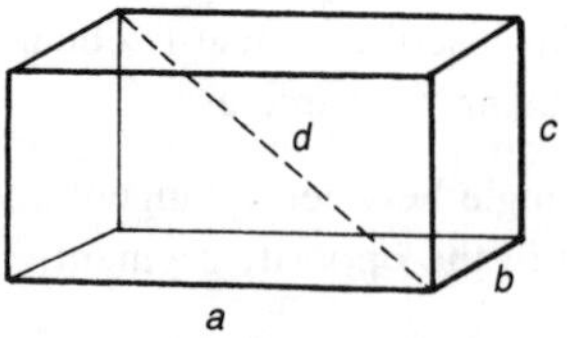

Prism

$V=Bh$

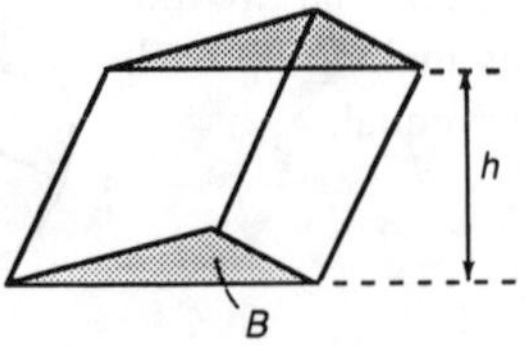

Pyramid

$$V=\frac{1}{3}Bh$$

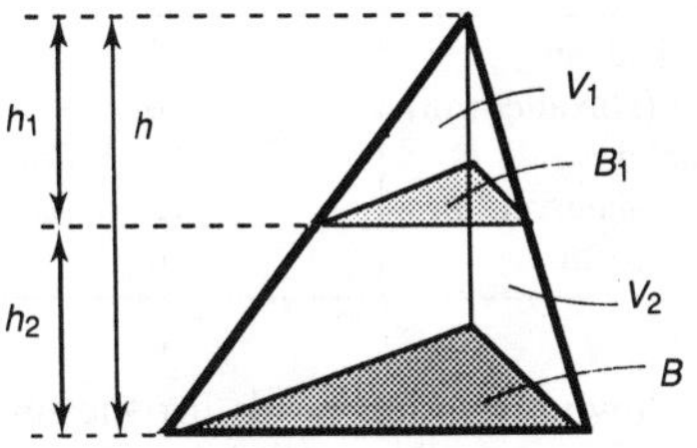

Frustum of pyramid

$$\frac{V_1}{V}=\left(\frac{B_1}{B}\right)^{3/2}=\left(\frac{h_1}{h}\right)^3$$

$$V_2=\frac{h_2}{3}\left(B+\sqrt{BB_1}+B_1\right)$$

Regular polyhedra

a=edge, V=volume, S=total surface area, R=circumradius, r=inradius

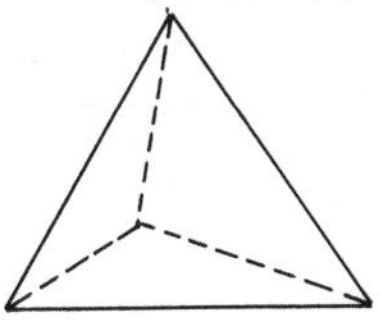

Tetrahedron

$V=\frac{a^3\sqrt{2}}{12}$ $S=a^2\sqrt{3}$

$R=\frac{a\sqrt{6}}{4}$ $r=\frac{a\sqrt{6}}{12}$

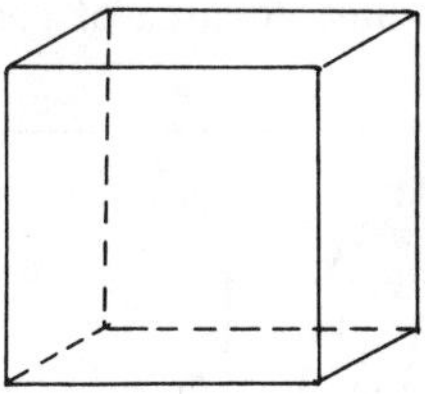

Cube

$V=a^3$ $S=6a^2$

$R=\frac{a\sqrt{3}}{2}$ $r=\frac{a}{2}$

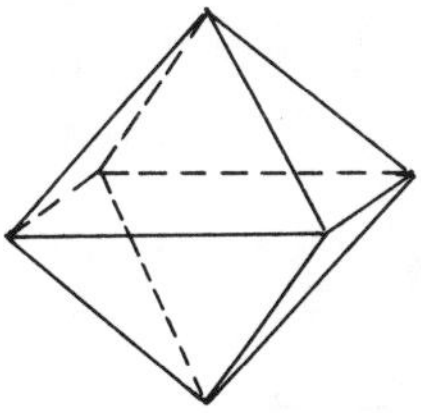

Octahedron

$V=\frac{a^3\sqrt{2}}{3}$ $S=2a^2\sqrt{3}$

$R=\frac{a}{\sqrt{2}}$ $r=\frac{a}{\sqrt{6}}$

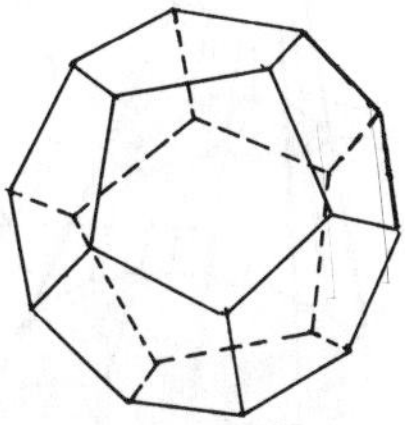

Dodecahedron

$$V=\frac{a^3(15+7\sqrt{5})}{4}$$

$$S=3a^2\sqrt{5(5+2\sqrt{5})}$$

$$R=\frac{a(1+\sqrt{5})\sqrt{3}}{4}$$

$$r=\frac{a}{4}\sqrt{\frac{50+22\sqrt{5}}{5}}$$

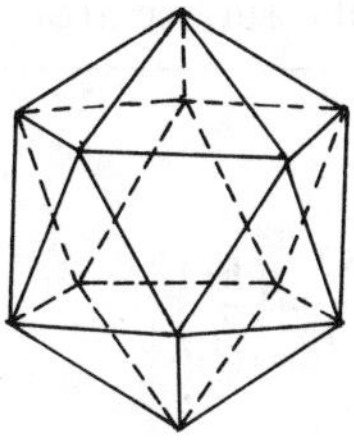

Icosahedron

$$V=\frac{5a^3}{12}(3+\sqrt{5})$$

$$S=5a^2\sqrt{3}$$

$$R=\frac{a}{4}\sqrt{2(5+\sqrt{5})}$$

$$r=\frac{a}{2}\sqrt{\frac{7+3\sqrt{5}}{6}}$$

Polyhedron	**Number of faces** F	**Number of edges** E	**Number of vertices** V
Tetrahedron	4	6	4
Cube (Hexahedron)	6	12	8
Octahedron	8	12	6
Dodecahedron	12	30	20
Icosahedron	20	30	12

The following relation holds for the numbers F, E and V.

$$F-E+V=2 \qquad \text{(The Euler relation)}$$

This relation also holds for nonregular polyhedra.

Cylinders

h=altitude, r=radius, B=area of base, A=lateral area, S=total surface area, V=volume

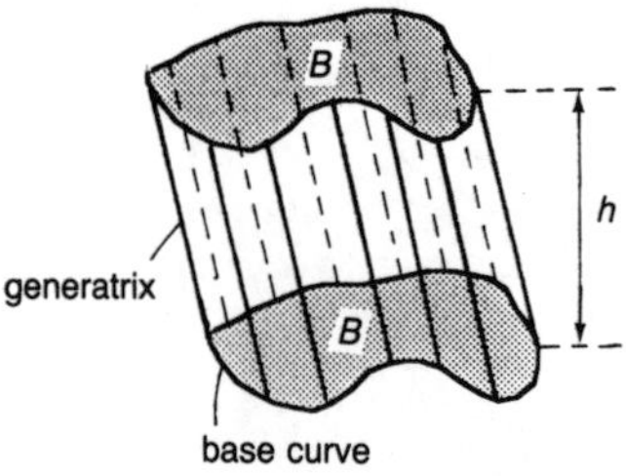

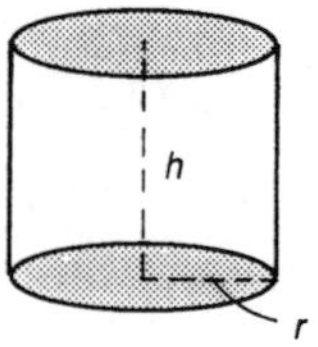

General cylinder

$V=Bh$

Right circular cylinder

$B=\pi r^2 \qquad A=2\pi rh,$
$S=2\pi r(r+h) \qquad V=\pi r^2 h$

Cones

h=altitude, r=radius, s=slant height, B=area of base, A=lateral area, S=total surface area, V=volume

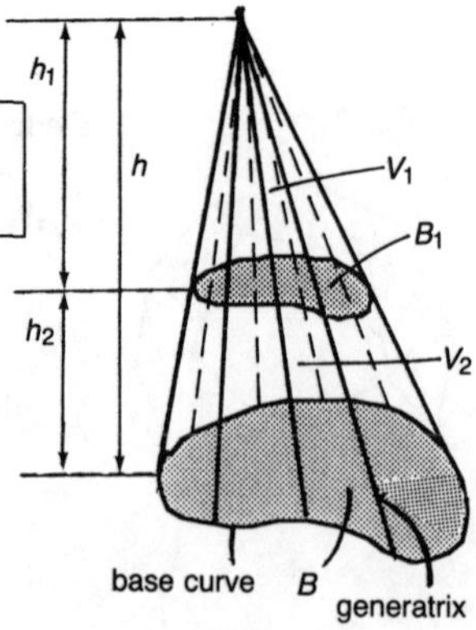

General cone

$$V=\frac{1}{3}Bh \qquad \frac{V_1}{V}=\left(\frac{B_1}{B}\right)^{3/2}=\left(\frac{h_1}{h}\right)^3$$

$$V_2=\frac{h_2}{3}(B+\sqrt{BB_1}+B_1)$$

Right circular cone

$s=\sqrt{r^2+h^2}$

$A=\pi rs$

$S=\pi r(r+s)$

$V=\frac{1}{3}\pi r^2 h$

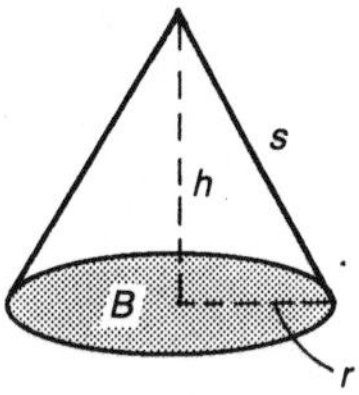

Frustum of right cone

$s=\sqrt{(r_1-r_2)^2+h^2}$

$A=\pi(r_1+r_2)s$

$S=\pi[r_1{}^2+(r_1+r_2)s+r_2{}^2]$

$V=\frac{\pi h}{3}(r_1{}^2+r_1r_2+r_2{}^2)$

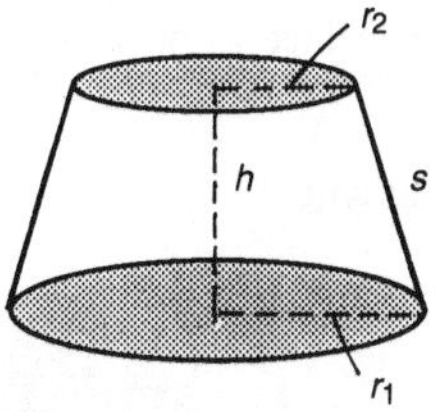

Spheres

r=radius, S=surface area, V=volume

Sphere

$S=4\pi r^2 \qquad V=\frac{4}{3}\pi r^3$

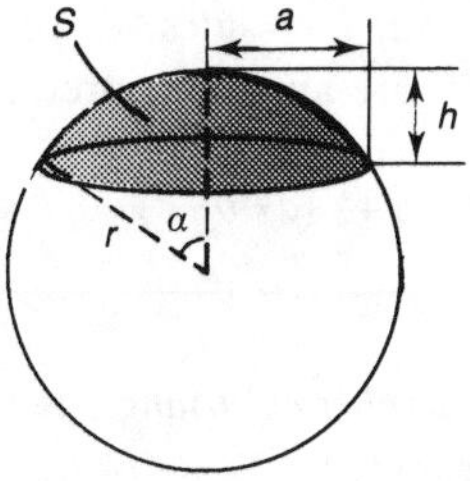

Segment (one base)

$a=r\sin\alpha \qquad h(2r-h)=a^2$

$h=r(1-\cos\alpha) \qquad h\approx\frac{a^2}{2r}\ (h<<r)$

$S=2\pi rh \qquad V=\frac{\pi}{3}h^2(3r-h)=\frac{\pi}{6}h(3a^2+h^2)$

Segment (two bases)

$S=2\pi rh$

$V=\frac{\pi}{6}h(3a^2+3b^2+h^2)$

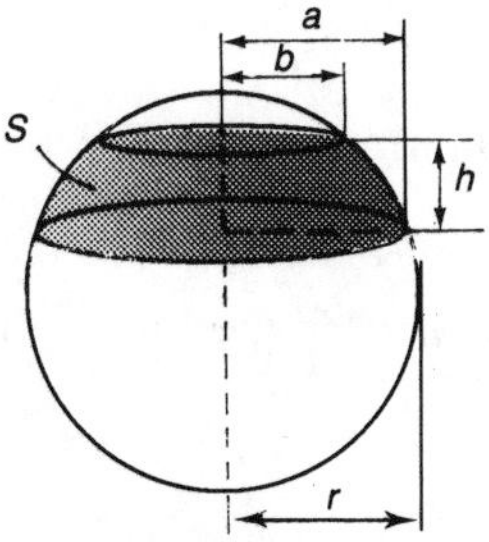

Spherical sector

$V=\frac{2\pi r^2 h}{3}$

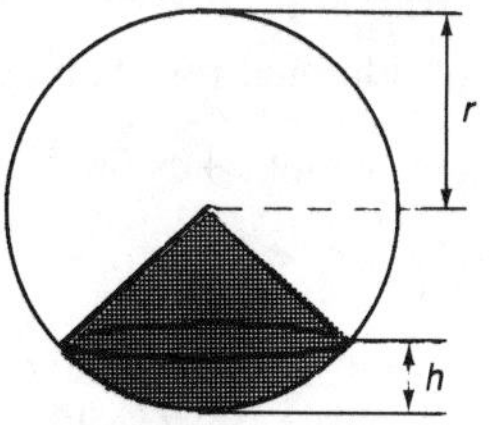

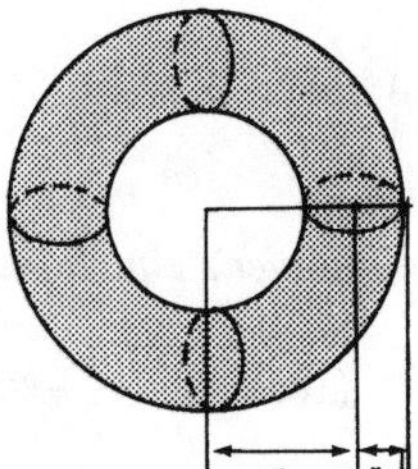

Circular torus

$S=4\pi^2 r_1 r_2$

$V=2\pi^2 r_1 r_2{}^2$

Solid angle (in radians)

ω=surface area on the unit sphere. ($\omega \leq 4\pi$)

Segment

$\omega = 4\pi \sin^2 \frac{\alpha}{2}$

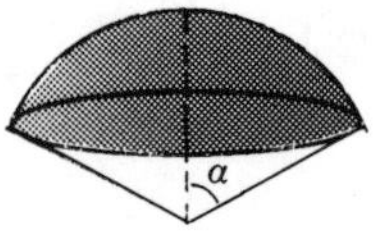

Spherical triangle

$\omega = \alpha + \beta + \gamma - \pi$

$$\tan \frac{\omega}{2} = \frac{|\boldsymbol{a} \cdot (\boldsymbol{b} \times \boldsymbol{c})|}{1 + \boldsymbol{a} \cdot \boldsymbol{b} + \boldsymbol{b} \cdot \boldsymbol{c} + \boldsymbol{c} \cdot \boldsymbol{a}}$$

(*Euler – Eriksson's* formula)

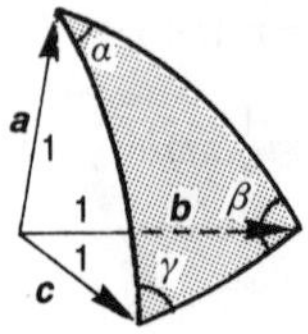

3.3 Spherical Trigonometry

Spherical triangles

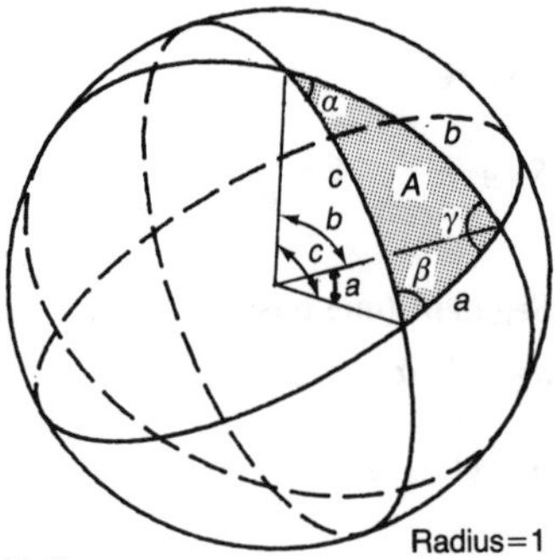

a, b,c=sides, α, β, γ=angles. All six elements are anglemeasured and less than 180°. $s = \frac{1}{2}(a+b+c)$ $\sigma = \frac{1}{2}(\alpha+\beta+\gamma)$

A *spherical triangle* is bounded by three great circles on a sphere.

General properties

(Additional results by cyclic permutation)

(1) $0° < a+b+c < 360°$ (2) $180° < \alpha+\beta+\gamma < 540°$

(3) $\alpha < \beta < \gamma \Leftrightarrow a < b < c$ (4) $a+b > c$ (5) $\alpha+\beta > \gamma + 180°$

(6) $\frac{\sin \alpha}{\sin a} = \frac{\sin \beta}{\sin b} = \frac{\sin \gamma}{\sin c}$ (*law of sines*)

(7) $\cos a = \cos b \cos c + \sin b \sin c \cos \alpha$

(8) $\cos \alpha = -\cos \beta \cos \gamma + \sin \beta \sin \gamma \cos a$

(*laws of cosines*)

Delambre's equations

(9a) $\sin \frac{\alpha}{2} \sin \frac{b+c}{2} = \sin \frac{a}{2} \cos \frac{\beta - \gamma}{2}$

Napier's equations

(10a) $\tan \frac{b+c}{2} \cos \frac{\beta+\gamma}{2} = \tan \frac{a}{2} \cos \frac{\beta - \gamma}{2}$

(9b) $\sin\frac{\alpha}{2}\cos\frac{b+c}{2}=\cos\frac{a}{2}\cos\frac{\beta+\gamma}{2}$

(10b) $\tan\frac{b-c}{2}\sin\frac{\beta+\gamma}{2}=\tan\frac{a}{2}\sin\frac{\beta-\gamma}{2}$

(9c) $\cos\frac{\alpha}{2}\sin\frac{b-c}{2}=\sin\frac{a}{2}\sin\frac{\beta-\gamma}{2}$

(10c) $\tan\frac{\beta+\gamma}{2}\cos\frac{b+c}{2}=\cot\frac{\alpha}{2}\cos\frac{b-c}{2}$

(9d) $\cos\frac{\alpha}{2}\cos\frac{b-c}{2}=\cos\frac{a}{2}\sin\frac{\beta+\gamma}{2}$

(10d) $\tan\frac{\beta-\gamma}{2}\sin\frac{b+c}{2}=\cot\frac{\alpha}{2}\sin\frac{b-c}{2}$

Half-angle formulas

(11a) $\sin^2\frac{\alpha}{2}=\frac{\sin(s-b)\sin(s-c)}{\sin b\sin c}$

(11b) $\cos^2\frac{\alpha}{2}=\frac{\sin s\sin(s-a)}{\sin b\sin c}$

(11c) $\sin^2\frac{a}{2}=-\frac{\cos\sigma\cos(\sigma-\alpha)}{\sin\beta\sin\gamma}$

(11d) $\cos^2\frac{a}{2}=\frac{\cos(\sigma-\beta)\cos(\sigma-\gamma)}{\sin\beta\sin\gamma}$

Area of spherical triangle

Excess $E=\alpha+\beta+\gamma-180°$,

$$\tan\frac{E}{4}=\left(\tan\frac{s}{2}\tan\frac{s-a}{2}\tan\frac{s-b}{2}\tan\frac{s-c}{2}\right)^{1/2}$$

Area $A=\pi R^2E/180$

Solution of spherical triangles

Note conditions (1)–(5) in determining the solution.

Given			Method: Find
1.	Three sides	a, b, c	α, β, γ from (7) or (11)
2.	Three angles	α, β, γ	a, b, c from (8) or (11)
3.	Two sides and the included angle	b, c, α	$\frac{\beta+\gamma}{2}$ and $\frac{\beta-\gamma}{2}$ from (10), hence β and γ; a from (8) or (11).
4.	Two angles and the included side	β, γ, a	$\frac{b+c}{2}$ and $\frac{b-c}{2}$ from (10), hence b and c; α from (7) or (11).
5.	Two sides and an opposite angle	b, c, β	γ from (6); α and a from (10) (Possibly two solutions)
6.	Two angles and an opposite side	β, γ, b	c from (6); α and a from (10) (Possibly two solutions)

Napier's rules for right spherical triangles

Let $\gamma=90°$

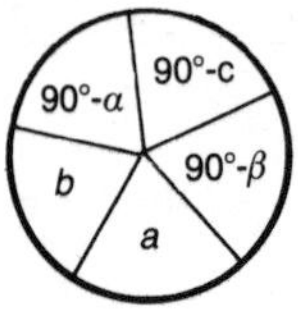

The sine of any angle is equal to

1. the product of the tangents of the two angles adjacent to it in the diagram.
2. the product of the cosines of the two angles opposite to it in the diagram.

3.4 Geometrical Vectors

Vectors

A (geometrical) *vector* in the plane or in space is characterized by *direction* and *length*. Notation:

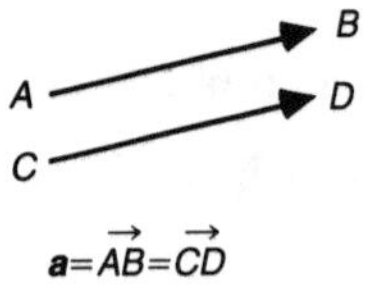

$\boldsymbol{a}, \boldsymbol{b}, \ldots, \overrightarrow{AB}, \overrightarrow{CD}, \ldots$

$\boldsymbol{a}=\boldsymbol{b} \Leftrightarrow \boldsymbol{a}$ and $\boldsymbol{b}$ coincide by a *parallel translation.*

$|\boldsymbol{a}|=|\overrightarrow{AB}|=$*length* of $\boldsymbol{a}$

Orthonormal basis

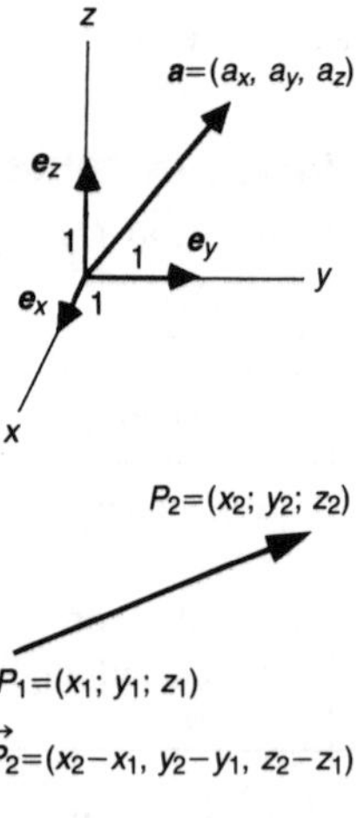

Let $\boldsymbol{e}_x$, $\boldsymbol{e}_y$ (and $\boldsymbol{e}_z$) be an *orthonormal basis*

[i.e. $\boldsymbol{e}_x \perp \boldsymbol{e}_y$, $\boldsymbol{e}_y \perp \boldsymbol{e}_z, \boldsymbol{e}_z \perp \boldsymbol{e}_x$, $|\boldsymbol{e}_x|=|\boldsymbol{e}_y|=|\boldsymbol{e}_z|=1$]

in the plane and in space (*right-hand system*) respectively (ONR-system).

$$\boldsymbol{a}=a_x\boldsymbol{e}_x+a_y\boldsymbol{e}_y+a_z\boldsymbol{e}_z=(a_x, a_y, a_z)$$

$$\boldsymbol{0}=(0, 0, 0),\ \boldsymbol{e}_x=(1, 0, 0),\ \boldsymbol{e}_y=(0, 1, 0),\ \boldsymbol{e}_z=(0, 0, 1)$$

$$|\boldsymbol{a}|=\sqrt{a_x^2+a_y^2+a_z^2}$$

$P_1=(x_1; y_1; z_1)$ and $P_2=(x_2; y_2; z_2)$ points $\Rightarrow$

$$\overrightarrow{P_1P_2}=(x_2-x_1, y_2-y_1, z_2-z_1)$$

Vector algebra

$\boldsymbol{a}=(a_x, a_y, a_z)$, $\boldsymbol{b}=(b_x, b_y, b_z)$, $\boldsymbol{c}=(c_x, c_y, c_z)$
s, t numbers, ONR-system

Addition and subtraction

$\boldsymbol{a}+\boldsymbol{b}=(a_x+b_x,\ a_y+b_y,\ a_z+b_z)$

$\boldsymbol{a}-\boldsymbol{b}=(a_x-b_x,\ a_y-b_y,\ a_z-b_z)$

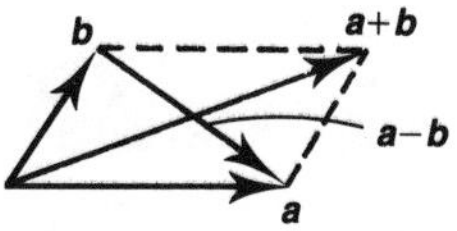

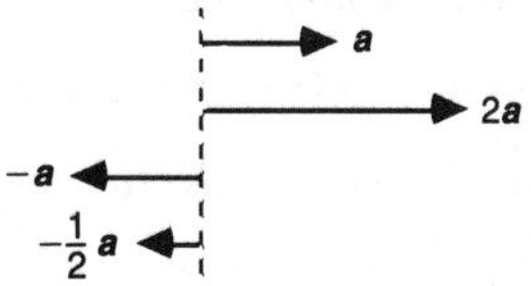

Multiplication by number

$t\boldsymbol{a}=(ta_x,\ ta_y,\ ta_z)$

$\boldsymbol{a}+\boldsymbol{b}=\boldsymbol{b}+\boldsymbol{a}$	$(\boldsymbol{a}+\boldsymbol{b})+\boldsymbol{c}=\boldsymbol{a}+(\boldsymbol{b}+\boldsymbol{c})$	$s(t\boldsymbol{a})=(st)\boldsymbol{a}$
$t(\boldsymbol{a}+\boldsymbol{b})=t\boldsymbol{a}+t\boldsymbol{b}$	$(s+t)\boldsymbol{a}=s\boldsymbol{a}+t\boldsymbol{a}$	

Scalar (dot) product

The number $\boldsymbol{a}\cdot\boldsymbol{b}$ is defined by $\boldsymbol{a}\cdot\boldsymbol{b}=|\boldsymbol{a}|\cdot|\boldsymbol{b}|\cos\theta=a_xb_x+a_yb_y+a_zb_z$

$\boldsymbol{a}\cdot\boldsymbol{b}=\boldsymbol{b}\cdot\boldsymbol{a}$	$(s\boldsymbol{a})\cdot(t\boldsymbol{b})=(st)(\boldsymbol{a}\cdot\boldsymbol{b})$
$\boldsymbol{a}\cdot(\boldsymbol{b}+\boldsymbol{c})=\boldsymbol{a}\cdot\boldsymbol{b}+\boldsymbol{a}\cdot\boldsymbol{c}$	
$\boldsymbol{a}\cdot\boldsymbol{a}=|\boldsymbol{a}|^2$	$\boldsymbol{a}\cdot\boldsymbol{b}=0\Leftrightarrow\boldsymbol{a}\perp\boldsymbol{b}$

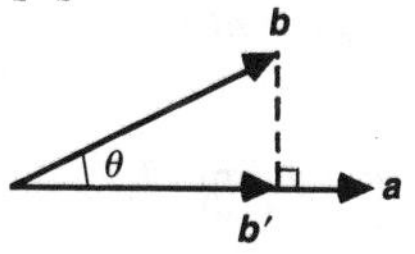

$\boldsymbol{a}\cdot\boldsymbol{b}=|\boldsymbol{a}|\cdot|\boldsymbol{b}'|$

Vector product

The vector $\boldsymbol{a}\times\boldsymbol{b}$ is defined by

(*i*) $|\boldsymbol{a}\times\boldsymbol{b}|=|\boldsymbol{a}|\cdot|\boldsymbol{b}|\cdot\sin\theta$

(*ii*) $\boldsymbol{a}\times\boldsymbol{b}$ is orthogonal to $\boldsymbol{a}$ and $\boldsymbol{b}$

(*iii*) $\boldsymbol{a}$, $\boldsymbol{b}$, $\boldsymbol{a}\times\boldsymbol{b}$ form a right-hand system

$$\boldsymbol{a}\times\boldsymbol{b}=(a_yb_z-a_zb_y,\ a_zb_x-a_xb_z,\ a_xb_y-a_yb_x)=\begin{vmatrix} \boldsymbol{e}_x & \boldsymbol{e}_y & \boldsymbol{e}_z \\ a_x & a_y & a_z \\ b_x & b_y & b_z \end{vmatrix}$$

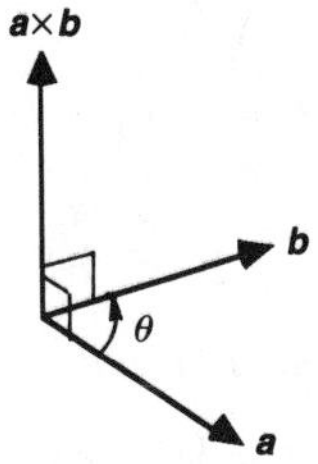

$\boldsymbol{a}\times\boldsymbol{b}=-\boldsymbol{b}\times\boldsymbol{a}$	$(\boldsymbol{a}\times\boldsymbol{b})\times\boldsymbol{c}\neq\boldsymbol{a}\times(\boldsymbol{b}\times\boldsymbol{c})$ in general		
$(s\boldsymbol{a})\times(t\boldsymbol{b})=(st)(\boldsymbol{a}\times\boldsymbol{b})$	$\boldsymbol{a}\times(\boldsymbol{b}+\boldsymbol{c})=\boldsymbol{a}\times\boldsymbol{b}+\boldsymbol{a}\times\boldsymbol{c}$		
$\boldsymbol{a}\times\boldsymbol{a}=\boldsymbol{0}$	$\boldsymbol{e}_x\times\boldsymbol{e}_y=\boldsymbol{e}_z$	$\boldsymbol{e}_y\times\boldsymbol{e}_z=\boldsymbol{e}_x$	$\boldsymbol{e}_z\times\boldsymbol{e}_x=\boldsymbol{e}_y$

Scalar triple product

$$[\boldsymbol{a},\boldsymbol{b},\boldsymbol{c}]=\boldsymbol{a}\cdot(\boldsymbol{b}\times\boldsymbol{c})=\begin{vmatrix} a_x & a_y & a_z \\ b_x & b_y & b_z \\ c_x & c_y & c_z \end{vmatrix}$$

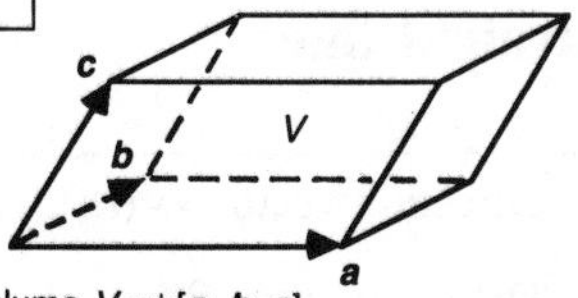

volume $V=\pm[\boldsymbol{a},\boldsymbol{b},\boldsymbol{c}]$

$[\boldsymbol{a},\boldsymbol{b},\boldsymbol{c}]=[\boldsymbol{b},\boldsymbol{c},\boldsymbol{a}]=[\boldsymbol{c},\boldsymbol{a},\boldsymbol{b}]=-[\boldsymbol{c},\boldsymbol{b},\boldsymbol{a}]=-[\boldsymbol{b},\boldsymbol{a},\boldsymbol{c}]=-[\boldsymbol{a},\boldsymbol{c},\boldsymbol{b}]$
$[\boldsymbol{a},\boldsymbol{b},\boldsymbol{c}]=0\Leftrightarrow\boldsymbol{a},\boldsymbol{b},\boldsymbol{c}$ are coplanar ($>0\Leftrightarrow\boldsymbol{a},\boldsymbol{b},\boldsymbol{c}$ are right-handed)

Vector triple product

$\boldsymbol{a}\times(\boldsymbol{b}\times\boldsymbol{c})=(\boldsymbol{a}\cdot\boldsymbol{c})\boldsymbol{b}-(\boldsymbol{a}\cdot\boldsymbol{b})\boldsymbol{c}$

$(\boldsymbol{a}\times\boldsymbol{b})\times\boldsymbol{c}=(\boldsymbol{a}\cdot\boldsymbol{c})\boldsymbol{b}-(\boldsymbol{b}\cdot\boldsymbol{c})\boldsymbol{a}$

3.5 Plane Analytic Geometry

$P=(x;\ y)$, $P_1=(x_1;\ y_1)$, ... points, $\boldsymbol{a}=(a_x,\ a_y)$ ... vectors

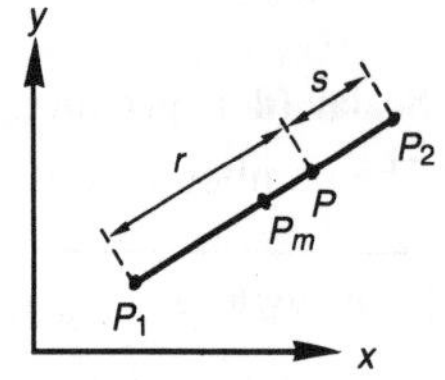

1. *Distance* between P_1 and $P_2=\sqrt{(x_1-x_2)^2+(y_1-y_2)^2}$

2. *Midpoint* $P_m=\left(\dfrac{x_1+x_2}{2};\ \dfrac{y_1+y_2}{2}\right)$

3. P dividing P_1P_2 into *ratio r/s*: $P=\left(\dfrac{rx_2+sx_1}{r+s};\ \dfrac{ry_2+sy_1}{r+s}\right)$

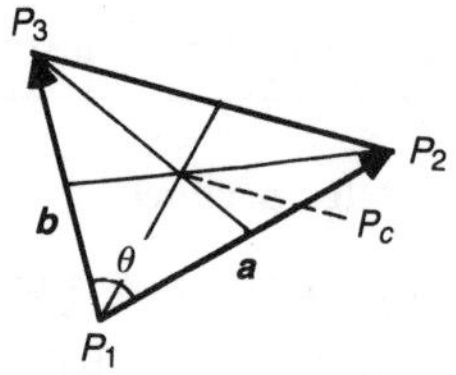

4. *Centroid* of triangle i.e. intersection of medians $P_c=$

$$=\left(\frac{x_1+x_2+x_3}{3};\ \frac{y_1+y_2+y_3}{3}\right)$$

5. *Area* of *triangle*=

$$=\pm\frac{1}{2}\begin{vmatrix} a_x & a_y \\ b_x & b_y \end{vmatrix}=\pm\frac{1}{2}\begin{vmatrix} x_2-x_1 & y_2-y_1 \\ x_3-x_1 & y_3-y_1 \end{vmatrix}$$

6. *Area* of *polygon* $P_1P_2\ \dots\ P_n=$

$$=\pm\frac{1}{2}\,(x_1y_2+x_2y_3+\ldots+x_{n-1}y_n+x_ny_1-x_2y_1-x_3y_2-\ldots-x_ny_{n-1}-x_1y_n)$$

7. *Angle* θ between vectors: $\cos\theta=\dfrac{\boldsymbol{a}\cdot\boldsymbol{b}}{|\boldsymbol{a}|\cdot|\boldsymbol{b}|}=\dfrac{a_xb_x+a_yb_y}{\sqrt{a_x^2+a_y^2}\ \sqrt{b_x^2+b_y^2}}$

Straight lines

Direction vector $\boldsymbol{v}=(\alpha,\ \beta)$
Direction angle θ
Normal vector $\boldsymbol{n}=(A,\ B)\,//\,(-\beta,\ \alpha)$
Slope $k=\tan\theta=\dfrac{\beta}{\alpha}=-\dfrac{A}{B}=\dfrac{y_2-y_1}{x_2-x_1}$

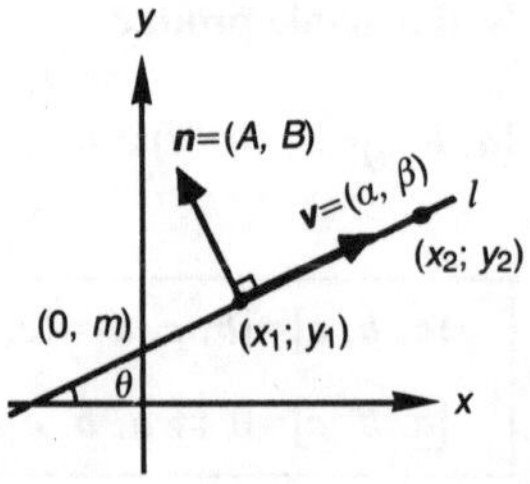

Equation forms

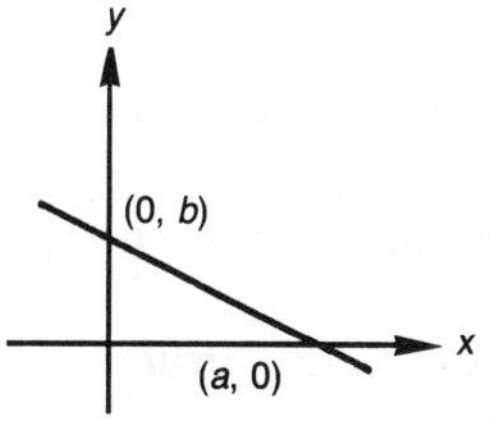

8. General form: $Ax+By+C=0$, $\boldsymbol{n}=(A, B)$, $k=-\dfrac{A}{B}$
9. Point-slope form: $y-y_1=k(x-x_1)$
10. Slope y-intercept form: $y=kx+m$
11. *Intercept* form: $\dfrac{x}{a}+\dfrac{y}{b}=1$
12. *Normal* form: $\dfrac{Ax+By+C}{\sqrt{A^2+B^2}}=0$

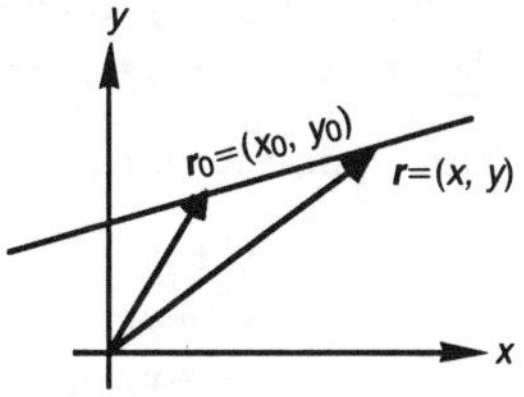

13. *Parametric* form: $\boldsymbol{r}=\boldsymbol{r}_0+t\boldsymbol{v} \Leftrightarrow \begin{cases} x=x_0+\alpha t \\ y=y_0+\beta t \end{cases}$
14. Angle θ between lines of slopes k_1 and k_2: $\tan\theta=\pm\dfrac{k_1-k_2}{1+k_1k_2}$
15. Two lines of slopes k_1 and k_2 are perpendicular $\Leftrightarrow k_1k_2=-1$.
16. Distance d from P_1 to $Ax+By+C=0$: $\quad d=\pm\dfrac{Ax_1+By_1+C}{\sqrt{A^2+B^2}}$

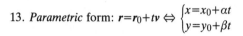

Second degree curves

General form

(3.1) $Ax^2+2Bxy+Cy^2+2Dx+2Ey+F=0$ (not all A, B, $C=0$)

Classification

(a) $AC-B^2>0$: *Ellipse case.* Possible geometrical meanings: ellipse, circle, one point, nothing.

(b) $AC-B^2=0$: *Parabola case.* Possible geometrical meanings: parabola, two parallel lines, one (double) line.

(c) $AC-B^2<0$: *Hyperbola case.* Possible geometrical meanings: hyperbola, two intersecting lines.

Analysis of (3.1):

(a) $B=0$: Completing the squares,

$$A\left(x+\frac{D}{A}\right)^2+C\left(y+\frac{E}{C}\right)^2=\frac{D^2}{A}+\frac{E^2}{C}-F \qquad \text{Center at}\left(-\frac{D}{A};-\frac{E}{C}\right).$$

Compare to the equations below.

(b) $B\neq0$: Spectral methods, see sec. 4.6.

Circle

a. Center at origin, radius R: $x^2+y^2=R^2$

Parameter form: $\begin{cases} x=R\cos t \\ y=R\sin t \end{cases}$, $0\leq t\leq 2\pi$

Area$=\pi R^2$, circumference$=2\pi R$.

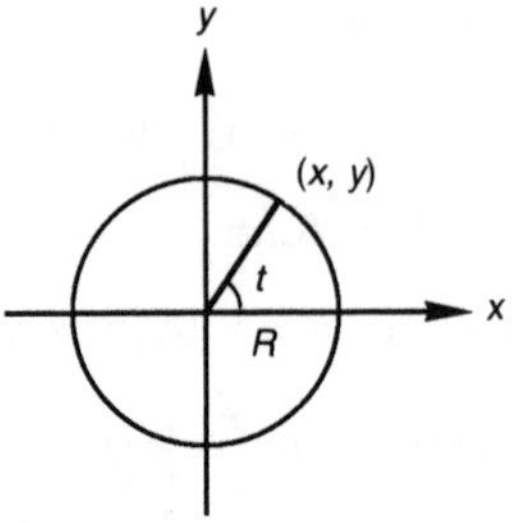

b. Center at (x_0, y_0), radius R: $(x-x_0)^2+(y-y_0)^2=R^2$
Parameter form $x=x_0+R\cos t$, $y=y_0+R\sin t$

c. Three point formula

$$\begin{vmatrix} x^2+y^2 & x & y & 1 \\ x_1^2+y_1^2 & x_1 & y_1 & 1 \\ x_2^2+y_2^2 & x_2 & y_2 & 1 \\ x_3^2+y_3^2 & x_3 & y_3 & 1 \end{vmatrix}=0$$

Ellipse

2a=major axis, 2b=minor axis

a. Center at origin, major axis along the x-axis:

$$\frac{x^2}{a^2}+\frac{y^2}{b^2}=1$$

Parameter form: $\begin{cases} x=a\cos t \\ y=b\sin t \end{cases}$, $0\leq t\leq 2\pi$

Foci at $(\pm c, 0)$ where $c=\sqrt{a^2-b^2}$

Excentricity $e=c/a$ $\quad(0\leq e<1)$

Area$=\pi ab$

Circumference$=4a\ E(k)$ with $k=\sqrt{a^2-b^2}/a$.

(See elliptic integrals sec. 12.5)

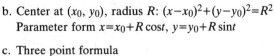

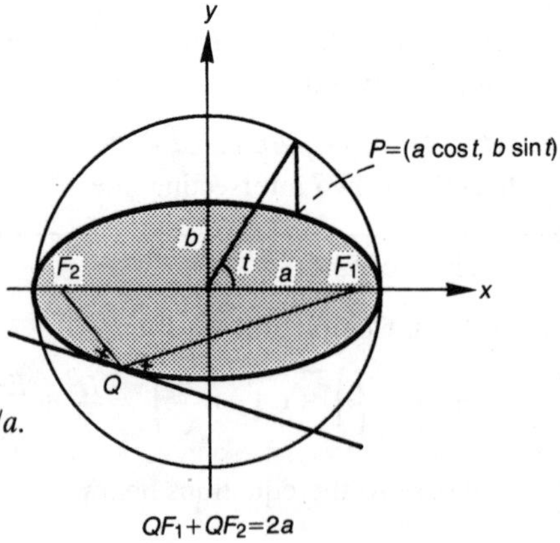

b. Center at (x_0, y_0): $\dfrac{(x-x_0)^2}{a^2}+\dfrac{(y-y_0)^2}{b^2}=1$

Parameter form: $x=x_0+a\cos t$, $y=y_0+b\sin t$.

Parabola

Vertex at the origin

Focus at $(p, 0)$

Directrix $x=-p$

Excentricity $e=1$

Equation: $y^2=4px$

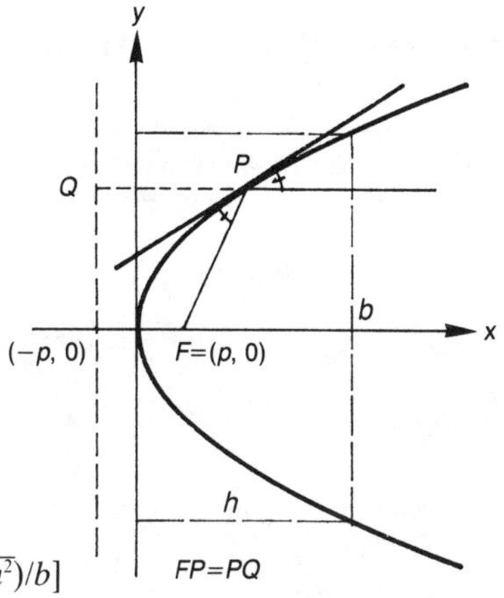

Segment
Area$=2bh/3$
Arc length$=\sqrt{b^2+16h^2}/2+(b^2/8h)\ln[4h+\sqrt{b^2+16h^2})/b]$

Rotation about x-axis
Volume$=\pi b^2h/8$
Area$=\pi b[(b^2+16h^2)^{3/2}-b^3]/96h^2$

Hyperbola

2a=transverse axis, 2b=conjugate axis

Center at the origin, transverse axis along the x-axis:

$$\frac{x^2}{a^2}-\frac{y^2}{b^2}=1$$

Parameter form of the right branch: $\begin{cases} x=a\cosh t \\ y=b\sinh t \end{cases}$ $\quad -\infty<t<\infty$

Foci at $(\pm c, 0)$ where $c=\sqrt{a^2+b^2}$

Excentricity $e=c/a \quad (e>1)$

Asymptotes $y=\pm bx/a$

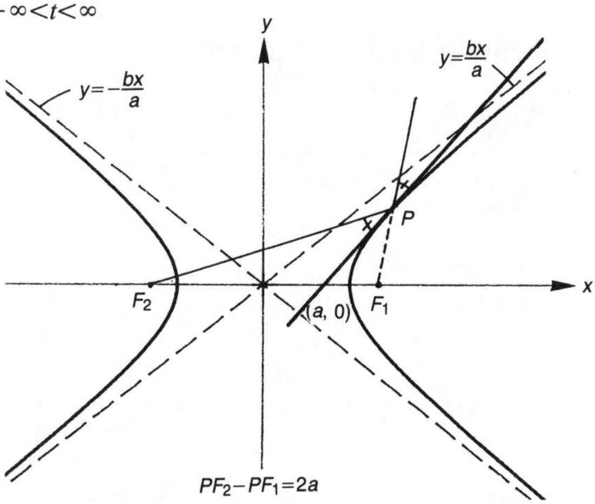

3.6 Analytic Geometry in Space

$P=(x; y; z)$, $P_1=(x_1; y_1; z_1)$ … points, $\boldsymbol{a}=(a_x, a_y, a_z)$ … vectors

1. *Distance* between P_1 and $P_2=$

$$=\sqrt{(x_1-x_2)^2+(y_1-y_2)^2+(z_1-z_2)^2}$$

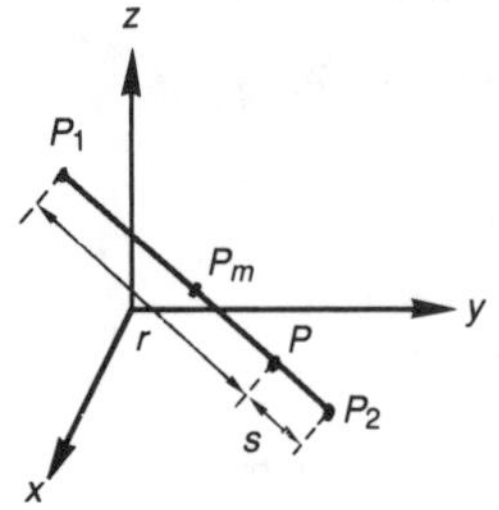

2. *Midpoint* $P_m=\left(\frac{x_1+x_2}{2}; \frac{y_1+y_2}{2}; \frac{z_1+z_2}{2}\right)$

3. P dividing P_1P_2 in *ratio* r/s:

$$P=\left(\frac{rx_2+sx_1}{r+s}; \frac{ry_2+sy_1}{r+s}; \frac{rz_2+sz_1}{r+s}\right)$$

4. *Centroid* of tetrahedron

$$P_c=\left(\frac{x_0+x_1+x_2+x_3}{4};\ldots;\ldots\right)$$

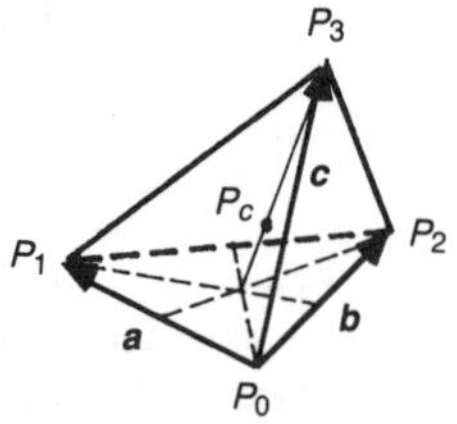

5. *Area* of *triangle* $P_1P_2P_3=\frac{1}{2}\left|\overrightarrow{P_1P_2}\times\overrightarrow{P_1P_3}\right|$

6. *Volume* of *tetrahedron* $=\pm\frac{1}{6}[\boldsymbol{a}, \boldsymbol{b}, \boldsymbol{c}]=\pm\frac{1}{6}\begin{vmatrix} a_x & a_y & a_z \\ b_x & b_y & b_z \\ c_x & c_y & c_z \end{vmatrix}$

7. Angle θ between vectors: $\cos\theta=\frac{\boldsymbol{a}\cdot\boldsymbol{b}}{|\boldsymbol{a}|\cdot|\boldsymbol{b}|}=\frac{a_xb_x+a_yb_y+a_zb_z}{\sqrt{a_x^2+a_y^2+a_z^2}\sqrt{b_x^2+b_y^2+b_z^2}}$

Straight lines and planes

Lines

Direction vector $\boldsymbol{v}=(\alpha, \beta, \gamma)$

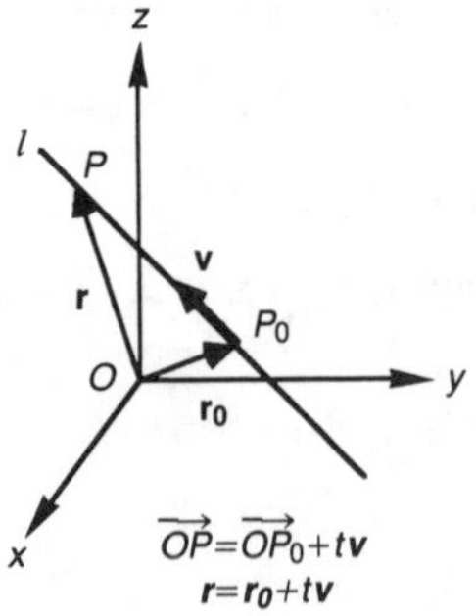

Line l given by

(1) point $P_0\in l$ and direction vector $\boldsymbol{v}$.

$$\text{Equation: }\begin{cases} x=x_0+\alpha t \\ y=y_0+\beta t \\ z=z_0+\gamma t \end{cases} \Leftrightarrow \frac{x-x_0}{\alpha}=\frac{y-y_0}{\beta}=\frac{z-z_0}{\gamma}$$

(2) two points P_1 and P_2. Set $\boldsymbol{v}=\overrightarrow{P_1P_2}=(x_2-x_1, y_2-y_1, z_2-z_1)$ and use (1)

Planes

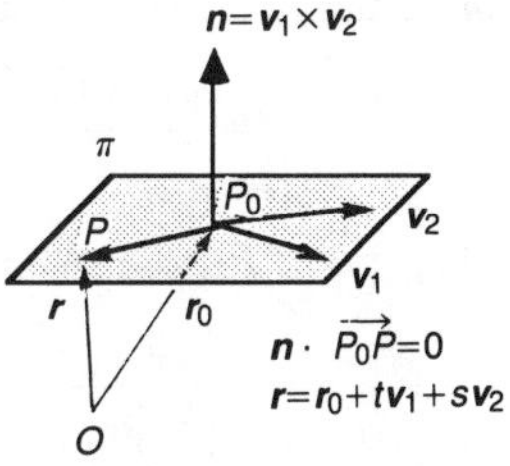

Normal vector $\boldsymbol{n}=(A, B, C)$
Spanning vectors $\boldsymbol{v}_1=(\alpha_1, \beta_1, \gamma_1)$, $\boldsymbol{v}_2=(\alpha_2, \beta_2, \gamma_2)$

General equation form: $Ax+By+Cz+D=0$

Plane π given by

(3) $P_0 \in \pi$ and normal vector $\boldsymbol{n}$:

$$A(x-x_0)+B(y-y_0)+C(z-z_0)=0$$

(4) $P_0 \in \pi$ and spanning vectors $\boldsymbol{v}_1$, $\boldsymbol{v}_2$ in the plane:

(a) Calculate $\boldsymbol{n}=\boldsymbol{v}_1 \times \boldsymbol{v}_2$ and use (3) or

(b) $\begin{vmatrix} x-x_0 & y-y_0 & z-z_0 \\ \alpha_1 & \beta_1 & \gamma_1 \\ \alpha_2 & \beta_2 & \gamma_2 \end{vmatrix}=0$ or

(c) $\begin{cases} x=x_0+\alpha_1 t+\alpha_2 s \\ y=y_0+\beta_1 t+\beta_2 s \\ z=z_0+\gamma_1 t+\gamma_2 s \end{cases}$ (parameter form)

(5) Three points $P_0, P_1, P_2 \in \pi$.

Calculate $\boldsymbol{n}=\overrightarrow{P_0P_1} \times \overrightarrow{P_0P_2}$ and use (3).

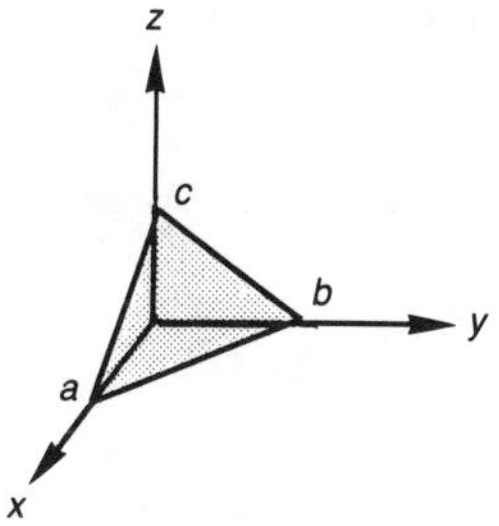

(6) Intersections with coordinate axis:

$$\frac{x}{a}+\frac{y}{b}+\frac{z}{c}=1 \quad \text{(intercept form)}$$

Angles

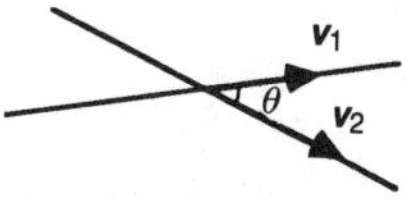

Between two lines: $\cos\theta=\dfrac{|\boldsymbol{v}_1 \cdot \boldsymbol{v}_2|}{|\boldsymbol{v}_1| \cdot |\boldsymbol{v}_2|}$

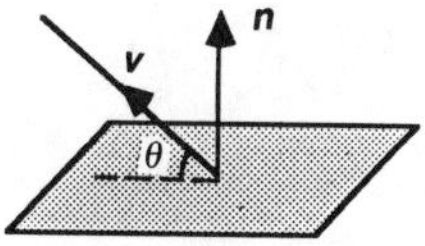

Between line and plane: $\cos(90°-\theta)=\dfrac{|\boldsymbol{v} \cdot \boldsymbol{n}|}{|\boldsymbol{v}| \cdot |\boldsymbol{n}|}$

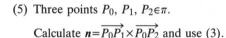

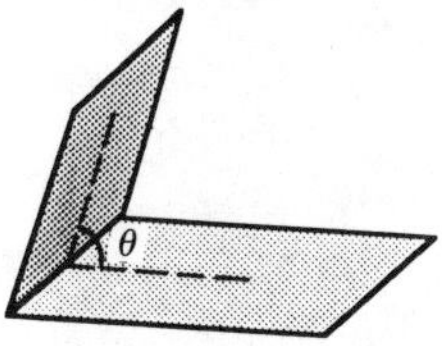

Between two planes: $\cos\theta=\dfrac{|\boldsymbol{n}_1 \cdot \boldsymbol{n}_2|}{|\boldsymbol{n}_1| \cdot |\boldsymbol{n}_2|}$

Distances

From a point P_1 to a line (P_0 arbitrary on the line)

$$d=\frac{|\boldsymbol{v}\times\overrightarrow{P_0P_1}|}{|\boldsymbol{v}|}$$

From a point P_1 to a plane (P_0 arbitrary in the plane)

$$d=\frac{|\boldsymbol{n}\cdot\overrightarrow{P_0P_1}|}{|\boldsymbol{n}|}=\frac{|Ax_1+By_1+Cz_1+D|}{\sqrt{A^2+B^2+C^2}}$$

Between two non-parallel lines (P_1, P_2 arbitrary on the lines respectively).

$$d=\frac{|\overrightarrow{P_1P_2}\cdot(\boldsymbol{v}_1\times\boldsymbol{v}_2)|}{|\boldsymbol{v}_1\times\boldsymbol{v}_2|}$$

Second degree surfaces

General form

(3.2) $$Ax^2+By^2+Cz^2+2Dxy+2Exz+2Fyz+2Gx+2Hy+2Kz+L=0$$

Analysis of (3.2):

(a) $D=E=F=0$: Complete the squares and compare to the standard forms below.

(b) Any D, E, $F\neq 0$: Spectral methods, see sec. 4.6.

Second degree surfaces in standard form

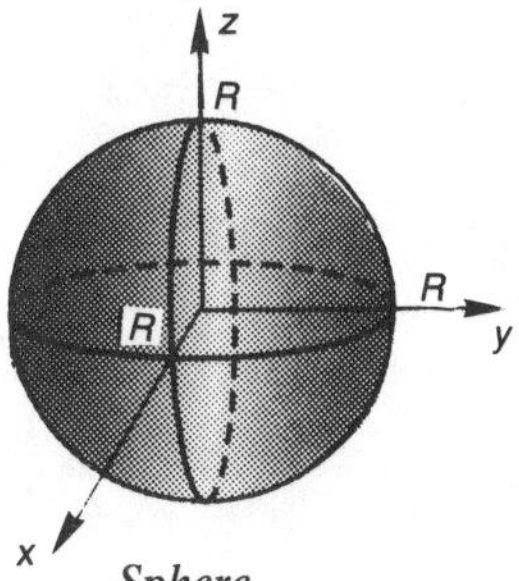

Sphere

$x^2+y^2+z^2=R^2$

$V=\dfrac{4\pi R^3}{3}$

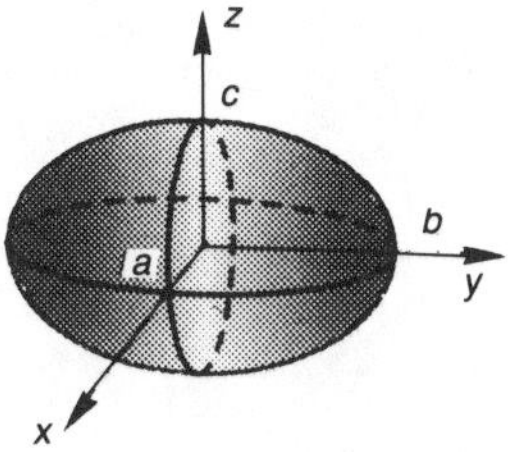

Ellipsoid

$\dfrac{x^2}{a^2}+\dfrac{y^2}{b^2}+\dfrac{z^2}{c^2}=1$

$V=\dfrac{4\pi abc}{3}$

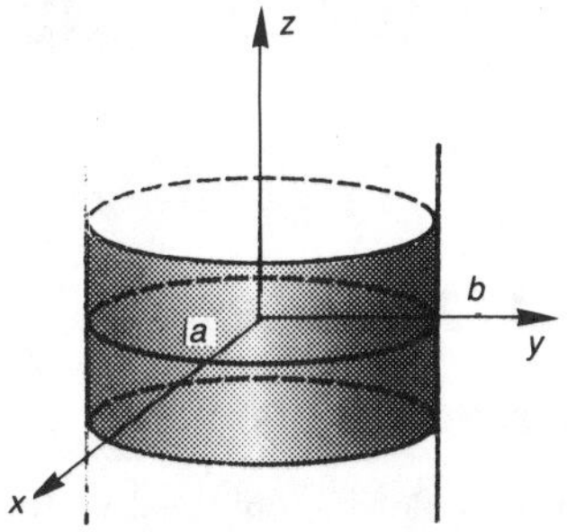

Elliptic cylinder

$\dfrac{x^2}{a^2}+\dfrac{y^2}{b^2}=1$

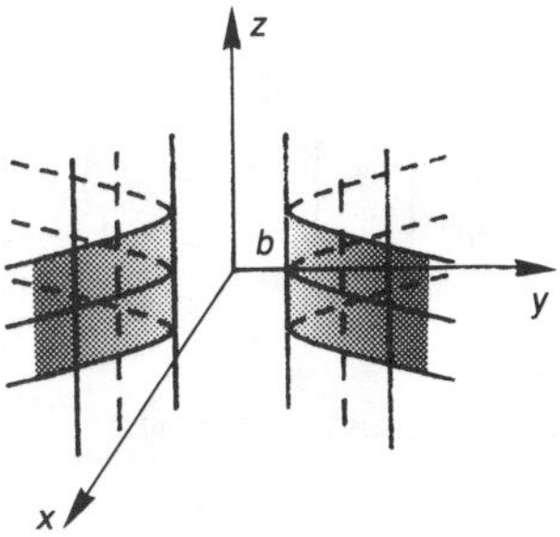

Hyperbolic Cylinder

$\dfrac{x^2}{a^2}-\dfrac{y^2}{b^2}=-1$

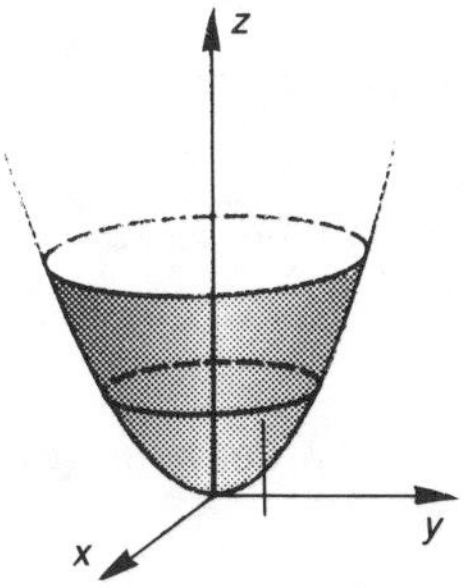

Elliptic Paraboloid

$z=\dfrac{x^2}{a^2}+\dfrac{y^2}{b^2}$

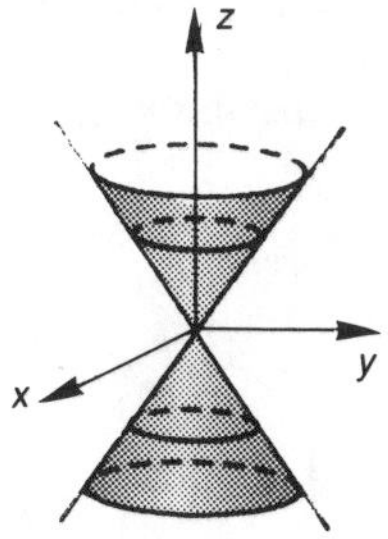

Elliptic Cone

$\dfrac{x^2}{a^2}+\dfrac{y^2}{b^2}-\dfrac{z^2}{c^2}=0$

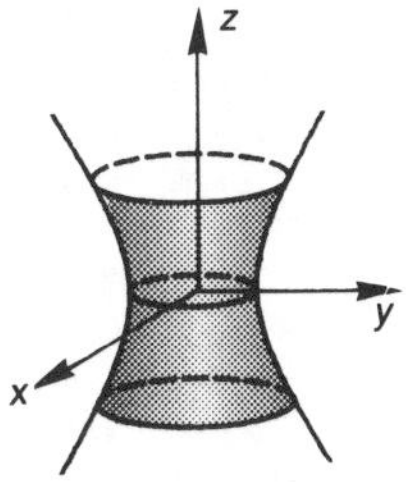

Elliptic Hyperboloid of one sheet

$\dfrac{x^2}{a^2}+\dfrac{y^2}{b^2}-\dfrac{z^2}{c^2}=1$

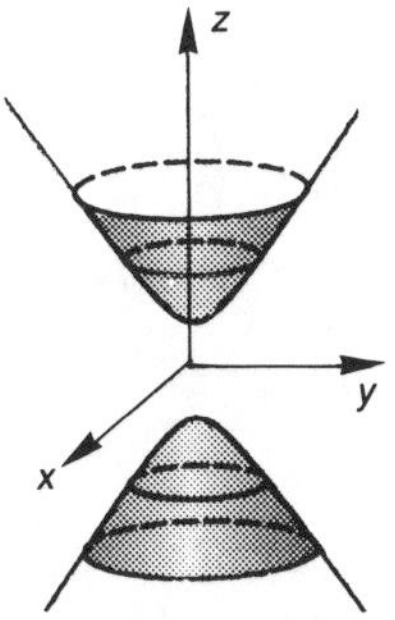

Elliptic Hyperboloid of two Sheets

$\dfrac{x^2}{a^2}+\dfrac{y^2}{b^2}-\dfrac{z^2}{c^2}=-1$

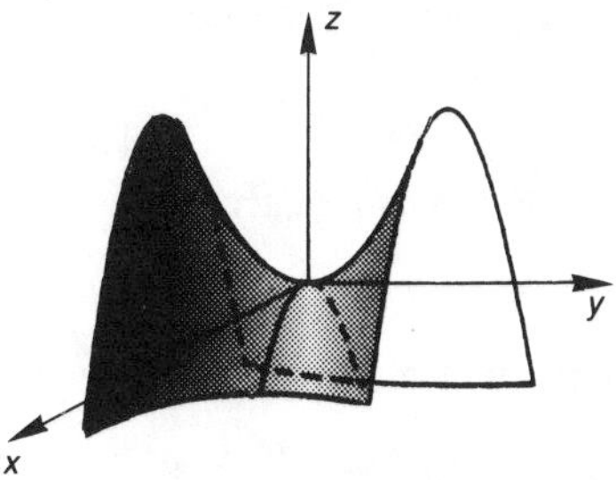

Hyperbolic Paraboloid

$z=\dfrac{y^2}{b^2}-\dfrac{x^2}{a^2}$

4 Linear Algebra

4.1 Matrices

Basic concepts

In the following only real vectors and matrices are considered.

Column vector $\boldsymbol{a}=\begin{bmatrix} a_1 \\ \dots \\ a_n \end{bmatrix} \in \boldsymbol{R}^n$ *Row vector* $\boldsymbol{a}^t=(a_1, \dots, a_n)$

Scalar product $\mathbf{a}^t\mathbf{b}=a_1b_1+\dots+a_nb_n$

Norm (length) $|\boldsymbol{a}|=\sqrt{\boldsymbol{a}^t\boldsymbol{a}}=\sqrt{a_1^2+\dots+a_n^2}$

Matrix of order $m\times n$: (A is *square* if $m=n$):

$$A=\begin{bmatrix} a_{11} \dots & a_{1n} \\ \dots & \\ a_{m1} \dots & a_{mn} \end{bmatrix}=(a_{ij})=[\boldsymbol{a}_1, \dots, \boldsymbol{a}_j, \dots, \boldsymbol{a}_n], \quad \boldsymbol{a}_j=\begin{bmatrix} a_{1j} \\ \dots \\ a_{mj} \end{bmatrix}$$

Transpose of $A: A^t=\begin{bmatrix} a_{11} \dots & a_{m1} \\ \dots & \\ a_{1n} \dots & a_{mn} \end{bmatrix}$ of order $n\times m$ (exchange rows and columns).

Diagonal matrix $D=\begin{bmatrix} a_{11} & & 0 \\ & a_{22} & \\ 0 & & \ddots \; a_{nn} \end{bmatrix}=\mathrm{diag}(a_{11}, \dots, a_{nn}) \quad (a_{ij}=0,\ i\neq j)$

Identity matrix $I=\mathrm{diag}(1, 1, \dots, 1)$ of order $n\times n$

Lower triangular matrix $T=\begin{bmatrix} a_{11} & & 0 \\ a_{21} & a_{22} & \\ \dots & & \ddots \\ a_{n1} & \dots & a_{nn} \end{bmatrix} \quad (a_{ij}=0,\ i<j)$

Upper triangular matrix analogously.

Exponential $e^A=\sum_{n=0}^{\infty}\frac{1}{n!}A^n \quad$ (A square)

Inverse of $A: A^{-1}$ satisfying $AA^{-1}=A^{-1}A=I \quad$ (A square)

A is *symmetric* if $A=A^t \Leftrightarrow a_{ij}=a_{ji}$

A is *orthogonal* if A is square and $A^tA=I \Leftrightarrow A^t=A^{-1}$

Matrix algebra

1. Addition $C=A+B: c_{ij}=a_{ij}+b_{ij}$ (A, B, C of the same order)
2. Subtraction $C=A-B: c_{ij}=a_{ij}-b_{ij}$ (A, B, C of the same order)
3. Multiplication by a number, $C=xA: c_{ij}=xa_{ij}$ (A, C of the same order)
4. Product AB of two matrices:

If order $(A)=m\times n$, order $(B)=n\times p$, then $C=AB$ is of order $m\times p$

and

$$c_{ij}=\sum_{k=1}^{n} a_{ik}b_{kj}$$

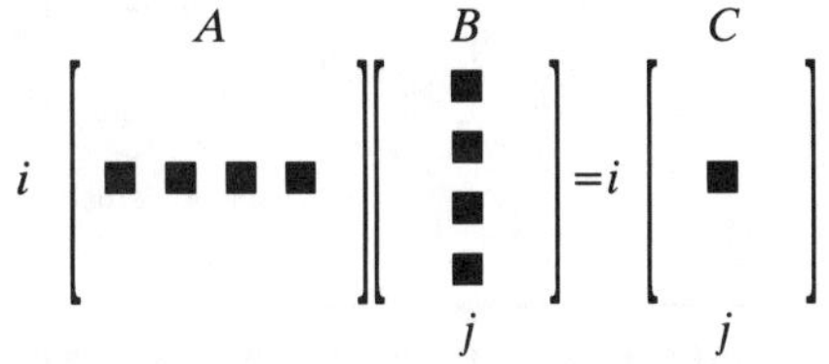

Note: $AB=A[\boldsymbol{b}_1, \ldots, \boldsymbol{b}_p]=[A\boldsymbol{b}_1, \ldots, A\boldsymbol{b}_p]$

$$A+B=B+A \quad (A+B)+C=A+(B+C) \quad x(A+B)=xA+xB$$
$$AB\neq BA \text{ (in general)} \quad (AB)C=A(BC) \quad IA=AI=A$$
$$A(B+C)=AB+AC \quad (A+B)C=AC+BC \quad (AB)^t=B^tA^t$$
$$(AB)^{-1}=B^{-1}A^{-1} \quad (A^{-1})^t=(A^t)^{-1} \quad (A+B)^t=A^t+B^t$$
$$e^{A+B}=e^Ae^B \text{ if } AB=BA \quad (e^A)^{-1}=e^{-A} \quad \frac{d}{dx}e^{xA}=Ae^{xA}$$

Differentiation

If $A=A(x)=(a_{ij}(x))$ and $B=B(x)$, then (*i*) $A'(x)=(a_{ij}'(x))$ (*ii*) $(A+B)'=A'+B'$ (*iii*) $(AB)'=AB'+A'B$ (*iv*) $(A^2)'=AA'+A'A$ (*v*) $(A^{-1})'=-A^{-1}A'A^{-1}$

Matrix norms, see sec. 16.2.

Rank

Elementary row operations

I. Exchange of two rows.
II. Multiplication of a row by a constant $\neq 0$.
III Addition of an arbitrary multiple of a row to another row.

Notation: $A\sim B$ if A can be transformed to B by a sequence of elementary row operations.

Example 1.

$$A=\begin{bmatrix} 1 & 1 & 1 \\ 2 & -1 & 5 \\ -1 & 0 & -2 \end{bmatrix} \begin{matrix} (-2) & (+) \end{matrix} \sim \begin{bmatrix} 1 & 1 & 1 \\ 0 & -3 & 3 \\ 0 & 1 & -1 \end{bmatrix} \left(\tfrac{1}{3}\right) \sim \begin{bmatrix} 1 & 1 & 1 \\ 0 & -1 & 1 \\ 0 & 1 & -1 \end{bmatrix} (+) \sim \begin{bmatrix} 1 & 1 & 1 \\ 0 & -1 & 1 \\ 0 & 0 & 0 \end{bmatrix}$$

The concepts *echelon form matrix* and *pivot element* are explained by the following figure:

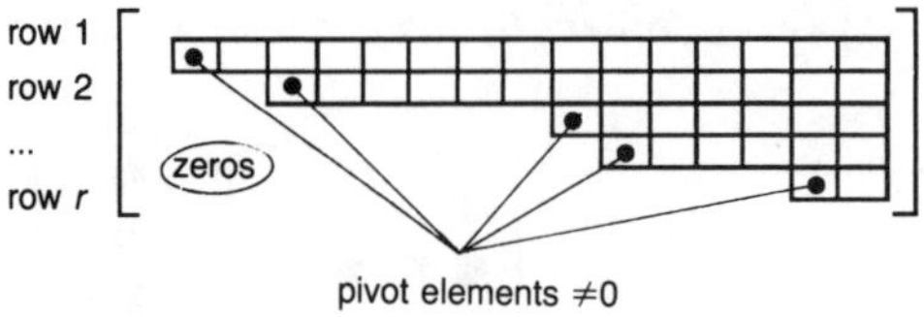

Any matrix can be transformed to an echelon form matrix by elementary row operations. The number r (the rank) is unique.

Definition of rank

The following characterizations of the *rank* $r(A)$ of a matrix A are equivalent.

1. The number of non-zero rows in an echolon form matrix $A' \sim A$.
2. The number of linearly independent columns or rows of A.
3. The greatest order of non-zero determinants of sub-matrices of A.

Example 2.

$r(A)=2$ for A in example 1.

$r(AB) \leq \min[r(A), r(B)] \qquad r(AA^t)=r(A^tA)=r(A)$

Trace

If A is square of order $n \times n$ then

$$tr\, A=\sum_{i=1}^{n} a_{ii} \quad \text{(i.e. the sum of the diagonal elements).}$$

$tr(x_1A+x_2B)=x_1 tr\, A+x_2 tr\, B \qquad (x_1, x_2 \text{ scalars})$

$tr(AB)=tr(BA) \qquad (\text{order } A=m \times n, \text{ order } B=n \times m)$

$tr\, A=\sum_{i=1}^{n} \lambda_i \qquad (\lambda_i \text{ eigenvalues of } A)$

4.2 Determinants

Definition

The nth *order determinant* of a square $n \times n$ matrix A is the number

$$D = \det A = \begin{vmatrix} a_{11} & \dots & a_{1n} \\ \dots & & \\ a_{n1} & \dots & a_{nn} \end{vmatrix} = \Sigma(-1)^{\alpha} a_{1p_1} \dots a_{np_n} \quad (n! \text{ terms})$$

The sum is taken over all *permutations* $(p_1, \dots, p_n)$ of $(1, \dots, n)$ and α is the number of all pairs (p_i, p_j) of the permutation such that $p_i > p_j$ if $i < j$. (Or α is the number of exchanges necessary to bring the sequence $(p_1, \dots, p_n)$ to $(1, \dots, n)$.)

Geometrical interpretation. $D = \pm$volume of the "n -dimensional parallelepiped" spanned by the column (or row) vectors of A.

Special cases

1. $n=2$: $\begin{vmatrix} a_{11} & a_{12} \\ a_{21} & a_{22} \end{vmatrix} = a_{11}a_{22} - a_{12}a_{21}$

2. $n=3$: $\begin{vmatrix} a_{11} & a_{12} & a_{13} \\ a_{21} & a_{22} & a_{23} \\ a_{31} & a_{32} & a_{33} \end{vmatrix} = a_{11}a_{22}a_{33} + a_{12}a_{23}a_{31} + a_{13}a_{21}a_{32} - a_{11}a_{23}a_{32} - a_{12}a_{21}a_{33} - a_{13}a_{22}a_{31}$

3. (Triangular determinant) $\begin{vmatrix} a_{11} & & & 0 \\ \cdot & a_{22} & & \\ \cdot & & \ddots & \\ \cdot & & & \\ a_{n1} & \dots & & a_{nn} \end{vmatrix} = a_{11}a_{22} \dots a_{nn}$

1. $\det A^t = \det A$	2. $\det(AB) = (\det A)(\det B)$	3. $\det A^{-1} = 1/\det A$
4. $\det I = 1$	5. $\det(xA) = x^n \det A$	6. $\det A = \lambda_1 \lambda_2 \dots \lambda_n$ (λ_i eigenvalues)

7. If all elements of a row (or a column) are multiplied by a constant c, then the determinant is multiplied by c.
8. Exchange of two rows (columns) changes the sign of the determinant.
9. A determinant does not change if one row (column) multiplied by a constant is added to another row (column).
10. A determinant equals zero if (*a*) all elements of a row (column) are zero, or (*b*) two rows (columns) coincide.

Development of a determinant. Cofactor

Subdeterminant D_{ij} of order $(n-1)\times(n-1)$ is formed by deleting the ith row and the jth column of D.

Cofactor $A_{ij}=(-1)^{i+j}\,D_{ij}$.

11. $\det A=a_{i1}A_{i1}+a_{i2}A_{i2}+\ldots+a_{in}A_{in}$ (development by i^{th} row)
12. $\det A=a_{1j}A_{1j}+a_{2j}A_{2j}+\ldots+a_{nj}A_{nj}$ (development by j^{th} column)

Example.

$$\begin{vmatrix} 1 & 2 & 3 & 4 \\ 5 & 0 & 2 & 0 \\ 3 & 4 & 1 & 1 \\ 2 & 3 & 4 & 5 \end{vmatrix} = [\text{development by } 2^{nd} \text{ row}] = 5\cdot(-1)^{2+1}\begin{vmatrix} 2 & 3 & 4 \\ 4 & 1 & 1 \\ 3 & 4 & 5 \end{vmatrix} + 2\cdot(-1)^{2+3}\begin{vmatrix} 1 & 2 & 4 \\ 3 & 4 & 1 \\ 2 & 3 & 5 \end{vmatrix}$$

Matrix inversion

13. A^{-1} exists $\Leftrightarrow \det A\neq 0$.

Calculation of A^{-1}

a. $[A^{-1}]_{ij}=\dfrac{1}{\det A}A_{ji}$ (*cofactor method*)

b. Assume, by elementary row operations,

$$[AI]=\begin{bmatrix} a_{11} \ldots a_{1n} & 1 \ldots 0 \\ \ldots & \\ a_{n1} \ldots a_{nn} & 0 \ldots 1 \end{bmatrix} \sim \begin{bmatrix} 1 \ldots 0 & b_{11} \ldots b_{1n} \\ \ldots & \\ 0 \ldots 1 & b_{n1} \ldots b_{nn} \end{bmatrix} = [IB].$$

Then $B=A^{-1}$. (*Jacobi's method*)

Special case, $n=2$: $\begin{bmatrix} a_{11} & a_{12} \\ a_{21} & a_{22} \end{bmatrix}^{-1} = \dfrac{1}{a_{11}a_{22}-a_{12}a_{21}}\begin{bmatrix} a_{22} & -a_{12} \\ -a_{21} & a_{11} \end{bmatrix}$.

Box inversion

$$\begin{matrix} & (p) & (q) \\ (p) & A & O \\ (q) & O & B \end{matrix}\ \Bigg]^{-1} \;\; \left[\begin{array}{c|c} A & O \\ \hline O & B \end{array}\right]^{-1} = \left[\begin{array}{c|c} A^{-1} & O \\ \hline O & B^{-1} \end{array}\right]$$

(rows and columns partitioned as (p), (q)) (order$(A)=p\times p$, order$(B)=q\times q$)

4.3 Systems of Linear Equations

A system of m linear equations and n unknowns $x_1, \ldots, x_n$ has the form

$$(ES)\begin{cases} a_{11}x_1+a_{12}x_2+\ldots+a_{1n}x_n=b_1 \\ a_{21}x_1+a_{22}x_2+\ldots+a_{2n}x_n=b_2 \\ \ldots \\ a_{m1}x_1+a_{m2}x_2+\ldots+a_{mn}x_n=b_m \end{cases}$$

$$\Leftrightarrow$$

$$(ES)\ A\boldsymbol{x}=\boldsymbol{b}, \quad A=\begin{bmatrix} a_{11} \ldots a_{1n} \\ \ldots \\ a_{m1} \ldots a_{mn} \end{bmatrix}, \boldsymbol{x}=\begin{bmatrix} x_1 \\ \ldots \\ x_n \end{bmatrix}, \boldsymbol{b}=\begin{bmatrix} b_1 \\ \ldots \\ b_m \end{bmatrix}$$

$$\Leftrightarrow$$

$$(ES)\ x_1\boldsymbol{a}_1+x_2\boldsymbol{a}_2+\ldots+x_n\boldsymbol{a}_n=\boldsymbol{b}, \quad \boldsymbol{a}_j=\begin{bmatrix} a_{1j} \\ \ldots \\ a_{mj} \end{bmatrix}$$

The (ES) is *homogeneous* if all $b_i=0$, otherwise *inhomogeneous.* The matrix A is called the *coefficient matrix* and

$$\text{the matrix } B=\begin{bmatrix} a_{11} \ldots & a_{1n}b_1 \\ \ldots & \\ a_{m1} \ldots & a_{mn}b_n \end{bmatrix}=[\boldsymbol{a}_1, \ldots, \boldsymbol{a}_n, \boldsymbol{b}]$$

the *augmented coefficient matrix.*

The number of solutions

Homogenous system			**Inhomogeneous system**	
Assumptions		Number of solutions*	Assumptions	Number of solutions*
$n<m$	$r(A)=n$	1	$r(A)<r(B)$	0
	$r(A)<n$	∞	$r(A)=r(B)=n$ $r(A)=r(B)<n$	1 ∞
$n=m$	$r(A)=n \Leftrightarrow$ $\det A\neq 0$	1	$r(A)=r(B)=n$ $\Leftrightarrow \det A\neq 0$	1
	$r(A)<n \Leftrightarrow$ $\det A=0$	∞	$r(A)<r(B)$ $r(A)=r(B)<n$	0 ∞
$n>m$		∞	$r(A)<r(B)$ $r(A)=r(B)$	0 ∞

* In the case of ∞ the number of free variables (indeterminates) is $n-r(A)$.

Gaussian elimination

By elementary row operations [i.e. 1. Exchange of equations, 2. Multiplication of an equation by a constant $\neq 0$, 3. Adding one equation to another] the system (*ES*) can be transformed into *echelon form*. From that form the unknowns are determined by backward substitution. Of these there are two groups:

1. *Basic variables*: corresponding to pivots.
2. *Free variables*: the others.

Example.

$$\begin{cases} x-\ y-2z+\ u=\ 0 \\ x-2y+\ z-\ u=-2 \\ 2x-\ y-\ z+3u=\ 2 \end{cases} \;\; \ominus \;\; (-2) \quad \Leftrightarrow \quad \begin{cases} x-y-2z+\ u=\ 0 \\ -y+3z-2u=-2 \\ y+3z+\ u=\ 2 \end{cases} \;\; \oplus \quad \Leftrightarrow$$

$$\Leftrightarrow \begin{cases} x-y-2z+\ u=\ 0 & (1) \\ -y+3z-2u=-2 & (2) \\ 6z-\ u=\ 0 & (3) \end{cases}$$

Solution: Set $\underline{z}=\underline{t}$ (arbitrary) $\overset{(3)}{\Rightarrow}$ $\underline{u}=6z=\underline{6t}$ $\overset{(2)}{\Rightarrow}$ $\underline{y}=2+3z-2u=\underline{2-9t}$ $\overset{(1)}{\Rightarrow}$ $\underline{x}=y+2z-u=\underline{2-13t}$

(x, y, z are basic variables, u free variable)

Quadratic systems

If $m=n$ in (*ES*) and if $\det A\neq 0$ the unique solution can be expressed explicitly by either

1. $\boldsymbol{x}=A^{-1}\boldsymbol{b}$ (*inverse matrix method*) or
2. $x_j=D_j/D$, $j=1, \ldots, n$, where $D=\det A$ and D_j=the determinant that arises when the jth column of D is replaced by the column elements $b_1, \ldots, b_n$. (*Cramer's rule*)

Least squares approximation

It may happen that (*ES*) has no exact solution. The problem of finding the approximate solution in mean consists in determining $x_1, \ldots, x_n$ such that the square mean error

$$\eta=\sqrt{\frac{1}{m}\left(\varepsilon_1{}^2+\ldots+\varepsilon_n{}^2\right)}=\frac{1}{\sqrt{m}}\,|A\boldsymbol{x}-\boldsymbol{b}|$$

where

$$\begin{cases} \varepsilon_1=a_{11}x_1+\ldots+a_{1n}x_n-b_1 \\ \ldots \\ \varepsilon_m=a_{m1}x_1+\ldots+a_{mn}x_n-b_m \end{cases}$$

is minimal.

Theorem

Any solution (there always exists at least one) of the $n \times n$ square system

$$A^tAx=A^tb$$

minimizes η.

4.4 Linear Coordinate Transformations

Orthogonal matrices

The vectors $\boldsymbol{a}$, $\boldsymbol{b}$ are *orthogonal* ($\boldsymbol{a} \perp \boldsymbol{b}$) if $\boldsymbol{a}^t\boldsymbol{b}=a_1b_1+\ldots+a_nb_n=0$.

Definition. A real square matrix P is *orthogonal* if $P^tP=I$.

Let $P=(p_{ij})=[\boldsymbol{p}_1, \boldsymbol{p}_2, \ldots, \boldsymbol{p}_n]$, where $\boldsymbol{p}_j$ are the column vectors of P. Then

1. P is orthogonal $\Leftrightarrow$ The column vectors of P are pairwise orthogonal and normed, i.e. $\boldsymbol{p}_i{}^t\boldsymbol{p}_j=0$ $(i \neq j)$, $|\boldsymbol{p}_i|=1$.
2. P orthogonal $\Rightarrow$

   (a) P^t is orthogonal (b) $P^{-1}=P^t$
   (c) $\det P=\pm 1$ (d) $(P\boldsymbol{a})^t(P\boldsymbol{b})=\boldsymbol{a}^t\boldsymbol{b}$
   (e) $|P\boldsymbol{a}|=|\boldsymbol{a}|$ (f) $\boldsymbol{a} \perp \boldsymbol{b} \Leftrightarrow P\boldsymbol{a} \perp P\boldsymbol{b}$
3. P, Q orthogonal $\Rightarrow$ PQ orthogonal.

Coordinate transformations

(The results are given in the 3-dimensional case. They can be generalized to arbitrary dimension in an obvious way. Cf. sec. 4.8.)

Given: Two parallel coordinate systems with bases $\boldsymbol{e}_x$, $\boldsymbol{e}_y$, $\boldsymbol{e}_z$ and $\overline{\boldsymbol{e}}_x$, $\overline{\boldsymbol{e}}_y$, $\overline{\boldsymbol{e}}_z$ respectively.

Relationship between bases (P=transformation matrix):

$$(4.1) \qquad \begin{cases} \overline{\boldsymbol{e}}_x=p_{11}\boldsymbol{e}_x+p_{21}\boldsymbol{e}_y+p_{31}\boldsymbol{e}_z \\ \overline{\boldsymbol{e}}_y=p_{12}\boldsymbol{e}_x+p_{22}\boldsymbol{e}_y+p_{32}\boldsymbol{e}_z \\ \overline{\boldsymbol{e}}_z=p_{13}\boldsymbol{e}_x+p_{23}\boldsymbol{e}_y+p_{33}\boldsymbol{e}_z \end{cases} \qquad P=\begin{bmatrix} p_{11} & p_{12} & p_{13} \\ p_{21} & p_{22} & p_{23} \\ p_{31} & p_{32} & p_{33} \end{bmatrix}=[\overline{\boldsymbol{e}}_x, \overline{\boldsymbol{e}}_y, \overline{\boldsymbol{e}}_z]$$

Components of a vector

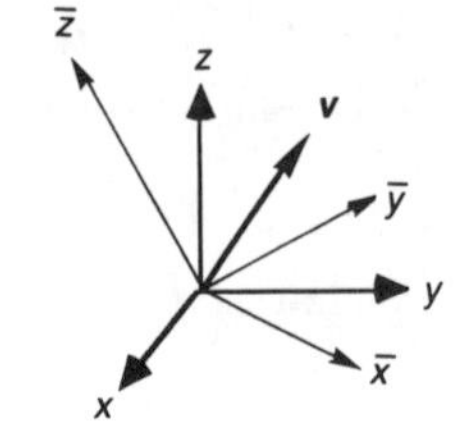

Let the vector $\boldsymbol{v}$ have components (v_x, v_y, v_z) and $(\bar{v}_x, \bar{v}_y, \bar{v}_z)$ with respect to the bases $\boldsymbol{e}_x, \boldsymbol{e}_y, \boldsymbol{e}_z$ and $\bar{\boldsymbol{e}}_x, \bar{\boldsymbol{e}}_y, \bar{\boldsymbol{e}}_z$ respectively. Then

(4.2) $$\begin{bmatrix} v_x \\ v_y \\ v_z \end{bmatrix} = P \begin{bmatrix} \bar{v}_x \\ \bar{v}_y \\ \bar{v}_z \end{bmatrix}$$

P is orthogonal if the coordinate systems are orthonormal (ON).

Coordinates of a point

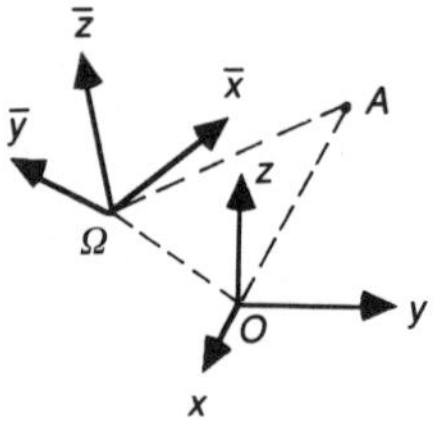

Let the point A have coordinates $(x; y; z)$ and $(\bar{x}; \bar{y}; \bar{z})$ with respect to the coordinate systems $(O, \boldsymbol{e}_x, \boldsymbol{e}_y, \boldsymbol{e}_z)$ and $(\Omega, \bar{\boldsymbol{e}}_x, \bar{\boldsymbol{e}}_y, \bar{\boldsymbol{e}}_z)$ respectively, where $\Omega=(x_0; y_0; z_0)$ in the former system. Then

(4.3) $$\begin{bmatrix} x \\ y \\ z \end{bmatrix} = \begin{bmatrix} x_0 \\ y_0 \\ z_0 \end{bmatrix} + P \begin{bmatrix} \bar{x} \\ \bar{y} \\ \bar{z} \end{bmatrix}$$

1. Both bases orthonormal $\Rightarrow$ P is orthogonal.
2. Rotation of an orthogonal coordinate system $\Rightarrow$ P is orthogonal.

Rotation of coordinate system

Coordinates of point A:

$$\begin{cases} x=\bar{x}\cos\alpha-\bar{y}\sin\alpha \\ y=\bar{x}\sin\alpha+\bar{y}\cos\alpha \end{cases} \Leftrightarrow \begin{cases} \bar{x}=x\cos\alpha+y\sin\alpha \\ \bar{y}=-x\sin\alpha+y\cos\alpha \end{cases}$$

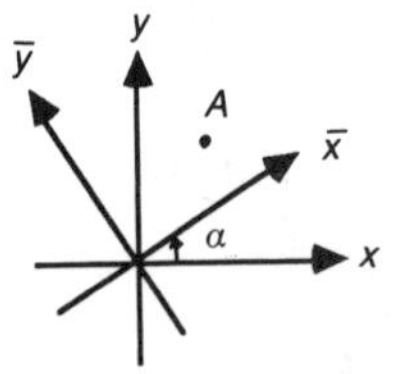

4.5 Eigenvalues. Diagonalization

Definition

The number λ is called an *eigenvalue* of a square matrix A and $\boldsymbol{g}\neq\boldsymbol{0}$ is a corresponding *eigenvector* if

(4.4) $$A\boldsymbol{g}=\lambda\boldsymbol{g}.$$

The characteristic equation

$$\lambda \text{ is an eigenvalue of } A \Leftrightarrow \det(A-\lambda I)=\begin{vmatrix} a_{11}-\lambda & a_{12} & \dots & a_{1n} \\ a_{21} & a_{22}-\lambda & & \cdot \\ \dots & & \ddots & \cdot \\ a_{n1} & \dots & & a_{nn}-\lambda \end{vmatrix}=0$$

$\det A=\lambda_1\lambda_2 \ldots \lambda_n$. (product of all eigenvalues).

Symmetric matrices

A real square matrix A is symmetric if $A^t=A$.

> A symmetric $\Rightarrow$ (a) all eigenvalues are real (b) eigenvectors corresponding to different eigenvalues are orthogonal (c) $\boldsymbol{a}^t(A\boldsymbol{b})=(A\boldsymbol{a})^t\boldsymbol{b}$

Spectral theorem for symmetric matrices

> Assume that (*i*) A is symmetric (*ii*) $\lambda_1, \ldots, \lambda_n$ are eigenvalues (incl. multiplicities).
>
> Then there exist corresponding eigenvectors $\mathbf{g}_1, \ldots, \mathbf{g}_n$ such that
>
> 1. $\mathbf{g}_i \perp \mathbf{g}_j$ $(i \neq j)$, $|\mathbf{g}_i|=1$ (pairwise orthogonal and normed)
>
>   Set $P=[g_1, g_2, \ldots, g_n]$. Then
> 2. P is an orthogonal matrix
> 3. $P^tAP=D=\text{diag}(\lambda_1, \lambda_2, \ldots, \lambda_n)$
> 4. $A=PDP^t$, $A^k=PD^kP^t=P \text{ diag}(\lambda_1{}^k, \ldots, \lambda_n{}^k)P^t$
> 5. $\lambda_{min} \leq \dfrac{\mathbf{x}^tA\mathbf{x}}{|\mathbf{x}|^2} \leq \lambda_{max}$, $(\mathbf{x}\neq\mathbf{0})$ (equality $\Leftrightarrow$ $\mathbf{x}$=corresponding eigenvector).

> ***Example.*** $A=\begin{bmatrix} 1 & 0 & -4 \\ 0 & 5 & 4 \\ -4 & 4 & 3 \end{bmatrix}$. Characteristic equation: $\begin{vmatrix} 1-\lambda & 0 & -4 \\ 0 & 5-\lambda & 4 \\ -4 & 4 & 3-\lambda \end{vmatrix}=-\lambda^3+9\lambda^2+9\lambda-81=0$
>
> with roots $\lambda_1=9$, $\lambda_2=3$, $\lambda_3=-3$. Corresponding eigenvectors:
>
> $\lambda_1=9$: $A\begin{bmatrix} x \\ y \\ z \end{bmatrix}=9\begin{bmatrix} x \\ y \\ z \end{bmatrix} \Rightarrow \left.\begin{array}{r} -8x-4z=0 \\ -4y+4z=0 \\ -4x+4y-6z=0 \end{array}\right\}$ with general solution
>
> $x=-t$, $y=2t$, $z=2t$. Choose $g_1=\frac{1}{3}\begin{bmatrix} -1 \\ 2 \\ 2 \end{bmatrix}$. Analogously $g_2=\frac{1}{3}\begin{bmatrix} 2 \\ 2 \\ -1 \end{bmatrix}$
>
> and $g_3=\frac{1}{3}\begin{bmatrix} 2 \\ -1 \\ 2 \end{bmatrix}$ so that $P=\frac{1}{3}\begin{bmatrix} -1 & 2 & 2 \\ 2 & 2 & -1 \\ 2 & -1 & 2 \end{bmatrix}$ and $D=P^tAP=\begin{bmatrix} 9 & 0 & 0 \\ 0 & 3 & 0 \\ 0 & 0 & -3 \end{bmatrix}$

Affine diagonalization

Assume that (the non-symmetric) A has n distinct real eigenvalues $\lambda_1, \ldots, \lambda_n$. Then

1. the corresponding eigenvectors $\boldsymbol{g}_1, \ldots, \boldsymbol{g}_n$ are linearly independent.
2. $L^{-1}AL = D = \text{diag}(\lambda_1, \ldots, \lambda_n)$, $\quad L = [\boldsymbol{g}_1, \ldots, \boldsymbol{g}_n]$.

(Also in the case of multiple eigenvalues there may exist linearly independent eigenvectors and hence 2.)

A generalized eigenvalue problem

Eigenvalue problem (A invertible):

(4.5) $\qquad B\boldsymbol{g} = \lambda A\boldsymbol{g}, \ \boldsymbol{g} \neq \boldsymbol{0}$

(This coincides with (4.4) if $A = I$).

The number λ is an eigenvalue of (4.5) $\Leftrightarrow \det(B - \lambda A) = 0$.

Generalized spectral theorem

Assume that (*i*) A, B symmetric, A positive definite (*ii*) $\lambda_1, \ldots, \lambda_n$ eigenvalues of (4.5). Then

1. All eigenvalues are real.
2. There exists a basis of corresponding eigenvectors $\boldsymbol{g}_1, \ldots, \boldsymbol{g}_n$ such that $\boldsymbol{g}_i^t A \boldsymbol{g}_j = \delta_{ij}$, $\boldsymbol{g}_i^t B \boldsymbol{g}_j = \lambda_j \delta_{ij}$.
3. If $L = [\boldsymbol{g}_1, \ldots, \boldsymbol{g}_n]$ then $\det L \neq 0$ and $L^t A L = I$, $\quad L^t B L = \text{diag}(\lambda_1, \ldots, \lambda_n)$ (*simultaneous diagonalization*).
4. $\lambda_{min} \leq \dfrac{\boldsymbol{x}^t B \boldsymbol{x}}{\boldsymbol{x}^t A \boldsymbol{x}} \leq \lambda_{max}$, $\boldsymbol{x} \neq \boldsymbol{0}$.

4.6 Quadratic Forms

Three dimensions

A *quadratic form* is a homogeneous polynomial of second degree,

$$Q = Q(\boldsymbol{x}) = Q(x, y, z) = a_{11}x^2 + a_{22}y^2 + a_{33}z^2 + 2a_{12}xy + 2a_{13}xz + 2a_{23}yz =$$

$$= [x \ y \ z] \begin{bmatrix} a_{11} & a_{12} & a_{13} \\ a_{12} & a_{22} & a_{23} \\ a_{13} & a_{23} & a_{33} \end{bmatrix} \begin{bmatrix} x \\ y \\ z \end{bmatrix} = \boldsymbol{x}^t A \boldsymbol{x},$$

where A is a symmetric matrix.

Theorem

Let (*i*) $\lambda_1, \lambda_2, \lambda_3$ be the eigenvalues of A

(*ii*) $\boldsymbol{g}_1, \boldsymbol{g}_2, \boldsymbol{g}_3$ correspondig eigenvectors, pairwise orthogonal and normed,

(*iii*) $P=[\boldsymbol{g}_1, \boldsymbol{g}_2, \boldsymbol{g}_3]$ orthogonal matrix

Then

1. the transformation $\begin{bmatrix} x \\ y \\ z \end{bmatrix} = P \begin{bmatrix} \bar{x} \\ \bar{y} \\ \bar{z} \end{bmatrix}$

brings Q into the canonical form

$$Q=\lambda_1\bar{x}^2+\lambda_2\bar{y}^2+\lambda_3\bar{z}^2$$

2. $\lambda_{min}(x^2+y^2+z^2) \leqslant Q \leqslant \lambda_{max}(x^2+y^2+z^2)$

(Equality $\Leftrightarrow$ $\boldsymbol{x}$=corresponding eigenvector)

Arbitrary dimension

Let 1. a_{ij} real, $a_{ij}=a_{ji}$, $A=(a_{ij})$ symmetric $n \times n$-matrix, $\boldsymbol{x}^t=(x_1, \ldots, x_n)$.

2. $\lambda_1, \ldots, \lambda_n$ eigenvalues of A, $\boldsymbol{g}_1, \ldots, \boldsymbol{g}_n$ corresponding eigenvectors, pairwise orthogonal and normed.
3. $P=[\boldsymbol{g}_1, \ldots, \boldsymbol{g}_n]$ orthogonal matrix.
4. $Q(\boldsymbol{x})=\boldsymbol{x}^t A \boldsymbol{x}$ quadratic form.

Then

1. setting $\bar{\boldsymbol{x}}^t=(\bar{x}_1, \ldots, \bar{x}_n)$, the transformation $\boldsymbol{x}=P\bar{\boldsymbol{x}}$ brings Q into the canonical form

$$Q=\lambda_1\bar{x}_1{}^2+\ldots+\lambda_n\bar{x}_n{}^2=\bar{\boldsymbol{x}}^t \operatorname{diag}(\lambda_1, \ldots, \lambda_n)\,\bar{\boldsymbol{x}}$$

2. $\lambda_{min}|\boldsymbol{x}|^2 \leqslant Q(\boldsymbol{x}) \leqslant \lambda_{max}|\boldsymbol{x}|^2$ (Equality $\Leftrightarrow$ $\boldsymbol{x}$=corresponding eigenvector)

Definiteness

Matrix $A=\begin{bmatrix} a_{11} \ldots a_{1k} \ldots a_{1n} \\ \ldots \\ a_{k1} \ldots a_{kk} \ldots \\ \ldots \\ a_{n1} \ldots \qquad\qquad a_{nn} \end{bmatrix}$ Submatrix $A_k=\begin{bmatrix} a_{11} \ldots a_{1k} \\ \ldots \\ a_{k1} \ldots a_{kk} \end{bmatrix}$

The symmetric matrix A and the corresponding quadratic form $Q(\boldsymbol{x})=\boldsymbol{x}^t A \boldsymbol{x}$ are said to be

1. positive definite if (any of the conditions)
   $Q(\boldsymbol{x})>0,\ \boldsymbol{x}\neq\boldsymbol{0} \Leftrightarrow$ all $\lambda_k>0 \Leftrightarrow$ all $\det A_k>0$
2. positive semidefinite if $Q(\boldsymbol{x})\geq 0 \Leftrightarrow$ all $\lambda_k\geq 0$
3. indefinite if
   Q assumes positive and negative values $\Leftrightarrow$ A has positive and negative eigenvalues.

A, Q are negative definite $\Leftrightarrow -A, -Q$ are positive definite.

Second degree curves (cf. sec. 3.5)

Example. (*ON*-system)

Classify the curve $6x^2+4xy+9y^2=1$.

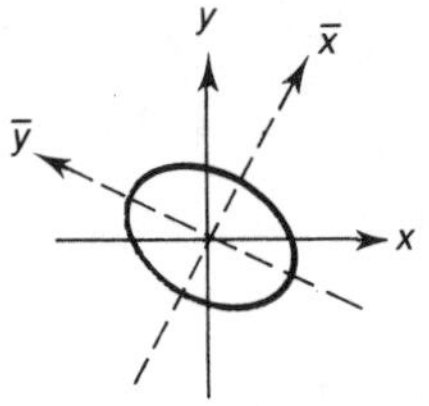

With notation as above:

$A=\begin{bmatrix}6 & 2\\ 2 & 9\end{bmatrix}$, $\lambda_1=10$, $\lambda_2=5$

$\boldsymbol{g}_1=\bar{\boldsymbol{e}}_x=\frac{1}{\sqrt{5}}\begin{bmatrix}1\\ 2\end{bmatrix}$, $\boldsymbol{g}_2=\bar{\boldsymbol{e}}_y=\frac{1}{\sqrt{5}}\begin{bmatrix}-2\\ 1\end{bmatrix}$

$P=\frac{1}{\sqrt{5}}\begin{bmatrix}1 & -2\\ 2 & 1\end{bmatrix}$. Coordinate transformation $\begin{bmatrix}x\\ y\end{bmatrix}=P\begin{bmatrix}\bar{x}\\ \bar{y}\end{bmatrix}$.

Canonical form: $10\bar{x}^2+5\bar{y}^2=1$

$\left(\text{Ellipse, semiaxis } \frac{1}{\sqrt{10}} \text{ and } \frac{1}{\sqrt{5}}\right)$.

$\bar{x}-$axis: $y=2x$, $\bar{y}-$axis: $y=-x/2$.

Second degree surfaces (cf. sec. 3.6)

Example. (*ON*-system)

Classify the surface

$$5x^2+5y^2+8z^2+8xy+4yz-4xz=1.$$

With notation as above:

$A=\begin{bmatrix}5 & 4 & -2\\ 4 & 5 & 2\\ -2 & 2 & 8\end{bmatrix}$ $\lambda_1=\lambda_2=9$, $\lambda_3=0$, $P=[\boldsymbol{g}_1, \boldsymbol{g}_2, \boldsymbol{g}_3]=\frac{1}{3}\begin{bmatrix}1 & 2 & 2\\ 2 & 1 & -2\\ 2 & -2 & 1\end{bmatrix}$

(Two eigenvalues coincide $\Rightarrow$ surface of revolution, axis along $\boldsymbol{g}_3$).

Coordinate transformation: $\begin{bmatrix}x\\ y\\ z\end{bmatrix}=P\begin{bmatrix}\bar{x}\\ \bar{y}\\ \bar{z}\end{bmatrix}$

Canonical form: $9\bar{x}^2+9\bar{y}^2=1$ (circular cylinder)

Axis of revolution: $(x, y, z)=t(2, -2, 1)$

Classification

Canonical form: $\lambda_1\bar{x}^2+\lambda_2\bar{y}^2+\lambda_3\bar{z}^2=c$

λ_1	λ_2	λ_3	c	Type
+	+	+	+	Ellipsoid
+	+	+	0	One point
+	+	−	+	Hyperboloid of one sheet
+	+	−	−	Hyperboloid of two sheets
+	+	−	0	Elliptic double cone
+	+	0	+	Elliptic cylinder
+	+	0	0	One line
+	−	0	+	Hyperbolic cylinder
+	−	0	0	Two intersecting planes
+	0	0	+	Two parallel planes
+	0	0	0	Two coincident planes

4.7 Linear Spaces

Vector spaces

A set L of elements $\boldsymbol{x}, \boldsymbol{y}, \boldsymbol{z}, \ldots$, is called a *linear space* or a *vector space* (over $\boldsymbol{R}$) if addition and multiplication by scalars are defined so that the following laws are satisfied for all $\boldsymbol{x}, \boldsymbol{y}, \boldsymbol{z} \in L$ and $\lambda, \mu \in \boldsymbol{R}$.

I. 1. $\boldsymbol{x}+\boldsymbol{y} \in L$ 2. $\boldsymbol{x}+\boldsymbol{y}=\boldsymbol{y}+\boldsymbol{x}$ 3. $(\boldsymbol{x}+\boldsymbol{y})+\boldsymbol{z}=\boldsymbol{x}+(\boldsymbol{y}+\boldsymbol{z})$
4. there exists $\boldsymbol{0}$ such that $\boldsymbol{x}+\boldsymbol{0}=\boldsymbol{x}$
5. there exists $-\boldsymbol{x}$ such that $\boldsymbol{x}+(-\boldsymbol{x})=\boldsymbol{0}$

II. 1. $\lambda\boldsymbol{x} \in L$ 2. $\lambda(\mu\boldsymbol{x})=(\lambda\mu)\boldsymbol{x}$ 3. $(\lambda+\mu)\boldsymbol{x}=\lambda\boldsymbol{x}+\mu\boldsymbol{x}$
4. $\lambda(\boldsymbol{x}+\boldsymbol{y})=\lambda\boldsymbol{x}+\lambda\boldsymbol{y}$ 5. $1\cdot\boldsymbol{x}=\boldsymbol{x}$ 6. $0\cdot\boldsymbol{x}=\boldsymbol{0}$ 7. $\lambda\boldsymbol{0}=\boldsymbol{0}$

Test for subspace

A nonempty subset M of L is a linear space itself if

1. $\boldsymbol{x}, \boldsymbol{y} \in M \Rightarrow \boldsymbol{x}+\boldsymbol{y} \in M$, 2. $\boldsymbol{x} \in M, \lambda \in \boldsymbol{R} \Rightarrow \lambda\boldsymbol{x} \in M$.

Linear combinations. Basis

1. The vector $\boldsymbol{y}$ is a linear combination of $\boldsymbol{x}_1, \ldots, \boldsymbol{x}_n$ if $\boldsymbol{y}=\lambda_1\boldsymbol{x}_1+\ldots+\lambda_n\boldsymbol{x}_n$.
2. The linear hull $LH(\boldsymbol{x}_1, \ldots, \boldsymbol{x}_n)$ is $\{\boldsymbol{y} : \boldsymbol{y}=\lambda_1\boldsymbol{x}_1+\ldots+\lambda_n\boldsymbol{x}_n, \lambda_i \in \boldsymbol{R}\}$.

3. $\boldsymbol{x}_1, \ldots, \boldsymbol{x}_n$ are

   (*i*) *linearly independent* if $\lambda_1\boldsymbol{x}_1+\ldots+\lambda_n\boldsymbol{x}_n=\boldsymbol{0} \Rightarrow \lambda_i=0$, all i
   (*ii*) *linearly dependent* if there exist $\lambda_1, \ldots, \lambda_n$, not all zero, such that $\lambda_1\boldsymbol{x}_1+\ldots+\lambda_n\boldsymbol{x}_n=\boldsymbol{0}$ ($\Leftrightarrow$ some $\boldsymbol{x}_i$ is a linear combination of the others.)

4. $\boldsymbol{e}_1, \ldots, \boldsymbol{e}_n$ is a basis of the linear space L and L is *n-dimensional* if

   (*i*) $\boldsymbol{e}_1, \ldots, \boldsymbol{e}_n$ are linearly independent
   (*ii*) every $\boldsymbol{x}\in L$ can be written (uniquely)

$$\boldsymbol{x}=x_1\boldsymbol{e}_1+\ldots+x_n\boldsymbol{e}_n$$

Scalar product

1. Let L be a linear space. A *scalar product* $(\boldsymbol{x}, \boldsymbol{y})$ [other notations $\boldsymbol{x}\cdot\boldsymbol{y}$, $(\boldsymbol{x}|\mathrm{y})$, $\langle\boldsymbol{x}, \boldsymbol{y}\rangle$ etc.] is a function $L\times L \to \boldsymbol{R}$ with the following properties holding for all $\boldsymbol{x}, \boldsymbol{y}, \boldsymbol{z}\in L$ and $\lambda, \mu\in\boldsymbol{R}$:

1. $(\boldsymbol{x}, \boldsymbol{y})=(\boldsymbol{y}, \boldsymbol{x})$ 2. $(\boldsymbol{x}, \lambda\boldsymbol{y}+\mu\boldsymbol{z})=\lambda(\boldsymbol{x}, \boldsymbol{y})+\mu(\boldsymbol{x}, \boldsymbol{z})$
3. $(\boldsymbol{x}, \boldsymbol{x})\geqslant 0$ (Equality $\Leftrightarrow \boldsymbol{x}=\boldsymbol{0}$)

2. *Length* of $\boldsymbol{x}$: $|\boldsymbol{x}|=\sqrt{(\boldsymbol{x}, \boldsymbol{x})}$, $|c\boldsymbol{x}|=|c|\cdot|\boldsymbol{x}|$ (c scalar).

3. $|(\boldsymbol{x}, \boldsymbol{y})|\leqslant|\boldsymbol{x}|\cdot|\boldsymbol{y}|$ (Cauchy-Schwarz' inequality) $|\boldsymbol{x}+\boldsymbol{y}|\leqslant|\boldsymbol{x}|+|\boldsymbol{y}|$ (Triangle ineq.)

Orthonormal basis

Let L be an n-dimensional linear space with scalar product (*Euclidean space*).

1. A basis $\boldsymbol{e}_1, \ldots, \boldsymbol{e}_n$ is called *orthonormal* (*ON*-basis) if

$$(\boldsymbol{e}_i, \boldsymbol{e}_j)=\begin{cases}1, & i=j\\ 0, & i\neq j\end{cases}$$

2. $\boldsymbol{e}_1, \ldots, \boldsymbol{e}_n$ *ON*-basis, $\boldsymbol{x}=\sum_{k=1}^{n} x_k\boldsymbol{e}_k$, $\boldsymbol{y}=\sum_{k=1}^{n} y_k\boldsymbol{e}_k \Rightarrow$
(*i*) $x_k=(\boldsymbol{x}, \boldsymbol{e}_k)$ (*ii*) $\lvert\boldsymbol{x}\rvert^2=\sum_{k=1}^{n} x_k^2$ (*iii*) $(\boldsymbol{x}, \boldsymbol{y})=\sum_{k=1}^{n} x_ky_k$

Orthogonal complement

M subspace of L:

$M^{\perp}=\{\boldsymbol{y}\in L\text{: } (\boldsymbol{x}, \boldsymbol{y})=0, \text{ all } \boldsymbol{x}\in M\}$

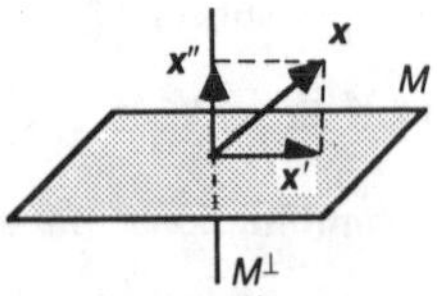

Orthogonal projection

M subspace, $\boldsymbol{e}_1, \ldots, \boldsymbol{e}_m$ *ON*-basis of M:

$\boldsymbol{x}'$ is the orthogonal projection of $\boldsymbol{x}$ on M if $\boldsymbol{x}=\boldsymbol{x}'+\boldsymbol{x}''$, $\boldsymbol{x}'\in M$, $\boldsymbol{x}''\in M^{\perp}$,

Moreover, $\boldsymbol{x}'=\sum_{k=1}^{m}(\boldsymbol{x}, \boldsymbol{e}_k)\boldsymbol{e}_k$

Gram – Schmidt orthogonalization

Given: $\boldsymbol{v}_1, \ldots, \boldsymbol{v}_n$ basis of linear space L with scalar product.

Sought: *ON*-basis $\boldsymbol{e}_1, \ldots, \boldsymbol{e}_n$ of L.

Construction: (Gram – Schmidt):

(1) $\boldsymbol{e}_1 = \dfrac{1}{|\boldsymbol{v}_1|}\boldsymbol{v}_1$

(2) $\boldsymbol{f}_2 = \boldsymbol{v}_2 - (\boldsymbol{v}_2, \boldsymbol{e}_1)\boldsymbol{e}_1, \quad \boldsymbol{e}_2 = \dfrac{1}{|\boldsymbol{f}_2|}\boldsymbol{f}_2$

...

(*k*) $\boldsymbol{f}_k = \boldsymbol{v}_k - (\boldsymbol{v}_k, \boldsymbol{e}_1)\boldsymbol{e}_1 - \ldots - (\boldsymbol{v}_k, \boldsymbol{e}_{k-1})\boldsymbol{e}_{k-1}; \quad \boldsymbol{e}_k = \dfrac{1}{|\boldsymbol{f}_k|}\boldsymbol{f}_k$

4.8 Linear Mappings

Let L, M be linear spaces. A function $F: L \to M$ is called a linear mapping if

$F(\lambda\boldsymbol{x}+\mu\boldsymbol{y}) = \lambda F(\boldsymbol{x}) + \mu F(\boldsymbol{y})$
for all $\boldsymbol{x}, \boldsymbol{y} \in L$, $\lambda, \mu \in \boldsymbol{R}$

Matrix representation

Assumption. (*i*) $\boldsymbol{e}_1, \ldots, \boldsymbol{e}_n$ basis of L, $\boldsymbol{f}_1, \ldots, \boldsymbol{f}_m$ basis of M.

$$(ii)\ F(\boldsymbol{e}_j) = \sum_{i=1}^{m} a_{ij}\boldsymbol{f}_i, \quad A = \begin{bmatrix} a_{11} \ldots & \boxed{a_{1j}} & \ldots a_{1n} \\ \ldots & & \\ a_{m1} \ldots & \boxed{a_{mj}} & \ldots a_{mn} \end{bmatrix} = [F(\boldsymbol{e}_1), \ldots, F(\boldsymbol{e}_n)]$$

$F(\boldsymbol{e}_j)$

Then if $\boldsymbol{x} = \sum_{j=1}^{n} x_j\boldsymbol{e}_j$, $F(\boldsymbol{x}) = \sum_{i=1}^{m} y_i\boldsymbol{f}_i$, the mapping $\boldsymbol{x} \to F(\boldsymbol{x})$

is represented by the matrix A, i.e.

$$\begin{bmatrix} y_1 \\ \ldots \\ y_m \end{bmatrix} = \begin{bmatrix} a_{11} \ldots a_{1n} \\ \ldots \\ a_{m1} \ldots a_{mn} \end{bmatrix} \begin{bmatrix} x_1 \\ \ldots \\ x_n \end{bmatrix} \Leftrightarrow Y = AX$$

Inverse mapping

If $F: L \to M$ is invertible then
$F^{-1}: M \to L$ is represented by A^{-1}

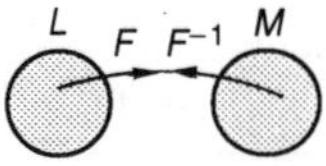

Composite mappings

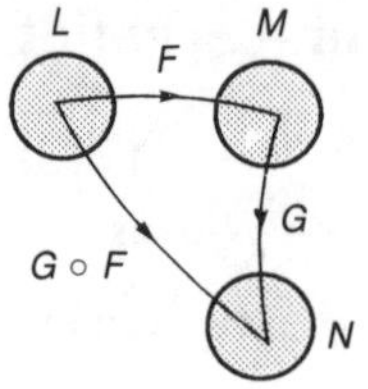

If $F : L \to M$, $G : M \to N$ and if F and G are represented by A and B respectively, then the mapping $G \circ F$ (defined by $(G \circ F)(\boldsymbol{x}) = G(F(\boldsymbol{x}))$ is represented by the matrix BA.

Symmetric mappings

Let L be a finite-dimensional Euclidean space and $F : L \to L$ a linear mapping.

1. F is called *symmetric* if $(F\boldsymbol{x}, \boldsymbol{y}) = (\boldsymbol{x}, F\boldsymbol{y})$, all $\boldsymbol{x}, \boldsymbol{y} \in L$.
2. The number λ is called *eigenvalue*, $\boldsymbol{x} \neq \boldsymbol{0}$ a corresponding *eigenvector*, if $F\boldsymbol{x} = \lambda\boldsymbol{x}$.
3. (*The spectral theorem.*) If F is a symmetric mapping then there is an orthonormal basis of L consisting of eigenvectors of F.
   The matrix representing F in this basis is diagonal with the eigenvalues along the diagonal.
4. (*Projections*) A matrix P represents an orthogonal projection $\Leftrightarrow$ $P^2 = P$ and $P^t = P \Leftrightarrow P^2 = P$ and $|P\boldsymbol{x}| \leq |\boldsymbol{x}|$, all $\boldsymbol{x}$.

   If a matrix A has linearly independent columns, then A^tA is invertible and $P = A(A^tA)^{-1}A^t$ represents the orthogonal projection onto the column space of A.

Change of basis

(Cf. sec. 4.4)

Let $\boldsymbol{e}_1, \ldots, \boldsymbol{e}_n$ and $\overline{\boldsymbol{e}}_1, \ldots, \overline{\boldsymbol{e}}_n$ be two bases of L with the relationships

$$\begin{cases} \overline{\boldsymbol{e}}_i = \sum_{j=1}^{n} p_{ji}\, \boldsymbol{e}_j, & P = (p_{ij}) = [\overline{\boldsymbol{e}}_1, \ldots, \overline{\boldsymbol{e}}_n] \\ \boldsymbol{e}_i = \sum_{j=1}^{n} q_{ji}\, \overline{\boldsymbol{e}}_j, & Q = (q_{ij}),\ Q = P^{-1} = [\boldsymbol{e}_1, \ldots, \boldsymbol{e}_n] \end{cases}$$

1. (*Relation between components*)

Let $\boldsymbol{x} = \sum_{i=1}^{n} x_i\, \boldsymbol{e}_i = \sum_{i=1}^{n} \overline{x}_i\, \overline{\boldsymbol{e}}_i$, $\quad \boldsymbol{X} = \begin{bmatrix} x_1 \\ x_2 \\ \cdots \\ x_n \end{bmatrix}$, $\overline{\boldsymbol{X}} = \begin{bmatrix} \overline{x}_1 \\ \overline{x}_2 \\ \cdots \\ \overline{x}_n \end{bmatrix}$. Then

$$\boldsymbol{X} = P\overline{\boldsymbol{X}} \quad x_i = \sum_{j=1}^{n} p_{ij}\, \overline{x}_j$$

$$\overline{\boldsymbol{X}} = Q\boldsymbol{X} \quad \overline{x}_i = \sum_{j=1}^{n} q_{ij}\, x_j$$

2. (*Relation between matrix representations.*)

If $F: L \to L$ is represented by the matrices A and $\bar{A}$ with respect to the two bases, respectively, then

$$\bar{A}=P^{-1}AP \quad \bar{a}_{ij}=\sum_{k,l} q_{il}\, a_{lk}\, p_{kj}$$

$$\bar{A}=P^{t}AP \quad \bar{a}_{ij}=\sum_{k,l} p_{li}\, a_{lk}\, p_{kj} \quad (P \text{ orthogonal})$$

Range and nullspace

Given $F: L \to M$, A the corresponding matrix.

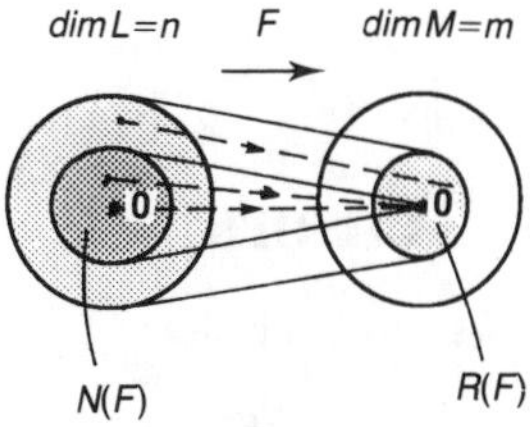

1. (a) *Nullspace*
   $N(F)=\{\mathbf{x}\in L : F(\mathbf{x})=\mathbf{0}\}\subset L$

   (b) $N(A)$=set of column vectors $\boldsymbol{X}\in \boldsymbol{R}^n$ such that $AX=\mathbf{0}$.

2. (a) *Range*
   $R(F)=\{F(x)\in M : x\in L\}\subset M$

   (b) $R(A)$=space spanned by the column vectors of A (subspace of R^m).

Theorem

1. $\dim L=\dim N(F)+\dim R(F)=\dim N(A)+\dim R(A)$
2. $\dim R(A)=\dim R(A^t)=\text{rank}(A)$
3. $\text{rank}\,(A)=\text{rank}\,(A^t)=\text{rank}\,(A^tA)=\text{rank}\,(AA^t)$
4. $\dim N(A)=n-\text{rank}\,(A)$
5. $N(A)=N(A^tA)$
6. $R(A^t)=R(A^tA)$
7. $N(A)=R(A^t)^{\perp}$

Example

Determine $R(A)$ and $N(A)$ of the matrix A below.

Solution: By elementary row operations (see sec. 4.1),

$$A=\begin{bmatrix} 1 & 2 & 1 & 1 & 3 \\ 1 & 2 & 2 & 3 & 6 \\ 2 & 4 & 2 & 3 & 8 \end{bmatrix} \sim \begin{bmatrix} 1 & 2 & 1 & 1 & 3 \\ 0 & 0 & 1 & 2 & 3 \\ 0 & 0 & 0 & 1 & 2 \end{bmatrix}=B$$
$$\;\;\uparrow \quad\;\; \uparrow\;\uparrow \qquad\qquad \uparrow \quad\;\; \uparrow\;\uparrow$$

(i) $R(A)$ is spanned by *those columns of A* which correspond to the columns (marked with arrows) with pivot elements of the echolon form matrix B. Thus the column vectors $(1,1,2)^t$, $(1,2,2)^t$ and $(1,3,3)^t$ constitute a basis for $R(A)$ and $\dim R(A)=3$.

(ii) Solving the system $A\mathbf{x}=\mathbf{0} \Leftrightarrow \begin{cases} x_1+2x_2+x_3+x_4+3x_5=0 \\ x_3+2x_4+3x_5=0 \\ x_4+2x_5=0 \end{cases}$

yields $x_5=t$, $x_4=-2t$, $x_3=t$, $x_2=s$, $x_1=-2t-2s$ or $(x_1, x_2, x_3, x_4, x_5)=t(-2, 0, 1, -2, 1)+s(-2, 1, 0, 0, 0)$.

Hence the column vectors $(-2, 0, 1, -2, 1)^t$ and $(-2, 1, 0, 0, 0)^t$ constitute a basis for $N(A)$ and $\dim N(A)=2$.

4.9 Tensors

Relations between bases:

$$\bar{e}_i=\sum_j p_i{}^j e_j, \quad e_i=\sum q_i{}^j \bar{e}_j, \quad Q=(q_i{}^j)=P^{-1}=(p_i{}^j)^{-1}$$

Tensors of order 1

1. $a=(a_i)$ *covariant* if $\bar{a}_i=\sum_j p_i{}^j a_j$ (transformation as basis vectors).

2. $b=(b^i)$ *contravariant* if $\bar{b}^i=\sum_j q_j{}^i b^j$ (transformation as vector components).

Tensors of order 2

3. $A=(a_{ij})$ *covariant* if $\bar{a}_{ij}=\sum_{k,l} p_i{}^k p_j{}^l a_{kl}$

4. $B=(b^{ij})$ *contravariant* if $\bar{b}^{ij}=\sum_{k,l} q_k{}^i q_l{}^j b^{kl}$

5. $C=(c_i{}^j)$ *mixed* if $\bar{c}_i{}^j=\sum_{k,l} p_i{}^k q_l{}^j c_k{}^l$

Tensors of higher orders are defined similarly.

The *stress tensor* (T_{ij}) and *the strain tensor* (ε_{ij}) are covariant tensors of second orders. The *metric tensor* (g_{ij}) where $g_{ij}=(e_i, e_j)$ is a covariant tensor.

Operations

1. Sum of tensors and multiplication by a scalar are defined component-wise.

2. Tensor product $C=A\otimes B$. E.g. if $A=(a_j{}^i)$, $B=(b_{ij})$ then $c^i{}_{jkl}=a_j{}^i b_{kl}$ (mixed tensor of order 4).

3. Contraction. E.g. $d_{jl}=\sum_k c^k{}_{jkl}=\sum_k a_j{}^k b_{kl}$.

4. Trace. $Tr(A)=\sum_i a_i{}^i$.

4.10 Complex matrices

In this section components of vectors and elements of matrices are complex numbers.

Vectors in $\mathbf{C}^n$

Let **a**, **b** be (column) vectors in $\mathbf{C}^n$

1. *Scalar product* $\mathbf{a}^*\mathbf{b}=\bar{a}_1b_1+\bar{a}_2b_2+\ldots+\bar{a}_nb_n$, $\mathbf{b}^*\mathbf{a}=\overline{\mathbf{a}^*\mathbf{b}}$
   $\mathbf{a}\perp\mathbf{b} \Leftrightarrow \mathbf{a}^*\mathbf{b}=0$ (**a** and **b** orthogonal) Here the bar denotes complex conjugate.
2. *Length (norm)* $|\mathbf{a}|^2=\mathbf{a}^*\mathbf{a}=|a_1|^2+|a_2|^2+\ldots+|a_n|^2$, $|c\mathbf{a}|=|c|\cdot|\mathbf{a}|$, c=complex number.
3. *Cauchy-Schwarz' inequality:* $|\mathbf{a}^*\mathbf{b}|\leq|\mathbf{a}|\cdot|\mathbf{b}|$.
4. *Triangle inequality:* $|\mathbf{a}+\mathbf{b}|\leq|\mathbf{a}|+|\mathbf{b}|$.
5. *Pythagorean relation:* $|\mathbf{a}+\mathbf{b}|^2=|\mathbf{a}|^2+|\mathbf{b}|^2 \Leftrightarrow \mathbf{a}\perp\mathbf{b}$.

Matrices

6. *Adjoint* A^* of $A=(a_{ij})$: $A^*=(\overline{a_{ji}})$.

$(A+B)^*=A^*+B^*$	$(cA)^*=\bar{c}A^*(c\in\mathbf{C})$
$(AB)^*=B^*A^*$	$A^{**}=A$
$(AB)^{-1}=B^{-1}A^{-1}$	$(A^{-1})^*=(A^*)^{-1}$

Inverse matrices

Set $C=A+iB$, A, B real matrices.

7. If A^{-1} exists then $C^{-1}=(A+BA^{-1}B)^{-1}-iA^{-1}B(A+BA^{-1}B)^{-1}$.
8. If B^{-1} exists then $C^{-1}=B^{-1}A(B+AB^{-1}A)^{-1}-i(B+AB^{-1}A)^{-1}$.
9. A, B singular, C regular: Let r be a real number such that $A+rB$ is regular. Set $F=A+rB$ and $G=A-rB$. Then $C^{-1}=(1-ir)(F+iG)^{-1}$. Continue as in 7.

Differentiation of matrices

10. If $A=A(x)=(a_{ij}(x))$ and $B=B(x)$, then

    (*i*) $A'(x)=(a_{ij}'(x))$ (*ii*) $(A+B)'=A'+B'$
    (*iii*) $(AB)'=AB'+A'B$ (*iv*) $(A^2)'=AA'+A'A$
    (*v*) $(A^{-1})'=-A^{-1}A'A^{-1}$

Matrix norms, see sec. 16.2.

Unitary matrices

11. A square $n\times n$-matrix U is *unitary* if $U^*U=UU^*=I$ (I=identity matrix).
12. Let $U=[\mathbf{u}_1, \mathbf{u}_2, \ldots, \mathbf{u}_n]$, where $\mathbf{u}_i$ are the column vectors of U. Then
    (*i*) U is unitary $\Leftrightarrow$ The column vectors of U are pairwise orthogonal and normed, i.e. $\mathbf{u}_i{}^*\mathbf{u}_j=\delta_{ij}$, $i, j=1, 2, \ldots, n$.

(*ii*) U unitary $\Rightarrow$

(*a*) U^* is unitary (*b*) $U^{-1}=U^*$

(*c*) $|\det U|=1$ (*d*) $(U\mathbf{a})^*(U\mathbf{b})=\mathbf{a}^*\mathbf{b}$

(*e*) $|U\mathbf{a}|=|\mathbf{a}|$ (*f*) $\mathbf{a}\perp\mathbf{b}\Leftrightarrow U\mathbf{a}\perp U\mathbf{b}$

(*iii*) U, V unitary of the same order $\Rightarrow UV$ unitary.

Normal matrices

13. A (square) matrix N is *normal* if $N^*N=NN^*$.
14. A (square) matrix H is *Hermitian* if $H^*=H$.
A (square) matrix H is *skew-Hermitian* if $H^*=-H$.

H is Hermitian $\Rightarrow iH$ is skew-Hermitian
H is skew-Hermitian $\Rightarrow iH$ is Hermitian
H is Hermitian $\Rightarrow H^{-1}$ is Hermitian (if it exists)
Hermitian, skew-Hermitian and unitary matrices are normal.

15. A normal matrix is
(*i*) Hermitian $\Leftrightarrow$ all eigenvalues are real
(*ii*) skew-Hermitian $\Leftrightarrow$ all eigenvalues are pure imaginary
(*iii*) unitary $\Leftrightarrow$ all eigenvalues have modulus 1.

Spectral theory, unitary transformations

16. λ is an eigenvalue of a square square $n\times n$-matrix A and $\mathbf{g}\neq\mathbf{0}$ is a corresponding eigenvector if $A\mathbf{g}=\lambda\mathbf{g}$.
17. λ is an eigenvalue of $A\Leftrightarrow\det(A-\lambda I)=0$.
18. *Gerschgorin's theorem.* Let C_i denote the cirular region
$$C_i=\{z: |z-a_{ii}|\leqslant \sum_{j=1 (j\neq i)}^{n} |a_{ij}|\}, \quad i=1, \ldots, n.$$
Then the eigenvalues of A are contained in the union of the C_i.
19. *Cayley-Hamilton's theorem:* The characteristic equation $P(\lambda)=\det(A-\lambda I)=(-1)^n\lambda^n+ +c_{n-1}\lambda^{n-1}+\ldots+c_0=0$ is satisfied by$A$, i.e. $P(A)=(-1)^nA^n+c_{n-1}A^{n-1}+\ldots+c_0I=0$.
20. *Schur's lemma:* For any square matrix A there exists a unitary matrix U such that $T=U^*AU$ is (upper) triangular. The elements t_{ii} of the diagonal of T are the eigenvalues (incl. multiplicities) of A.
21. **Spectral theorem.** (*Unitary transformation*)

Assume that (*i*) N is normal (*ii*) $\lambda_1, \lambda_2, \ldots, \lambda_n$ are eigenvalues (incl. multiplicities) of N.

Then there exist corresponding eigenvectors $\mathbf{g}_1, \mathbf{g}_2, \ldots, \mathbf{g}_n$ such that

(*a*) $\mathbf{g}_i\perp\mathbf{g}_j$, $i\neq j$, $|\mathbf{g}_i|=1$ (pairwise orthogonal and normed)

(*b*) $U=[\mathbf{g}_1, \mathbf{g}_2, \ldots, \mathbf{g}_n]$ is a unitary matrix

(*c*) $U^*NU=D=\text{diag}(\lambda_1, \lambda_2, \ldots, \lambda_n)$

22. If N is normal then eigenvectors corresponding to different eigenvalues are orthogonal.

Hermitian forms

A hermitian form is a homogeneous polynomial of second degree of the form

$$h=h(\mathbf{z})=\mathbf{z}^*H\mathbf{z}=\sum_{ij=1}^{n} h_{ij}z_i\bar{z}_j \quad (h_{ji}=\overline{h_{ij}})$$

23. The values $h(\mathbf{z})$ are real, all $\mathbf{z}$.
24. By a suitable unitary transformation (see above) $\mathbf{z}=U\zeta$, $h=\sum_{k=1}^{n} \lambda_k|\zeta_k|^2$.
25. $\lambda_{\min}|\mathbf{z}|^2 \leq h(\mathbf{z}) \leq \lambda_{\max}|\mathbf{z}|^2$ (Equality $\Leftrightarrow$ $\mathbf{z}$=corresponding eigenvector).
26. *Law of inertia.* If a Hermitian form is written as a sum of squares (e.g. by completing squares or by a unitary transformation) in two different ways, then the numbers of positive coefficients as well as negative and zero coefficients are the same in both cases.
27. A Hermitian form is positive definite if

$$h(\mathbf{z})>0,\ \mathbf{z}\neq 0 \text{ or all } \lambda_k>0.$$

Decomposition of matrices

28. For any square matrix A there exist unique Hermitian matrices H_1 and H_2 [$H_1=(A+A^*)/2$ and $H_1=(A-A^*)/2i$] such that $A=H_1+iH_2$.
29. N is normal $\Leftrightarrow N=H_1+iH_2$ with commuting Hermitian matrices H_1 and H_2 (i.e. $H_1H_2=H_2H_1$).
30. Let H_1 and H_2 be Hermitian. Then there exist a unitary matrix U simultaneously diagonalizing H_1 and H_2 (i.e. U^*H_1U and U^*H_2U are diagonal) $\Leftrightarrow$ $H_1H_2=H_2H_1$.

Non-unitary transformations

31. Assume that the square $n\times n$-matrix A has n linear independent eigenvectors $\mathbf{g}_1, \mathbf{g}_2, \ldots, \mathbf{g}_n$ (e.g. this is the case if the n eigenvalues $\lambda_1, \lambda_2, \ldots, \lambda_n$ are distinct). Then with $L=[\mathbf{g}_1, \mathbf{g}_2, \ldots, \mathbf{g}_n]$, $L^{-1}AL=D=\text{diag}(\lambda_1, \lambda_2, \ldots, \lambda_n)$.
32. *Jordan form:* For any matrix A there exists a non-singular matrix S such that

$$S^{-1}AS=J=\begin{bmatrix} J_1 & 0 & \cdots & & 0 \\ 0 & J_2 & 0 & \cdots & 0 \\ \cdots & & & & \\ \cdots & & & & \\ \cdots & & & 0 & J_m \end{bmatrix}, \qquad J_i=\begin{bmatrix} \lambda_i & 1 & 0 & \cdots & 0 \\ 0 & \lambda_i & 1 & 0 \cdots & 0 \\ \cdots & & & & \\ 0 & \cdots & 0 & \lambda_i & 1 \\ 0 & \cdots & & 0 & \lambda_i \end{bmatrix}$$

where J_i are the Jordan blocks. The same eigenvalue λ_i may appear in several blocks if it corresponds to several independent eigenvectors.

5 The Elementary Functions

5.1 A Survey of the Elementary Functions

Function $y=f(x)$	Domain D_f	Range R_f	Inverse function $x=f^{-1}(y)$	Derivative $f'(x)$	Primitive function $\int f(x)dx$
$y=x^n,\ n\in\mathbf{Z}^+$				nx^{n-1}	$\frac{x^{n+1}}{n+1}$
n even	all x	$y\geqslant 0$	$x=\sqrt[n]{y},\ x\geqslant 0$		
n odd	all x	all y	$x=\sqrt[n]{y}$		
$y=x^{-n},\ n\in\mathbf{Z}^+$				$-\frac{n}{x^{n+1}}$	$\frac{x^{1-n}}{1-n},\ n\neq 1$
n even	$x\neq 0$	$y>0$	$x=1/\sqrt[n]{y},\ x>0$		
n odd	$x\neq 0$	$y\neq 0$	$x=1/\sqrt[n]{y}$		$\ln\lvert x\rvert,\ n=1$
$y=x^a,\ a\notin\mathbf{Z}$				ax^{a-1}	$\frac{x^{a+1}}{a+1}$
$a>0$	$x\geqslant 0$	$y\geqslant 0$	$x=y^{1/a}$		
$a<0$	$x>0$	$y>0$			
$y=e^x$	all x	$y>0$	$x=\ln y$	e^x	e^x
$y=a^x (a>0,\ a\neq 1)$	all x	$y>0$	$x={}^a\log y=\frac{\ln y}{\ln a}$	$a^x \ln a$	$a^x/\ln a$
$y=\ln x$	$x>0$	all y	$x=e^y$	$1/x$	$x \ln x-x$
$y={}^a\log x\ (a>0,\ a\neq 1)$	$x>0$	all y	$x=a^y$	$1/(x \ln a)$	$(x \ln x-x)/\ln a$
$y=\sinh x$	all x	all y	$x=\ln(y+\sqrt{y^2+1})$	$\cosh x$	$\cosh x$
$y=\cosh x$	all x	$y\geqslant 1$	$x=\ln(y+\sqrt{y^2-1}),\ (x\geqslant 0)$	$\sinh x$	$\sinh x$
$y=\tanh x$	all x	$\lvert y\rvert<1$	$x=\frac{1}{2}\ln\frac{1+y}{1-y}$	$1/\cosh^2 x$	$\ln(\cosh x)$
$y=\coth x$	$x\neq 0$	$\lvert y\rvert>1$	$x=\frac{1}{2}\ln\frac{y+1}{y-1}$	$-1/\sinh^2 x$	$\ln\lvert\sinh x\rvert$
$y=\sin x$	all x	$-1\leqslant y\leqslant 1$	$x=\arcsin y\ \left(-\frac{\pi}{2}\leqslant x\leqslant\frac{\pi}{2}\right)$	$\cos x$	$-\cos x$
$y=\cos x$	all x	$-1\leqslant y\leqslant 1$	$x=\arccos y\ (0\leqslant x\leqslant\pi)$	$-\sin x$	$\sin x$
$y=\tan x$	$x\neq\frac{\pi}{2}+n\pi$	all y	$x=\arctan y\ \left(-\frac{\pi}{2}<x<\frac{\pi}{2}\right)$	$1/\cos^2 x$	$-\ln\lvert\cos x\rvert$
$y=\cot x$	$x\neq n\pi$	all y	$x=\operatorname{arccot} y\ (0<x<\pi)$	$-1/\sin^2 x$	$\ln\lvert\sin x\rvert$
$y=\sec x$	$x\neq\frac{\pi}{2}+n\pi$	$\lvert y\rvert\geqslant 1$	$x=\arccos\frac{1}{y}\ (0\leqslant x\leqslant\pi)$	$\sin x\ \sec^2 x$	$\ln\left\lvert\tan\left(\frac{x}{2}+\frac{\pi}{4}\right)\right\rvert$
$y=\csc x$	$x\neq n\pi$	$\lvert y\rvert\geqslant 1$	$x=\arcsin\frac{1}{y}\ \left(-\frac{\pi}{2}\leqslant x\leqslant\frac{\pi}{2}\right)$	$-\cos x\ \csc^2 x$	$\ln\left\lvert\tan\frac{x}{2}\right\rvert$
$y=\arcsin x$	$-1\leqslant x\leqslant 1$	$-\frac{\pi}{2}\leqslant y\leqslant\frac{\pi}{2}$	$x=\sin y$	$\frac{1}{\sqrt{1-x^2}}$	$x\arcsin x+\sqrt{1-x^2}$
$y=\arccos x$	$-1\leqslant x\leqslant 1$	$0\leqslant y\leqslant\pi$	$x=\cos y$	$-\frac{1}{\sqrt{1-x^2}}$	$x\arccos x-\sqrt{1-x^2}$
$y=\arctan x$	all x	$-\frac{\pi}{2}<y<\frac{\pi}{2}$	$x=\tan y$	$\frac{1}{1+x^2}$	$x\arctan x-\frac{1}{2}\ln(1+x^2)$
$y=\operatorname{arccot} x$	all x	$0<y<\pi$	$x=\cot y$	$-\frac{1}{1+x^2}$	$x\operatorname{arccot} x+\frac{1}{2}\ln(1+x^2)$

5.2 Polynomials and Rational Functions

Polynomials

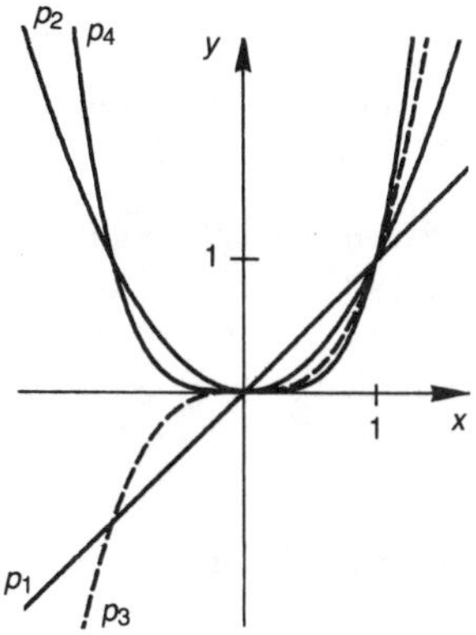

$P_k(x)=x^k$, $k=1, 2, 3, 4$

$P(x)=a_nx^n+a_{n-1}x^{n-1}+\ldots+a_1x+a_0$
degree $P(x)=n$ if $a_n\neq 0$.

Asymptotic behaviour

$P(x)\sim a_nx^n$ as $x\to\pm\infty$ $(a_n\neq 0)$

Algebraic equations and the factor theorem, see sec. 2.4

Polynomial division

If $\deg P_1(x)\geqslant\deg P_2(x)$ then uniquely

$$\frac{P_1(x)}{P_2(x)}=Q(x)+\frac{R(x)}{P_2(x)} \quad \text{or } P_1(x)=Q(x)P_2(x)+R(x),$$

where $\deg Q(x)=\deg P_1(x)-\deg P_2(x)$, $\deg R(x)<\deg P_2(x)$.

Here, $Q(x)$ is the *quotient polynomial* and $R(x)$ the *remainder polynomial.*

The following example illustrates division of polynomials.

$$\frac{2x^3-3x^2+9x+5}{x^2-x+3}=2x-1+\frac{2x+8}{x^2-x+3}:$$

$$\begin{array}{r|l} & 2x-1 \\ \hline x^2-x+3 & 2x^3-3x^2+9x+5 \\ & -(2x^3-2x^2+6x) \\ \cline{2-2} & -x^2+3x+5 \\ & -(-x^2+x-3) \\ \cline{2-2} & 2x+8 \end{array}$$

Greatest common divisor

An algorithm (*Euclid's algorithm*) for determining a polynomial of maximal degree dividing two given polynomials $P(x)$ and $Q(x)$ is the following:

Assume that $\deg P(x)\geqslant\deg Q(x)$. Dividing $P(x)$ by $Q(x)$ yields $P(x)=K_1(x)Q(x)+R_1(x)$ where $\deg R_1(x)<\deg Q(x)$. If $R_1(x)\neq 0$ then dividing $Q(x)$ by $R_1(x)$ gives $Q(x)=K_2(x)R_1(x)+R_2(x)$ where $\deg R_2(x)<\deg R_1(x)$. Continuing like this, let $R_j(x)$ be the first remainder polynomial which is zero. Then $R_{j-1}(x)$ is a polynomial of maximal degree that divides both $P(x)$ and $Q(x)$.

Rational functions

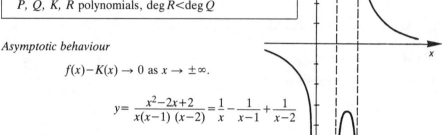

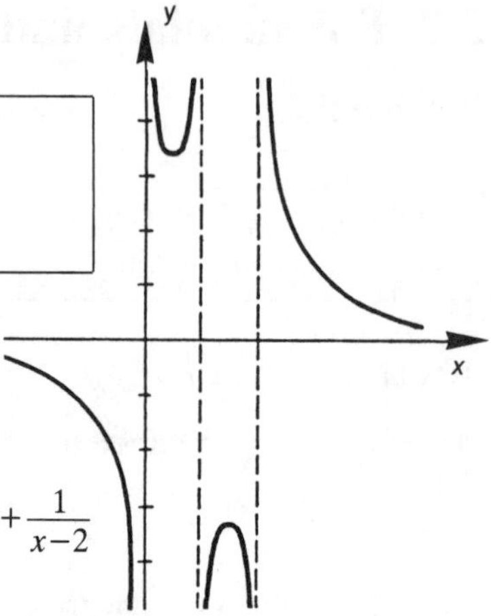

$$f(x)=\frac{P(x)}{Q(x)}=K(x)+\frac{R(x)}{Q(x)},$$

P, Q, K, R polynomials, $\deg R<\deg Q$

Asymptotic behaviour

$$f(x)-K(x)\to 0 \text{ as } x\to\pm\infty.$$

$$y=\frac{x^2-2x+2}{x(x-1)(x-2)}=\frac{1}{x}-\frac{1}{x-1}+\frac{1}{x-2}$$

Partial fractions

Assume that

(*i*) $R(x)$, $Q(x)$ are real polynomials, $\deg R<\deg Q$,
(*ii*) $Q(x)$ is decomposed in real factors of degree $\leqslant 2$, i.e.
$Q(x)=C(x-r)^m(x-s)^n\ \ldots\ (x^2+2ax+b)^p(x^2+2cx+d)^q\ \ldots,\ (a^2<b,\ c^2<d)$

Then

$$\frac{R(x)}{Q(x)}=\frac{R_1}{x-r}+\frac{R_2}{(x-r)^2}+\ldots+\frac{R_m}{(x-r)^m}+$$
$$+\frac{S_1}{x-s}+\frac{S_2}{(x-s)^2}+\ldots+\frac{S_n}{(x-s)^n}+\ldots+$$
$$+\frac{A_1x+B_1}{(x^2+2ax+b)}+\ldots+\frac{A_px+B_p}{(x^2+2ax+b)^p}+$$
$$+\frac{C_1x+D_1}{x^2+2cx+d}+\ldots+\frac{C_qx+D_q}{(x^2+2cx+d)^q}+\ldots,$$

where the capitals in the numerators are uniquely determined constants.

The following example illustrates how to find the constants.

$$\frac{1}{(x^2+1)(x+1)^2}=\frac{A}{x+1}+\frac{B}{(x+1)^2}+\frac{Cx+D}{x^2+1}=$$
$$=\frac{A(x+1)(x^2+1)+B(x^2+1)+(Cx+D)(x+1)^2}{(x+1)^2(x^2+1)}=$$
$$=\frac{(A+C)x^3+(A+B+2C+D)x^2+(A+C+2D)x+(A+B+D)}{(x+1)^2(x^2+1)}$$

Identification of coefficients:

$A+C=0,\ A+B+2C+D=0,\ A+C+2D=0,\ A+B+D=1$

$\Rightarrow A=B=-C=1/2,\ D=0$. Thus

$$\frac{1}{(x^2+1)(x+1)^2}=\frac{1}{2(x+1)}+\frac{1}{2(x+1)^2}-\frac{x}{2(x^2+1)}$$

5.3 Logarithmic, Exponential, Power and Hyperbolic Functions

Logarithmic functions

$y=\ln x,\quad y'=\frac{1}{x}\quad (x>0)$

$y={}^a\log x,\quad y'=\frac{1}{x\ln a}\quad (a>0,\ a\neq 1)$

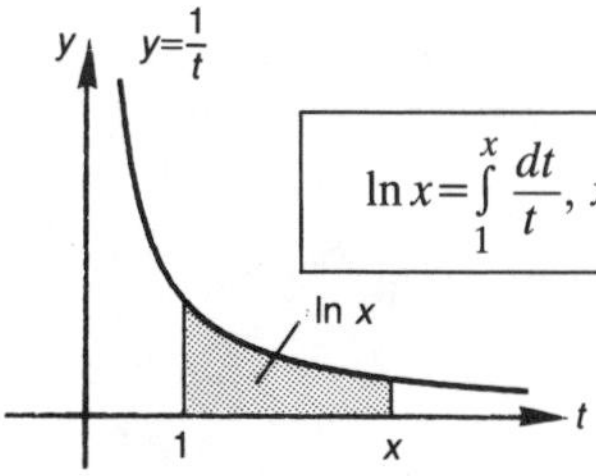

$\ln x=\int_1^x \frac{dt}{t},\ x>0$

$\ln 1=0,\ \ln e=1,\ \lim_{x\to 0^+}\ln x=-\infty,\ \lim_{x\to\infty}\ln x=\infty$

${}^a\log x+{}^a\log y={}^a\log xy\quad {}^a\log x-{}^a\log y={}^a\log\frac{x}{y}\quad {}^a\log x^p=p\ {}^a\log x$

${}^a\log\frac{1}{x}=-{}^a\log x\quad {}^a\log x=\frac{{}^b\log x}{{}^b\log a}=\frac{\ln x}{\ln a}$

Complex case: ${}^e\log z=\ln|z|+i\arg z$

Inverses

$$y=\ln x \Leftrightarrow x=e^y\quad y={}^a\log x\Leftrightarrow x=a^y=e^{y\ln a}$$

Exponential functions

Natural base $e=\lim_{n\to\infty}\left(1+\frac{1}{n}\right)^n\approx 2.71828\ 18285$

$y=e^x=\exp(x),\quad y'=e^x$

$y=a^x,\quad y'=a^x\ln a\ (a>0)$

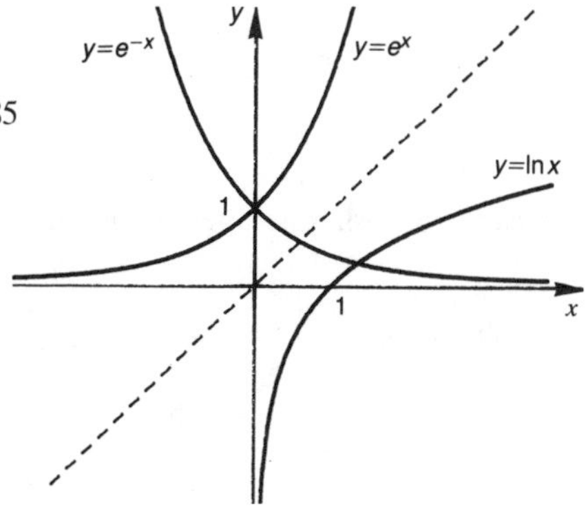

$a^0=1,\ \lim_{x\to-\infty}e^x=0,\ \lim_{x\to\infty}e^x=\infty$

$a^x a^y = a^{x+y}$ $\quad \dfrac{a^x}{a^y} = a^{x-y}$ $\quad (a^x)^y = a^{xy}$
$a^x b^x = (ab)^x$ $\quad a^{-x} = \dfrac{1}{a^x}$ $\quad a^x = e^{x \ln a}$
Complex case: $e^z = e^{x+iy} = e^x e^{iy} = e^x\ (\cos y + i \sin y)$

Inverses

$$y = e^x \Leftrightarrow x = \ln y, \quad y = a^x \Leftrightarrow x = {}^a\!\log y = \frac{\ln y}{\ln a}$$

Power functions

$y = x^a, \quad y' = a x^{a-1} \quad (x > 0)$

Complex case: $z^a = e^{a\,{}^e\!\log z}$

Inverses

$$y = x^a \Leftrightarrow x = y^{1/a}$$

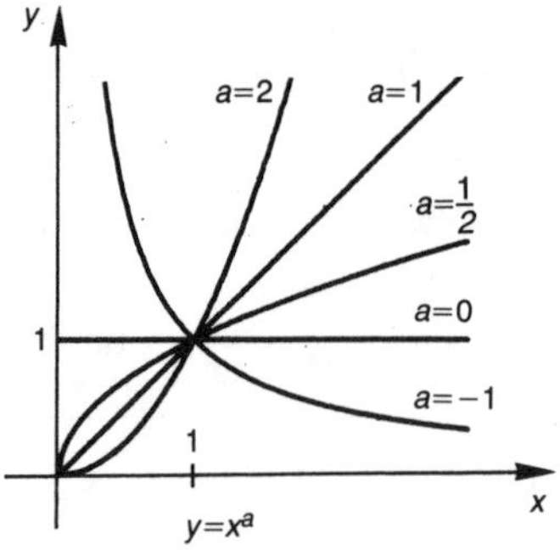

$y=x^a$

Hyperbolic functions

The hyperbolic functions are defined as follows.

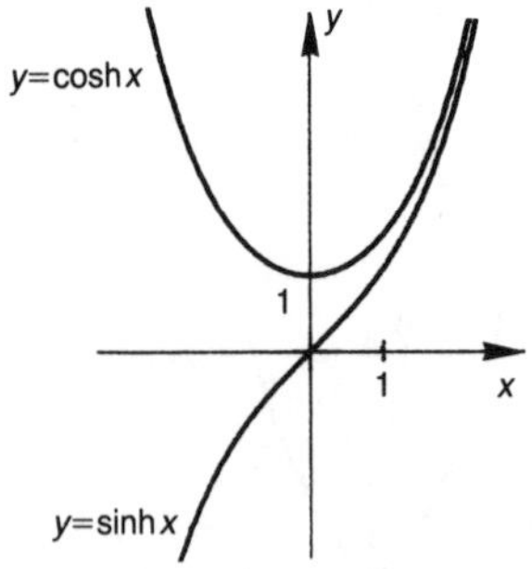

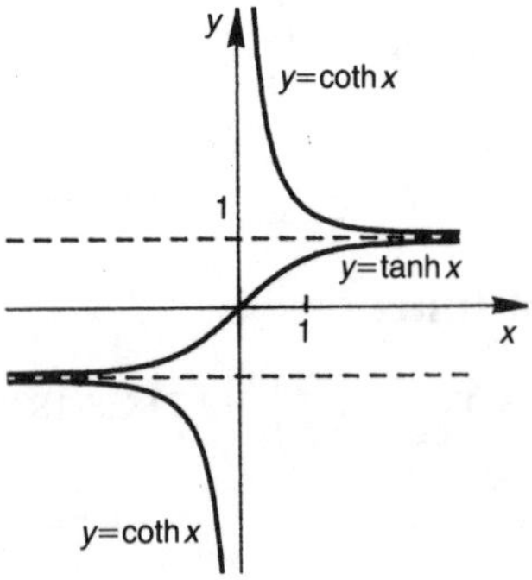

The curve y=cosh x is called *catenary*.

$y = \sinh x = \dfrac{e^x - e^{-x}}{2}$	$y = \cosh x = \dfrac{e^x + e^{-x}}{2}$	$y = \tanh x = \dfrac{e^x - e^{-x}}{e^x + e^{-x}}$	$y = \coth x = \dfrac{e^x + e^{-x}}{e^x - e^{-x}}$
$y' = \cosh x$	$y' = \sinh x$	$y' = 1/\cosh^2 x$	$y' = -1/\sinh^2 x$

Transformation table

	$\sinh x$	$\cosh x$	$\tanh x$	$\coth x$
$\sinh x=$	–	$\pm\sqrt{\cosh^2 x-1}$	$\dfrac{\tanh x}{\sqrt{1-\tanh^2 x}}$	$\pm\dfrac{1}{\sqrt{\coth^2 x-1}}$
$\cosh x=$	$\sqrt{1+\sinh^2 x}$	–	$\dfrac{1}{\sqrt{1-\tanh^2 x}}$	$\dfrac{\lvert\coth x\rvert}{\sqrt{\coth^2 x-1}}$
$\tanh x=$	$\dfrac{\sinh x}{\sqrt{1+\sinh^2 x}}$	$\pm\dfrac{\sqrt{\cosh^2 x-1}}{\cosh x}$	–	$\dfrac{1}{\coth x}$
$\coth x=$	$\dfrac{\sqrt{1+\sinh^2 x}}{\sinh x}$	$\pm\dfrac{\cosh x}{\sqrt{\cosh^2 x-1}}$	$\dfrac{1}{\tanh x}$	–

Geometrical interpretation

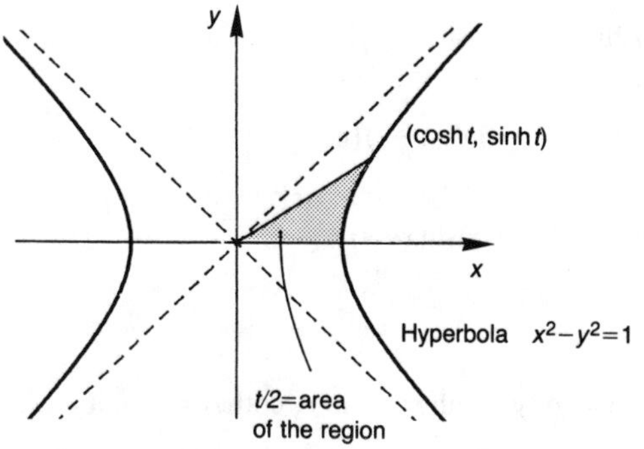

1. $\sinh(-x)=-\sinh x$
   $\cosh(-x)=\cosh x$
   $\tanh(-x)=-\tanh x$
   $\coth(-x)=-\coth x$

2. $\cosh^2 x-\sinh^2 x=1$

   $\tanh x=\dfrac{\sinh x}{\cosh x}$

   $\coth x=\dfrac{\cosh x}{\sinh x}=\dfrac{1}{\tanh x}$

3. $\sinh(x\pm y)=\sinh x\cosh y\pm\cosh x\sinh y$
   $\cosh(x\pm y)=\cosh x\cosh y\pm\sinh x\sinh y$

   $\tanh(x\pm y)=\dfrac{\tanh x\pm\tanh y}{1\pm\tanh x\tanh y}$

   $\coth(x\pm y)=\dfrac{1\pm\coth x\coth y}{\coth x\pm\coth y}$

4. $$\sinh 2x = 2\sinh x\cosh y \qquad \sinh\frac{x}{2} = \pm\sqrt{\frac{\cosh x - 1}{2}}$$

$$\cosh 2x = \sinh^2 x + \cosh^2 x \qquad \cosh\frac{x}{2} = \sqrt{\frac{\cosh x + 1}{2}}$$

$$\tanh 2x = \frac{2\tanh x}{1 + \tanh^2 x} \qquad \tanh\frac{x}{2} = \pm\sqrt{\frac{\cosh x - 1}{\cosh x + 1}} = \frac{\sinh x}{\cosh x + 1}$$

$$\coth 2x = \frac{\coth^2 x + 1}{2\coth x} \qquad \coth\frac{x}{2} = \pm\sqrt{\frac{\cosh x + 1}{\cosh x - 1}} = \frac{\sinh x}{\cosh x - 1}$$

5. $$\sinh x + \sinh y = 2\sinh\frac{x+y}{2}\cosh\frac{x-y}{2} \qquad \sinh x - \sinh y = 2\cosh\frac{x+y}{2}\sinh\frac{x-y}{2}$$

$$\cosh x + \cosh y = 2\cosh\frac{x+y}{2}\cosh\frac{x-y}{2} \qquad \cosh x - \cosh y = 2\sinh\frac{x+y}{2}\sinh\frac{x-y}{2}$$

$$\tanh x \pm \tanh y = \frac{\sinh(x \pm y)}{\cosh x\cosh y} \qquad \coth x \pm \coth y = \frac{\sinh(x \pm y)}{\sinh x\sinh y}$$

6. $$\sinh x\sinh y = \frac{1}{2}[\cosh(x+y) - \cosh(x-y)]$$

$$\sinh x\cosh y = \frac{1}{2}[\sinh(x+y) + \sinh(x-y)]$$

$$\cosh x\cosh y = \frac{1}{2}[\cosh(x+y) + \cosh(x-y)]$$

Complex case:

7. $\sinh iy = i\sin y$, $\cosh iy = \cos y$, $\tanh iy = i\tan y$, $\coth iy = -i\cot y$

8. $\sinh(x+iy)$, $\cosh(x+iy)$, $\tanh(x+iy)$, $\coth(x+iy)$: Use 3 and 7.

Inverses

$$y = \sinh x \Leftrightarrow x = \operatorname{arsinh} y = \ln(y + \sqrt{y^2+1})$$

$$y = \cosh x,\ x \geqslant 0 \Leftrightarrow x = \operatorname{arcosh} y = \ln(y + \sqrt{y^2-1}),\ y \geqslant 1$$

$$y = \tanh x \Leftrightarrow x = \operatorname{artanh} y = \frac{1}{2}\ln\frac{1+y}{1-y},\ |y| < 1$$

$$y = \coth x \Leftrightarrow x = \operatorname{arcoth} y = \frac{1}{2}\ln\frac{y+1}{y-1},\ |y| > 1$$

5.4 Trigonometric and Inverse Trigonometric Functions

Trigonometric functions

The trigonometric functions are defined with the aid of the unit circle. The angle α is measured in degrees or radians. One rotation or 360° is 2π radians.

$$1° = \frac{\pi}{180} \text{ radians} \approx 0.017453 \text{ rad.} \quad 1 \text{ radian} = \frac{180°}{\pi}$$

Degrees:	30	45	60	90	120	135	150	180	210	225	240	270	300	315	330	360
Radians:	$\frac{\pi}{6}$	$\frac{\pi}{4}$	$\frac{\pi}{3}$	$\frac{\pi}{2}$	$\frac{2\pi}{3}$	$\frac{3\pi}{4}$	$\frac{5\pi}{6}$	π	$\frac{7\pi}{6}$	$\frac{5\pi}{4}$	$\frac{4\pi}{3}$	$\frac{3\pi}{2}$	$\frac{5\pi}{3}$	$\frac{7\pi}{4}$	$\frac{11\pi}{6}$	2π

Definitions

$$\sin\alpha = y \qquad \cos\alpha = x$$

$$\tan\alpha = \frac{y}{x} \qquad \cot\alpha = \frac{x}{y}$$

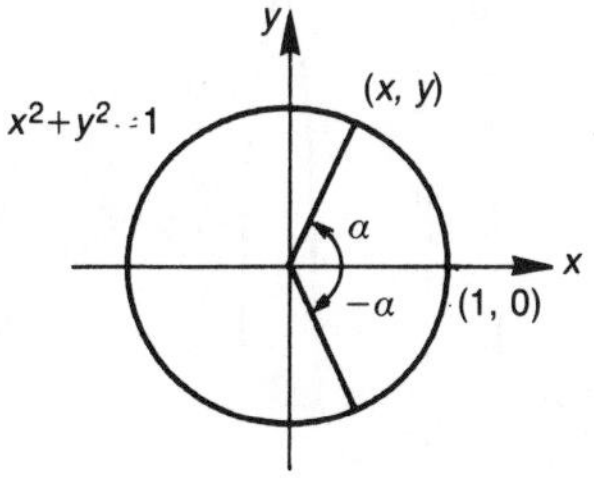

Derivatives. $D\sin x = \cos x \qquad D\cos x = -\sin x$

$$D\tan x = 1 + \tan^2 x = \frac{1}{\cos^2 x} \qquad D\cot x = -1 - \cot^2 x = -\frac{1}{\sin^2 x}$$

Related functions: $\sec x = \frac{1}{\cos x}, \quad \csc x = \frac{1}{\sin x}$

Trigonometric functions for some angles of the first quadrant.

Angle	sin	cos	tan	cot
$0° = 0$	0	1	0	–
$15° = \pi/12$	$\frac{1}{4}(\sqrt{6}-\sqrt{2})$	$\frac{1}{4}(\sqrt{6}+\sqrt{2})$	$2-\sqrt{3}$	$2+\sqrt{3}$
$30° = \pi/6$	$1/2$	$\sqrt{3}/2$	$1/\sqrt{3}$	$\sqrt{3}$
$45° = \pi/4$	$1/\sqrt{2}$	$1/\sqrt{2}$	1	1
$60° = \pi/3$	$\sqrt{3}/2$	$1/2$	$\sqrt{3}$	$1/\sqrt{3}$
$75° = 5\pi/12$	$\frac{1}{4}(\sqrt{6}+\sqrt{2})$	$\frac{1}{4}(\sqrt{6}-\sqrt{2})$	$2+\sqrt{3}$	$2-\sqrt{3}$
$90° = \pi/2$	1	0	–	0

Graphs

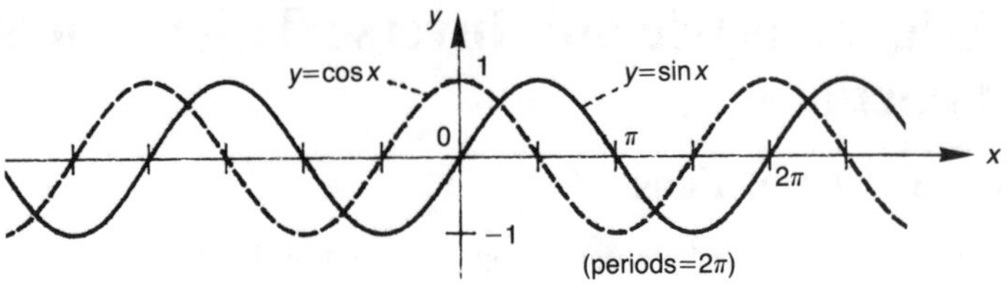

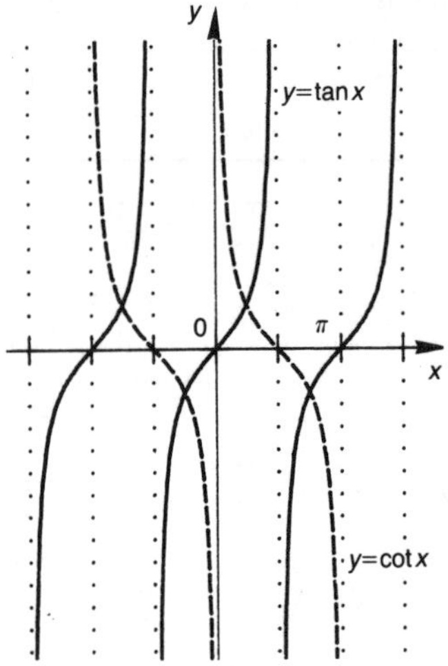

Geometrical interpretations

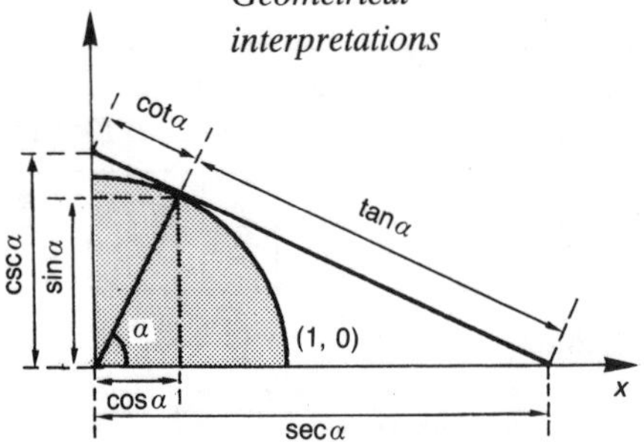

$$\lim_{x\to 0}\frac{\sin x}{x}=1 \quad (x \text{ in radians})$$

$\sin x < x < \tan x \ , 0 < x < \pi/2$

Transformation table

	$\sin\alpha$	$\cos\alpha$	$\tan\alpha$	$\cot\alpha$	$\sec\alpha$	$\csc\alpha$
$\sin\alpha=$	–	$\pm\sqrt{1-\cos^2\alpha}$	$\frac{\tan\alpha}{\pm\sqrt{1+\tan^2\alpha}}$	$\frac{1}{\pm\sqrt{1+\cot^2\alpha}}$	$\frac{\pm\sqrt{\sec^2\alpha-1}}{\sec\alpha}$	$\frac{1}{\csc\alpha}$
$\cos\alpha=$	$\pm\sqrt{1-\sin^2\alpha}$	–	$\frac{1}{\pm\sqrt{1+\tan^2\alpha}}$	$\frac{\cot\alpha}{\pm\sqrt{1+\cot^2\alpha}}$	$\frac{1}{\sec\alpha}$	$\frac{\pm\sqrt{\csc^2\alpha-1}}{\csc\alpha}$
$\tan\alpha=$	$\frac{\sin\alpha}{\pm\sqrt{1-\sin^2\alpha}}$	$\frac{\pm\sqrt{1-\cos^2\alpha}}{\cos\alpha}$	–	$\frac{1}{\cot\alpha}$	$\pm\sqrt{\sec^2\alpha-1}$	$\frac{1}{\pm\sqrt{\csc^2\alpha-1}}$
$\cot\alpha=$	$\frac{\pm\sqrt{1-\sin^2\alpha}}{\sin\alpha}$	$\frac{\cos\alpha}{\pm\sqrt{1-\cos^2\alpha}}$	$\frac{1}{\tan\alpha}$	–	$\frac{1}{\pm\sqrt{\sec^2\alpha-1}}$	$\pm\sqrt{\csc^2\alpha-1}$
$\sec\alpha=$	$\frac{1}{\pm\sqrt{1-\sin^2\alpha}}$	$\frac{1}{\cos\alpha}$	$\pm\sqrt{1+\tan^2\alpha}$	$\frac{\pm\sqrt{1+\cot^2\alpha}}{\cot\alpha}$	–	$\frac{\csc\alpha}{\pm\sqrt{\csc^2\alpha-1}}$
$\csc\alpha=$	$\frac{1}{\sin\alpha}$	$\frac{1}{\pm\sqrt{1-\cos^2\alpha}}$	$\frac{\pm\sqrt{1+\tan^2\alpha}}{\tan\alpha}$	$\pm\sqrt{1+\cot^2\alpha}$	$\frac{\sec\alpha}{\pm\sqrt{\sec^2\alpha-1}}$	–

Reductions

$x=$	$-\alpha$	$\frac{\pi}{2}-\alpha$	$\alpha\pm\frac{\pi}{2}$	$\pi-\alpha$	$\alpha\pm\pi$	$2\pi-\alpha$	$\alpha\pm 2\pi$
$\sin x=$	$-\sin\alpha$	$\cos\alpha$	$\pm\cos\alpha$	$\sin\alpha$	$-\sin\alpha$	$-\sin\alpha$	$\sin\alpha$
$\cos x=$	$\cos\alpha$	$\sin\alpha$	$\mp\sin\alpha$	$-\cos\alpha$	$-\cos\alpha$	$\cos\alpha$	$\cos\alpha$
$\tan x=$	$-\tan\alpha$	$\cot\alpha$	$-\cot\alpha$	$-\tan\alpha$	$\tan\alpha$	$-\tan\alpha$	$\tan\alpha$
$\cot x=$	$-\cot\alpha$	$\tan\alpha$	$-\tan\alpha$	$-\cot\alpha$	$\cot\alpha$	$-\cot\alpha$	$\cot\alpha$

E.g. $\cos(\pi-\alpha)=-\cos\alpha$

1. $\sin^2\alpha+\cos^2\alpha=1 \qquad \tan\alpha=\frac{\sin\alpha}{\cos\alpha} \qquad \cot\alpha=\frac{\cos\alpha}{\sin\alpha}=\frac{1}{\tan\alpha}$

$$\frac{1}{\cos^2\alpha}=1+\tan^2\alpha \qquad \frac{1}{\sin^2\alpha}=1+\cot^2\alpha$$

2. $\sin(\alpha+\beta)=\sin\alpha\cos\beta+\cos\alpha\sin\beta \qquad \sin(\alpha-\beta)=\sin\alpha\cos\beta-\cos\alpha\sin\beta$

$$\cos(\alpha+\beta)=\cos\alpha\cos\beta-\sin\alpha\sin\beta \qquad \cos(\alpha-\beta)=\cos\alpha\cos\beta+\sin\alpha\sin\beta$$

$$\tan(\alpha+\beta)=\frac{\tan\alpha+\tan\beta}{1-\tan\alpha\tan\beta} \qquad \tan(\alpha-\beta)=\frac{\tan\alpha-\tan\beta}{1+\tan\alpha\tan\beta}$$

$$\cot(\alpha+\beta)=\frac{\cot\alpha\cot\beta-1}{\cot\alpha+\cot\beta} \qquad \cot(\alpha-\beta)=-\frac{\cot\alpha\cot\beta+1}{\cot\alpha-\cot\beta}$$

3. $\sin\alpha+\sin\beta=2\sin\frac{\alpha+\beta}{2}\cos\frac{\alpha-\beta}{2} \qquad \sin\alpha-\sin\beta=2\sin\frac{\alpha-\beta}{2}\cos\frac{\alpha+\beta}{2}$

$$\cos\alpha+\cos\beta=2\cos\frac{\alpha+\beta}{2}\cos\frac{\alpha-\beta}{2} \qquad \cos\alpha-\cos\beta=-2\sin\frac{\alpha-\beta}{2}\sin\frac{\alpha+\beta}{2}$$

4. $\sin\alpha\sin\beta=\frac{1}{2}[\cos(\alpha-\beta)-\cos(\alpha+\beta)]$

$$\sin\alpha\cos\beta=\frac{1}{2}[\sin(\alpha-\beta)+\sin(\alpha+\beta)]$$

$$\cos\alpha\cos\beta=\frac{1}{2}[\cos(\alpha-\beta)+\cos(\alpha+\beta)]$$

5. $\sin 2\alpha=2\sin\alpha\cos\alpha \qquad \sin^2\frac{\alpha}{2}=\frac{1-\cos\alpha}{2}$

$$\cos 2\alpha=\cos^2\alpha-\sin^2\alpha= =2\cos^2\alpha-1=1-2\sin^2\alpha \qquad \cos^2\frac{\alpha}{2}=\frac{1+\cos\alpha}{2}$$

$$\tan 2\alpha=\frac{2\tan\alpha}{1-\tan^2\alpha} \qquad \tan\frac{\alpha}{2}=\frac{\sin\alpha}{1+\cos\alpha}=\pm\sqrt{\frac{1-\cos\alpha}{1+\cos\alpha}}$$

$$\cot 2\alpha=\frac{\cot^2\alpha-1}{2\cot\alpha} \qquad \cot\frac{\alpha}{2}=\frac{\sin\alpha}{1-\cos\alpha}=\pm\sqrt{\frac{1+\cos\alpha}{1-\cos\alpha}}$$

6. $\sin 3\alpha=3\sin\alpha-4\sin^3\alpha$ $\quad$ $\sin 4\alpha=4\sin\alpha\cos\alpha-8\sin^3\alpha\cos\alpha$

$\cos 3\alpha=4\cos^3\alpha-3\cos\alpha$ $\quad$ $\cos 4\alpha=8\cos^4\alpha-8\cos^2\alpha+1$

$\tan 3\alpha=\frac{3\tan\alpha-\tan^3\alpha}{1-3\tan^2\alpha}$ $\quad$ $\tan 4\alpha=\frac{4\tan\alpha-4\tan^3\alpha}{1-6\tan^2\alpha+\tan^4\alpha}$

7. $\sin n\alpha=n\sin\alpha\cos^{n-1}\alpha-\binom{n}{3}\sin^3\alpha\cos^{n-3}\alpha+\binom{n}{5}\sin^5\alpha\cos^{n-5}\alpha-\ldots$

$\cos n\alpha=\cos^n\alpha-\binom{n}{2}\sin^2\alpha\cos^{n-2}\alpha+\binom{n}{4}\sin^4\alpha\cos^{n-4}\alpha-\ldots$

8. $$\sin^{2n}\alpha=\frac{(-1)^n}{2^{2n-1}}\cdot$$
$$\cdot\left[\cos 2n\alpha-\binom{2n}{1}\cos(2n-2)\alpha+\ldots+(-1)^{n-1}\binom{2n}{n-1}\cos 2\alpha\right]+\binom{2n}{n}\frac{1}{2^{2n}}$$

$$\sin^{2n-1}\alpha=\frac{(-1)^{n-1}}{2^{2n-2}}\cdot$$
$$\cdot\left[\sin(2n-1)\alpha-\binom{2n-1}{1}\sin(2n-3)\alpha+\ldots+(-1)^{n-1}\binom{2n-1}{n-1}\sin\alpha\right]$$

$$\cos^{2n}\alpha=\frac{1}{2^{2n-1}}\left[\cos 2n\alpha+\binom{2n}{1}\cos(2n-2)\alpha+\ldots+\binom{2n}{n-1}\cos 2\alpha\right]+\binom{2n}{n}\frac{1}{2^{2n}}$$

$$\cos^{2n-1}\alpha=\frac{1}{2^{2n-2}}\left[\cos(2n-1)\alpha+\binom{2n-1}{1}\cos(2n-3)\alpha+\ldots+\binom{2n-1}{n-1}\cos\alpha\right]$$

Complex case:

9. $\sin iy=i\sinh y \quad \cos iy=\cosh y,$
$\tan iy=i\tanh y \quad \cot iy=-i\coth y$

10. $\sin(x+iy)$, $\cos(x+iy)$, $\tan(x+iy)$, $\cot(x+iy)$: Use 2 and 9.

11. $\cos x=\frac{1}{2}(e^{ix}+e^{-ix}) \quad \sin x=\frac{1}{2i}(e^{ix}-e^{-ix})$

The expression $a\cos x+b\sin x$

Let $a>0$:

(5.1) $$\begin{cases} a\cos x+b\sin x=r\cos(x-\varphi)\\ a\sin x+b\cos x=r\sin(x+\varphi)\end{cases}$$

where $r=\sqrt{a^2+b^2}$, $\varphi=\arctan\frac{b}{a}$. *(Amplitude-phase angle form)*

Basic trigonometric equations

1. Let x_0 denote an (arbitrary) particular solution of each of the following equations.

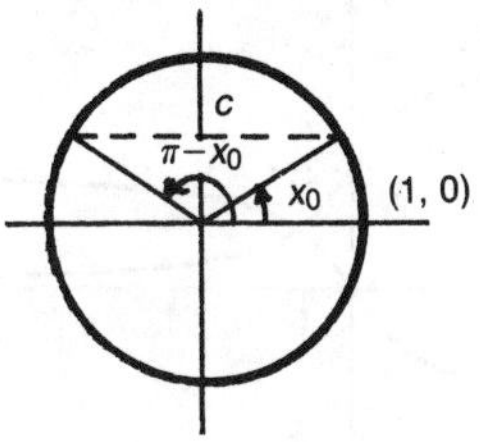

a.

$\sin x = c \ (-1 \leqslant c \leqslant 1)$

$x = \begin{cases} x_0 + 2n\pi \\ (\pi - x_0) + 2n\pi \end{cases} \quad (n \in \mathbf{Z})$

$(x_0 = \arcsin c)$

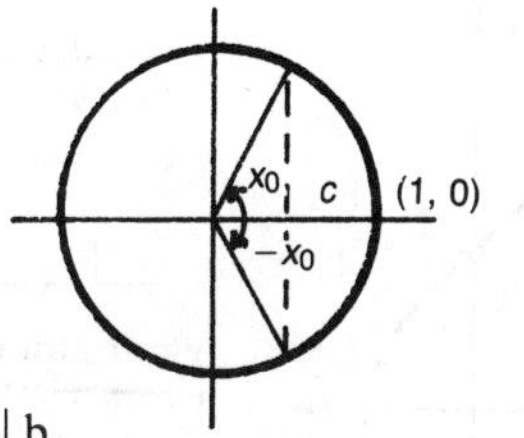

b.

$\cos x = c \ (-1 \leqslant c \leqslant 1)$

$x = \pm x_0 + 2n\pi \quad (n \in \mathbf{Z})$

$(x_0 = \arccos c)$

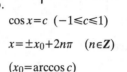

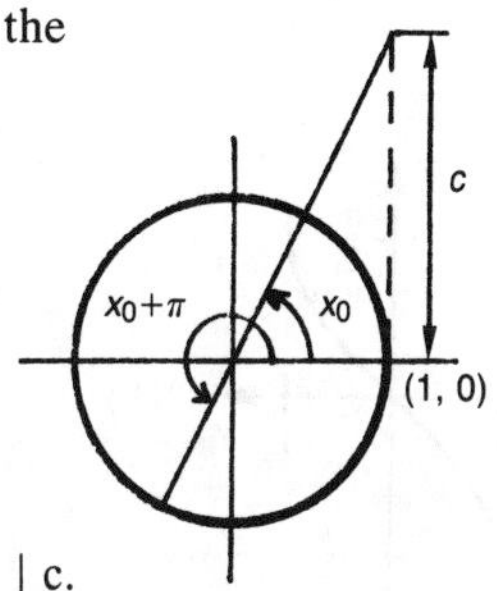

c.

$\tan x = c$

$x = x_0 + n\pi \quad (n \in \mathbf{Z})$

$(x_0 = \arctan c)$

2. a. $a \cos x + b \sin x = 0$

   Dividing by $\cos x$: $a + b \tan x = 0$
   etc. by 1c.

   b. $a \cos x + b \sin x = c \ (a > 0)$

   By(5.1): $r \cos(x - \varphi) = c$
   etc. by 1b.

Triangle theorems

See sec. 3.1.

Inverse trigonometric functions

The inverse trigonometric functions are defined as follows. (These definitions are sometimes called Principal values of the inverse trigonometric functions.)

$$y = \arcsin x \Leftrightarrow x = \sin y, \ -1 \leqslant x \leqslant 1, \quad -\frac{\pi}{2} \leqslant y \leqslant \frac{\pi}{2}$$

$$y = \arccos x \Leftrightarrow x = \cos y, \ -1 \leqslant x \leqslant 1, \quad 0 \leqslant y \leqslant \pi$$

$$y = \arctan x \Leftrightarrow x = \tan y, \ -\infty < x < \infty, \ -\frac{\pi}{2} < y < \frac{\pi}{2}$$

$$y = \operatorname{arccot} x \Leftrightarrow x = \cot y, \ -\infty < x < \infty, \quad 0 < y < \pi$$

(Alternative notation: $\sin^{-1} x$, $\cos^{-1} x$, $\tan^{-1} x$ and $\cot^{-1} x$).

Derivatives: $D \arcsin x = \dfrac{1}{\sqrt{1-x^2}}$ $\qquad D \arccos x = -\dfrac{1}{\sqrt{1-x^2}}$

$D \arctan x = \dfrac{1}{1+x^2}$ $\qquad D \operatorname{arccot} x = -\dfrac{1}{1+x^2}$

Graphs

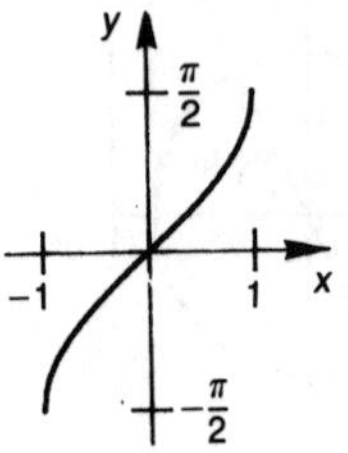

$y = \arcsin x$

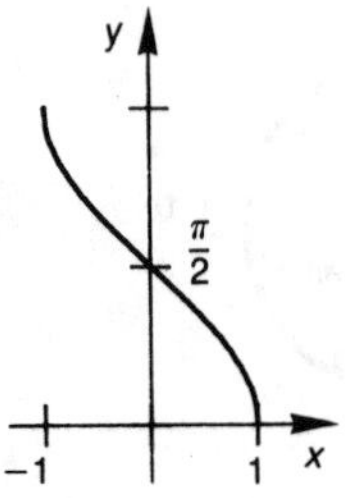

$y = \arccos x$

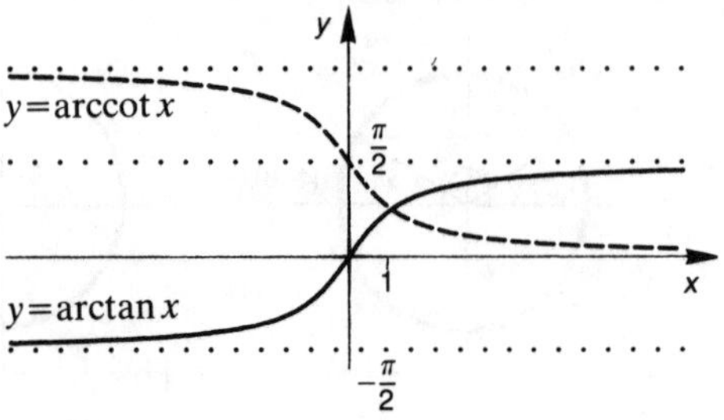

Special values

x	$\arcsin x$	$\arccos x$	$\arctan x$	$\text{arccot}\, x$
$-\infty$	–	–	$-\pi/2$	π
$-\sqrt{3}$	–	–	$-\pi/3$	$5\pi/6$
-1	$-\pi/2$	π	$-\pi/4$	$3\pi/4$
$-\sqrt{3}/2$	$-\pi/3$	$5\pi/6$		
$-1/\sqrt{2}$	$-\pi/4$	$3\pi/4$		
$-1/\sqrt{3}$			$-\pi/6$	$2\pi/3$
$-1/2$	$-\pi/6$	$2\pi/3$		
0	0	$\pi/2$	0	$\pi/2$
$1/2$	$\pi/6$	$\pi/3$		
$1/\sqrt{3}$			$\pi/6$	$\pi/3$
$1/\sqrt{2}$	$\pi/4$	$\pi/4$		
$\sqrt{3}/2$	$\pi/3$	$\pi/6$		
1	$\pi/2$	0	$\pi/4$	$\pi/4$
$\sqrt{3}$	–	–	$\pi/3$	$\pi/6$
∞	–	–	$\pi/2$	0

1. $\arcsin(-x) = -\arcsin x$ $\qquad$ $\arccos(-x) = \pi - \arccos x$

   $\arctan(-x) = -\arctan x$ $\qquad$ $\operatorname{arccot}(-x) = \pi - \operatorname{arccot} x$

2. $\arcsin x + \arccos x = \frac{\pi}{2}$ $\qquad$ $\arctan x + \operatorname{arccot} x = \frac{\pi}{2}$

$$\arctan \frac{1}{x} = \begin{cases} \frac{\pi}{2} - \arctan x, & x > 0 \\ -\frac{\pi}{2} - \arctan x, & x < 0 \end{cases}$$

3. $\arctan x + \arctan y = \arctan \frac{x+y}{1-xy} + \begin{cases} \pi, & xy > 1,\ x > 0 \\ 0, & xy < 1 \\ -\pi, & xy > 1,\ x < 0 \end{cases}$

Transformation table ($x>0$)

$x>0$	arcsin	arccos	arctan	arccot
$\arcsin x =$	–	$\arccos \sqrt{1-x^2}$	$\arctan \frac{x}{\sqrt{1-x^2}}$	$\operatorname{arccot} \frac{\sqrt{1-x^2}}{x}$
$\arccos x =$	$\arcsin \sqrt{1-x^2}$	–	$\arctan \frac{\sqrt{1-x^2}}{x}$	$\operatorname{arccot} \frac{x}{\sqrt{1-x^2}}$
$\arctan x =$	$\arcsin \frac{x}{\sqrt{1+x^2}}$	$\arccos \frac{1}{\sqrt{1+x^2}}$	–	$\operatorname{arccot} \frac{1}{x}$
$\operatorname{arccot} x =$	$\arcsin \frac{1}{\sqrt{1+x^2}}$	$\arccos \frac{x}{\sqrt{1+x^2}}$	$\arctan \frac{1}{x}$	–

6. Differential Calculus (one variable)

6.1 Some Basic Concepts

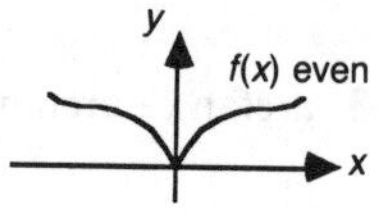

Intervals

$[a, b]=\{x : a \leqslant x \leqslant b\}$, *closed interval*

$(a, b)=]a, b[=\{x : a<x<b\}$, *open interval*

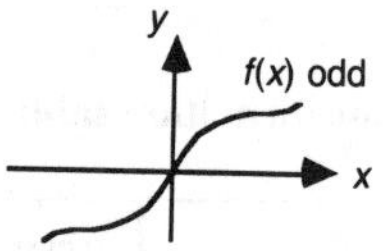

$[a, b)=\{x : a \leqslant x<b\}$, *half-open interval*

$(a, \infty)=\{x : x>a\}$, *infinite interval*

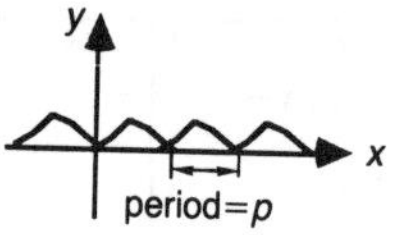

A function $y=f(x)$

1. is *even* if $f(-x)=f(x)$
2. is *odd* if $f(-x)=-f(x)$
3. has *period p* if $f(x+p)=f(x)$, all x.

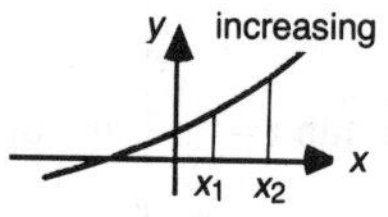

4. is *increasing* [*decreasing*] if $x_1<x_2 \Rightarrow f(x_1) \leqslant f(x_2)$ $[f(x_1) \geqslant f(x_2)]$.

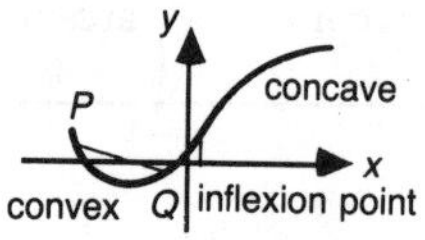

5. is *convex* [*concave*] if for any two points P and Q of the curve $y=f(x)$, the chord lies above [below] the curve.

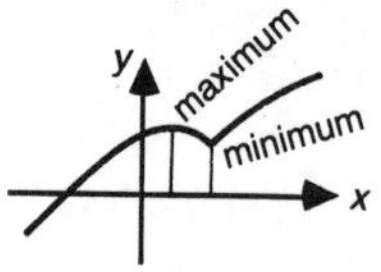

6. has an *inflexion point* at $x=a$ if the curve changes from convex to concave (or vice versa) at a.
7. has a *local maximum* [*minimum*] at $x=a$ if there is a neighborhood U of a such that $f(x) \leqslant f(a)$ $[f(x) \geqslant f(a)]$ for all $x \in U \cap D_f$.

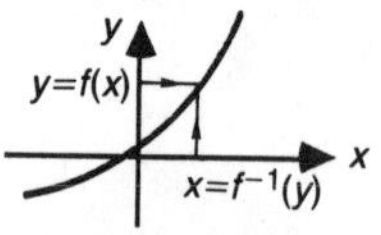

8. has an *inverse* f^{-1} if for any $y \in R_f$ there exists a unique $x \in D_f$ such that $f(x)=y$.

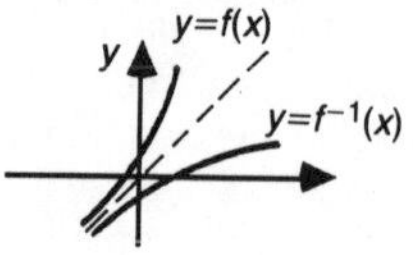

$$y=f(x) \Leftrightarrow x=f^{-1}(y)$$

$$D_{f^{-1}}=R_f, \quad R_{f^{-1}}=D_f$$

6.2 Limits and Continuity

Limits of a function

Definitions

1. $\lim_{x\to a} f(x)=A$ [$\lim_{x\to a^+} f(x)=A$]:

   For any $\varepsilon>0$ there exists $\delta>0$ such that $|f(x)-A|<\varepsilon$ for all $x\in D_f$ such that $|x-a|<\delta$ [$a<x<a+\delta$]. (Sometimes $|x-a|<\delta$ is replaced by the deleted neighborhood $0<|x-a|<\delta$).

2. $\lim_{x\to\infty} f(x)=A$: For any $\varepsilon>0$ there exists ω such that $|f(x)-A|<\varepsilon$ for all $x>\omega$, $x\in D_f$.

3. $\lim_{x\to a} f(x)=\infty$: For any M there exists $\delta>0$ such that $f(x)>M$ for all $x\in D_f$ such that $|x-a|<\delta$.

Alternative notation: $f(x)\to A$ as $x\to a$ etc.

Assume $\lim_{x\to a} f(x)=A$, $\lim_{x\to a} g(x)=B$ (finite limits) exist.

Then

1. $\lim\,(f(x)\pm g(x))=A\pm B$
2. $\lim f(x)g(x)=AB$
3. $\lim \dfrac{f(x)}{g(x)}=\dfrac{A}{B}$ $(B\neq 0)$
4. $\lim[f(x)]^{g(x)}=A^B$ $(A>0)$
5. $\lim h(f(x))=h(A)$ $(h(t)$ continuous$)$
6. $f(x)\leqslant g(x) \Rightarrow A\leqslant B$
7. $A=B,\ f(x)\leqslant h(x)\leqslant g(x) \Rightarrow \lim h(x)=A$

l'Hospital's rules

Assume that

(*i*) $g(x)\neq 0$ and $f(x)$, $g(x)$ have continuous derivatives in a deleted neighborhood of a (or ∞)

(*ii*) $f(x)$ and $g(x)\to 0$ (or ∞) as $x\to a$ (or $x\to\infty$)

Then

$$\lim_{\substack{x\to a\\(x\to\infty)}} \frac{f(x)}{g(x)} = \lim_{\substack{x\to a\\(x\to\infty)}} \frac{f'(x)}{g'(x)} \quad \text{if the latter limit exists.}$$

Indeterminate forms

Functions $f(x)$ of the forms $\dfrac{u(x)}{v(x)}$, $u(x)v(x)$, $[u(x)]^{v(x)}$ and $u(x)-v(x)$ are not well defined at a (or at ∞) if $f(a)$ has formally any of the forms

(1) $\dfrac{0}{0}$ (2) $\dfrac{\infty}{\infty}$ (3) $0\cdot\infty$ (4) $[0^+]^0$ (5) ∞^0 (6) 1^∞ (7) $\infty-\infty$

However the limit $\lim_{x\to a} f(x)$ may exist and may be found by the use of l'Hospital's rules, rewriting the functions as follows:

(3) $uv=\frac{u}{\frac{1}{v}}$. (4), (5), (6) $u^v=e^{v \ln u}$. Here the exponent will be of the form (3).

Some standard limits

$$\lim_{x\to\pm\infty}\left(1+\frac{1}{x}\right)^x=e \qquad \lim_{x\to\pm\infty}\left(1+\frac{t}{x}\right)^x=e^t \qquad \lim_{x\to\infty} x^{1/x}=1$$

$$\lim_{x\to\infty}\frac{x^p}{a^x}=\lim_{x\to\infty}x^pe^{-qx}=0\ (a>1,\ q>0) \qquad \lim_{x\to\infty}\frac{(\ln x)^p}{x^q}=0\ (q>0)$$

$$\lim_{x\to 0^+} x^p|\ln x|^q=0\ (p>0) \qquad \lim_{m\to\infty}\frac{a^m}{m!}=0$$

$$\lim_{x\to 0}\frac{\sin ax}{x}=a \qquad \lim_{x\to 0}\frac{a^x-1}{x}=\ln a \qquad \lim_{x\to 0}\frac{\ln(1+x)}{x}=1$$

$$\lim_{x\to\infty}\frac{a_mx^m+\ldots+a_0}{b_nx^n+\ldots+b_0}=\begin{cases}0 \text{ if } m<n\\ \dfrac{a_m}{b_n} \text{ if } m=n\\ \pm\infty \text{ if } m>n\end{cases} \qquad (a_m,\ b_n\neq 0)$$

Examples

I. (l'Hospital's rule)

(a) $$\lim_{x\to 0}\frac{1-\cos x}{x^2}=\left[\frac{0}{0}\right]=\lim_{x\to 0}\frac{\sin x}{2x}=\left[\frac{0}{0}\right]=\lim_{x\to 0}\frac{\cos x}{2}=\frac{1}{2}$$

(b) $$\lim_{x\to\infty}\frac{\ln x}{\sqrt{x}}=\left[\frac{\infty}{\infty}\right]=\lim_{x\to\infty}\frac{\frac{1}{x}}{\frac{1}{2\sqrt{x}}}=\lim_{x\to\infty}\frac{2}{\sqrt{x}}=0.$$

II. (Taylor expansion) $$\lim_{x\to 0}\frac{1-\cos x}{x^2}=\lim_{x\to 0}\frac{1-\left(1-\frac{x^2}{2}+O(x^4)\right)}{x^2}=\lim_{x\to 0}\left(\frac{1}{2}+O(x^2)\right)=\frac{1}{2}$$

Continuity

Definitions

A function $y=f(x)$ is said to be

1. continuous at x_o if $x_o\in D_f$ and $\lim_{x\to x_o} f(x)=f(x_o)$,
2. continuous in an interval I if $f(x)$ is continuous at every point of I,
3. uniformly continuous in an interval I if for any $\varepsilon>0$ there exists a $\delta>0$ such that

$$|f(x_1)-f(x_2)|<\varepsilon \text{ for all } x_1,\ x_2\in I \text{ such that } |x_1-x_2|<\delta.$$

Theorems

1. f, g continuous $\Rightarrow f\pm g$, fg, f/g, $f\circ g$ continuous (where they are defined).
2. Any composition of the elementary functions is continuous where it is defined.
3. $f(x)$ continuous on a closed interval $[a, b] \Rightarrow$
   (a) $f(x)$ assumes every value between $f(a)$ and $f(b)$.
   (b) $f(x)$ assumes its supremum (greatest value) and its infimum (least value) in $[a, b]$.
   (c) $f(x)$ is bounded in $[a, b]$
   (d) $f(x)$ is uniformly continuous in $[a, b]$.
4. $f'(x)$ bounded in an interval $I \Rightarrow f(x)$ uniformly continuous in I.

6.3 Derivatives

The *derivative* $f'(x)$ of a function $y=f(x)$ is defined by

$$f'(x)=\lim_{\Delta x\to 0}\frac{f(x+\Delta x)-f(x)}{\Delta x}=\lim_{\Delta x\to 0}\frac{\Delta y}{\Delta x}$$

Alternative notation:

$$y'=f'(x)=\frac{dy}{dx}=\frac{d}{dx}f(x)=Df(x),$$

$$\dot{y}=\frac{dy}{dt}\ (y \text{ a function of time})$$

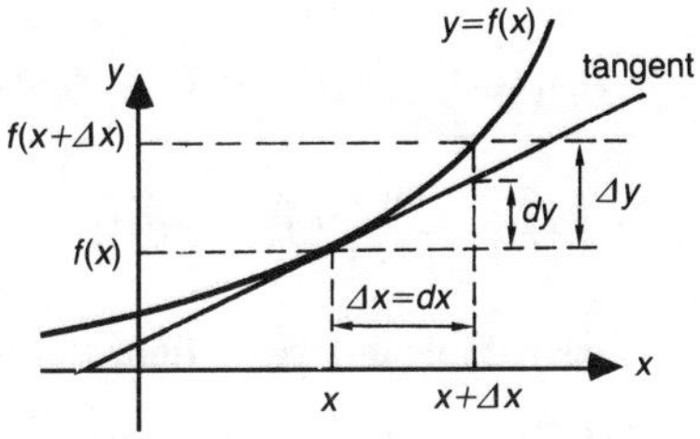

Higher derivatives

$$y''=f''(x)=\frac{d^2y}{dx^2}=D^2f(x),$$

$$y^{(n)}=f^{(n)}(x)=\frac{d^ny}{dx^n}=D^nf(x)=[\text{def: } D\{D^{n-1}f(x)\}].$$

Differential

Difference $\Delta f=f(x+\Delta x)-f(x)$
Differential $df=f'(x)\Delta x=f'(x)dx$

$f(x)$ differentiable $\Rightarrow \Delta f=df+\varepsilon(x)\Delta x$, where $\varepsilon(x)\to 0$ as $\Delta x\to 0$

Equation of *tangent* at $(a, f(a))$: $y-f(a)=f'(a)(x-a)$

Equation of *normal* at $(a, f(a))$: $y-f(a)=-\dfrac{1}{f'(a)}(x-a)$

Lagrange's mean value theorem

Assume that $f(x)$ is (*i*) continuous in $[a, b]$, (*ii*) differentiable in (a, b).

Then there exists $\xi \in (a, b)$ such that $\quad f(b)-f(a)=(b-a)f'(\xi)$

Cauchy's mean value theorem

Assume that $f(x)$ and $g(x)$ are (*i*) continuous in $[a, b]$, (*ii*) differentiable in (a, b).

Then there exists $\xi \in (a, b)$ such that $\quad [f(b)-f(a)]g'(\xi)=[g(b)-g(a)]f'(\xi)$

Differential formulas

$f=f(x)$, $g=g(x)$, $D=d/dx$

Sum, product, quotient

$$D(f+g)=f'+g' \quad D(fg)=f'g+fg' \quad D\left(\frac{f}{g}\right)=\frac{f'g-fg'}{g^2}$$

$$D^n(fg)=\sum_{k=0}^{n}\binom{n}{k}f^{(n-k)}g^{(k)} \quad D(f^n g^m)=f^{n-1}g^{m-1}(ngf'+mfg')$$

$$D\left(\frac{f^n}{g^m}\right)=\frac{f^{n-1}}{g^{m+1}}(ngf'-mfg')$$

Logarithmic differentiation. Powers

$f(x)=u(x)^a\, v(x)^b\, w(x)^c \ldots \Rightarrow$

(*i*) $\ln|f|=a\ln|u|+b\ln|v|+c\ln|w|+\ldots$

(*ii*) $\dfrac{df}{f}=a\dfrac{du}{u}+b\cdot\dfrac{dv}{v}+c\dfrac{dw}{w}+\ldots$

(*iii*) $\dfrac{f'(x)}{f(x)}=a\dfrac{u'(x)}{u(x)}+b\dfrac{v'(x)}{v(x)}+c\cdot\dfrac{w'(x)}{w(x)}+\ldots$

$$D(f^g)=De^{g\ln f}=f^g\left(\frac{f'g}{f}+g'\ln f\right)$$

Composite functions. (Chain rule)

$$Df(g(x))=f'(g(x))g'(x) \qquad \frac{dz}{dx}=\frac{dz}{dy}\cdot\frac{dy}{dx}$$

$$D^2f(g(x))=f''(g(x))\,[g'(x)]^2+f'(g(x))g''(x) \quad \frac{d^2z}{dx^2}=\frac{d^2z}{dy^2}\left(\frac{dy}{dx}\right)^2+\frac{dz}{dy}\cdot\frac{d^2y}{dx^2}$$

Integrals

$$\frac{d}{dx}\int_a^x f(t)dt=f(x) \qquad \frac{d}{dx}\int_x^a f(t)dt=-f(x)$$

$$\frac{d}{dx}\int_{u(x)}^{v(x)} f(t)dt=f(v(x))v'(x)-f(u(x))u'(x)$$

$$\frac{d}{dx}\int_{u(x)}^{v(x)} F(x,\ t)dt=F(x,\ v)\ \frac{dv}{dx}-F(x,\ u)\ \frac{du}{dx}+\int_{u(x)}^{v(x)}\frac{\partial}{\partial x}F(x,\ t)dt$$

Inverse function

$$\frac{dx}{dy}=\left(\frac{dy}{dx}\right)^{-1} \qquad \frac{d^2x}{dy^2}=-\frac{d^2y}{dx^2}\bigg/\left(\frac{dy}{dx}\right)^3$$

Implicit function

$y=y(x)$ given implicitly by $F(x,\ y)=0$:

$$\frac{dy}{dx}=-\frac{F_x}{F_y} \qquad \frac{d^2y}{dx^2}=-\frac{1}{F_y{}^3}[F_{xx}F_y{}^2-2F_{xy}F_xF_y+F_{yy}F_x{}^2]$$

Basic derivatives

$f(x)$	$f'(x)$	$f(x)$	$f'(x)$	$f(x)$	$f'(x)$
x^a	ax^{a-1}	$\sinh x$	$\cosh x$	$\sin x$	$\cos x$
$\frac{1}{x^a}$	$-\frac{a}{x^{a+1}}$	$\cosh x$	$\sinh x$	$\cos x$	$-\sin x$
$\sqrt{x}$	$\frac{1}{2\sqrt{x}}$	$\tanh x$	$\frac{1}{\cosh^2 x}=1-\tanh^2 x$	$\tan x$	$\frac{1}{\cos^2 x}=1+\tan^2 x$
$\frac{1}{x}$	$-\frac{1}{x^2}$	$\coth x$	$-\frac{1}{\sinh^2 x}=1-\coth^2 x$	$\cot x$	$-\frac{1}{\sin^2 x}=-1-\cot^2 x$
$\frac{1}{x^2}$	$-\frac{2}{x^3}$	$\operatorname{arsinh} x$	$\frac{1}{\sqrt{x^2+1}}$	$\sec x$	$\sin x\ \sec^2 x$
e^x	e^x	$\operatorname{arcosh} x$	$\frac{1}{\sqrt{x^2-1}}$	$\csc x$	$-\cos x\ \csc^2 x$
a^x	$a^x \ln a$	$\operatorname{artanh} x$	$\frac{1}{1-x^2}$	$\arcsin x$	$\frac{1}{\sqrt{1-x^2}}$
$\ln \lvert x\rvert$	$\frac{1}{x}$	$\operatorname{arcoth} x$	$\frac{1}{1-x^2}$	$\arccos x$	$-\frac{1}{\sqrt{1-x^2}}$
${}^a\log\lvert x\rvert$	$\frac{1}{x \ln a}$			$\arctan x$	$\frac{1}{1+x^2}$
				$\operatorname{arccot} x$	$-\frac{1}{1+x^2}$

Derivatives of composite functions ($u=u(x)$, $u'=u'(x)$, $D=d/dx$)

$$Df(u)=u'f'(u) \quad Du^a=au^{a-1}u' \quad D\frac{1}{u^a}=-\frac{au'}{u^{a+1}} \quad De^u=u'e^u$$

$$D\ln|u|=\frac{u'}{u} \quad D\sin u=u'\cos u \quad D\sin^k u=ku'\sin^{k-1}u\cos u$$

$$D\cos u=-u'\sin u \quad D\cos^k u=-ku'\sin u\cos^{k-1}u \quad D\tan u=\frac{u'}{\cos^2 u}$$

$$D\cot u=-\frac{u'}{\sin^2 u} \quad D\arcsin u=\frac{u'}{\sqrt{1-u^2}} \quad D\arctan u=\frac{u'}{1+u^2}$$

Higher derivatives

$f(x)$	$f^{(k)}(x)$
$(x-a)^n$, n positive integer	$\begin{cases} n(n-1)\ldots(n-k+1)(x-a)^{n-k}, & k<n \\ n!, & k=n \\ 0, & k>n \end{cases}$
$\frac{1}{(x-a)^n}$, n positive integer	$(-1)^k\cdot\frac{n(n+1)\ldots(n+k-1)}{(x-a)^{n+k}}$
x^a	$a(a-1)\ldots(a-k+1)x^{a-k}$
e^{ax}	$a^k e^{ax}$
$\ln\|x\|$	$(-1)^{k-1}(k-1)!\,x^{-k}$
$\sin ax$	$a^k\sin\left(ax+\frac{k\pi}{2}\right)$
$\cos ax$	$a^k\cos\left(ax+\frac{k\pi}{2}\right)$

6.4 Monotonicity. Extremes of Functions

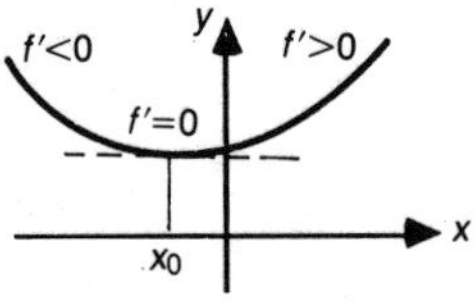

Let $f(x)$ be differentiable in an interval I. Then (in I):

$f'(x)>0 \Rightarrow f(x)$ *strictly increasing*
$f'(x)\geq 0 \Rightarrow f(x)$ *increasing*
$f'(x)=0 \Rightarrow f(x)$ *constant*
$f'(x)\leq 0 \Rightarrow f(x)$ *decreasing*
$f'(x)<0 \Rightarrow f(x)$ *strictly decreasing*

A point x_0 is a *stationary (critical) point* if $f'(x_0)=0$.

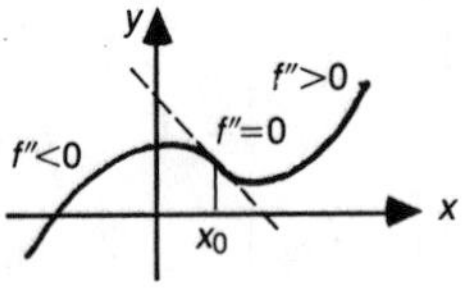

Let $f(x)$ be twice differentiable in an interval I. Then (in I)

$f''(x)\geq 0 \Rightarrow f(x)$ *convex*
$f''(x)\leq 0 \Rightarrow f(x)$ *concave*

A point x_0 is an *inflexion point* if (*i*) $f''(x_0)=0$ (*ii*) $f''(x)$ changes sign at x_0.

Jensen's inequality
If $f(x)$ is convex and $\lambda_1+\ldots+\lambda_n=1$, $\lambda_i>0$, then

$$f(\lambda_1x_1+\ldots+\lambda_nx_n)\leq\lambda_1f(x_1)+\ldots+\lambda_nf(x_n)$$

Necessary and sufficient conditions for extremum

Assume that $f(x)$ is differentiable.

Necessary condition

x_0 local extremum (maximum or minimum) of $f(x) \Rightarrow f'(x_0)=0$.

Sufficient conditions

1. *Sign changes of the derivative*:

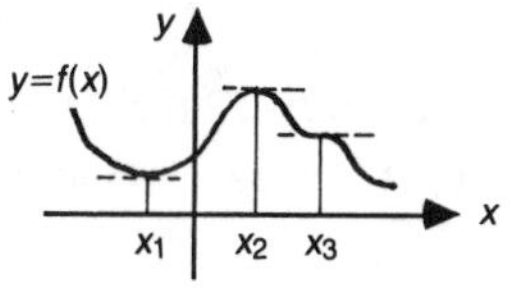

x		x_1		x_2		x_3	
$f'(x)$	−	0	+	0	−	0	−
$f(x)$	↘	↓ minimum	↗	↓ maximum	↘	↓ terrace	↘

2. *Higher derivatives*

A. $f(x)$ has local maximum [minimum] at x_0 if

(*i*) $f'(x_0)=0$ and (*ii*) $f''(x_0)<0$ [>0] or $f''(x_0)=\ldots=f^{(n-1)}(x_0)=0$, $f^{(n)}(x_0)<0$ [>0], n even

B. x_0 is a terrace point of $f(x)$ if

(*i*) $f'(x_0)=0$ and (*ii*) $f''(x_0)=\ldots=f^{(n-1)}(x_0)=0$, $f^{(n)}(x_0)\neq 0$, n odd.

Global (absolute) extremum

Global extremum of a function $f(x)$ in an interval occurs (if it exists) in one of the following points:

1. Points where $f'(x)=0$.
2. Points where $f'(x)$ does not exist.
3. Endpoints of the interval.

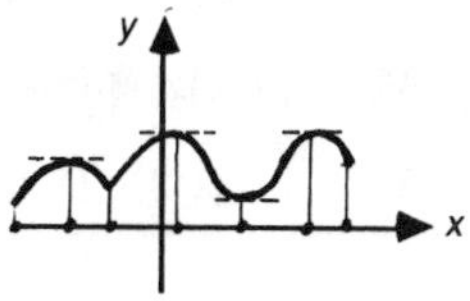

7. Integral Calculus

7.1 Indefinite Integrals

Primitive function

A function $F(x)$ is called a *primitive function* of $f(x)$ on an interval I if $F'(x)=f(x)$ for all $x \in I$. Any primitive function of $f(x)$ can be written $F(x)+C$ where C is an arbitrary constant.

Notation: $F(x)=\int f(x)dx$. The function $f(x)$ is called *integrand*. (Note: A primitive function of a (composite) elementary function is not always an elementary function. For example, $\int e^{-x^2}dx$ and $\int \frac{\sin x}{x}dx$ are not elementary.)

Methods of integration

General properties

A1.	$\int [af(x)+bg(x)]dx=a\int f(x)dx+b\int g(x)dx$	*(linearity)*
A2.	$\int f(x)g(x)dx=F(x)g(x)-\int F(x)g'(x)dx$	*(integration by parts)*
A3.	$\int f(g(x))g'(x)dx=\int f(t)dt,\ [t=g(x)]$	*(substitution)*

Example

$$\int \sin\sqrt{x}\,dx=[\text{substitution: } \sqrt{x}=t \Leftrightarrow x=t^2;\ dx=2tdt]=$$
$$=\int 2t\sin t\,dt=[\text{integration by parts}]=-2t\cos t+2\int \cos t\,dt=$$
$$=-2t\cos t+2\sin t+C=-2\sqrt{x}\cos\sqrt{x}+2\sin\sqrt{x}+C$$

A4.	$\int f(g(x))g'(x)dx=F(g(x))$
A4a.	$\int f(ax+b)dx=\frac{1}{a}F(ax+b)$
A4b.	$\int \frac{f'(x)}{f(x)}dx=\ln\lvert f(x)\rvert$
A5a.	$f(x)$ odd $\Rightarrow F(x)$ even
A5b.	$f(x)$ even $\Rightarrow F(x)$ odd (if $F(0)=0$)

Basic primitive functions

B1. $\int x^a dx = \frac{x^{a+1}}{a+1}$ $(a \neq -1)$	B2. $\int \frac{dx}{x} = \ln\|x\|$
B3. $\int e^x dx = e^x$	B4. $\int \sin x\, dx = -\cos x$
B5. $\int \cos x\, dx = \sin x$	B6. $\int \frac{dx}{\sin^2 x} = -\cot x$
B7. $\int \frac{dx}{\cos^2 x} = \tan x$	B8. $\int \frac{dx}{a^2+x^2} = \frac{1}{a} \arctan \frac{x}{a}$
B9. $\int \frac{dx}{\sqrt{a^2-x^2}} = \arcsin \frac{x}{a}$ $(a>0)$	B10. $\int \frac{dx}{\sqrt{x^2+a}} = \ln\|x+\sqrt{x^2+a}\|$
B11. $\int \sinh x\, dx = \cosh x$	B12. $\int \cosh x\, dx = \sinh x$

Some methods for determining primitive functions of certain classes of functions are given below. The methods essentially consists in reducing the integrals to one of the basic integrals above. (Sometimes simpler methods can be used than those generally recommended.)

Rational functions

Integral	Method
C1. $\int \frac{P(x)}{Q(x)} dx$	By partial fraction decomposition (see sec. 5.2) C1 is reduced to integration of a polynomial and C2 and C3.
C2. $\int \frac{A dx}{(x-a)^n}$	$= \begin{cases} A \ln\|x-a\|, & n=1 \\ -\frac{A}{(n-1)(x-a)^{n-1}}, & n \geq 2 \end{cases}$
C3. $\int \frac{Ax+B}{(x^2+2ax+b)^n} dx$ $(a^2<b)$	The substitution $x+a=t$ transforms to C4 and C5.
C4. $\int \frac{Ct}{(t^2+\alpha^2)^n} dt$	$= \begin{cases} \frac{C}{2} \ln(t^2+\alpha^2), & n=1 \\ -\frac{C}{2(n-1)(t^2+\alpha^2)^{n-1}}, & n \geq 2 \end{cases}$
C5a. $\int \frac{D}{(t^2+\alpha^2)} dt$	$= \frac{D}{\alpha} \arctan \frac{t}{\alpha}$
C5b. $I_n = \int \frac{D}{(t^2+\alpha^2)^n} dt$ $(n \geq 2)$	Recursively, $I_{n+1} = \frac{Dt}{2n\alpha^2(t^2+\alpha^2)^n} + \frac{2n-1}{2n\alpha^2} I_n$ (Or differentiate I_{n-1} with respect to α).

Algebraic functions

Integral	Method
D1. $\int R\left(x, \sqrt[n]{\frac{ax+b}{cx+d}}\right) dx$ (*R* rational function)	Substitution: $t=\sqrt[n]{\frac{ax+b}{cx+d}}$, $x=\frac{dt^n-b}{a-ct^n}$ brings the integrand to a rational function of *t*.
D2. $\int \frac{dx}{\sqrt{ax^2+bx+c}}$ $(a\neq 0)$	Completing the square: $ax^2+bx+c=a\left(x+\frac{b}{2a}\right)^2+$ $+c-\frac{b^2}{4a}$ and the substitution $t=x+\frac{b}{2a}$ brings D2 to B9 (if $a<0$) and to B10 (if $a>0$)
D3. $\int \frac{P(x)}{\sqrt{ax^2+bx+c}} dx$ ($P(x)$ polynomial)	$=Q(x)\sqrt{ax^2+bx+c}+K\cdot\int \frac{dx}{\sqrt{ax^2+bx+c}}$ $(\deg Q<\deg P)$ $Q(x)$ and K will be determined by differentiation and identifying coefficients. Use D2 for the last integral.
D4. $\int R(x, \sqrt{a^2-x^2})\, dx$	$x=a\sin t$, $dx=a\cos t\, dt$, $\sqrt{a^2-x^2}=a\cos t$
D5. $\int R(x, \sqrt{x^2+a^2})\, dx$	$x=a\tan t$, $dx=\frac{a\,dt}{\cos^2 t}$, $\sqrt{x^2+a^2}=\frac{a}{\cos t}$ or D7
D6. $\int R(x, \sqrt{x^2-a^2})\, dx$	$x=a\cosh t$, $dx=a\sinh t\, dt$, $\sqrt{x^2-a^2}=a\sinh t$ or D7
D7. $\int R(x, \sqrt{ax^2+bx+c})dx$	$a>0$: Subst: $\sqrt{ax^2+bx+c}=t+x\sqrt{a}$ $a<0$: Write $ax^2+bx+c=a(x-p)(x-q)=a(x-p)^2\,\frac{x-q}{x-p}$ and use D1. Alternative: Complete the square and use D4 or D6.

Exponential and logarithmic functions

Integral	Method
E1. $\int R(e^{ax})\,dx$	Substitution: $e^{ax}=t,\ x=\frac{1}{a}\ln t,\ dx=\frac{dt}{at}$
E2. $\int P(x)\, e^{ax}\,dx$ ($P(x)$ polynomial)	$=$[integration by parts]$=\frac{1}{a}P(x)\,e^{ax}-\frac{1}{a}\int P'(x)\,e^{ax}\,dx$ etc.
E3. $\int x^a(\ln x)^n\,dx$	$=\begin{cases}\text{[integration by parts]}=\frac{x^{a+1}}{a+1}(\ln x)^n- \\ \frac{n}{a+1}\int x^a(\ln x)^{n-1}\,dx \text{ etc } (a\neq -1) \\ \frac{(\ln x)^{n+1}}{n+1}\ (a=-1)\end{cases}$ or $t=\ln x$ and use E2.
E4. $\int \frac{1}{x} f(\ln x)\,dx$	Substitution $\ln x=t,\ \frac{dx}{x}=dt$

Trigonometric and inverse trigonometric functions

Note: Trigonometric integrands may be rewritten by means of suitable formulas (cf. sec. 5.4) and then be integrated.

Integral	Method
F1. $\int f(\sin x)\cos x\,dx$	Substitution: $\sin x = t$, $\cos x\,dx = dt$
F2. $\int f(\cos x)\sin x\,dx$	Substitution: $\cos x = t$, $-\sin x\,dx = dt$
F3. $\int f(\tan x)\,dx$	Substitution: $\tan x = t$, $dx = \frac{dt}{1+t^2}$
F4. $\int R(\cos x, \sin x)\,dx$	Substitution: $\tan\frac{x}{2} = t$, $\sin x = \frac{2t}{1+t^2}$, $\cos x = \frac{1-t^2}{1+t^2}$, $dx = \frac{2dt}{1+t^2}$
F5. $\int \sin^n x\,dx$ $(n \geq 1)$	n odd: Use $\sin^2 x = 1-\cos^2 x$ and F2. n even: Use $\sin^2 x = \frac{1}{2}(1-\cos 2x)$ etc. Or use 149, 173 sec. 7.4
F6. $\int \cos^n x\,dx$ $(n \geq 1)$	n odd: Use $\cos^2 x = 1-\sin^2 x$ and F1. n even: Use $\cos^2 x = \frac{1}{2}(1+\cos 2x)$ etc. Or use 150, 174 sec. 7.4
F7. $\int P(x)\begin{Bmatrix}\cos x\\ \sin x\end{Bmatrix}dx$ ($P(x)$ polynomial)	Integrations by parts, differentiating the polynomial (cf. E2).
F8a. $\int P(x)e^{ax}\cos bx\,dx$	$= \text{Re}\int P(x)e^{(a+ib)x}dx$. Use E2.
F8b. $\int P(x)e^{ax}\sin bx\,dx$	$= \text{Im}\int P(x)e^{(a+ib)x}dx$. Use E2.
F9. $\int x^n \arctan x\,dx$	=[integration by parts]= $= \frac{x^{n+1}}{n+1}\arctan x - \frac{1}{n+1}\int\frac{x^{n+1}}{1+x^2}dx$. Use C1.
F10. $\int x^n \arcsin x\,dx$	=[integration by parts]= $= \frac{x^{n+1}}{n+1}\arcsin x - \frac{1}{n+1}\int\frac{x^{n+1}}{\sqrt{1-x^2}}dx$. Use D3.
F11. $\int f(\arcsin x)\,dx$	Substitution: $\arcsin x = t$, $x = \sin t$
F12. $\int f(\arctan x)\,dx$	Substitution: $\arctan x = t$, $x = \tan t$

7.2 Definite Integrals

Riemann integrals

If (*i*) $f(x)$ is continuous in $[a, b]$, (*ii*) $F(x)$ is a primitive function of $f(x)$, then

$$\int_a^b f(x)\,dx=[F(x)]_a^b=F(b)-F(a)$$

(Remark. If $f(x)$ has a finite number of discontinuities, then the integral is the sum of the integrals over those subintervals, where $f(x)$ is continuous.)

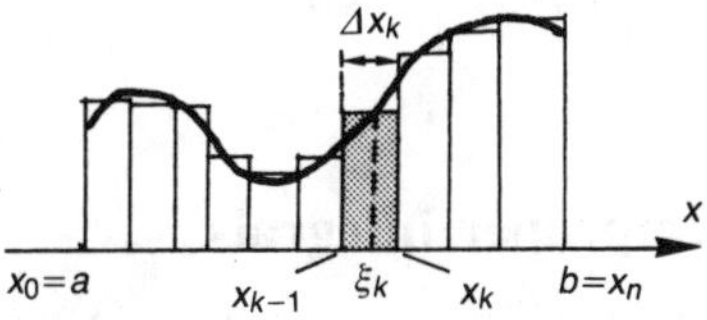

Riemann sums

Let $a=x_0<x_1<\ldots<x_n=b$, $\Delta x_k=x_k-x_{k-1}$, $x_{k-1}\leq\xi_k\leq x_k$, $\bar{d}=\max \Delta x_k$. Then

$$\int_a^b f(x)\,dx=\lim_{\bar{d}\to 0}\sum_{k=1}^{n} f(\xi_k)\Delta x_k=\lim_{n\to\infty}\frac{b-a}{n}\sum_{k=1}^{n} f\left(a+\frac{k}{n}(b-a)\right)$$

Properties of integrals

1. $\int_b^a f(x)\,dx=-\int_a^b f(x)\,dx$ 2. $\int_a^a f(x)\,dx=0$

3. $\int_a^b f(x)\,dx+\int_b^c f(x)\,dx=\int_a^c f(x)\,dx$ (*additivity*)

4. $\int_a^b [\alpha f(x)+\beta g(x)]\,dx=\alpha\int_a^b f(x)\,dx+\beta\int_a^b g(x)\,dx$ (*linearity*)

5. $\int_a^b f(x)g(x)\,dx=[F(x)g(x)]_a^b-\int_a^b F(x)g'(x)\,dx$ (*integration by parts*)

6. $\int_a^b f(g(x))\,g'(x)\,dx=[t=g(x)]=\int_{g(a)}^{g(b)} f(t)\,dt$ (*substitution*)

7. $f(x)\leq g(x) \Rightarrow \int_a^b f(x)\,dx\leq \int_a^b g(x)\,dx$

8. $\left|\int_a^b f(x)\,dx\right|\leq \int_a^b |f(x)|\,dx\leq M(b-a),\ M=\max_{[a,\,b]}|f(x)|$

Mean value theorems

Assume that (*i*) $f(x)$, $g(x)$ are continuous in $[a, b]$, (*ii*) $g(x)$ does not change sign. Then there exists a $\xi\in(a, b)$ such that

9. $\int_a^b f(x)\,dx=f(\xi)\,(b-a)$ 10. $\int_a^b f(x)g(x)\,dx=f(\xi)\int_a^b g(x)\,dx$

Inequalities

11. $\int_a^b |fg| \leq \left[\int_a^b |f|^p\right]^{\frac{1}{p}} \left[\int_a^b |g|^q\right]^{\frac{1}{q}}$, $\frac{1}{p}+\frac{1}{q}=1,\ p,\ q>1$ (*Hölder's inequality*)

12. $\int_a^b |fg| \leq \left[\int_a^b f^2\right]^{\frac{1}{2}} \left[\int_a^b g^2\right]^{\frac{1}{2}}$ (*Schwarz' inequality*)

13. $\left[\int_a^b |f+g|^p\right]^{\frac{1}{p}} \leq \left[\int_a^b |f|^p\right]^{\frac{1}{p}} + \left[\int_a^b |g|^p\right]^{\frac{1}{p}}$, $p \geq 1$ (*Minkowski's inequality*)

Improper integrals

The following integrals are said to be *convergent* if the limit exists, otherwise *divergent*.

Infinite interval:

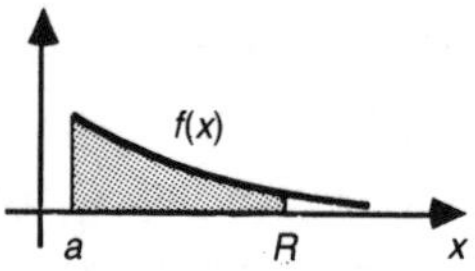

(a) $\int_a^\infty f(x)\,dx = \lim_{R\to\infty} \int_a^R f(x)\,dx$

(b) *Cauchy principal value*: $(CPV) \int_{-\infty}^{\infty} f(x)\,dx = \lim_{R\to\infty} \int_{-R}^{R} f(x)\,dx$

Unbounded function

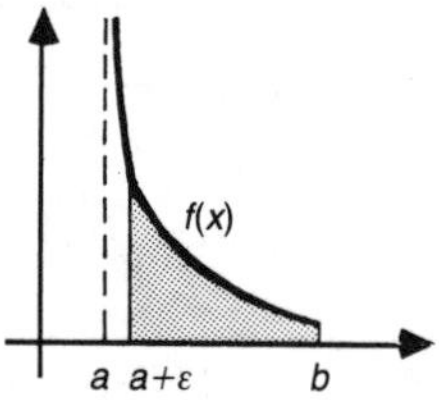

(a) $\int_a^b f(x)\,dx = \lim_{\varepsilon\to 0^+} \int_{a+\varepsilon}^{b} f(x)\,dx$

(b) *Cauchy principal value*: $(CPV) \int_a^b f(x)\,dx = \lim_{\varepsilon\to 0^+} \left(\int_a^{c-\varepsilon} f(x)\,dx + \int_{c+\varepsilon}^{b} f(x)\,dx \right)$

$\int_1^\infty \frac{dx}{x^p}$ and $\int_2^\infty \frac{dx}{x(\ln x)^p}$ are $\begin{cases} \textit{convergent} \text{ if } p>1 \\ \textit{divergent} \text{ if } p\leq 1 \end{cases}$

$\int_0^1 \frac{dx}{x^p}$ is $\begin{cases} \textit{convergent} \text{ if } p<1 \\ \textit{divergent} \text{ if } p\geq 1 \end{cases}$

Convergence tests ($-a$ or b may be ∞)

(a) $0 \leq f(x) \leq g(x)$, $\int_a^b g(x)\,dx$ convergent $\Rightarrow \int_a^b f(x)\,dx$ convergent

(b) $\int_a^b |f(x)|\,dx$ convergent $\Rightarrow \int_a^b f(x)\,dx$ convergent

Uniform convergence

$\int_a^\infty f(x, t)\,dt$ *converges uniformly* for $x \in I$ if $\quad \sup_{x \in I} \left| \int_R^\infty f(x, t)\,dt \right| \to 0$ as $R \to \infty$

Test

(*i*) $|f(x, t)| \leq g(t)$, $x \in I$, (*ii*) $\int_a^\infty g(t)\,dt$ convergent $\Rightarrow$

$\int_a^\infty f(x, t)\,dt$ uniformly convergent for $x \in I$

7.3 Applications of Differential and Integral Calculus

Plane curves

(Curves in space, see sec. 11.1)

A=area, l=arc length, $\varkappa$=curvature, $\varrho = \frac{1}{|\varkappa|}$=radius of curvature,

(ξ, η)=centre of curvature

Curves in parametric form

$\left(\text{Dot denotes differentiation with respect to } t\text{, e.g. } \dot{x} = \frac{dx}{dt}\right)$

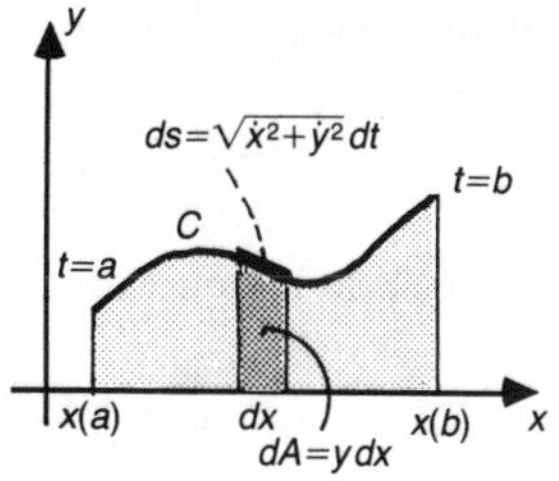

$$\text{Curve } C: \begin{cases} x = x(t) \\ y = y(t) \end{cases}, \quad a \leq t \leq b$$

$$A = \int_{x(a)}^{x(b)} y\,dx = \int_a^b y(t)\dot{x}(t)dt \quad (y \geq 0)$$

$$l = \int_C ds = \int_a^b \sqrt{\dot{x}(t)^2 + \dot{y}(t)^2}\,dt, \qquad \frac{dy}{dx} = \frac{\dot{y}(t)}{\dot{x}(t)}, \qquad \frac{d^2y}{dx^2} = \frac{\dot{x}\ddot{y} - \ddot{x}\dot{y}}{\dot{x}^3}$$

Asymptotes

(*i*) $y = kx + m$ if $\lim_{t \to t_o} x(t) = \pm\infty$ and $k = \lim_{t \to t_o} \frac{y(t)}{x(t)}$, $m = \lim_{t \to t_o} [y(t) - kx(t)]$

(*ii*) vertical $x = x_o$ if $\lim_{t \to t_o} x(t) = x_o$, $\lim_{t \to t_o} y(t) = \pm\infty$

$$\varkappa = \frac{\dot{x}\ddot{y} - \ddot{x}\dot{y}}{(\dot{x}^2 + \dot{y}^2)^{3/2}} = \frac{u}{v^{3/2}} \qquad \begin{cases} \xi = x - \frac{\dot{y}v}{u} \\ \eta = y + \frac{\dot{x}v}{u} \end{cases}$$

The *evolute* is the curve consisting of all centres of curvature of the given curve.

Curves in function form. $y=y(x)$

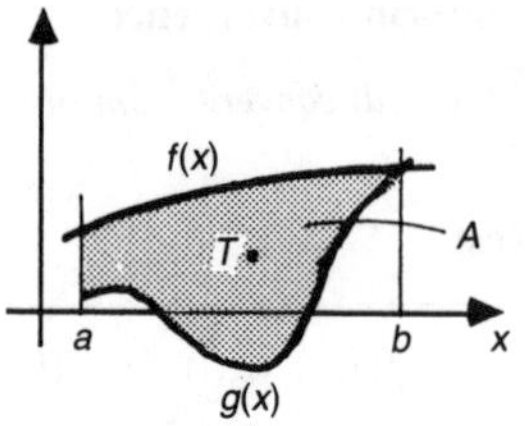

$$A=\int_a^b [f(x)-g(x)]\,dx \quad (f(x)\geq g(x))$$

$$l=\int_a^b \sqrt{1+f'(x)^2}\,dx$$

Asymptote $y=kx+m$: $k=\lim_{x\to\pm\infty}\frac{y(x)}{x}$, $m=\lim_{x\to\pm\infty}(y(x)-kx)$

$$\varkappa=\frac{y''(x)}{[1+y'(x)^2]^{3/2}} \qquad \begin{cases} \xi=x-\dfrac{y'(1+y'^2)}{y''} \\ \eta=y+\dfrac{1+y'^2}{y''} \end{cases}$$

Centroid $T=(T_x,\ T_y)$ $\begin{cases} T_x=\dfrac{1}{A}\displaystyle\int_a^b x[f(x)-g(x)]dx \\ T_y=\dfrac{1}{2A}\displaystyle\int_a^b [f^2(x)-g^2(x)]dx \end{cases}$

Moments of inertia about the y-axis / *about the x-axis*

(*i*) of curve $y=f(x)$ with density $\varrho(x)$

$$I_y=\int_a^b x^2\,\varrho(x)\sqrt{1+f'(x)^2}\,dx \qquad I_x=\int_a^b f(x)^2\,\varrho(x)\sqrt{1+f'(x)^2}\,dx$$

(*ii*) of plane region with constant density ϱ_0

$$I_y=\varrho_0\int_a^b x^2[f(x)-g(x)]\,dx \quad (f(x)\geq g(x)) \qquad I_x=\varrho_0/3\int_a^b [f(x)^3-g(x)^3]\,dx$$

Curves in implicit form

$$C:\ F(x,\ y)=0, \quad \frac{dy}{dx}=-\frac{F_x}{F_y}, \quad \frac{d^2y}{dx^2}=\frac{-F_y^2F_{xx}+2F_xF_yF_{xy}-F_x^2F_{yy}}{F_y^3}$$

$$\varkappa=\frac{-F_y^2F_{xx}+2F_xF_yF_{xy}-F_x^2F_{yy}}{(F_x^2+F_y^2)^{3/2}}=\frac{u}{v^{3/2}}$$

$$\begin{cases} \xi=x+\dfrac{F_xv}{u} \\ \eta=y+\dfrac{F_yv}{u} \end{cases}$$

Curves in polar coordinates, $x=r\cos\theta$, $y=r\sin\theta$

$$C: r=r(\theta),\ \alpha\leqslant\theta\leqslant\beta$$

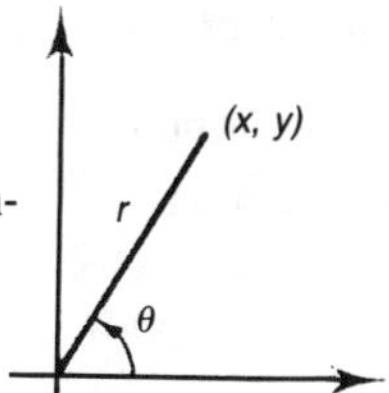

$$\begin{cases} x=r\cos\theta \\ y=r\sin\theta \end{cases}$$

Remark. The curve $r=r(\theta)$ may be transformed to rectangular parametric form by

$$\begin{cases} x=r(\theta)\cos\theta \\ y=r(\theta)\sin\theta \end{cases},\ \alpha\leqslant\theta\leqslant\beta$$

$$A=\frac{1}{2}\int_{\alpha}^{\beta} r^2(\theta)d\theta \qquad l=\int_{\alpha}^{\beta}\sqrt{r(\theta)^2+r'(\theta)^2}\,d\theta$$

$$\tan\mu=\frac{r(\theta)}{r'(\theta)} \qquad \varkappa=\frac{r^2+2r'^2-rr''}{[r^2+r'^2]^{3/2}}=\frac{u}{v^{3/2}}$$

$$\begin{cases} \xi=x-\dfrac{v(r\cos\theta+r'\sin\theta)}{u} \\ \eta=y-\dfrac{v(r\sin\theta-r'\cos\theta)}{u} \end{cases}$$

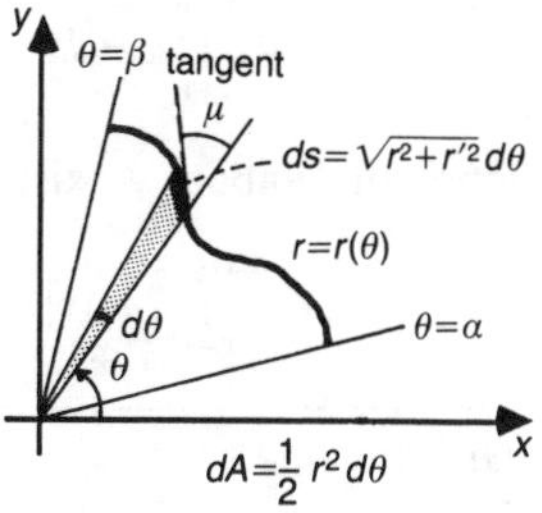

Family of curves

Family given by $F(x, y, \lambda)=0$, λ=parameter:

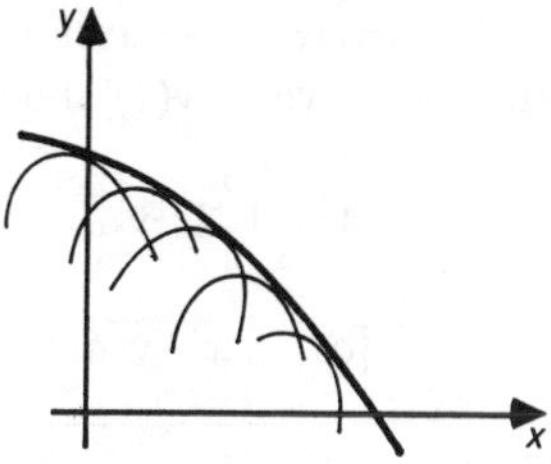

Envelope: Tangent to every curve of the family.

Envelope

The *envelope* is the solution (eliminating λ) of the system

$$\begin{cases} F(x, y, \lambda)=0 \\ F'_{\lambda}(x, y, \lambda)=0 \end{cases}$$

Orthogonal trajectories

Differential equation of

Family of curves	Family of orthogonal trajectories
Rectangular coord.	
$F(x, y, y')=0$	$F(x, y, -1/y')=0$
Polar coord.	
$F(\theta, r, r')=0$	$F(\theta, r, -r^2/r')=0$

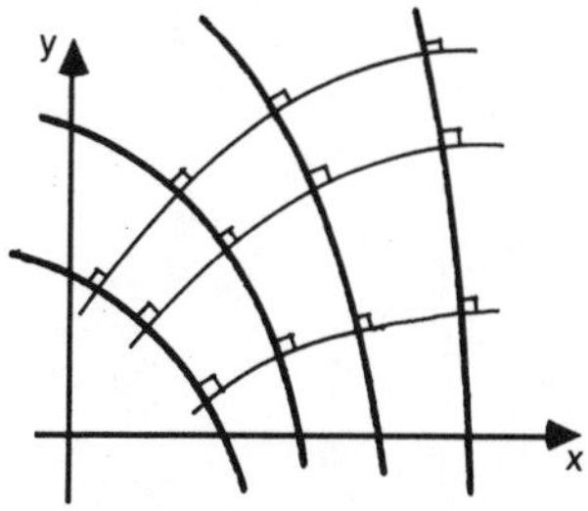

Orthogonal trajectories: Intersect the curves of the family perpendicularly.

Solids and surfaces of revolution

General volume formula

$A(x)$=area of section

$$V=\int_a^b A(x)dx$$

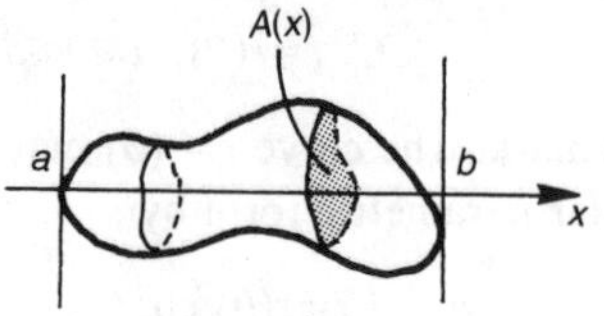

Volume of a solid of revolution

Rotation of region D about x-axis:

$$V=\int_{x=a}^{x=b} \pi y^2\, dx$$

Rotation of D about y-axis:

$$V=\int_{x=a}^{x=b} 2\pi x|y|\, dx \quad (0<a<b)$$

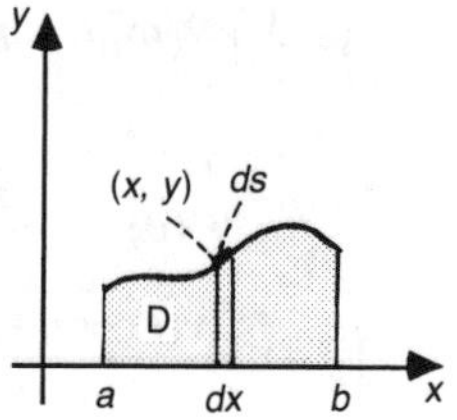

> *Guldin's rule*:
> V=product of area of generating region and distance traversed by its centre of mass.

Area of a surface of revolution

Rotation of curve $y=y(x)$ about x-axis:

$$A=\int_{x=a}^{x=b} 2\pi|y|\, ds$$

Rotation of curve $y=y(x)$ about y-axis:

$$A=\int_{x=a}^{x=b} 2\pi x\, ds \quad (0<a<b)$$

$$\left[ds=\sqrt{\dot{x}^2+\dot{y}^2}\,dt, \quad ds=\sqrt{1+y'(x)^2}\,dx, \quad ds=\sqrt{r^2+(r')^2}\,d\theta\right]$$

> *Guldin's rule*:
> A=product of length of generating curve and distance traversed by its centre of mass.

Examples of plane curves

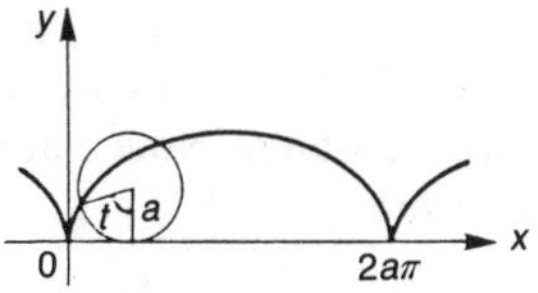

Cycloid

$$\begin{cases} x=a(t-\sin t) \\ y=a(1-\cos t) \end{cases}$$

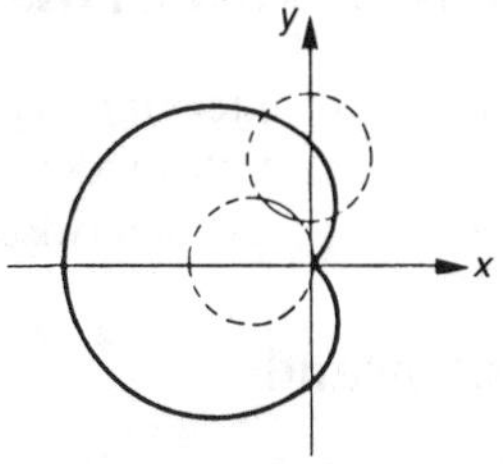

Cardioid

$r=2a(1-\cos t)$

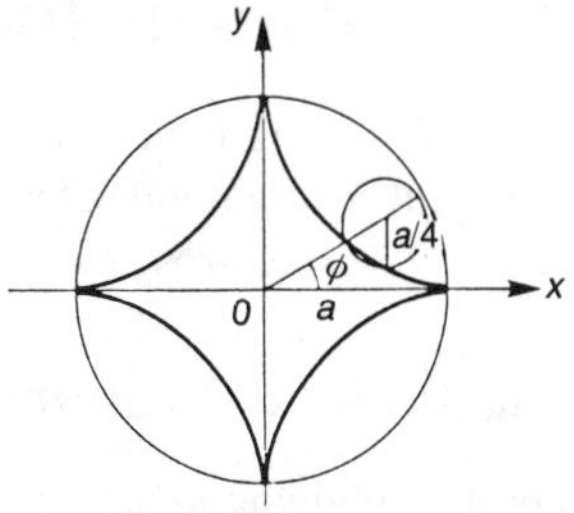

Astroid

$x^{2/3}+y^{2/3}=a^{2/3}$

$x=a\cos^3\phi$

$y=a\sin^3\phi$

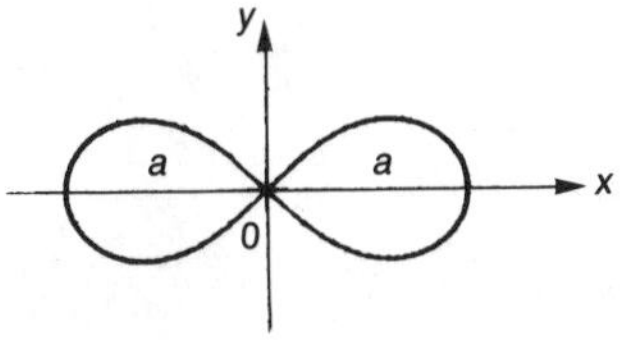

Lemniscate

$(x^2+y^2)^2=a^2(x^2-y^2)$

$r^2=a^2\cos 2\theta$

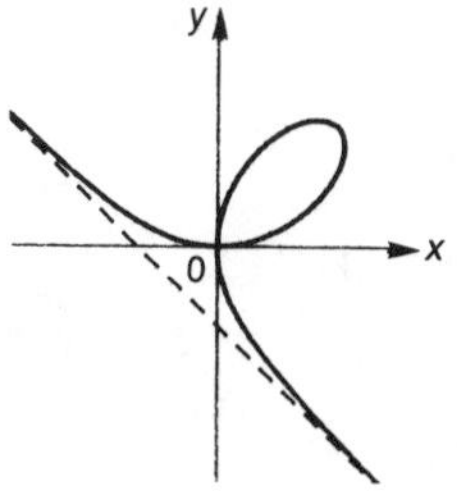

Folium of Descartes

$x^3+y^3-3axy=0$

$$\begin{cases} x=3a\phi/(1+\phi^3) \\ y=3a\phi^2/(1+\phi^3) \end{cases}$$

$$r=\frac{3a\sin\theta\cos\theta}{\sin^3\theta+\cos^3\theta}$$

[asymptote: $x+y+a=0$]

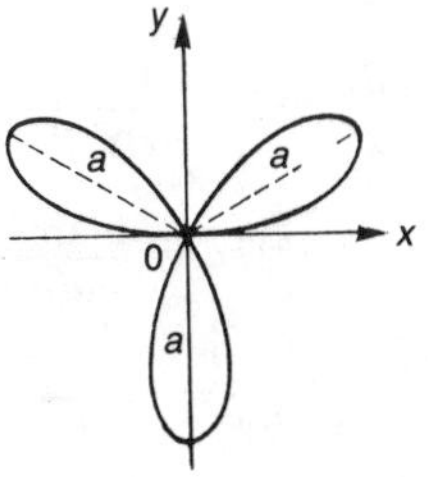

Three-leaved rose

$r=a\sin 3\phi$

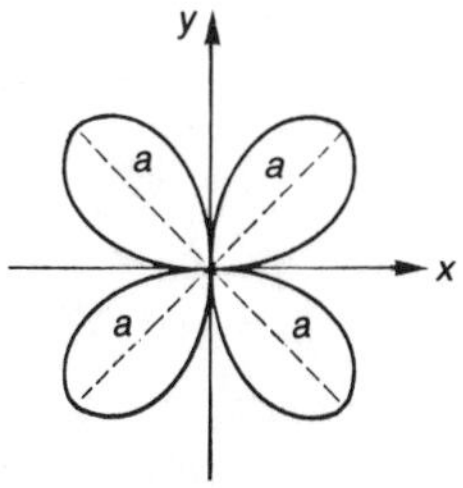

Four-leaved rose

$r=a\,|\sin 2\phi|$

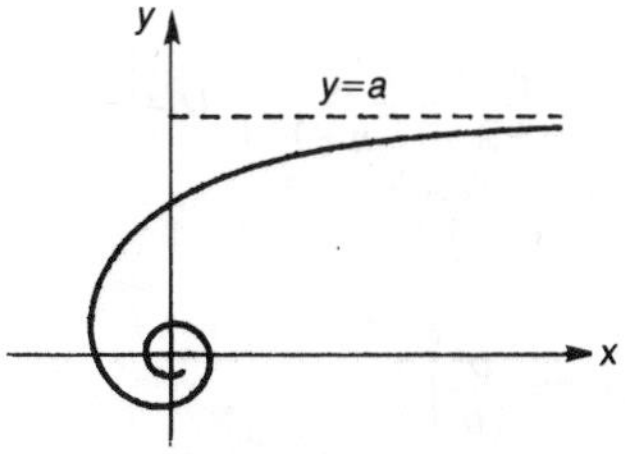

Hyperbolic spiral

$r=\dfrac{a}{\theta}$

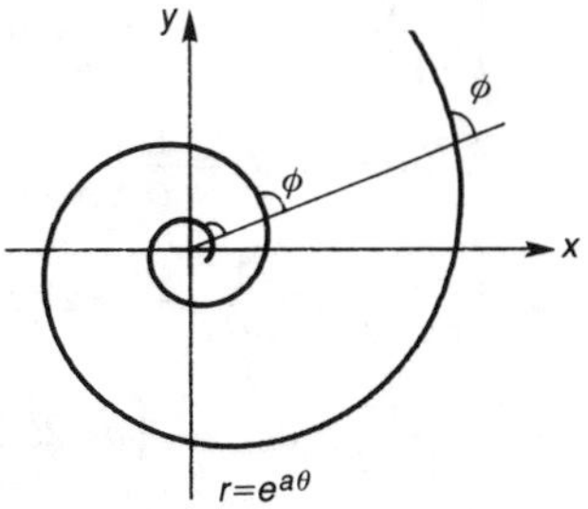

Logarithmic spiral

$\phi=\arctan\dfrac{1}{a}$

7.4 Tables of Indefinite Integrals

In the following table, the constant C of integration is omitted. The letters u and v represent different indicated functions of x. Other letters represent constants. Note that m and n denote *arbitrary real constants* unless otherwise is indicated.

Algebraic and rational integrands

Forms containing $u=ax+b$, $v=\sqrt{x}+c$

1. $\int c\,dx=cx$

2. $\int x^n dx=\frac{x^{n+1}}{n+1} \quad (n\neq -1)$

3. $\int \frac{dx}{x}=\ln|x|$

4. $\int \frac{dx}{x^n}=-\frac{1}{(n-1)x^{n-1}} \quad (n\neq 1)$

5. $\int (ax+b)^n dx=\frac{(ax+b)^{n+1}}{a(n+1)} \quad (n\neq -1)$

6. $\int \frac{dx}{(ax+b)^n}=-\frac{1}{a(n-1)(ax+b)^{n-1}} \quad (n\neq 1)$

7. $\int \frac{dx}{ax+b}=\frac{1}{a}\ln|ax+b|$

8. $\int x(ax+b)^n dx=\frac{1}{a^2}\left(\frac{u^{n+2}}{n+2}-\frac{bu^{n+1}}{n+1}\right) \quad (n\neq -1,\ -2)$

9. $\int \frac{x}{ax+b}dx=\frac{x}{a}-\frac{b}{a^2}\ln|u|$

10. $\int \frac{x^2}{ax+b}dx=\frac{x^2}{2a}-\frac{bx}{a^2}+\frac{b^2}{a^3}\ln|u|$

11. $\int \frac{x}{(ax+b)^2}dx=\frac{b}{a^2u}+\frac{1}{a^2}\ln|u|$

12. $\int x^2(ax+b)^n dx=\frac{1}{a^3}\left(\frac{u^{n+3}}{n+3}-\frac{2bu^{n+2}}{n+2}+\frac{b^2u^{n+1}}{n+1}\right) \quad (n\neq -1,\ -2,\ -3)$

13. $\int \frac{x^2}{ax+b}dx=\frac{1}{a^3}\left(\frac{u^2}{2}-2bu+b^2\ln|u|\right)$

14. $\int \frac{x^2}{(ax+b)^2}dx=\frac{1}{a^3}\left(u-2b\ln|u|-\frac{b^2}{u}\right)$

15. $\int \frac{x^2}{(ax+b)^3}dx=\frac{1}{a^3}\left(\ln|u|+\frac{2b}{u}-\frac{b^2}{2u^2}\right)$

16. $\int \frac{dx}{x(ax+b)}=-\frac{1}{b}\ln\left|\frac{u}{x}\right|$

17. $\int \frac{dx}{x^2(ax+b)} = -\frac{1}{bx} + \frac{a}{b^2} \ln \left| \frac{u}{x} \right|$

18. $\int \frac{dx}{x^3(ax+b)} = \frac{2ax-b}{2b^2x^2} - \frac{a^2}{b^3} \ln \left| \frac{u}{x} \right|$

19. $\int \frac{dx}{x(ax+b)^2} = \frac{1}{bu} - \frac{1}{b^2} \ln \left| \frac{u}{x} \right|$

20. $\int \frac{dx}{x(ax+b)^n} = \frac{1}{b(n-1)u^{n-1}} + \frac{1}{b} \int \frac{dx}{xu^{n-1}} \quad (n \neq 1)$

21. $\int \frac{dx}{x^2(ax+b)^2} = -\frac{2ax+b}{b^2xu} + \frac{2a}{b^3} \ln \left| \frac{u}{x} \right|$

22. $\int \sqrt{ax+b}\, dx = \frac{2u^{3/2}}{3a}$

23. $\int x\sqrt{ax+b}\, dx = \frac{2u^{3/2}}{15a^2}(3ax-2b)$

24. $\int x^2\sqrt{ax+b}\, dx = \frac{2u^{3/2}}{105a^3}(15a^2x^2-12abx+8b^2)$

25. $\int x^n\sqrt{ax+b}\, dx = \frac{2}{a(2n+3)}(x^n u^{3/2} - bn \int x^{n-1}\sqrt{u}\, dx)$

26. $\int \frac{\sqrt{ax+b}}{x}\, dx = 2\sqrt{u} + b \int \frac{dx}{x\sqrt{u}}$

27. $\int \frac{\sqrt{ax+b}}{x^n}\, dx = -\frac{1}{b(n-1)} \left(\frac{u^{3/2}}{x^{n-1}} + \frac{(2n-5)a}{2} \int \frac{\sqrt{u}}{x^{n-1}}\, dx \right)$

28. $\int \frac{dx}{\sqrt{ax+b}} = \frac{2\sqrt{u}}{a}$

29. $\int \frac{x^n}{\sqrt{ax+b}} dx = \frac{2}{a(2n+1)}(x^n\sqrt{u} - bn \int \frac{x^{n-1}}{\sqrt{u}}\, dx)$

30. $\int \frac{dx}{x^n\sqrt{ax+b}} = -\frac{\sqrt{u}}{(n-1)bx^{n-1}} - \frac{(2n-3)a}{(2n-2)b} \int \frac{dx}{x^{n-1}\sqrt{u}} \quad (n \neq 1)$

31. $$\int \frac{dx}{x\sqrt{ax+b}} = \begin{cases} \frac{1}{\sqrt{b}} \ln \left| \frac{\sqrt{u}-\sqrt{b}}{\sqrt{u}+\sqrt{b}} \right| & (b>0) \\ \frac{2}{\sqrt{-b}} \arctan \sqrt{\frac{u}{-b}} & (b<0) \end{cases}$$

32. $\int \frac{dx}{\sqrt{x}+c} = 2\sqrt{x} - 2c \ln |v|$

33. $\int \frac{\sqrt{x}}{\sqrt{x}+c}\, dx = x - 2c\sqrt{x} + 2c^2 \ln |v|$

34. $\int \frac{x}{\sqrt{x}+c} dx = \frac{2x\sqrt{x}}{3} - cx + 2c^2\sqrt{x} - 2c^3 \ln |v|$

35. $\int \frac{dx}{\sqrt{x}(\sqrt{x}+c)} = 2\ln|v|$

36 $\int \frac{dx}{x(\sqrt{x}+c)} = \frac{2}{c}\ln\frac{\sqrt{x}}{|v|}$

37. $\int \frac{dx}{(\sqrt{x}+c)^2} = 2\ln|v| + \frac{2c}{v}$

38. $\int \frac{\sqrt{x}}{(\sqrt{x}+c)^2}dx = 2\sqrt{x} - 4c\ln|v| - \frac{2c^2}{v}$

39. $\int \frac{x}{(\sqrt{x}+c)^2}dx = x - 4c\sqrt{x} + \frac{2c^3}{v} + 6c^2\ln|v|$

40. $\int \frac{dx}{\sqrt{x}(\sqrt{x}+c)^2} = -\frac{2}{v}$

41. $\int \frac{dx}{x(\sqrt{x}+c)^2} = \frac{2}{cv} + \frac{2}{c^2}\ln\frac{\sqrt{x}}{|v|}$

Forms containing $u=ax+b$, $v=cx+d$, $k=ad-bc\neq 0$

42. $\int \frac{dx}{(ax+b)^n(cx+d)^m} = \frac{1}{k(m-1)}\left[\frac{1}{u^{n-1}v^{m-1}} + a(m+n-2)\int\frac{dx}{u^n v^{m-1}}\right]$ $(m>1,\ n>0)$

43. $\int \frac{dx}{(ax+b)(cx+d)} = \frac{1}{k}\ln\left|\frac{u}{v}\right|$

44. $\int \frac{x}{(ax+b)(cx+d)}dx = -\frac{1}{k}\left(\frac{b}{a}\ln|u| - \frac{d}{c}\ln|v|\right)$

45. $\int \frac{dx}{(ax+b)^2(cx+d)} = -\frac{1}{k}\left(\frac{1}{u} + \frac{c}{k}\ln\left|\frac{u}{v}\right|\right)$

46. $\int \frac{x}{(ax+b)^2(cx+d)}dx = \frac{b}{aku} + \frac{d}{k^2}\ln\left|\frac{u}{v}\right|$

47. $\int \frac{x^2}{(ax+b)^2(cx+d)}dx = -\frac{b^2}{a^2ku} + \frac{1}{k^2}\left(\frac{d^2}{c}\ln|v| - \frac{b(k+ad)}{a^2}\ln|u|\right)$

48. $\int \frac{dx}{x(ax+b)(cx+d)} = \frac{1}{bd}\ln|x| - \frac{a}{bk}\ln|u| + \frac{c}{dk}\ln|v|$

49. $\int \frac{dx}{x^2(ax+b)(cx+d)} = -\frac{a^2d^2+b^2c^2}{b^2d^2k}\ln|x| - \frac{1}{bdx} + \frac{a^2}{b^2k}\ln|u| + \frac{c^2}{d^2k}\ln|v|$

50. $\int \frac{ax+b}{cx+d}dx = \frac{ax}{c} - \frac{k}{c^2}\ln|v|$

51. $\int \frac{(ax+b)^n}{(cx+d)^m}dx = \frac{1}{k(m-1)}\left[\frac{u^{n+1}}{v^{m-1}} + (m-n-2)a\int\frac{u^n}{v^{m-1}}dx\right] =$

$= -\frac{1}{(m-n-1)c}\left[\frac{u^n}{v^{m-1}} - kn\int\frac{u^{n-1}}{v^m}dx\right]$

52. $\int \sqrt{\frac{x+b}{x+d}}\,dx = \sqrt{x+b}\,\sqrt{x+d} + (b-d)\ln\left(\sqrt{x+b}+\sqrt{x+d}\right)$

53. $\int \sqrt{\frac{b-x}{d+x}}\,dx = \sqrt{b-x}\,\sqrt{d+x} + (b+d)\arcsin\sqrt{\frac{d+x}{b+d}}$

54. $$\int \frac{dx}{(cx+d)\sqrt{ax+b}} = \begin{cases} \frac{1}{\sqrt{-kc}}\ln\left|\frac{\sqrt{cu}-\sqrt{-k}}{\sqrt{cu}+\sqrt{-k}}\right| & (c>0,\ k<0) \\ \frac{2}{\sqrt{kc}}\arctan\sqrt{\frac{cu}{k}} & (c,\ k>0) \end{cases}$$

Forms containing $u=ax^2+b$, $v=b-ax^2$

55. $$I_1 = \int \frac{dx}{ax^2+b} = \begin{cases} \frac{1}{\sqrt{ab}}\arctan\left(x\sqrt{\frac{a}{b}}\right) & (a>0,\ b>0) \\ \frac{1}{2\sqrt{-ab}}\ln\left|\frac{x\sqrt{a}-\sqrt{-b}}{x\sqrt{a}+\sqrt{-b}}\right| & (a>0,\ b<0) \end{cases}$$

56. $I_2 = \int \frac{dx}{(ax^2+b)^2} = \frac{x}{2bu} + \frac{1}{2b}I_1$

57. $I_n = \int \frac{dx}{(ax^2+b)^n} = \frac{x}{2(n-1)bu^{n-1}} + \frac{2n-3}{2(n-1)b}I_{n-1}$

58. $\int x(ax^2+b)^n\,dx = \frac{u^{n+1}}{2a(n+1)} \quad (n\neq -1)$

59. $\int \frac{x}{ax^2+b}\,dx = \frac{1}{2a}\ln|u|$

60. $\int \frac{x\,dx}{(ax^2+b)^n} = -\frac{1}{2a(n-1)u^{n-1}} \quad (n\neq 1)$

61. $\int \frac{dx}{x(ax^2+b)} = \frac{1}{2b}\ln\left|\frac{x^2}{u}\right|$

62. $\int \frac{dx}{x^2(ax^2+b)} = -\frac{1}{bx} - \frac{a}{b}I_1$ (see 55.)

63. $\int \frac{x^2}{ax^2+b}\,dx = \frac{x}{a} - \frac{b}{a}I_1$ (see 55.)

64. $\int \frac{x^n}{ax^2+b}\,dx = \frac{x^{n-1}}{a(n-1)} - \frac{b}{a}\int \frac{x^{n-2}}{u}\,dx \quad (n\neq 1)$

65. $\int \frac{x^2}{(ax^2+b)^n}\,dx = -\frac{x}{2(n-1)au^{n-1}} + \frac{1}{2(n-1)a}I_{n-1}$ (see 57.)

66. $\int \frac{dx}{x(ax^2+b)^n} = \frac{1}{2b(n-1)u^{n-1}} + \frac{1}{b}\int \frac{dx}{x\,u^{n-1}}$

67. $\int \frac{dx}{x^2(ax^2+b)^n} = \frac{1}{b}\int \frac{dx}{x^2u^{n-1}} - \frac{a}{b}I_n$ (see 57.)

68. $\displaystyle\int \frac{dx}{(cx+d)(ax^2+b)} = \frac{1}{k}\left[\frac{c}{2}\ln\frac{(cx+d)^2}{|u|} + ad\,I_1\right]$ $(k=ad^2+bc^2)$ (see 55)

69. $\displaystyle\int \frac{x\,dx}{(cx+d)(ax^2+b)} = \frac{1}{k}\left[-\frac{d}{2}\ln\frac{(cx+d)^2}{|u|} + bc\,I_1\right]$ $(k=ad^2+bc^2)$ (see 55)

70. $\displaystyle\int \frac{x^2dx}{(cx+d)(ax^2+b)} = \frac{1}{k}\left[\frac{d^2}{c}\ln|cx+d| + \frac{bc}{2a}\ln|u| - bd\,I_1\right]$ $(k=ad^2+bc^2)$ (see 55)

71. $\displaystyle\int \sqrt{ax^2+b}\,dx = \frac{x\sqrt{u}}{2} + \frac{b}{2\sqrt{a}}\ln|x\sqrt{a}+\sqrt{u}|$ $(a>0)$

72. $\displaystyle\int \sqrt{b-ax^2}\,dx = \frac{x\sqrt{v}}{2} + \frac{b}{2\sqrt{a}}\arcsin\left(x\sqrt{\frac{a}{b}}\right)$ $(a>0)$

73. $\displaystyle\int \frac{dx}{\sqrt{ax^2+b}} = \frac{1}{\sqrt{a}}\ln|x\sqrt{a}+\sqrt{u}|$ $(a>0)$

74. $\displaystyle\int \frac{dx}{\sqrt{b-ax^2}} = \frac{1}{\sqrt{a}}\arcsin\left(x\sqrt{\frac{a}{b}}\right)$ $(a>0)$

75. $\displaystyle\int \frac{x}{\sqrt{ax^2+b}}\,dx = \frac{\sqrt{u}}{a}$

76. $$\int \frac{dx}{x\sqrt{ax^2+b}} = \begin{cases} \dfrac{1}{\sqrt{b}}\ln\left|\dfrac{\sqrt{u}-\sqrt{b}}{x}\right| & (b>0) \\ \dfrac{1}{\sqrt{-b}}\arctan\sqrt{\dfrac{u}{-b}} & (b<0) \end{cases}$$

77. $\displaystyle\int x\sqrt{ax^2+b}\,dx = \frac{u^{3/2}}{3a}$

78. $\displaystyle\int x^2\sqrt{ax^2+b}\,dx = \frac{x}{4a}u^{3/2} - \frac{bx}{8a}\sqrt{u} - \frac{b^2}{8a^{3/2}}\ln|x\sqrt{a}+\sqrt{u}|$ $(a>0)$

79. $\displaystyle\int x^2\sqrt{b-ax^2}\,dx = -\frac{x}{4a}v^{3/2} + \frac{bx}{8a}\sqrt{v} + \frac{b^2}{8a^{3/2}}\arcsin\left(x\sqrt{\frac{a}{b}}\right)$ $(a>0)$

80. $$\int \frac{\sqrt{ax^2+b}}{x}\,dx = \begin{cases} \sqrt{u}+\sqrt{b}\ln\left|\dfrac{\sqrt{u}-\sqrt{b}}{x}\right| & (b>0) \\ \sqrt{u}-\sqrt{-b}\arctan\sqrt{\dfrac{u}{-b}} & (b<0) \end{cases}$$

81. $\displaystyle\int \frac{dx}{x^2\sqrt{ax^2+b}} = -\frac{\sqrt{u}}{bx}$

82. $\displaystyle\int \frac{x^n}{\sqrt{ax^2+b}}\,dx = \frac{x^{n-1}\sqrt{u}}{na} - \frac{(n-1)b}{na}\int \frac{x^{n-2}}{\sqrt{u}}\,dx$ $(n>0)$

83. $\displaystyle\int x^n\sqrt{ax^2+b}\,dx = \frac{x^{n-1}u^{3/2}}{(n+2)a} - \frac{(n-1)b}{(n+2)a}\int x^{n-2}\sqrt{u}\,dx$ $(n>0)$

84. $\displaystyle\int \frac{\sqrt{ax^2+b}}{x^n}\,dx = -\frac{u^{3/2}}{b(n-1)x^{n-1}} - \frac{(n-4)a}{(n-1)b}\int \frac{\sqrt{u}}{x^{n-2}}\,dx$ $(n>1)$

85. $\displaystyle\int \frac{dx}{x^n\sqrt{ax^2+b}} = -\frac{\sqrt{u}}{b(n-1)x^{n-1}} - \frac{(n-2)a}{(n-1)b}\int \frac{dx}{x^{n-2}\sqrt{u}} \quad (n>1)$

86. $\displaystyle\int \frac{dx}{(x-c)\sqrt{x^2-c^2}} = -\frac{\sqrt{x^2-c^2}}{c(x-c)}$ (c has arbitrary sign)

87. $\displaystyle\int \frac{dx}{(x-c)\sqrt{ax^2+b}} = \left[x-c=\frac{1}{t}\right] = \mp\int \frac{dt}{\sqrt{(ac^2+b)t^2+2act+a}}$

(− if $x>c$, + if $x<c$, see 104 or 27)

88. $\displaystyle\int \frac{dx}{(x-c)^n\sqrt{ax^2+b}} = \left[x-c=\frac{1}{t}\right] = \mp\int \frac{t^{n-1}}{\sqrt{(ac^2+b)t^2+2act+a}}\,dt$

(− if $x>c$, + if $x<c$, see 105, 106, 115 or 28)

89. $\displaystyle\int (ax^2+b)^{3/2}dx = \frac{x}{8}(2ax^2+5b)\sqrt{u} + \frac{3b^2}{8\sqrt{a}}\ln\left|x\sqrt{a}+\sqrt{u}\right| \quad (a>0)$

90. $\displaystyle\int (b-ax^2)^{3/2}dx = \frac{x}{8}(-2ax^2+5b)\sqrt{v} + \frac{3b^2}{8\sqrt{a}}\arcsin\left(x\sqrt{\frac{a}{b}}\right) \quad (a>0)$

91. $\displaystyle\int x(ax^2+b)^{3/2}dx = \frac{1}{5a}u^{5/2}$

92. $\displaystyle\int x^2(ax^2+b)^{3/2}dx = \frac{1}{6}x^3u^{3/2} + \frac{b}{2}\int x^2\sqrt{u}\,dx$ (see 78, 79)

93. $\displaystyle\int x^3(ax^2+b)^{3/2}dx = \frac{1}{7a^2}u^{7/2} - \frac{b}{5a^2}u^{5/2}$

94. $\displaystyle\int \frac{(ax^2+b)^{3/2}}{x}dx = \frac{1}{3}u^{3/2} + b\int \frac{\sqrt{u}}{x}dx$ (see 80)

95. $\displaystyle\int \frac{dx}{(ax^2+b)^{3/2}} = \frac{x}{b\sqrt{u}}$

96. $\displaystyle\int \frac{x}{(ax^2+b)^{3/2}}dx = -\frac{1}{a\sqrt{u}}$

97. $\displaystyle\int \frac{x^2}{(ax^2+b)^{3/2}}dx = -\frac{x}{a\sqrt{u}} + \frac{1}{a}\int \frac{dx}{\sqrt{u}}$ (see 72, 73)

98. $\displaystyle\int \frac{x^3}{(ax^2+b)^{3/2}}dx = -\frac{x^2}{a\sqrt{u}} + \frac{2\sqrt{u}}{a^2}$

99. $\displaystyle\int \frac{dx}{x(ax^2+b)^{3/2}} = \frac{1}{b\sqrt{u}} + \frac{1}{b}\int \frac{dx}{x\sqrt{u}}$ (see 76)

100. $\displaystyle\int \frac{dx}{x^2(ax^2+b)^{3/2}} = -\frac{1}{b^2}\left(\frac{\sqrt{u}}{x} + \frac{ax}{\sqrt{u}}\right)$

101. $\displaystyle\int \frac{dx}{x^3(ax^2+b)^{3/2}} = -\frac{1}{2b}\left(\frac{1}{x^2\sqrt{u}} + 3a\int \frac{dx}{x\,u^{3/2}}\right)$ (see 99)

Forms containing $u=ax^2+bx+c$, ($k=4ac-b^2$)

> Remark. Writing $ax^2+bx+c=a\left(x+\frac{b}{2a}\right)^2+c-\frac{b^2}{4a}$ this form is transformed by the substitution $t=x+\frac{b}{2a}$ to the form at^2+b_1.

102. $$\int\frac{dx}{u}=\begin{cases}\frac{1}{\sqrt{-k}}\ln\left|\frac{2ax+b-\sqrt{-k}}{2ax+b+\sqrt{-k}}\right| & (4ac<b^2)\\ \frac{2}{\sqrt{k}}\arctan\frac{2ax+b}{\sqrt{k}} & (4ac>b^2)\\ -\frac{2}{2ax+b} & (4ac=b^2)\end{cases}$$

103. $$\int\frac{x}{u}dx=\frac{1}{2a}\ln|u|-\frac{b}{2a}\int\frac{dx}{u}$$

104. $$\int\frac{dx}{u^2}=\frac{2ax+b}{ku}+\frac{2a}{k}\int\frac{dx}{u}$$

105. $$\int\frac{x}{u^2}dx=-\frac{bx+2c}{ku}-\frac{b}{k}\int\frac{dx}{u}$$

106. $$\int\frac{dx}{xu}=\frac{1}{2c}\ln\left|\frac{x^2}{u}\right|-\frac{b}{2c}\int\frac{dx}{u}$$

107. $$\int\frac{dx}{x^2u}=\frac{b}{2c^2}\ln\left|\frac{u}{x^2}\right|-\frac{1}{cx}+\left(\frac{b^2}{2c^2}-\frac{a}{c}\right)\int\frac{dx}{u}$$

108. $$\int\frac{dx}{\sqrt{u}}=\begin{cases}\frac{1}{\sqrt{a}}\ln|2ax+b+2\sqrt{au}| & (a>0)\\ \frac{1}{\sqrt{-a}}\arcsin\frac{-2ax-b}{\sqrt{-k}} & (a<0)\end{cases}$$

109. $$\int\frac{x}{\sqrt{u}}dx=\frac{\sqrt{u}}{a}-\frac{b}{2a}\int\frac{dx}{\sqrt{u}}$$

110. $$\int\frac{x^2}{\sqrt{u}}dx=\frac{x\sqrt{u}}{2a}-\frac{3b}{4a}\int\frac{x}{\sqrt{u}}dx-\frac{c}{2a}\int\frac{dx}{\sqrt{u}}$$

111. $$\int\frac{dx}{x\sqrt{u}}=\begin{cases}-\frac{1}{\sqrt{c}}\ln\left|\frac{\sqrt{u}+\sqrt{c}}{x}+\frac{b}{2\sqrt{c}}\right| & (c>0)\\ \frac{1}{\sqrt{-c}}\arcsin\frac{bx+2c}{x\sqrt{-k}} & (c<0)\\ -\frac{2\sqrt{u}}{bx} & (c=0)\end{cases}$$

112. $$\int\frac{dx}{x^2\sqrt{u}}=-\frac{\sqrt{u}}{cx}-\frac{b}{2c}\int\frac{dx}{x\sqrt{u}}$$

113. $\displaystyle\int \frac{dx}{(x-d)^n\sqrt{u}} = \left[x-d=\frac{1}{t}\right] = \mp\int \frac{t^{n-1}}{\sqrt{(ad^2+bd+c)t^2+(2ad+b)t+a}}dt$
(− if $x>d$, + if $x<d$)

114. $\displaystyle\int \sqrt{u}\,dx = \frac{2ax+b}{4a}\sqrt{u} + \frac{k}{8a}\int\frac{dx}{\sqrt{u}}$

115. $\displaystyle\int x\sqrt{u}\,dx = \frac{u^{3/2}}{3a} - \frac{b}{2a}\int\sqrt{u}\,dx$

116. $\displaystyle\int x^2\sqrt{u}\,dx = \left(x-\frac{5b}{6a}\right)\frac{u^{3/2}}{4a} + \frac{5b^2-4ac}{16a^2}\int\sqrt{u}\,dx$

117. $\displaystyle\int\frac{dx}{u^{3/2}} = \frac{2(2ax+b)}{k\sqrt{u}}$

118. $\displaystyle\int\frac{dx}{u^{n+1}} = \frac{2ax+b}{knu^n} + \frac{2(2n-1)a}{kn}\int\frac{dx}{u^n}$

119. $\displaystyle\int\frac{x}{u^{n+1}}dx = -\frac{bx+2c}{knu^n} - \frac{b(2n-1)}{kn}\int\frac{dx}{u^n}$

120. $$\int\frac{x^m}{u^n}dx = \begin{cases} -\dfrac{x^{m-1}}{a(2n-m-1)u^n} - \dfrac{(n-m)b}{(2n-m-1)a}\displaystyle\int\frac{x^{m-1}}{u^n}\,dx + \\ \qquad + \dfrac{(m-1)c}{(2n-m-1)a}\displaystyle\int\frac{x^{m-2}}{u^n}\,dx \quad (m\neq 2n-1) \\ \dfrac{1}{a}\displaystyle\int\frac{x^{m-2}}{u^{n-1}}dx - \frac{b}{a}\int\frac{x^{m-1}}{u^n}dx - \frac{c}{a}\int\frac{x^{m-2}}{u^n}\,dx \quad (\text{all } m,n) \end{cases}$$

121. $\displaystyle\int\frac{dx}{x^m u^n} = -\frac{1}{(m-1)cx^{m-1}u^{n-1}} - \frac{(n+m-2)b}{(m-1)c}\int\frac{dx}{x^{m-1}u^n} -$
$\displaystyle\qquad -\frac{(2n+m-3)a}{(m-1)c}\int\frac{dx}{x^{m-2}u^n}$

122. $\displaystyle\int\frac{dx}{x\,u^n} = \frac{1}{2c(n-1)u^{n-1}} - \frac{b}{2c}\int\frac{dx}{u^n} + \frac{1}{c}\int\frac{dx}{x\,u^{n-1}}$

123. $\displaystyle\int u^n\,dx = \frac{(2ax+b)u^n}{2(2n+1)a} + \frac{nk}{2(2n+1)a}\int u^{n-1}\,dx$

124. $\displaystyle\int xu^n\,dx = \frac{u^{n+1}}{2(n+1)a} - \frac{b}{2a}\int u^n\,dx$

Forms containing $u=x^3+a^3$

(For forms x^3-a^3 change a to $-a$)

125. $\displaystyle\int\frac{dx}{x^3+a^3} = \frac{1}{3a^2}\left[\frac{1}{2}\ln\frac{(x+a)^2}{x^2-ax+a^2} + \sqrt{3}\arctan\frac{2x-a}{a\sqrt{3}}\right]$

126. $\displaystyle\int\frac{dx}{(x^3+a^3)^2} = \frac{x}{3a^3u} + \frac{2}{3a^3}\int\frac{dx}{u}$

127. $\int \frac{x}{x^3+a^3}dx = \frac{1}{3a}\left[-\frac{1}{2}\ln\frac{(x+a)^2}{x^2-ax+a^2} + \sqrt{3}\arctan\frac{2x-a}{a\sqrt{3}}\right]$

128. $\int \frac{x^2}{x^3+a^3}dx = \frac{1}{3}\ln|u|$

129. $\int \frac{dx}{x(x^3+a^3)} = \frac{1}{3a^3}\ln\left|\frac{x^3}{u}\right|$

130. $\int \frac{dx}{x^2(x^3+a^3)} = -\frac{1}{a^3x} - \frac{1}{a^3}\int\frac{x}{u}dx$

Forms containing $u=x^4 \pm a^4$

131. $\int \frac{dx}{x^4+a^4} = \frac{1}{2\sqrt{2}a^3}\left[\frac{1}{2}\ln\frac{x^2+ax\sqrt{2}+a^2}{x^2-ax\sqrt{2}+a^2} + \arctan\frac{ax\sqrt{2}}{a^2-x^2} + (n\pi)^*\right]$

132. $\int \frac{dx}{x^4-a^4} = \frac{1}{2a^3}\left[\frac{1}{2}\ln\left|\frac{x-a}{x+a}\right| - \arctan\frac{x}{a}\right]$

133. $\int \frac{x}{x^4+a^4}dx = \frac{1}{2a^2}\arctan\frac{x^2}{a^2}$

134. $\int \frac{x}{x^4-a^4}dx = \frac{1}{4a^2}\ln\left|\frac{x^2-a^2}{x^2+a^2}\right|$

135. $\int \frac{x^2}{x^4+a^4}dx = \frac{1}{2\sqrt{2}a}\left[\frac{1}{2}\ln\frac{x^2-ax\sqrt{2}+a^2}{x^2+ax\sqrt{2}+a^2} + \arctan\frac{ax\sqrt{2}}{a^2-x^2} + (n\pi)^*\right]$

136. $\int \frac{x^2}{x^4-a^4}dx = \frac{1}{2a}\left[\frac{1}{2}\ln\left|\frac{x-a}{x+a}\right| + \arctan\frac{x}{a}\right]$

137. $\int \frac{x^3}{x^4 \pm a^4}dx = \frac{1}{4}\ln|x^4 \pm a^4|$

Forms containing $u=ax^n+b$

138. $\int \frac{dx}{x(ax^n+b)} = \frac{1}{bn}\ln\left|\frac{x^n}{u}\right|$

139. $\int \frac{dx}{x\sqrt{ax^n+b}} = \begin{cases} \frac{1}{n\sqrt{b}}\ln\left|\frac{\sqrt{u}-\sqrt{b}}{\sqrt{u}+\sqrt{b}}\right| & (b>0) \\ \frac{2}{n\sqrt{-b}}\arctan\sqrt{\frac{u}{-b}} & (b<0) \end{cases}$

140. $\int x^m(ax^n+b)^p dx =$

$= \frac{1}{m+np+1}[x^{m+1}u^p + npb\int x^m u^{p-1}dx] =$

$= \frac{1}{bn(p+1)}[-x^{m+1}u^{p+1} + (m+np+n+1)\int x^m u^{p+1}dx] =$

* Add π when going past $-a$ and a.

$$= \frac{1}{a(m+np+1)} \left[x^{m-n+1}u^{p+1} - (m-n+1)b \int x^{m-n}u^p dx\right] =$$

$$= \frac{1}{b(m+1)} \left[x^{m+1}u^{p+1} - (m+np+n+1)b \int x^{m+n}u^p dx\right]$$

(m, n, p arbitrarily real)

Transcendental integrands

Forms containing sin *ax*, cos *ax*

$\left(\text{For forms involving } \sec x,\ \csc x, \text{ use } \sec x = \frac{1}{\cos x},\ \csc x = \frac{1}{\sin x}\right).$

141. $\int \sin ax\, dx = -\frac{1}{a}\cos ax$

142. $\int \cos ax\, dx = \frac{1}{a}\sin ax$

143. $\int \sin^2 ax\, dx = \frac{x}{2} - \frac{\sin 2ax}{4a}$

144. $\int \cos^2 ax\, dx = \frac{x}{2} + \frac{\sin 2ax}{4a}$

145. $\int \sin^3 ax\, dx = -\frac{1}{a}\cos ax + \frac{1}{3a}\cos^3 ax$

146. $\int \cos^3 ax\, dx = \frac{1}{a}\sin ax - \frac{1}{3a}\sin^3 ax$

147. $\int \sin^4 ax\, dx = \frac{3x}{8} - \frac{\sin 2ax}{4a} + \frac{\sin 4ax}{32a}$

148. $\int \cos^4 ax\, dx = \frac{3x}{8} + \frac{\sin 2ax}{4a} + \frac{\sin 4ax}{32a}$

149. $\int \sin^n ax\, dx = -\frac{1}{na}\sin^{n-1} ax\, \cos ax + \frac{n-1}{n}\int \sin^{n-2} ax\, dx$

150. $\int \cos^n ax\, dx = \frac{1}{na}\sin ax\, \cos^{n-1} ax + \frac{n-1}{n}\int \cos^{n-2} ax\, dx$

151. $\int \sin ax \cos^n ax\, dx = \begin{cases} -\frac{\cos^{n+1} ax}{a(n+1)} & (n \neq -1) \\ -\frac{1}{a}\ln|\cos ax| & (n = -1) \end{cases}$

152. $\int \sin^n ax \cos ax\, dx = \begin{cases} \frac{\sin^{n+1} ax}{a(n+1)} & (n \neq -1) \\ \frac{1}{a}\ln|\sin ax| & (n = -1) \end{cases}$

153. $\int \sin^m x \cos^n x\, dx = \begin{cases} \dfrac{\sin^{m+1}x \cos^{n-1}x}{m+n} + \dfrac{n-1}{m+n} \int \sin^m x \cos^{n-2} x\, dx, \text{ or} \\ -\dfrac{\sin^{m-1}x \cos^{n+1}x}{m+n} + \dfrac{m-1}{m+n} \int \sin^{m-2} x \cos^n x\, dx \end{cases}$

154. $\int x \sin ax\, dx = \frac{1}{a^2}(\sin ax - ax \cos ax)$

155. $\int x \cos ax\, dx = \frac{1}{a^2}(\cos ax + ax \sin ax)$

156. $\int x^2 \sin ax\, dx = \frac{1}{a^3}(2\cos ax + 2ax \sin ax - a^2x^2 \cos ax)$

157. $\int x^2 \cos ax\, dx = \frac{1}{a^3}(-2\sin ax + 2ax \cos ax + a^2x^2 \sin ax)$

158. $\int x^n \sin ax\, dx = -\frac{1}{a} x^n \cos ax + \frac{n}{a} \int x^{n-1} \cos ax\, dx$

159. $\int x^n \cos ax\, dx = \frac{1}{a} x^n \sin ax - \frac{n}{a} \int x^{n-1} \sin ax\, dx$

160. $\int x \sin^2 ax\, dx = \frac{x^2}{4} - \frac{x \sin 2ax}{4a} - \frac{\cos 2ax}{8a^2}$

161. $\int x \cos^2 ax\, dx = \frac{x^2}{4} + \frac{x \sin 2ax}{4a} + \frac{\cos 2ax}{8a^2}$

162. $\int \sin mx \sin nx\, dx = \frac{\sin(m-n)x}{2(m-n)} - \frac{\sin(m+n)x}{2(m+n)} \quad (m^2 \neq n^2)$

163. $\int \sin mx \cos nx\, dx = -\frac{\cos(m-n)x}{2(m-n)} - \frac{\cos(m+n)x}{2(m+n)} \quad (m^2 \neq n^2)$

164. $\int \cos mx \cos nx\, dx = \frac{\sin(m-n)x}{2(m-n)} + \frac{\sin(m+n)x}{2(m+n)} \quad (m^2 \neq n^2)$

165. $\int \csc ax\, dx = \int \frac{dx}{\sin ax} = \frac{1}{a} \ln \left| \tan \frac{ax}{2} \right|$

166. $\int \csc^2 ax\, dx = -\frac{1}{a} \cot ax$

167. $\int \sec ax\, dx = \int \frac{dx}{\cos ax} = \frac{1}{a} \ln \left| \tan \left(\frac{ax}{2} + \frac{\pi}{4} \right) \right|$

168. $\int \sec^2 ax\, dx = \frac{1}{a} \tan ax$

169. $\int \frac{dx}{\sin^2 ax} = -\frac{1}{a} \cot ax$

170. $\int \sec ax \tan ax\, dx = \int \frac{\sin ax}{\cos^2 ax} dx = \frac{1}{a} \sec ax = \frac{1}{a \cos ax}$

171. $\int \frac{dx}{\cos^2 ax} = \frac{1}{a} \tan ax$

172. $\int \csc ax \cot ax\, dx = \int \frac{\cos ax}{\sin^2 ax} dx = -\frac{1}{a}\csc ax = -\frac{1}{a \sin ax}$

173. $\int \frac{dx}{\sin^n ax} = -\frac{\cos ax}{a(n-1)\sin^{n-1} ax} + \frac{n-2}{n-1}\int \frac{dx}{\sin^{n-2} ax}$

174. $\int \frac{dx}{\cos^n ax} = \frac{\sin ax}{a(n-1)\cos^{n-1} ax} + \frac{n-2}{n-1}\int \frac{dx}{\cos^{n-2} ax}$

175. $\int \frac{dx}{\sin ax \cos ax} = \frac{1}{a}\ln|\tan ax|$

176. $\int \frac{dx}{\sin ax \cos^2 ax} = \frac{1}{a}\left(\frac{1}{\cos ax} + \ln\left|\tan\frac{ax}{2}\right|\right)$

177. $\int \frac{dx}{\sin^2 ax \cos ax} = \frac{1}{a}\left(-\frac{1}{\sin ax} + \ln\left|\tan\left(\frac{ax}{2} + \frac{\pi}{4}\right)\right|\right)$

178. $$\int \frac{dx}{\sin^m x \cos^n x} = \begin{cases} -\frac{1}{(m-1)\sin^{\mathrm{m}-1} x \cos^{n-1} x} + \frac{m+n-2}{\mathrm{m}-1}\int \frac{dx}{\sin^{m-2} x \cos^n x}, \text{ or} \\ \frac{1}{(n-1)\sin^{m-1} x \cos^{n-1} x} - \frac{m+n-2}{n-1}\int \frac{dx}{\sin^m x \cos^{n-2} x} \end{cases}$$

179. $$\int \frac{\sin^m x}{\cos^n x} dx = \begin{cases} \frac{\sin^{m+1} x}{(n-1)\cos^{n-1} x} - \frac{m-n+2}{n-1}\int \frac{\sin^m x}{\cos^{n-2} x} dx, \text{ or} \\ -\frac{\sin^{m-1} x}{(m-n)\cos^{n-1} x} + \frac{m-1}{m-n}\int \frac{\sin^{m-2} x}{\cos^n x} dx \end{cases}$$

180. $$\int \frac{\cos^n x}{\sin^m x} dx = \begin{cases} -\frac{\cos^{n+1} x}{(m-1)\sin^{m-1} x} - \frac{n-m+2}{m-1}\int \frac{\cos^n x}{\sin^{m-2} x} dx, \text{ or} \\ \frac{\cos^{n-1} x}{(n-m)\sin^{m-1} x} + \frac{n-1}{n-m}\int \frac{\cos^{n-2} x}{\sin^m x} dx \end{cases}$$

181. $\int \frac{x}{\sin^2 ax} dx = -\frac{x}{a}\cot ax + \frac{1}{a^2}\ln|\sin ax|$

182. $\int \frac{x}{\cos^2 ax} dx = \frac{x}{a}\tan ax + \frac{1}{a^2}\ln|\cos ax|$

183. $\int \frac{1}{1+\sin ax} dx = -\frac{1}{a}\tan\left(\frac{\pi}{4} - \frac{ax}{2}\right)$

184. $\int \frac{dx}{1-\sin ax} = \frac{1}{a}\tan\left(\frac{\pi}{4} + \frac{ax}{2}\right)$

185. $\int \frac{1}{1+\cos ax} dx = \frac{1}{a}\tan\frac{ax}{2}$

186. $\int \frac{dx}{1-\cos ax} = -\frac{1}{a}\cot\frac{ax}{2}$

187. $$\int \frac{1}{b+c\sin ax}dx=\begin{cases}\dfrac{2}{a\sqrt{b^2-c^2}}\arctan\dfrac{b\tan\dfrac{ax}{2}+c}{\sqrt{b^2-c^2}} & (b^2>c^2)\\[2ex] \dfrac{1}{a\sqrt{c^2-b^2}}\ln\left|\dfrac{b\tan\dfrac{ax}{2}+c-\sqrt{c^2-b^2}}{b\tan\dfrac{ax}{2}+c+\sqrt{c^2-b^2}}\right| & (b^2<c^2)\end{cases}$$

188. $$\int \frac{dx}{\sin x(b+c\sin x)}=\frac{1}{b}\ln\left|\tan\frac{x}{2}\right|-\frac{c}{b}\int\frac{dx}{b+c\sin x}$$

189. $$\int \frac{dx}{\sin x(1+\sin x)}=\ln\left|\tan\frac{x}{2}\right|-\tan\left(\frac{x}{2}-\frac{\pi}{4}\right)$$

190. $$\int \frac{dx}{\sin x(1-\sin x)}=\ln\left|\tan\frac{x}{2}\right|+\tan\left(\frac{x}{2}+\frac{\pi}{4}\right)$$

191. $$\int \frac{dx}{(b+c\sin x)^2}=\frac{c\cos x}{(b^2-c^2)(b+c\sin x)}+\frac{b}{b^2-c^2}\int\frac{dx}{b+c\sin x}$$

192. $$\int \frac{\sin x\,dx}{(b+c\sin x)^2}=\frac{b\cos x}{(c^2-b^2)(b+c\sin x)}+\frac{c}{c^2-b^2}\int\frac{dx}{b+c\sin x}$$

193. $$\int \frac{\cos x\,dx}{(b+c\sin x)^2}=-\frac{1}{c(b+c\sin x)}$$

194. $$\int \frac{1}{b+c\cos ax}dx=\begin{cases}\dfrac{2}{a\sqrt{b^2-c^2}}\arctan\dfrac{(b-c)\tan\dfrac{ax}{2}}{\sqrt{b^2-c^2}} & (b^2>c^2)\\[2ex] \dfrac{1}{a\sqrt{c^2-b^2}}\ln\left|\dfrac{(c-b)\tan\dfrac{ax}{2}+\sqrt{c^2-b^2}}{(c-b)\tan\dfrac{ax}{2}-\sqrt{c^2-b^2}}\right| & (b^2<c^2)\end{cases}$$

195. $$\int \frac{dx}{\cos x(b+c\cos x)}=\frac{1}{b}\ln\left|\tan\left(\frac{x}{2}+\frac{\pi}{4}\right)\right|-\frac{c}{b}\int\frac{dx}{b+c\cos x}$$

196. $$\int \frac{dx}{\cos x(1+\cos x)}=\ln\left|\tan\left(\frac{x}{2}+\frac{\pi}{4}\right)\right|-\tan\frac{x}{2}$$

197. $$\int \frac{dx}{\cos x(1-\cos x)}=\ln\left|\tan\left(\frac{x}{2}+\frac{\pi}{4}\right)\right|-\cot\frac{x}{2}$$

198. $$\int \frac{dx}{(b+c\cos x)^2}=\frac{c\sin x}{(c^2-b^2)(b+c\cos x)}-\frac{b}{c^2-b^2}\int\frac{dx}{b+c\cos x}$$

199. $$\int \frac{\cos x\,dx}{(b+c\cos x)^2}=\frac{b\sin x}{(b^2-c^2)(b+c\cos x)}-\frac{c}{b^2-c^2}\int\frac{dx}{b+c\cos x}$$

200. $$\int \frac{\sin x\,dx}{(b+c\cos x)^2}=\frac{1}{c(b+c\cos x)}$$

201. $$\int \frac{dx}{b\cos x+c\sin x}=\frac{1}{r}\ln\left|\tan\frac{x+\alpha}{2}\right|$$

$r=\sqrt{b^2+c^2}$, $\alpha=\arctan\dfrac{b}{c}$ $(c>0)$

202. $\displaystyle\int \frac{dx}{a+b\cos x+c\sin x}=[x+\alpha=t]=\int \frac{dt}{a+r\sin t}$ (r, α as in 201) ($c>0$)

203. $\displaystyle\int \frac{\sin ax}{b+c\cos ax}\,dx=-\frac{1}{ac}\ln|b+c\cos ax|$

204. $\displaystyle\int \frac{\cos ax}{b+c\sin ax}\,dx=\frac{1}{ac}\ln|b+c\sin ax|$

205. $\displaystyle\int \frac{\sin ax}{b+c\sin ax}\,dx=\frac{x}{c}-\frac{b}{c}\int \frac{dx}{b+c\sin ax}$

206. $\displaystyle\int \frac{\cos ax}{b+c\cos ax}\,dx=\frac{x}{c}-\frac{b}{c}\int \frac{dx}{b+c\cos ax}$

207. $\displaystyle\int \frac{dx}{a\sin^2 x+b}=\int \frac{dx}{(a+b)\sin^2 x+b\cos^2 x}$ (see 209 or 210)

208. $\displaystyle\int \frac{dx}{a\cos^2 x+b}=\int \frac{dx}{(a+b)\cos^2 x+b\sin^2 x}$ (see 209 or 210)

209. $\displaystyle\int \frac{dx}{a^2\cos^2 x+b^2\sin^2 x}=\frac{1}{ab}\arctan\left(\frac{b}{a}\tan x\right)$

210. $\displaystyle\int \frac{dx}{a^2\cos^2 x-b^2\sin^2 x}=\frac{1}{2ab}\ln\left|\frac{b\tan x+a}{b\tan x-a}\right|$

211. $\displaystyle\int \frac{\sin x}{a\cos^2 x+b}\,dx=[t=\cos x]=-\int \frac{dt}{at^2+b}$ (see 55)

212. $\displaystyle\int \frac{\cos x}{a\sin^2 x+b}\,dx=[t=\sin x]=\int \frac{dt}{at^2+b}$ (see 55)

For forms containing $\cos^2 x$ instead of $\sin^2 x$ in 213–220, use $\cos^2 x=1-\sin^2 x$.

213. $\displaystyle\int \sin x\sqrt{a\sin^2 x+b}\,dx=-\frac{\cos x}{2}\sqrt{a\sin^2 x+b}-$

$$-\frac{a+b}{2\sqrt{a}}\arcsin\frac{\sqrt{a}\cos x}{\sqrt{a+b}} \quad (a>0)$$

214. $\displaystyle\int \sin x\sqrt{b-a\sin^2 x}\,dx=-\frac{\cos x}{2}\sqrt{b-a\sin^2 x}-$

$$-\frac{a-b}{2\sqrt{a}}\ln\left|\sqrt{a}\cos x+\sqrt{b-a\sin^2 x}\right| \quad (a>0)$$

215. $\displaystyle\int \frac{\sin x}{\sqrt{a\sin^2 x+b}}\,dx=-\frac{1}{\sqrt{a}}\arcsin\frac{\sqrt{a}\cos x}{\sqrt{a+b}}$ ($a>0$)

216. $\displaystyle\int \frac{\sin x}{\sqrt{b-a\sin^2 x}}\,dx=-\frac{1}{\sqrt{a}}\ln\left|\sqrt{a}\cos x+\sqrt{b-a\sin^2 x}\right|$ ($a>0$)

217. $\displaystyle\int \cos x\sqrt{a\sin^2 x+b}\,dx=\frac{\sin x}{2}\sqrt{a\sin^2 x+b}+$

$$+\frac{b}{2\sqrt{a}}\ln\left|\sqrt{a}\sin x+\sqrt{a\sin^2 x+b}\right| \quad (a>0)$$

218. $\int \cos x\sqrt{b-a\sin^2 x}\, dx = \frac{\sin x}{2}\sqrt{b-a\sin^2 x} + \frac{b}{2\sqrt{a}}\arcsin\left(\sqrt{\frac{a}{b}}\sin x\right) \quad (a>0)$

219. $\int \frac{\cos x}{\sqrt{a\sin^2 x+b}}\, dx = \frac{1}{\sqrt{a}} \ln|\sqrt{a}\sin x + \sqrt{a\sin^2 x+b}| \quad (a>0)$

220. $\int \frac{\cos x}{\sqrt{b-a\sin^2 x}}\, dx = \frac{1}{\sqrt{a}}\arcsin\left(\sqrt{\frac{a}{b}}\sin x\right) \quad (a>0)$

Forms containing $\tan ax$ and $\cot ax = \frac{1}{\tan ax}$

221. $\int \tan ax\, dx = -\frac{1}{a}\ln|\cos ax|$

222. $\int \tan^2 ax\, dx = \frac{1}{a}\tan ax - x$

223. $\int \tan^3 ax\, dx = \frac{1}{2a}\tan^2 ax + \frac{1}{a}\ln|\cos ax|$

224. $\int \tan^n ax\, dx = \frac{1}{a(n-1)}\tan^{n-1} ax - \int \tan^{n-2} ax\, dx$

225. $\int \tan^n ax \sec^2 ax\, dx = \int \frac{\tan^n ax}{\cos^2 ax}\, dx = \frac{1}{a(n+1)}\tan^{n+1} ax \quad (n \neq -1)$

226. $\int \frac{\sec^2 ax}{\tan ax}\, dx = \int \frac{dx}{\cos^2 ax \tan ax} = \frac{1}{a}\ln|\tan ax|$

227. $\int \cot ax\, dx = \frac{1}{a}\ln|\sin ax|$

228. $\int \cot^2 ax\, dx = -\frac{1}{a}\cot ax - x$

229. $\int \cot^3 ax\, dx = -\frac{1}{2a}\cot^2 ax - \frac{1}{a}\ln|\sin ax|$

230. $\int \cot^n ax\, dx = -\frac{1}{a(n-1)}\cot^{n-1} ax - \int \cot^{n-2} ax\, dx$

231. $\int \cot^n ax \csc^2 ax\, dx = \int \frac{\cot^n ax}{\sin^2 ax}\, dx = -\frac{1}{a(n+1)}\cot^{n+1} ax \quad (n \neq -1)$

232. $\int \frac{\csc^2 ax}{\cot ax}\, dx = \int \frac{dx}{\sin^2 ax \cot ax} = -\frac{1}{a}\ln|\cot ax|$

233. $\int \frac{dx}{b+c\tan x} = \frac{1}{b^2+c^2}(bx + c\ln|b\cos x + c\sin x|)$

234. $\int \frac{dx}{\sqrt{b+c\tan^2 x}} = \frac{1}{\sqrt{b-c}}\arcsin\left(\sqrt{\frac{b-c}{b}}\sin x\right) \quad (b>0,\ b^2>c^2)$

Forms containing inverse trigonometric functions

235. $\int \arcsin ax\, dx = x \arcsin ax + \frac{1}{a}\sqrt{1-a^2x^2}$

236. $\int (\arcsin ax)^2 dx = x(\arcsin ax)^2 - 2x + \frac{2}{a}\sqrt{1-a^2x^2} \arcsin ax$

237. $\int x \arcsin ax\, dx = \frac{1}{4a^2}(2a^2x^2 \arcsin ax - \arcsin ax + ax\sqrt{1-a^2x^2})$

238. $\int x^2 \arcsin ax\, dx = \frac{1}{9a^3}(3a^3x^3 \arcsin ax + (a^2x^2+2)\sqrt{1-a^2x^2})$

239. $\int \frac{\arcsin ax}{x^2} dx = -\frac{1}{x}\arcsin ax - a \ln\left|\frac{1+\sqrt{1-a^2x^2}}{ax}\right|$

240. $\int \arccos ax\, dx = x \arccos ax - \frac{1}{a}\sqrt{1-a^2x^2}$

241. $\int (\arccos ax)^2 dx = x(\arccos ax)^2 - 2x - \frac{2}{a}\sqrt{1-a^2x^2} \arccos ax$

242. $\int x \arccos ax\, dx = \frac{1}{4a^2}(2a^2x^2 \arccos ax - \arccos ax - ax\sqrt{1-a^2x^2})$

243. $\int x^2 \arccos ax\, dx = \frac{1}{9a^3}(3a^3x^3 \arccos ax - (a^2x^2+2)\sqrt{1-a^2x^2})$

244. $\int \frac{\arccos ax}{x^2} dx = -\frac{1}{x}\arccos ax + a \ln\left|\frac{1+\sqrt{1-a^2x^2}}{ax}\right|$

245. $\int \arctan ax\, dx = \frac{1}{2a}[2ax \arctan ax - \ln(1+a^2x^2)]$

246. $\int \operatorname{arccot} ax\, dx = \frac{1}{2a}[2ax \operatorname{arccot} ax + \ln(1+a^2x^2)]$

247. $\int x \arctan ax\, dx = \frac{1}{2a^2}[(1+a^2x^2) \arctan ax - ax]$

248. $\int x^2 \arctan ax\, dx = \frac{1}{6a^3}[2a^3x^3 \arctan ax - a^2x^2 + \ln(1+a^2x^2)]$

249. $\int \frac{\arctan ax}{x^2} dx = -\frac{1}{x}\arctan ax - \frac{a}{2}\ln\frac{1+a^2x^2}{a^2x^2}$

250. $\int \sec^{-1} ax\, dx = x \sec^{-1} ax - \frac{1}{a}\ln\left|ax + \sqrt{a^2x^2-1}\right|$

251. $\int \csc^{-1} ax\, dx = x \csc^{-1} ax + \frac{1}{a}\ln\left|ax + \sqrt{a^2x^2-1}\right|$

252. $\int x \sec^{-1} ax\, dx = \frac{x^2}{2}\sec^{-1} ax - \frac{1}{2a^2}\sqrt{a^2x^2-1}$

253. $\int x \csc^{-1} ax\, dx = \frac{x^2}{2}\csc^{-1} ax + \frac{1}{2a^2}\sqrt{a^2x^2-1}$

Forms containing exponential, hyperbolic and logarithmic functions (also combined with sin and cos)

254. $\int e^{ax}dx=\frac{1}{a}\,e^{ax}$

255. $\int a^x dx=\int e^{x\ln a}dx=\frac{a^x}{\ln a}$

256. $\int xe^{ax}dx=\frac{e^{ax}}{a^2}\,(ax-1)$

257. $\int x^2e^{ax}dx=\frac{e^{ax}}{a^3}\,(a^2x^2-2ax+2)$

258. $\int x^ne^{ax}dx=\frac{e^{ax}}{a^{n+1}}\,[(ax)^n-n(ax)^{n-1}+n(n-1)(ax)^{n-2}-\ldots+(-1)^n n!]$ (n pos. integer)

259. $\int\frac{dx}{b+ce^{ax}}=\frac{1}{ab}\,(ax-\ln|b+ce^{ax}|)$

260. $\int\frac{e^{ax}}{b+ce^{ax}}dx=\frac{1}{ac}\ln|b+ce^{ax}|$

261. $\int xe^{ax^2}dx=\frac{1}{2a}\,e^{ax^2}$

262. $\int x^{2n+1}e^{ax^2}dx=\frac{1}{2}\int t^ne^{at}dt.\ \ t=x^2$. See 258.

263. $\int\frac{xe^{ax}}{(1+ax)^2}dx=\frac{e^{ax}}{a^2(1+ax)}$

264. $\int \sinh ax\;dx=\frac{1}{a}\cosh ax$

265. $\int \cosh ax\;dx=\frac{1}{a}\sinh ax$

266. $\int \tanh ax\;dx=\frac{1}{a}\ln(\cosh ax)$

267. $\int \coth ax\;dx=\frac{1}{a}\ln|\sinh ax|$

268. $\int \sinh^2 ax\;dx=\frac{1}{4a}\,(\sinh 2ax-2ax)$

269. $\int \sinh^n ax\,dx=\frac{1}{an}\sinh^{n-1}ax\cosh ax-\frac{n-1}{n}\int\sinh^{n-2}ax\,dx$

270. $\int \operatorname{csch} ax\,dx=\int\frac{dx}{\sinh ax}=\frac{1}{a}\ln\left|\tanh\frac{ax}{2}\right|$

271. $\int \operatorname{sech}^2 ax\,dx=\int\frac{dx}{\cosh^2 ax}=\frac{1}{a}\tanh ax$

272. $\int \operatorname{sech} ax\tanh ax\,dx=\int\frac{\sinh ax}{\cosh^2 ax}\,dx=-\frac{1}{a}\operatorname{sech} ax$

273. $\int \cosh^2 ax\ dx = \frac{1}{4a}(\sinh 2ax + 2ax)$

274. $\int \cosh^n ax\, dx = \frac{1}{an}\cosh^{n-1} ax \sinh ax + \frac{n-1}{n}\int \cosh^{n-2} ax\, dx$

275. $\int \operatorname{sech} ax\, dx = \int \frac{dx}{\cosh ax} = \frac{2}{a}\arctan e^{ax}$

276. $\int \operatorname{csch}^2 ax\, dx = \int \frac{dx}{\sinh^2 ax} = -\frac{1}{a}\coth ax$

277. $\int \operatorname{csch} ax \coth ax\, dx = \int \frac{\cosh ax}{\sinh^2 ax}\, dx = -\operatorname{csch} ax$

278. $\int \tanh^2 ax\ dx = x - \frac{1}{a}\tanh ax$

279. $\int \coth^2 ax\ dx = x - \frac{1}{a}\coth ax$

280. $\int \ln ax\ dx = x \ln ax - x$

281. $\int x^n \ln ax\, dx = x^{n+1}\left[\frac{\ln ax}{n+1} - \frac{1}{(n+1)^2}\right] \quad (n \neq -1)$

282. $\int x^n (\ln ax)^m dx = \frac{x^{n+1}}{n+1}(\ln ax)^m - \frac{m}{n+1}\int x^n (\ln ax)^{m-1} dx \ (n \neq -1)$

283. $\int \frac{(\ln ax)^n}{x} dx = \frac{(\ln ax)^{n+1}}{n+1} \quad (n \neq -1)$

284. $\int \frac{dx}{x \ln ax} = \ln(\ln ax)$

285. $\int (\ln ax)^2 dx = x(\ln ax)^2 - 2x \ln ax + 2x$

286. $\int (\ln ax)^n dx = x(\ln ax)^n - n\int (\ln ax)^{n-1}\, dx$

287. $\int e^{ax} \sin bx\ dx = \frac{e^{ax}}{a^2+b^2}(a \sin bx - b \cos bx)$

288. $\int e^{ax} \sin^n bx\, dx = \frac{e^{ax}\sin^{n-1} bx}{a^2+n^2b^2}(a \sin bx - nb \cos bx) +$

$$+\frac{n(n-1)b^2}{a^2+n^2b^2}\int e^{ax}\sin^{n-2} bx\, dx$$

289. $\int e^{ax} \cos bx\, dx = \frac{e^{ax}}{a^2+b^2}(a \cos bx + b \sin bx)$

290. $\int e^{ax} \cos^n bx\, dx = \frac{e^{ax}\cos^{n-1} bx}{a^2+n^2b^2}(a \cos bx + nb \sin bx) +$

$$+\frac{n(n-1)b^2}{a^2+n^2b^2}\int e^{ax}\cos^{n-2} bx\, dx$$

291. $\int xe^{ax} \sin bx\ dx = \frac{xe^{ax}}{a^2+b^2}(a \sin bx - b \cos bx) -$

$$-\frac{e^{ax}}{(a^2+b^2)^2}\,[(a^2-b^2)\,\sin bx-2ab\,\cos bx]$$

292. $\int xe^{ax}\cos bx\,dx=\frac{xe^{ax}}{a^2+b^2}\,(a\,\cos bx+b\,\sin bx)-$

$$-\frac{e^{ax}}{(a^2+b^2)^2}\,[(a^2-b^2)\,\cos bx+2ab\,\sin bx]$$

293. $\int\,\sin(\ln ax)dx=\frac{x}{2}\,[\sin(\ln ax)-\cos(\ln ax)]$

294. $\int\,\cos(\ln ax)dx=\frac{x}{2}\,[\sin(\ln ax)+\cos(\ln ax)]$

295. $\int\,\ln(ax+b)\,dx=\frac{ax+b}{a}\,\ln(ax+b)-x$

296. $\int\,\ln(x^2+a^2)\,dx=x\,\ln(x^2+a^2)-2x+2a\,\arctan\frac{x}{a}$

297. $\int\,\ln(x^2-a^2)\,dx=x\,\ln(x^2-a^2)-2x+a\,\ln\frac{x+a}{x-a}$

298. $\int\,x\,\ln(x^2\pm a^2)\,dx=\frac{1}{2}(x^2\pm a^2)\,\ln(x^2\pm a^2)-\frac{1}{2}x^2$

299. $\int\ln|x+\sqrt{x^2+a}|\,dx=x\ln|x+\sqrt{x^2+a}|-\sqrt{x^2+a}$

300. $\int x\ln|x+\sqrt{x^2+a}|\,dx=\left(\frac{x^2}{2}+\frac{a}{4}\right)\,\ln|x+\sqrt{x^2+a}|-\frac{x\sqrt{x^2+a}}{4}$

301. $\int\,\operatorname{arsinh}x\,dx=\int\,\ln(x+\sqrt{x^2+1})\,dx=x\,\operatorname{arsinh}x-\sqrt{x^2+1}$

302. $\int\,\operatorname{arcosh}x\,dx=\int\,\ln(x+\sqrt{x^2-1})\,dx=x\,\operatorname{arcosh}x-\sqrt{x^2-1}$

303. $\int\,\operatorname{artanh}x\,dx=x\,\operatorname{artanh}x+\frac{1}{2}\,\ln(x^2-1)$

304. $\int\,\operatorname{arcoth}x\,dx=x\,\operatorname{arcoth}x+\frac{1}{2}\,\ln(x^2-1)$

305. $\int\,\operatorname{sech}^{-1}x\,dx=x\,\operatorname{sech}^{-1}x+\sin^{-1}x$

306. $\int x\operatorname{sech}^{-1}x\,dx=\frac{x^2}{2}\operatorname{sech}^{-1}x-\frac{1}{2}\sqrt{1-x^2}$

307. $\int\,\operatorname{csch}^{-1}x\,dx=x\,\operatorname{csch}^{-1}x+\operatorname{sgn}x\,\sinh^{-1}x$

308. $\int x\operatorname{csch}^{-1}x\,dx=\frac{x^2}{2}\,\operatorname{csch}^{-1}x+\frac{1}{2}\,\operatorname{sgn}x\sqrt{1+x^2}$

7.5 Tables of Definite Integrals

$\Gamma(x)$ is the Gamma function (see sec. 12.6.)

$\gamma=0.5772156649$ ... is the Euler constant.

Elliptic integrals, see sec. 12.5.

Integrands containing algebraic functions

1. $\int_0^1 x^{m-1}(1-x)^{n-1}dx=\frac{\Gamma(m)\Gamma(n)}{\Gamma(m+n)}$ $(m,\ n>0)$

2. $\int_a^b (x-a)^{m-1}(b-x)^{n-1}dx=(b-a)^{m+n-1}\ \frac{\Gamma(m)\Gamma(n)}{\Gamma(m+n)}$ $(a<b,\ m,\ n>0)$

3. $\int_0^1 \frac{x^n}{1+x}dx=(-1)^n\left[\ln 2-1+\frac{1}{2}-\ldots+\frac{(-1)^n}{n}\right]$ $(n=1,\ 2,\ 3,\ \ldots)$

4. $\int_0^1 \frac{dx}{(1-x^n)^{1/n}}=\frac{\pi}{n\sin\frac{\pi}{n}}$ $(n>1)$

5. $\int_0^1 \frac{x^a}{\sqrt{1-x^2}}dx=\frac{\sqrt{\pi}\ \Gamma((a+1)/2)}{2\ \Gamma((a+2)/2)}$ $(a>-1)$

6. $\int_0^1 \frac{x^{a-1}}{(1-x)^a}=\frac{\pi}{\sin a\pi}$ $(0<a<1)$

7. $\int_0^1 \frac{dx}{\sqrt{1-x^\alpha}}=\frac{\sqrt{\pi}\ \Gamma\left(\frac{1}{\alpha}\right)}{\alpha\ \Gamma\left(\frac{1}{\alpha}+\frac{1}{2}\right)}$

8. $\int_0^\infty \frac{dx}{1+x^a}=\frac{\pi}{a\sin\frac{\pi}{a}}$ $(a>1)$

9. $\int_0^\infty \frac{dx}{x^\alpha(1+x)}=\frac{\pi}{\sin\alpha\pi}$ $(0<\alpha<1)$

10. $\int_0^\infty \frac{x^{\alpha-1}}{1+x^\beta}dx=\frac{\pi}{\beta\sin\left(\frac{\alpha\pi}{\beta}\right)}$ $(0<\alpha<\beta)$

11. $\int_0^\infty \frac{dx}{a^2+x^2}=\frac{\pi}{2a}$ $(a>0)$

12. $\int_0^\infty \frac{dx}{(a^2+x^2)^n}=\frac{\pi(2n-3)!!}{2a^{2n-1}(2n-2)!!}$ $(a>0,\ n=2,\ 3,\ \ldots)$

13. $\int_0^\infty \frac{dx}{(a^2+x^2)(b^2+x^2)}=\frac{\pi}{2ab(a+b)}$ $(a,\ b>0)$

14. $\int_0^\infty \frac{x^{m-1}}{(ax+b)^{m+n}}\,dx = \frac{\Gamma(m)\Gamma(n)}{a^m b^n \Gamma(m+n)} \quad (a,\ b,\ m,\ n>0)$

15. $\int_0^\infty \frac{dx}{ax^2+2bx+c} = \frac{1}{\sqrt{ac-b^2}}\left[\frac{\pi}{2} - \arctan\frac{b}{\sqrt{ac-b^2}}\right] \quad (a,\ ac-b^2>0)$

16. $\int_0^\infty \frac{dx}{ax^4+2bx^2+c} = \frac{\pi}{2\sqrt{cd}},\ d=2(b+\sqrt{ac}) \quad (a,\ c,\ d>0)$

Integrands containing trigonometric functions, (combined with algebraic functions)

17. $\int_0^{\pi/2} \sin^n x\,dx = \int_0^{\pi/2} \cos^n x\,dx = \begin{cases} \frac{(n-1)!!}{n!!}, & n=1, 3, 5, \ldots \\ \frac{(n-1)!!}{n!!}\cdot\frac{\pi}{2}, & n=2, 4, 6, \ldots \end{cases}$

18. $\int_0^{\pi/2} \sin^a x\,dx = \int_0^{\pi/2} \cos^a x\,dx = \frac{\sqrt{\pi}}{2}\,\frac{\Gamma\left(\frac{a+1}{2}\right)}{\Gamma\left(\frac{a+2}{2}\right)} \quad (a>-1)$

19. $\int_0^{\pi} x\sin^n x\,dx = \begin{cases} \frac{(n-1)!!}{n!!}\,\pi, & n=1, 3, 5, \ldots \\ \frac{(n-1)!!}{n!!}\cdot\frac{\pi^2}{2}, & n=2, 4, 6, \ldots \\ \frac{\pi^{3/2}}{2}\cdot\frac{\Gamma\left(\frac{n+1}{2}\right)}{\Gamma\left(\frac{n+2}{2}\right)}, & n>-1 \end{cases}$

20. $\int_0^{\pi/2} \sin^{2\alpha+1}x\cos^{2\beta+1}x\,dx = \frac{\Gamma(\alpha+1)\,\Gamma(\beta+1)}{2\,\Gamma(\alpha+\beta+2)}$

21. $\int_0^{\pi} \sin mx\,\sin nx\,dx = \begin{cases} 0 & (m\neq n \text{ integers}) \\ \frac{\pi}{2} & (m=n \text{ integers}) \end{cases}$

22. $\int_0^{\pi} \cos mx\,\cos nx\,dx = \begin{cases} 0 & (m\neq n \text{ integers}) \\ \frac{\pi}{2} & (m=n \text{ integers}) \end{cases}$

23. $\int_0^{\pi} \sin mx\,\cos nx\,dx = \begin{cases} 0 & (m,\ n \text{ integers, } m+n \text{ even}) \\ \frac{2m}{m^2-n^2} & (m,\ n \text{ integers, } m+n \text{ odd}) \end{cases}$

24. $\int_0^{\pi/2} \frac{dx}{1+a\cos x} = \int_0^{\pi/2} \frac{dx}{1+a\sin x} = \frac{\arccos a}{\sqrt{1-a^2}} \quad (|a|<1)$

25. $\int_0^{\pi} \frac{dx}{1+a \sin x} = \frac{\pi - 2\arcsin a}{\sqrt{1-a^2}} \quad (-1<a<1)$

26. $\int_0^{\pi} \frac{dx}{1+a \cos x} = \frac{\pi}{\sqrt{1-a^2}} \quad (-1<a<1)$

27. $\int_0^{\pi/2} \frac{dx}{a^2\cos^2 x + b^2 \sin^2 x} = \frac{\pi}{2ab} \quad (a, b>0)$

28a. $\int_0^{\infty} \sin x^2 dx = \int_0^{\infty} \cos x^2 dx = \frac{\sqrt{2\pi}}{4}$

28b. $\int_0^{\infty} \sin x^a dx = \Gamma\left(1+\frac{1}{a}\right) \sin \frac{\pi}{2a} \qquad \int_0^{\infty} \cos x^a dx = \Gamma\left(1+\frac{1}{a}\right) \cos \frac{\pi}{2a} \qquad (a>1)$

29. $\int_0^{\infty} \frac{\sin ax}{x} dx = \frac{\pi}{2} \quad (a>0)$

30. $\int_0^{\infty} \frac{\sin x}{\sqrt{x}} dx = \int_0^{\infty} \frac{\cos x}{\sqrt{x}} = \sqrt{\frac{\pi}{2}}$

31. $\int_0^{\infty} \frac{\sin^2 x}{x^2} dx = \frac{\pi}{2}$

32. $\int_0^{\infty} \frac{\sin^3 x}{x^3} dx = \frac{3\pi}{8}$

33. $\int_0^{\infty} \frac{\sin^4 x}{x^4} dx = \frac{\pi}{3}$

34. $\int_0^{\infty} \frac{\sin x}{x^{\alpha}} dx = \frac{\pi}{2\,\Gamma(\alpha) \sin \alpha\pi/2} \quad (0<\alpha<2)$

35. $\int_0^{\infty} \frac{\cos x}{x^{\alpha}} dx = \frac{\pi}{2\,\Gamma(\alpha) \cos \alpha\pi/2} \quad (0<\alpha<1)$

36. $\int_0^{\infty} \frac{\cos ax - \cos bx}{x} dx = \ln \frac{b}{a}$

37. $\int_0^{\infty} \frac{x \sin ax}{b^2+x^2} dx = \frac{\pi}{2} e^{-ab} \qquad (a, b>0)$

38. $\int_0^{\infty} \frac{\cos ax}{b^2+x^2} dx = \frac{\pi}{2b} e^{-ab} \qquad (a, b>0)$

Integrals containing exponential and logarithmic functions (combined with algebraic and trigonometric functions)

39. $\int_0^{\infty} x^n e^{-x} dx = n! \quad (n=0, 1, 2, \ldots)$

40. $\int_0^{\infty} x^n e^{-ax} dx = \begin{cases} \frac{\Gamma(n+1)}{a^{n+1}} & (n>-1,\ a>0) \\ \frac{n!}{a^{n+1}} & (n=0, 1, 2, \ldots, a>0) \end{cases}$

41. $\int_0^{\infty} e^{-ax^2}dx=\frac{1}{2}\sqrt{\frac{\pi}{a}} \qquad \int_{-\infty}^{\infty} e^{2bx-ax^2}dx=\sqrt{\frac{\pi}{a}}\, e^{b^2/a} \qquad (a>0)$

42. $$\int_0^{\infty} x^n e^{-ax^2}dx=\begin{cases} \frac{1}{2}\,\Gamma\left(\frac{n+1}{2}\right) \Big/ a^{\frac{n+1}{2}} & (n>-1,\ a>0) \\ \frac{(2k-1)!!}{2^{k+1}a^k}\sqrt{\frac{\pi}{a}} & (n=2k,\ k \text{ integer},\ a>0) \\ \frac{k!}{2a^{k+1}} & (n=2k+1,\ k \text{ integer},\ a>0) \end{cases}$$

43. $\int_0^{\infty} e^{-ax}\sin bx\, dx=\frac{b}{a^2+b^2} \qquad (a>0)$

44. $\int_0^{\infty} e^{-ax}\cos bx\, dx=\frac{a}{a^2+b^2} \qquad (a>0)$

45. $\int_0^{\infty} xe^{-ax}\sin bx\, dx=\frac{2ab}{(a^2+b^2)^2} \qquad (a>0)$

46. $\int_0^{\infty} xe^{-ax}\cos bx\, dx=\frac{a^2-b^2}{(a^2+b^2)^2} \qquad (a>0)$

47. $\int_0^{\infty} \frac{e^{-ax}\sin bx}{x}\, dx=\arctan\frac{b}{a} \qquad (a>0)$

48. $\int_0^{1} (\ln x)^n dx=(-1)^n n! \qquad (n=1, 2, 3, \ldots)$

49. $\int_0^{1} \ln|\ln x|\, dx=\int_0^{\infty} e^{-x}\ln x\, dx=-\gamma$

50. $\int_0^{1} \frac{\ln x}{x-1}\, dx=\frac{\pi^2}{6}$

51. $\int_0^{1} \frac{\ln x}{x+1}\, dx=-\frac{\pi^2}{12}$

52. $\int_0^{1} \frac{\ln x}{\sqrt{1-x^2}}\, dx=-\frac{\pi}{2}\ln 2$

53. $\int_0^{1} x^m\left(\ln\frac{1}{x}\right)^n dx=\frac{\Gamma(n+1)}{(m+1)^{n+1}} \qquad (m>-1,\ n>-1)$

54. $\int_0^{\infty} \frac{\sin x}{x}\ln x\, dx=-\frac{\pi}{2}\gamma$

55. $\int_0^{\pi/2} \ln(\sin x)dx=\int_0^{\pi/2} \ln(\cos x)\, dx=-\frac{\pi}{2}\ln 2$

56. $\int_0^{\pi/4} \ln(1+\tan x)dx=\frac{\pi}{8}\ln 2$

8. Sequences and Series

8.1 Sequences of Numbers

Notation: $\{a_n\}_1^\infty$ or $a_1, a_2, a_3, \ldots, a_n, \ldots$

Limit: The sequence $\{a_n\}_1^\infty$ has the limit A, $\lim_{n\to\infty} a_n = A$ or $a_n \to A$ as $n \to \infty$, if *for any* number $\varepsilon > 0$ *there exists* an integer N such that

$$|a_n - A| < \varepsilon \text{ for all } n > N.$$

If the limit exists the sequence is *convergent,* otherwise *divergent.*

(Laws of limits and rules for determining them, cf. the corresponding laws and rules for functions in section 6.2.)

$\limsup_{n\to\infty} a_n = \lim_{n\to\infty} (\sup_{k\geq n} a_k)$ exists for all sequences (possibly $\pm\infty$).

Theorems

1. $\{a_n\}_1^\infty$ monotone and bounded $\Rightarrow \lim_{n\to\infty} a_n$ exists (finite).
2. $\lim_{n\to\infty} a_n$ exists $\Leftrightarrow \lim_{\substack{m\to\infty \\ n\to\infty}} |a_m - a_n| = 0$ (*Cauchy condition*)

Examples

1. $\lim_{n\to\infty} a^n = \begin{cases} 0 & \text{if } |a| < 1 \\ \infty & \text{if } a > 1 \end{cases}$
2. $\lim_{n\to\infty} \sqrt[n]{a} = \lim_{n\to\infty} \sqrt[n]{n^a} = \lim_{n\to\infty} \sqrt[n]{p(n)} = 1$ (a=positive constant, $p(n)$ polynomial)
3. $\lim_{n\to\infty} \dfrac{a^n}{n!} = 0$ (a constant)
4. $a_n = \left(1 + \dfrac{1}{n}\right)^n$ is increasing and $a_n \to e$ as $n \to \infty$.
5. $a_{n+1} = \sqrt{\dfrac{1+a_n}{2}}$, $a_1 = 0$ *recursively* given sequence.

   $\{a_n\}_1^\infty$ is increasing and bounded

   $\Rightarrow \lim_{n\to\infty} a_n = A$ exists and

   $A = \sqrt{\dfrac{1+A}{2}} \Rightarrow A = 1$

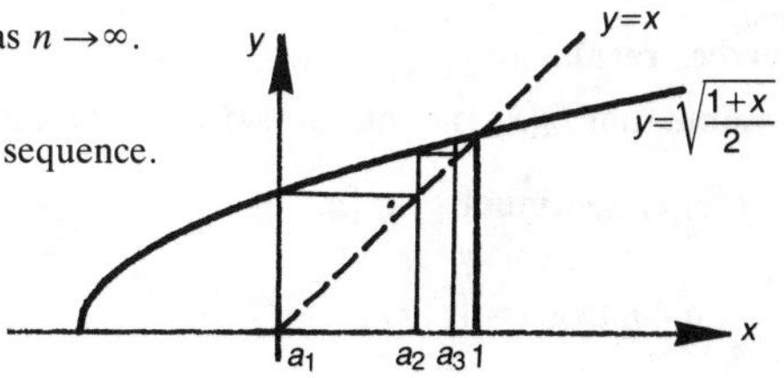

8.2 Sequences of Functions

Pointwise convergence

The sequence $\{f_n(x)\}$ is said to *converge pointwise* to $f(x)$ on the interval I if for every *fixed* $x \in I$: $\lim_{n\to\infty} f_n(x)=f(x)$ [$f(x)$=limit function].

1. Arzelà's theorem

Assume (*i*) $f_n(x) \to f(x)$ pointwise on $[a, b]$
(*ii*) $|f_n(x)|<M$, all n and $x \in [a, b]$
(*iii*) $f_n(x)$, $f(x)$ integrable

Then

$$\lim_{n\to\infty} \int_a^b f_n(x)dx = \int_a^b f(x)dx$$

Uniform convergence

The sequence $\{f_n(x)\}$ is said to *converge uniformly* to $f(x)$ in the interval I if $\sup_{x\in I}|f_n(x)-f(x)| \to 0$ as $n \to \infty$.

Example. $f_n(x)=\dfrac{x}{nx+1} \to 0$ uniformly for $x \in [0, 1]$ because

$$\sup_{x\in[0,1]} |f_n(x)| = \frac{1}{n+1} \to 0 \text{ as } n \to \infty.$$

2. Dini's theorem

Assume (*i*) $\{f_n(x)\}_1^\infty$ increasing, i.e. $f_n(x) \leq f_{n+1}(x)$, all n, x (or decreasing) (*ii*) $f_n(x) \to f(x)$ pointwise on $[a, b]$ (*iii*) $f_n(x)$, $f(x)$ continuous on $[a, b]$. Then the convergence is uniform.

Further results

Assume that $f_n(x)$ is continuous for each n and $f_n(x) \to f(x)$ uniformly on $[a, b]$. Then

3. $f(x)$ is continuous on $[a, b]$.

4. $\lim_{n\to\infty} \int_a^b f_n(x)dx = \int_a^b f(x)dx$

5. $\{f_n'(x)\}_1^\infty$ converges uniformly $\Rightarrow f'(x)$ exists and is equal to $\lim_{n\to\infty} f_n'(x)$.

8.3 Series of Constant Terms

An *infinite series* $\sum_{k=1}^{\infty} a_k = a_1 + a_2 + a_3 + \ldots$ is *convergent* with sum s if the sequence of *partial sums* $s_n = \sum_{k=1}^{n} a_k$ is *convergent* with limit s. (Otherwise the series is *divergent*.)

1. $\sum_{n=1}^{\infty} a_n$ convergent $\Rightarrow a_n \to 0$ as $n \to \infty$.

Summation by parts (*Abel*)

2. $\sum_{k=1}^{n} a_k b_k = A_n b_{n+1} - \sum_{k=1}^{n} A_k (b_{k+1} - b_k)$, where $A_n = \sum_{k=1}^{n} a_k$

Integral estimates

3. $f(x)$ increasing: $\int_{m-1}^{n} f(x)dx \leq \sum_{k=m}^{n} f(k) \leq \int_{m}^{n+1} f(x)dx$
4. $f(x)$ decreasing: $\int_{m}^{n+1} f(x)dx \leq \sum_{k=m}^{n} f(k) \leq \int_{m-1}^{n} f(x)dx$

Convergence tests

Series with non-negative terms

5. (*Comparison test*). Assume $0 \leq a_n \leq b_n$. Then

   (a) $\sum_{1}^{\infty} b_n$ *convergent* $\Rightarrow \sum_{1}^{\infty} a_n$ *convergent*

   (b) $\sum_{1}^{\infty} a_n$ *divergent* $\Rightarrow \sum_{1}^{\infty} b_n$ *divergent*

6. Assume $a_n, b_n > 0$, $\lim_{n \to \infty} \frac{a_n}{b_n} = c \neq 0, \infty$ (or $a_n \sim b_n$). Then

$$\sum_{1}^{\infty} a_n \text{ convergent} \Leftrightarrow \sum_{1}^{\infty} b_n \text{ convergent}$$

7. (*Integral test*). Assume that $f(x)$ is positive and decreasing for $x \geq N$. Then

$$\sum_{n=N}^{\infty} f(n) \text{ convergent} \Leftrightarrow \int_{N}^{\infty} f(x)dx \text{ convergent}$$

Series with arbitrary (complex) terms

8. $\sum_{1}^{\infty} |a_n|$ *convergent* $\Rightarrow \sum_{1}^{\infty} a_n$ *convergent*

9. (*Ratio test*). Assume $\lim_{n\to\infty}\left|\frac{a_{n+1}}{a_n}\right|=c$. Then

(a) $c<1 \Rightarrow \sum_1^\infty |a_n|$ *convergent* (b) $c>1 \Rightarrow \sum_1^\infty a_n$ *divergent*

10. (*Root test*). Assume $\lim_{n\to\infty}\sqrt[n]{|a_n|}=c$. Then

(a) $c<1 \Rightarrow \sum_1^\infty |a_n|$ *convergent* (b) $c>1 \Rightarrow \sum_1^\infty a_n$ *divergent*

11. (*Leibniz' test*). Assume (*i*) $\{a_n\}$ decreasing (*ii*) $a_n \to 0$, $n\to\infty$. Then

$$\sum_1^\infty (-1)^n a_n \textit{ convergent}$$

12. (*Dirichlet's test*). Assume (*i*) $A_n=\sum_{k=1}^n a_k$ is bounded (complex) sequence, (*ii*) $b_n \searrow 0$, $n\to\infty$. Then

$$\sum_1^\infty a_n b_n \textit{ convergent}$$

13. (*Abel's test*). Assume (*i*) $\sum_1^\infty a_n$ *convergent* (*ii*) $\{b_n\}$ monotone and convergent. Then

$$\sum_1^\infty a_n b_n \textit{ convergent}$$

Examples

1. $\sum_{n=1}^\infty \left(\frac{x}{2}\right)^n$ $\begin{cases}\text{conv. for } |x|<2\\ \text{div. for } |x|>2\end{cases}$ (Root test)
2. $\sum_{n=1}^\infty \frac{1}{n^p}$ and $\sum_{n=1}^\infty \frac{1}{n(\ln n)^p}$ $\begin{cases}\text{conv. for } p>1\\ \text{div. for } p\leq 1\end{cases}$ (integral test)
3. $\sum_{n=1}^\infty \left(1-\cos\frac{1}{n}\right)$ conv.

   $\left(\text{Comparison test: } 1-\cos\frac{1}{n}=\frac{1}{2n^2}+O\left(\frac{1}{n^4}\right) \text{ and } \sum_1^\infty \frac{1}{n^2} \text{ conv.}\right)$
4. $\sum_1^\infty \frac{(-1)^n}{\sqrt{n}}$ conv. (Leibniz' test)
5. $\sum_{k=1}^\infty \frac{1}{\sqrt{k}} e^{ikx}$ conv. for $x\neq 2m\pi$ (Dirichlet's test, cf. 8.6.18)

Infinite products

An *infinite product* $\prod_{k=1}^\infty (1+a_k)=(1+a_1)(1+a_2)\ldots$ with $1+a_k\neq 0$ *converges* to $p\neq 0$ if

$$\lim_{n\to\infty}\prod_{k=1}^n (1+a_k)=p.$$

1. $\prod_1^\infty (1+|a_k|)$ *convergent* $\Rightarrow \prod_1^\infty (1+a_k)$ *convergent*
2. $\prod_1^\infty (1+|a_k|)$ *convergent* $\Leftrightarrow \sum_1^\infty |a_k|$ *convergent*

8.4 Series of Functions

Uniform convergence

The series $\sum_{k=1}^{\infty} f_k(x)$ $[=s(x)]$ is *uniformly convergent* for $x\in I$ if the sequence $s_n(x)= = \sum_{k=1}^{n} f_k(x)$ converges uniformly to $s(x)$, $x\in I$, i.e. $\sup_{x\in I} |\sum_{k=n}^{\infty} f_k(x)| \to 0$ as $n \to \infty$.

Tests

1. (*Weierstrass' majorant test*). Assume (*i*) $|f_n(x)|\leq M_n$, $x\in I$, (*ii*) $\sum_1^{\infty} M_n$ *convergent*. Then
$$\sum_1^{\infty} f_n(x) \text{ is \textit{uniformly convergent}.}$$

2. (*Dirichlet's test*). Assume (*i*) $F_n(x)= \sum_{k=1}^{n} f_k(x)$, $|F_n(x)|\leq M$, $x\in I$ (*ii*) $g_{n+1}(x)\leq g_n(x)$, $x\in I$ (*iii*) $g_n(x)\to 0$ uniformly, $x\in I$. Then
$$\sum_1^{\infty} f_n(x)g_n(x) \text{ is \textit{uniformly convergent}, } x\in I.$$

Theorems

Assume $s(x)= \sum_{n=1}^{\infty} f_n(x)$ ($f_n(x)$ continuous), is uniformly convergent in $[a, b]$. Then

3. $s(x)$ is continuous.

4. $\int_a^b s(x)dx= \sum_{n=1}^{\infty} \int_a^b f_n(x)dx$

5. $\sum_1^{\infty} f_n'(x)$ uniformly convergent $\Rightarrow s'(x)= \sum_1^{\infty} f_n'(x)$.

Power series

A power series of a real (or complex) variable x is of the form (a_n may be complex)

$$f(x)= \sum_{n=0}^{\infty} a_n(x-x_0)^n, \qquad a_n=\frac{f^{(n)}(x_0)}{n!}$$

In particular if $x_0=0$:

(8.1) $$f(x)= \sum_{n=0}^{\infty} a_nx^n=a_0+a_1x+a_2x^2+\ldots, \qquad a_n=\frac{f^n(0)}{n!}$$

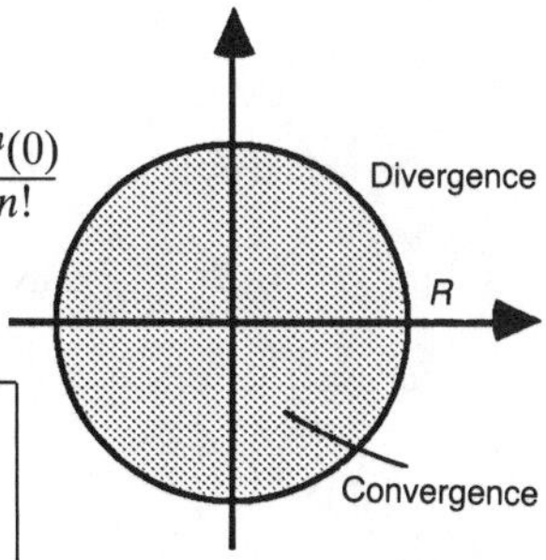

Radius of convergence *R*

$$\frac{1}{R}= \limsup_{n\to\infty} \sqrt[n]{|a_n|}=[\lim_{n\to\infty} \sqrt[n]{|a_n|}= \lim_{n\to\infty} \left|\frac{a_{n+1}}{a_n}\right|$$

if the latter limits exist.]

Properties. (Assume $R>0$)

1. The power series (8.1)

   (i) *converges* for $|x|<R$ and *converges uniformly* for $|x|\leqslant R-\varepsilon$, any $\varepsilon>0$.

   (ii) diverges for $|x|>R$.

   (Convergence investigations for $|x|=R$: Look for tests in sec. 8.3).

2. The series (8.1) may be differentiated or integrated term by term arbitrary many times and the received new series also have radius of convergence$=R$, i.e.

$$f'(x)=\sum_{n=1}^{\infty} na_nx^{n-1},\ |x|<R$$

$$\int f(x)dx=\sum_{n=0}^{\infty}\frac{a_nx^{n+1}}{n+1}+C,\ |x|<R$$

3. (Uniqueness theorem)

$$\sum_0^{\infty} a_nx^n=\sum_0^{\infty} b_nx^n \Rightarrow a_n=b_n,\ \text{all } n.$$

4. (Multiplication of power series)

$$\sum_0^{\infty} a_nx^n,\ |x|<R_1;\ \sum_0^{\infty} b_nx^n,\ |x|<R_2 \Rightarrow$$

$$\left(\sum_0^{\infty} a_nx^n\right)\left(\sum_0^{\infty} b_nx^n\right)=\sum_0^{\infty} c_nx^n,$$

$$c_n=\sum_{k=0}^{n} a_kb_{n-k},\ |x|<\min(R_1, R_2)$$

5. (Abel's limit theorem)

$$(i)\ f(x)=\sum_0^{\infty} a_nx^n,\ -R<x<R \qquad (ii)\ \sum_0^{\infty} a_nR^n=s\ \text{(convergent)} \Rightarrow \lim_{x\to R^-} f(x)=s.$$

Example

For which (complex) x does $\sum_1^{\infty}\frac{x^n}{n}$ converge?

Solution:

A. $\frac{1}{R}=\lim_{n\to\infty}\frac{1}{\sqrt[n]{n}}=1 \Rightarrow R=1.$

B. The boundary $|x|=1$: (a) $x=1$: $\sum_1^{\infty}\frac{1}{n}$ div.

(b) $x=-1$: $\sum_1^{\infty}\frac{(-1)^n}{n}$ conv. (Leibniz' test 8.3.11).

(c) $x\neq 1$: set $x=e^{i\varphi}$: $\sum_1^{\infty}\frac{1}{n}e^{in\varphi}$ convergent by Dirichlet's test (8.3.12) because

(i) $\frac{1}{n}\searrow 0$ (ii) $\left|\sum_1^{N} e^{in\varphi}\right|\leqslant\frac{1}{\sin\varphi/2}$ (8.6.18)

Answer: $|x|\leqslant 1,\ x\neq 1.$

8.5 Taylor Series

Taylor's formula

Let $f(x)$ and its $n+1$ first derivatives be continuous in an interval about $x=a$. Then in this interval:

(8.2)
$$f(x)=f(a)+\frac{f'(a)}{1!}(x-a)+\frac{f''(a)}{2!}(x-a)^2+\ldots+\frac{f^{(n)}(a)}{n!}(x-a)^n+R_{n+1}(x),$$

where $R_{n+1}(x)=\int_a^x \frac{(x-t)^n}{n!}f^{(n+1)}(t)dt=\frac{f^{(n+1)}(\xi)}{(n+1)!}(x-a)^{n+1}$,

(ξ between a and x)

Maclaurin's formula

$$f(x)=f(0)+\frac{f'(0)}{1!}x+\frac{f''(0)}{2!}x^2+\ldots+\frac{f^{(n)}(0)}{n!}x^n+\frac{x^{n+1}}{(n+1)!}f^{(n+1)}(\theta x),\ (0<\theta<1)$$

Note: $f(x)$ is $\begin{cases}\text{odd: only odd powers of } x\\ \text{even: only even powers of } x.\end{cases}$

Taylor series

If $R_n(x)\to 0$ as $n\to\infty$ then

$$f(x)=\sum_{k=0}^{\infty}\frac{f^{(k)}(a)}{k!}(x-a)^k \qquad \text{[Taylor series]}$$

$$f(x)=\sum_{k=0}^{\infty}\frac{f^{(k)}(0)}{k!}x^k \qquad \text{[Maclaurin series]}$$

The Ordo concept (Big O and Little o)

1. $f(x)=O(x^a)$ as $x\to 0$ means: $f(x)=x^aH(x)$, where $H(x)$ is bounded in a neighbourhood of $x=0$.
2. $f(x)=o(x^a)$ as $x\to 0$ means: $f(x)/x^a\to 0$ as $x\to 0$.

1. $O(x^4) \pm O(x^4) = O(x^4)$
2. $O(x^3) \pm O(x^4) = O(x^3)$
3. $x^2 O(x^3) = O(x^5) = x^5 O(1)$
4. $O(x^2) O(x^3) = O(x^5)$

Corresponding rules for little o.

Example.

$$e^x = 1 + x + \frac{x^2}{2} + \begin{cases} O(x^3) \\ o(x^2) \end{cases} = 1 + x + \frac{x^2}{2} + \begin{cases} x^3 O(1) \\ x^2 o(1) \end{cases}$$

Asymptotic equivalence

$$f(x) \sim g(x) \text{ as } x \to a \text{ means: } \frac{f(x)}{g(x)} \to 1 \text{ as } x \to a$$

Methods of deriving Taylor series

Examples of other methods than a direct use of formula (8.2) are given below. In each case, the problem is to determine the Taylor series expansion about the given point a with the given order n of the remainder term.

1. (*Substitution*): $f(x) = e^{-2x^2}$, $a = 0$, $n = 6$.

$$f(x) = [t = -2x^2] = e^t = 1 + t + \frac{t^2}{2} + O(t^3) = 1 - 2x^2 + 2x^4 + O(x^6).$$

2. (*Multiplication*): $f(x) = e^x \sin x$, $a = 0$, $n = 5$.

$$f(x) = \left(1 + x + \frac{x^2}{2} + \frac{x^3}{6} + O(x^4)\right)\left(x - \frac{x^3}{6} + O(x^5)\right) = x + x^2 + \frac{x^3}{3} + O(x^5)$$

3. (*Division, using the geometrical series*): $f(x) = \dfrac{x}{\arctan x}$, $a = 0$, $n = 6$.

$$f(x) = \frac{x}{x - \frac{x^3}{3} + \frac{x^5}{5} + O(x^7)} = \frac{1}{1 - \left(\frac{x^2}{3} - \frac{x^4}{5} + O(x^6)\right)} = \left[t = \frac{x^2}{3} - \frac{x^4}{5} + O(x^6)\right] =$$

$$= 1 + t + t^2 + O(t^3) = 1 + \frac{x^2}{3} - \frac{4x^4}{45} + O(x^6)$$

(*using "long division"*):

$$\begin{array}{r|l}
 & 1 + x^2/3 - 4x^4/45 + \ldots \\
\hline
x - x^3/3 + x^5/5 & x \\
 & -(x - x^3/3 + x^5/5) \\
\hline
 & x^3/3 - x^5/5 \\
 & -(x^3/3 - x^5/9 + x^7/15) \\
\hline
 & -4x^5/45 - x^7/15 \\
 & -(-4x^5/45 + \ldots) \\
\hline
\end{array}$$

4. (*Composite function*): $f(x) = \ln(\cos x)$, $a = 0$, $n = 6$.

$$f(x) = \ln\left[1 + \left(-\frac{x^2}{2} + \frac{x^4}{24} + O(x^6)\right)\right] = \left[t = -\frac{x^2}{2} + \frac{x^4}{24} + O(x^6)\right] =$$

$$= t - \frac{t^2}{2} + O(t^3) = -\frac{x^2}{2} - \frac{x^4}{12} + O(x^6)$$

5. (*Rewrite*)

(a) $f(x)=\sqrt{9+x}$, $a=0$, $n=3$.

$$f(x)=3\sqrt{1+\frac{x}{9}}=3\left(1+\frac{1}{2}\cdot\frac{x}{9}-\frac{1}{8}\cdot\frac{x^2}{81}+O(x^3)\right)=3+\frac{x}{6}-\frac{x^2}{216}+O(x^3)$$

(b) $f(x)=e^{(x-1)^2}$, $a=0$, $n=3$.

$$f(x)=e^{x^2-2x+1}=e\cdot e^{x^2-2x}=[t=x^2-2x]=e\cdot e^t=$$
$$=e\left(1+t+\frac{t^2}{2}+O(t^3)\right)=e(1-2x+3x^2)+O(x^3).$$

6. $f(x)=\tan x$, $a=\frac{\pi}{4}$, $n=3$.

Method 1.

$$f\left(\frac{\pi}{4}\right)=1, f'(x)=1+\tan^2 x, f'\left(\frac{\pi}{4}\right)=2 \text{ etc}, f''\left(\frac{\pi}{4}\right)=4\cdot$$
$$\Rightarrow f(x)=1+2\left(x-\frac{\pi}{4}\right)+2\left(x-\frac{\pi}{4}\right)^2+O\left[\left(x-\frac{\pi}{4}\right)^3\right].$$

Method 2.

$$f(x)=\left[x=\frac{\pi}{4}+t\right]=\tan\left(\frac{\pi}{4}+t\right)=\frac{1+\tan t}{1-\tan t}=\frac{1+t+O(t^3)}{1-(t+O(t^3))}=$$
$$=[\text{geom.ser.}]=(1+t+O(t^3))(1+t+t^2+O(t^3))=1+2t+2t^2+O(t^3).$$

7. (*Asymptotic behavior*) $a_n=\sqrt[n]{n}$, $n\to\infty$

$$a_n=e^{\frac{1}{n}\ln n}\sim\left[\text{note: }\frac{\ln n}{n}\to 0 \text{ as } n\to\infty\right]\sim 1+\frac{\ln n}{n}+\frac{\ln^2 n}{2n^2}$$

8. Differentiation and integration of a given series, e.g.

$$\frac{1}{(1-x)^2}=\frac{d}{dx}\left[\frac{1}{1-x}\right]=\frac{d}{dx}(1+x+x^2+\ldots)=1+2x+3x^2+\ldots$$

Operations with series

Let $s=a_1x+a_2x^2+a_3x^3+\ldots$

$t=b_0+b_1x+b_2x^2+\ldots$

$t=$	b_0	b_1	b_2	b_3	b_4
$\frac{1}{1-s}$	1	a_1	$a_2+a_1b_1$	$a_3+a_2b_1+a_1b_2$	$a_4+a_3b_1+a_2b_2+a_1b_3$
$(1+s)^{1/2}$	1	$\frac{1}{2}a_1$	$\frac{1}{2}a_2-\frac{1}{8}a_1^2$	$\frac{1}{2}a_3-\frac{1}{4}a_1a_2+\frac{1}{16}a_1^3$	$\frac{1}{2}a_4-\frac{1}{4}a_1a_3-\frac{1}{8}a_2^2+\frac{3}{16}a_1^2a_2-\frac{5}{128}a_1^4$
$(1+s)^{-1/2}$	1	$-\frac{1}{2}a_1$	$\frac{3}{8}a_1^2-\frac{1}{2}a_2$	$\frac{3}{4}a_1a_2-\frac{1}{2}a_3-\frac{5}{16}a_1^3$	$\frac{3}{4}a_1a_3+\frac{3}{8}a_2^2-\frac{1}{2}a_4-\frac{15}{16}a_1^2a_2+\frac{35}{128}a_1^4$
e^s	1	a_1	$a_2+\frac{1}{2}a_1^2$	$a_3+a_1a_2+\frac{1}{6}a_1^3$	$a_4+a_1a_3+\frac{1}{2}a_2^2+\frac{1}{2}a_1^2a_2+\frac{1}{24}a_1^4$
$\ln(1+s)$	0	a_1	$a_2-\frac{1}{2}a_1^2$	$a_3-a_1a_2+\frac{1}{3}a_1^2$	$a_4-\frac{1}{2}a_2^2-a_1a_3+a_1^2a_2-\frac{1}{4}a_1^4$
$\cos s$	1	0	$-\frac{1}{2}a_1^2$	$-a_1a_2$	$-\frac{1}{2}a_2^2-a_1a_3+\frac{1}{24}a_1^4$
$\sin s$	0	a_1	a_2	$a_3-\frac{1}{6}a_1^3$	$a_4-\frac{1}{2}a_1^2a_2$

8.6 Special Sums and Series

Euler-Maclaurin summation formula, see sec. 16.6.

Miscellaneous sums and series

Arithmetic series: $a_n=a_{n-1}+d=a_1+(n-1)d$ (d=difference)

1. $\sum_{k=1}^{n} a_k=\sum_{k=1}^{n}[a_1+(k-1)d]=\frac{n(a_1+a_n)}{2}=\frac{n[2a_1+(n-1)d]}{2}$

Geometric series: $a_n=a_{n-1}\cdot x=a_0x^n$ (x=quotient)

2. $\sum_{k=0}^{n-1} ax^k=a+ax+\ldots+ax^{n-1}=a\cdot\frac{x^n-1}{x-1}=a\cdot\frac{1-x^n}{1-x},\ (x\neq 1)$

3. $\sum_{k=0}^{\infty} ax^k=a+ax+ax^2+\ldots=\frac{a}{1-x}\qquad(-1<x<1)$

4. $\sum_{k=1}^{\infty} k^m x^k=(1-x)^{-m-1}\sum_{j=1}^{m} a_j^{(m)}x^j\qquad(m=1, 2, 3, \ldots),\ -1<x<1$

$a_1^{(m)}=a_m^{(m)}=1,\ a_j^{(m)}=ja_j^{(m-1)}+(m-j+1)a_{j-1}^{(m-1)},\ j=2, \ldots, m-1$

Table of $a_j^{(m)}$

m \ j	1	2	3	4	5	6	7	8	9	10
1	1									
2	1	1								
3	1	4	1							
4	1	11	11	1						
5	1	26	66	26	1					
6	1	57	302	302	57	1				
7	1	120	1191	2416	1191	120	1			
8	1	247	4293	15619	15619	4293	247	1		
9	1	502	14608	88234	156190	88234	14608	502	1	
10	1	1013	47840	455192	1310354	1310354	455192	47840	1013	1

E.g. $\sum_{k=1}^{\infty} k^5x^k=(1-x)^{-6}(x+26x^2+66x^3+26x^4+x^5)$

$\sum_{k=1}^{\infty} kx^k=\frac{x}{(1-x)^2},\ (-1<x<1)$

5. $\sum_{k=1}^{\infty}\frac{x^k}{k}=-\ln(1-x)\qquad(-1\leq x<1)$

Binomic terms (cf. sec. 2.1)

6. $\sum_{k=0}^{n} \binom{n}{k} a^{n-k} b^k = (a+b)^n$

7. $\sum_{k=0}^{n} \binom{n}{k} = 2^n$

8. $\sum_{k=0}^{n} k \binom{n}{k} = n2^{n-1}$

9. $\sum_{k=0}^{n} k^2 \binom{n}{k} = (n^2+n)2^{n-2}$

Sums of powers

10. $\sum_{k=1}^{n} k = \frac{n(n+1)}{2}$

11. $\sum_{k=1}^{n} k^2 = \frac{n(n+1)(2n+1)}{6}$

12. $\sum_{k=1}^{n} k^3 = \frac{n^2(n+1)^2}{4}$

13. $\sum_{k=1}^{n} k^m = \frac{n^{m+1}}{m+1} + \frac{n^m}{2} + \frac{1}{2}\binom{m}{1} B_2 n^{m-1} + \frac{1}{4}\binom{m}{3} B_4 n^{m-3} + \frac{1}{6}\binom{m}{5} B_6 n^{m-5} + \ldots$

(positive powers of n), where B_k are the Bernoulli numbers (sec. 12.3)

Series of reciprocal powers

14. $\sum_{k=1}^{n} \frac{1}{k} - \ln n \to \gamma \approx 0.5772$ as $n \to \infty$ (γ=Euler's constant)

Partial sums $H_n = 1 + \frac{1}{2} + \frac{1}{3} + \ldots + \frac{1}{n}$ *of the harmonic series*

n	H_n
10	2.92896 82539 68253 96825 39683
100	5.18737 75176 39620 26080 51177
1000	7.48547 08605 50344 91265 65182
10000	9.78760 60360 44382 26417 84779
100000	12.09014 61298 63427 94736 32194
10^6	14.39272 67228 65723 63138 11275
10^7	16.69531 13658 59851 81539 91189
10^8	18.99789 64138 53898 32441 71104
10^9	21.30048 15023 47944 01668 51018

15. $s_m = \sum_{k=1}^{\infty} \frac{1}{k^m}$, $m \geq 2$: $s_2 = \frac{\pi^2}{6}$, $s_3 \approx 1.2021$, $s_4 = \frac{\pi^4}{90}$

$s_5 \approx 1.0369$, $s_6 = \frac{\pi^6}{945}$, $s_{2n} = \frac{2^{2n-1}\pi^{2n}}{(2n)!}(-1)^{n-1}B_{2n}$

(B_i=Bernoulli numbers, sec. 12.3)

Exponential terms

16. $\sum_{k=1}^{n} e^{kx}=e^{x}\cdot\frac{e^{nx}-1}{e^{x}-1}=\frac{\sinh\frac{nx}{2}}{\sinh\frac{x}{2}}\,e^{(n+1)x/2}\quad(x\neq 0)$

17. $\sum_{k=0}^{\infty} e^{-kx}=\frac{1}{1-e^{-x}}\quad(x>0)$

18. $\sum_{k=1}^{n} e^{ikx}=e^{ix}\frac{1-e^{inx}}{1-e^{ix}}=\frac{\sin\frac{nx}{2}}{\sin\frac{x}{2}}\cdot e^{i(n+1)x/2}\quad(x\neq 2m\pi)$

Trigonometric terms

19. $\sum_{k=1}^{n}\sin kx=\operatorname{Im}\sum_{k=0}^{n} e^{ikx}=\frac{\sin\frac{nx}{2}\sin\frac{(n+1)x}{2}}{\sin\frac{x}{2}}$ (cf. 18)

20. $\sum_{k=0}^{n}\cos kx=\operatorname{Re}\sum_{k=0}^{n} e^{ikx}=\frac{\cos\frac{nx}{2}\sin\frac{(n+1)x}{2}}{\sin\frac{x}{2}}$ (cf. 18)

21. $\sum_{k=1}^{n-1} r^{k}\sin kx=\operatorname{Im}\sum_{k=0}^{n-1}(re^{ix})^{k}=\frac{r\sin x(1-r^{n}\cos nx)-(1-r\cos x)r^{n}\sin nx}{1-2r\cos x+r^{2}}$

22. $\sum_{k=0}^{n-1} r^{k}\cos kx=\operatorname{Re}\sum_{k=0}^{n-1}(re^{ix})^{k}=\frac{(1-r\cos x)(1-r^{n}\cos nx)+r^{n+1}\sin x\sin nx}{1-2r\cos x+r^{2}}$

23. $\sum_{k=1}^{n-1}\sin\frac{k\pi}{n}=\cot\frac{\pi}{2n}$

Some special numbers

24. $e=\sum_{k=0}^{\infty}\frac{1}{k!}=2.7182818284\ldots$ (transcendental)

25. $\pi=4\arctan 1=4\sum_{k=0}^{\infty}\frac{(-1)^{k}}{2k+1}=3.1415926535\ldots$ (transcendental)

26. $\ln 2=\sum_{k=1}^{\infty}\frac{(-1)^{k-1}}{k}=\sum_{k=1}^{\infty}\frac{1}{k\cdot 2^{k}}=0.69315\ldots$ (transcendental)

27. $\gamma=\lim_{n\to\infty}\left(\sum_{k=1}^{n}\frac{1}{k}-\ln n\right)=0.577215665\ldots$ (Euler's constant, irrational, transcendental?)

Table of power series expansions

(In some cases the remainder term $R_n(x)$ is given)

(B_n=Bernoulli numbers, E_n=Euler numbers, see sec. 12.3)

Function	Power series expansion	Interval of convergence
	Algebraic functions $\binom{\alpha}{n}=\frac{\alpha(\alpha-1)\ldots(\alpha-n+1)}{n!}$, α real number	
$(1+x)^{\alpha}$	$1+\alpha x+\frac{\alpha(\alpha-1)}{2!}x^2+\frac{\alpha(\alpha-1)(\alpha-2)}{3!}x^3+\ldots+\binom{\alpha}{n}x^n+\ldots,$	$-1<x<1$
	$R_n(x)=\binom{\alpha}{n}(1+\theta x)^{\alpha-n}x^n,\ 0<\theta<1$	
$\frac{1}{1-x}$	$1+x+x^2+x^3+\ldots+x^n+\ldots$	$-1<x<1$
$\frac{1}{1+x}$	$1-x+x^2-x^3+\ldots+(-1)^n x^n+\ldots$	$-1<x<1$
$\frac{1}{a-bx}$	$\frac{1}{a}\left[1+\frac{bx}{a}+\left(\frac{bx}{a}\right)^2+\ldots+\left(\frac{bx}{a}\right)^n+\ldots\right]$ or	$\|x\|<\left\|\frac{a}{b}\right\|$
	$-\frac{1}{bx}\left[1+\frac{a}{bx}+\left(\frac{a}{bx}\right)^2+\ldots+\left(\frac{a}{bx}\right)^n+\ldots\right]$	$\|x\|>\left\|\frac{a}{b}\right\|$
$\frac{1}{(1-x)^2}$	$1+2x+3x^2+\ldots+(n+1)x^n+\ldots$	$-1<x<1$
$\sqrt{1+x}$	$1+\frac{x}{2}-\frac{x^2}{8}+\frac{x^3}{16}-\frac{5x^4}{128}+\ldots+\binom{1/2}{n}x^n+\ldots$	$-1\leq x\leq 1$
$\frac{1}{\sqrt{1+x}}$	$1-\frac{x}{2}+\frac{3x^2}{8}-\frac{5x^3}{16}+\frac{35x^4}{128}-\ldots+\binom{-1/2}{n}x^n+\ldots$	$-1<x\leq 1$

Note: $\left|\binom{\alpha}{n}\right|\sim C_{\alpha}n^{-\alpha-1},\ n\to\infty$

Table of fractional binomial coefficients, see below.

	Exponential, hyperbolic, logarithmic and inverse hyperbolic functions	
e^x	$1+x+\frac{x^2}{2!}+\frac{x^3}{3!}+\ldots+\frac{x^n}{n!}+\ldots$	$-\infty<x<\infty$
	$R_n(x)=\frac{e^{\theta x}}{n!}x^n,\ 0<\theta<1$	
a^x	$1+x\ln a+\frac{(x\ln a)^2}{2!}+\ldots+\frac{(x\ln a)^n}{n!}+\ldots$	$-\infty<x<\infty$

$\frac{1}{e^x-1}$	$\frac{1}{x}-\frac{1}{2}+\frac{x}{12}-\frac{x^3}{30\cdot 4!}+\ldots+\frac{B_{2n}\,x^{2n-1}}{(2n)!}+\ldots$	$-2\pi<x<2\pi,\ x\neq 0$
$\sinh x$	$x+\frac{x^3}{3!}+\frac{x^5}{5!}+\ldots+\frac{x^{2n+1}}{(2n+1)!}+\ldots$	$-\infty<x<\infty$
$\cosh x$	$1+\frac{x^2}{2!}+\frac{x^4}{4!}+\ldots+\frac{x^{2n}}{(2n)!}+\ldots$	$-\infty<x<\infty$
$\tanh x$	$x-\frac{x^3}{3}+\frac{2x^5}{15}-\frac{17x^7}{315}+\ldots+\frac{2^{2n}(2^{2n}-1)}{(2n)!}B_{2n}x^{2n-1}+\ldots$	$-\frac{\pi}{2}<x<\frac{\pi}{2}$
$\coth x$	$\frac{1}{x}+\frac{x}{3}-\frac{x^3}{45}+\frac{2x^5}{945}+\ldots+\frac{2^{2n}\,B_{2n}}{(2n)!}\,x^{2n-1}+\ldots$	$-\pi<x<\pi,\ x\neq 0$
$\frac{1}{\sinh x}$	$\frac{1}{x}-\frac{x}{6}+\frac{7x^3}{360}-\ldots-\frac{2^{2n}-2}{(2n)!}\,B_{2n}x^{2n-1}+\ldots$	$-\pi<x<\pi,\ x\neq 0$
$\frac{1}{\cosh x}$	$1-\frac{x^2}{2}+\frac{5x^4}{24}-\ldots+\frac{E_{2n}}{(2n)!}\,x^{2n}+\ldots$	$-\frac{\pi}{2}<x<\frac{\pi}{2}$
$\ln(1+x)$	$x-\frac{x^2}{2}+\frac{x^3}{3}-\frac{x^4}{4}+\ldots+(-1)^{n-1}\,\frac{x^n}{n}+\ldots$	$-1<x\leqslant 1$
	$R_n(x)=\frac{(-1)^{n-1}}{1+\theta x}\cdot\frac{x^n}{n},\ 0<\theta<1$	
$\ln(a+x)$	$\ln a+\frac{x}{a}-\frac{1}{2}\left(\frac{x}{a}\right)^2+\frac{1}{3}\left(\frac{x}{a}\right)^3-\ldots+\frac{(-1)^{n-1}}{n}\left(\frac{x}{a}\right)^n+\ldots$	$-a<x\leqslant a$
$\ln(1+x)$	$\frac{x}{1+x}+\frac{1}{2}\left(\frac{x}{1+x}\right)^2+\ldots+\frac{1}{n}\left(\frac{x}{1+x}\right)^n+\ldots$	$x>-\frac{1}{2}$
$\operatorname{arsinh} x$	$x-\frac{x^3}{6}+\frac{3x^5}{40}-\ldots+(-1)^n\cdot\frac{(2n-1)!!}{(2n)!!}\,\frac{x^{2n+1}}{2n+1}+\ldots$	$-1<x<1$
arcoshx	$\ln\|2x\|-\frac{1}{4x^2}-\frac{3}{32x^4}-\ldots-\frac{(2n-1)!!}{(2n)!!}\cdot\frac{1}{2nx^{2n}}-\ldots$	$\|x\|>1$
$\operatorname{artanh} x$	$x+\frac{x^3}{3}+\frac{x^5}{5}+\ldots+\frac{x^{2n+1}}{2n+1}+\ldots$	$-1<x<1$
$\operatorname{arcoth} x$	$\frac{1}{x}+\frac{1}{3x^3}+\frac{1}{5x^5}+\ldots+\frac{1}{(2n+1)x^{2n+1}}+\ldots$	$\|x\|>1$

	Trigonometric and inverse trigonometric functions	
$\sin x$	$x-\frac{x^3}{3!}+\frac{x^5}{5!}-\frac{x^7}{7!}+\ldots+(-1)^n\,\frac{x^{2n+1}}{(2n+1)!}+\ldots$	$-\infty<x<\infty$
	$R_{2n+1}(x)=(-1)^n\,\frac{\cos\theta x}{(2n+1)!}\,x^{2n+1},\ 0<\theta<1$	
$\cos x$	$1-\frac{x^2}{2!}+\frac{x^4}{4!}-\frac{x^6}{6!}+\ldots+(-1)^n\,\frac{x^{2n}}{(2n)!}+\ldots$	$-\infty<x<\infty$
	$R_{2n}(x)=(-1)^n\,\frac{\cos\theta x}{(2n)!}\,x^{2n},\ 0<\theta<1$	

$\tan x$	$x+\frac{x^3}{3}+\frac{2x^5}{15}+\frac{17x^7}{315}+\ldots+(-1)^{n-1}\frac{2^{2n}(2^{2n}-1)}{(2n)!}B_{2n}x^{2n-1}+\ldots$	$-\frac{\pi}{2}<x<\frac{\pi}{2}$
$\cot x$	$\frac{1}{x}-\frac{x}{3}-\frac{x^3}{45}-\frac{2x^5}{945}-\ldots+(-1)^n\frac{2^{2n}}{(2n)!}B_{2n}x^{2n-1}+\ldots$	$-\pi<x<\pi$, $x\neq 0$
$\sec x=\frac{1}{\cos x}$	$1+\frac{x^2}{2!}+\frac{5x^4}{4!}+\frac{61x^6}{6!}+\ldots+(-1)^n\frac{E_{2n}}{(2n)!}x^{2n}+\ldots$	$-\frac{\pi}{2}<x<\frac{\pi}{2}$
$\csc x=\frac{1}{\sin x}$	$\frac{1}{x}+\frac{x}{3!}+\frac{7x^3}{3\cdot 5!}+\frac{31x^5}{3\cdot 7!}+\ldots+(-1)^{n-1}\cdot\frac{2^{2n}-2}{(2n)!}B_{2n}x^{2n-1}+\ldots$	$-\pi<x<\pi$, $x\neq 0$
$\arcsin x$	$x+\frac{x^3}{6}+\frac{3x^5}{40}+\ldots+\frac{(2n-1)!!}{(2n)!!}\frac{x^{2n+1}}{2n+1}+\ldots$	$-1<x<1$
$\arctan x$	$x-\frac{x^3}{3}+\frac{x^5}{5}-\frac{x^7}{7}+\ldots+(-1)^n\frac{x^{2n+1}}{2n+1}+\ldots$	$-1\leqslant x\leqslant 1$
	$R_{2n+1}(x)=(-1)^n\frac{1}{1+\theta^2x^2}\cdot\frac{x^{2n+1}}{2n+1}$, $0<\theta<1$	
$\arccos x$	$=\frac{\pi}{2}-\arcsin x$	
$\operatorname{arccot} x$	$=\frac{\pi}{2}-\arctan x$	

Graphs of some Taylor polynomials $P_n(x)$ of degree n

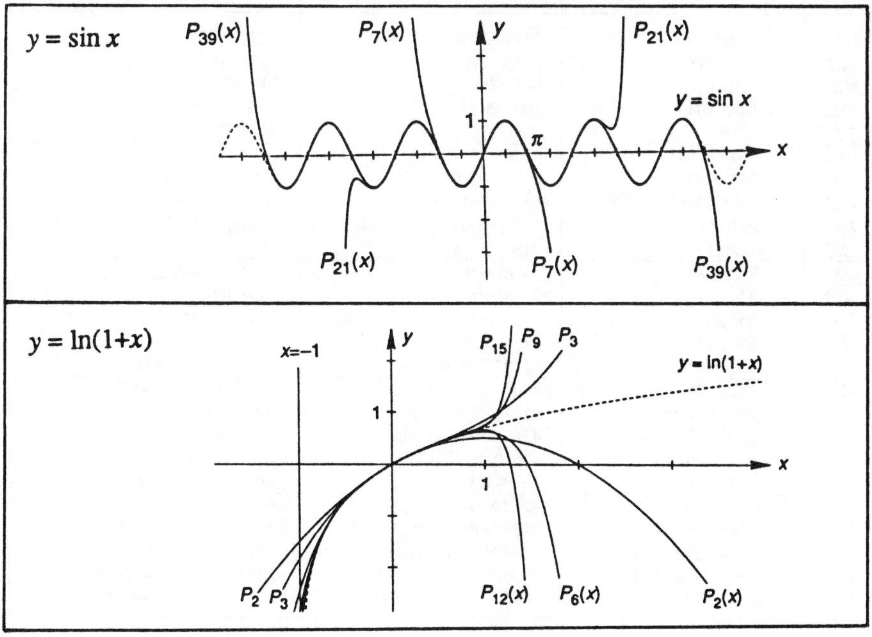

Fractional Binomial Coefficients $\binom{a}{k}$

$a \backslash k$	0	1	2	3	4	5	6
1/2	1	1/2	–1/8	1/16	–5/128	7/256	–21/1024
–1/2	1	–1/2	3/8	–5/16	35/128	–63/256	231/1024
3/2	1	3/2	3/8	–1/16	3/128	–3/256	7/1024
–3/2	1	–3/2	15/8	–35/16	315/128	–693/256	3003/1024
5/2	1	5/2	15/8	5/16	–5/128	3/256	–5/1024
–5/2	1	–5/2	35/8	–105/16	1155/128	–3003/256	15015/1024
1/3	1	1/3	–1/9	5/81	–10/243	22/729	–154/6561
–1/3	1	–1/3	2/9	–14/81	35/243	–91/729	728/6561
2/3	1	2/3	–1/9	4/81	–7/243	14/729	–91/6561
–2/3	1	–2/3	5/9	–40/81	110/243	–308/729	2618/6561
4/3	1	4/3	2/9	–4/81	5/243	–8/729	44/6561
–4/3	1	–4/3	14/9	–140/81	455/243	–1456/729	13832/6561
5/3	1	5/3	5/9	–5/81	5/243	–7/729	35/6561
–5/3	1	–5/3	20/9	–220/81	770/243	–2618/729	26180/6561
1/4	1	1/4	–3/32	7/128	–77/2048	231/8192	–1463/65536
–1/4	1	–1/4	5/32	–15/128	195/2048	–663/8192	4641/65536
3/4	1	3/4	–3/32	5/128	–45/2048	117/8192	–663/65536
–3/4	1	–3/4	21/32	–77/128	1155/2048	–4389/8192	33649/65536
1/5	1	1/5	–2/25	6/125	–21/625	399/15625	–1596/78125
–1/5	1	–1/5	3/25	–11/125	44/625	–924/15625	4004/78125
1/6	1	1/6	–5/72	55/1296	–935/31104	4301/186624	–124729/6,718464
1/7	1	1/7	–3/49	13/343	–65/2401	351/16807	–1989/117649
1/8	1	1/8	–7/128	35/1024	–805/32768	4991/262144	–64883/4,194304
1/9	1	1/9	–4/81	68/2187	–442/19683	3094/177147	–68068/4,782969
1/10	1	1/10	–9/200	57/2000	–1653/80000	64467/4,000000	–1,052961/80,000000

$a \backslash k$	7	8	9
1/2	33/2048	–429/32768	715/65536
–1/2	–429/2048	6435/32768	–12155/65536
3/2	–9/2048	99/32768	–143/65536
–3/2	–6435/2048	109395/32768	–230945/65536
5/2	5/2048	–45/32768	55/65536
–5/2	–36465/2048	692835/32768	–1,616615/65536
1/3	374/19683	–935/59049	21505/1,594323
–1/3	–1976/19683	5434/59049	–135850/1,594323
2/3	208/19683	–494/59049	10868/1,594323
–2/3	–7480/19683	21505/59049	–559130/1,594323
4/3	–88/19683	187/59049	–3740/1,594323
–4/3	–43472/19683	135850/59049	–3,803800/1,594323
5/3	–65/19683	130/59049	–2470/1,594323
–5/3	–86020/19683	279565/59049	–8,107385/1,594323
1/4	4807/262144	–129789/8,388608	447051/33,554432
–1/4	–16575/262144	480675/8,388608	–1,762475/33,554432
3/4	1989/262144	–49725/8,388608	160225/33,554432
–3/4	–129789/262144	4,023459/8,388608	–15,646785/33,554432
1/5	6612/390625	–28101/1,953125	121771/9,765625
–1/5	–17732/390625	79794/1,953125	–363506/9,765625
1/6	623645/40,310784	–25,569445/1934,917632	1201,763915/104485,552128
1/7	81549/5,764801	–489294/40,353607	2,990130/282,475249
1/8	435643/33,554432	–23,960365/2147,483648	167,722555/17179,869184
1/9	515372/43,046721	–3,994133/387,420489	283,583443/31381,059609
1/10	8,874957/800,000000	–612,372033/64000,000000	5375,265623/64000,000000

9 Ordinary Differential Equations (ODE)

General considerations and terminology, see sec. 9.4.

9.1 Differential Equations of the First Order

Special types	**Solution or method of solution**
1. $y'=f(x)$	$y=\int f(x)dx+C$
2. $f(y)\dfrac{dy}{dx}=g(x)$ (*separable*)	$\Leftrightarrow f(y)dy=g(x)dx$; $\int f(y)dy=\int g(x)dx+C$; $F(y)=G(x)+C$ (Test of separation: $f(x, y)=g(x)h(y) \Leftrightarrow f f''_{xy}=f'_x f'_y$)
3. $y'+f(x)y=g(x)$ (*linear*)	$y(x)=e^{-F(x)}(\int e^{F(x)}g(x)dx+C)$, where $F(x)=\int f(x)dx$

Table of some special cases
Equation: $y'-ay=g(x)$, a=constant

$g(x)=$	General solution $y(x)=$
$P(x)$, polynomial of degree n	$-\dfrac{P(x)}{a}-\dfrac{P'(x)}{a^2}-\ldots-\dfrac{P^{(n)}(x)}{a^{n+1}}+Ce^{ax}$
Ae^{kx}	$\dfrac{Ae^{kx}}{k-a}+Ce^{ax}$, $k\neq a$ $(Ax+C)e^{ax}$, $k=a$
$A\cos\omega x+B\sin\omega x$	$-\dfrac{Aa+B\omega}{a^2+\omega^2}\cos\omega x+\dfrac{A\omega-Ba}{a^2+\omega^2}\sin\omega x+Ce^{ax}$
$e^{kx}(A\cos\omega x+B\sin\omega x)$ $\alpha=k-a$	$(\alpha^2+\omega^2)^{-1}e^{kx}[(A\alpha-B\omega)\cos\omega x+(B\alpha+A\omega)\sin\omega x]+Ce^{ax}$

4. $y'+y\,f(x)=y^a g(x)$ $(a\neq 0,\ a\neq 1)$ (*Bernoulli's equation*)	Substitution: $z=y^{1-a}$, $z'=(1-a)y^{-a}\cdot y'$ gives $z'+(1-a)zf(x)=(1-a)g(x)$ of type 3.
5. $y'=f\left(\frac{y}{x}\right)$	Substitution: $y=xz$, $y'=xz'+z$ gives $\frac{dz}{f(z)-z}=\frac{dx}{x}$ of type 2
6. $y'=f(ax+by)$	Substitution: $z=ax+by$, $\frac{dz}{dx}=a+b\,\frac{dy}{dx}$ gives a separable equation of type 2
7. $y'=f\left(\frac{ax+by+c}{px+qy+r}\right)$	(*i*) $c=r=0$: Type 5. (*ii*) $ax+by$ and $px+qy$ proportional (i.e. $a/p=b/q$): Type 6. (*iii*) Subst. $x=u+\alpha$, $y=v+\beta$, $\frac{dy}{dx}=\frac{dv}{du}$ where α, β solution of $\begin{cases} a\alpha+b\beta+c=0 \\ p\alpha+q\beta+r=0 \end{cases}$ gives an equation of type 5 in u and v.
8. $\frac{dy}{dx}=-\frac{P(x,\ y)}{Q(x,\ y)} \Leftrightarrow$ $P(x,\ y)dx+Q(x,\ y)dy=0$ with $P_y=Q_x$ (*Exact equation*)	There exists $F(x,\ y)$ such that $F_x=P$, $F_y=Q$. $F(x,\ y)$ may be determined by these equations. General solution: $F(x,\ y)=C$

Examples

Type 2. $(x^2+1)\,\frac{dy}{dx}-xy=0;\ \int\frac{dy}{y}=\int\frac{x\,dx}{x^2+1};\ \ln|y|=\frac{1}{2}\ln(x^2+1)+C_1;$

$|y|=C_2\sqrt{x^2+1},\ y=C\sqrt{x^2+1}$

Type 3. $y'+\frac{3}{x}y=x(x>0),\ F(x)=\int\frac{3}{x}dx=3\ln x \Rightarrow e^{F(x)}=x^3$

$y(x)=\frac{1}{x^3}\left(\int x^3\cdot x\,dx+C\right)=\frac{x^2}{5}+\frac{C}{x^3}$

Type 8. $(4x^3+2xy+y^2)dx+(x^2+2xy-4y^3)dy=0 \quad (P\,dx+Q\,dy=0)$

$P_y=Q_x=2x+2y$ (exact)

$F_x=P \Rightarrow F=x^4+x^2y+xy^2+\varphi(y) \Rightarrow$

$F_y=x^2+2xy+\varphi'(y)=Q \Rightarrow \varphi'(y)=-4y^3 \Rightarrow \varphi(y)=-y^4+C$

Solution: $x^4+x^2y+xy^2-y^4=C$

9.2 Differential Equations of the Second Order

Special types	Solution or method of solution
1. $y''=f(x, y')$	Substitution: $z=y'$ gives $z'=f(x,z)$ i.e. ODE of first order.
2. $y''=f(y, y')$	Substitution: $p=\frac{dy}{dx}$, $\frac{d^2y}{dx^2}=p\cdot\frac{dp}{dy}$ gives $p\frac{dp}{dy}-f(y, p)=0$, i.e. ODE of 1st order. Given $p=p(y)$ then $p=\frac{dy}{dx}$ is separable.
3. $y''+f(x)y'+g(x)y=R(x)$	Let $\varphi(x)$ be a particular solution of the homogeneous equation $y''+f(x)y'+g(x)y=0$. Then the substitution $y=z\varphi(x)$ gives $\varphi z''+(2\varphi'+f\varphi)z'=R$, i.e. an ODE of type 1 in $z(x)$.
4. $y''+ay'+by=0$ (a, b real constants) *Characteristic equation:* (*) $f(r)=r^2+ar+b=0$	Let r_1, r_2 be the roots of (*). (*i*) $r_1\neq r_2$ real: $y=C_1e^{r_1x}+C_2e^{r_2x}$ (*ii*) $r=r_1=r_2$: $y=(C_1x+C_2)\,e^{rx}$ (*iii*) $r_1=\alpha+i\beta$, $r_2=\alpha-i\beta$: $y=e^{\alpha x}(C_1\cos\beta x+C_2\sin\beta x)=$ $=e^{\alpha x}C\cos(\beta x+\theta)$
5. $y''+ay'+by=R(x)$ (a, b real constants)	y_p+y_h=particular solution + solution of homogeneous equation. Refering to 4, above, $y_p=$ (*i*) $\frac{1}{r_1-r_2}\left[e^{r_1x}\int e^{-r_1x}R(x)dx-e^{r_2x}\int e^{-r_2x}R(x)dx\right]$ (*ii*) $xe^{rx}\int e^{-rx}R(x)dx-e^{rx}\int xe^{-rx}R(x)dx$ (*iii*) $\frac{1}{\beta}\,e^{\alpha x}[\sin\beta x\int e^{-\alpha x}\cos\beta x\,R(x)dx-$ $-\cos\beta x\int e^{-\alpha x}\sin\beta x\,R(x)dx]$

Table of some special cases

$R(x)=$	$y_p(x)=$
$P(x)=Ax^2+Bx+C$	$\frac{1}{b}\left[P(x)-\frac{a}{b}P'(x)+\frac{a^2-b}{b^2}P''(x)\right]$, $b\neq0$ $\frac{1}{a}\left[\int P(x)dx-\frac{1}{a}P(x)+\frac{1}{a^2}P'(x)\right]$, $b=0$, $a\neq0$

Ae^{kx}	$\frac{A}{f(k)}e^{kx}$, k not root of (*), $f(k)=k^2+ak+b$ $\frac{Ax}{f'(k)}e^{kx}$, k simple root of (*) $\frac{Ax^2}{f''(k)}e^{kx}$, k double root of (*)
$A\cos\omega x+B\sin\omega x$ $(\alpha=b-\omega^2,\ \beta=a\omega)$	$\frac{(A\alpha-B\beta)\cos\omega x+(A\beta+B\alpha)\sin\omega x}{\alpha^2+\beta^2}$, $i\omega$ not root of (*) $\frac{x}{2\omega}(-B\cos\omega x+A\sin\omega x)$, $i\omega$ root of (*)
$e^{kx}(A\cos\omega x+B\sin\omega x)$ $(\alpha=b+ak+k^2-\omega^2,$ $\beta=a\omega+2k\omega)$	$e^{kx}\frac{(A\alpha-B\beta)\cos\omega x+(A\beta+B\alpha)\sin\omega x}{\alpha^2+\beta^2}$, $k+i\omega$ not root of (*) $\frac{x}{2\omega}e^{kx}(A\sin\omega x-B\cos\omega x)$, $k+i\omega$ root of (*)
$\delta(x-c)$	In analogy to 5. above, (i) $\frac{1}{r_1-r_2}(e^{r_1(x-c)}-e^{r_2(x-c)})\,\theta(x-c)$ (ii) $(x-c)e^{r(x-c)}\,\theta(x-c)$ (iii) $\frac{1}{\beta}e^{\alpha(x-c)}\sin(\beta(x-c))\,\theta(x-c)$

6. $x^2y''+axy'+by=R(x)$ a, b constants, $x>0$ (*Euler's equation*)	The substitution $x=e^t$ gives $\frac{d^2y}{dt^2}+(a-1)\frac{dy}{dt}+by=R(e^t)$, i.e. an equation of type 5.

Further examples of 2nd order differential equations, see chapter 12.

Solution by power series expansion

7. *Gauss' hypergeometric differential equation*:

$$x(1-x)y''-[(a+b+1)x-c]y'-aby=0 \quad (a,\ b,\ c \text{ constants})$$

Power series solution for $|x|<1$: $y=C_1y_1+C_2y_2$, where

$$y_1=F(a,b,c,x)=1+\frac{ab}{c}\cdot\frac{x}{1!}+\frac{a(a+1)b(b+1)}{c(c+1)}\cdot\frac{x^2}{2!}+\ldots$$

$$y_2=x^{1-c}F(a-c+1,\ b-c+1,\ 2-c,\ x),\quad c\neq 0,1,2,\ldots$$

8. *Kummer's confluent hypergeometric differential equation*:

$$xy''+(c-x)y'-by=0 \quad (b,\ c \text{ constants})$$

Solution for $|x|<\infty$: $y=C_1y_1+C_2y_2$, where

$$y_1=F(b,c,x)=1+\frac{b}{c}\cdot\frac{x}{1!}+\frac{b(b+1)}{c(c+1)}\cdot\frac{x^2}{2!}+\ldots$$

$$y_2=x^{1-c}F(b-c+1,\ 2-c,\ x),\quad c\neq 0,1,2,\ldots$$

9.3 Linear Differential Equations

Equations with constant coefficients

Differential equation (a_i real constants):

(9.1) $y^{(n)}+a_{n-1}y^{(n-1)}+\ldots+a_1y'+a_0y=R(x) \quad [y=y(x)]$

or

$P(D)y=R(x)$, where $P(D)=D^n+a_{n-1}D^{n-1}+\ldots+a_1D+a_0$

General solution of (9.1): $y(x)=y_p(x)+y_h(x)=$ =Particular solution + Solution of homogeneous equation.

Homogeneous equations

(9.2) $y^{(n)}+a_{n-1}y^{(n-1)}+\ldots+a_1y'+a_0y=0$ or $P(D)y=0$

Characteristic equation:

(9.3) $r^n+a_{n-1}r^{n-1}+\ldots+a_1r+a_0=0$

with roots $r_1, \ldots, r_k$ of multiplicity $m_1, \ldots, m_k$ respectively.

General solution of (9.2)

$y_h(x)=P_1(x)e^{r_1x}+\ldots+P_k(x)e^{r_kx}$,
where $P_j(x)$=polynomial of degree $\leqslant m_j-1$.

In particular, if all roots are *simple,* then

$$y_h(x)=C_1e^{r_1x}+\ldots+C_ne^{r_nx},\quad C_j=\text{constant}$$

Note. If (e.g.) $r_1=\alpha+i\beta$, $r_2=\alpha-i\beta$, $m=m_1=m_2$, then

$$P_1(x)e^{r_1x}+P_2(x)e^{r_2x}=e^{\alpha x}\,(Q_1(x)\cos\beta x+Q_2(x)\sin\beta x)$$

where $Q_1(x)$, $Q_2(x)$=arbitrary polynomials of degree $\leqslant m-1$.

Example

1. $y'''-y=0$. Characteristic equation $r^3-1=0 \Rightarrow r_1=1, r_{2,3}=\frac{1}{2}(-1\pm i\sqrt{3})$

$$y_h=Ae^x+e^{-x/2}\,(B\cos\sqrt{3}x/2+C\sin\sqrt{3}x/2)$$

2. $y'''+3y''+3y'+y=0 \Leftrightarrow (D+1)^3y=0$. Characteristic roots $r=-1$ (triple root)

$$y_h=(Ax^2+Bx+C)\,e^{-x}$$

Particular solutions

The right hand side $R(x)$ of (9.1) is

1. Polynomial

(9.4) $$P(D)y=Q(x) \qquad (\deg Q(x)=m)$$

(*i*) $a_0 \neq 0$: Substitute $y_p=k_m x^m+\ldots+k_1 x+k_0$. The coefficients $k_0, \ldots, k_m$ will be determined by substituting y_p in (9.4) and identifying coefficients.
(*ii*) $a_0=\ldots=a_{k-1}=0$, $a_k \neq 0$. Substitute $y_p=k_m x^{m+k}+\ldots+k_0 x^k$.

2. Exponential function

(9.5) $$P(D)y=Q(x)\, e^{kx} \; (k=\text{constant},\; Q(x) \text{ polynomial})$$

The substitution $y(x)=e^{kx}z(x)$ will transform (9.5) to $P(D+k)z=Q(x)$, i.e. an equation of the form (9.4) in $z(x)$.

3. Trigonometric functions

$$\begin{aligned} &P(D)y_1=Q(x)\, e^{kx} \cos \omega x \qquad (k,\ \omega=\text{real constants},\ Q(x) \text{ real polynomial}) \\ &P(D)y_2=Q(x)\, e^{kx} \sin \omega x \end{aligned}$$

Considering $P(D)y=Q(x)\, e^{(k+i\omega)x}$ and determining y_p as in 2, yields

$$y_{1p}=\operatorname{Re} y_p,\quad y_{2p}=\operatorname{Im} y_p$$

Special case $P(D)y=Q(x) \begin{Bmatrix} \cos \omega x \\ \sin \omega x \end{Bmatrix}$.

Substitute $y_p=x^s\,[Q_1(x) \cos \omega x+Q_2(x) \sin \omega x]$ where $\deg Q_1=\deg Q_2=\deg Q$, $s=0$ if $i\omega$ is not a root of (9.3), otherwise $s=$ the multiplicity of $i\omega$ as a root of (9.3).

4. Discontinuous functions,

such as the unit step function

$$\theta(x-a)=\begin{cases} 1, & x>a \\ 0, & x<a \end{cases}$$

and the impulse function $\delta(x-a)=\theta'(x-a)$ etc.

Try the substitution $y_p(x)=u(x)\theta(x-a)+A\,\delta(x-a)+B\delta'(x-a)+\ldots$

5. Arbitrary function

(*i*) Let $h(x)$ be the specific solution of $P(D)y=0$ which satisfies $h(0)=h'(0)=\ldots= =h^{(n-2)}(0)=0$, $h^{(n-1)}(0)=1$. Then

$$y_p(x)=\int_a^x h(x-t)f(t)dt, \qquad a \text{ suitably chosen.}$$

(*ii*) *Factorization of the operator*

E.g. $(D-a)(D-b)y=R(x)$ leads to two equations of 1st order:

$$(D-a)z=R(x) \text{ and } (D-b)y=z$$

The Euler equation

(9.6) $$a_n x^n D^n y + a_{n-1} x^{n-1} D^{n-1} y + \ldots + a_1 x D y + a_0 y = R(x),$$

$$a_i=\text{constants}, D=\frac{d}{dx}, \ (x>0)$$

Substitution: $x=e^t$, $t=\ln x$, $\theta=\dfrac{d}{dt}$ gives

$$\boxed{xD=\theta,\ x^2D^2=\theta(\theta-1),\ \ldots,\ x^kD^k=\theta(\theta-1)\ \ldots\ (\theta-k+1)}$$

Hence, (9.6) will be transformed into an equation with constant coefficients.

Systems with constant coefficients

System of two unknown functions $y(x)$, $z(x)$:

(9.7) $$\begin{cases} P_{11}(D)y+P_{12}(D)z=R_1 & [P_{22}(D)] \\ P_{21}(D)y+P_{22}(D)z=R_2 & [-P_{12}(D)] \end{cases} \quad \text{(operator multiplication)}$$

Eliminating $z(x)$ gives

$$\{P_{11}(D)P_{22}(D)-P_{12}(D)P_{21}(D)\}y(x)=P_{22}(D)\,R_1(x)-P_{12}(D)\,R_2(x),$$

which is solved by the methods above. Analogously for $z(x)$, or $y(x)$ is substituted into (9.7).

Note.

1. In general the solutions have to be tested in (9.7). The total number of integration constants equals the order of the operator $P_{11}P_{22}-P_{12}P_{21}$ assuming no common differential factors of P_{11} and P_{21} or P_{12} and P_{22}.
2. Systems with more unknown functions are treated similarly.
3. Systems combined with initial values may be solved by Laplace transformation. (Cf. below).

Matrix method

Consider a *homogeneous* system of differential equations in vector form,

(9.8) $$\mathbf{y}'(t) = A\mathbf{y}(t), \quad \text{where} \quad \mathbf{y}(t) = \begin{bmatrix} y_1(t) \\ y_2(t) \\ \dots \\ y_n(t) \end{bmatrix} \quad \text{and } A \text{ is a constant } n \times n\text{-matrix.}$$

1. If λ is an eigenvalue of A (i.e. $\det(A - \lambda I) = 0$), and $\mathbf{v}$ is a corresponding eigenvector (i.e. $A\mathbf{v} = \lambda\mathbf{v}$, $\mathbf{v} \neq 0$), then $\mathbf{y}(t) = e^{\lambda t}\mathbf{v}$ is a particular solution of (9.8).

2. If A has n linearly independent eigenvectors $\mathbf{v}_1, \mathbf{v}_2, \dots, \mathbf{v}_n$ corresponding to the eigenvalues $\lambda_1, \lambda_2, \dots, \lambda_n$ (not necessarily distinct), then the general solution of (9.8) is

$$\mathbf{y}(t) = C_1 e^{\lambda_1 t}\mathbf{v}_1 + C_2 e^{\lambda_2 t}\mathbf{v}_2 + \dots + C_n e^{\lambda_n t}\mathbf{v}_n$$

where $C_1, C_2, \dots, C_n$ are arbitrary constants.

Initial value problems. (Laplace transformation)

Problem

(9.10), (9.11)
$$\begin{cases} y^{(n)} + a_{n-1}y^{(n-1)} + \dots + a_1 y' + a_0 y = f(t), \ t>0 \ (y=y(t)) & (9.10) \\ y(0)=y_0, \ y'(0)=y_1, \ \dots, \ y^{(n-1)}(0)=y_{n-1} & (9.11) \end{cases}$$

Method 1

(*i*) Find the general solution of (9.10).
(*ii*) Determine the integration constants by the conditions (9.11).

Method 2

Apply Laplace transformation to (9.10) using (9.11). Cf. sec. 13.5.

Example

$$\begin{cases} y''+y=f(t) \\ y(0)=0, \ y'(0)=1 \end{cases}$$

Laplace transformation gives

$$s^2Y(s)-1+Y(s)= \frac{1}{s^2} - \frac{e^{-s}}{s^2} \Rightarrow$$

$$Y(s)= \frac{1}{s^2} - \left(\frac{1}{s^2} - \frac{1}{s^2+1}\right) e^{-s} \Rightarrow$$

$$y(t)=t-[(t-1)-\sin(t-1)]\,\theta(t-1)=\begin{cases} t, & 0<t<1 \\ 1+\sin(t-1), & t\geq 1 \end{cases}$$

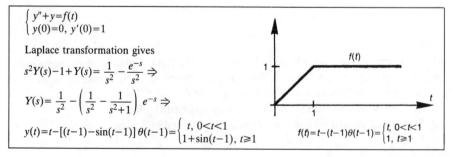

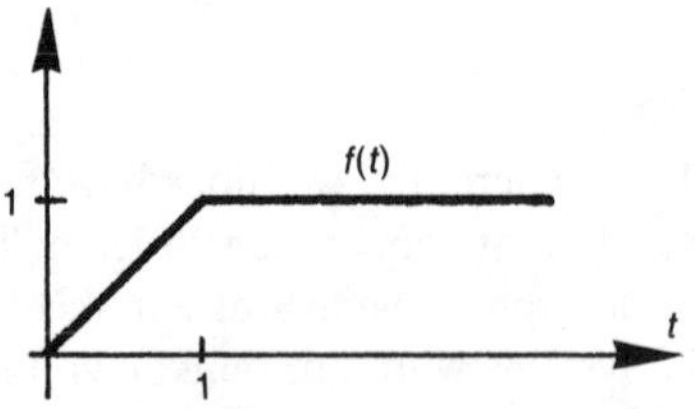

$$f(t)=t-(t-1)\theta(t-1)=\begin{cases} t, & 0<t<1 \\ 1, & t\geq 1 \end{cases}$$

9.4 General Concepts and Results

Terminology

An *ordinary differential equation* (ODE) is an equation involving derivatives of an unknown function of one variable. The *order* of the equation is the order of the highest derivative of the unknown function. An (explicit) ODE of order n has the form

(9.12) $$y^{(n)}=f(x, y, y', \dots, y^{(n-1)}), \qquad y=y(x),$$

with the *general solution*

(9.13) $$y=y(x, C_1, C_2, \dots, C_n),$$

where C_i are arbitrary (and independent) constants. Each particular choice of the n constants yields a *particular solution* of (9.12).

The constants C_i in (9.13) may be uniquely determined by for example

(*i*) *n initial conditions* $y(x_0)=a_0$, $y'(x_0)=a_1$, …, $y^{(n-1)}(x_0)=a_{n-1}$, or
(*ii*) *n boundary conditions* of $y(x)$ and its derivatives at two distinct points x_1 and x_2.

An ODE is *linear* if it has the form

(9.14) $$L[y]=y^{(n)}+a_{n-1}\,y^{(n-1)}+\dots+a_0y=g(x), \qquad y=y(x),\ a_i=a_i(x)$$

This equation is (*i*) *homogeneous* if $g(x)\equiv 0$, (*ii*) *inhomogeneous* otherwise.

Existence and uniqueness theorem

Notation: $\mathbf{y}=\begin{bmatrix} y_1 \\ y_2 \\ \dots \\ y_n \end{bmatrix}$, $\mathbf{f}(x, \mathbf{y})=\begin{bmatrix} f_1(x, y_1, \dots, y_n) \\ \dots \\ f_n(x, y_1, \dots, y_n) \end{bmatrix}$ $\qquad D=\{(x, \mathbf{y}):\ |x-x_0|\leq a,\ |y_i-y_{0,i}|\leq b\}\subset \mathbf{R}^{n+1}$

The Cauchy problem

(9.15) $$\begin{cases} \mathbf{y}'(x)=\mathbf{f}(x, \mathbf{y}),\ (x, \mathbf{y})\in D \\ \mathbf{y}(x_0)=\mathbf{y}_0 \end{cases}$$

> *Theorem*
>
> Assume (*i*) $f_i(x, \mathbf{y})$ are continuous in D.
> (*ii*) $f_i(x, \mathbf{y})$ satisfy a (uniform) Lipschitz condition in D,
> i.e. $|f_i(x, \mathbf{y})-f_i(x, \bar{\mathbf{y}})|<c|\mathbf{y}-\bar{\mathbf{y}}|$, ($c$=constant).
>
> Then there exists a unique solution $\mathbf{y}(x)$ of (9.15) in the interval $|x-x_0|<d$, where $d=\min(a, b/B)$, $|f_i(x, \mathbf{y})|<B$ in D.

Remark. By setting $y_1=y$, $y_2=y'$, …, $y_n=y^{(n-1)}$, the equation (9.12) may be written as the system $y_1'=y_2$, $y_2'=y_3$, …, $y_n'=f(x, y_1, \dots, y_n)$.

Systems of linear differential equations

Given the system

(9.16) $$\mathbf{y}'(x)=A(x)\mathbf{y}(x)+\mathbf{g}(x), \qquad A(x)\ n\times n\text{-matrix},$$

and the corresponding homogeneous equation

(9.17) $\quad y'(x)=A(x)y(x)$

Definition. $y_1(x), \ldots, y_n(x)$ is a *fundamental system* (*basis*) of solutions of (9.17) if they are linearly independent solutions of (9.17).

Fundamental matrix $Y(x)=[y_1(x), \ldots, y_n(x)]$.

1. $Y(x)$ is a fundamental matrix of (9.17) $\Leftrightarrow$

$$Y'(x)=A(x)Y(x) \text{ and } \det Y(x)\neq 0$$

2. $Y(x)$ a fundamental matrix of (9.17) $\Rightarrow$

$$y(x)=Y(x)C+Y(x)\int_{x_0}^{x} Y^{-1}(t)g(t)dt, \qquad C=\text{constant vector},$$

is the general solution of (9.16).

3. If $A(x)=A$ is constant then

$$y(x)=e^{Ax}C+e^{Ax}\int_{x_0}^{x} e^{-At}g(t)dt$$

is the general solution of (9.16). [Also, cf. sec. 9.3]

Asymptotic behavior

Let A be a constant matrix with eigenvalues λ_k and set $\sigma(A)=\max \operatorname{Re}\lambda_k$.

1. If $y(x)$ is any solution of

$$y'(x)=Ay(x),$$

then

(*i*) $|y(x)|$ is bounded as $x\to\infty$ if $\sigma(A)\leqslant 0$.
(*ii*) $|y(x)|\to 0$ as $x\to\infty$ if $\sigma(A)<0$.

2. If $y(x)$ is any solution of

$$y'(x)=Ay(x)+g(x),$$

then

(*i*) $|y(x)|$ is bounded as $x\to\infty$ if $\sigma(A)\leqslant 0$, $\int_{x_0}^{\infty}|g(t)|dt<\infty$

or if $\sigma(A)<0$, $\int_{x-1}^{x}|g(t)|dt$ bounded.

(*ii*) $|y(x)|\to 0$ as $x\to\infty$ if $\sigma(A)<0$, $\int_{x-1}^{x}|g(t)|dt\to 0,\ x\to\infty$

Linear equations of higher order

($y_1(x), \ldots, y_n(x)$ are linearly independent if $\sum_{i=1}^{n}\lambda_i y_i(x)\equiv 0 \Rightarrow \lambda_i=0$, all i)

Given the linear ODE

(9.18) $\quad L[y](x)=a_n(x)y^{(n)}(x)+a_{n-1}(x)y^{(n-1)}(x)+\ldots+a_0(x)y(x)=g(x)$

and the corresponding homogeneous equation

(9.19) $\quad L[y]=0$

Definition. The set of functions $y_1(x), \ldots, y_n(x)$ is a fundamental system (basis) of solutions of (9.19) if they are linearly independent solutions of (9.19).

Wronski's determinant $W(y_1, \ldots, y_n)=\begin{vmatrix} y_1 & y_2 & \ldots & y_n \\ y_1' & y_2' & \ldots & y_n' \\ \ldots \\ y_1^{(n-1)} & y_2^{(n-1)} & \ldots & y_n^{(n-1)} \end{vmatrix}$

4. The equation (9.19) has at most n linearly independent solutions.
5. $y_1(x), \ldots, y_n(x)$ is a fundamental system of solutions of (9.19) $\Leftrightarrow$ $W(y_1, \ldots, y_n)\neq 0$.
6. If $y_h(x)$ is the general solution of (9.19) and $y_p(x)$ any particular solution of (9.18) then $y(x)=y_p(x)+y_h(x)$ is the general solution of (9.18).

Boundary value problems

Consider the problem

(9.20) $$\begin{cases} L[y](x)=\sum_{k=0}^{n} a_k(x)y^{(k)}(x)=h(x), \ a<x<b \\ B_k y\equiv \sum_{i=0}^{n-1}[\alpha_{ik}y^{(i)}(a)+\beta_{ik}y^{(i)}(b)]=c_k, \ k=1, \ldots, n \ (\alpha_{ik}, \beta_{ik}, c_k \text{ constants}) \end{cases}$$

where $h(x)$, $a_k(x)\in C[a, b]$, $a_n(x)\neq 0$.

Let $y_1, \ldots, y_n$ be a basis of solutions of $L[y]=0$.

Theorem. The above problem is uniquely solvable $\Leftrightarrow$

(9.21) $\det(B_i y_j)\neq 0$.

Green's function

Green's function $G(x, \xi)$, which is continuous in the square $a\leq x, \xi\leq b$ *(if* $n\geq 2$), is defined as the solution of

$$\begin{cases} L[G](x, \xi)=\delta(x-\xi) & , a<\xi<b \\ B_k G(x, \xi)=0 & , a<\xi<b, \ k=1, \ldots, n \end{cases}$$

Theorem. Assume that (9.21) holds. Then the solution of (9.20) with $c_k=0$ can be written

$$y(x)=\int_a^b G(x, \xi)h(\xi)d\xi$$

Example

$$\begin{cases} (1+x)y''+y'=h(x), \ 0<x<1 \\ y'(0)=y(1)=0 \end{cases}$$

Determination of the Green's function:

$$(1+x)y''+y'=\delta(x-\xi) \Leftrightarrow \frac{d}{dx}\{(1+x)y'\}=\delta(x-\xi).$$

Thus, $(1+x)y'=\theta(x-\xi)+A$. $y'(0)=0 \Rightarrow A=0$.

$$y'=\frac{\theta(x-\xi)}{1+x} \Rightarrow y=[\ln(1+x)-\ln(1+\xi)]\,\theta(x-\xi)+B \qquad \text{(cf sec. 12.6).}$$

$y(1)=0 \Rightarrow B=\ln(1+\xi)-\ln 2$. Hence,

$$G(x, \xi)=\begin{cases} \ln(1+x)-\ln 2, & 0\leq\xi\leq x\leq 1 \\ \ln(1+\xi)-\ln 2, & 0\leq x\leq\xi\leq 1 \end{cases}$$

Table of Green's functions

For the general solution of the differential equation $y''+ay'+by=\delta(x-\xi)$, see sec. 9.2, "*Table of some special cases*".

In the following examples the corresponding boundary value problem is self-adjoint (see sec. 12.1) so that $G(x, \xi)$ is symmetric, i.e. $G(x, \xi)=G(\xi, x)$.

Differential operator $L[y]$ *in the interval* $(0, a)$	*Boundary contitions*	$G(x, \xi)$, $(x\leq\xi)$ $[G(\xi, x), (\xi\leq x)]$
D^2y	$y(0)=y(a)=0$	$(\xi a-1)x \quad [(xa-1)\xi, \ \xi\leq x]$
	$y(0)=y'(a)=0$	$-x$
	$y'(0)=y(a)=0$	$\xi-a$
	$\begin{cases} y(0)+y(a)=0 \\ y'(0)+y'(a)=0 \end{cases}$	$(\xi-x)/2-a/4$
$(D^2-k^2)y$	$y(0)=y(a)=0$	$-\sinh kx \sinh k(a-\xi)/(k\sinh ka)$
	$y(0)=y'(a)=0$	$-\sinh kx \cosh k(a-\xi)/(k\cosh ka)$
	$y'(0)=y(a)=0$	$-\cosh kx \sinh k(a-\xi)/(k\cosh ka)$
$(D^2+k^2)y$	$y(0)=y(a)=0$	$-\sin kx \sin k(a-\xi)/(k\sin ka)$
	$y(0)=y'(a)=0$	$-\sin kx \cos k(a-\xi)/(k\cos ka)$
	$y'(0)=y(a)=0$	$-\cos kx \sin k(a-\xi)/(k\cos ka)$
D^4y	$\begin{cases} y(0)=y'(0)= \\ =y(a)=y'(a)=0 \end{cases}$	$x^2(a-\xi)^2(3a\xi-(2\xi+a)x)/6a^3$

Integral equations

1. The Cauchy problem (9.15) is equivalent to the integral equation

$$y(x)=\int_{x_0}^{x} f(t, y(t))dt+y_0$$

Example. $y(x)=\int_0^x t^2y(t)dt+x^2+1 \Leftrightarrow$ [by differentiating]

$$\begin{cases} y'(x)=x^2y(x)+2x \\ y(0)=1 \end{cases}$$

2. *Fredholm equations.*

First kind $\int_a^b K(x, t)y(t)dt=h(x)$ Second kind $y(x)-\int_a^b K(x, t)y(t)dt=h(x)$ (Cf. sec. 12.7).

3. *Volterra equations.*

First kind $\int_a^x K(x, t)y(t)dt=h(x)$ Second kind $y(x)-\int_a^x K(x, t)y(t)dt=h(x)$

9.5 Linear Difference Equations

Difference (or *recurrence*) *equation* of *order* N (a_i real constants):

(9.22) $x(n+N)+a_{N-1}x(n+N-1)+\ldots+a_0x(n)=R(n),\ n=0,\ 1,\ 2,\ \ldots$

or

$P(T)x(n)=R(n)$,

where the *translation* operator $P(T)=T^N+a_{N-1}T^{N-1}+\ldots+a_1T+a_0$, $T^kx(n)=x(n+k)$.

General solution:

$x(n)=x_p(n)+x_h(n)=$Particular solution+Solution of homogeneous equation.

Homogeneous equations

(9.23) $x(n+N)+a_{N-1}x(n+N-1)+\ldots+a_0x(n)=0$ or $P(T)x(n)=0$.

(9.24) *Characteristic equation:* $r^N+a_{N-1}r^{N-1}+\ldots+a_0=0$

with roots $r_1, \ldots, r_k$ of multiplicity $m_1, \ldots, m_k$, respectively.

General solution of (9.23)

$x_h(n)=P_1(n)r_1{}^n+\ldots+P_k(n)r_k{}^n$

where $P_j(n)$ are polynomials (in n) of degree $\leqslant m_j-1$.

In particular, if all roots are simple, then

$$x_h(n)=C_1r_1{}^n+\ldots+C_Nr_N{}^n,\ C_j=\text{constant}$$

If $r_1=\varrho e^{i\theta}$, $r_2=\varrho e^{-i\theta}$, $m=m_1=m_2$, then

$$P_1(n)r_1{}^n+P_2(n)r_k{}^n=\varrho^n(Q_1(n)\cos n\theta+Q_2(n)\sin n\theta),$$

where $Q_1(n)$ and $Q_2(n)$ are polynomials of degree $\leqslant m-1$.

Example. The Fibonacci numbers $x(n)=F_{n+1}$ (cf. sec. 2.2) are defined by

(9.25) $x(n+2)=x(n)+x(n+1),\ n\geqslant 0$

(9.26) $x(0)=x(1)=1$

The characteristic equation of (9.25) is $r^2-r-1=0$ with roots $a=(1+\sqrt{5})/2$ and $b=(1-\sqrt{5})/2$. Thus $x(n)=Aa^n+Bb^n$. Then (9.26) gives $A=a/\sqrt{5}$ and $B=-b/\sqrt{5}$. Thus $F_n=(a^n-b^n)/\sqrt{5},\ n\geqslant 1$.

Particular solutions

The right hand side $R(n)$ of (9.22) is

1. Polynomial

$$R(n)=b_q n^q+b_{q-1}n^{q-1}+\dots+b_0 \text{ a polynomial in } n \text{ of degree } q.$$

If $r=1$ is a root of multiplicity m of (9.24) ($m=0$ if $r=1$ is not a root) then substitute

$$x_p(n)=n^m(k_q n^q+k_{q-1}n^{q-1}+\dots+k_0).$$

The coefficients $k_0, \dots, k_q$ will be determined by substituting x_p in (9.22) and identifying coefficients.

2. Exponential function

$$R(n)=Q(n)c^n \ (c=\text{constant}, \ Q(n)=\text{polynomial of degree } q)$$

The substitution $x(n)=c^n y(n)$ will transform (9.22) to $P(cT)y(n)=Q(n)$. Continue as in **1.**

[Or substitute $x(n)=n^m(k_q n^q+k_{q-1}n^{q-1}+\dots+k_0)c^n$ if $r=c$ is a root of multiplicty m of (9.24).]

Example.
The substitution $x(n)=2^n y(n)$ transforms the difference equation

$$x(n+2)-4x(n)=n\,2^n \Leftrightarrow (T^2-4)x(n)=n\,2^n \text{ to}$$
$$((2T)^2-4)y(n)=n \Leftrightarrow (T^2-1)y(n)=n/4 \Leftrightarrow y(n+2)-y(n)=n/4$$

3. Trigonometric functions

$$R_1(n)=Q(n)c^n \cos n\theta \text{ or } R_2(n)=Q(n)\, c^n \sin n\theta$$
$$(c,\ \theta=\text{real constants}, \ Q(n)=\text{real polynomial})$$

Replace $R(n)$ in (9.22) by $R^*(n)=Q(n)(ce^{i\theta})^n$. Determining $x_p^*(n)$ as in **2** yields $x_{1p}(n)=\text{Re } x_p^*(n)$ and $x_{2p}(n)=\text{Im } x_p^*(n)$.

4. Arbitrary function

For $x(0), \dots, x(N-1)$ given (e.g. $=0$) then $x(n)$, $n \geq N$ are recursively uniquely determined by (9.22). For example, $x(n+1)-ax(n)=R(n)$, $n=0, 1, 2, \dots$ has a particular solution

$$x(0)=0, \ x(n)=\sum_{k=0}^{n-1} a^{n-k-1} R(k), \ n=1, 2, 3, \dots$$

See also sec. 13.4 (z-transformation).

10 Multidimensional Calculus

10.1 The Space R^n

The Euclidean space R^n

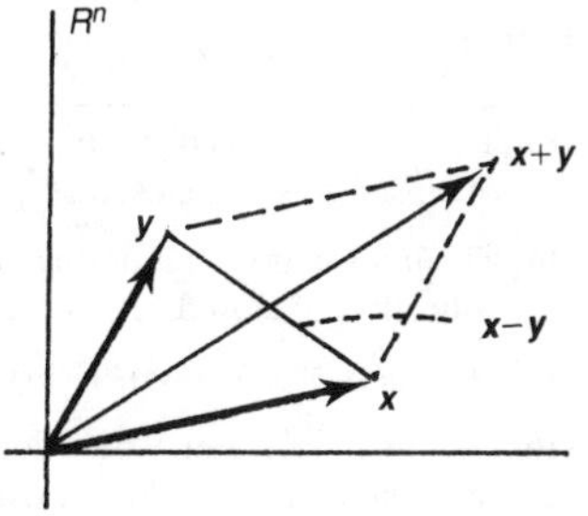

The Euclidean space R^n has the following characteristics.

1. R^n is the set of (real) n-tuples $\boldsymbol{x}=(x_1, \ldots, x_n)$
2. Addition and multiplication with a scalar λ are defined:

$$\boldsymbol{x}+\boldsymbol{y}=(x_1+y_1, \ldots, x_n+y_n), \qquad \lambda\boldsymbol{x}=(\lambda x_1, \ldots, \lambda x_n)$$

3. Scalar product: $\boldsymbol{x}\cdot\boldsymbol{y}=x_1y_1+x_2y_2+\ldots+x_ny_n$
4. Norm and distance:

$$|\boldsymbol{x}|=\sqrt{\boldsymbol{x}\cdot\boldsymbol{x}}=\sqrt{x_1^2+\ldots+x_n^2}, \quad |\boldsymbol{x}-\boldsymbol{y}|=\sqrt{(x_1-y_1)^2+\ldots+(x_n-y_n)^2}$$

5. Angle θ between $\boldsymbol{x}$ and $\boldsymbol{y}$: $\cos\theta=\dfrac{\boldsymbol{x}\cdot\boldsymbol{y}}{|\boldsymbol{x}|\cdot|\boldsymbol{y}|}$
6. Cauchy-Schwarz' inequality: $|\boldsymbol{x}\cdot\boldsymbol{y}|\leq|\boldsymbol{x}|\cdot|\boldsymbol{y}|$
7. The triangle inequality: $||\boldsymbol{x}|-|\boldsymbol{y}||\leq|\boldsymbol{x}+\boldsymbol{y}|\leq|\boldsymbol{x}|+|\boldsymbol{y}|$

Topological concepts

Neighborhood	Interior point	Exterior point	Boundary point	Limit point	Open set
Closed set	Boundary	Closure	Bounded set	Compact set	Connected set	Domain
Region					

Let $\boldsymbol{a}, \boldsymbol{b}, \boldsymbol{c}, \ldots$ be points (vectors) in R^n and let S be a set of points.

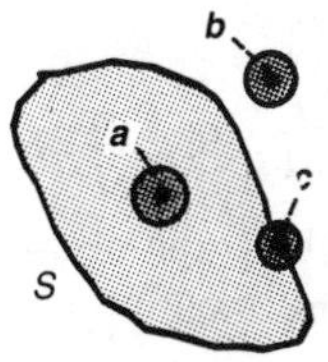

Definitions

1. A *neighborhood* of a point $\boldsymbol{p}$ is any ball B with center $\boldsymbol{p}$, i.e. $B=\{\boldsymbol{x}: |\boldsymbol{x}-\boldsymbol{p}|<\delta, \delta>0\}$. The neighborhood is *deleted* if $\boldsymbol{p}$ is omitted.
2. $\boldsymbol{a}\in S$ is an interior point of S if there is a neighborhood of $\boldsymbol{a}$ all of whose points belong to S.
3. $\boldsymbol{b}\notin S$ is an *exterior point* of S if there exists a neighborhood of $\boldsymbol{b}$ all of whose points do not belong to S.
4. $\boldsymbol{c}$ is a *boundary point* of S if every neighborhood of $\boldsymbol{c}$ contains at least one point in S and at least one point outside S. The *boundary* $\partial S=\{\text{boundary points of } S\}$.
5. $\boldsymbol{p}$ is called a *limit point* (or a *cluster point* or a *point of accumulation*) of S if every deleted neighborhood of $\boldsymbol{p}$ contains at least one point of S.

6. S is said to be *open* if S contains only interior points.
7. S is *closed* if every boundary point belongs to S (i.e. the complement of S is open). The *closure* $\bar{S}$ of S is defined by $\bar{S}=S \cup \partial S$.
8. S is *bounded* if there exists a constant M such that $|\boldsymbol{x}|<M$ for every $\boldsymbol{x} \in S$.
9. S is *compact* if S is closed and bounded.
10. S is *connected* if any two points of S can be joined by a continuous path contained in S.
11. S is *simply connected* if any closed curve lying in S can be shrunk to a point without leaving S.
12. S is a *domain* or an *open region* if S is open and connected.

Theorems

13. The union of any collection of open sets is an open set. The intersection of any collection of closed sets is a closed set.
14. The intersection of a finite collection of open sets is an open set. The union of a finite collection of closed sets is a closed set.
15. (*Bolzano–Weierstrass*). Every bounded infinite set has at least one limit point.
16. (*Heine–Borel*). Let S be a compact set and assume that S is covered by a family $\{A_i\}$ of open sets. Then there exists a finite number of the sets A_i which cover S.

10.2 Surfaces. Tangent Planes

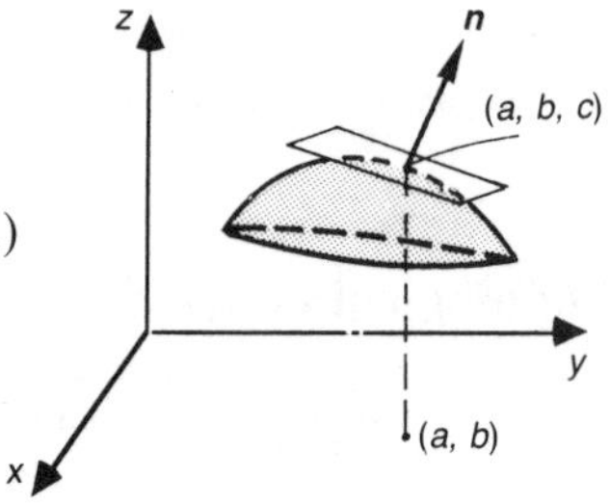

Graph of a function

$$z=f(x, y), \text{ normal vector } \boldsymbol{n}=(-f_x', -f_y', 1)$$

Tangent plane at (a, b, c), $c=f(a, b)$:

$$z-c=f_x'(a, b)(x-a)+f_y'(a, b)(y-b)$$

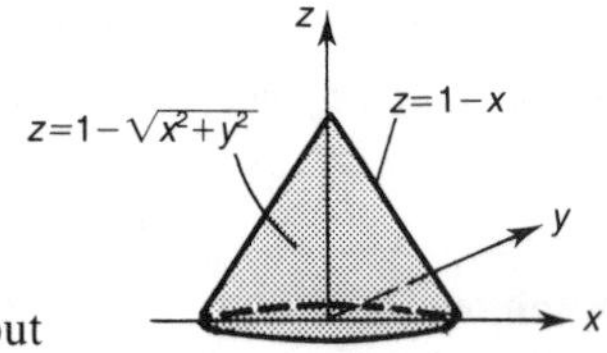

Remarks

(*i*) The graph of $z=f(x, y)$ is a *surface of revolution* about the z-axis $\Leftrightarrow f(x, y)$ depends only on (x^2+y^2).

(*ii*) The equation of the surface arising when $z=f(x)$ [or $f(y)$] rotates about the z-axis is

$$z=f(r)=f(\sqrt{x^2+y^2}) \quad \text{(cf. figure)}$$

Level surfaces

$$F(x, y, z)=C, \ \boldsymbol{n}=\text{grad } F=(F_x', F_y', F_z')$$

Tangent plane:

$$F_x'(a, b, c)(x-a)+F_y'(a, b, c)(y-b)+F_z'(a, b, c)(z-c)=0$$

Level curves. $F(x, y)=C, \ \boldsymbol{n}=\text{grad } F=(F_x', F_y')$

Parameterized surfaces

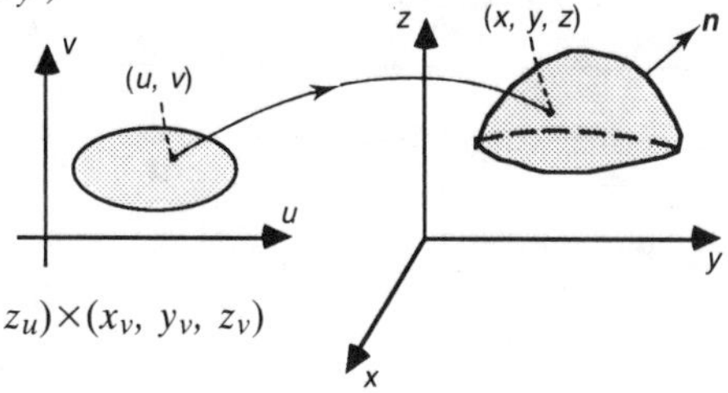

$$\begin{cases} x=x(u, v) \\ y=y(u, v), \\ z=z(u, v) \end{cases} \qquad \boldsymbol{r}=(x, y, z)$$

$$\boldsymbol{n}=(A, B, C)=\boldsymbol{r}_u\times\boldsymbol{r}_v=(x_u, y_u, z_u)\times(x_v, y_v, z_v)$$

Tangent plane

$$A(x-a)+B(y-b)+C(z-c)=0$$

10.3 Limits and Continuity

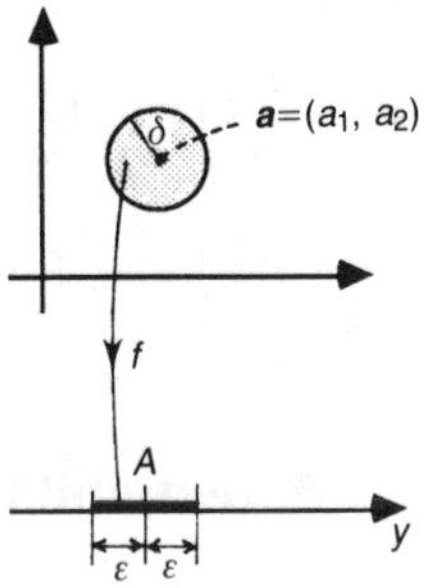

Functions $f:R^n \to R$. $y=f(\boldsymbol{x})=f(x_1, x_2, \ldots, x_n)$

Definition

$\lim\limits_{\boldsymbol{x}\to\boldsymbol{a}} f(\boldsymbol{x})=A$ means:

For any $\varepsilon>0$ there exists a $\delta>0$ such that $|f(\boldsymbol{x})-A|<\varepsilon$ for all $\boldsymbol{x}\in D_f$ such that $|\boldsymbol{x}-\boldsymbol{a}|=\sqrt{(x_1-a_1)^2+\ldots+(x_n-a_n)^2}<\delta$

Example

1. $\lim\limits_{(x,y)\to(0,0)} \dfrac{x^2y^2}{x^2+y^2}=[x=r\cos\theta, \ y=r\sin\theta]= \lim\limits_{\substack{r\to0^+ \\ \theta \text{ arbitrary}}} (r^2\cos^2\theta\sin^2\theta)=0$

2. $\lim\limits_{(x,y)\to(0,0)} \dfrac{xy}{x^2+y^2}$ does not exist, because (*i*) $f(x, 0)\to 0$ as $x\to 0$

   (*ii*) $f(x, x)\to \frac{1}{2}$ as $x\to 0$.

The concept of continuity is defined similarly as for functions of a single variable (cf. sec. 6.2).

Theorem

If $f(\boldsymbol{x})=f(x_1, \ldots, x_n)$ is continuous on a compact (i.e. bounded and closed) set D, then

(*i*) $f(\boldsymbol{x})$ is bounded on D

(*ii*) $f(\boldsymbol{x})$ assumes its supremum (maximum) and infimum (minimum) on D.

10.4 Partial Derivatives

Partial derivative

Consider functions $f\colon R^2 \to R$ (analogously for $f\colon R^n \to R$).

Definition

$$f_x'(a,b)=\lim_{h\to 0}\frac{f(a+h,\ b)-f(a,\ b)}{h} \qquad f_y'(a,b)=\lim_{k\to 0}\frac{f(a,\ b+k)-f(a,\ b)}{k}$$

Alternative notations: $f_x=D_xf=\frac{\partial f}{\partial x}=\left(\frac{\partial f}{\partial x}\right)_{y(=\text{constant})}$

Higher derivatives:

$$f_{xx}=f''_{xx}=\frac{\partial^2 f}{\partial x^2}=D_xf_x,\ f_{yx}=f''_{yx}=\frac{\partial^2 f}{\partial x\partial y}=D_xf_y,\ f_{yy}=f''_{yy}=\frac{\partial^2 f}{\partial y^2}=D_yf_y \text{ etc.}$$

Remark. $f_{xy}=f_{yx}$ if these functions are continuous.

Notation: $f\in C^k \Leftrightarrow f$ has continuous partial derivatives of order $\leqslant k$.

$u=u(x,\ y),\ v=v(x,\ y),\ f=f(x,\ y)$:

$$\left(\frac{\partial f}{\partial u}\right)_v(P)=\lim_{Q\to P}\frac{f(Q)-f(P)}{u(Q)-u(P)}=$$

$$=\left(\frac{\partial f}{\partial x}\right)_y\left(\frac{\partial x}{\partial u}\right)_v+\left(\frac{\partial f}{\partial y}\right)_x\left(\frac{\partial y}{\partial u}\right)_v$$

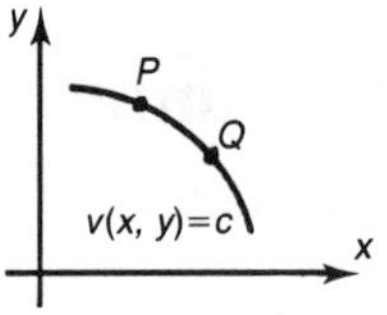

Differentiability (Linear approximation)

$f(x,\ y)$ is *differentiable* at $(x,\ y)$ if

$$\Delta f=f(x+h,\ y+k)-f(x,\ y)=hf_x'(x,\ y)+kf_y'(x,\ y)+\sqrt{h^2+k^2}\ \varepsilon(h,\ k)$$

where $\varepsilon(h,\ k)\to 0$ as $(h,\ k)\to(0,\ 0)$.

(Analogously for $f\colon R^n\to R$)

Differential: $df=\frac{\partial f}{\partial x}dx+\frac{\partial f}{\partial y}dy$

$f(x,\ y)$ is differentiable at $(x,\ y)$ if the partial derivatives of $f(x,\ y)$ exist in a neighborhood of $(x,\ y)$ and are continuous at $(x,\ y)$.

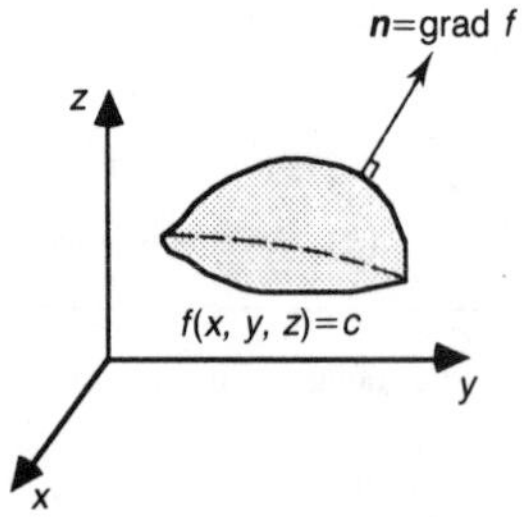

Gradient

Given $f\colon R^3\to R$ (analogously for $f\colon R^n\to R$). The vector $\nabla f=\text{grad } f=(f_x',\ f_y',\ f_z')$ is orthogonal to the corresponding level surface $f(x,\ y,\ z)=C$.

Directional derivative

Given the directional vector $\boldsymbol{e}=(e_x, e_y, e_z)$ of length 1.

$$f_{\boldsymbol{e}}'(a, b, c)=\lim_{t\to 0}\frac{1}{t}[f(a+te_x, b+te_y, c+te_z)-f(a, b, c)]=$$
$$=\frac{d}{dt}f(a+te_x, b+te_y, c+te_z)\big|_{t=0}=\boldsymbol{e}\cdot \operatorname{grad} f(a, b, c)$$

(if grad f is continuous).

1. $f_{\boldsymbol{e}}'$ is maximal in the direction $\boldsymbol{e}=\dfrac{\operatorname{grad} f}{|\operatorname{grad} f|}$.
2. $\max f_{\boldsymbol{e}}'=|\operatorname{grad} f|$.

The chain rule

1. $z=z(x, y)$, $x=x(t)$, $y=y(t)$:

$$\frac{dz}{dt}=\frac{\partial z}{\partial x}\cdot\frac{dx}{dt}+\frac{\partial z}{\partial y}\cdot\frac{dy}{dt}$$

$$\frac{d^2z}{dt^2}=\frac{\partial z}{\partial x}\cdot\frac{d^2x}{dt^2}+\frac{dx}{dt}\left(\frac{\partial^2 z}{\partial x^2}\cdot\frac{dx}{dt}+\frac{\partial^2 z}{\partial x\partial y}\cdot\frac{dy}{dt}\right)+$$
$$+\frac{\partial z}{\partial y}\cdot\frac{d^2y}{dt^2}+\frac{dy}{dt}\left(\frac{\partial^2 z}{\partial x\partial y}\cdot\frac{dx}{dt}+\frac{\partial^2 z}{\partial y^2}\cdot\frac{dy}{dt}\right).$$

2. $z=z(x, y)$, $x=x(u, v)$, $y=y(u, v)$:

$$\begin{cases}\dfrac{\partial z}{\partial u}=\dfrac{\partial z}{\partial x}\cdot\dfrac{\partial x}{\partial u}+\dfrac{\partial z}{\partial y}\cdot\dfrac{\partial y}{\partial u}\\[2ex]\dfrac{\partial z}{\partial v}=\dfrac{\partial z}{\partial x}\cdot\dfrac{\partial x}{\partial v}+\dfrac{\partial z}{\partial y}\cdot\dfrac{\partial y}{\partial v}\end{cases}$$

3. $f=f(x_1, \dots, x_n)$, $x_k=x_k(u_1, \dots, u_m)$, $k=1, \dots, n$

$$\frac{\partial f}{\partial u_j}=\sum_{k=1}^{n}\frac{\partial f}{\partial x_k}\cdot\frac{\partial x_k}{\partial u_j},\qquad j=1, \dots, m.$$

4. x, y, z depending on each other:

(i) $\left(\dfrac{\partial x}{\partial y}\right)_z=1/\left(\dfrac{\partial y}{\partial x}\right)_z$ etc.

(ii) $\left(\dfrac{\partial x}{\partial y}\right)_z\left(\dfrac{\partial y}{\partial z}\right)_x\left(\dfrac{\partial z}{\partial x}\right)_y=-1$

The mean value theorem

If f: $R^n\to R$ has continuous partial derivatives, then

$$f(\boldsymbol{x}+\boldsymbol{h})-f(\boldsymbol{x})=\boldsymbol{h}\cdot\operatorname{grad} f(\boldsymbol{x}+\theta\boldsymbol{h}),\ 0<\theta<1$$

Taylor's formula

$f: R^2 \to R$:

Assume that $f(x, y)$ has continuous partial derivatives of order $\leq n$ in a region containing the straight line segment between (a, b) and $(a+h, b+k)$. Then

(10.1) $$f(a+h, b+k)=f(a, b)+ +hf_x'(a, b)+kf_y'(a, b)+ +\frac{1}{2}[h^2f''_{xx}(a, b)+2hkf''_{xy}(a, b)+k^2f''_{yy}(a, b)]+\ldots+R_n=$$

$$=\sum_{j=0}^{n-1}\frac{1}{j!}\left(h\frac{\partial}{\partial x}+k\frac{\partial}{\partial y}\right)^j f(a, b)+R_n$$

where $R_n=\frac{1}{n!}\left(h\frac{\partial}{\partial x}+k\frac{\partial}{\partial y}\right)^n f(a+\theta h, b+\theta k)$, $0<\theta<1$

or $R_n=(h^2+k^2)^{n/2}\,B(h, k)$, where $B(h, k)$ is bounded in a neighborhood of $(0, 0)$.

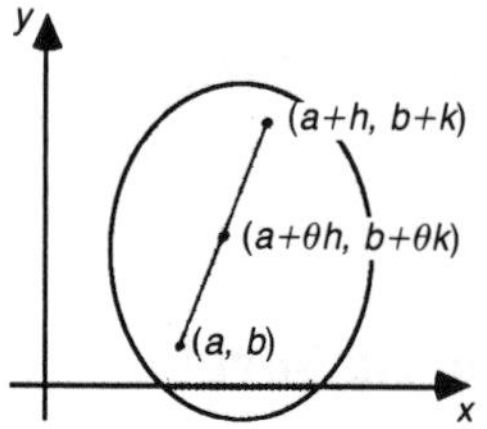

General case $f: R^n \to R$: *Multi-index* $\boldsymbol{j}=(j_1, \ldots, j_n)$. $D_{\boldsymbol{j}}f=D_{j_1}\ldots D_{j_n}f$, $\boldsymbol{h}^{\boldsymbol{j}}=h_1^{j_1}\ldots h_n^{j_n}$, $\boldsymbol{j}!=j_1!\ldots j_n!$

$$f(\boldsymbol{a}+\boldsymbol{h})=\sum_{j_1+\ldots+j_n\leq m-1}\frac{1}{\boldsymbol{j}!}(D_{\boldsymbol{j}}f)\boldsymbol{h}^{\boldsymbol{j}}+R_m, \text{ where}$$

$R_m=|\boldsymbol{h}|^m\,B(\boldsymbol{h})$, $B(\boldsymbol{h})$ bounded around $\boldsymbol{0}$

Example

Find the *Taylor polynomial* of order 2 at $(1, 0)$ of $f(x, y)=\ln(x-y)$.

Method 1. Using (10.1): $f(1, 0)=0$; $f_x'=\frac{1}{x-y}\Rightarrow f_x'(1, 0)=1$ etc. gives

$$P_2(x, y)=0+1\cdot(x-1)-1\cdot y+\frac{1}{2}[-1(x-1)^2+2(x-1)y-1\cdot y^2]$$

Method 2. Using standard Maclaurin expansions:

$$f(x, y)=[x=1+h, y=k]=\ln(1+h-k)=$$

$$=[t=h-k]=\ln(1+t)=t-\frac{t^2}{2}+\ldots=h-k-\frac{1}{2}(h-k)^2+\ldots=\text{etc.}$$

The implicit function theorem

Assume

(*i*) $F: R^3 \to R$ is continuously differentiable

(*ii*) $F(a, b, c)=0$

(*iii*) $F_z'(a, b, c)\neq 0$

Then $F(x, y, z)=0$ defines a function $z=z(x, y)$ in a neighborhood of (a, b) [i.e. $F(x, y, z(x, y))=0$] and

$$\frac{\partial z}{\partial x}=-\frac{F_x'}{F_z'} \qquad \frac{\partial z}{\partial y}=-\frac{F_y'}{F_z'}$$

Example

Find the Taylor polynomial of order 2 at origin of $z=z(x, y)$ with $z(0, 0)=0$ which satisfies $(*)$ $xy+xz+\sin z=0$.

Solution. $\frac{\partial}{\partial z}\{xy+xz+\sin z\}=x+\cos z=1$ at $(0, 0, 0) \Rightarrow z=z(x, y)$ exists in a neighborhood of $(x, y)=(0, 0)$. Differentiating $(*)$:

$D_x(*)$: $y+xz_x'+z+z_x' \cos z=0 \Rightarrow z_x'(0, 0)=0$
$D_y(*)$: $x+xz_y'+z_y' \cos z=0 \Rightarrow z_y'(0, 0)=0$
$D_{xy}(*)$: $1+xz''_{xy}+z_y'+z''_{xy} \cos z-z_x'z_y' \sin z=0 \Rightarrow z''_{xy}(0, 0)=-1$
etc.: By (10.1): $P_2(x, y)=-xy$

10.5 Extremes of Functions

All functions assumed to be differentiable.

Extremum in the interior of a domain

Necessary condition

a. $f: R^2 \to R$. Assume that $f(x, y)$ has a local maximum or minimum at an interior point (a, b) of D_f. Then

$$f_x'(a, b)=f_y'(a, b)=0$$

i.e. (a, b) is a *stationary point.*

b. $f: R^n \to R$. All partial derivatives$=0$ at an interior extremum point.

Sufficient conditions

a. $f: R^2 \to R$. Assume (a, b) is a stationary point.

Set $D=f''_{xx} f''_{yy}-(f''_{xy})^2$. Then

(*i*) $D(a, b)>0, f''_{xx}(a, b)>0 \Rightarrow (a, b)$ is a minimum point
(*ii*) $D(a, b)>0, f''_{xx}(a, b)<0 \Rightarrow (a, b)$ is a maximum point
(*iii*) $D(a, b)<0 \Rightarrow (a, b)$ is a saddle point
(*iv*) $D(a, b)=0$ tells nothing. Try Taylor expansion or consider the function directly.

b. $f: R^n \to R$. Assume $P=(a_1, \ldots, a_n)$ is a stationary point.

Set matrix $A=(a_{ij})$ where $a_{ij}=\dfrac{\partial^2}{\partial x_i \partial x_j} f(P)$. Then

(*i*) if $Q=\mathbf{x}^t A\mathbf{x}$ is positive (negative) definite [cf. sec. 4.6] then P is a minimum (maximum) point.

(*ii*) if $Q=\mathbf{x}^t A\mathbf{x}$ is indefinite [cf. sec. 4.6] then P is a saddle point.

Extremum with side conditions (constraints)

Problem. Find extrema of $f(x, y, z)$ with side condition $g(x, y, z)=0$.

1. Substitution

Solving for (e.g.) z in the last relation gives the problem of finding interior extremas of $h(x, y)=f(x, y, z(x, y))$.

2. Lagrange's multipliers

a. *Special case.*

Necessary condition for maximum or minimum of $f(x, y)$ with side condition $g(x, y)=0$:

$$\begin{cases} f_x'+\lambda g_x'=0 \\ f_y'+\lambda g_y'=0 \\ g=0 \end{cases} \quad \text{or} \quad \begin{cases} g_x'=0 & \text{(degenerate case)} \\ g_y'=0 \\ g=0 \end{cases}$$

b. *General case.*

Necessary condition for maximum or minimum of $f(x_1, \ldots, x_n)$ with side conditions

$$g_1(x_1, \ldots, x_n)=0, \ldots, g_k(x_1, \ldots, x_n)=0, \; k<n$$

$$\begin{cases} \dfrac{\partial}{\partial x_i}(f+\lambda_1 g_1+\ldots+\lambda_k g_k)=0, \; i=1, \ldots, n \\ g_j=0, \; j=1, \ldots, k \end{cases}$$

or $\left\{ \dfrac{\partial(g_1, \ldots, g_k)}{\partial(x_{i_1}, \ldots, x_{i_k})}=\ldots=0 \right.$ [all $\binom{n}{k}$ functional determinants of $g_1, \ldots, g_k$ with respect to k of the variables $x_1, \ldots, x_n$].

(degenerate case)

3. Method of functional determinants

Problem as 2b. Solution from

$\left\{ \dfrac{\partial(f, g_1, \ldots, g_k)}{\partial(x_{i_1}, \ldots, x_{i_{k+1}})}=\ldots=0 \right.$ [all $\binom{n}{k+1}$ functional determinants of $f, g_1, \ldots, g_k$ with respect to $k+1$ of the variables $x_1, \ldots, x_n$].

10.6 Functions f: $R^n \to R^m$ ($R^n \to R^n$)

Functions f: $R^n \to R^m$

Notation. $\boldsymbol{y}=\boldsymbol{f}(\boldsymbol{x})=(f_1(\boldsymbol{x}), \ldots, f_m(\boldsymbol{x}))^t$, $\boldsymbol{x}=(x_1, \ldots, x_n)\in \boldsymbol{R}^n$, $\boldsymbol{y}=(y_1, \ldots, y_m)\in \boldsymbol{R}^m$.

Limit. $\lim\limits_{\boldsymbol{x}\to\boldsymbol{a}} \boldsymbol{f}(\boldsymbol{x})=\boldsymbol{A}=(A_1, \ldots, A_m) \Leftrightarrow \lim\limits_{\boldsymbol{x}\to\boldsymbol{a}} f_k(\boldsymbol{x})=A_k,\ k=1, \ldots, m.$

Continuity. $\boldsymbol{f}(\boldsymbol{x})$ continuous at $\boldsymbol{a}\in D_f$ if $\lim\limits_{\boldsymbol{x}\to\boldsymbol{a}} \boldsymbol{f}(\boldsymbol{x})=\boldsymbol{f}(\boldsymbol{a})$.

Total derivative. $D\boldsymbol{f}(\boldsymbol{x})=\boldsymbol{f}'(\boldsymbol{x})=\begin{bmatrix} \dfrac{\partial f_1}{\partial x_1} & \cdots & \dfrac{\partial f_1}{\partial x_n} \\ \cdots & & \\ \dfrac{\partial f_m}{\partial x_1} & \cdots & \dfrac{\partial f_m}{\partial x_n} \end{bmatrix}$ ($m\times n$−matrix)

Differentiability (*Linear Approximation*). $\dfrac{\partial f_i}{\partial x_j}$ continuous $\Rightarrow$

$$\boldsymbol{f}(\boldsymbol{a}+\boldsymbol{h})-\boldsymbol{f}(\boldsymbol{a})=\boldsymbol{f}'(\boldsymbol{a})\boldsymbol{h}+|\boldsymbol{h}|\varepsilon(\boldsymbol{h}) \text{ where } |\varepsilon(\boldsymbol{h})| \to 0 \text{ as } \boldsymbol{h} \to \boldsymbol{0}.$$

Differential. $d\boldsymbol{f}=\boldsymbol{f}'(\boldsymbol{a})\boldsymbol{h}$, $\boldsymbol{h}=(h_1, \ldots, h_n)^t$

Chain rule. $\boldsymbol{g}$: $\boldsymbol{R}^n \to \boldsymbol{R}^m$, $\boldsymbol{f}$: $\boldsymbol{R}^m \to \boldsymbol{R}^p$:

$$D\boldsymbol{f}(\boldsymbol{g}(\boldsymbol{x}))=\boldsymbol{f}'(\boldsymbol{g}(\boldsymbol{x}))\boldsymbol{g}'(\boldsymbol{x}) \text{ [matrix multiplication].}$$

Mean value theorem. ($\boldsymbol{f}$ differentiable). For any $\boldsymbol{v}\in \boldsymbol{R}^m$ there exists $0<\theta<1$ such that

$$[\boldsymbol{f}(\boldsymbol{a}+\boldsymbol{h})-\boldsymbol{f}(\boldsymbol{a})]\cdot \boldsymbol{v}=\boldsymbol{f}'(\boldsymbol{a}+\theta\boldsymbol{h})\boldsymbol{h}\cdot \boldsymbol{v} \quad \text{(scalar product)}$$

The implicit function theorem

Assume that

(*i*) $f_k(x_1, \ldots, x_n, y_1, \ldots, y_m)$, $k=1, \ldots, m$, have continuous partial derivatives with respect to $y_1, \ldots, y_m$ in a neighborhood of the point $\boldsymbol{p}=(a_1, \ldots, a_n, b_1, \ldots, b_m)\in \boldsymbol{R}^{n+m}$,

(*ii*) $f_k(\boldsymbol{p})=0$, $k=1, \ldots, m$,

(*iii*) $\dfrac{\partial(f_1, \ldots, f_m)}{\partial(y_1, \ldots, y_m)} \neq 0$ at $\boldsymbol{p}$.

Then

(*i*) the system

$$(10.2) \quad \begin{cases} f_1(x_1, \ldots, x_n, y_1, \ldots, y_m)=0 \\ \cdots \\ f_m(x_1, \ldots, x_n, y_1, \ldots, y_m)=0 \end{cases}$$

defines m functions $y_k=y_k(x_1, \ldots, x_n)$, $k=1, \ldots, m$ in a neighborhood of $(a_1, \ldots, a_n)$,

(*ii*) $\dfrac{\partial y_k}{\partial x_j}$, $k=1, \ldots, m$, may be calculated by solving the linear system arising when differentiating (10.2) with respect to x_j.

Functions $f: R^n \to R^n$

Notation. $\boldsymbol{y}=\boldsymbol{y}(\boldsymbol{x})$: $\begin{cases} y_1=y_1(x_1, \ldots, x_n) \\ \ldots \\ y_n=y_n(x_1, \ldots, x_n) \end{cases}$

Functional (Jacobian) matrix. $\left(\dfrac{\partial y_i}{\partial x_j}\right) = \begin{bmatrix} \dfrac{\partial y_1}{\partial x_1} & \cdots & \dfrac{\partial y_1}{\partial x_n} \\ \cdots & & \\ \dfrac{\partial y_n}{\partial x_1} & \cdots & \dfrac{\partial y_n}{\partial x_n} \end{bmatrix}$

Functional (Jacobian) determinant. $J=\dfrac{\partial(y_1, \ldots, y_n)}{\partial(x_1, \ldots, x_n)}=\det\left(\dfrac{\partial y_i}{\partial x_j}\right) = \begin{vmatrix} \dfrac{\partial y_1}{\partial x_1} & \cdots & \dfrac{\partial y_1}{\partial x_n} \\ \cdots & & \\ \dfrac{\partial y_n}{\partial x_1} & \cdots & \dfrac{\partial y_n}{\partial x_n} \end{vmatrix}$

1. *Chain rule.* $\left(\dfrac{\partial z_i}{\partial x_j}\right) = \left(\dfrac{\partial z_i}{\partial y_j}\right)\left(\dfrac{\partial y_i}{\partial x_j}\right)$, matrix multiplication
2. $\left(\dfrac{\partial x_i}{\partial y_j}\right) = \left(\dfrac{\partial y_i}{\partial x_j}\right)^{-1}$, matrix inversion
3. $\dfrac{\partial(z_1, \ldots, z_n)}{\partial(x_1, \ldots, x_n)}=\dfrac{\partial(z_1, \ldots, z_n)}{\partial(y_1, \ldots, y_n)} \cdot \dfrac{\partial(y_1, \ldots, y_n)}{\partial(x_1, \ldots, x_n)}$
4. $\dfrac{\partial(x_1, \ldots, x_n)}{\partial(y_1, \ldots, y_n)}=1/\dfrac{\partial(y_1, \ldots, y_n)}{\partial(x_1, \ldots, x_n)}$

Examples.

1. $\begin{cases} x=r\cos\theta \\ y=r\sin\theta \end{cases} \Rightarrow \dfrac{\partial(x, y)}{\partial(r, \theta)}=r$
2. $\begin{cases} x=r\sin\theta\cos\varphi \\ y=r\sin\theta\sin\varphi \\ z=r\cos\theta \end{cases} \Rightarrow \dfrac{\partial(x, y, z)}{\partial(r, \theta, \varphi)}=r^2\sin\theta$
3. $\begin{cases} x=au+bv \\ y=cu+dv \end{cases} \Rightarrow \dfrac{\partial(x, y)}{\partial(u, v)}=ad-bc, \quad \dfrac{\partial(u, v)}{\partial(x, y)}=\dfrac{1}{ad-bc}$

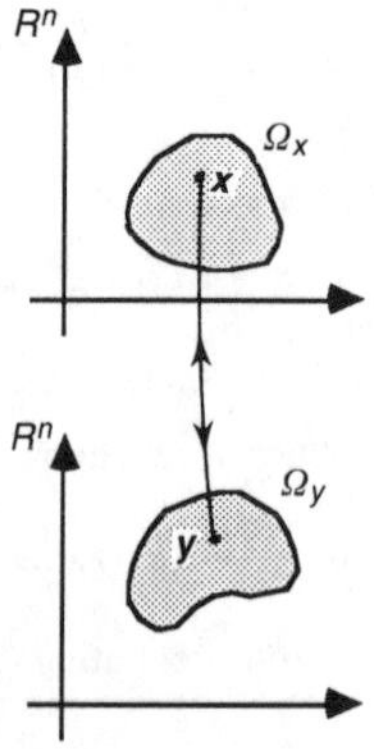

Local volume (area) scale

Set $m(\Omega)$=volume (or area) of Ω and assume that Ω_x and Ω_y are in one-to-one correspondence by

$$\begin{cases} y_1=y_1(x_1, \ldots, x_n) \\ \ldots \\ y_n=y_n(x_1, \ldots, x_n). \text{ Then} \end{cases}$$

1. $\dfrac{m(\Omega_y)}{m(\Omega_x)} \to \left|\dfrac{\partial(y_1, \ldots, y_n)}{\partial(x_1, \ldots, x_n)}\right|$ as $m(\Omega_x) \to 0$

2. $m(\Omega_y) \approx \left|\dfrac{\partial(y_1, \ldots, y_n)}{\partial(x_1, \ldots, x_n)}\right| \cdot m(\Omega_x)$ for "small" Ω_x and Ω_y.

Gradient. f: $\boldsymbol{R}^n \to \boldsymbol{R}$: grad $f = \nabla f = \left(\dfrac{\partial f}{\partial x_1}, \ldots, \dfrac{\partial f}{\partial x_n}\right)$

Divergence. $\boldsymbol{f}$: $\boldsymbol{R}^n \to \boldsymbol{R}^n$: div$\boldsymbol{f} = \nabla \cdot \boldsymbol{f} = \dfrac{\partial f_1}{\partial x_1} + \ldots + \dfrac{\partial f_n}{\partial x_n}$

Rotation. $\boldsymbol{f}$: $\boldsymbol{R}^3 \to \boldsymbol{R}^3$: curl$\boldsymbol{f}$=rot$\boldsymbol{f} = \nabla \times \boldsymbol{f} = \left(\dfrac{\partial f_3}{\partial x_2} - \dfrac{\partial f_2}{\partial x_3}, \dfrac{\partial f_1}{\partial x_3} - \dfrac{\partial f_3}{\partial x_1}, \dfrac{\partial f_2}{\partial x_1} - \dfrac{\partial f_1}{\partial x_2}\right)$

Formulas involving grad, div, curl, see sec. 11.2.

The inverse function theorem

Assume (*i*) $\boldsymbol{y}=\boldsymbol{f}(\boldsymbol{x})$ is continuously differentiable

(*ii*) $J = \dfrac{\partial(y_1, \ldots, y_n)}{\partial(x_1, \ldots, x_n)} \neq 0$ at $\boldsymbol{a}$.

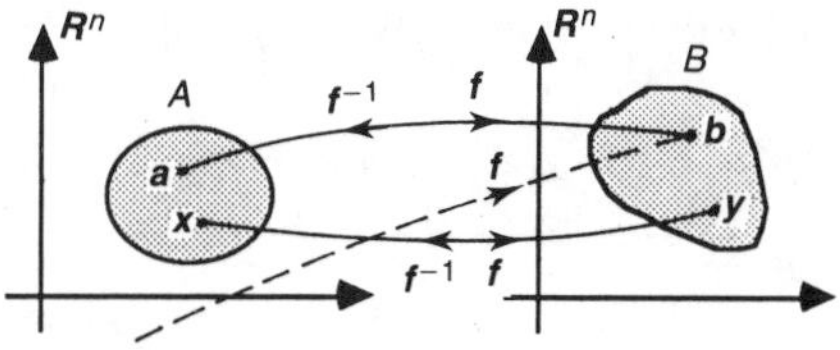

Then there exist open sets A, B ($\boldsymbol{a} \in A$, $\boldsymbol{b}=\boldsymbol{f}(\boldsymbol{a}) \in B$) and a uniquely determined inverse function $\boldsymbol{f}^{-1}$: $B \to A$ such that

$\boldsymbol{f}$: $\begin{cases} y_1 = y_1(x_1, \ldots, x_n) \\ \ldots \\ y_n = y_n(x_1, \ldots, x_n) \end{cases}$ $\quad \boldsymbol{f}^{-1}$: $\begin{cases} x_1 = x_1(y_1, \ldots, y_n) \\ \ldots \\ x_n = x_n(y_1, \ldots, y_n) \end{cases}$

(*i*) $\boldsymbol{f}$, ($\boldsymbol{f}^{-1}$) are one-to-one on A, (B).
(*ii*) $\boldsymbol{f}^{-1}$ is continuously differentiable and

(a) $\left(\dfrac{\partial x_i}{\partial y_j}\right) = \left(\dfrac{\partial y_i}{\partial x_j}\right)^{-1}$ (matrix inversion)

or $D\boldsymbol{f}^{-1}(\boldsymbol{y}) = [D\boldsymbol{f}(\boldsymbol{x})]^{-1}$

(b) $\dfrac{\partial(x_1, \ldots, x_n)}{\partial(y_1, \ldots, y_n)} = 1/\dfrac{\partial(y_1, \ldots, y_n)}{\partial(x_1, \ldots, x_n)}$

10.7 Double Integrals

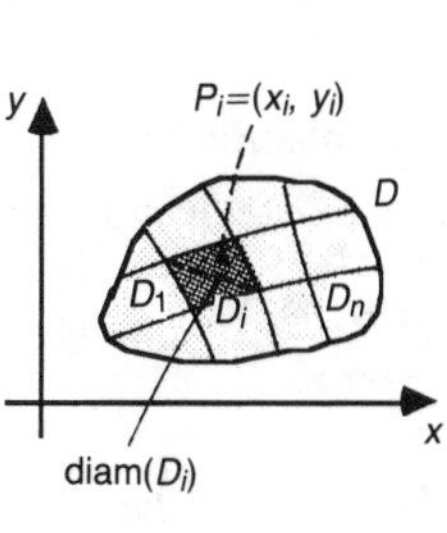

Let D be a bounded (measurable) set, $f(x, y)$ continuous and bounded on D, $A(D)$=area of D.

Riemann sum

$$s_n = \sum_{i=1}^{n} f(x_i, y_i)\, A(D_i) \to \iint_D f(x, y)\, dxdy \text{ as } \max_i \text{diam}\,(D_i) \to 0$$

Iterated integration

$$\iint_D f(x, y)\, dxdy = \int_a^b \left[\int_{\alpha(x)}^{\beta(x)} f(x, y)\, dy \right] dx =$$

$$= \int_a^b dx \int_{\alpha(x)}^{\beta(x)} f(x, y)\, dy$$

(Analogously with variables changed).

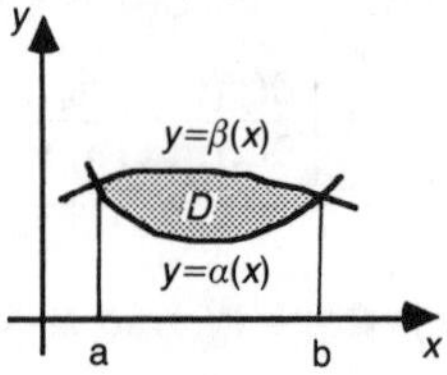

Example

$$\iint_D 6x^2y\, dxdy = \int_0^1 x^2dx \int_{x^2}^{x} 6y\, dy = \int_0^1 x^2dx[3y^2]_{x^2}^{x} =$$

$$= 3\int_0^1 x^2(x^2 - x^4)dx = 6/35$$

$$\iint_D 6x^2y\, dxdy = \int_0^1 y\, dy \int_y^{\sqrt{y}} 6x^2dx = \int_0^1 y\, dy[2x^3]_y^{\sqrt{y}} =$$

$$= 2\int_0^1 y(y^{3/2} - y^3)dy = 6/35$$

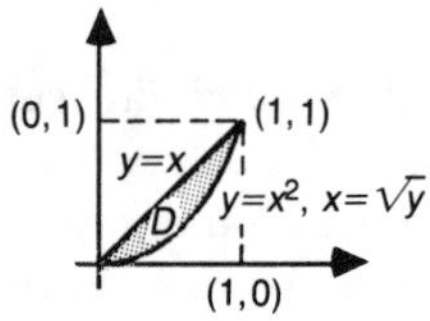

Substitution

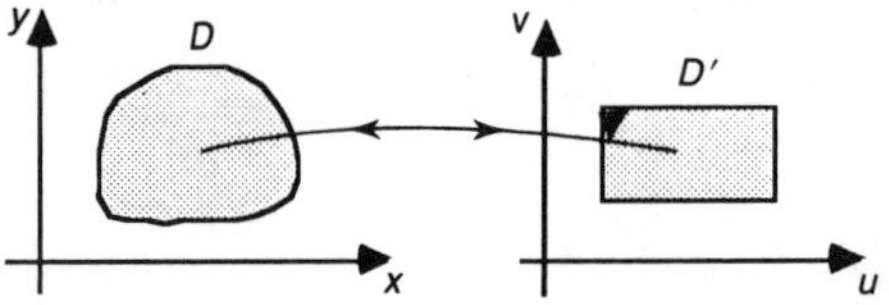

$D \leftrightarrow D'$ one-to-one by

$$\begin{cases} x = x(u, v) \\ y = y(u, v) \end{cases} \Leftrightarrow \begin{cases} u = u(x, y) \\ v = v(x, y) \end{cases}$$

$$J = \frac{\partial(x, y)}{\partial(u, v)} = \begin{vmatrix} x_u' & x_v' \\ y_u' & y_v' \end{vmatrix} \neq 0$$

$$\iint_D f(x, y)dxdy = \iint_{D'} f(x(u, v), y(u, v)) \left| \frac{\partial(x, y)}{\partial(u, v)} \right| dudv$$

Special substitutions

1. *Polar coordinates*

$$\begin{cases} x = r\cos\theta \\ y = r\sin\theta \end{cases}, \quad dxdy = rdrd\theta$$

$$\iint_D f(x, y)\, dxdy = \int_\alpha^\beta d\theta \int_0^{r(\theta)} f(r\cos\theta, r\sin\theta)\, r\, dr$$

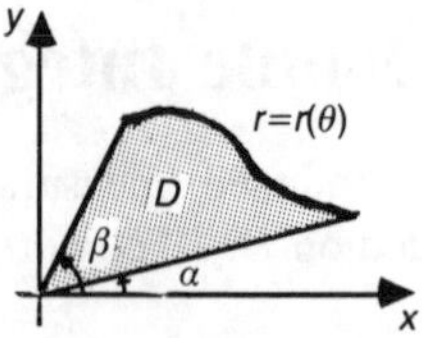

2. *Linear transformation*

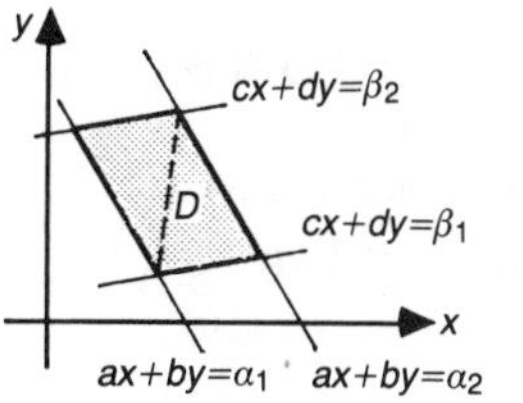

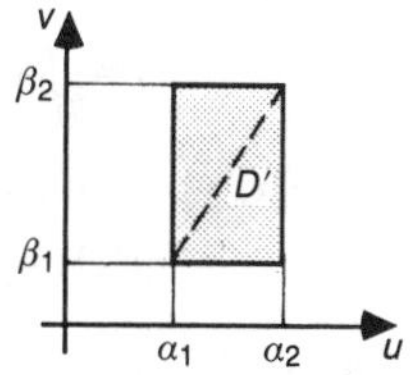

$$\begin{cases} u=ax+by \\ v=cx+dy \end{cases}, \quad \frac{\partial(x,\ y)}{\partial(u,\ v)}=\frac{1}{ad-bc}$$

$$\iint_D f(x,\ y)\ dxdy=\frac{1}{|ad-bc|}\iint_{D'} f(x(u,\ v),\ y(u,\ v))\ dudv$$

Improper integrals

Assume $f(x,\ y)\geqslant 0$ and f unbounded on D or D unbounded. Let $\{D_n\}_1^{\infty}$ satisfy: (*i*) D_n bounded and f bounded on D_n (*ii*) $D_n\subset D_{n+1}$ (*iii*) $D=\bigcup_{n=1}^{\infty} D_n$. Then (definition)

$$\iint_D f(x,\ y)dxdy=\lim_{n\to\infty}\iint_{D_n} f(x,\ y)dxdy$$

Fubini's theorem

Assume $f(x,\ y)\geqslant 0$ in D: $a<x<b,\ c<x<d$, where $a,\ b,\ c,\ d$ are finite or infinite. Then, if one of the integrals $I=\iint_D f\,dxdy$, $I_1=\int_a^b dx\int_c^d f\ dy$, $I_2=\int_c^d dy\int_a^b f\,dx$ is convergent, so are all of them and $I=I_1=I_2$.

Example

$$I=\int_{-\infty}^{\infty} e^{-x^2}dx=\sqrt{\pi},\text{ because}$$

$$I^2=\int_{-\infty}^{\infty} e^{-x^2}dx\int_{-\infty}^{\infty} e^{-y^2}dy=\iint_{R^2} e^{-x^2-y^2}dxdy=$$

$$=\int_0^{2\pi} d\theta\int_0^{\infty} re^{-r^2}dr=2\pi\left[-\frac{1}{2}e^{-r^2}\right]_0^{\infty}=\pi$$

Applications

Geometry

1. *Area.* $A(D)=\iint_D dxdy$

2. *Volume V*

$$V=\iint_D [f(x,\ y)-g(x,\ y)]\ dxdy$$

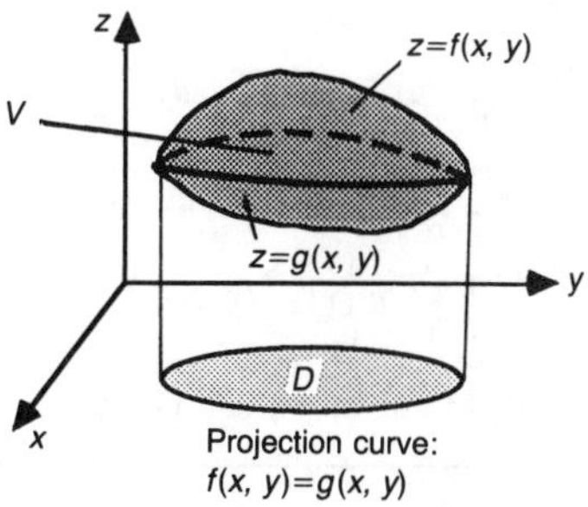

3. *Area of surface* $z=f(x, y)$ over D:

$$A=\iint_D \sqrt{1+f_x^2+f_y^2}\, dxdy$$

Mass $\varrho(x, y)$=density of plane region D

$$m(D)=\iint_D \varrho(x, y)\, dxdy$$

Centroid $C=(x_c, y_c)$

(*i*) Homogeneous body.

$$\begin{cases} x_c=\dfrac{1}{A(D)}\displaystyle\iint_D x\, dxdy \\ y_c=\dfrac{1}{A(D)}\displaystyle\iint_D y\, dxdy \end{cases}$$

(*ii*) Density distribution $\varrho(x, y)$.

$$\begin{cases} x_c=\dfrac{1}{m(D)}\displaystyle\iint_D x\varrho(x, y)\, dxdy \\ y_c=\dfrac{1}{m(D)}\displaystyle\iint_D y\varrho(x, y)\, dxdy \end{cases}$$

Moments of inertia ($\varrho(x, y)$=density)

(*i*) about the x-axis: $I_x=\iint_D y^2\varrho(x, y)\, dxdy$

(*ii*) about the y-axis: $I_y=\iint_D x^2\varrho(x, y)\, dxdy$

(*iii*) polar moment about the origin: $I_0=I_x+I_y=\iint_D r^2\varrho(x, y)\, dxdy$

10.8 Triple Integrals

Iterated integration

There are two possibilities (cf. the example below):

1. $\Omega=\{(x, y, z): \varphi(x, y)\leqslant z\leqslant\psi(x, y), (x, y)\in D\}$

$$\iiint_\Omega f d\Omega=\iiint_\Omega f(x, y, z)\, dxdydz=\iint_D dxdy\int_{\varphi(x, y)}^{\psi(x, y)} f(x, y, z)dz$$

2. $\Omega=\{(x, y, z): (x, y)\in D_z, a\leqslant z\leqslant b\}$.

$$\iiint_\Omega f d\Omega=\iiint_\Omega f(x, y, z)\, dxdydz=\int_a^b dz\iint_{D_z} f(x; y, z)\, dxdy$$

Example

Ω=cone: $\sqrt{x^2+y^2}\leqslant z\leqslant 1$

1. $\iiint_\Omega z\, dxdydz=\iint_D dxdy\int_{\sqrt{x^2+y^2}}^{1} z\, dz=$

$=\frac{1}{2}\iint_D(1-x^2-y^2)\, dxdy$=[pol. coord.]=

$=\frac{1}{2}\int_0^{2\pi} d\theta\int_0^1(1-r^2)r\, dr=\frac{\pi}{4}$

2. $\iiint_\Omega z\, dxdydz=\int_0^1 zdz\iint_{D_z} dxdy=\int_0^1 z\cdot\pi z^2dz=\frac{\pi}{4}$

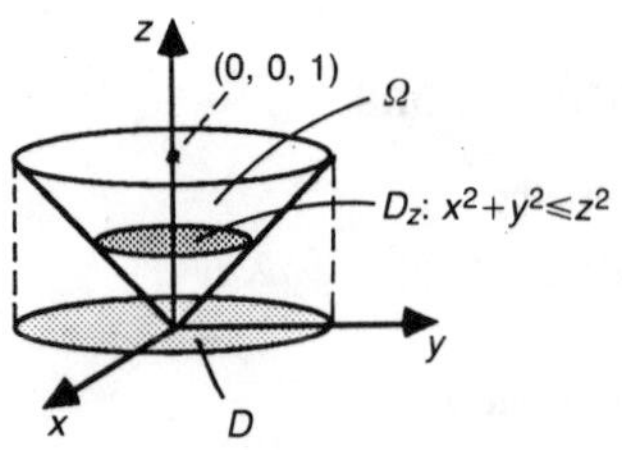

Substitution

Assume that there is a one-to-one correspondence between Ω in the (x, y, z)-space and Ω' in the (u, v, w)-space, and that

$$\frac{\partial(x, y, z)}{\partial(u, v, w)} \neq 0.$$

Then

$$\iiint_{\Omega} f(x, y, z)\, dxdydz = \iiint_{\Omega'} f(x, y, z) \left| \frac{\partial(x, y, z)}{\partial(u, v, w)} \right| dudvdw$$

Special substitutions

1. *Spherical coordinates*

$$\begin{cases} x = r \sin\theta \cos\varphi \\ y = r \sin\theta \sin\varphi \\ z = r \cos\theta \end{cases} \quad dxdydz = r^2 \sin\theta\, dr d\theta d\varphi$$

$$x^2 + y^2 + z^2 = r^2$$

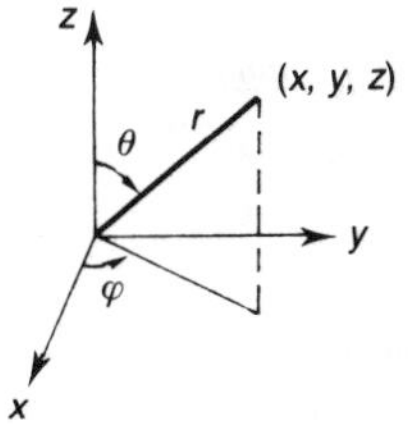

2. *Cylindrical coordinates*

$$\begin{cases} x = r \cos\varphi \\ y = r \sin\varphi \\ z = z \end{cases} \quad dxdydz = r dr d\varphi dz$$

$$x^2 + y^2 = r^2$$

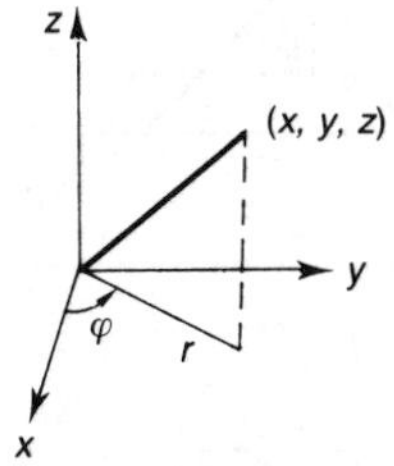

Applications

Geometry

Volume $V(\Omega) = \iiint_{\Omega} dxdydz$

Mass. $\varrho(x, y, z)$=density distribution of body Ω.

$$m(\Omega) = \iiint_{\Omega} \varrho(x, y, z)\, dxdydz$$

Centroid $C = (x_c, y_c, z_c)$

(*i*) Homogeneous body: $x_c = \frac{1}{V(\Omega)} \iiint_{\Omega} x\, dxdydz,$ $\quad y_c, z_c$ similarly.

(*ii*) Density $\varrho(x, y, z)$: $x_c = \frac{1}{m(\Omega)} \iiint_{\Omega} x\, \varrho(x, y, z)\, dxdydz,$ $\quad y_c, z_c$ similarly.

Moments of inertia ($\varrho = \varrho(x, y, z)$=density, $I = \int r^2 dm$)

(*i*) about the x-axis: $I_x = \iiint_{\Omega} \varrho(y^2 + z^2)\, dxdydz$

(*ii*) about the y-axis: $I_y = \iiint_{\Omega} \varrho(z^2 + x^2)\, dxdydz$

(*iii*) about the z-axis: $I_z = \iiint_{\Omega} \varrho(x^2. + y^2.)\, dxdydz$

(*iv*) polar moment about the origin: $I_0 = \frac{1}{2}(I_x + I_y + I_z)$

Table of centroids and moments of inertia

Body (homogeneous)	**Centroid**	**Moments of inertia** m=**mass of body**
Straight bar 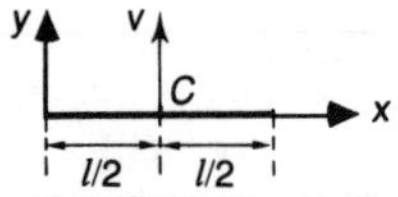	$x_c = l/2$ $y_c = 0$	$I_x = 0$ $I_y = \frac{ml^2}{3}$ $I_v = \frac{ml^2}{12}$
Rectangle 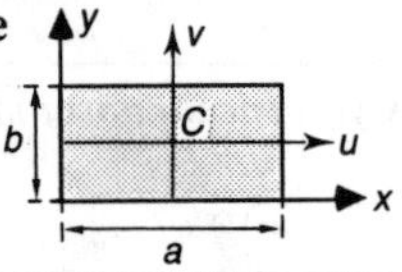	$x_c = \frac{a}{2}$ $y_c = \frac{b}{2}$	$I_x = \frac{mb^2}{3}$ $I_y = \frac{ma^2}{3}$ $I_u = \frac{mb^2}{12}$ $I_v = \frac{ma^2}{12}$
Triangle 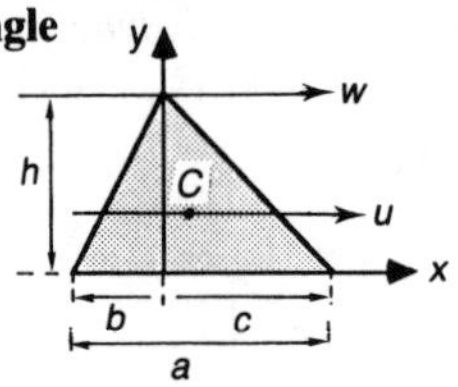	$x_c = \frac{c-b}{3}$ $y_c = \frac{h}{3}$	$I_x = \frac{mh^2}{6}$ $I_y = \frac{m(b^3+c^3)}{6a}$ $I_u = \frac{mh^2}{18}$ $I_w = \frac{mh^2}{2}$
Circle 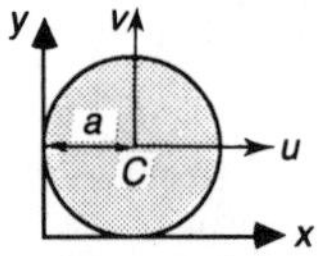	$x_c = a$ $y_c = a$	$I_x = I_y = \frac{5ma^2}{4}$ $I_u = I_v = \frac{ma^2}{4}$
Circular sector 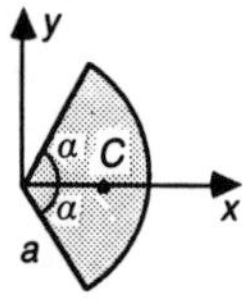	$x_c = \frac{2a \sin \alpha}{3\alpha}$ $y_c = 0$	$I_x = \frac{ma^2}{4}\left(1 - \frac{\sin 2\alpha}{2\alpha}\right)$ $I_y = \frac{ma^2}{4}\left(1 + \frac{\sin 2\alpha}{2\alpha}\right)$
Hollow circle 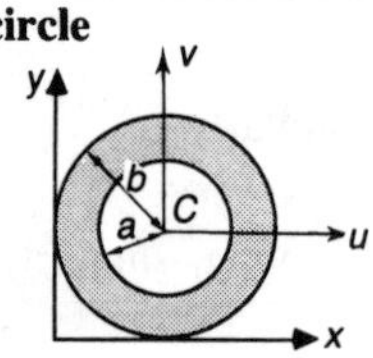	$x_c = b$ $y_c = b$	$I_x = I_y = I_u + mb^2$ $I_u = I_v = \frac{m(a^2+b^2)}{4}$
Prism 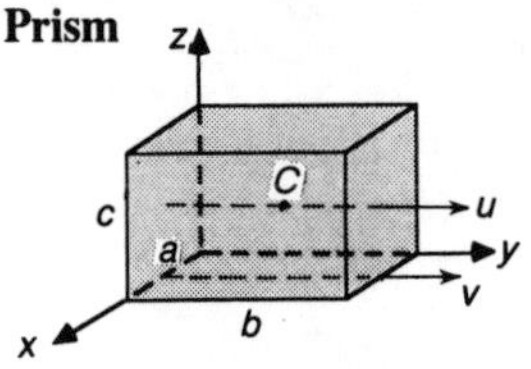	$x_c = a/2$ $y_c = b/2$ $z_c = c/2$	$I_y = \frac{m}{3}(a^2+c^2)$ $I_u = \frac{m}{12}(a^2+c^2)$ $I_v = \frac{m}{12}(a^2+4c^2)$

Body (homogeneous)	**Centroid**	**Moments of inertia** **m=mass of body**
Sphere 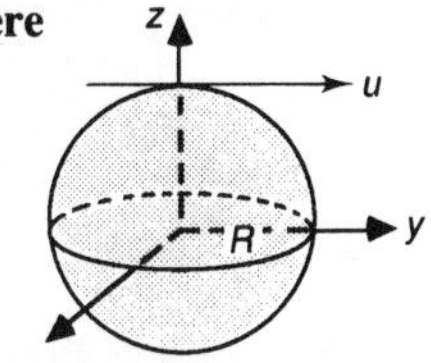	$x_c=y_c=z_c=0$	$I_x=I_y=I_z=\frac{2mR^2}{5}$ $I_u=\frac{7mR^2}{5}$
Spherical shell	$x_c=y_c=z_c=0$	$I_x=I_y=I_z=\frac{2mR^2}{3}$ $I_u=\frac{5mR^2}{3}$
Cylinder 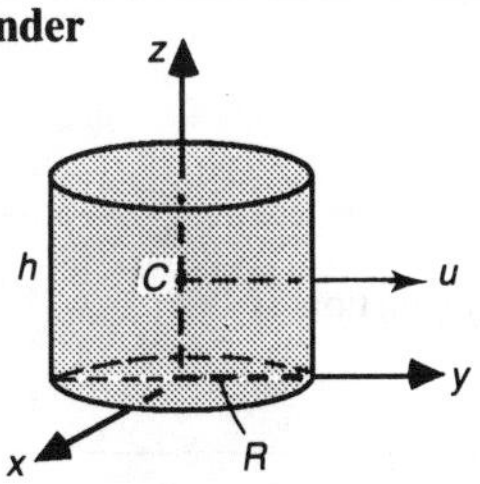	$x_c=y_c=0$ $z_c=h/2$	$I_x=I_y=\frac{m}{12}(3R^2+4h^2)$ $I_z=\frac{mR^2}{2}$ $I_u=\frac{m}{12}(3R^2+h^2)$
Cylindrical shell (open)	$x_c=y_c=0$ $z_c=h/2$	$I_x=I_y=\frac{m}{6}(3R^2+2h^2)$ $I_z=mR^2$ $I_u=\frac{m}{12}(6R^2+h^2)$
Cone 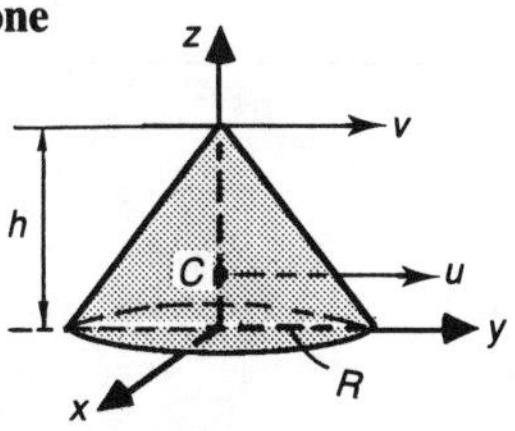	$x_c=y_c=0$ $z_c=h/4$	$I_x=I_y=\frac{m}{20}(3R^2+2h^2)$ $I_z=\frac{3mR^2}{10}$ $I_u=\frac{3m}{80}(4R^2+h^2)$ $I_v=\frac{3m}{20}(R^2+4h^2)$
Conical shell (open)	$x_c=y_c=0$ $z_c=2h/3$	$I_v=\frac{m}{4}(R^2+2h^2)$ $I_u=\frac{m}{18}(9R^2+10h^2)$ $I_z=\frac{mR^2}{2}$

10.9 Partial Differential Equations

A partial differential equation (*PDE*) is an equation involving partial derivatives of an unknown function of two or more independent variables.

First order (quasi) linear equations

(10.3) $\quad a(x, y)u_x'+b(x, y)u_y'=c(x, y, u), \qquad u=u(x, y)$

For general solution:

(*i*) Find the characteristic curves, $\dfrac{dy}{dx}=\dfrac{b(x, y)}{a(x, y)}$ with the general solution $\xi(x, y)=C$.

(*ii*) Make the coordinate transformation

$$\begin{cases}\xi=\xi(x, y)\\ \eta=\text{a suitable function of } x, y \text{ (e.g. } \eta=x \text{ or } \eta=y)\end{cases}$$

(*iii*) The equation (10.3) will take the form

$$(a\eta_x'+b\eta_y')\frac{\partial u}{\partial \eta}=c,$$

which may be solved as an ordinary differential equation.

Remark. The general solution $u=u(\xi, \eta)$ will contain an arbitrary function of ξ.

Example.

$$xu_x'+yu_y'=u.$$

Characteristics: $\dfrac{dy}{dx}=\dfrac{y}{x} \Rightarrow \displaystyle\int\frac{dy}{y}=\int\frac{dx}{x} \Rightarrow \frac{y}{x}=C$

Setting $\xi=y/x$, $\eta=x$, the given *PDE* will be transformed to

$$\eta\frac{\partial u}{\partial \eta}=u.$$

Separation of variables gives $u=\eta f(\xi)=xf\left(\dfrac{y}{x}\right)$.

Second order (quasi) linear equations

(10.4) $\quad a(x, y)u''_{xx}+2b(x, y)u''_{xy}+c(x, y)u''_{yy}=f(x, y, u, u_x', u_y')$

Classification of (10.4)

1. Elliptic if $ac-b^2>0$ (e.g. $\Delta u=u''_{xx}+u''_{yy}=0$, the Laplace equation)
2. Parabolic if $ac-b^2=0$ (e.g. $u_t'=\alpha^2u''_{xx}$, the heat equation)
3. Hyperbolic if $ac-b^2<0$ (e.g. $u''_{tt}=c^2u''_{xx}$, the wave equation)

Characteristics

$$a\left(\frac{dy}{dx}\right)^2-2b\frac{dy}{dx}+c=0 \Rightarrow \frac{dy}{dx}=\frac{1}{a}\left(b\pm\sqrt{b^2-ac}\right)$$

Thus, if (10.4) is (*i*) elliptic, there are no real characteristics, (*ii*) parabolic, one family, (*iii*) hyperbolic, two families of characteristic curves.

Examples of initial and boundary value problems

The wave equation

Example 1. $u''_{tt}-c^2u''_{xx}=0$, c=constant

The transformation $\xi=x+ct$, $\eta=x-ct$ gives $u_{\xi\eta}=0$ with general solution $u=f(\xi)+g(\eta)=$ $=f(x+ct)+g(x-ct)$.

The initial value problem

$$\begin{cases} u''_{tt}=c^2u''_{xx}, & t>0, \ -\infty<x<\infty \\ u(x,0)=\varphi(x), & -\infty<x<\infty \\ u_t'(x,0)=\psi(x), & -\infty<x<\infty \end{cases}$$

has the solution

$$u(x,t)=\frac{1}{2}\,[\varphi(x+ct)+\varphi(x-ct)]+\frac{1}{2c}\int_{x-ct}^{x+ct}\psi(s)ds$$

(*d'Alembert's formula*)

The Dirichlet problem

The problem (u continuous in $\bar{\Omega}$)

$$\begin{cases} \Delta u=u''_{xx}+u''_{yy}=0 & \text{in } \Omega \\ u=f & \text{on } \partial\Omega \ (f \text{ continuous}) \end{cases}$$

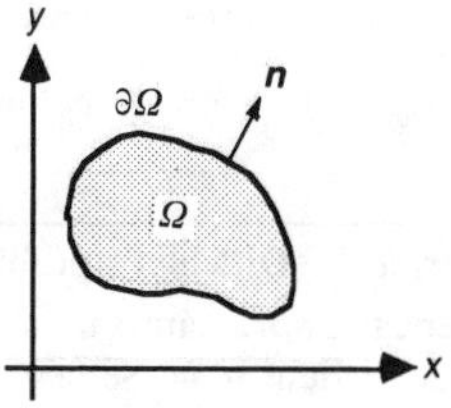

admits a unique solution.

Poisson's integral formulas

1. Ω: The unit disc. Solution:

$$u=u(r,\theta)=\frac{1}{2\pi}\int_0^{2\pi}\frac{(1-r^2)\,f(\varphi)d\varphi}{1-2r\cos(\theta-\varphi)+r^2}$$

2. Ω: The upper half plane. Solution:

$$u(x,y)=\frac{1}{\pi}\int_{-\infty}^{\infty}\frac{yf(t)}{y^2+(x-t)^2}\,dt$$

3. Ω arbitrary: Cf. conformal mapping (sec. 14.5).

The Neumann problem

$$\text{(10.5)} \qquad \begin{cases} \Delta u=0 \text{ in } \Omega \\ \dfrac{\partial u}{\partial n}=g \text{ on } \partial\Omega \end{cases}$$

Necessary for solution is $\oint_{\partial\Omega} g(s)ds=0$ and in that case the problem admits a unique solution (up to an additive constant).

Equivalent Dirichlet problem

u solves (10.5) $\Leftrightarrow$ v, the conjugate of u, solves

$$\begin{cases} \Delta v=0 \text{ in } \Omega \\ v=\int_a^s g(s)ds \text{ on } \partial\Omega \ (s=\text{arc length}). \end{cases}$$

Orthogonal series representation of solutions by separation of variables (Fourier's method)

Example 2. (Heat conduction in a bar)

$$\begin{cases} (PDE) & u_t'=\alpha^2 u''_{xx},\ t>0,\ 0<x<L \\ (BC) & u(0,t)=u(L,t)=0,\ t>0 \\ (IC) & u(x,0)=\varphi(x),\ 0\leq x\leq L \end{cases}$$

(*i*) Separation of variables: $u(x,t)=X(x)T(t) \Rightarrow$ [by (PDE)]

(10.6) $$\frac{T'(t)}{\alpha^2 T(t)}=\frac{X''(x)}{X(x)}=\lambda \qquad (\lambda=\text{separation constant})$$

(*ii*) $(BC) \Rightarrow \left.\begin{matrix} X''-\lambda X=0 \\ X(0)=X(L)=0 \end{matrix}\right\} \Rightarrow \begin{cases} X_n(x)=\sin\dfrac{n\pi x}{L},\ n=1,2,3,\ldots \text{ (eigenfunctions)} \\ \lambda_n=-\dfrac{n^2\pi^2}{L^2},\ n=1,2,3,\ldots \text{ (eigenvalues)} \end{cases}$

(*iii*) By (10.6), $T'+\dfrac{\alpha^2 n^2\pi^2}{L^2}T=0 \Rightarrow T_n(t)=c_n\, e^{-\alpha^2 n^2\pi^2 t/L^2}$

(*iv*) Set $u(x,t)=\sum_{n=1}^{\infty} T_n(t)X_n(x)=\sum_{n=1}^{\infty} c_n\, e^{-\alpha^2 n^2\pi^2 t/L^2}\sin\dfrac{n\pi x}{L}$

(*v*) By (IC), $\varphi(x)=\sum_{n=1}^{\infty} c_n \sin\dfrac{n\pi x}{L} \Rightarrow c_n=\dfrac{2}{L}\int_0^L \varphi(x)\sin\dfrac{n\pi x}{L}\,dx$

Example 3. (Dirichlet's problem for a sphere.)

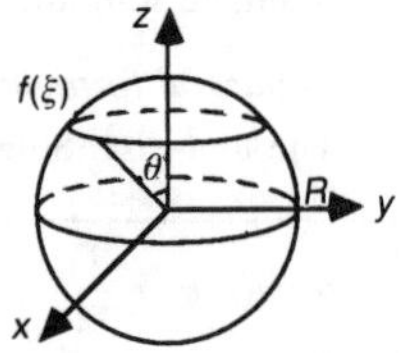

Spherical Coordinates: $u=u(r,\theta,\varphi)=u(r,\theta)$ assuming that u is independent of φ. Set $\xi=\cos\theta$:

$$\begin{cases} (PDE)\ \triangle u=\dfrac{1}{r^2}\dfrac{\partial}{\partial r}\left(r^2\dfrac{\partial u}{\partial r}\right)+\dfrac{1}{r^2}\dfrac{\partial}{\partial\xi}\left((1-\xi^2)\dfrac{\partial u}{\partial\xi}\right)=0,\ 0<r<R \\ (BC)\ \ u(R,\xi)=f(\xi) \qquad\qquad ,\ -1<\xi<1 \end{cases}$$

General solution of (PDE): $u(r,\xi)=\sum_{n=0}^{\infty}(A_n r^n+B_n r^{-n-1})\,P_n(\xi)$, where $P_n(\xi)$ are Legendre's polynomials (cf. sec. 12.2).

$A_n=\dfrac{2n+1}{2R^n}\int_{-1}^{1} f(\xi)P_n(\xi)d\xi$, B_n ($B_n=0$ if u is bounded for $r=0$) are determined as Legendre-Fourier coefficients by (BC). [cf. sec. 12.1]

Example 4. (Oscillations of a Circular Membrane.)

Polar Coordinates: $u=u(r,\varphi,t)=u(r,t)$ assuming that u is independent of φ:

$$\begin{cases} (PDE)\ \triangle u=u''_{rr}+\dfrac{1}{r}u_r'=\dfrac{1}{c^2}u''_{tt},\ 0<r<R,\ t>0 \\ (BC)\ \ u(R,t)=0,\ t>0 \\ (IC\ 1)\ u(r,0)=f(r),\quad 0\leq r\leq R \\ (IC\ 2)\ u_t'(r,0)=0,\quad 0\leq r\leq R \end{cases}$$

Separation of variables (α_n zeros of $J_0(x)$, see sec. 12.4) $\Rightarrow$

$$u(r,t)=\sum_{n=1}^{\infty}\left(A_n\cos\frac{c\alpha_n}{R}t+B_n\sin\frac{c\alpha_n}{R}t\right)J_0\left(\frac{\alpha_n}{R}r\right),$$

where J_0 is a Bessel function and $A_n=\dfrac{2}{R^2J_1(\alpha_n)^2}\int_0^R rf(r)J_0\left(\dfrac{\alpha_n r}{R}\right)dr$, $(B_n=0)$ are determined as Fourier-Bessel coefficients by (IC).

Transform representation of solutions

Example 5. ($U(\omega)$ Fourier transform of $u(x)$)

$$\begin{cases} (PDE)\ \triangle u = u''_{xx}+u''_{yy}=0, & -\infty<x<\infty,\ 0<y<1 \\ (BC)\ \ u(x,0)=f(x),\ u(x,1)=0, & -\infty<x<\infty \end{cases}$$

Set $u(x, y)=\frac{1}{2\pi}\int_{-\infty}^{\infty} U(\omega, y)e^{i\omega x}d\omega \Rightarrow$ [by PDE]

$$U_{yy}-\omega^2U=0 \Rightarrow U=A(\omega)\cosh\omega y+B(\omega)\sinh\omega y \Rightarrow$$

$$u(x, y)=\frac{1}{2\pi}\int_{-\infty}^{\infty}(A(\omega)\cosh\omega y+B(\omega)\sinh\omega y)e^{i\omega x}d\omega$$

$(BC1) \Rightarrow A(\omega)=F(\omega)$

$(BC2) \Rightarrow B(\omega)=-A(\omega)\dfrac{\cosh\omega}{\sinh\omega}=-F(\omega)\dfrac{\cosh\omega}{\sinh\omega}$

$$\Rightarrow u(x, y)=\frac{1}{2\pi}\int_{-\infty}^{\infty}F(\omega)\frac{\sinh\omega(1-y)}{\sinh\omega}e^{i\omega x}d\omega$$

Example 6. ($U(s)$ Laplace transform of $u(t)$)

$$\begin{cases} (PDE)\ u''_{xx}=u_t',\ x>0,\ t>0 \\ (BC)\ \ u_x'(0, t)=f(t),\ \lim\limits_{x\to\infty}u(x, t)=0,\ t>0 \\ (IC)\ \ u(x, 0)=0,\ x>0 \end{cases}$$

Laplace Transformation of $(PDE) \Rightarrow$

$$\frac{\partial^2}{\partial x^2}U(x, s)=sU(x, s) \Rightarrow U(x, s)=A(s)e^{x\sqrt{s}}+B(s)e^{-x\sqrt{s}}$$

$(BC2) \Rightarrow A(s)=0; \dfrac{\partial U}{\partial x}=-B(s)\sqrt{s}\,e^{-x\sqrt{s}}\cdot (BC1) \Rightarrow B(s)=-\dfrac{1}{\sqrt{s}}F(s) \Rightarrow$

$$U(x, s)=-\frac{F(s)}{\sqrt{s}}e^{-x\sqrt{s}} \Rightarrow$$

$$u(x, t)=-\int_0^t \frac{1}{\sqrt{\pi\tau}}e^{-x^2/4\tau}f(t-\tau)d\tau$$

11 Vector Analysis

11.1 Curves

Vector-valued functions $\boldsymbol{R} \rightarrow \boldsymbol{R}^m$

$$\boldsymbol{f}(t)=[f_1(t), \ldots, f_m(t)], \qquad \dot{\boldsymbol{f}}(t)=\frac{d\boldsymbol{f}}{dt}=[\dot{f}_1(t), \ldots, \dot{f}_m(t)]$$

$$\frac{d}{dt}\{a\boldsymbol{f}+b\boldsymbol{g}\}=a\dot{\boldsymbol{f}}+b\dot{\boldsymbol{g}} \qquad \frac{d}{dt}\{h(t)\boldsymbol{f}(t)\}=h(t)\dot{\boldsymbol{f}}(t)+\dot{h}(t)\boldsymbol{f}(t)$$

$$\frac{d}{dt}\{\boldsymbol{f}\cdot\boldsymbol{g}\}=\dot{\boldsymbol{f}}\cdot\boldsymbol{g}+\boldsymbol{f}\cdot\dot{\boldsymbol{g}} \qquad \frac{d}{dt}(\boldsymbol{f}\times\boldsymbol{g})=\dot{\boldsymbol{f}}\times\boldsymbol{g}+\boldsymbol{f}\times\dot{\boldsymbol{g}} \quad (m=3)$$

$$\frac{d}{dt}[\boldsymbol{f}, \boldsymbol{g}, \boldsymbol{h}]=[\dot{\boldsymbol{f}}, \boldsymbol{g}, \boldsymbol{h}]+[\boldsymbol{f}, \dot{\boldsymbol{g}}, \boldsymbol{h}]+[\boldsymbol{f}, \boldsymbol{g}, \dot{\boldsymbol{h}}] \qquad \frac{d}{dt}h(\boldsymbol{f}(t)) = \nabla h(\boldsymbol{f}(t))\cdot\dot{\boldsymbol{f}}(t)$$

$$\boldsymbol{f}(t+h)=\boldsymbol{f}(t)+h\dot{\boldsymbol{f}}(t)+\frac{h^2}{2!}\ddot{\boldsymbol{f}}(t)+\ldots+\frac{h^n}{n!}\boldsymbol{f}^{(n)}(t)+\ldots$$

Curves C: $\boldsymbol{r}=\boldsymbol{r}(t)$: $\boldsymbol{R} \rightarrow \boldsymbol{R}^3$

Curve $\boldsymbol{r}=\boldsymbol{r}(t)=(x(t), y(t), z(t)), a\leq t\leq b$

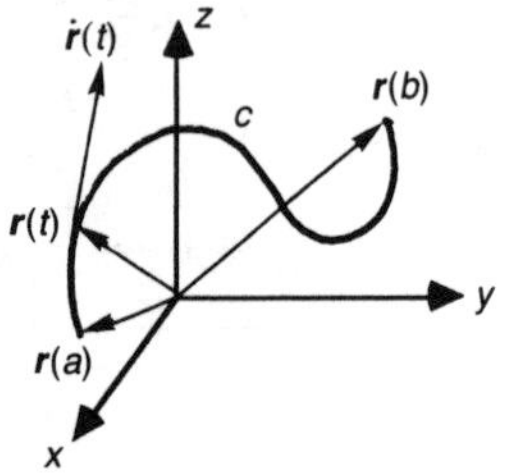

Tangent vector

$$\dot{\boldsymbol{r}}(t)=\lim_{\Delta t\to 0}\frac{\boldsymbol{r}(t+\Delta t)-\boldsymbol{r}(t)}{\Delta t}=(\dot{x}(t), \dot{y}(t), \dot{z}(t))$$

Length of curve $=s=\int_C ds=\int_a^b|\dot{\boldsymbol{r}}(t)|dt=\int_a^b\sqrt{\dot{x}^2+\dot{y}^2+\dot{z}^2}\,dt$

Arc length element $ds=|d\boldsymbol{r}|=|\dot{\boldsymbol{r}}|dt=v\,dt$

Motion of particle

Below, $\boldsymbol{r}=\boldsymbol{r}(t)$ = position vector, $\boldsymbol{t}$ = unit tangent vector, $\boldsymbol{n}$ = unit principal normal vector, a_t = tangential component, a_n = normal component:

Velocity $\boldsymbol{v}=\dot{\boldsymbol{r}}=v\boldsymbol{t}$ *Speed* $v=\dot{s}=|\dot{\boldsymbol{r}}|$ *Acceleration* $\boldsymbol{a}=\dot{\boldsymbol{v}}=\ddot{\boldsymbol{r}}=a_t\boldsymbol{t}+a_n\boldsymbol{n}$

$$a_t=\dot{v}=\frac{\boldsymbol{v}\cdot\boldsymbol{a}}{v}=\frac{\dot{\boldsymbol{r}}\cdot\ddot{\boldsymbol{r}}}{|\dot{\boldsymbol{r}}|} \qquad a_n=\kappa v^2=\frac{|\boldsymbol{v}\times\boldsymbol{a}|}{v}=\frac{|\dot{\boldsymbol{r}}\times\ddot{\boldsymbol{r}}|}{|\dot{\boldsymbol{r}}|}$$

Rotation round an axis:

$$\boldsymbol{v}=\dot{\boldsymbol{r}}=\boldsymbol{\omega}\times\boldsymbol{r}$$

Differential geometry

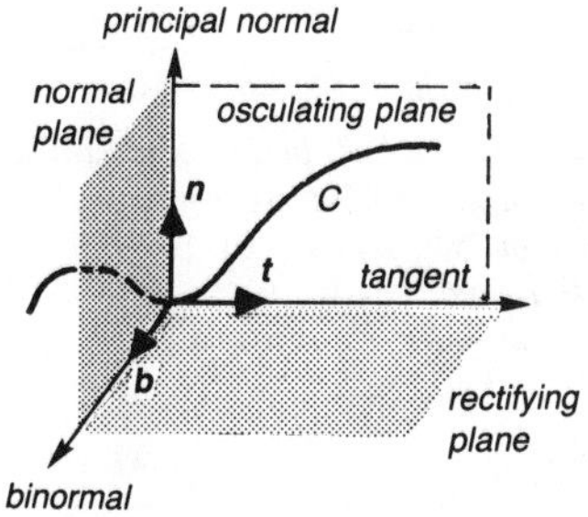

Think of C, $\boldsymbol{t}$, $\boldsymbol{n}$ in the plane of paper

Concept	Arbitrary parameter t	Arc length $s=\int_C^t \sqrt{\dot{x}^2+\dot{y}^2+\dot{z}^2}\,dt$ as parameter
Unit tangent vector	$\boldsymbol{t}=\frac{\dot{\boldsymbol{r}}}{\lvert\dot{\boldsymbol{r}}\rvert}=\frac{\dot{\boldsymbol{r}}}{v}$	$\boldsymbol{t}=\boldsymbol{r}'$
Unit principal normal vector	$\boldsymbol{n}=\frac{\ddot{\boldsymbol{r}}-\dot{v}\boldsymbol{t}}{\lvert\ddot{\boldsymbol{r}}-\dot{v}\boldsymbol{t}\rvert}$	$\boldsymbol{n}=\frac{\boldsymbol{r}''}{\lvert\boldsymbol{r}''\rvert}$
Unit binormal vector	$\boldsymbol{b}=\boldsymbol{t}\times\boldsymbol{n}$	$\boldsymbol{b}=\boldsymbol{t}\times\boldsymbol{n}$
Curvature	$\varkappa=\frac{\lvert\dot{\boldsymbol{r}}\times\ddot{\boldsymbol{r}}\rvert}{\lvert\dot{\boldsymbol{r}}\rvert^3}$	$\varkappa=\lvert\boldsymbol{r}''\rvert$
Radius of Curvature	$\varrho_\varkappa=\frac{1}{\varkappa}$	$\varrho_\varkappa=\frac{1}{\varkappa}$
Torsion	$\tau=\frac{[\dot{\boldsymbol{r}}, \ddot{\boldsymbol{r}}, \dddot{\boldsymbol{r}}]}{\lvert\dot{\boldsymbol{r}}\times\ddot{\boldsymbol{r}}\rvert^2}$	$\tau=\frac{[\boldsymbol{r}', \boldsymbol{r}'', \boldsymbol{r}''']}{\lvert\boldsymbol{r}''\rvert^2}$
Radius of torsion	$\varrho_\tau=\frac{1}{\tau}$	$\varrho_\tau=\frac{1}{\tau}$

Frenet's formulas

$\dot{\boldsymbol{t}}=\varkappa v\boldsymbol{n}$, $\quad \dot{\boldsymbol{n}}=-\varkappa v\boldsymbol{t}+\tau v\boldsymbol{b}$, $\quad \dot{\boldsymbol{b}}=-\tau v\boldsymbol{n}$

($v=1$ for s=arc length as parameter)

Remark. C straight line $\Leftrightarrow \varkappa\equiv 0$.
C plane curve $\Leftrightarrow \tau\equiv 0$.

Example. (*Helix*)

$\boldsymbol{r}\,(t)=(a\cos t,\ a\sin t,\ bt),\ c=\sqrt{a^2+b^2}$

$\dot{\boldsymbol{r}}\,(t)=(-a\sin t,\ a\cos t,\ \mathrm{b}),\ v=c$

$\ddot{\boldsymbol{r}}\,(t)=(-a\cos t,\ -a\sin t,\ 0)$

$\dddot{\boldsymbol{r}}\,(t)=(a\sin t,\ -a\cos t,\ 0)$

$\dot{\boldsymbol{r}}\times\ddot{\boldsymbol{r}}(t)=(ab\sin t,\ -ab\cos t,\ a^2)$

$\varkappa=\dfrac{a}{c^2},\ \tau=\dfrac{b}{c^2}$

$s=\int_0^{t_0}|\dot{\boldsymbol{r}}(t)|dt=ct_0$

$\boldsymbol{t}=\dfrac{1}{c}(-a\sin t,\ a\cos t,\ b),\ \boldsymbol{n}=-(\cos t,\ \sin t,\ 0),\ \boldsymbol{b}=\dfrac{1}{c}\ (b\sin t,\ -b\cos t,\ a)$

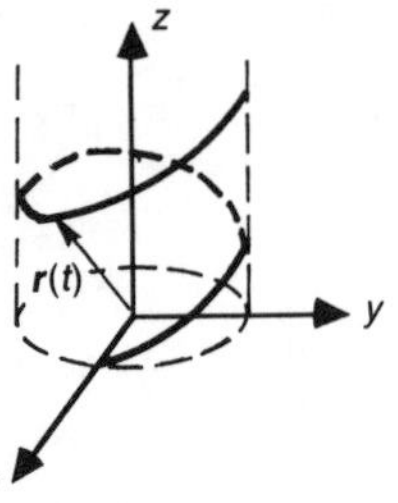

11.2 Vector Fields

$\boldsymbol{r}=(x,\ y,\ z),\ \boldsymbol{F}=(P,\ Q,\ R),\ \hat{\boldsymbol{x}},\ \hat{\boldsymbol{y}},\ \hat{\boldsymbol{z}}$ *ONR*-basis.

Vector field. $\boldsymbol{F}(\boldsymbol{r})=(P(x,\ y,\ z),\ Q(x,\ y,z),\ R(x,\ y,\ z)):\ R^3\to R^3$

Scalar field. $\phi(\boldsymbol{r})=\phi(x,\ y,\ z):\ R^3\to R$

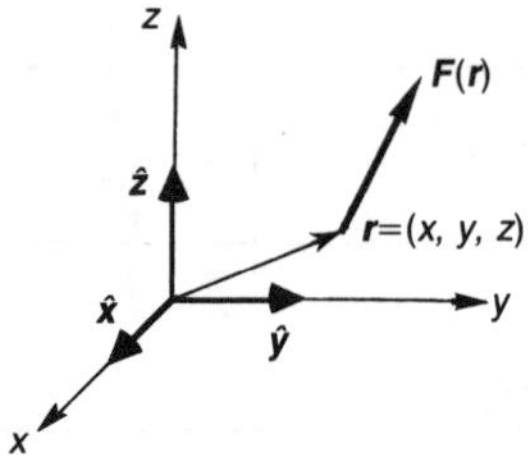

Every point is assigned a vector

1. grad $\phi=\nabla\phi=(\phi_x{}',\ \phi_y{}',\ \phi_z{}')$

   Vector operator $\nabla=\left(\dfrac{\partial}{\partial x},\dfrac{\partial}{\partial y},\dfrac{\partial}{\partial z}\right)=$

   $=\hat{\boldsymbol{x}}\dfrac{\partial}{\partial x}+\hat{\boldsymbol{y}}\dfrac{\partial}{\partial y}+\hat{\boldsymbol{z}}\dfrac{\partial}{\partial z}$

2. $\boldsymbol{F}$ is a *gradient field* $\Leftrightarrow$ there exists ϕ such that $\boldsymbol{F}=\text{grad }\phi$ [ϕ is *potential* of $\boldsymbol{F}$]
3. div $\boldsymbol{F}=\nabla\cdot\boldsymbol{F}=P_x{}'+Q_y{}'+R_z{}'$
4. curl $\boldsymbol{F}=\text{rot }\boldsymbol{F}=\nabla\times\boldsymbol{F}=(R_y{}'-Q_z{}',\ P_z{}'-R_x{}',\ Q_x{}'-P_y{}')$
5. Laplacian $\triangle\phi=\nabla\cdot\nabla\ \phi=\text{div grad }\phi=\phi''_{xx}+\phi''_{yy}+\phi''_{zz}$
6. $\boldsymbol{F}=\text{grad }\phi\Leftrightarrow\text{curl }\boldsymbol{F}=\boldsymbol{0}$ (in a simply-connected domain)
7. $\boldsymbol{F}=\text{curl }\boldsymbol{G}\Leftrightarrow\text{div }\boldsymbol{F}=0$ (in a simply-connected domain)

Laws for operations with the operator ∇

Linearity	
1. $\nabla(\alpha\phi+\beta\psi)=\alpha\nabla\phi+\beta\nabla\psi$	$\text{grad}(\alpha\phi+\beta\psi)=\alpha\,\text{grad}\,\phi+\beta\,\text{grad}\,\psi$
2. $\nabla\cdot(\alpha\boldsymbol{F}+\beta\boldsymbol{G})=\alpha\nabla\cdot\boldsymbol{F}+\beta\nabla\cdot\boldsymbol{G}$	$\text{div}(\alpha\boldsymbol{F}+\beta\boldsymbol{G})=\alpha\,\text{div}\,\boldsymbol{F}+\beta\,\text{div}\,\boldsymbol{G}$
3. $\nabla\times(\alpha\boldsymbol{F}+\beta\boldsymbol{G})=\alpha\nabla\times\boldsymbol{F}+\beta\nabla\times\boldsymbol{G}$	$\text{curl}(\alpha\boldsymbol{F}+\beta\boldsymbol{G})=\alpha\,\text{curl}\,\boldsymbol{F}+\beta\,\text{curl}\,\boldsymbol{G}$
Operations on products	
4. $\nabla(\phi\psi)=\phi\nabla\psi+\psi\nabla\phi$	$\text{grad}(\phi\psi)=\phi\,\text{grad}\,\psi+\psi\,\text{grad}\,\phi$
5. $\nabla(\boldsymbol{F}\cdot\boldsymbol{G})=(\boldsymbol{F}\cdot\nabla)\boldsymbol{G}+(\boldsymbol{G}\cdot\nabla)\boldsymbol{F}+$ $+\boldsymbol{F}\times(\nabla\times\boldsymbol{G})+\boldsymbol{G}\times(\nabla\times\boldsymbol{F})$	$\text{grad}(\boldsymbol{F}\cdot\boldsymbol{G})=(\boldsymbol{F}\cdot\text{grad})\boldsymbol{G}+$ $+(\boldsymbol{G}\cdot\text{grad})\boldsymbol{F}+\boldsymbol{F}\times\text{curl}\,\boldsymbol{G}+\boldsymbol{G}\times\text{curl}\,\boldsymbol{F}$
6. $\nabla\cdot(\phi\boldsymbol{F})=\phi\nabla\cdot\boldsymbol{F}+(\nabla\phi)\cdot\boldsymbol{F}$	$\text{div}(\phi\boldsymbol{F})=\phi\text{div}\boldsymbol{F}+\boldsymbol{F}\cdot\text{grad}\,\phi$
7. $\nabla\cdot(\boldsymbol{F}\times\boldsymbol{G})=\boldsymbol{G}\cdot\nabla\times\boldsymbol{F}-\boldsymbol{F}\cdot\nabla\times\boldsymbol{G}$	$\text{div}(\boldsymbol{F}\times\boldsymbol{G})=\boldsymbol{G}\cdot\text{curl}\,\boldsymbol{F}-\boldsymbol{F}\cdot\text{curl}\,\boldsymbol{G}$
8. $\nabla\times(\phi\boldsymbol{F})=\phi\nabla\times\boldsymbol{F}+(\nabla\phi)\times\boldsymbol{F}$	$\text{curl}(\phi\boldsymbol{F})=\phi\,\text{curl}\,\boldsymbol{F}+(\text{grad}\,\phi)\times\boldsymbol{F}$
9. $\nabla\times(\boldsymbol{F}\times\boldsymbol{G})=(\boldsymbol{G}\cdot\nabla)\boldsymbol{F}-(\boldsymbol{F}\cdot\nabla)\boldsymbol{G}+$ $+\boldsymbol{F}(\nabla\cdot\boldsymbol{G})-\boldsymbol{G}(\nabla\cdot\boldsymbol{F})$	$\text{curl}(\boldsymbol{F}\times\boldsymbol{G})=(\boldsymbol{G}\cdot\text{grad})\boldsymbol{F}-$ $-(\boldsymbol{F}\cdot\text{grad})\boldsymbol{G}+\boldsymbol{F}\,\text{div}\,\boldsymbol{G}-\boldsymbol{G}\,\text{div}\,\boldsymbol{F}$
Double application of ∇	
10. $\nabla\cdot(\nabla\times\boldsymbol{F})=0$	$\text{div}\,\text{curl}\,\boldsymbol{F}=0$
11. $\nabla\times(\nabla\phi)=\boldsymbol{0}$	$\text{curl}\,\text{grad}\,\phi=\boldsymbol{0}$
12. $\nabla\times(\nabla\times\boldsymbol{F})=\nabla(\nabla\cdot\boldsymbol{F})-\nabla^2\boldsymbol{F}$	$\text{curl}\,\text{curl}\,\boldsymbol{F}=\text{grad}\,\text{div}\,\boldsymbol{F}-\triangle\boldsymbol{F}$

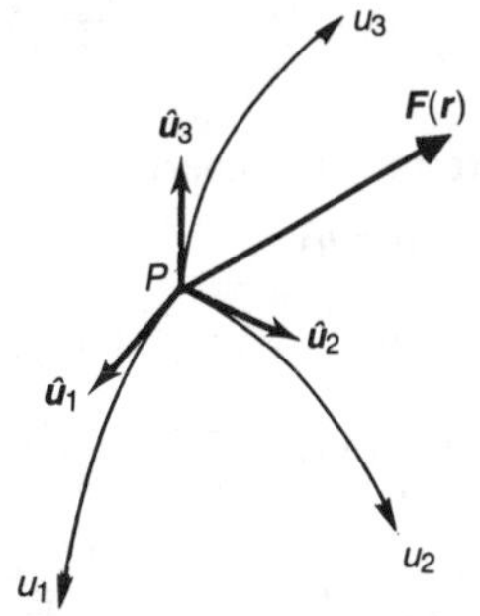

Orthogonal curvilinear coordinates

Assume that, in a domain of a Cartesian (x, y, z)-coordinate system with unit basis vectors $\hat{\boldsymbol{x}}, \hat{\boldsymbol{y}}, \hat{\boldsymbol{z}}$, a one-to-one correspondence between (x, y, z) and (u_1, u_2, u_3) is given by

$$\text{(11.1)}\qquad \begin{cases} x=x(u_1, u_2, u_3) \\ y=y(u_1, u_2, u_3) \\ z=z(u_1, u_2, u_3) \end{cases}$$

$$\boldsymbol{F}(\boldsymbol{r})=F_x\hat{\boldsymbol{x}}+F_y\hat{\boldsymbol{y}}+F_z\hat{\boldsymbol{z}}=$$
$$=F_{u_1}\hat{\boldsymbol{u}}_1+F_{u_2}\hat{\boldsymbol{u}}_2+F_{u_3}\hat{\boldsymbol{u}}_3$$

Given $P=(x; y; z)$, assume that the surfaces u_i=constant, $i=1, 2, 3$, intersect at P under right angles. The intersection curves are called *coordinate lines*. The *local unit basis vectors* $\hat{\boldsymbol{u}}_1$, $\hat{\boldsymbol{u}}_2$, $\hat{\boldsymbol{u}}_3$ form a *local ON-system*.

Condition for (11.1) to define an orthogonal (u_1, u_2, u_3)-system:

$$\frac{\partial \boldsymbol{r}}{\partial u_i}\cdot\frac{\partial \boldsymbol{r}}{\partial u_j}=0,\ i\neq j,\ \boldsymbol{r}=(x, y, z)$$

The system is right-handed $\Leftrightarrow \dfrac{\partial(x, y, z)}{\partial(u_1, u_2, u_3)}>0.$

Differential formulas in orthogonal coordinate systems

General coordinates (u_1, u_2, u_3).

Let $\boldsymbol{F}(\boldsymbol{r})=F_x\hat{\boldsymbol{x}}+F_y\hat{\boldsymbol{y}}+F_z\hat{\boldsymbol{z}}=F_{u_1}\hat{\boldsymbol{u}}_1+F_{u_2}\hat{\boldsymbol{u}}_2+F_{u_3}\hat{\boldsymbol{u}}_3$

Set $h_i=\left|\dfrac{\partial \boldsymbol{r}}{\partial u_i}\right|=\sqrt{\left(\dfrac{\partial x}{\partial u_i}\right)^2+\left(\dfrac{\partial y}{\partial u_i}\right)^2+\left(\dfrac{\partial z}{\partial u_i}\right)^2}$, $i=1, 2, 3.$

Then

(*i*) $$F_{u_i}=h_i^{-1}\left(F_x\frac{\partial x}{\partial u_i}+F_y\frac{\partial y}{\partial u_i}+F_z\frac{\partial z}{\partial u_i}\right),\ i=1, 2, 3$$

(*vector component relationship*)

(*iia*) $ds^2=h_1^2du_1^2+h_2^2du_2^2+h_3^2du_3^2$ (*arc length element*)

(*b*) $h_1h_2\,du_1du_2,\quad h_2h_3\,du_2du_3,\quad h_3h_1\,du_3du_1$ (*surface elements*)

(*c*) $h_1h_2h_3\,du_1du_2du_3$ (*volume element*)

(*iii*) $\operatorname{grad}\phi=\nabla\phi=\displaystyle\sum_{i=1}^{3}\frac{1}{h_i}\frac{\partial\phi}{\partial u_i}\hat{\boldsymbol{u}}_i$

(*iv*) $\operatorname{div}\boldsymbol{F}=\nabla\cdot\boldsymbol{F}=\displaystyle\frac{1}{h_1h_2h_3}\sum_{i=1}^{3}\frac{\partial}{\partial u_i}\left(\frac{h_1h_2h_3}{h_i}F_{u_i}\right)$

(*v*) $$\operatorname{curl}\boldsymbol{F}=\nabla\times\boldsymbol{F}=\frac{1}{h_1h_2h_3}\begin{vmatrix} h_1\hat{\boldsymbol{u}}_1 & h_2\hat{\boldsymbol{u}}_2 & h_3\hat{\boldsymbol{u}}_3 \\ \dfrac{\partial}{\partial u_1} & \dfrac{\partial}{\partial u_2} & \dfrac{\partial}{\partial u_3} \\ h_1F_{u_1} & h_2F_{u_2} & h_3F_{u_3}\end{vmatrix}$$

(*vi*) $$\Delta\phi=\nabla^2\phi=\frac{1}{h_1h_2h_3}\sum_{i=1}^{3}\frac{\partial}{\partial u_i}\left(\frac{h_1h_2h_3}{h_i^2}\frac{\partial\phi}{\partial u_i}\right)$$

Rectangular coordinates (x, y, z)

$h_1=h_2=h_3=1$

(*ii*) $ds^2=dx^2+dy^2+dz^2$

(*iii*) $\operatorname{grad}\phi=\dfrac{\partial\phi}{\partial x}\hat{\boldsymbol{x}}+\dfrac{\partial\phi}{\partial y}\hat{\boldsymbol{y}}+\dfrac{\partial\phi}{\partial z}\hat{\boldsymbol{z}}$ $\quad(\hat{\boldsymbol{x}}=\mathbf{i}, \hat{\boldsymbol{y}}=\mathbf{j}, \hat{\boldsymbol{z}}=\mathbf{k})$

(*iv*) $\operatorname{div}\boldsymbol{F}=\dfrac{\partial F_x}{\partial x}+\dfrac{\partial F_y}{\partial y}+\dfrac{\partial F_z}{\partial z}$

(*v*) $$\operatorname{curl}\boldsymbol{F}=\left(\frac{\partial F_z}{\partial y}-\frac{\partial F_y}{\partial z}\right)\hat{\boldsymbol{x}}+\left(\frac{\partial F_x}{\partial z}-\frac{\partial F_z}{\partial x}\right)\hat{\boldsymbol{y}}+\left(\frac{\partial F_y}{\partial x}-\frac{\partial F_x}{\partial y}\right)\hat{\boldsymbol{z}}$$

(*vi*) $\Delta\phi=\dfrac{\partial^2\phi}{\partial x^2}+\dfrac{\partial^2\phi}{\partial y^2}+\dfrac{\partial^2\phi}{\partial z^2}$

Translated and rotated coordinates (ξ, η, ζ)

$$\begin{cases} x=\xi_0+a_{11}\xi+a_{12}\eta+a_{13}\zeta \\ y=\eta_0+a_{21}\xi+a_{22}\eta+a_{23}\zeta \\ z=\zeta_0+a_{31}\xi+a_{32}\eta+a_{33}\zeta \end{cases}\qquad (a_{ij})\text{ orthogonal matrix}$$

$h_1=h_2=h_3=1$

$$(i)\ \begin{cases} F_\xi = a_{11}F_x + a_{21}F_y + a_{31}F_z \\ F_\eta = a_{12}F_x + a_{22}F_y + a_{32}F_z \\ F_\zeta = a_{13}F_x + a_{23}F_y + a_{33}F_z \end{cases}$$

$$(ii)\ ds^2 = d\xi^2 + d\eta^2 + d\zeta^2$$

$$(iii)\ \operatorname{grad}\phi = \frac{\partial\phi}{\partial\xi}\hat{\xi} + \frac{\partial\phi}{\partial\eta}\hat{\eta} + \frac{\partial\phi}{\partial\zeta}\hat{\zeta}$$

$$(iv)\ \operatorname{div}\boldsymbol{F} = \frac{\partial F_\xi}{\partial\xi} + \frac{\partial F_\eta}{\partial\eta} + \frac{\partial F_\zeta}{\partial\zeta}$$

$$(v)\ \operatorname{curl}\boldsymbol{F} = \left(\frac{\partial F_\zeta}{\partial\eta} - \frac{\partial F_\eta}{\partial\zeta}\right)\hat{\xi} + \left(\frac{\partial F_\xi}{\partial\zeta} - \frac{\partial F_\zeta}{\partial\xi}\right)\hat{\eta} + \left(\frac{\partial F_\eta}{\partial\xi} - \frac{\partial F_\xi}{\partial\eta}\right)\hat{\zeta}$$

$$(vi)\ \triangle\phi = \frac{\partial^2\phi}{\partial\xi^2} + \frac{\partial^2\phi}{\partial\eta^2} + \frac{\partial^2\phi}{\partial\zeta^2}$$

Cylindrical coordinates (r, φ, z)

$x = r\cos\varphi,\ y = r\sin\varphi,\ z = z$

$h_1 = 1,\ h_2 = r,\ h_3 = 1$

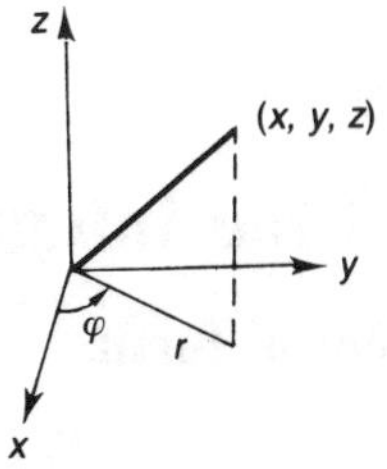

$$(i)\ \begin{cases} F_r = F_x\cos\varphi + F_y\sin\varphi \\ F_\varphi = F_y\cos\varphi - F_x\sin\varphi \\ F_z = F_z \end{cases}$$

$$(ii)\ ds^2 = dr^2 + r^2 d\varphi^2 + dz^2$$

$$(iii)\ \operatorname{grad}\phi = \frac{\partial\phi}{\partial r}\hat{\boldsymbol{r}} + \frac{1}{r}\frac{\partial\phi}{\partial\varphi}\hat{\varphi} + \frac{\partial\phi}{\partial z}\hat{\boldsymbol{z}}$$

$$(iv)\ \operatorname{div}\boldsymbol{F} = \frac{1}{r}\frac{\partial(rF_r)}{\partial r} + \frac{1}{r}\frac{\partial F_\varphi}{\partial\varphi} + \frac{\partial F_z}{\partial z}$$

$$(v)\ \operatorname{curl}\boldsymbol{F} = \left(\frac{1}{r}\frac{\partial F_z}{\partial\varphi} - \frac{\partial F_\varphi}{\partial z}\right)\hat{\boldsymbol{r}} + \left(\frac{\partial F_r}{\partial z} - \frac{\partial F_z}{\partial r}\right)\hat{\varphi} + \frac{1}{r}\left(\frac{\partial(rF_\varphi)}{\partial r} - \frac{\partial F_r}{\partial\varphi}\right)\hat{\boldsymbol{z}}$$

$$(vi)\ \triangle\phi = \frac{1}{r}\frac{\partial}{\partial r}\left(r\frac{\partial\phi}{\partial r}\right) + \frac{1}{r^2}\frac{\partial^2\phi}{\partial\varphi^2} + \frac{\partial^2\phi}{\partial z^2}$$

Spherical coordinates (r, θ, φ)

$x = r\sin\theta\cos\varphi,\ y = r\sin\theta\sin\varphi,\ z = r\cos\theta$

$h_1 = 1,\ h_2 = r,\ h_3 = r\sin\theta$

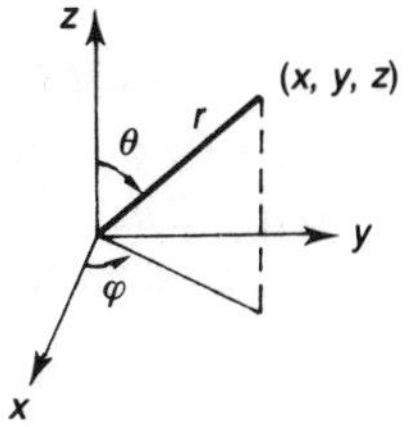

$$(i)\ \begin{cases} F_r = F_x\sin\theta\cos\varphi + F_y\sin\theta\sin\varphi + F_z\cos\theta \\ F_\theta = F_x\cos\theta\cos\varphi + F_y\cos\theta\sin\varphi - F_z\sin\theta \\ F_\varphi = -F_x\sin\varphi + F_y\cos\varphi \end{cases}$$

$$(ii)\ ds^2 = dr^2 + r^2\sin^2\theta\, d\varphi^2 + r^2\, d\theta^2$$

$$(iii)\ \operatorname{grad}\phi = \frac{\partial\phi}{\partial r}\hat{\boldsymbol{r}} + \frac{1}{r}\frac{\partial\phi}{\partial\theta}\hat{\boldsymbol{\theta}} + \frac{1}{r\sin\theta}\frac{\partial\phi}{\partial\varphi}\hat{\varphi}$$

(*iv*) $\operatorname{div}\boldsymbol{F}=\frac{1}{r^2}\frac{\partial(r^2F_r)}{\partial r}+\frac{1}{r\sin\theta}\frac{\partial(F_\theta\sin\theta)}{\partial\theta}+\frac{1}{r\sin\theta}\frac{\partial F_\varphi}{\partial\varphi}$

(*v*) $\operatorname{curl}\boldsymbol{F}=\frac{1}{r\sin\theta}\left[\frac{\partial(F_\varphi\sin\theta)}{\partial\theta}-\frac{\partial F_\theta}{\partial\varphi}\right]\hat{\boldsymbol{r}}+\frac{1}{r\sin\theta}\left[\frac{\partial F_r}{\partial\varphi}-\sin\theta\frac{\partial(rF_\varphi)}{\partial r}\right]\hat{\boldsymbol{\theta}}+$

$+\frac{1}{r}\left[\frac{\partial(rF_\theta)}{\partial r}-\frac{\partial F_r}{\partial\theta}\right]\hat{\boldsymbol{\varphi}}$

(*vi*) $\Delta\phi=\frac{1}{r^2}\frac{\partial}{\partial r}\left(r^2\frac{\partial\phi}{\partial r}\right)+\frac{1}{r^2\sin\theta}\frac{\partial}{\partial\theta}\left(\sin\theta\frac{\partial\phi}{\partial\theta}\right)+\frac{1}{r^2\sin^2\theta}\frac{\partial^2\phi}{\partial\varphi^2}=$

$=\frac{\partial^2\phi}{\partial r^2}+\frac{2}{r}\frac{\partial\phi}{\partial r}+\frac{1}{r^2}\left[\frac{\partial}{\partial\xi}\left((1-\xi^2)\frac{\partial\phi}{\partial\xi}\right)+\frac{1}{1-\xi^2}\frac{\partial^2\phi}{\partial\varphi^2}\right]$ if $\xi=\cos\theta$

11.3 Line Integrals

Differential forms

$f, g: R^n\to R$, $h: R\to R$. Differential form:

$$\omega=f\,dg=f\left(\frac{\partial g}{\partial x_1}dx_1+\ldots+\frac{\partial g}{\partial x_n}dx_n\right)$$

1. $d(af+bg)=a\,df+b\,dg$	2. $d(fg)=f\,dg+g\,df$
3. $d\left(\frac{f}{g}\right)=\frac{g\,df-f\,dg}{g^2}$	4. $d(h(f))=h'(f)df$

Exact differential forms

In R^2: $\omega=P\,dx+Q\,dy$ is *exact* if there exists $\phi(x, y)$ called *primitive function* of ω such that $\phi_x'=P$, $\phi_y'=Q$, (i.e. $\omega=d\,\phi$).
Test: $P\,dx+Q\,dy$ exact $\Leftrightarrow$ $P_y'=Q_x'$, (in a simply connected domain).

In R^3: $\omega=P\,dx+Q\,dy+R\,dz$ *is exact* if there exists $\phi(x, y, z)$ [*primitive function*] such that $\phi_x'=P$, $\phi_y'=Q$, $\phi_z'=R$, (i.e. $\omega=d\,\phi$).
Test: $P\,dx+Q\,dy+R\,dz$ exact $\Leftrightarrow$ $\operatorname{curl}(P, Q, R)=\mathbf{0}$ $\Leftrightarrow$ $P_y'=Q_x'$, $P_z'=R_x'$, $Q_z'=R_y'$, (in a simply connected domain).

Line integrals

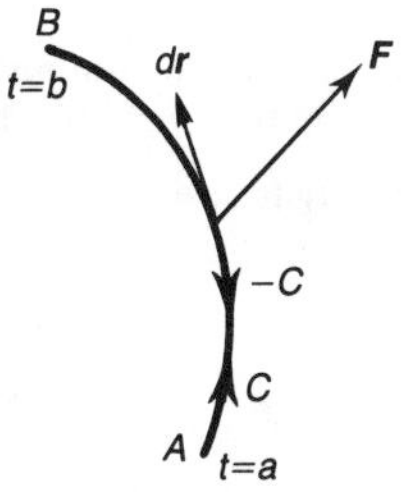

Given: Curve C: $\boldsymbol{r}=\boldsymbol{r}(t)=(x(t), y(t), z(t)),\ a\leq t\leq b$

$d\boldsymbol{r}=(dx, dy, dz),\ ds=|d\boldsymbol{r}|=\sqrt{\dot{x}^2+\dot{y}^2+\dot{z}^2}\ dt$

Vector field $\boldsymbol{F}=\boldsymbol{F}(\boldsymbol{r})=$
$=(P(x, y, z), Q(x, y, z), R(x, y, z))$

Scalar field $\phi(\boldsymbol{r})=\phi(x, y, z)$

There are four kinds of line integrals:

1. $\int_C \boldsymbol{F}\cdot d\boldsymbol{r}$	2. $\int_C \phi\lvert d\boldsymbol{r}\rvert$	3. $\int_C \boldsymbol{F}\times d\boldsymbol{r}$	4. $\int_C \phi d\boldsymbol{r}$

1. Tangent line integral

$$\int_C \omega=\int_C \boldsymbol{F}\cdot d\boldsymbol{r}=\int_C P\,dx+Q\,dy+R\,dz=$$
$$=\int_a^b \left(P\frac{dx}{dt}+Q\frac{dy}{dt}+R\frac{dz}{dt}\right)dt$$

Properties: (*i*) $\int_{-C}\omega=-\int_C\omega$ (*ii*) $\int_{C_1+C_2}\omega=\int_{C_1}\omega+\int_{C_2}\omega$

(*iii*) $\int_C d\phi=\int_C \nabla\phi\cdot d\boldsymbol{r}=\phi(B)-\phi(A)$ (*iv*) ω exact $\Rightarrow$

$\int_C\omega=\phi(B)-\phi(A)$, ϕ primitive function of ω

Example. Calculate $I=\int_C y\,dx+z\,dy-x^2dz,\ C\colon (x, y, z)=(\cos t, \sin t, t),\ 0\leq t\leq 2\pi$. Solution: $dx=-\sin t\,dt,\ dy=\cos t\,dt,\ dz=dt$. Thus, $I=\int_0^{2\pi}(-\sin^2 t+t\cos t-\cos^2 t)dt=-2\pi$

Green's formula in the plane

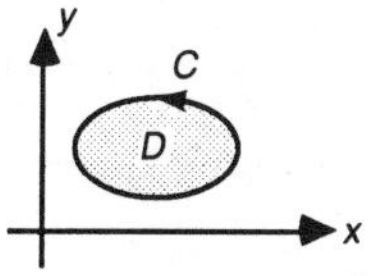

Assume (*i*) C closed curve with positive orientation, (*ii*) P, Q continuously differentiable on C and in D. Then

$$\oint_C P\,dx+Q\,dy=\iint_D\left(\frac{\partial Q}{\partial x}-\frac{\partial P}{\partial y}\right)dxdy$$

Area of $D=$
$=\int_C x\,dy=-\int_C y\,dx=$
$=\frac{1}{2}\int_C x\,dy-y\,dx$

Theorem

Assume that P, Q are continuously differentiable in a simply connected domain D. Then the following conditions are equivalent.

(*i*) $P\,dx+Q\,dy$ is exact

(*ii*) $\dfrac{\partial P}{\partial y}=\dfrac{\partial Q}{\partial x}$

(*iii*) $\oint_C P\,dx+Q\,dy=0$ (any closed C in D)

(*iv*) $\int_{C_1} P\,dx+Q\,dy$ depends only on the initial point and the end point of the curve C_1 (lying in D).

Stokes' theorem (see sec. 11.4)

2. $$\int_C \phi|d\boldsymbol{r}|=\int_C \phi ds=\int_a^b \phi[\boldsymbol{r}(t)]\left|\frac{d\boldsymbol{r}}{dt}\right|dt=$$

$$=\int_a^b \phi(x(t),y(t),z(t))\sqrt{\left(\frac{dx}{dt}\right)^2+\left(\frac{dy}{dt}\right)^2+\left(\frac{dz}{dt}\right)^2}\,dt$$

In particular, $l=\int_C ds$ (*arc length*)

3. $$\int_C \boldsymbol{F}\times d\boldsymbol{r}=\int_a^b \boldsymbol{F}(\boldsymbol{r}(t))\times\frac{d\boldsymbol{r}}{dt}\,dt \qquad \text{(vector valued)}$$

4. $$\int_C \phi d\boldsymbol{r}=\int_a^b \phi(\boldsymbol{r}(t))\frac{d\boldsymbol{r}}{dt}\,dt \qquad \text{(vector valued)}$$

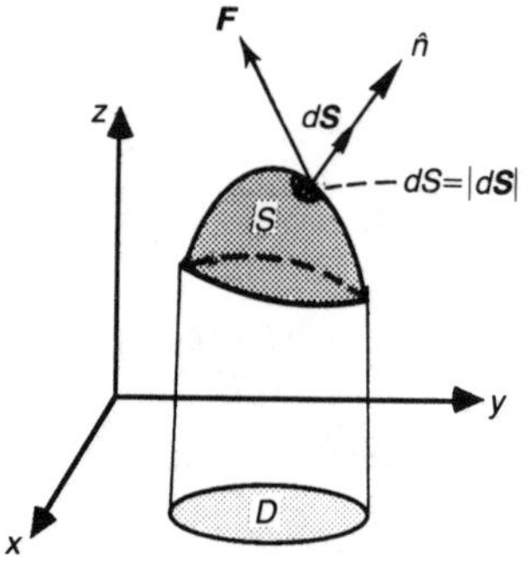

11.4 Surface Integrals

Surface

Surfaces are given in one of the following two ways:

A. *Graph of a function S:* $z=z(x,y)$, $(x,y)\in D$

$$\hat{\boldsymbol{n}}=\pm\frac{(z_x', z_y', -1)}{\sqrt{1+(z_x')^2+(z_y')^2}},\ d\boldsymbol{S}=\hat{\boldsymbol{n}}\,dS=$$

$$=\pm(z_x', z_y', -1)dxdy,\ dS=\sqrt{1+(z_x')^2+(z_y')^2}\,dxdy$$

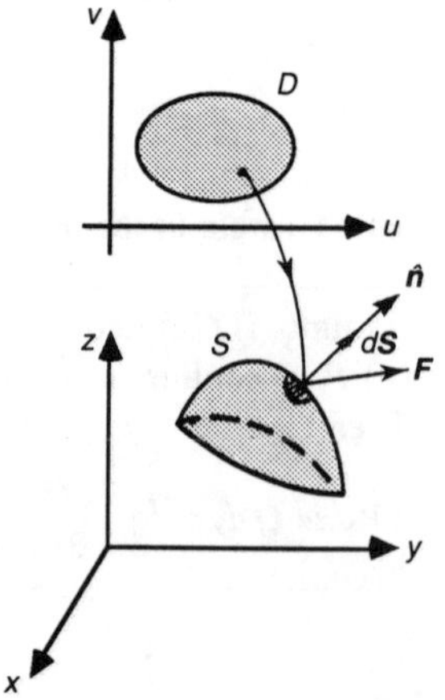

B. *Parametrization*

S: $\boldsymbol{r}=\boldsymbol{r}(u,v)=(x(u,v), y(u,v), z(u,v))$, $(u,v)\in D$

$$\hat{\boldsymbol{n}}=\pm\frac{\boldsymbol{r}_u'\times\boldsymbol{r}_v'}{|\boldsymbol{r}_u'\times\boldsymbol{r}_v'|},\ d\boldsymbol{S}=\hat{\boldsymbol{n}}\,dS=\pm\boldsymbol{r}_u'\times\boldsymbol{r}_v'\,dudv$$

$$dS=|\boldsymbol{r}_u'\times\boldsymbol{r}_v'|\,dudv$$

Surface integrals

Given: Surface S as above.

Vector field $\boldsymbol{F}=\boldsymbol{F}(\boldsymbol{r})=(P, Q, R)$

Scalar field $\phi=\phi(\boldsymbol{r})$.

There are four kinds of surface integrals:

1. $\iint_S \boldsymbol{F}\cdot d\boldsymbol{S}$ 2. $\iint_S \phi dS$ 3. $\iint_S \boldsymbol{F}\times d\boldsymbol{S}$ 4. $\iint_S \phi d\boldsymbol{S}$

1. Normal surface integral

A. $$\iint_S \boldsymbol{F}\cdot d\boldsymbol{S}=\iint_S \boldsymbol{F}\cdot\hat{\boldsymbol{n}}\,dS=\pm\iint_D\left(-P\frac{\partial z}{\partial x}-Q\frac{\partial z}{\partial y}+R\right)dxdy$$

(+ refers to the "upwards" direction of the normal vector)

B. $$\iint_S \boldsymbol{F}\cdot d\boldsymbol{S}=\iint_S \boldsymbol{F}\cdot\hat{\boldsymbol{n}}\,dS=\iint_D \boldsymbol{F}(\boldsymbol{r}(u, v))\cdot(\boldsymbol{r}_u'\times\boldsymbol{r}_v')\,dudv$$

Alternative notation:

$$\iint_S \boldsymbol{F}\cdot d\boldsymbol{S}=\iint_S P\,dydz+Q\,dzdx+R\,dxdy=$$

$$=\pm\iint_{D_{yz}} P[x(y, z), y, z]\,dydz\pm\iint_{D_{xz}} Q[x, y(x, z), z]\,dxdz\pm$$

$$\pm\iint_{D_{xy}} R[x, y, z(x, y)]\,dxdy,\qquad (D_{xy}=\text{projection of } S \text{ in } xy\text{-plane, etc.})$$

where the last integrals are "ordinary" double integrals. The + sign is chosen if the normal vector of S is in the "direction of" the positive x-axis, y-axis and z-axis, respectively (i.e. the first, second and third components of the normal vector of S is positive in respective cases).

Remark. $\iint_S P\,dydz=-\iint_S P\,dzdy$ etc.

Example

$S=S_1+S_2+S_3$ closed cylinder

S_1: $\boldsymbol{r}=(\cos u, \sin u, v)$, D: $0\leq u\leq 2\pi$, $0\leq v\leq 1$

$\boldsymbol{r}_u'=(-\sin u, \cos u, 0)$, $\boldsymbol{r}_v'=(0, 0, 1)$

$\boldsymbol{r}_u'\times\boldsymbol{r}_v'=(\cos u, \sin u, 0)$

S_2: $z=1$, S_3: $z=0$, D_{xy}: $x^2+y^2\leq 1$

$$\iint_S \boldsymbol{r}\cdot d\boldsymbol{S}=\iint_S x\,dydz+y\,dzdx+z\,dxdy$$

$$\iint_{S_1}=\iint_D(\cos u, \sin u, v)\cdot(\cos u, \sin u, 0)\,dudv=$$

$$=\iint_D(\cos^2u+\sin^2u)\,dudv=\iint_D dudv=2\pi$$

$$\iint_{S_2}=+\iint_{D_{xy}} z\,dxdy=\iint_{D_{xy}} 1\cdot dxdy=\pi,\ \iint_{S_3}=-\iint_{D_{xy}} z\,dxdy=-\iint_{D_{xy}} 0\cdot dxdy=0$$

Thus, $\iint_S \boldsymbol{r}\cdot d\boldsymbol{S}=3\pi$.

Alternative (and simpler) solution: Use Gauss' theorem.

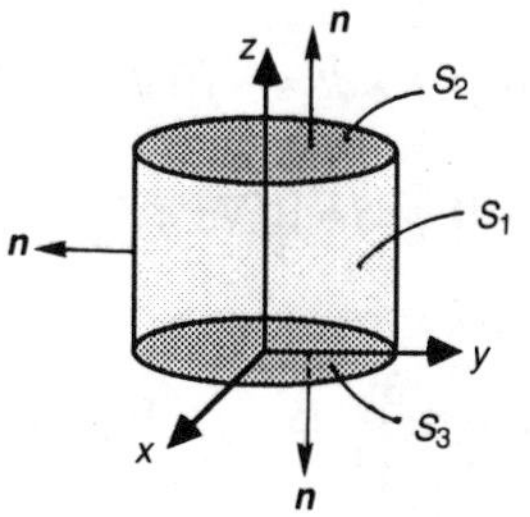

S_1: $x^2+y^2=1$
S_2: $z=1$
S_3: $z=0$

2. A. $\iint_S \phi dS = \iint_D \phi(x, y, z(x, y))\sqrt{1+(z_x')^2+(z_y')^2}\, dxdy$

B. $\iint_S \phi dS = \iint_D \phi(\boldsymbol{r}(u, v))|\boldsymbol{r}_u' \times \boldsymbol{r}_v'|\, dudv$

Area of a surface

Area of $S = \iint_S dS \overset{(A)}{=} \iint_D \sqrt{1+(z_x')^2+(z_y')^2}\, dxdy \overset{(B)}{=} \iint_D |\boldsymbol{r}_u' \times \boldsymbol{r}_v'|\, dudv$

3. $\iint_S \boldsymbol{F} \times d\boldsymbol{S} = \iint_D \boldsymbol{F}(\boldsymbol{r}(u, v)) \times (\boldsymbol{r}_u' \times \boldsymbol{r}_v')\, dudv$ (vector valued)

4. $\iint_S \phi d\boldsymbol{S} = \iint_D \phi(\boldsymbol{r}(u, v))(\boldsymbol{r}_u' \times \boldsymbol{r}_v')\, dudv$ (vector valued)

Integral theorems

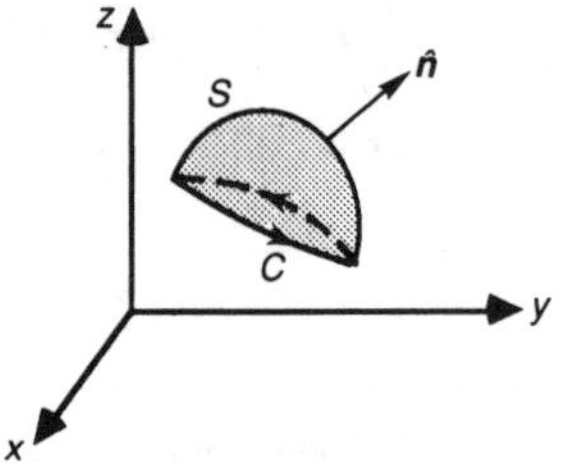

(Functions assumed to be continuously differentiable.)

1. *Stokes' theorem.* ($C=\partial S$ closed curve in space.)

(a) $\oint_C \boldsymbol{F} \cdot d\boldsymbol{r} = \iint_S (\nabla \times \boldsymbol{F}) \cdot d\boldsymbol{S} = \iint_S \operatorname{curl} \boldsymbol{F} \cdot d\boldsymbol{S}$, or

$\oint_C P\,dx + Q\,dy + R\,dz = \iint_S (R_y' - Q_z')\, dydz + (P_z' - R_x')\, dzdx + (Q_x' - P_y')\, dxdy$

(b) $\oint_C \boldsymbol{F} \times d\boldsymbol{r} = -\iint_S (d\boldsymbol{S} \times \nabla) \times \boldsymbol{F}$

(c) $\oint_C \phi d\boldsymbol{r} = \iint_S (d\boldsymbol{S} \times \nabla)\phi$

2. *Gauss' theorem.* ($S=\partial V$ closed surface)

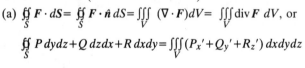

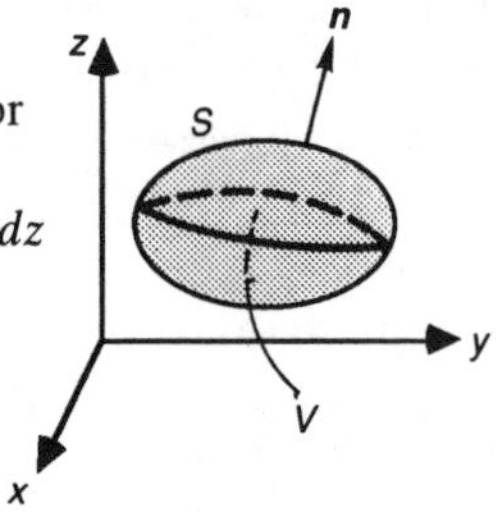

(a) $\oiint_S \boldsymbol{F} \cdot d\boldsymbol{S} = \oiint_S \boldsymbol{F} \cdot \hat{\boldsymbol{n}}\, dS = \iiint_V (\nabla \cdot \boldsymbol{F}) dV = \iiint_V \operatorname{div} \boldsymbol{F}\, dV$, or

$\oiint_S P\,dydz + Q\,dzdx + R\,dxdy = \iiint_V (P_x' + Q_y' + R_z')\, dxdydz$

(b) $\oiint_S \boldsymbol{F} \times d\boldsymbol{S} = -\iiint_V \nabla \times \boldsymbol{F}\, dV$

(c) $\oiint_S \phi d\boldsymbol{S} = \iiint_V \nabla\phi\, dV$

3. *Green's formulas*

(a) $\oiint_S \phi \dfrac{\partial \psi}{\partial n}\, dS = \iiint_V \nabla\phi \cdot \nabla\psi\, dV + \iiint_V \phi\, \Delta\psi\, dV$

(b) $\oiint_S \left(\phi \dfrac{\partial \psi}{\partial n} - \psi \dfrac{\partial \phi}{\partial n}\right) dS = \iiint_V (\phi\Delta\psi - \psi\Delta\phi)\, dV$

(c) Green's formula in the plane (see sec. 11.3).

Remark. The *normal derivative* $\dfrac{\partial \phi}{\partial n}$ is the *directional derivative* (see sec. 10.4) of ϕ in the direction of the outward normal.

12 Orthogonal Series and Special Functions

12.1 Orthogonal Systems

Assumptions. $\varphi_k(x)$, $w(x)$ are real and continuous, $f(x)$ real and piece-wise continuous.

Definition.

The set $\{\varphi_k(x)\}_1^\infty$ is an *orthogonal system* on (a, b) [finite or infinite interval] with respect to the *weight function* $w(x)\geq 0$ if

(*i*) $\int_a^b \varphi_k(x)\, \varphi_n(x) w(x) dx=0,\ n\neq k,$ (φ_k, φ_n *w-orthogonal*)

(*ii*) $N_k=\int_a^b \varphi_k^2(x) w(x) dx>0$

General Fourier Series

(12.1)
$$f(x)\sim \sum_{k=1}^{\infty} c_k\varphi_k(x), \qquad c_k=\frac{1}{N_k}\int_a^b f(x)\varphi_k(x)w(x)dx$$

Here, c_k are the (general) Fourier coefficients of $f(x)$.

Approximation in mean

If $s_n(x)=a_1\varphi_1(x)+\ldots+a_n\varphi_n(x)$, then

1. $Q_n=\int_a^b [f(x)-s_n(x)]^2 w(x) dx$ is *minimal* $\Leftrightarrow$ $a_k=c_k$, $k=1, \ldots, n$

2. $\min Q_n=\int_a^b f^2(x)w(x)dx-\sum_{k=1}^{n} N_k c_k^2$

The system $\{\varphi_k(x)\}_1^\infty$ is *complete* if $\lim_{n\to\infty}\int_a^b [f(x)-s_n(x)]^2 w(x)dx=0$ (with $a_k=c_k$), i.e. $s_n \to f$ *in mean* as $n \to\infty$.

Theorem.

3. $\{\varphi_k(x)\}_1^\infty$ orthogonal system $\Rightarrow \sum_{k=1}^{\infty} N_k c_k^2 \leqslant \int_a^b f^2(x)w(x)dx$
   (Bessel's inequality)

4. $\{\varphi_k(x)\}_1^\infty$ complete orthogonal system $\Rightarrow$

   (i) $\sum_{k=1}^{\infty} N_k c_k^2 = \int_a^b f^2(x)w(x)dx$ (Parseval's identity)

   (ii) $\sum_{k=1}^{\infty} N_k c_k d_k = \int_a^b f(x)g(x)w(x)dx$ (d_k Fourier coefficients of $g(x)$)

The Sturm–Liouville eigenvalue problem

Assumptions.

(i) $p(x)$, $p'(x)$, $q(x)$, $w(x)$ real, continuous in $[a, b]$
(ii) $p(x)>0$, $q(x)\geqslant 0$ in $[a, b]$, $w(x)>0$ in (a, b)
(iii) A, B, C, D real non-negative constants, $A^2+B^2>0$, $C^2+D^2>0$

The Sturm–Liouville eigenvalue problem

Find eigenvalues λ and corresponding eigenfunctions $\varphi(x)\not\equiv 0$ of the boundary value problem

$$\begin{cases} -\dfrac{d}{dx}\{p(x)\varphi'(x)\}+q(x)\varphi(x)=\lambda\, w(x)\,\varphi(x) \\ A\,\varphi'(a)-B\,\varphi(a)=0 \\ C\,\varphi'(b)+D\,\varphi(b)=0 \end{cases}$$

5. The eigenvalues λ_k satisfy $0\leqslant\lambda_1<\lambda_2<\lambda_3<\ldots$,
   $\lim_{n\to\infty}\dfrac{n^2}{\lambda_n}=\dfrac{1}{\pi^2}\left(\int_a^b\sqrt{\dfrac{w(x)}{p(x)}}dx\right)^2.$
6. Eigenfunctions corresponding to different eigenvalues are w-orthogonal.
7. The set $\{\varphi_k(x)\}_1^\infty$ of eigenfunctions is a complete orthogonal system.

Examples of complete trigonometric systems on the interval [0, a]

Differential equation: $\varphi''(x)+\lambda\varphi(x)=0,\ 0<x<a$

Boundary conditions	*Eigenvalues*	*Eigenfunctions*	$\beta_k=$	$N_k=$
$\varphi(0)=\varphi(a)=0$	$\lambda_k=\beta_k^2$	$\varphi_k(x)=\sin\beta_k x$	$k\pi/a,\ k=1, 2, 3, \ldots$	$a/2$
$\varphi'(0)=\varphi(a)=0$	$\lambda_k=\beta_k^2$	$\varphi_k(x)=\cos\beta_k x$	$(k+1/2)\pi/a,\ k=0, 1, 2, \ldots$	$a/2$
$\varphi(0)=\varphi'(a)=0$	$\lambda_k=\beta_k^2$	$\varphi_k(x)=\sin\beta_k x$	$(k+1/2)\pi/a,\ k=0, 1, 2, \ldots$	$a/2$
$\varphi'(0)=\varphi'(a)=0$	$\lambda_k=\beta_k^2$	$\varphi_k(x)=\cos\beta_k x$	$k\pi/a,\ k=0, 1, 2, \ldots$	a if $k=0$, $a/2$ if $k\geq 1$
$\begin{cases}\varphi(0)=0,\\ \varphi'(a)+c\varphi(a)=0\\ (c>0 \text{ constant})\end{cases}$	$\lambda_k=\beta_k^2$	$\varphi_k(x)=\sin\beta_k x$	β_k are the positive roots of $\tan a\beta=-\beta/c$	$[a+c/(c^2+\beta_k^2)]/2$
$\begin{cases}\varphi'(0)=0,\\ \varphi'(a)+c\varphi(a)=0\\ (c>0 \text{ constant})\end{cases}$	$\lambda_k=\beta_k^2$	$\varphi_k(x)=\cos\beta_k x$	β_k are the positive roots of $\tan a\beta=c/\beta$	$[a+c/(c^2+\beta_k^2)]/2$

A generalized eigenvalue problem

Notation: $(u|v)=\int_a^b u(x)v(x)dx$

Let A, B be real linear ordinary differential operators and consider the eigenvalue problem

$$(EP)\ \begin{cases}(DE)\ B\,v(x)=\lambda A\,v(x),\ a<x<b\\ (BC)\ \text{Real linear and homogeneous boundary conditions on } v \text{ and derivatives of } v \text{ at } a \text{ and}\end{cases}$$

Test function space $V=\{\text{functions satisfying } (BC)\}$.

The operator A is

(*i*) symmetric if $(Au|v)=(u|Av)$, all $u, v\in V$
(*ii*) positive definite [semidefinite] if

$$(Av|v)>0\ [(Av|v)\geq 0] \text{ all } v\in V,\ v\not\equiv 0$$

The eigenvalue problem (EP) is

(*i*) *self-adjoint* if A and B are symmetric
(*ii*) *totally definite* if A and B are positive definite
(*iii*) *positive semidefinite* if A is positive definite and B is positive semidefinite.

Remark. Sturm–Liouville's (EP) is self-adjoint and positive semidefinite.

Assume that (EP) is self-adjoint.

8. If u, v are eigenfunctions of (EP) corresponding to different eigenvalues, then $(Au|v)=(Bu|v)=0$.
9. If (EP) is totally definite [positive semidefinite], then every eigenvalue is >0 [≥ 0].

Rayleigh's quotient $R(v)=\dfrac{(Bv|v)}{(Av|v)}$.

Assume that

(*i*) (EP) is self-adjoint and positive semidefinite and order (B) > order (A)
(*ii*) $0 \leq \lambda_1 \leq \lambda_2 \leq \ldots$ and $v_1, v_2, \ldots$ are eigenvalues and eigenfunctions respectively of (EP).

Then

$$\lambda_n = \min R(v) = R(v_n),$$

where the minimum is taken over $v \in V$ such that

$$(Av|v_k) = 0,\ k = 1, \ldots, n-1$$

and the minimum is assumed by the eigenfunctions corresponding to λ_n.

Iteration

Assume that (EP) is self-adjoint and total definite, order (B) > order (A).

Iteration for λ_1 and v_1:

$$\begin{cases} \text{Choose } u_0 \in V \\ u_k\text{: } Bu_k = Au_{k-1} \text{ with } (BC),\ v_1 \approx u_k\ (k \text{ big enough})\ \lambda_1 \approx u_k(x)/u_{k+1}(x). \end{cases}$$

Schwarz' constants $a_k = (u_{k-m}|Au_m)$ independent of m.

Schwarz' quotients $\mu_k = a_{k-1}/a_k \searrow \lambda_1$.

$$\mu_k - \lambda_1 \leq \frac{\mu_k}{\lambda_2 - \mu_k}(\mu_{k-1} - \mu_k) \text{ if } \lambda_2 > \mu_k$$

12.2 Orthogonal Polynomials

Legendre polynomials $P_n(x)$

Explicit form. $P_n(x) = 2^{-n} \sum_{k=0}^{[n/2]} (-1)^k \binom{n}{k} \binom{2n-2k}{n} x^{n-2k},\ n = 0, 1, 2, \ldots$

Rodrigues' formula: $P_n(x) = \dfrac{1}{n!\,2^n} D^n\{(x^2-1)^n\}$

$$P_n(-x) = (-1)^n P_n(x); \qquad P_n(1) = 1,\ |P_n(x)| \leq 1,\ -1 \leq x \leq 1$$

Orthogonality: $\int_{-1}^{1} P_k(x)P_n(x)dx = \dfrac{2}{2n+1}\delta_{kn}$

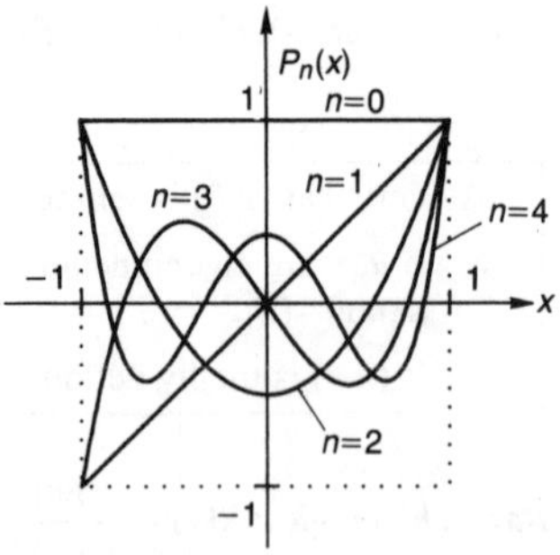

Recurrence formulas

$$(n+1)P_{n+1}(x) = (2n+1)xP_n(x) - nP_{n-1}(x)$$
$$(x^2-1)P_n'(x) = nxP_n(x) - nP_{n-1}(x)$$

Generating function

$$(1-2xt+t^2)^{-1/2} = \sum_{n=0}^{\infty} P_n(x)t^n,\ |t| < 1,\ |x| \leq 1$$

Differential equation. $y = P_n(x)$ satisfies

$$(1-x^2)y'' - 2xy' + n(n+1)y = 0$$

$P_n=P_n(x)$:	
$P_0=1$	$x^{10}=(256P_{10}+2176P_8+7904P_6+1550P_4+16150P_2+4199P_0)/46189$
$P_1=x$	$x^9=(128P_9+960P_7+2992P_5+4760P_3+3315P_1)/12155$
$P_2=(3x^2-1)/2$	$x^8=(128P_8+832P_6+2160P_4+2600P_2+715P_0)/6435$
$P_3=(5x^3-3x)/2$	$x^7=(16P_7+88P_5+182P_3+143P_1)/429$
$P_4=(35x^4-30x^2+3)/8$	$x^6=(16P_6+72P_4+110P_2+33P_0)/231$
$P_5=(63x^5-70x^3+15x)/8$	$x^5=(8P_5+28P_3+27P_1)/63$
$P_6=(231x^6-315x^4+105x^2-5)/16$	$x^4=(8P_4+20P_2+7P_0)/35$
$P_7=(429x^7-693x^5+315x^3-35x)/16$	$x^3=(2P_3+3P_1)/5$
$P_8=(6435x^8-12012x^6+6930x^4-1260x^2+35)/128$	$x^2=(2P_2+P_0)/3$
$P_9=(12155x^9-25740x^7+18018x^5-4620x^3+315x)/128$	$x=P_1$
$P_{10}=(46189x^{10}-109395x^8+90090x^6-30030x^4+3465x^2-63)/256$	$1=P_0$

The associated Legendre functions $P_n^m(x)$

$y=P_n^m(x)=(1-x^2)^{m/2}D^mP_n(x)$, $0\leq m\leq n$, satisfy the *differential equation*

$$(1-x^2)y''-2xy'+\left[n(n+1)-\frac{m^2}{1-x^2}\right]y=0$$

Orthogonality. $\int_{-1}^{1} P_k^m(x)\,P_n^m(x)dx=\frac{(n+m)!}{(n-m)!}\cdot\frac{2}{2n+1}\,\delta_{kn}$

Spherical harmonics

$\cos m\varphi\ P_n^m(\cos\theta)$ and $\sin m\varphi\ P_n^m(\cos\theta)$ satisfy the *partial differential equation*

$$\frac{1}{\sin\theta}\frac{\partial}{\partial\theta}\left(\sin\theta\,\frac{\partial y}{\partial\theta}\right)+\frac{1}{\sin^2\theta}\frac{\partial^2 y}{\partial\varphi^2}+n(n+1)y=0$$

Chebyshev's polynomials $T_n(x)$ and $U_n(x)$

$T_n(x)=\cos(n\arccos x)$ $\quad U_n(x)=\frac{\sin[(n+1)\arccos x]}{\sqrt{1-x^2}}$, $-1<x<1$, $n=0, 1, 2, \ldots$

$T_n(\cos\theta)=\cos n\theta$ $\quad U_n(\cos\theta)=\frac{\sin(n+1)\theta}{\sin\theta}$

$T_n(-x)=(-1)^nT_n(x)$, $\quad |T_n(x)|\leq 1$

$U_n(-x)=(-1)^nU_n(x)$, $\quad |U_n(x)|\leq n+1$

Orthogonality

$$\int_{-1}^{1} T_k(x)T_n(x)\frac{dx}{\sqrt{1-x^2}}=\begin{cases}0, & k\neq n\\ \pi, & k=n=0\\ \pi/2, & k=n\neq 0\end{cases}\qquad \int_{-1}^{1} U_k(x)U_n(x)\sqrt{1-x^2}\,dx=\frac{\pi}{2}\,\delta_{kn}$$

Recurrence formulas

$T_{n+1}(x)=2x\ T_n(x)-T_{n-1}(x)$ $\quad U_{n+1}(x)=2x\ U_n(x)-U_{n-1}(x)$

Generating functions

$$\frac{1-xt}{1-2xt+t^2}=\sum_{n=0}^{\infty}T_n(x)t^n,\qquad |x|<1,\ |t|<1$$

$$\frac{1}{1-2xt+t^2}=\sum_{n=0}^{\infty}U_n(x)t^n,\qquad |x|<1,\ |t|<1$$

Differential equation

$$(1-x^2)y''-xy'+n^2y=0,\ -1<x<1$$

General solution: $y=AT_n(x)+BU_{n-1}(x)\sqrt{1-x^2}$

$U_n(x)$ solves the equation $(1-x^2)y''-3xy'+n(n+2)y=0$

$T_n(x)=0 \Leftrightarrow x=x_k=\cos((k+1/2)\pi/n),\ k=0,\ 1,\ \ldots,\ n-1.$
$|T_n(x)|=1 \Leftrightarrow x=x_k'=\cos(k\pi/n),\ k=0,\ 1,\ \ldots,\ n.$
$T_n=T_n(x)$:

$T_0=1$	$x^{10}=(T_{10}+10T_8+45T_6+120T_4+210T_2+126T_0)/512$
$T_1=x$	$x^9=(T_9+9T_7+36T_5+84T_3+126T_1)/256$
$T_2=2x^2-1$	$x^8=(T_8+8T_6+28T_4+56T_2+35T_0)/128$
$T_3=4x^3-3x$	$x^7=(T_7+7T_5+21T_3+35T_1)/64$
$T_4=8x^4-8x^2+1$	$x^6=(T_6+6T_4+15T_2+10T_0)/32$
$T_5=16x^5-20x^3+5x$	$x^5=(T_5+5T_3+10T_1)/16$
$T_6=32x^6-48x^4+18x^2-1$	$x^4=(T_4+4T_2+3T_0)/8$
$T_7=64x^7-112x^5+56x^3-7x$	$x^3=(T_3+3T_1)/4$
$T_8=128x^8-256x^6+160x^4-32x^2+1$	$x^2=(T_2+T_0)/2$
$T_9=256x^9-576x^7+432x^5-120x^3+9x$	$x=T_1$
$T_{10}=512x^{10}-1280x^8+1120x^6-400x^4+50x^2-1$	$1=T_0$

$U_n=U_n(x)$:

$U_0=1$	$x^{10}=(U_{10}+9U_8+35U_6+75U_4+90U_2+42U_0)/1024$
$U_1=2x$	$x^9=(U_9+8U_7+27U_5+48U_3+42U_1)/512$
$U_2=4x^2-1$	$x^8=(U_8+7U_6+20U_4+28U_2+14U_0)/256$
$U_3=8x^3-4x$	$x^7=(U_7+6U_5+14U_3+14U_1)/128$
$U_4=16x^4-12x^2+1$	$x^6=(U_6+5U_4+9U_2+5U_0)/64$
$U_5=32x^5-32x^3+6x$	$x^5=(U_5+4U_3+5U_1)/32$
$U_6=64x^6-80x^4+24x^2-1$	$x^4=(U_4+3U_2+2U_0)/16$
$U_7=128x^7-192x^5+80x^3-8x$	$x^3=(U_3+2U_1)/8$
$U_8=256x^8-448x^6+240x^4-40x^2+1$	$x^2=(U_2+U_0)/4$
$U_9=512x^9-1024x^7+672x^5-160x^3+10x$	$x=U_1/2$
$U_{10}=1024x^{10}-2304x^8+1792x^6-560x^4+60x^2-1$	$1=U_0$

Shifted Chebyshev's polynomials $T_n^*(x)$ and $U_n^*(x)$

$T_n^*(x)=T_n(2x-1)=T_{2n}(\sqrt{x}),\ 0\leq x\leq 1$

$U_n^*(x)=U_n(2x-1),\ 0\leq x\leq 1$

Orthogonality

$$\int_0^1 T_k^*(x)T_n^*(x)(x-x^2)^{-1/2}dx=\begin{cases}0, & k\neq n\\ \pi, & k=n\\ \pi/2, & k=n\neq 0\end{cases}$$

$$\int_0^1 U_k^*(x)U_n^*(x)(x-x^2)^{1/2}dx=(\pi/8)\delta_{kn}$$

Recurrence formulas

$T_{n+1}^*(x)=(4x-2)T_n^*(x)-T_{n-1}^*(x)$

$U_{n+1}^*(x)=(4x-2)U_n^*(x)-U_{n-1}^*(x)$

Differential equations

$T_n^*(x)$ solves the equation
$(x-x^2)y''-(x-1/2)y'+n^2y=0$

$U_n^*(x)$ solves the equation
$(x-x^2)y''-3(x-1/2)y'+n(n+2)y=0$

$T_n^* = T_n^*(x)$:

$T_0^*=1$	$x^7=(T_7^*+14T_6^*+91T_5^*+364T_4^*+1001T_3^*+2002T_2^*+3003T_1^*+1716T_0^*)/8192$
$T_1^*=2x-1$	$x^6=(T_6^*+12T_5^*+66T_4^*+220T_3^*+495T_2^*+792T_1^*+462T_0^*)/2048$
$T_2^*=8x^2-8x+1$	$x^5=(T_5^*+10T_4^*+45T_3^*+120T_2^*+210T_1^*+126T_0^*)/512$
$T_3^*=32x^3-48x^2+18x-1$	$x^4=(T_4^*+8T_3^*+28T_2^*+56T_1^*+35T_0^*)/128$
$T_4^*=128x^4-256x^3+160x^2-32x+1$	$x^3=(T_3^*+6T_2^*+15T_1^*+10T_0^*)/32$
$T_5^*=512x^5-1280x^4+1120x^3-400x^2+50x-1$	$x^2=(T_2^*+4T_1^*+3T_0^*)/8$
$T_6^*=2048x^6-6144x^5+6912x^4-3584x^3+840x^2-72x+1$	$x=(T_1^*+T_0^*)/2$
$T_7^*=8192x^7-28672x^6+39424x^5-26880x^4+9408x^3-1568x^2+98x-1$	$1=T_0^*$

$U_n^* = U_n^*(x)$:

$U_0^*=1$	$x^7=(U_7^*+14U_6^*+90U_5^*+350U_4^*+910U_3^*+1638U_2^*+2002U_1^*+1430\,U_0^*)/16384$
$U_1^*=4x-2$	$x^6=(U_6^*+12U_5^*+65U_4^*+208U_3^*+429U_2^*+572U_1^*+429U_0^*)/4096$
$U_2^*=16x^2-16x+3$	$x^5=(U_5^*+10U_4^*+44U_3^*+110U_2^*+165U_1^*+132U_0^*)/1024$
$U_3^*=64x^3-96x^2+40x-4$	$x^4=(U_4^*+8U_3^*+27U_2^*+48U_1^*+42U_0^*)/256$
$U_4^*=256x^4-512x^3+336x^2-80x+5$	$x^3=(U_3^*+6U_2^*+14U_1^*+14U_0^*)/64$
$U_5^*=1024x^5-2560x^4+2304x^3-896x^2+140x-6$	$x^2=(U_2^*+4U_1^*+5U_0^*)/16$
$U_6^*=4096x^6-12288x^5+14080x^4-7680x^3+2016x^2-224x+7$	$x=(U_1^*+2U_0^*)/4$
$U_7^*=16384x^7-57344x^6+79872x^5-56320x^4+21120x^3-4032x^2+336x-8$	$1=U_0^*$

Hermite's polynomials $H_n(x)$

Rodrigues' formula. $H_n(x)=(-1)^n e^{x^2} D^n(e^{-x^2})$, $n=0, 1, 2, \dots$

Hermite's functions $h_n(x)=e^{-x^2/2}H_n(x)$

Orthogonality

$$\int_{-\infty}^{\infty} H_k(x)H_n(x)e^{-x^2}dx = \int_{-\infty}^{\infty} h_k(x)h_n(x)dx = n!\ 2^n\sqrt{\pi}\ \delta_{kn}$$

Recurrence formulas

$H_{n+1}(x)=2x\ H_n(x)-2n\ H_{n-1}(x)$

$H_n'(x)=2n\ H_{n-1}(x)$

Generating function

$$e^{2tx-t^2}=\sum_{n=0}^{\infty} H_n(x)\,\frac{t^n}{n!}, \qquad -\infty<x<\infty,\ -\infty<t<\infty$$

Differential equation

$y=H_n(x)$ satisfies $\quad y''-2xy'+2ny=0$

$y=h_n(x)$ satisfies $\quad y''+(2n+1-x^2)y=0$

$H_n = H_n(x)$:

$H_0=1$	$x^{10}=(H_{10}+90H_8+2520H_6+25200H_4+75600H_2+30240H_0)/1024$
$H_1=2x$	$x^9=(H_9+72H_7+1512H_5+10080H_3+15120H_1)/512$
$H_2=4x^2-2$	$x^8=(H_8+56H_6+840H_4+3360H_2+1680H_0)/256$
$H_3=8x^3-12x$	$x^7=(H_7+42H_5+420H_3+840H_1)/128$
$H_4=16x^4-48x^2+12$	$x^6=(H_6+30H_4+180H_2+120H_0)/64$
$H_5=32x^5-160x^3+120x$	$x^5=(H_5+20H_3+60H_1)/32$
$H_6=64x^6-480x^4+720x^2-120$	$x^4=(H_4+12H_2+12H_0)/16$
$H_7=128x^7-1344x^5+3360x^3-1680x$	$x^3=(H_3+6H_1)/8$
$H_8=256x^8-3584x^6+13440x^4-13440x^2+1680$	$x^2=(H_2+2H_0)/4$
$H_9=512x^9-9216x^7+48384x^5-80640x^3+30240x$	$x=H_1/2$
$H_{10}=1024x^{10}-23040x^8+161280x^6-403200x^4+302400x^2-30240$	$1=H_0$

Laguerre's polynomials $L_n(x)$, $L_n^{(\alpha)}(x)$

Rodrigues' formula

$$L_n^{(\alpha)}(x)=\frac{x^{-\alpha}e^x}{n!}D^n(x^{n+\alpha}e^{-x}),\qquad L_n(x)=L_n^{(0)}(x),\ n=0,\ 1,\ 2,\ \dots$$

Laguerre's function. $l_n(x)=e^{-x/2}L_n(x)$

Orthogonality

$$\int_0^\infty L_k^{(\alpha)}(x)\,L_n^{(\alpha)}(x)\,x^\alpha e^{-x}dx=\frac{\Gamma(1+\alpha+n)}{n!}\,\delta_{kn}\quad(\alpha>-1)$$

$$\int_0^\infty L_k(x)\,L_n(x)\,e^{-x}dx=\delta_{kn}$$

Recurrence formulas

$$(n+1)\,L_{n+1}^{(\alpha)}(x)=(2n+\alpha+1-x)\,L_n^{(\alpha)}(x)-(n+\alpha)\,L_{n-1}^{(\alpha)}(x)$$

$$\frac{d}{dx}\,L_n^{(\alpha)}(x)=-L_{n-1}^{(\alpha+1)}(x)$$

Generating function

$$(1-t)^{-\alpha-1}\exp\left(-\frac{xt}{1-t}\right)=\sum_{n=0}^{\infty}L_n^{(\alpha)}(x)t^n,\qquad |t|<1$$

Differential equation

$y=L_n^{(\alpha)}(x)$ satisfies

$$xy''+(1+\alpha-x)y'+ny=0$$

$L_n=L_n(x)$:

$L_0=1$	$x^7=5040L_0-35280L_1+105840L_2-176400L_3+176400L_4-105840L_5+35280L_6-5040L_7$
$L_1=1-x$	$x^6=720L_0-4320L_1+10800L_2-14400L_3+10800L_4-4320L_5+720L_6$
$L_2=1-2x+x^2/2$	$x^5=120L_0-600L_1+1200L_2-1200L_3+600L_4-120L_5$
$L_3=1-3x+3x^2/2-x^3/6$	$x^4=24L_0-96L_1+144L_2-96L_3+24L_4$
$L_4=1-4x+3x^2-2x^3/3+x^4/24$	$x^3=6L_0-18L_1+18L_2-6L_3$
$L_5=1-5x+5x^2-5x^3/3+5x^4/24-x^5/120$	$x^2=2L_0-4L_1+2L_2$
$L_6=1-6x+15x^2/2-10x^3/3+5x^4/8-x^5/20+x^6/720$	$x=L_0-L_1$
$L_7=1-7x+21x^2/2-35x^3/6+35x^4/24-7x^5/40+7x^6/720-x^7/5040$	$1=L_0$

Jacobi's polynomials $P_n^{(\alpha,\beta)}(x)$

Rodrigues' formula

$$P_n^{(\alpha,\beta)}(x)=\frac{(-1)^n}{2^n n!}(1-x)^{-\alpha}(1+x)^{-\beta}D^n\{(1-x)^{n+\alpha}(1+x)^{n+\beta}\}$$

Explicit form

$$P_n^{(\alpha,\beta)}(x)=2^{-n}\sum_{k=0}^{n}\binom{n+\alpha}{k}\binom{n+\beta}{n-k}(x-1)^{n-k}(x+1)^k$$

Orthogonality

$$\int_{-1}^{1}P_k^{(\alpha,\beta)}(x)\,P_n^{(\alpha,\beta)}(x)\,(1-x)^\alpha(1+x)^\beta dx=\frac{2^{\alpha+\beta+1}\Gamma(n+\alpha+1)\,\Gamma(n+\beta+1)}{(2n+\alpha+\beta+1)n!\,\Gamma(n+\alpha+\beta+1)}\,\delta_{kn}$$

Generating function

$$u^{-1}(1-t+u)^{-\alpha}(1+t+u)^{-\beta}=\sum_{n=0}^{\infty}2^{-\alpha-\beta}P_n^{(\alpha,\beta)}(x)t^n,\qquad |x|<1,\ |t|<1$$

$$u=\sqrt{1-2xt+t^2}$$

Differential equation

$y=P_n^{(\alpha,\beta)}(x)$ satisfies

$$(1-x^2)y''+[\beta-\alpha-(\alpha+\beta+2)x]y'+n(n+\alpha+\beta+1)y=0$$

12.3 Bernoulli and Euler Polynomials

Bernoulli polynomials $B_n(x)$

Generating function

$$\frac{te^{xt}}{e^t-1}=\sum_{n=0}^{\infty}B_n(x)\frac{t^n}{n!},\qquad |t|<2\pi$$

Relations. $B_n'(x)=nB_{n-1}(x)$; $B_n(x+1)-B_n(x)=nx^{n-1}$

$B_0(x)=1$	$B_1(x)=x-\frac{1}{2}$	$B_2(x)=x^2-x+\frac{1}{6}$	$B_3(x)=x^3-\frac{3}{2}x^2+\frac{1}{2}x$
$B_4(x)=x^4-2x^3+x^2-\frac{1}{30}$	$B_5(x)=x^5-\frac{5}{2}x^4+\frac{5}{3}x^3-\frac{1}{6}x$		

Bernoulli numbers B_n

$$B_n=B_n(0),\ B_{2n+1}=0,\ B_{2n}=(-1)^{n+1}\frac{2(2n)!}{\pi^{2n}(2^{2n}-1)}\sum_{k=0}^{\infty}(2k+1)^{-2n},\ n\geq 1$$

n	B_n	n	B_n	n	B_n
0	1	6	1/42	14	7/6
1	−1/2	8	−1/30	16	−3617/510
2	1/6	10	5/66	18	43867/798
4	−1/30	12	−691/2730	20	−174611/330

Euler polynomials $E_n(x)$

Generating function

$$\frac{2e^{xt}}{e^t+1} = \sum_{n=0}^{\infty} E_n(x) \frac{t^n}{n!}, \quad |t|<\pi$$

Relations. $E_n'(x)=nE_{n-1}(x);\quad E_n(x+1)+E_n(x)=2x^n$

$E_0(x)=1$	$E_1(x)=x-\frac{1}{2}$	$E_2(x)=x^2-x$	$E_3(x)=x^3-\frac{3}{2}x^2+\frac{1}{4}$
$E_4(x)=x^4-2x^3+x$	$E_5(x)=x^5-\frac{5}{2}x^4+\frac{5}{2}x^2-\frac{1}{2}$		

Euler numbers E_n

$E_n=2^n E_n\left(\frac{1}{2}\right)$=integer. $E_{2n+1}=0$, $E_{2n}=(-1)^n\frac{(2n)!\,2^{2n+2}}{\pi^{2n+1}}\sum_{k=0}^{\infty}(-1)^k(2k+1)^{-2n-1}$, $n\geq 0$

Generating function

$$\frac{2e^{t/2}}{e^t+1} = \sum_{n=0}^{\infty} 2^{-n} E_n \frac{t^n}{n!}$$

n	E_n	n	E_n	n	E_n
0	1	8	1385	16	193915 12145
2	−1	10	−50521	18	−240 48796 75441
4	5	12	27 02765	20	37037 11882 37525
6	−61	14	−1993 60981		

12.4 Bessel Functions

Bessel functions $J_p(x)$

$$\begin{cases} J_p(x)=\displaystyle\sum_{k=0}^{\infty}\frac{(-1)^k}{k!\,\Gamma(p+k+1)}\left(\frac{x}{2}\right)^{p+2k}, \quad (p\neq\text{integer})\ (0<x<\infty) \\ J_n(x)=\displaystyle\sum_{k=0}^{\infty}\frac{(-1)^k}{k!(n+k)!}\left(\frac{x}{2}\right)^{n+2k}=\frac{1}{\pi}\int_0^{\pi}\cos(x\sin\varphi-n\varphi)d\varphi=\frac{1}{2\pi}\int_{-\pi}^{\pi}e^{i(x\sin\varphi-n\varphi)}d\varphi, \\ \qquad n=0,1,2,\ldots \\ J_{-n}(x)=(-1)^n J_n(x),\ n=1,2,3,\ldots \end{cases}$$

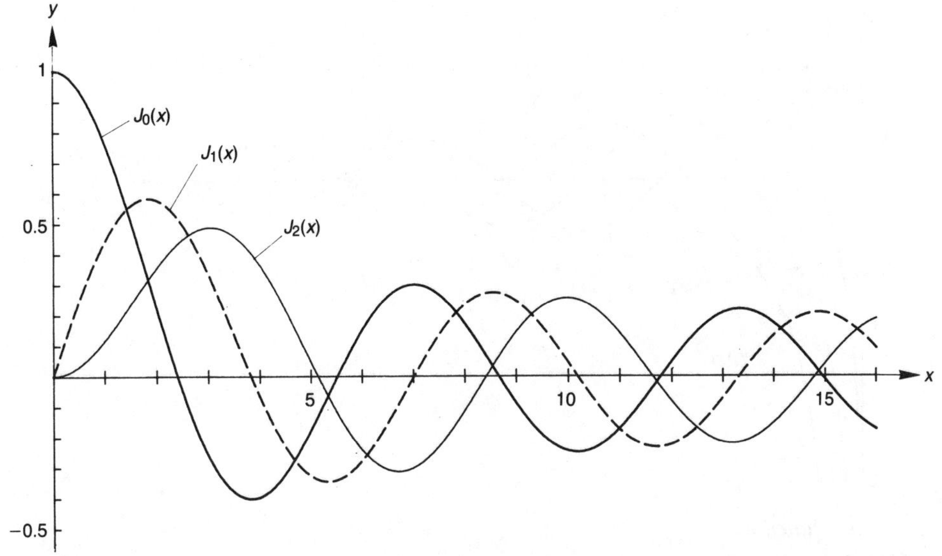

Related functions

Weber (Neumann) functions $Y_p(x)$ (or $N_p(x)$)

$$
\begin{cases}
Y_p(x)=\dfrac{J_p(x)\cos p\pi - J_{-p}(x)}{\sin p\pi}, \qquad p\neq\text{integer} \\[2ex]
Y_n(x)=\lim\limits_{p\to n} Y_p(x)=\dfrac{2}{\pi}\left(\gamma+\ln\dfrac{x}{2}\right)J_n(x)-\dfrac{1}{\pi}\sum\limits_{k=0}^{n-1}\dfrac{(n-k-1)!}{k!}\left(\dfrac{x}{2}\right)^{2k-n}- \\[2ex]
\qquad -\dfrac{1}{\pi}\sum\limits_{k=0}^{\infty}(H_k+H_{k+n})\dfrac{(-1)^k}{k!(n+k)!}\left(\dfrac{x}{2}\right)^{2k+n}=\dfrac{1}{\pi}\int\limits_0^{\pi}\sin(x\sin t-nt)dt- \\[2ex]
\qquad -\dfrac{1}{\pi}\int\limits_0^{\infty}[e^{nt}+(-1)^n e^{-nt}]e^{-x\sinh t}dt, \\[1ex]
\qquad\qquad\qquad n=0,\ 1,\ 2,\ \ldots, \\[1ex]
\text{where } H_m=\sum\limits_{j=1}^{m}\dfrac{1}{j}, \qquad \gamma=\text{Euler's constant.} \\[2ex]
Y_{-n}(x)=(-1)^n\, Y_n(x),\ n=1,\ 2,\ 3,\ \ldots
\end{cases}
$$

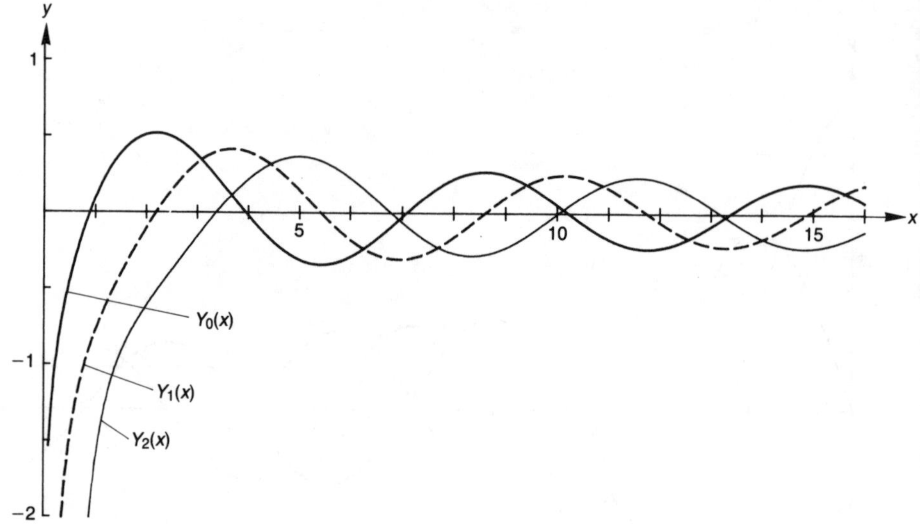

Hankel functions

$H_p^{(1)}(x)=J_p(x)+iY_p(x)$ $\qquad$ $H_p^{(2)}(x)=J_p(x)-iY_p(x)$

The modified Bessel functions

$$I_n(x)=i^{-n}\,J_n(ix)=\sum_{k=0}^{\infty}\frac{1}{k!\,(n+k)!}\left(\frac{x}{2}\right)^{n+2k}=\frac{1}{\pi}\int_0^{\pi}e^{x\cos t}\cos nt\,dt$$

$$K_n(x)=\frac{\pi}{2}\,i^{n+1}\,H_n^{(1)}(ix)=\frac{\pi}{2}\,i^{n+1}\,[J_n(ix)+iY_n(ix)]=\int_0^{\infty}e^{-x\cosh t}\cosh nt\,dt$$

The functions ber(x), bei(x), ker(x) and kei(x) are defined by

$$\text{ber}(x)+i\,\text{bei}(x)=J_0(i^{3/2}x)$$

$$\text{ker}(x)+i\,\text{kei}(x)=\frac{i\pi}{2}\,H_0^{(1)}(i^{3/2}x)$$

Differential equations

1. $x^2y''+xy'+(a^2x^2-p^2)y=0$, $(a>0)$
   Solution: $y=AJ_p(ax)+BY_p(ax)$
   [Observe that $|Y_p(x)|\to\infty$ as $x\to 0^+$.]

2. $x^2y''+xy'-(a^2x^2+n^2)y=0$
   Solution: $y=AI_n(ax)+BK_n(ax)$

Transformed Bessel differential equations

Equation	*Solution* $(C_p=AJ_p+BY_p)$
$(xy')'+(a^2x-p^2/x)y=0$	$y=C_p(ax)$
$x^2y''+(a^2x^2-p^2+1/4)y=0$	$y=x^{1/2}C_p(ax)$
$xy''-(2p-1)y'+a^2xy=0$	$y=x^pC_p(ax)$
$y''+a^2x^{p-2}y=0$	$y=x^{1/2}C_{1/p}(2ax^{p/2}/p))$
$y''+(a^2e^{2x}-p^2)y=0$	$y=C_p(ae^x)$
$x^2y''+(2p+1)xy'+(b^2x^{2r}+c^2)y=0$	$y=x^{-p}\{C_{q/r}(bx^r/r)\}$, $q=\sqrt{p^2-q^2}$

If a^2 is replaced by $-a^2$ in the equations, then in the solutions J_p and Y_p are replaced by I_p and K_p respectively.

Generating function

$$\exp\left[\frac{x}{2}\left(t-\frac{1}{t}\right)\right] = \sum_{-\infty}^{\infty} J_n(x)t^n \quad e^{ix\sin\varphi} = \sum_{-\infty}^{\infty} J_n(x)\, e^{in\varphi}$$

In particular $\cos x = J_0(x) - 2J_2(x) + 2J_4(x) - 2J_6(x) + \dots$

$\sin x = 2J_1(x) - 2J_3(x) + 2J_5(x) - \dots$

Zeros

The functions $J_n(x)$ and $J_n'(x)$, $n=0, 1, 2, \dots$ have inifinitely many simple zeros. These are denoted by ξ_{nj} and η_{nj}, $j=1, 2, 3, \dots$ respectively and are tabulated below for $0 \leqslant n \leqslant 7$, $1 \leqslant j \leqslant 10$.

Eigenvalue problems

1. Dirichlet condition. (Singular) eigenvalue problem:

$$\begin{cases} (xy'(x))' + (\lambda x - n^2/x)y(x) = 0, \; 0<x<a, \; n=0, 1, 2, \dots \\ y(0) \text{ bounded} \\ y(a)=0 \end{cases}$$

Eigenvalues: $\lambda = (\xi_{nj}/a)^2$, $j=1, 2, 3, \dots$

Eigenfunctions: $J_n(\xi_{nj}x/a)$, $j=1, 2, 3, \dots$

Orthogonality: $\int_0^a J_n(\xi_{nj}x/a)\, J_n(\xi_{nk}x/a)x\, dx = (a^2/2)[J_n'(\xi_{nj})]^2\delta_{jk}$

2. Neumann condition. (Singular) eigenvalue problem:

$$\begin{cases} (xy'(x))' + (\lambda x - n^2/x)y(x) = 0, \; 0<x<a, \; n=0, 1, 2, \dots \\ y(0) \text{ bounded} \\ y'(a)=0 \end{cases}$$

Eigenvalues: $\lambda = (\eta_{nj}/a)^2$, $j=1, 2, 3, \dots$

Eigenfunctions: $J_n(\eta_{nj}x/a)$, $j=1, 2, 3, \dots$

Orthogonality: $\int_0^a J_n(\eta_{nj}x/a)\, J_n(\eta_{nk}x/a)x\, dx = (a^2/2)(1-n^2/\eta_{nj}^2)[J_n(\eta_{nj})]^2\delta_{jk}$

Roots α_n of $J_0(x)=0$ and the corresponding values of $J_1(x)$

n	α_n	$J_1(\alpha_n)$	n	α_n	$J_1(\alpha_n)$
1	2.40	+0.5191	21	65.19	+0.0988
2	5.52	−0.3403	22	68.33	−0.0965
3	8.65	+0.2715	23	71.47	+0.0944
4	11.79	−0.2325	24	74.61	−0.0924
5	14.93	+0.2065	25	77.76	+0.0905
6	18.07	−0.1877	26	80.90	−0.0887
7	21.21	+0.1733	27	84.04	+0.0870
8	24.35	−0.1617	28	87.18	−0.0854
9	27.49	+0.1522	29	90.32	+0.0839
10	30.63	−0.1442	30	93.46	−0.0825
11	33.78	+0.1373	31	96.61	+0.0812
12	36.92	−0.1313	32	99.75	−0.0799
13	40.06	+0.1261	33	102.89	+0.0787
14	43.20	−0.1214	34	106.03	−0.0775
15	46.34	+0.1172	35	109.17	+0.0764
16	49.48	−0.1134	36	112.31	−0.0753
17	52.62	+0.1100	37	115.45	+0.0743
18	55.77	−0.1068	38	118.60	−0.0733
19	58.91	+0.1040	39	121.74	+0.0723
20	62.05	−0.1013	40	124.88	−0.0714

Asymptotic behavior

1. $x\to\infty$

$$J_n(x)=\sqrt{\frac{2}{\pi x}}\left[\cos\left(x-\frac{\pi}{4}-\frac{n\pi}{2}\right)+O\left(\frac{1}{x}\right)\right] \qquad Y_n(x)=\sqrt{\frac{2}{\pi x}}\left[\sin\left(x-\frac{\pi}{4}-\frac{n\pi}{2}\right)+O\left(\frac{1}{x}\right)\right]$$

$$I_n(x)=\frac{e^x}{\sqrt{2\pi x}}\left[1+O\left(\frac{1}{x}\right)\right] \qquad K_n(x)=\sqrt{\frac{\pi}{2x}}e^{-x}\left[1+O\left(\frac{1}{x}\right)\right]$$

2. $x\to 0^+$

$$J_n(x)\sim\frac{1}{n!}\left(\frac{x}{2}\right)^n \qquad Y_0(x)\sim\frac{2}{\pi}\ln x \qquad Y_n(x)\sim-\frac{(n-1)!}{\pi}\left(\frac{2}{x}\right)^n,\ n>0$$

$$I_n(x)\sim\frac{1}{n!}\left(\frac{x}{2}\right)^n \qquad K_0(x)\sim-\ln x \qquad K_n(x)\sim\frac{(n-1)!}{2}\left(\frac{2}{x}\right)^n,\ n>0$$

Recurrence formulas

For any cylinder function $C_p(x)=J_p(x)$, $Y_p(x)$, $H_p^{(1)}(x)$ or $H_p^{(2)}(x)$:

$$C_{p-1}(x)+C_{p+1}(x)=\frac{2p}{x}C_p(x) \qquad C_{p-1}(x)-C_{p+1}(x)=2C_p'(x)$$

$$\frac{d}{dx}\{x^pC_p(x)\}=x^pC_{p-1}(x) \qquad \frac{d}{dx}\{x^{-p}C_p(x)\}=-x^{-p}C_{p+1}(x)$$

In particular: $J_0'(x)=-J_1(x)$, $Y_0'(x)=-Y_1(x)$

$$\int C_n^2(x)x\,dx=\frac{1}{2}\,x^2[C_n'(x)]^2+\frac{1}{2}(x^2-n^2)C_n^2(x)$$

$$\int x^{1+n}C_n(x)dx=x^{1+n}C_{n+1}(x)=-x^{1-n}\left[C_n'(x)-\frac{n}{x}\,C_n(x)\right]$$

$$\int x^{1-n}C_n(x)dx=-x^{1-n}C_{n-1}(x)=-x^{1-n}\left[C_n'(x)+\frac{n}{x}\,C_n(x)\right]$$

$$I_{n+1}(x)=I_{n-1}(x)-\frac{2n}{x}\,I_n(x)=2I_n'(x)-I_{n-1}(x)$$

$$K_{n+1}(x)=K_{n-1}(x)+\frac{2n}{x}\,K_n(x)=-2K_n'(x)-K_{n-1}(x)$$

Half order functions

$$J_{1/2}(x)=\sqrt{\frac{2}{\pi x}}\,\sin x \qquad J_{-1/2}(x)=\sqrt{\frac{2}{\pi x}}\,\cos x$$

$$I_{1/2}(x)=\sqrt{\frac{2}{\pi x}}\,\sinh x \qquad I_{-1/2}(x)=\sqrt{\frac{2}{\pi x}}\,\cosh x$$

Further half order Bessel functions may be obtained by the recurrence formulas above.

Spherical Bessel functions

$$j_n(x)=\sqrt{\frac{\pi}{2x}}\,J_{n+1/2}(x)=x^n\left(-\frac{1}{x}\frac{d}{dx}\right)^n\frac{\sin x}{x}=\frac{x^n}{2^{n+1}n!}\int_0^{\pi}\cos(x\cos t)\sin^{2n+1}t\,dt$$

$$y_n(x)=(-1)^{n+1}\sqrt{\frac{\pi}{2x}}\,J_{-n-1/2}(x)=\sqrt{\frac{\pi}{2x}}\,Y_{n+1/2}(x)=-x^n\left(-\frac{1}{x}\frac{d}{dx}\right)^n\frac{\cos x}{x}$$

$$h_n^{(1)}(x)=j_n(x)+iy_n(x)=\sqrt{\frac{\pi}{2x}}\,H_{n+1/2}^{(1)}(x)$$

$$h_n^{(2)}(x)=j_n(x)-iy_n(x)=\sqrt{\frac{\pi}{2x}}\,H_{n+1/2}^{(2)}(x)$$

$$j_0(x)=\frac{\sin x}{x} \qquad y_0(x)=-\frac{\cos x}{x}$$

$$j_1(x)=\frac{\sin x}{x^2}-\frac{\cos x}{x} \qquad y_1(x)=-\frac{\sin x}{x}-\frac{\cos x}{x^2}$$

$$j_2(x)=\left(\frac{3}{x^3}-\frac{1}{x}\right)\sin x-\frac{3}{x^2}\cos x \qquad y_2(x)=\left(-\frac{3}{x^3}+\frac{1}{x}\right)\cos x-\frac{3}{x^2}\sin x$$

Reccurence formulas

For any function $f_n(x)=j_n(x)$, $y_n(x)$, $h_n^{(1)}(x)$ or $h_n^{(2)}(x)$:

$$f_{n+1}(x)=(2n+1)f_n(x)/x-f_{n-1}(x)$$

$$f_n'(x)=\frac{n}{2n+1}f_{n-1}(x)-\frac{n+1}{2n+1}f_{n+1}(x)=f_{n-1}(x)-\frac{n+1}{x}f_n(x)=$$
$$=\frac{n}{x}f_n(x)-f_{n+1}(x)$$

Differential equation

$$x^2y''+2xy'+(a^2x^2-n(n+1))y=0$$

Solution: $y=Aj_n(ax)+By_n(ax)$

Zeros

The functions $j_n(x)$ and $j_n'(x)$, $n=0, 1, 2, \dots$ have infinitely many simple zeros. These are denoted by $\xi_{n+1/2,j}$ and ζ_{nj}, $j=1, 2, 3, \dots$ respectively and are tabulated below for $0\leqslant n\leqslant 7$, $1\leqslant j\leqslant 10$.

Eigenvalue problems

1. Dirichlet condition. (Singular) eigenvalue problem:

$$\begin{cases}(x^2y'(x))'+(\lambda x^2-n(n+1))y(x)=0, \ 0<x<a, \ n=0, 1, 2, \dots \\ y(0) \text{ bounded} \\ y(a)=0\end{cases}$$

Eigenvalues: $\lambda=(\xi_{n+1/2,j}/a)^2$, $j=1, 2, 3, \dots$

Eigenfunctions: $j_n(\xi_{n+1/2,j}x/a)$, $j=1, 2, 3, \dots$

Orthogonality: $\int_0^a j_n(\xi_{n+1/2,j}x/a)\,j_n(\xi_{n+1/2,k}x/a)x^2\,dx=(a^3/2)[j_n'(\xi_{n+1/2,j})]^2\delta_{jk}$

2. Neumann condition. (Singular) eigenvalue problem:

$$\begin{cases}(x^2y'(x))'+(\lambda x^2-n(n+1))y(x)=0, \ 0<x<a, \ n=0, 1, 2, \dots \\ y(0) \text{ bounded} \\ y'(a)=0\end{cases}$$

Eigenvalues: $\lambda=(\zeta_{nj}/a)^2$, $j=1, 2, 3, \dots$

Eigenfunctions: $j_n(\zeta_{nj}x/a)$, $j=1, 2, 3, \dots$

Orthogonality: $\int_0^a j_n(\zeta_{nj}x/a)\,j_n(\zeta_{nk}x/a)x^2\,dx=(a^3/2)[1-n(n+1)/\zeta_{nj}^2)][j_n(\zeta_{nj})]^2\delta_{jk}$

Numerical tables of Bessel and Modified Bessel Functions

x	$J_0(x)$	$J_1(x)$	$Y_0(x)$	$Y_1(x)$	$e^{-x}I_0(x)$	$e^{-x}I_1(x)$	$e^xK_0(x)$	$e^xK_1(x)$
0.0	+ 1.000000	+ 0.000000	- ∞	- ∞	1.0000	0.0000	∞	∞
0.1	+ 0.997502	+ 0.049938	- 1.5342	- 6.4590	0.9071	0.0453	2.6823	10.890
0.2	+ 0.990025	+ 0.099501	- 1.0811	- 3.3238	0.8269	0.0823	2.1408	5.8334
0.3	+ 0.977626	+ 0.148319	- 0.8073	- 2.2931	0.7576	0.1124	1.8526	4.1252
0.4	+ 0.960398	+ 0.196027	- 0.6060	- 1.7809	0.6974	0.1368	1.6627	3.2587
0.5	+ 0.938470	+ 0.242268	- 0.4445	- 1.4715	0.6450	0.1564	1.5241	2.7310
0.6	+ 0.912005	+ 0.286701	- 0.3085	- 1.2604	0.5993	0.1722	1.4167	2.3739
0.7	+ 0.881201	+ 0.328996	- 0.1907	- 1.1032	0.5593	0.1847	1.3301	2.1150
0.8	+ 0.846287	+ 0.368842	- 0.0868	- 0.9781	0.5241	0.1945	1.2582	1.9179
0.9	+ 0.807524	+ 0.405950	+ 0.0056	- 0.8731	0.4932	0.2021	1.1972	1.7624
1.0	+ 0.765198	+ 0.440051	+ 0.0883	- 0.7812	0.4658	0.2079	1.1445	1.6362
1.1	+ 0.719622	+ 0.470902	+ 0.1622	- 0.6981	0.4414	0.2122	1.0983	1.5314
1.2	+ 0.671133	+ 0.498289	+ 0.2281	- 0.6211	0.4198	0.2153	1.0575	1.4429
1.3	+ 0.620086	+ 0.522023	+ 0.2865	- 0.5485	0.4004	0.2173	1.0210	1.3670
1.4	+ 0.566855	+ 0.541948	+ 0.3379	- 0.4791	0.3831	0.2185	0.9881	1.3011
1.5	+ 0.511828	+ 0.557937	+ 0.3824	- 0.4123	0.3674	0.2190	0.9582	1.2432
1.6	+ 0.455402	+ 0.569896	+ 0.4204	- 0.3476	0.3533	0.2190	0.9309	1.1919
1.7	+ 0.397985	+ 0.577765	+ 0.4520	- 0.2847	0.3405	0.2186	0.9059	1.1460
1.8	+ 0.339986	+ 0.581517	+ 0.4774	- 0.2237	0.3289	0.2177	0.8828	1.1048
1.9	+ 0.281819	+ 0.581157	+ 0.4968	- 0.1644	0.3182	0.2166	0.8615	1.0675
2.0	+ 0.223891	+ 0.576725	+ 0.5104	- 0.1070	0.3085	0.2153	0.8416	1.0335
2.1	+ 0.166607	+ 0.568292	+ 0.5183	- 0.0517	0.2996	0.2137	0.8230	1.0024
2.2	+ 0.110362	+ 0.555963	+ 0.5208	+ 0.0015	0.2913	0.2121	0.8057	0.9738
2.3	+ 0.055540	+ 0.539873	+ 0.5181	+ 0.0523	0.2837	0.2103	0.7894	0.9474
2.4	+ 0.002508	+ 0.520185	+ 0.5104	+ 0.1005	0.2766	0.2085	0.7740	0.9229
2.5	- 0.048384	+ 0.497094	+ 0.4981	+ 0.1459	0.2700	0.2066	0.7595	0.9002
2.6	- 0.096805	+ 0.470818	+ 0.4813	+ 0.1884	0.2639	0.2047	0.7459	0.8790
2.7	- 0.142449	+ 0.441601	+ 0.4605	+ 0.2276	0.2582	0.2027	0.7329	0.8591
2.8	- 0.185036	+ 0.409709	+ 0.4359	+ 0.2635	0.2528	0.2007	0.7206	0.8405
2.9	- 0.224312	+ 0.375427	+ 0.4079	+ 0.2959	0.2478	0.1988	0.7089	0.8230
3.0	- 0.260052	+ 0.339059	+ 0.3769	+ 0.3247	0.2430	0.1968	0.6978	0.8066
3.1	- 0.292064	+ 0.300921	+ 0.3431	+ 0.3496	0.2385	0.1949	0.6871	0.7910
3.2	- 0.320188	+ 0.261343	+ 0.3071	+ 0.3707	0.2343	0.1930	0.6770	0.7763
3.3	- 0.344296	+ 0.220663	+ 0.2691	+ 0.3879	0.2302	0.1911	0.6673	0.7623
3.4	- 0.364296	+ 0.179226	+ 0.2296	+ 0.4010	0.2264	0.1892	0.6580	0.7491
3.5	- 0.380128	+ 0.137378	+ 0.1890	+ 0.4102	0.2228	0.1874	0.6490	0.7365
3.6	- 0.391769	+ 0.095466	+ 0.1477	+ 0.4154	0.2193	0.1856	0.6405	0.7245
3.7	- 0.399230	+ 0.053834	+ 0.1061	+ 0.4167	0.2160	0.1838	0.6322	0.7130
3.8	- 0.402556	+ 0.012821	+ 0.0645	+ 0.4141	0.2129	0.1821	0.6243	0.7021
3.9	- 0.401826	- 0.027244	+ 0.0234	+ 0.4078	0.2099	0.1804	0.6167	0.6916

x	$J_0(x)$	$J_1(x)$	$Y_0(x)$	$Y_1(x)$	$e^{-x}I_0(x)$	$e^{-x}I_1(x)$	$e^xK_0(x)$	$e^xK_1(x)$
4.0	- 0.397150	- 0.066043	- 0.0169	+ 0.3979	0.2070	0.1788	0.6093	0.6816
4.1	- 0.388670	- 0.103273	- 0.0561	+ 0.3846	0.2042	0.1771	0.6022	0.6720
4.2	- 0.376557	- 0.138647	- 0.0938	+ 0.3680	0.2016	0.1755	0.5953	0.6627
4.3	- 0.361011	- 0.171897	- 0.1296	+ 0.3484	0.1990	0.1740	0.5887	0.6539
4.4	- 0.342257	- 0.202776	- 0.1633	+ 0.3260	0.1966	0.1725	0.5823	0.6454
4.5	- 0.320543	- 0.231060	- 0.1947	+ 0.3010	0.1942	0.1710	0.5761	0.6371
4.6	- 0.296138	- 0.256553	- 0.2235	+ 0.2737	0.1919	0.1695	0.5701	0.6292
4.7	- 0.269331	- 0.279081	- 0.2494	+ 0.2445	0.1897	0.1681	0.5643	0.6216
4.8	- 0.240425	- 0.298500	- 0.2723	+ 0.2136	0.1876	0.1667	0.5586	0.6143
4.9	- 0.209738	- 0.314695	- 0.2921	+ 0.1812	0.1855	0.1653	0.5531	0.6071
5.0	- 0.177597	- 0.327579	- 0.3085	+ 0.1479	0.1835	0.1640	0.5478	0.6003
5.1	- 0.144335	- 0.337097	- 0.3216	+ 0.1137	0.1816	0.1627	0.5426	0.5936
5.2	- 0.110290	- 0.343223	- 0.3313	+ 0.0792	0.1797	0.1614	0.5376	0.5872
5.3	- 0.075803	- 0.345961	- 0.3374	+ 0.0445	0.1779	0.1601	0.5327	0.5810
5.4	- 0.041210	- 0.345345	- 0.3402	+ 0.0101	0.1762	0.1589	0.5280	0.5749
5.5	- 0.006844	- 0.341438	- 0.3395	- 0.0238	0.1745	0.1577	0.5233	0.5690
5.6	+ 0.026971	- 0.334333	- 0.3354	- 0.0568	0.1728	0.1565	0.5188	0.5634
5.7	+ 0.059920	- 0.324148	- 0.3282	- 0.0887	0.1712	0.1554	0.5144	0.5578
5.8	+ 0.091703	- 0.311028	- 0.3177	- 0.1192	0.1697	0.1542	0.5101	0.5525
5.9	+ 0.122033	- 0.295142	- 0.3044	- 0.1481	0.1681	0.1531	0.5059	0.5473
6.0	+ 0.150645	- 0.276684	- 0.2882	- 0.1750	0.1667	0.1521	0.5019	0.5422
6.1	+ 0.177291	- 0.255865	- 0.2694	- 0.1998	0.1652	0.1510	0.4979	0.5372
6.2	+ 0.201747	- 0.232917	- 0.2483	- 0.2223	0.1638	0.1499	0.4940	0.5324
6.3	+ 0.223812	- 0.208087	- 0.2251	- 0.2422	0.1624	0.1489	0.4902	0.5277
6.4	+ 0.243311	- 0.181638	- 0.1999	- 0.2596	0.1611	0.1479	0.4865	0.5232
6.5	+ 0.260095	- 0.153841	- 0.1732	- 0.2741	0.1598	0.1469	0.4828	0.5187
6.6	+ 0.274043	- 0.124980	- 0.1452	- 0.2857	0.1585	0.1460	0.4793	0.5144
6.7	+ 0.285065	- 0.095342	- 0.1162	- 0.2945	0.1573	0.1450	0.4758	0.5102
6.8	+ 0.293096	- 0.065219	- 0.0864	- 0.3002	0.1561	0.1441	0.4724	0.5060
6.9	+ 0.298102	- 0.034902	- 0.0563	- 0.3029	0.1549	0.1432	0.4691	0.5020
7.0	+ 0.300079	- 0.004683	- 0.0259	- 0.3027	0.1537	0.1423	0.4658	0.4981
7.1	+ 0.299051	+ 0.025153	+ 0.0042	- 0.2995	0.1526	0.1414	0.4627	0.4942
7.2	+ 0.295071	+ 0.054327	+ 0.0339	- 0.2934	0.1515	0.1405	0.4595	0.4905
7.3	+ 0.288217	+ 0.082570	+ 0.0628	- 0.2846	0.1504	0.1397	0.4565	0.4868
7.4	+ 0.278596	+ 0.109625	+ 0.0907	- 0.2731	0.1494	0.1389	0.4535	0.4832
7.5	+ 0.266340	+ 0.135248	+ 0.1173	- 0.2591	0.1483	0.1380	0.4505	0.4797
7.6	+ 0.251602	+ 0.159214	+ 0.1424	- 0.2428	0.1473	0.1372	0.4476	0.4762
7.7	+ 0.234559	+ 0.181313	+ 0.1658	- 0.2243	0.1463	0.1364	0.4448	0.4729
7.8	+ 0.215408	+ 0.201357	+ 0.1872	- 0.2039	0.1453	0.1357	0.4420	0.4696
7.9	+ 0.194362	+ 0.219179	+ 0.2065	- 0.1817	0.1444	0.1349	0.4393	0.4663
8.0	+ 0.171651	+ 0.234636	+ 0.2235	- 0.1581	0.1434	0.1341	0.4366	0.4631
8.1	+ 0.147517	+ 0.247608	+ 0.2381	- 0.1331	0.1425	0.1334	0.4340	0.4600
8.2	+ 0.122215	+ 0.257999	+ 0.2501	- 0.1072	0.1416	0.1327	0.4314	0.4570
8.3	+ 0.096006	+ 0.265739	+ 0.2595	- 0.0806	0.1407	0.1320	0.4289	0.4540
8.4	+ 0.069157	+ 0.270786	+ 0.2662	- 0.0535	0.1399	0.1312	0.4264	0.4511
8.5	+ 0.041939	+ 0.273122	+ 0.2702	- 0.0262	0.1390	0.1305	0.4239	0.4482
8.6	+ 0.014623	+ 0.272755	+ 0.2715	+ 0.0011	0.1382	0.1299	0.4215	0.4454
8.7	- 0.012523	+ 0.269719	+ 0.2700	+ 0.0280	0.1373	0.1292	0.4192	0.4426
8.8	- 0.039234	+ 0.264074	+ 0.2659	+ 0.0544	0.1365	0.1285	0.4168	0.4399
8.9	- 0.065253	+ 0.255902	+ 0.2592	+ 0.0799	0.1357	0.1279	0.4145	0.4372

x	$J_0(x)$	$J_1(x)$	$Y_0(x)$	$Y_1(x)$	$e^{-x}I_0(x)$	$e^{-x}I_1(x)$	$e^xK_0(x)$	$e^xK_1(x)$
9.0	- 0.090334	+ 0.245312	+ 0.2499	+ 0.1043	0.1350	0.1272	0.4123	0.4346
9.1	- 0.114239	+ 0.232431	+ 0.2383	+ 0.1275	0.1342	0.1266	0.4101	0.4321
9.2	- 0.136748	+ 0.217409	+ 0.2245	+ 0.1491	0.1334	0.1260	0.4079	0.4295
9.3	- 0.157655	+ 0.200414	+ 0.2086	+ 0.1691	0.1327	0.1253	0.4058	0.4270
9.4	- 0.176772	+ 0.181632	+ 0.1907	+ 0.1871	0.1320	0.1247	0.4036	0.4246
9.5	- 0.193929	+ 0.161264	+ 0.1712	+ 0.2032	0.1313	0.1241	0.4016	0.4222
9.6	- 0.208979	+ 0.139525	+ 0.1502	+ 0.2171	0.1305	0.1235	0.3995	0.4198
9.7	- 0.221795	+ 0.116639	+ 0.1279	+ 0.2287	0.1299	0.1230	0.3975	0.4175
9.8	- 0.232276	+ 0.092840	+ 0.1045	+ 0.2379	0.1292	0.1224	0.3955	0.4152
9.9	- 0.240341	+ 0.068370	+ 0.0804	+ 0.2447	0.1285	0.1218	0.3936	0.4130
10.0	- 0.245936	+ 0.043473	+ 0.0557	+ 0.2490	0.1278	0.1213	0.3916	0.4108
10.1	- 0.249030	+ 0.018396	+ 0.0307	+ 0.2508	0.1272	0.1207	0.3897	0.4086
10.2	- 0.249617	- 0.006616	+ 0.0056	+ 0.2502	0.1265	0.1202	0.3879	0.4064
10.3	- 0.247717	- 0.031318	- 0.0193	+ 0.2471	0.1259	0.1196	0.3860	0.4043
10.4	- 0.243372	- 0.055473	- 0.0437	+ 0.2416	0.1253	0.1191	0.3842	0.4023
10.5	- 0.236648	- 0.078850	- 0.0675	+ 0.2337	0.1247	0.1186	0.3824	0.4002
10.6	- 0.227635	- 0.101229	- 0.0904	+ 0.2236	0.1241	0.1181	0.3806	0.3982
10.7	- 0.216443	- 0.122399	- 0.1122	+ 0.2114	0.1235	0.1175	0.3789	0.3962
10.8	- 0.203202	- 0.142167	- 0.1326	+ 0.1973	0.1229	0.1170	0.3772	0.3943
10.9	- 0.188062	- 0.160350	- 0.1516	+ 0.1813	0.1223	0.1165	0.3755	0.3923
11.0	- 0.171190	- 0.176785	- 0.1688	+ 0.1637	0.1217	0.1161	0.3738	0.3904
11.1	- 0.152768	- 0.191328	- 0.1843	+ 0.1446	0.1212	0.1156	0.3721	0.3886
11.2	- 0.132992	- 0.203853	- 0.1977	+ 0.1243	0.1206	0.1151	0.3705	0.3867
11.3	- 0.112068	- 0.214255	- 0.2091	+ 0.1029	0.1201	0.1146	0.3689	0.3849
11.4	- 0.090215	- 0.222451	- 0.2183	+ 0.0807	0.1195	0.1142	0.3673	0.3831
11.5	- 0.067654	- 0.228379	- 0.2252	+ 0.0579	0.1190	0.1137	0.3657	0.3813
11.6	- 0.044616	- 0.232000	- 0.2299	+ 0.0348	0.1185	0.1132	0.3642	0.3796
11.7	- 0.021331	- 0.233300	- 0.2322	+ 0.0114	0.1179	0.1128	0.3627	0.3779
11.8	+ 0.001967	- 0.232285	- 0.2322	- 0.0118	0.1174	0.1123	0.3612	0.3762
11.9	+ 0.025049	- 0.228983	- 0.2298	- 0.0347	0.1169	0.1119	0.3597	0.3745
12.0	+ 0.047689	- 0.223447	- 0.2252	- 0.0571	0.1164	0.1115	0.3582	0.3728
12.1	+ 0.069667	- 0.215749	- 0.2184	- 0.0787	0.1159	0.1110	0.3567	0.3712
12.2	+ 0.090770	- 0.205982	- 0.2095	- 0.0994	0.1154	0.1106	0.3553	0.3696
12.3	+ 0.110798	- 0.194259	- 0.1986	- 0.1189	0.1150	0.1102	0.3539	0.3680
12.4	+ 0.129561	- 0.180710	- 0.1858	- 0.1371	0.1145	0.1098	0.3525	0.3664
12.5	+ 0.146884	- 0.165484	- 0.1712	- 0.1538	0.1140	0.1094	0.3511	0.3649
12.6	+ 0.162607	- 0.148742	- 0.1551	- 0.1689	0.1136	0.1090	0.3497	0.3633
12.7	+ 0.176588	- 0.130662	- 0.1375	- 0.1821	0.1131	0.1086	0.3484	0.3618
12.8	+ 0.188701	- 0.111432	- 01187	- 0.1935	0.1126	0.1082	0.3470	0.3603
12.9	+ 0.198842	- 0.091248	- 0.0989	- 0.2028	0.1122	0.1078	0.3457	0.3589
13.0	+ 0.206926	- 0.070318	- 0.0782	- 0.2101	0.1118	0.1074	0.3444	0.3574
13.1	+ 0.212888	- 0.048852	- 0.0569	- 0.2152	0.1113	0.1070	0.3431	0.3560
13.2	+ 0.216686	- 0.027067	- 0.0352	- 0.2182	0.1109	0.1066	0.3418	0.3545
13.3	+ 0.218298	- 0.005177	- 0.0134	- 0.2190	0.1105	0.1062	0.3406	0.3531
13.4	+ 0.217725	+ 0.016599	+ 0.0085	- 0.2176	0.1100	0.1059	0.3393	0.3518
13.5	+ 0.214989	+ 0.038049	+ 0.0301	- 0.2140	0.1096	0.1055	0.3381	0.3504
13.6	+ 0.210133	+ 0.058965	+ 0.0512	- 0.2084	0.1092	0.1051	0.3368	0.3490
13.7	+ 0.203221	+ 0.079143	+ 0.0717	- 0.2007	0.1088	0.1048	0.3356	0.3477
13.8	+ 0.194336	+ 0.098391	+ 0.0913	- 0.1912	0.1084	0.1044	0.3344	0.3464
13.9	+ 0.183580	+ 0.116525	+ 0.1099	- 0.1798	0.1080	0.1040	0.3333	0.3450

x	$J_0(x)$	$J_1(x)$	$Y_0(x)$	$Y_1(x)$	$e^{-x}I_0(x)$	$e^{-x}I_1(x)$	$e^{x}K_0(x)$	$e^{x}K_1(x)$
14.0	+ 0.1711	+ 0.1334	+ 0.1272	- 0.1666	0.1076	0.1037	0.3321	0.3437
14.1	+ 0.1570	+ 0.1488	+ 0.1431	- 0.1520	0.1072	0.1034	0.3309	0.3425
14.2	+ 0.1414	+ 0.1626	+ 0.1575	- 0.1359	0.1068	0.1030	0.3298	0.3412
14.3	+ 0.1245	+ 0.1747	+ 0.1703	- 0.1186	0.1065	0.1027	0.3286	0.3399
14.4	+ 0.1065	+ 0.1850	+ 0.1812	- 0.1003	0.1061	0.1023	0.3275	0.3387
14.5	+ 0.0875	+ 0.1934	+ 0.1903	- 0.0810	0.1057	0.1020	0.3264	0.3375
14.6	+ 0.0679	+ 0.1999	+ 0.1974	- 0.0612	0.1053	0.1017	0.3253	0.3363
14.7	+ 0.0476	+ 0.2043	+ 0.2025	- 0.0408	0.1050	0.1013	0.3242	0.3351
14.8	+ 0.0271	+ 0.2066	+ 0.2056	- 0.0202	0.1046	0.1010	0.3231	0.3339
14.9	+ 0.0064	+ 0.2069	+ 0.2065	+ 0.0005	0.1043	0.1007	0.3221	0.3327
15.0	- 0.0142	+ 0.2051	+ 0.2055	+ 0.0211	0.1039	0.1004	0.3210	0.3315
15.1	- 0.0346	+ 0.2013	+ 0.2023	+ 0.0413	0.1035	0.1001	0.3200	0.3304
15.2	- 0.0544	+ 0.1955	+ 0.1972	+ 0.0609	0.1032	0.0997	0.3189	0.3292
15.3	- 0.0736	+ 0.1879	+ 0.1902	+ 0.0799	0.1029	0.0994	0.3179	0.3281
15.4	- 0.0919	+ 0.1784	+ 0.1813	+ 0.0979	0.1025	0.0991	0.3169	0.3270
15.5	- 0.1092	+ 0.1672	+ 0.1706	+ 0.1148	0.1022	0.0988	0.3159	0.3259
15.6	- 0.1253	+ 0.1544	+ 0.1584	+ 0.1305	0.1018	0.0985	0.3149	0.3248
15.7	- 0.1401	+ 0.1402	+ 0.1446	+ 0.1447	0.1015	0.0982	0.3139	0.3237
15.8	- 0.1533	+ 0.1247	+ 0.1295	+ 0.1575	0.1012	0.0979	0.3129	0.3226
15.9	- 0.1650	+ 0.1080	+ 0.1132	+ 0.1686	0.1009	0.0976	0.3119	0.3216
16.0	- 0.1749	+ 0.0904	+ 0.0958	+ 0.1780	0.1005	0.0973	0.3110	0.3205
16.1	- 0.1830	+ 0.0720	+ 0.0776	+ 0.1855	0.1002	0.0971	0.3100	0.3195
16.2	- 0.1893	+ 0.0530	+ 0.0588	+ 0.1912	0.0999	0.0968	0.3091	0.3185
16.3	- 0.1936	+ 0.0335	+ 0.0394	+ 0.1949	0.0996	0.0965	0.3081	0.3174
16.4	- 0.1960	+ 0.0139	+ 0.0199	+ 0.1967	0.0993	0.0962	0.3072	0.3164
16.5	- 0.1964	- 0.0058	+ 0.0002	+ 0.1965	0.0990	0.0959	0.3063	0.3154
16.6	- 0.1948	- 0.0252	- 0.0194	+ 0.1943	0.0987	0.0957	0.3054	0.3144
16.7	- 0.1913	- 0.0444	- 0.0386	+ 0.1903	0.0984	0.0954	0.3045	0.3135
16.8	- 0.1860	- 0.0629	- 0.0574	+ 0.1843	0.0981	0.0951	0.3036	0.3125
16.9	- 0.1788	- 0.0807	- 0.0754	+ 0.1766	0.0978	0.0948	0.3027	0.3115
17.0	- 0.1699	- 0.0977	- 0.0926	+ 0.1672	0.0975	0.0946	0.3018	0.3106
17.1	- 0.1593	- 0.1135	- 0.1088	+ 0.1562	0.0972	0.0943	0.3009	0.3096
17.2	- 0.1472	- 0.1281	- 0.1238	+ 0.1437	0.0969	0.0941	0.3001	0.3087
17.3	- 0.1337	- 0.1414	- 0.1375	+ 0.1298	0.0966	0.0938	0.2992	0.3077
17.4	- 0.1190	- 0.1532	- 0.1497	+ 0.1147	0.0963	0.0935	0.2984	0.3068
17.5	- 0.1031	- 0.1634	- 0.1604	+ 0.0986	0.0961	0.0933	0.2975	0.3059
17.6	- 0.0863	- 0.1719	- 0.1694	+ 0.0816	0.0958	0.0930	0.2967	0.3050
17.7	- 0.0688	- 0.1787	- 0.1767	+ 0.0638	0.0955	0.0928	0.2959	0.3041
17.8	- 0.0506	- 0.1837	- 0.1822	+ 0.0456	0.0952	0.0925	0.2950	0.3032
17.9	- 0.0321	- 0.1868	- 0.1858	+ 0.0269	0.0950	0.0923	0.2942	0.3023
18.0	- 0.0134	- 0.1880	- 0.1876	+ 0.0081	0.0947	0.0920	0.2934	0.3015
18.1	+ 0.0054	- 0.1874	- 0.1874	- 0.0106	0.0944	0.0918	0.2926	0.3006
18.2	+ 0.0241	- 0.1848	- 0.1854	- 0.0291	0.0942	0.0916	0.2918	0.2997
18.3	+ 0.0423	- 0.1805	- 0.1816	- 0.0473	0.0939	0.0913	0.2910	0.2989
18.4	+ 0.0601	- 0.1744	- 0.1760	- 0.0649	0.0937	0.0911	0.2903	0.2980
18.5	+ 0.0772	- 0.1666	- 0.1687	- 0.0818	0.0934	0.0908	0.2895	0.2972
18.6	+ 0.0934	- 0.1572	- 0.1597	- 0.0977	0.0931	0.0906	0.2887	0.2964
18.7	+ 0.1086	- 0.1463	- 0.1491	- 0.1126	0.0929	0.0904	0.2879	0.2955
18.8	+ 0.1226	- 0.1340	- 0.1372	- 0.1263	0.0926	0.0901	0.2872	0.2947
18.9	+ 0.1353	- 0.1204	- 0.1239	- 0.1386	0.0924	0.0899	0.2864	0.2939

x	$J_0(x)$	$J_1(x)$	$Y_0(x)$	$Y_1(x)$	$e^{-x}I_0(x)$	$e^{-x}I_1(x)$	$e^xK_0(x)$	$e^xK_1(x)$
19.0	+ 0.1466	- 0.1057	- 0.1095	- 0.1496	0.0921	0.0897	0.2857	0.2931
19.1	+ 0.1564	- 0.0900	- 0.0941	- 0.1590	0.0919	0.0895	0.2850	0.2923
19.2	+ 0.1646	- 0.0735	- 0.0778	- 0.1667	0.0917	0.0892	0.2842	0.2915
19.3	+ 0.1711	- 0.0564	- 0.0608	- 0.1727	0.0914	0.0890	0.2835	0.2907
19.4	+ 0.1759	- 0.0388	- 0.0433	- 0.1771	0.0912	0.0888	0.2828	0.2900
19.5	+ 0.1789	- 0.0209	- 0.0255	- 0.1796	0.0909	0.0886	0.2821	0.2892
19.6	+ 0.1800	- 0.0029	- 0.0075	- 0.1803	0.0907	0.0884	0.2813	0.2884
19.7	+ 0.1794	+ 0.0151	+ 0.0105	- 0.1792	0.0905	0.0881	0.2806	0.2877
19.8	+ 0.1770	+ 0.0328	+ 0.0283	- 0.1764	0.0902	0.0879	0.2799	0.2869
19.9	+ 0.1729	+ 0.0501	+ 0.0457	- 0.1718	0.0900	0.0877	0.2792	0.2862
20.0	+ 0.1670	+ 0.0668	+ 0.0626	- 0.1655	0.0898	0.0875	0.2785	0.2854

$$J_0(x)\approx\sqrt{\frac{2}{\pi x}}\left[\cos\left(x-\frac{\pi}{4}\right)+\frac{1}{8x}\sin\left(x-\frac{\pi}{4}\right)\right],\ \text{error}<0.0001 \text{ for } x>20$$

$$J_1(x)\approx\sqrt{\frac{2}{\pi x}}\left[\sin\left(x-\frac{3\pi}{4}\right)-\frac{3}{8x}\cos\left(x-\frac{3\pi}{4}\right)\right],\ \text{error}<0.0001 \text{ for } x>20$$

$$Y_0(x)\approx\sqrt{\frac{2}{\pi x}}\left[\sin\left(x-\frac{\pi}{4}\right)-\frac{1}{8x}\cos\left(x-\frac{\pi}{4}\right)\right],\ \text{error}<0.0001 \text{ for } x>20$$

$$Y_1(x)\approx\sqrt{\frac{2}{\pi x}}\left[\sin\left(x-\frac{3\pi}{4}\right)+\frac{3}{8x}\cos\left(x-\frac{3\pi}{4}\right)\right],\ \text{error}<0.0001 \text{ for } x>20$$

$$J_n(x)=\frac{2(n-1)}{x}J_{n-1}(x)-J_{n-2}(x) \qquad J_0'(x)=-J_1(x)$$

$$Y_n(x)=\frac{2(n-1)}{x}Y_{n-1}(x)-Y_{n-2}(x) \qquad Y_0'(x)=-Y_1(x)$$

For $x>20$, $I_0(x)\approx e^x\left(1+\frac{1}{8x}\right)/\sqrt{2\pi x}$ $\qquad I_1(x)\approx e^x\left(1-\frac{3}{8x}\right)/\sqrt{2\pi x}$

$$K_0(x)\approx e^{-x}\sqrt{\frac{\pi}{2x}}\left(1-\frac{1}{8x}\right) \qquad K_1(x)\approx e^{-x}\sqrt{\frac{\pi}{2x}}\left(1+\frac{3}{8x}\right)$$

$$I_n(x)=I_{n-2}(x)-\frac{2(n-1)}{x}I_{n-1}(x) \qquad I_0'(x)=I_1(x)$$

$$K_n(x)=K_{n-2}(x)+\frac{2(n-1)}{x}K_{n-1}(x) \qquad K_0'(x)=-K_1(x)$$

Zeros ξ_{nj} of $J_n(x)$ and associated values $J_n'(\xi_{nj})$

j	ξ_{0j}	$J_0'(\xi_{0j})$	ξ_{1j}	$J_1'(\xi_{1j})$	ξ_{2j}	$J_2'(\xi_{2j})$	ξ_{3j}	$J_3'(\xi_{3j})$
1	2.4048	- 0.5191	3.8317	- 0.4028	5.1356	- 0.3397	6.3802	- 0.2983
2	5.5201	0.3403	7.0156	0.3001	8.4172	0.2714	9.7610	0.2494
3	8.6537	- 0.2715	10.1735	- 0.2497	11.6198	- 0.2324	13.0152	- 0.2183
4	11.7915	0.2325	13.3237	0.2184	14.7960	0.2065	16.2235	0.1964
5	14.9309	- 0.2065	16.4706	- 0.1965	17.9598	- 0.1877	19.4094	- 0.1800
6	18.0711	0.1877	19.6159	0.1801	21.1170	0.1733	22.5827	0.1672
7	21.2116	- 0.1733	22.7601	- 0.1672	24.2701	- 0.1617	25.7482	- 0.1567
8	24.3525	0.1617	25.9037	0.1567	27.4206	0.1522	28.9084	0.1480
9	27.4935	- 0.1522	29.0468	- 0.1480	30.5692	- 0.1442	32.0649	- 0.1406
10	30.6346	0.1442	32.1897	0.1406	33.7165	0.1373	35.2187	0.1342

j	ξ_{4j}	$J_4'(\xi_{4j})$	ξ_{5j}	$J_5'(\xi_{5j})$	ξ_{6j}	$J_6'(\xi_{6j})$	ξ_{7j}	$J_7'(\xi_{7j})$
1	7.5883	- 0.2684	8.7715	- 0.2454	9.9361	- 0.2271	11.0864	- 0.2121
2	11.0647	0.2319	12.3386	0.2174	13.5893	0.2052	14.8213	0.1948
3	14.3725	- 0.2064	15.7002	- 0.1961	17.0038	- 0.1873	18.2876	- 0.1794
4	17.6160	0.1877	18.9801	0.1799	20.3208	0.1731	21.6415	0.1669
5	20.8269	- 0.1732	22.2178	- 0.1671	23.5861	- 0.1616	24.9349	- 0.1566
6	24.0190	0.1617	25.4303	0.1567	26.8202	0.1521	28.1912	0.1479
7	27.1991	- 0.1522	28.6266	- 0.1480	30.0337	- 0.1441	31.4228	- 0.1405
8	30.3710	0.1442	31.8117	0.1406	33.2330	0.1373	34.6371	0.1342
9	33.5371	- 0.1373	34.9888	- 0.1342	36.4220	- 0.1313	37.8387	- 0.1286
10	36.6990	0.1313	38.1599	0.1286	39.6032	0.1261	41.0308	0.1237

Zeros η_{nj} of $J_n'(x)$ and associated values $J_n(\eta_{nj})$

j	η_{0j}	$J_0(\eta_{0j})$	η_{1j}	$J_1(\eta_{1j})$	η_{2j}	$J_2(\eta_{2j})$	η_{3j}	$J_3(\eta_{3j})$
1	0.0000	1.0000	1.8412	0.5819	3.0542	0.4865	4.2012	0.4344
2	3.8317	- 0.4028	5.3314	- 0.3461	6.7061	- 0.3135	8.0152	- 0.2912
3	7.0156	0.3001	8.5363	0.2733	9.9695	0.2547	11.3459	0.2407
4	10.1735	- 0.2497	11.7060	- 0.2333	13.1704	- 0.2209	14.5858	- 0.2110
5	13.3237	0.2184	14.8636	0.2070	16.3475	0.1979	17.7887	0.1904
6	16.4706	- 0.1965	18.0155	- 0.1880	19.5129	- 0.1810	20.9725	- 0.1750
7	19.6159	0.1801	21.1644	0.1735	22.6716	0.1678	24.1449	0.1630
8	22.7601	- 0.1672	24.3113	- 0.1618	25.8260	- 0.1572	27.3101	- 0.1531
9	25.9037	0.1567	27.4571	0.1523	28.9777	0.1484	30.4703	0.1449
10	29.0468	- 0.1480	30.6019	- 0.1442	32.1273	- 0.1409	33.6269	- 0.1378

j	η_{4j}	$J_4(\eta_{4j})$	η_{5j}	$J_5(\eta_{5j})$	η_{6j}	$J_6(\eta_{6j})$	η_{7j}	$J_7(\eta_{7j})$
1	5.3176	0.3997	6.4156	0.3741	7.5013	0.3541	8.5778	0.3379
2	9.2824	- 0.2744	10.5199	- 0.2611	11.7349	- 0.2502	12.9324	- 0.2410
3	12.6819	0.2296	13.9872	0.2204	15.2682	0.2126	16.5294	0.2059
4	15.9641	- 0.2028	17.3128	- 0.1958	18.6374	- 0.1898	19.9419	- 0.1845
5	19.1960	0.1840	20.5755	0.1785	21.9317	0.1736	23.2681	0.1693
6	22.4010	- 0.1699	23.8036	- 0.1653	25.1839	- 0.1613	26.5450	- 0.1576
7	25.5898	0.1587	27.0103	0.1548	28.4098	0.1514	29.7907	0.1482
8	28.7678	- 0.1495	30.2028	- 0.1462	31.6179	- 0.1432	33.0152	- 0.1404
9	31.9385	0.1417	33.3854	0.1388	34.8134	0.1362	36.2244	0.1338
10	35.1039	- 0.1351	36.5608	- 0.1326	37.9996	- 0.1302	39.4223	- 0.1281

Zeros $\chi_{nj} = \xi_{n+1/2,j}$ of $j_n(x)$ and associated values $j_n'(\chi_{nj,j})$

j	χ_{0j}	$j_0'(\chi_{0j})$	χ_{1j}	$j_1'(\chi_{1j})$	χ_{2j}	$j_2'(\chi_{2j})$	χ_{3j}	$j_3'(\chi_{3j})$
1	3.1416	- 0.3183	4.4934	- 0.2172	5.7635	- 0.1655	6.9879	- 0.1338
2	6.2832	0.1592	7.7253	0.1284	9.0950	0.1079	10.4171	0.0933
3	9.4248	- 0.1061	10.9041	- 0.0913	12.3229	- 0.0803	13.6980	- 0.0718
4	12.5664	0.0796	14.0662	0.0709	15.5146	0.0641	16.9236	0.0585
5	15.7080	- 0.0637	17.2208	- 0.0580	18.6890	- 0.0533	20.1218	- 0.0493
6	18.8496	0.0531	20.3713	0.0490	21.8539	0.0456	23.3042	0.0427
7	21.9911	- 0.0455	23.5195	- 0.0425	25.0128	- 0.0399	26.4768	- 0.0376
8	25.1327	0.0398	26.6661	0.0375	28.1678	0.0354	29.6426	0.0336
9	28.2743	- 0.0354	29.8116	- 0.0335	31.3201	- 0.0319	32.8037	- 0.0304
10	31.4159	0.0318	32.9564	0.0303	34.4705	0.0290	35.9614	0.0277

j	χ_{4j}	$j_4'(\chi_{4j})$	χ_{5j}	$j_5'(\chi_{5j})$	χ_{6j}	$j_6'(\chi_{6j})$	χ_{7j}	$j_7'(\chi_{7j})$
1	8.1820	- 0.1123	9.3558	- 0.0966	10.5128	- 0.0848	11.6570	- 0.0754
2	11.7049	0.0822	12.9665	0.0735	14.2074	0.0664	15.4313	0.0606
3	15.0397	- 0.0650	16.3547	- 0.0594	17.6480	- 0.0547	18.9230	- 0.0507
4	18.3013	0.0538	19.6532	0.0499	20.9835	0.0465	22.2953	0.0435
5	21.5254	- 0.0459	22.9046	- 0.0430	24.2628	- 0.0405	25.6029	- 0.0382
6	24.7276	0.0401	26.1278	0.0378	27.5079	0.0358	28.8704	0.0340
7	27.9156	- 0.0356	29.3326	- 0.0338	30.7304	- 0.0322	32.1112	- 0.0307
8	31.0939	0.0320	32.5247	0.0305	33.9371	0.0292	35.3332	0.0280
9	34.2654	- 0.0291	35.7076	- 0.0278	37.1323	- 0.0267	38.5414	- 0.0257
10	37.4317	0.0266	38.8836	0.0256	40.3189	0.0246	41.7391	0.0238

Zeros ζ_{nj} of $j_n'(x)$ and associated values $j_n(\zeta_{nj})$

j	ζ_{0j}	$j_0(\zeta_{0j})$	ζ_{1j}	$j_1(\zeta_{1j})$	ζ_{2j}	$j_2(\zeta_{2j})$	ζ_{3j}	$j_3(\zeta_{3j})$
1	0.0000	1.0000	2.0816	0.4362	3.3421	0.3068	4.5141	0.2417
2	4.4934	- 0.2172	5.9404	- 0.1681	7.2899	- 0.1396	8.5838	- 0.1205
3	7.7253	0.1284	9.2058	0.1086	10.6139	0.0950	11.9727	0.0850
4	10.9041	- 0.0913	12.4044	- 0.0806	13.8461	- 0.0726	15.2445	- 0.0663
5	14.0662	0.0709	15.5792	0.0642	17.0429	0.0589	18.4681	0.0545
6	17.2208	- 0.0580	18.7426	- 0.0534	20.2219	- 0.0496	21.6666	- 0.0464
7	20.3713	0.0490	21.8997	0.0457	23.3905	0.0428	24.8501	0.0404
8	23.5195	- 0.0425	25.0528	- 0.0399	26.5526	- 0.0377	28.0239	- 0.0358
9	26.6661	0.0375	28.2034	0.0355	29.7103	0.0337	31.1910	0.0321
10	29.8116	- 0.0335	31.3521	- 0.0319	32.8649	- 0.0305	34.3534	- 0.0292

j	ζ_{4j}	$j_4(4_{0j})$	ζ_{5j}	$j_5(\zeta_{5j})$	ζ_{6j}	$j_6(\zeta_{6j})$	ζ_{7j}	$j_7(\zeta_{7j})$
1	5.6467	0.2016	6.7565	0.1740	7.8511	0.1537	8.9348	0.1380
2	9.8404	- 0.1067	11.0702	- 0.0961	12.2793	- 0.0877	13.4720	- 0.0808
3	13.2956	0.0772	14.5906	0.0709	15.8631	0.0658	17.1175	0.0614
4	16.6093	- 0.0612	17.9472	- 0.0570	19.2627	- 0.0534	20.5594	- 0.0503
5	19.8624	0.0509	21.2311	0.0479	22.5781	0.0452	23.9064	0.0429
6	23.0828	- 0.0437	24.4748	- 0.0413	25.8461	- 0.0393	27.1992	- 0.0375
7	26.2833	0.0383	27.6937	0.0364	29.0843	0.0348	30.4575	0.0333
8	29.4706	- 0.0341	30.8960	- 0.0326	32.3025	- 0.0313	33.6922	- 0.0300
9	32.6489	0.0308	34.0866	0.0295	35.5063	0.0284	36.9099	0.0274
10	35.8205	- 0.0280	37.2686	- 0.0270	38.6996	- 0.0260	40.1151	- 0.0251

12.5 Functions Defined by Transcendental Integrals

The Gamma function $\Gamma(z)$

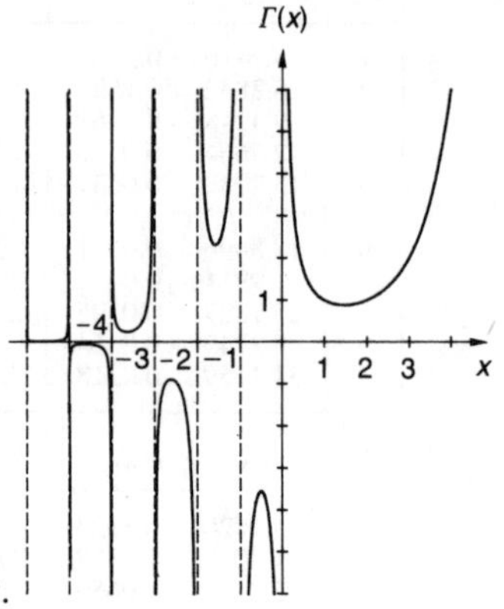

$\Gamma(z)$ is an analytic function in the whole plane except for simple poles at $0, -1, -2, \ldots$

$$\Gamma(z) = \lim_{n\to\infty} \frac{n^z n!}{z(z+1)(z+2)\ldots(z+n)}$$

$$\Gamma(z) = \int_0^\infty t^{z-1}e^{-t}dt, \ \mathrm{Re}\, z>0$$

$$\Gamma(z) = \int_1^\infty t^{z-1}e^{-t}dt + \sum_{n=0}^{\infty} \frac{(-1)^n}{n!\,(z+n)}, \qquad z\neq 0, -1, -2, \ldots$$

$$\frac{\Gamma'(z)}{\Gamma(z)} = -\frac{1}{z} - \gamma + \sum_{n=1}^{\infty}\left(\frac{1}{n} - \frac{1}{z+n}\right), \qquad \gamma = \text{Euler's constant.}$$

$$\Gamma(z+1) = z\Gamma(z) \quad \Gamma(n) = (n-1)!,\ n=1,2,3,\ldots \quad \Gamma\left(\frac{1}{2}\right) = \sqrt{\pi}$$

$$\Gamma(z)\,\Gamma(1-z) = \frac{\pi}{\sin \pi z} \qquad \Gamma\left(\frac{1}{2}+z\right)\Gamma\left(\frac{1}{2}-z\right) = \frac{\pi}{\cos \pi z}$$

$$\Gamma(2z) = \frac{1}{\sqrt{\pi}}\, 2^{2z-1}\Gamma(z)\,\Gamma\left(z+\frac{1}{2}\right) \qquad \Gamma'(1) = -\gamma$$

Asymptotic behavior (Stirling's formula)

$$\Gamma(z) = \sqrt{2\pi}\, e^{-z} z^{z-1/2}\left[1 + \frac{1}{12z} + \frac{1}{288z^2} - \frac{139}{51840z^3} + O\left(\frac{1}{z^4}\right)\right], \ |\arg z| < \pi,\ |z| \to \infty$$

The Beta function $B(p, q)$

$$B(p,q) = \frac{\Gamma(p)\,\Gamma(q)}{\Gamma(p+q)} = \int_0^1 t^{p-1}(1-t)^{q-1}dt = 2\int_0^{\pi/2} \sin^{2p-1}t\, \cos^{2q-1}t\, dt \qquad (\mathrm{Re}\, p,\ \mathrm{Re}\, q > 0)$$

Numerical table of $\Gamma(x)$

Values of $\Gamma(x)$ outside the interval $1 \leqslant x \leqslant 2$ may be obtained using $\Gamma(x+1) = x\Gamma(x)$ recursively.

x	$\Gamma(x)$	x	$\Gamma(x)$	x	$\Gamma(x)$	x	$\Gamma(x)$
1.00	1.00000	1.25	0.90640	1.50	0.88623	1.75	0.91906
.01	0.99433	.26	0.90440	.51	0.88659	.76	0.92137
.02	0.98884	.27	0.90250	.52	0.88704	.77	0.92376
.03	0.98355	.28	0.90072	.53	0.88757	.78	0.92623
.04	0.97844	.29	0.89904	.54	0.88818	.79	0.92877
1.05	0.97350	1.30	0.89747	1.55	0.88887	1.80	0.93138
.06	0.96874	.31	0.89600	.56	0.88964	.81	0.93408
.07	0.96415	.32	0.89464	.57	0.89049	.82	0.93685
.08	0.95973	.33	0.89338	.58	0.89142	.83	0.93969
.09	0.95546	.34	0.89222	.59	0.89243	.84	0.94261
1.10	0.95135	1.35	0.89115	1.60	0.89352	1.85	0.94561
.11	0.94740	.36	0.89018	.61	0.89468	.86	0.94869
.12	0.94359	.37	0.88931	.62	0.89592	.87	0.95184
.13	0.93993	.38	0.88854	.63	0.89724	.88	0.95507
.14	0.93642	.39	0.88785	.64	0.89864	.89	0.95838
1.15	0.93304	1.40	0.88726	1.65	0.90012	1.90	0.96177
.16	0.92980	.41	0.88676	.66	0.90167	.91	0.96523
.17	0.92670	.42	0.88636	.67	0.90330	.92	0.96877
.18	0.92373	.43	0.88604	.68	0.90500	.93	0.97240
.19	0.92089	.44	0.88581	.69	0.90678	.94	0.97610
1.20	0.91817	1.45	0.88566	1.70	0.90864	1.95	0.97988
.21	0.91558	.46	0.88560	.71	0.91057	.96	0.98374
.22	0.91311	.47	0.88563	.72	0.91258	.97	0.98768
.23	0.91075	.48	0.88575	.73	0.91467	.98	0.99171
.24	0.90852	.49	0.88595	.74	0.91683	.99	0.99581
						2.00	1.00000

Elliptic integrals

Elliptic integrals of the first kind

$$F(k,\varphi)=\int_0^{\varphi}\frac{d\theta}{\sqrt{1-k^2\sin^2\theta}}=\int_0^{x}\frac{dt}{\sqrt{(1-t^2)(1-k^2t^2)}}\quad (k^2<1,\ x=\sin\varphi)$$

Elliptic integrals of the second kind

$$E(k,\varphi)=\int_0^{\varphi}\sqrt{1-k^2\sin^2\theta}\,d\theta=\int_0^{x}\sqrt{\frac{1-k^2t^2}{1-t^2}}\,dt\quad (k^2<1,\ x=\sin\varphi)$$

Elliptic integrals of the third kind

$$\pi(k,n,\varphi)=\int_0^{\varphi}\frac{d\theta}{(1+n\sin^2\theta)\sqrt{1-k^2\sin^2\theta}}=$$

$$=\int_0^{x}\frac{dt}{(1+nt^2)\sqrt{1-t^2)(1-k^2t^2)}}\quad (k^2<1,\ x=\sin\varphi)$$

Complete elliptic integrals

$$K=K(k)=F\left(k,\frac{\pi}{2}\right)=\int_0^{\pi/2}\frac{d\theta}{\sqrt{1-k^2\sin^2\theta}}\quad (k^2<1)$$

$$E=E(k)=E\left(k,\frac{\pi}{2}\right)=\int_0^{\pi/2}\sqrt{1-k^2\sin^2\theta}\,d\theta\quad (k^2<1)$$

Legendre's relation ($k'=\sqrt{1-k^2}$)

$$E(k)K(k')+E(k')K(k)-K(k)K(k')=\frac{\pi}{2}$$

Differential equations

$$k(1-k^2)\frac{d^2K}{dk^2}+(1-3k^2)\frac{dK}{dk}-kK=0$$

$$k(1-k^2)\frac{d^2E}{dk^2}+(1-k^2)\frac{dE}{dk}+kE=0$$

Numerical tables of complete elliptic integrals

$k=\sin\alpha$ (α in degrees)

α	K	E
0°	1.5708	1.5708
1	1.5709	1.5707
2	1.5713	1.5703
3	1.5719	1.5697
4	1.5727	1.5689
5	1.5738	1.5678
6	1.5751	1.5665
7	1.5767	1.5649
8	1.5785	1.5632
9	1.5805	1.5611
10	1.5828	1.5589
11	1.5854	1.5564
12	1.5882	1.5537
13	1.5913	1.5507
14	1.5946	1.5476
15	1.5981	1.5442
16	1.6020	1.5405
17	1.6061	1.5367
18	1.6105	1.5326
19	1.6151	1.5283
20	1.6200	1.5238
21	1.6252	1.5191
22	1.6307	1.5141
23	1.6365	1.5090
24	1.6426	1.5037
25	1.6490	1.4981
26	1.6557	1.4924
27	1.6627	1.4864
28	1.6701	1.4803
29	1.6777	1.4740
30	1.6858	1.4675
31	1.6941	1.4608
32	1.7028	1.4539
33	1.7119	1.4469
34	1.7214	1.4397
35	1.7312	1.4323
36	1.7415	1.4248
37	1.7522	1.4171
38	1.7633	1.4092
39	1.7748	1.4013
40	1.7868	1.3931
41	1.7992	1.3849
42	1.8122	1.3765
43	1.8256	1.3680
44	1.8396	1.3594
45	1.8541	1.3506
46	1.8691	1.3418
47	1.8848	1.3329
48	1.9011	1.3238
49	1.9180	1.3147
50°	1.9356	1.3055
51	1.9539	1.2963
52	1.9729	1.2870
53	1.9927	1.2776
54	2.0133	1.2681
55	2.0347	1.2587
56	2.0571	1.2492
57	2.0804	1.2397
58	2.1047	1.2301
59	2.1300	1.2206
60	2.1565	1.2111
61	2.1842	1.2015
62	2.2132	1.1920
63	2.2435	1.1826
64	2.2754	1.1732
65	2.3088	1.1638
65.5	2.3261	1.1592
66.0	2.3439	1.1545
66.5	2.3622	1.1499
67.0	2.3809	1.1453
67.5	2.4001	1.1408
68.0	2.4198	1.1362
68.5	2.4401	1.1317
69.0	2.4610	1.1272
69.5	2.4825	1.1228
70.0	2.5046	1.1184
70.5	2.5273	1.1140
71.0	2.5507	1.1096
71.5	2.5749	1.1053
72.0	2.5998	1.1011
72.5	2.6256	1.0968
73.0	2.6521	1.0927
73.5	2.6796	1.0885
74.0	2.7081	1.0844
74.5	2.7375	1.0804
75.0	2.7681	1.0764
75.5	2.7998	1.0725
76.0	2.8327	1.0686
76.5	2.8669	1.0648
77.0	2.9026	1.0611
77.5	2.9397	1.0574
78.0	2.9786	1.0538
78.5	3.0192	1.0502
79.0	3.0617	1.0468
79.5	3.1064	1.0434
80.0	3.1534	1.0401
80.2	3.1729	1.0388
80.4	3.1928	1.0375
80.6	3.2132	1.0363
80.8	3.2340	1.0350
81°.0	3.2553	1.0338
81.2	3.2771	1.0326
81.4	3.2995	1.0314
81.6	3.3223	1.0302
81.8	3.3458	1.0290
82.0	3.3699	1.0278
82.2	3.3946	1.0267
82.4	3.4199	1.0256
82.6	3.4460	1.0245
82.8	3.4728	1.0234
83.0	3.5004	1.0223
83.2	3.5288	1.0213
83.4	3.5581	1.0202
83.6	3.5884	1.0192
83.8	3.6196	1.0182
84.0	3.6519	1.0172
84.2	3.6852	1.0163
84.4	3.7198	1.0153
84.6	3.7557	1.0144
84.8	3.7930	1.0135
85.0	3.8317	1.0127
85.2	3.8721	1.0118
85.4	3.9142	1.0110
85.6	3.9583	1.0102
85.8	4.0044	1.0094
86.0	4.0528	1.0086
86.2	4.1037	1.0079
86.4	4.1574	1.0072
86.6	4.2142	1.0065
86.8	4.2744	1.0059
87.0	4.3387	1.0053
87.2	4.4073	1.0047
87.4	4.4811	1.0041
87.6	4.5609	1.0036
87.8	4.6477	1.0031
88.0	4.7427	1.0026
88.2	4.8478	1.0021
88.4	4.9654	1.0017
88.6	5.0988	1.0014
88.8	5.2527	1.0010
89.0	5.4349	1.0008
89.1	5.5402	1.0006
89.2	5.6579	1.0005
89.3	5.7914	1.0004
89.4	5.9455	1.0003
89.5	6.1278	1.0002
89.6	6.3509	1.0001
89.7	6.6385	1.0001
89.8	7.0440	1.0000
89.9	7.7371	1.0000

Exponential integrals

1. $E_1(x)=\int_x^{\infty}\frac{e^{-t}}{t}dt=\int_1^{\infty}\frac{e^{-xt}}{t}dt=-\gamma-\ln x-\sum_{n=1}^{\infty}\frac{(-1)^n x^n}{n\,n!}\quad (x>0)$

2. $E_n(x)=\int_1^{\infty}\frac{e^{-xt}}{t^n}dt\quad (x>0,\ n=0,\ 1,\ 2,\ \ldots)\quad E_{n+1}(x)=\frac{1}{n}\left[e^{-x}-x\,E_n(x)\right],\ n=1,\ 2,\ \ldots$

3. $\mathrm{Ei}(x)=\int_{-\infty}^{x}\frac{e^t}{t}dt=\gamma+\ln x+\sum_{n=1}^{\infty}\frac{x^n}{n n!}$ ($x>0$, Cauchy P.V.), γ=Euler's constant

4. $\mathrm{li}(x)=\int_0^x\frac{dt}{\ln t}=\mathrm{Ei}(\ln x)$, ($x>1$, Cauchy P.V.)

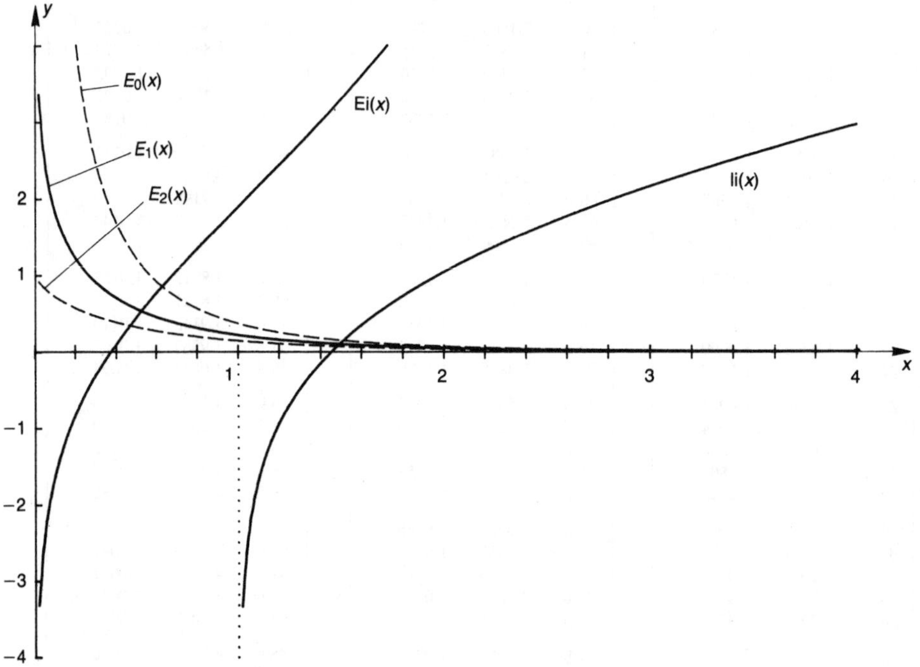

Asymptotic behaviour ($x\to\infty$):

$E_n(x)\sim\frac{e^{-x}}{x}\qquad \mathrm{Ei}(x)\sim\frac{e^x}{x}\qquad \mathrm{li}(x)\sim\frac{x}{\ln x}$

Error function

5. $\mathrm{erf}(x)=\frac{2}{\sqrt{\pi}}\int_0^x e^{-t^2}dt=\frac{2}{\sqrt{\pi}}\sum_{n=0}^{\infty}\frac{(-1)^n}{n!(2n+1)}x^{2n+1}$

Cf. table, sec, 17.8.

Sine and cosine integrals

6. $\mathrm{Si}(x)=\int_0^x \frac{\sin t}{t}\,dt=\sum_{n=0}^{\infty}\frac{(-1)^n x^{2n+1}}{(2n+1)(2n+1)!}\quad (x\geqslant 0)$

7. $\mathrm{Ci}(x)=-\int_x^{\infty}\frac{\cos t}{t}\,dt=\gamma+\ln x+\sum_{n=1}^{\infty}\frac{(-1)^n x^{2n}}{2n(2n)!}\quad (x>0)$

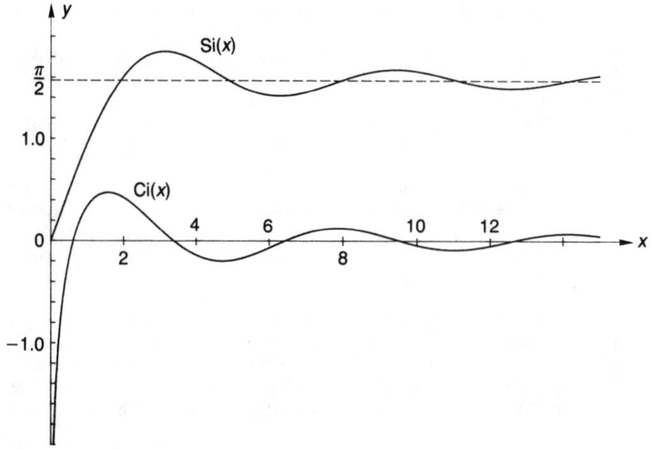

Fresnel integrals

8. $C(x)=\int_0^x \cos\frac{\pi t^2}{2}\,dt=$

$=\sum_{n=0}^{\infty}\frac{(-1)^n(\pi/2)^{2n}}{(2n)!(4n+1)}\,x^{4n+1}$

9. $S(x)=\int_0^x \sin\frac{\pi t^2}{2}\,dt=$

$=\sum_{n=0}^{\infty}\frac{(-1)^n(\pi/2)^{2n+1}}{(2n+1)!(4n+3)}\,x^{4n+3}$

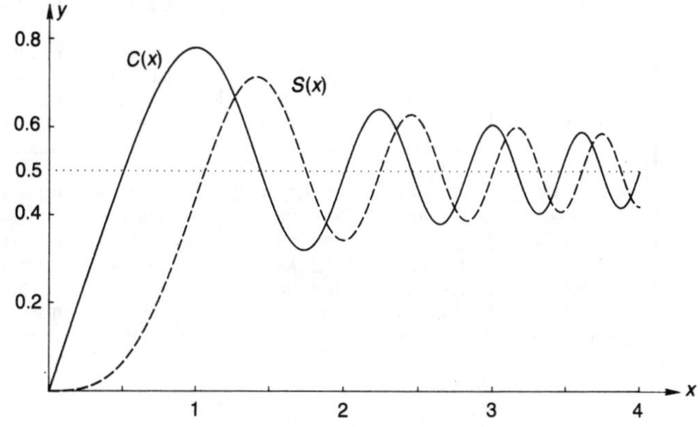

Numerical tables of Exponential, Sine, Cosine, and Fresnel Integrals

x	$E_1(x)$	$\mathrm{Ei}(x)$	$\mathrm{Si}(x)$	$\mathrm{Ci}(x)$	$C(x)$	$S(x)$
0.00	∞	- ∞	0.000000	- ∞	0.00000	0.00000
0.02	3.354708	- 3.314707	0.020000	- 3.334907	0.02000	0.00000
0.04	2.681264	- 2.601257	0.039996	- 2.642060	0.04000	0.00003
0.06	2.295307	- 2.175283	0.059988	- 2.237095	0.06000	0.00011
0.08	2.026941	- 1.866884	0.079972	- 1.950113	0.08000	0.00027
0.10	1.822924	- 1.622813	0.099944	- 1.727868	0.10000	0.00052
0.12	1.659542	- 1.419350	0.119904	- 1.546646	0.11999	0.00090
0.14	1.524146	- 1.243841	0.139848	- 1.393793	0.13999	0.00144
0.16	1.409187	- 1.088731	0.159773	- 1.261759	0.15997	0.00214
0.18	1.309796	- 0.949148	0.179676	- 1.145672	0.17995	0.00305
0.20	1.222651	- 0.821761	0.199556	- 1.042206	0.19992	0.00419
0.22	1.145380	- 0.704195	0.219409	- 0.948988	0.21987	0.00557
0.24	1.076235	- 0.594697	0.239233	- 0.864266	0.23980	0.00723
0.26	1.013889	- 0.491932	0.259026	- 0.786710	0.25971	0.00920
0.28	0.957308	- 0.394863	0.278783	- 0.715286	0.27958	0.01148
0.30	0.905677	- 0.302669	0.298504	- 0.649173	0.29940	0.01412
0.32	0.858335	- 0.214683	0.318185	- 0.587710	0.31917	0.01713
0.34	0.814746	- 0.130363	0.337824	- 0.530355	0.33888	0.02053
0.36	0.774462	- 0.049258	0.357418	- 0.476661	0.35851	0.02436
0.38	0.737112	+0.029011	0.376965	- 0.426252	0.37805	0.02863
0.40	0.702380	0.104765	0.396461	- 0.378809	0.39748	0.03336
0.42	0.669997	0.178278	0.415906	- 0.334062	0.41679	0.03858
0.44	0.639733	0.249787	0.435295	- 0.291776	0.43595	0.04431
0.46	0.611387	0.319497	0.454627	- 0.251749	0.45494	0.05056
0.48	0.584784	0.387589	0.473898	- 0.213803	0.47375	0.05737
0.50	0.559774	0.454220	0.493107	- 0.177784	0.49234	0.06473
0.52	0.536220	0.519531	0.512252	- 0.143554	0.51070	0.07268
0.54	0.514004	0.583646	0.531328	- 0.110990	0.52878	0.08122
0.56	0.493020	0.646677	0.550335	- 0.079986	0.54656	0.09037
0.58	0.473173	0.708726	0.569269	- 0.050441	0.56401	0.10014
0.60	0.454380	0.769881	0.588129	- 0.022271	0.58110	0.11054
0.62	0.436562	0.830226	0.606911	+0.004606	0.59777	0.12158
0.64	0.419652	0.889836	0.625614	+0.030260	0.61401	0.13325
0.66	0.403586	0.948778	0.644235	+0.054758	0.62976	0.14557
0.68	0.388309	1.007116	0.662772	+0.078158	0.64499	0.15854
0.70	0.373769	1.064907	0.681222	0.100515	0.65965	0.17214
0.72	0.359918	1.122205	0.699584	0.121879	0.67370	0.18637
0.74	0.346713	1.179058	0.717854	0.142296	0.68709	0.20122
0.76	0.334115	1.235513	0.736031	0.161810	0.69978	0.21668
0.78	0.322088	1.291613	0.754112	0.180458	0.71171	0.23273
0.80	0.310597	1.347397	0.772096	0.198279	0.72284	0.24934
0.82	0.299611	1.402902	0.789979	0.215305	0.73313	0.26649
0.84	0.289103	1.458164	0.807761	0.231568	0.74252	0.28415
0.86	0.279045	1.513216	0.825438	0.247098	0.75096	0.30228
0.88	0.269413	1.568089	0.843009	0.261923	0.75841	0.32084
0.90	0.260184	1.622812	0.860471	0.276068	0.76482	0.33978

x	$E_1(x)$	Ei(x)	Si(x)	Ci(x)	$C(x)$	$S(x)$
1.00	0.219384	1.895118	0.946083	0.337404	0.77989	0.43826
1.05	0.201873	2.031087	0.987775	0.362737	0.77591	0.48805
1.10	0.185991	2.167378	1.028685	0.384873	0.76381	0.53650
1.15	0.171555	2.304288	1.068785	0.404045	0.74356	0.58214
1.20	0.158408	2.442092	1.108047	0.420459	0.71544	0.62340
1.25	0.146413	2.581048	1.146446	0.434301	0.68009	0.65866
1.30	0.135451	2.721399	1.183958	0.445739	0.63855	0.68633
1.35	0.125417	2.863377	1.220559	0.454927	0.59227	0.70501
1.40	0.116219	3.007207	1.256227	0.462007	0.54310	0.71353
1.45	0.107777	3.153106	1.290941	0.467109	0.49326	0.71111
1.50	0.100020	3.301285	1.324684	0.470356	0.44526	0.69750
1.55	0.092882	3.451955	1.357435	0.471862	0.40177	0.67308
1.60	0.086308	3.605320	1.389180	0.471733	0.36546	0.63889
1.65	0.080248	3.761588	1.419904	0.470070	0.33880	0.59675
1.70	0.074655	3.920963	1.449592	0.466968	0.32383	0.54920
1.75	0.069489	4.083654	1.478233	0.462520	0.32193	0.49938
1.80	0.064713	4.249868	1.505817	0.456811	0.33363	0.45094
1.85	0.060295	4.419816	1.532333	0.449925	0.35838	0.40769
1.90	0.056204	4.593714	1.557775	0.441940	0.39447	0.37335
1.95	0.052414	4.771779	1.582137	0.432934	0.43906	0.35114
2.0	0.722657	1.340965	1.605413	0.422981	0.48825	0.34342
2.1	0.730792	1.371487	1.648699	0.400512	0.58156	0.37427
2.2	0.738431	1.397422	1.687625	0.375075	0.63629	0.45570
2.3	0.745622	1.419172	1.722207	0.347176	062656	0.55315
2.4	0.752405	1.437118	1.752486	0.317292	0.55496	0.61969
2.5	0.758815	1.451625	1.778520	0.285871	0.45741	0.61918
2.6	0.764883	1.463033	1.800394	0.253337	0.38894	0.54999
2.7	0.770637	1.471662	1.818212	0.220085	0.39249	0.45292
2.8	0.776102	1.477808	1.832097	0.186488	0.46749	0.39153
2.9	0.781300	1.481746	1.842190	0.152895	0.56238	0.41014
3.0	0.786251	1.483729	1.848653	+0.119630	0.60572	0.49631
3.1	0.790973	1.483990	1.851659	+0.086992	0.56159	0.58182
3.2	0.795481	1.482740	1.851401	+0.055257	0.46632	0.59335
3.3	0.799791	1.480174	1.848081	+0.024678	0.40569	0.51929
3.4	0.803916	1.476469	1.841914	- 0.004518	0.43849	0.42965
3.5	0.807868	1.471782	1.833125	- 0.032129	0.53257	0.41525
3.6	0.811657	1.466260	1.821948	- 0.057974	0.58795	0.49231
3.7	0.815294	1.460030	1.808622	- 0.081901	0.54195	0.57498
3.8	0.818789	1.453211	1.793390	- 0.103778	0.44809	0.56562
3.9	0.822149	1.445906	1.776501	- 0.123499	0.42233	0.47520
4.0	0.825383	1.438208	1.758203	- 0.140982	0.49843	0.42052
4.1	0.828497	1.430201	1.738744	- 0.156165	0.57370	0.47580
4.2	0.831499	1.421957	1.718369	- 0.169013	0.54172	0.56320
4.3	0.834394	1.413542	1.697320	- 0.179510	0.44944	0.55400
4.4	0.837188	1.405012	1.675834	- 0.187660	0.43833	0.46227
4.5	0.839887	1.396419	1.654140	- 0.193491	0.52603	0.43427
4.6	0.842496	1.387805	1.632460	- 0.197047	0.56724	0.51619
4.7	0.845018	1.379209	1.611005	- 0.198391	0.49143	0.56715
4.8	0.847459	1.370663	1.589975	- 0.197604	0.43380	0.49675
4.9	0.849822	1.362196	1.569559	- 0.194780	0.50016	0.43507

x	$xe^xE_1(x)$	$xe^{-x}\text{Ei}(x)$	Si(x)	Ci(x)	$C(x)$	$S(x)$
5.0	0.852111	1.353831	1.549931	- 0.190030	0.56363	0.49919
5.1	0.854330	1.345589	1.531253	- 0.183476		
5.2	0.856481	1.337487	1.513671	- 0.175254		
5.3	0.858568	1.329538	1.497315	- 0.165506		
5.4	0.860594	1.321754	1.482300	- 0.154386		
5.5	0.862562	1.314144	1.468724	- 0.142053		
5.6	0.864473	1.306714	1.456668	- 0.128672		
5.7	0.866331	1.299471	1.446198	- 0.114411		
5.8	0.868138	1.292416	1.437359	- 0.099441		
5.9	0.869895	1.285554	1.430184	- 0.083933		
6.0	0.871606	1.278884	1.424688	- 0.068057		
6.1	0.873271	1.272406	1.420867	- 0.051983		
6.2	0.874892	1.266120	1.418707	- 0.035873		
6.3	0.876472	1.260024	1.418174	- 0.019888		
6.4	0.878012	1.254115	1.419223	- 0.004181		
6.5	0:879513	1.248391	1.421794	+0.011102		
6.6	0.880977	1.242848	1.425816	+0.025823		
6.7	0.882405	1.237482	1.431205	+0.039855		
6.8	0.883799	1.232290	1.437868	+0.053081		
6.9	0.885159	1.227267	1.445702	+0.065392		
7.0	0.886488	1.222408	1.454597	0.076695		
7.1	0.887785	1.217709	1.464433	0.086907		
7.2	0.889053	1.213166	1.475089	0.095957		
7.3	0.890292	1.208774	1.486436	0.103789		
7.4	0.891503	1.204527	1.498345	0.110358		
7.5	0.892688	1.200421	1.510682	0.115633		
7.6	0.893846	1.196452	1.523314	0.119598		
7.7	0.894980	1.192615	1.536109	0.122246		
7.8	0.896089	1.188905	1.548937	0.123586		
7.9	0.897174	1.185317	1.561671	0.123638		
8.0	0.898237	1.181848	1.574187	0.122434		
8.1	0.899278	1.178493	1.586367	0.120017		
8.2	0.900297	1.175247	1.598099	0.116440		
8.3	0.901296	1.172106	1.609278	0.111767		
8.4	0.902275	1.169068	1.619807	0.106071		
8.5	0.903234	1.166127	1.629597	0.099431		
8.6	0.904174	1.163279	1.638570	0.091936		
8.7	0.905096	1.160522	1.646655	0.083679		
8.8	0.906000	1.157852	1.653792	0.074760		
8.9	0.906887	1.155266	1.659934	0.065280		
9.0	0.907758	1.152759	1.665040	0.055348		
9.1	0.908611	1.150330	1.669084	0.045069		
9.2	0.909450	1.147974	1.672049	0.034555		
9.3	0.910272	1.145690	1.673930	0.023913		
9.4	0.911080	1.143474	1.674729	0.013252		
9.5	0.911873	1.141323	1.674463	+0.002678		
9.6	0.912652	1.139236	1.673157	- 0.007707		
9.7	0.913417	1.137210	1.670845	- 0.017804		
9.8	0.914169	1.135241	1.667570	- 0.027519		
9.9	0.914907	1.133329	1.663384	- 0.036764		

For $x>5$,

$$C(x)=0.5+\left(\frac{1}{\pi x}-\frac{3}{\pi^3x^5}\right)\sin\frac{\pi x^2}{2}-\left(\frac{1}{\pi^2x^3}-\frac{15}{\pi^4x^7}\right)\cos\frac{\pi x^2}{2}+\varepsilon(x),\text{ where }|\varepsilon(x)|<x^{-9}$$

$$S(x)=0.5-\left(\frac{1}{\pi x}-\frac{3}{\pi^3x^5}\right)\cos\frac{\pi x^2}{2}-\left(\frac{1}{\pi^2x^3}-\frac{15}{\pi^4x^7}\right)\sin\frac{\pi x^2}{2}+\varepsilon(x),\text{ where }|\varepsilon(x)|<x^{-9}$$

x	$xe^xE_1(x)$	$xe^{-x}\mathrm{Ei}(x)$	$\mathrm{Si}(x)$	$\mathrm{Ci}(x)$
10.0	0.915633	1.131470	1.658348	- 0.045456
10.2	0.917048	1.127906	1.645995	- 0.060892
10.4	0.918416	1.124534	1.631117	- 0.073320
10.6	0.919739	1.121340	1.614391	- 0.082368
10.8	0.921020	1.118309	1.596541	- 0.087809
11.0	0.922260	1.115431	1.578307	- 0.089563
11.2	0.923461	1.112694	1.560416	- 0.087693
11.4	0.924625	1.110089	1.543557	- 0.082402
11.6	0.925754	1.107606	1.528354	- 0.074015
11.8	0.926850	1.105237	1.515347	- 0.062967
12.0	0.927914	1.102975	1.504971	- 0.049780
12.2	0.928946	1.100811	1.497547	- 0.035042
12.4	0.929950	1.098740	1.493270	- 0.019383
12.6	0.930925	1.096756	1.492206	- 0.003444
12.8	0.931874	1.094854	1.494297	+0.012138
13.0	0.932796	1.093027	1.499362	0.026764
13.2	0.933694	1.091273	1.507111	0.039889
13.4	0.934567	1.089586	1.517161	0.051043
13.6	0.935418	1.087962	1.529047	0.059845
13.8	0.936247	1.086399	1.542249	0.066018
14.0	0.937055	1.084892	1.556211	0.069397
14.2	0.937843	1.083438	1.570362	0.069926
14.4	0.938611	1.082035	1.584141	0.067666
14.6	0.939360	1.080680	1.597016	0.062781
14.8	0.940091	1.079370	1.608505	0.055536
15.0	0.940804	1.078103	1.618194	+0.046279
15.2	0.941501	1.076877		
15.4	0.942181	1.075690		
15.6	0.942846	1.074541		
15.8	0.943495	1.073426		
16.0	0.944130	1.072345		
16.2	0.944750	1.071296		
16.4	0.945357	1.070278		
16.6	0.945951	1.069289		
16.8	0.946532	1.068328		
17.0	0.947100	1.067394		
17.2	0.947656	1.066485		
17.4	0.948201	1.065601		
17.6	0.948735	1.064741		
17.8	0.949257	1.063903		
18.0	0.949769	1.063087		
19.0	0.952181	1.059305		
20.0	0.954371	1.055956		
50	0.980755	1.020852		
100	0.990194	1.010206		
200	0.995049	1.005051		
∞	1.000000	1.000000		

12.6 Step and Impulse Functions

Heaviside's step function $\theta(t)$

$$\theta(t)=\begin{cases}1, & t>0\\ 0, & t<0\end{cases}\qquad \theta(t-a)=\begin{cases}1, & t>a\\ 0, & t<a\end{cases}$$

$$\text{sgn}(t)=2\theta(t)-1=\begin{cases}1, & t>0\\ -1, & t<0\end{cases}$$

$$f(t)[\theta(t-a)-\theta(t-b)]=\begin{cases}0, & t<a\\ f(t), & a<t<b\\ 0, & t>b\end{cases}$$

$\int f(t)\theta(t-a)dt=(F(t)-F(a))\,\theta(t-a)+C$

F primitive function of f.

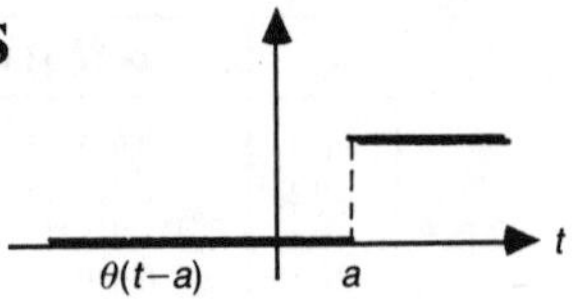

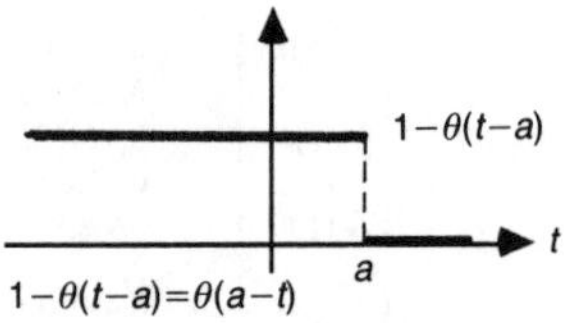

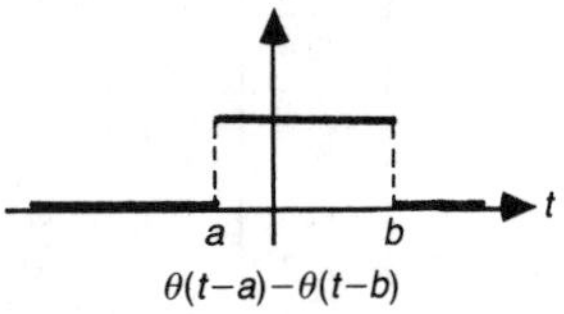

Dirac's delta function $\delta(t)$

The generalized function $\delta(t)$ has the following properties. (Cf. sec. 12.9).

1. $\delta(t)=0,\ (t\neq 0),\ \delta(0)=+\infty$
2. $\int_{-\infty}^{\infty}\delta(t)dt=1 \qquad \int_{-\infty}^{\infty}|\delta^{(n)}(t)|dt=\infty,\ n\geq 1$
3. $\delta(-t)=\delta(t)$ (δ is even)
4. $\delta\left(\frac{t}{a}\right)=a\,\delta(t),\ a>0$
5. $\theta'(t)=\delta(t), \qquad \frac{d}{dt}\text{sgn}(t)=2\delta(t)$

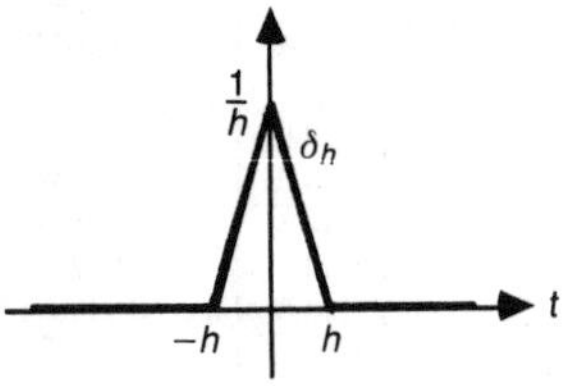

$\delta(t)=$" $\lim_{h\to 0}\delta_h(t)$"

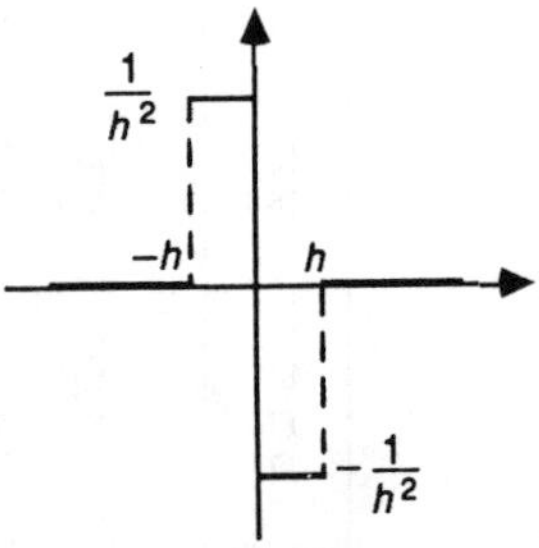

$\delta'(x)=$" $\lim_{h\to 0}\delta_h'(x)$"

$f(t)$ continuous:

6. $f(t)\delta(t-a)=f(a)\delta(t-a)$
7. $\int_{-\infty}^{\infty}f(t)\delta(t-a)dt=f(a) \qquad \int\delta(t-a)dt=\theta(t-a)+C$

$f'(t)$ continuous:

8. $f(t)\delta'(t-a)=f(a)\delta'(t-a)-f'(a)\delta(t-a)$
9. $\int_{-\infty}^{\infty}f(t)\delta'(t-a)dt=-\int_{-\infty}^{\infty}f'(t)\delta(t-a)dt=-f'(a)$

$f^{(n)}(t)$ continuous:

10. $\int_{-\infty}^{\infty}f(t)\delta^{(n)}(t-a)dt=(-1)^n f^{(n)}(a)$
11. $f*\delta^{(n)}(t)=f^{(n)}(t),\ n=0, 1, 2, \ldots$

---

12. $t^n f(t)=0\Rightarrow f(t)=C_0\delta(t)+C_1\delta'(t)+\ldots+C_{n-1}\delta^{(n-1)}(t),\ n=1, 2, 3, \ldots$

Example

Let $f(t)=e^{-|t|}$. Calculate the (distributional) third derivative $f'''(t)$. Solution:

$$f'(t)=-\operatorname{sgn} t\, e^{-|t|}$$
$$f''(t)=-2\delta(t)e^{-|t|}+(\operatorname{sgn} t)^2e^{-|t|}=-2\delta(t)+e^{-|t|}$$
$$f'''(t)=-2\delta'(t)-\operatorname{sgn} t\, e^{-|t|}$$

12.7 Functional Analysis

Spaces

1. *Linear space* or *vector space*.
2. *Metric space*: Set with a distance function.
3. *Normed space*: Linear space with a norm (distance).
4. *Banach space*: Complete normed space.
5. *Hilbert space*: Banach space with an inner product.

1. Linear spaces. See sec. 4.7.

2. Metric spaces

A set M of points $u, v, \ldots$ is a *metric space* if there is a distance function $d(u, v)$ on M satisfying for all $u, v, w \in M$:

(*i*) $d(u, v) \geqslant 0$ $[=0 \Leftrightarrow u=v]$
(*ii*) $d(v, u)=d(u, v)$
(*iii*) $d(u, v) \leqslant d(u, w)+d(w, v)$

Convergence

1. $u_n \to u$ as $n \to \infty$ if $d(u_n, u) \to 0$ as $n \to \infty$ $(u, u_n \in M)$
2. A sequence $\{u_n\}_1^\infty$ in M is a *Cauchy sequence* if $\lim_{m, n \to \infty} d(u_m, u_n)=0$, i.e. for any $\varepsilon>0$ there exist an N such that $d(u_m, u_n)<\varepsilon$, all $m, n>N$.
3. M is *complete* if every Cauchy sequence in M has a limit in M.

Topology

Let S be a subset of M.

1. $u_0 \in S$ is an *interior point* of S if there exists a $\delta>0$ such that $u \in S$ for all $u \in M$ satisfying $d(u, u_0)<\delta$.
2. $u_0 \in M$ is an *accumulation point* of S if there is a sequence $\{u_n\}$, $u_n \in S$, $u_n \neq u_0$ with $\lim_{n \to \infty} u_n=u_0$.
3. S is *open* if every point of S is an interior point.
4. The *closure* $\overline{S}$ (of S) is formed by adding to S all its accumulation points.
5. S is *closed* if $S=\overline{S}$, or equivalently, if the complement $M \setminus S$ is open.
6. S is *compact* if any sequence in S contains a subsequence that converges in S.
7. S is *dense* in M if every point of M is the limit of some sequence in S.
8. M is *separable* if it contains a countable dense subset.

3. Normed spaces

A linear space L of elements $u, v, \ldots$ is a *normed space* if there is a norm $\|\cdot\|$ on L satisfying for all $u, v \in L$:

(*i*) $\|u\| \geqslant 0$ $[=0 \Leftrightarrow u=0]$
(*ii*) $\|\alpha u\|=|\alpha| \cdot \|u\|$ (α scalar)
(*iii*) $\|u+v\| \leqslant \|u\|+\|v\|$

With the distance $d(u, v)=\|u-v\|$ the space is a metric space.

4. Banach spaces

A complete normed space is called a *Banach space.*

> *Examples of Banach spaces*
>
> 1. Every finite-dimensional normed space (e.g. $\boldsymbol{R}^n$).
>
> 2. $C([a, b])=\{$continuous functions on $[a, b]\}$ with norm $\|f\|= \max_{a\leqslant x\leqslant b} |f(x)|$
>
> 3. The $L_p(\Omega)-$spaces, $1\leqslant p\leqslant\infty$, with ordinary L_p-norm.

5. Hilbert spaces

A linear space L is a real [complex] inner (scalar) product space if there is defined an inner product $(u|v)$ on L satisfying for all $u, v\in L$:

(*i*) $(u|u)\geqslant 0$ $(=0 \Leftrightarrow u=0)$
(*ii*) $(v|u)=(u|v)$ $[(v|u)=\overline{(u|v)}]$
(*iii*) $(\alpha_1 u_1+\alpha_2 u_2|v)=\alpha_1(u_1|v)+\alpha_2(u_2|v)$

The elements u and v are orthogonal if $(u|v)=0$.

A norm is defined by $\|u\|^2=(u|u)$.

A Banach space with inner product is called a *Hilbert space.*

> *Examples of Hilbert spaces*
>
> 4. $\boldsymbol{R}^n$ with inner product $(\boldsymbol{x}, \boldsymbol{y})= \sum_{k=1}^{n} x_k y_k$.
>
> 5. $L_2(\Omega)$ with inner product $(f, g)= \int_\Omega f(x)g(x)dx$.

Let H be a Hilbert space.

Orthonormal basis

An orthonormal family $\{e_n\}_1^\infty$, i.e. $(e_i|e_j)=\delta_{ij}$, is an *orthonormal basis* (ON-basis) of H if $u=\sum_1^\infty (u|e_n)e_n$, all $u\in H$.

1. H is finite dimensional: See sec. 4.7.
2. H is infinite dimensional: The following conditions are equivalent:

   (*i*) $\{e_n\}_1^\infty$ is an ON-basis
   (*ii*) $(u|e_n)=0$, all $n \Rightarrow u=0$
   (*iii*) $\|u\|^2=\sum_1^\infty (u|e_n)^2$, all $u\in H$. (*Parseval's identity*)

Gram-Schmidt orthogonalization process, cf. sec. 4.7.

Orthogonal complement

Two subspaces U and V of H are orthogonal ($U\perp V$) if $(u|v)=0$, all $u\in U$, $v\in V$.

Orthogonal complement of a set $U\subset H$: $U^\perp=\{v\in H\text{: } (u|v)=0, \forall u\in U\}$, which is a closed linear subspace of H (i.e. $U^\perp$ is a Hilbert space).

Linear operators

Let T be a *linear operator* [i.e. $T(\alpha u+\beta v)=\alpha Tu+\beta Tv$] on a Banach space B or a Hilbert space H (or from one such space to another).

The operator T is

(*i*) *continuous* at u_0 if $\|Tu-Tu_0\| \to 0$ as $u \to u_0$
(*ii*) *bounded* if there exists a constant C such that

$$\|Tu\| \leq C\|u\|, \text{ all } u \in B$$

The smallest bound C is called the *norm* of T, denoted $\|T\|$, i.e.

$$\|T\| = \sup_{\substack{u \in B \\ u \neq 0}} \frac{\|Tu\|}{\|u\|} = \sup_{\|u\|=1} \|Tu\|$$

$\|Tu\| \leq \|T\| \cdot \|u\|$	$\|T_1 T_2\| \leq \|T_1\| \cdot \|T_2\|$
$\|T_1+T_2\| \leq \|T_1\|+\|T_2\|$	$\|\alpha T\| = \|\alpha\| \cdot \|T\|$

(With this norm the space consisting of all bounded linear operators on B is itself a Banach space.)

T is continuous at $u_0 \Leftrightarrow T$ is continuous at $0 \Leftrightarrow T$ is bounded

A linear operator T is *compact* (or *completely continuous*) if for every bounded sequence $\{u_n\}$ the sequence $\{Tu_n\}$ contains a convergent subsequence.

1. If T, T_n compact and S bounded linear operators, then
   (*i*) T is bounded,
   (*ii*) T_1+T_2, TS and ST are compact,
   (*iii*) $T_n \to A$, i.e. $\|T_n-A\| \to 0 \Rightarrow A$ is compact.
2. If T is linear and the range of T is finite dimensional, then T is compact.

Examples

6. If B is infinite dimensional, then the identity operator $Iu=u$ is *not* compact.
7. Let $K(x, y)$ be an *integral operator* on $[a, b]$, i.e.

$$Kf(x) = \int_a^b K(x, y) f(y)\, dy.$$

Then

(*i*) $K(x, y)$ is continuous in the square $a \leq x, y \leq b \Rightarrow$
K is compact on $C[a, b]$ and $\|K\| \leq \max|K(x, y)|$
(*ii*) $K(x, y) \in L_2$ in the square $a \leq x, y \leq b \Rightarrow$
K is compact on $L_2[a, b]$ and $\|K\| \leq \|K(x, y)\|_{L_2}$

Inverse operator T^{-1}: $v=Tu \Leftrightarrow u=T^{-1}v$; $\quad TT^{-1}=T^{-1}T=I$.

T^{-1} exists if

(*i*) T is surjective and $\|Tu\| \geq c\|u\|$, $c>0$. Then $\|T^{-1}\| \leq 1/c$.
(*ii*) $T=I-K$, $\|K\|<1$. Then $T^{-1}=(I-K)^{-1}=I+K+K^2+\ldots$

Symmetric operator

Let $T: H \to H$ be a linear operator. The *adjoint operator* T^* is defined by $(Tu|v)=(u|T^*v)$

$T^{**}=T \quad (TS)^*=S^*T^* \quad (T^*)^{-1}=(T^{-1})^* \quad \|T^*\|=\|T\|$

T is *symmetric* (*hermitian* in the complex case) if $T^*=T$, i.e. $(Tu|v)=(u|Tv)$.

Example 8. The integral operator in example 7 above is symmetric if $K(x, y)=K(y, x)$.

Projection

Let H_1 be a subspace of H and Pu the orthogonal projection of u onto H_1, i.e. $(u-Pu|v)=0$, all $v \in H_1$

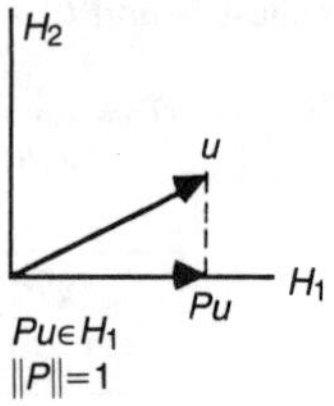

$\Leftrightarrow$ P is an orthogonal projection $P^2=P$ and $P^*=P$

Spectrum

Let T be a linear operator on H. The number λ is an *eigenvalue* of T and u the corresponding *eigenvector* if $Tu=\lambda u$, $u \neq 0$.

1. *Eigenspace* $H_\lambda=\{u \in H: Tu=\lambda u\}$.
2. T *symmetric* $\Rightarrow$ all eigenvalues are real and eigenvectors corresponding to different eigenvalues are orthogonal. $(H_{\lambda_1} \perp H_{\lambda_2}, \lambda_1 \neq \lambda_2)$.
3. T *bounded* $\Rightarrow$ $|\lambda| \leq \|T\|$.
4. T *symmetric* and *compact* $\Rightarrow$

   (*i*) H_λ is finite dimensional if $\lambda \neq 0$

   (*ii*) $\|T\|$ or $-\|T\|$ is an eigenvalue.

A spectral theorem. (Hilbert) Assume that $T \neq 0$ is a linear, symmetric and compact operator on H. Let $\|\lambda_1\| \geq \|\lambda_2\| \geq \|\lambda_3\| \geq \ldots > 0$ be the non-zero eigenvalues of T and let $\{e_i\}$ be a corresponding orthonormal family of eigenvectors. Then (*i*) $u=\sum_i (u\|e_i)e_i+u_0$, where $u_0 \in H_0$, i.e. $Tu_0=0$ (*ii*) $Tu=\sum_i \lambda_i(u\|e_i)e_i$

The Fredholm alternative in Banach spaces

Let K be a compact (integral) operator on B. Then the equation

$$f-Kf=g$$

has a unique solution $f \in B$ for any $g \in B$ if the equation $f-Kf=0$ admits only the trivial solution $f=0$.

The Fredholm alternative in Hilbert spaces

Let K be a compact (integral) operator on H. Then the equation

$$f-Kf=g$$

has a solution $f \in H$ if and only if g is orthogonal to every solution of the equation

$$f-K^*f=0$$

The solution is unique if this equation admits only the trivial solution $f=0$.

Linear functionals

A linear operator F: B (or H) $\to R$ (or C) is called a *linear functional* (or *linear form*).

> *Theorems*
> 1. (*Hahn-Banach*): Let F_1 be a bounded linear functional on a linear subspace B_1 of B. Then F_1 can be extended to a linear functional F on B such that $\|F\|=\|F_1\|$.
> 2. (*Riesz*): Every bounded linear functional F on H has the form $Fu=(u|v)$ for a unique $v\in H$, and $\|F\|=\|v\|$.

Bilinear forms

An operator $A(u, v)$: $H\times H\to C$ (or R) is a *bilinear form* if

$$\begin{cases} A(\alpha_1u_1+\alpha_2u_2, v)=\underline{\alpha}_1A(u_1, v)+\underline{\alpha}_2A(u_2, v), \\ A(u, \beta_1v_1+\beta_2v_2)=\bar{\beta}_1A(u, v_1)+\bar{\beta}_2 A(u, v_2). \end{cases}$$

The form is bounded if $|A(u, v)|\leqslant C\|u\|\cdot\|v\|$. The norm $\|A\|=\sup\limits_{\|u\|=\|v\|=1}|A(u, v)|$

$A(u, v)$ is *elliptic* if $|A(u, u)|\geqslant c\|u\|^2$, $c>0$.

> *Theorem* (*Lax-Milgram*)
> (*i*) Every bounded bilinear form A on H has the form $A(u, v)=(Tu|v)$ for a unique linear operator T, and $\|A\|=\|T\|$.
> (*ii*) Let A be a bounded and elliptic bilinear form on a linear space V and let F be a bounded linear form on V. Then the variational problem $A(u, v)=F(v)$, all $v\in V$ has a unique solution $u\in V$.

12.8 Lebesgue Integrals

Lebesgue measure

Set measure

Let S be a subset of an interval $I=[a, b]$ and let $S'=I\setminus S$. Set $m(I)=b-a$.

Exterior Lebesgue measure $m_e(S)=\inf\sum\limits_1^\infty m(I_n)$, $S\subset\bigcup\limits_1^\infty I_n$, I_n interval.

Interior Lebesgue measure $m_i(S)=(b-a)-m_e(S')$.

If $m_e(S)=m_i(S)$, then the set S is *Lebesgue measurable* and the *Lebesgue measure* is

$$\boxed{m(S)=m_e(S)=m_i(S)}$$

An *unbounded* set S is measurable if $S\cap(-c, c)$ is measurable for any $c>0$ and

$$m(S)=\lim_{c\to\infty} m[S\cap(-c, c)]$$

Sets of measure zero

A countable set $S=\{p_n\}_1^\infty$ e.g. $\mathbf{Q}$, has measure zero, because, for any $\varepsilon>0$, letting p_n belong to an interval of length $\varepsilon\cdot 2^{-n}$ it follows

$$m_e(S)<\sum_1^\infty \varepsilon\cdot 2^{-n}=\varepsilon$$

On the other hand, a set of measure zero does not have to be countable.

Measurable functions

A function $f(x)$ is *measurable* if the set $\{x: f(x)\geq c\}$ is measurable for every c.

A property which holds for every point (of a set) with the exception of a set of measure zero is said to hold *almost everywhere (a.e.).*

The Lebesgue integral

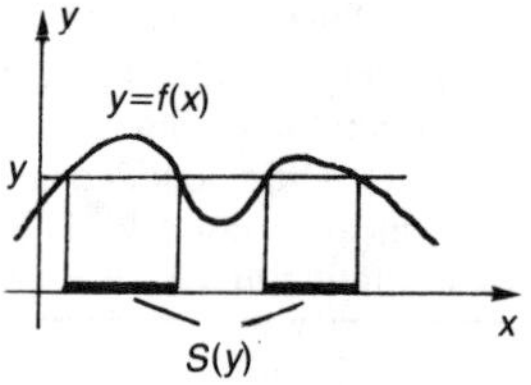

1. Assume that (*i*) $f(x)\geq 0$ and measurable (*ii*) S measurable.

   Set $S(y)=\{x: f(x)\geq y\}$. Then $m(S(y))$ is decreasing and Riemann integrable:

   Definition: $$\int_S f(x)dx=\int_0^\infty m(S(y))dy$$

   If this value is finite, then $f(x)$ is *summable* over S.
2. $f(x)$ arbitrary sign. Set $f_+(x)=\max[f(x), 0]\geq 0$ and $f_-(x)=\max[-f(x), 0]\geq 0$.

   Then $f(x)=f_+(x)-f_-(x)$ and

$$\int_S f(x)dx=\int_S f_+(x)dx-\int_S f_-(x)dx$$

Fubini's theorem

If $f(x, y)$ is summable over $\mathbf{R}^2$, then $\iint_{\mathbf{R}^2} f(x, y)dxdy=\int_{\mathbf{R}} dx\int_{\mathbf{R}} f(x, y)dy$.

(The formulation implies that the right hand side makes sense.)

Limit theorems

Fatou's lemma

If (*i*) $f_n(x)\geq 0$ summable over Ω
 (*ii*) $f_n(x)\to f(x)$ a.e.
 (*iii*) $\int_\Omega f_n(x)dx\leq C$,

then $f(x)$ is summable and $\int_\Omega f(x)dx\leq C$

Lebesgue's theorem

If (*i*) $f_n(x)$, $g(x)$ are summable over Ω
(*ii*) $|f_n(x)| \leq g(x)$
(*iii*) $f_n(x) \to f(x)$ a.e.,

then $f(x)$ is summable and

$$\int_\Omega f(x)dx = \lim_{n\to\infty} \int_\Omega f_n(x)dx$$

Approximation

If $f(x)$ is summable, then for any $\varepsilon > 0$ there exists a piece-wise constant function (with a finite number of steps) $\varphi(x)$ such that $\int_\Omega |f(x) - \varphi(x)|dx < \varepsilon$.

Some function spaces in $\mathbf{R}^n$

Notation.

a. Below $\Omega \subset \mathbf{R}^n$ denotes a domain (open and connected). All functions are complex valued and defined on a subset of $\mathbf{R}^n$

b. Multi-index $\alpha = (\alpha_1, \ldots, \alpha_n)$, $|\alpha| = \alpha_1 + \ldots + \alpha_n$, $\beta \leq \alpha \Leftrightarrow \beta_i \leq \alpha_i$, all i.

c. $D^\alpha = D_1^{\alpha_1} \ldots D_n^{\alpha_n}$, $D_i = \partial/\partial x_i$.

Leibniz' formula:

$$D^\alpha(uv) = \sum_{\beta \leq \alpha} \binom{\alpha}{\beta} D^\beta u D^{\alpha-\beta} v, \quad \binom{\alpha}{\beta} = \binom{\alpha_1}{\beta_1} \cdots \binom{\alpha_n}{\beta_n}$$

d. $\Omega_1 \subset\subset \Omega$ means that $\bar{\Omega}_1 \subset \Omega$ and Ω_1 is bounded.

e. Let u be a function. The *support* of u is the set supp $u = \overline{\{x \in \mathbf{R}^n : u(x) \neq 0\}}$.

f. u has *compact support* in Ω if supp $u \subset\subset \Omega$.

Spaces of continuous functions

$C^0(\Omega) = C(\Omega) = \{\phi : \phi \text{ is continuous in } \Omega\}$

$C^m(\Omega) = \{\phi : D^\alpha\phi \text{ is continuous in } \Omega, |\alpha| \leq m\}$

$C^\infty(\Omega) = \{\phi : D^\alpha\phi \text{ is continuous in } \Omega, \text{ all } \alpha\}$

$C_0{}^m(\Omega) = \{\text{Functions in } C^m(\Omega) \text{ with compact support in } \Omega\}$, $0 \leq m \leq \infty$.

$C^m(\bar{\Omega}) = \{\phi \in C^m(\Omega) : D^\alpha\phi \text{ is bounded and uniformly continuous in } \Omega, |\alpha| \leq m\}$

$C_B{}^m(\Omega) = \{\phi \in C^m(\Omega) : D^\alpha\phi \text{ is bounded in } \Omega, |\alpha| \leq m\}$

1. With the norm $\|\phi\| = \max_{0 \leq |\alpha| \leq m} \sup_{x \in \Omega} |D^\alpha\phi(x)|$,

   $C^m(\bar{\Omega})$ is (*i*) a Banach space (*ii*) separable if Ω is bounded.

2. Let $\Omega_2 \subset\subset \Omega_1 \subset\subset \Omega$. Then there exists a $\phi \in C_0{}^\infty(\Omega)$ such that $\phi(x) = 1$ in Ω_2 and $\phi(x) = 0$ outside Ω_1.

L^p-spaces

A function $f(x)$ belongs to $L^p(\Omega)$ if $\int_\Omega |f(x)|^p dx<\infty$. $L^p(\Omega)$ is a Banach space with the norm

$$\|f\|_p=\|f\|_{L^p(\Omega)}=\left(\int_\Omega |f(x)|^p dx\right)^{1/p}$$

A function $f(x)\in L^\infty(\Omega)$ if $f(x)$ is measurable over Ω and if there is a constant C such that $|f(x)|\leq C$ a.e. The smallest possible value of C, $\operatorname{essup}_{x\in\Omega}|f(x)|$, is the L^∞-norm of $f(x)$, i.e.

$$\|f\|_\infty=\operatorname{essup}_{x\in\Omega}|f(x)|$$

Also $L^\infty(\Omega)$ is a Banach space.

Furthermore, $L^2(\Omega)$ is a Hilbert space with the inner product $(u, v)=\int_\Omega u(x)\overline{v(x)}dx$.

Approximation
a. $L^p(\Omega)$ is separable if $1\leq p<\infty$ [but not so for $p=\infty$].
b. $C_0^\infty(\Omega)$ is dense in $L^p(\Omega)$ if $1\leq p<\infty$ [but not so for $p=\infty$].

Weak derivatives
The function $D^\alpha u\in L^1(\Omega)$ is the *weak (distributional) derivative* of $u\in L^1(\Omega)$ if

$$\int_\Omega D^\alpha u(x)\,\phi(x)dx=(-1)^{|\alpha|}\int_\Omega u(x)D^\alpha\phi(x)dx,\ \text{all } \phi\in C_0^\infty(\Omega).$$

Remark. (*i*) If $D^\alpha u(x)$ exists, then it is unique up to a set of measure zero. (*ii*) If $D^\alpha u(x)$ exists in the classical sense, then it coincides with the weak derivative a.e.

Dual space
If $\frac{1}{p}+\frac{1}{q}=1$, $p, q\geq 1$, L^p and L^q are said to be dual.

Inequalities. ($1\leq p\leq\infty$)

$\|f+g\|_p\leq\|f\|_p+\|g\|_p$ [*Minkowski's inequality*]

$\|fg\|_1\leq\|f\|_p\|g\|_q$, $\frac{1}{p}+\frac{1}{q}=1$, $p, q\geq 1$ [*Hölder's inequality*]

$\|f*g\|_r\leq\|f\|_p\|g\|_q$, $\frac{1}{p}+\frac{1}{q}-\frac{1}{r}=1$, $p, q, r\geq 1$, $f, g\geq 0$ [*Young's inequality*]

Convolutions (one dimension)

$$f*g(x)=\int_{-\infty}^{\infty} f(x-y)g(y)dy$$

1. $f\in L^1$, g bounded $\Rightarrow f*g$ bounded
2. $f\in L^1$, $g\in L^p$ $(1\leq p\leq\infty)\Rightarrow f*g\in L^p$
3. $f(x)$, $g(x)=0$, $x<0\Rightarrow f*g(x)=0$, $x<0$
4. $f*g=g*f$
5. $f*(g+h)=f*g+f*h$
6. $(f*g)'=f'*g=f*g'$ if $f, f', g, g'\in L^1$.

Sobolev spaces

1. Norms: (*i*) $\|u\|_{m,p}=\|u\|_{m,p,\Omega}=\left\{\sum_{|\alpha|\leqslant m}\|D^\alpha u\|_p^p\right\}^{1/p}$, $1\leqslant p<\infty$

   (*ii*) $\|u\|_{m,\infty}=\|u\|_{m,\infty,\Omega}=\max_{0\leqslant|\alpha|\leqslant m}\|D^\alpha u\|_\infty$

2. The *Sobolev spaces* $W^{m,p}(\Omega)=\{u\in L^p(\Omega)\colon D^\alpha u\in L^p(\Omega)$ for $|\alpha|\leqslant m\}$=The completion of $\{u\in C^m(\Omega)\colon \|u\|_{m,p}<\infty\}$ with respect to the norm $\|\bullet\|_{m,p}$.

3. $W_0^{m,p}(\Omega)$=the closure of $C_0^\infty(\Omega)$ in the space $W^{m,p}(\Omega)$.

4. $W^{m,p}(\Omega)$ is a Banach space (and separable if $1\leqslant p<\infty$).

5. $H^m(\Omega)=W^{m,2}(\Omega)$ $[H_0^m(\Omega)=W_0^{m,2}(\Omega)]$ is a separable Hilbert space with inner product

   $(u,v)=\sum_{|\alpha|\leqslant m}(D^\alpha u, D^\alpha v)$, $(u,v)=\int_\Omega u(x)\overline{v(x)}dx$.

6. *Characterizations:* Let Ω be bounded. $H_0^1(\Omega)=\{u\in H^1(\Omega)\colon u=0 \text{ on } \partial\Omega\}$, $H_0^2(\Omega)=$ $=\{u\in H^2(\Omega)\colon u=\partial u/\partial n=0 \text{ on } \partial\Omega\}$

7. *Interpolation type inequalities.* Introduce the semi-norm

   $|u|_{m,p}=|u|_{m,p,\Omega}=\left\{\sum_{|\alpha|=m}\|D^\alpha u\|_p^p\right\}^{1/p}$,

   Let Ω be "sufficiently regular" (e.g. convex). Then for any $\varepsilon>0$, m, and p there exists a constant C such that for $0\leqslant j\leqslant m-1$,

   $$|u|_{j,p}\leqslant\varepsilon|u|_{m,p}+C|u|_{0,p}, \text{ all } u\in W^{m,p}(\Omega).$$

8. *Poincaré-Friedrich's inequality:* If Ω is bounded then there exists a constant $C=C(\Omega)$ such that $|u|_{0,2}\leqslant C|u|_{1,2}$, all $u\in H_0^1(\Omega)$.

9. *Sobolev imbedding theorem.*

   Notation. A normed space X with norm $\|\bullet\|_X$ is *imbedded* in another normed space Y, written $X\to Y$, if (*i*) X is a vector subspace of Y (*ii*) There exists a constant C such that

   $$\|x\|_Y\leqslant C\|x\|_X, \text{all } x\in X.$$

   Let Ω be "sufficiently regular" (e.g. convex). Then

   A. If $mp<n$, then $W^{j+m,p}(\Omega)\to W^{j,q}(\Omega)$, $p\leqslant q\leqslant np/(n-mp)$, $j\geqslant 0$.

   B. If $mp=n$, then (*i*) $W^{j+m,p}(\Omega)$, $\to W^{j,q}(\Omega)$, $p\leqslant q<\infty$ (*ii*) $W^{j+n,1}(\Omega)\to C_B^j(\Omega)$. (If Ω is bounded $W^{j+m,p}(\Omega)\to W^{j,q}(\Omega)$, $p\leqslant q\leqslant\infty$. In particular, $\|u\|_{0,\infty}\leqslant C\|u\|_{[n/2]+1,2}$, $[\bullet]$=integer part.)

   C. If $mp>n$, then $W^{j+m,p}(\Omega)\to C_B^j(\Omega)$.

12.9 Generalized functions (Distributions)*

Test functions. (The class S)

1. A *test function* is a complex-valued function $\varphi(t)$ on $\boldsymbol{R}=(-\infty, \infty)$ satisfying:

   (*i*) $\varphi(t)$ is infinitely differentiable, i.e. $\varphi \in C^{\infty}(\boldsymbol{R})$
   (*ii*) $\lim_{|t| \to \infty} t^p \varphi^{(q)}(t)=0$ for all integers p, $q \geq 0$.

   The class of all test functions is denoted S. (Another class of test functions is D, consisting of all infinitely differentiable functions, which are zero outside a bounded subset of $\boldsymbol{R}$. Note that $D \subset S$.)

   [Example: $\varphi(t)=e^{-t^2} \in S$]

2. A sequence $\varphi_n \in S$ is a *zero sequence* if

$$\lim_{n \to \infty} \max_{t \in R} |t^p \varphi_n^{(q)}(t)|=0 \text{ for all } p,\ q \geq 0$$

   [Example: $\varphi \in S \Rightarrow \varphi_n(t)=\varphi\left(t+\frac{1}{n}\right)-\varphi(t)$ is a zero sequence.]

Generalized functions. (The set S')

3. A functional on S is a function f which maps $\varphi \in S$ to a complex number, denoted $(f|\varphi)$ or $f(\varphi)$.
4. A *generalized function* (g.f.) (or *temperate distribution*) is a continuous linear functional f on S, i.e.

   (*i*) $(f|\alpha\varphi+\beta\psi)=\alpha(f|\varphi)+\beta(f|\psi)$, α, $\beta \in C$; φ, $\psi \in S$.
   (*ii*) $\lim_{n \to \infty} (f|\varphi_n)=0$ for any zero sequence $\varphi_n \in S$.

   The set of all temperate distributions is denoted S'. (The corresponding functionals on D are called *distributions*.)
5. $f=g \Leftrightarrow (f|\varphi)=(g|\varphi)$ for all $\varphi \in S$.
6. The *support* of $\varphi \in S$ is the smallest closed set outside of which $\varphi(t)=0$.
7. Let $A \subset R$ be an open set. Then $f=g$ in A if $(f|\varphi)=(g|\varphi)$, all $\varphi \in S$ with support in A.
8. The *support* of $f \in S'$ is the smallest closed set outside of which $f=0$.
9. Let $f(t)$ be a piece-wise continuous function such that

$$\int_{-\infty}^{\infty} (1+t^2)^{-m} |f(t)| dt < \infty$$

   for some integer m. Then

$$(f|\varphi)= \int_{-\infty}^{\infty} f(t)\varphi(t) dt$$

   defines a *regular* g.f. A non-regular g.f. is called *singular*. Also for singular g.fs. the notation $f(t)$ (instead of f) is used as well as

$$(f|\varphi)= \oint_{-\infty}^{\infty} f(t)\varphi(t) dt$$

10. Dirac's delta function $\delta(t)$ is a singular g.f., defined by

$$(\delta|\varphi)=\varphi(0)$$

11. The g.f. $f(at+b)$, $a \neq 0$, is defined by

$$\oint_{-\infty}^{\infty} f(at+b)\varphi(t) dt = \frac{1}{|a|} \oint_{-\infty}^{\infty} f(t)\varphi\left(\frac{t-b}{a}\right) dt$$

* Essentially listed from Jan Petersson: Fourieranalys, Chalmers University of Technology, Göteborg, 1987.

12. The g.fs. $f+g$, cf, $\bar{f}$ and ψf (where $\psi \in C^\infty$ and for each integer $q \geq 0$ there exists an integer p such that $t^{-p}\,\psi^{(q)}(t) \to 0$ as $|t| \to \infty$) are defined by

$(f+g\vert\varphi)=(f\vert\varphi)+(g\vert\varphi)$	$(cf\vert\varphi)=c(f\vert\varphi)$
$(\bar{f}\vert\varphi)=\overline{(f\vert\bar{\varphi})}$	$(\psi f\vert\varphi)=(f\vert\psi\varphi)$

13. $\psi(t)\delta(t-a)=\psi(a)\delta(t-a)$
14. $tf(t)=0 \Leftrightarrow f(t)=c\delta(t)$

Derivatives

15. The derivative $f'(t)$ is defined by $(f'|\varphi)=-(f|\varphi')$

16. $(\psi f)'=\psi f'+\psi' f$

17. $\theta'(t)=\delta(t)$

18. The singular g.fs. t^{-n}, $n=1,2,3,\ldots$ are defined by $t^{-n}=\dfrac{(-1)^{n-1}}{(n-1)!}\,D^n \ln|t|$

19. $tt^{-1}=1$

Fourier transforms

20. The Fourier transform $\hat{\varphi}(\omega)$ of a test function $\varphi(t)$ is defined by

$$\hat{\varphi}(\omega)=\int_{-\infty}^{\infty}\varphi(t)e^{-i\omega t}dt$$

21. $\varphi \in S \Rightarrow \hat{\varphi} \in S$
22. φ_n is a zero sequence $\Rightarrow \hat{\varphi}_n$ is a zero sequence.
23. The Fourier transform $\hat{f}(\omega)$ of a g.f. $f(t)$ is defined by

$$\boxed{(\hat{f}|\varphi)=(f|\hat{\varphi})}$$

24. $f \in S' \Rightarrow \hat{f} \in S'$, $\hat{\hat{f}}(t)=2\pi f(-t)$, $\hat{f}=\hat{g} \Rightarrow f=g$.
25. The laws F3–F11, sec. 13.2, hold for g.fs.
26. $\hat{\delta}(\omega)=1$, $\hat{\theta}(\omega)=\pi\delta(\omega)-i\omega^{-1}$.

Convolutions

27. $f \in S'$, $\varphi \in S \Rightarrow f*\varphi(t)=\oint_{-\infty}^{\infty}\varphi(t-\tau)f(\tau)d\tau \in C^\infty(\boldsymbol{R})$

28. The convolution $h=f*g$, $\quad f, g \in S'$ is defined if $\hat{f}\hat{g} \in S'$, by

$$\hat{h}=\hat{f}\hat{g}$$

29. If the convolutions exist:

$$f*g=g*f, \qquad f*(g+h)=f*g+f*h, \qquad (f*g)'=f'*g=f*g'$$

30. $f*\delta^{(n)}=f^{(n)}$, all $f \in S'$.

13 Transforms*

13.1 Trigonometric Fourier Series

Fourier series of periodic functions

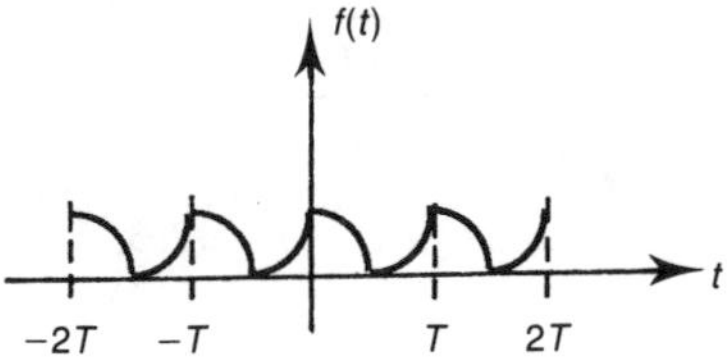

Assume that $f(t)$

(i) is bounded and has period T

(ii) is piece-wise differentiable

(iii) $=\frac{1}{2}[f(t^+)+f(t^-)]$ at jumps

$$\Omega=\frac{2\pi}{T}$$

Cosine – sine – form

Orthogonality

$$\int_0^T \cos k\Omega t \cos n\Omega t\, dt=\begin{cases} 0, & n\neq k \\ T, & n=k=0 \\ T/2, & n=k>0 \end{cases}$$

$$\int_0^T \sin k\Omega t \sin n\Omega t\, dt=\frac{T}{2}\,\delta_{kn}\ (k,\ n>0)$$

$$\int_0^T \sin k\Omega t \cos n\Omega t\, dt=0$$

$$f(t)=\frac{a_0}{2}+\sum_{n=1}^{\infty}(a_n \cos n\Omega t+b_n \sin n\Omega t)$$

$$a_n=\frac{2}{T}\int_a^{a+T} f(t)\cos n\Omega t\, dt\ (n\geq 0) \qquad b_n=\frac{2}{T}\int_a^{a+T} f(t)\sin n\Omega t\, dt\ (n\geq 1)$$

Special case. Period $T=2\pi$

$$\begin{cases} f(t)=\frac{a_0}{2}+\sum_{n=1}^{\infty}(a_n \cos nt+b_n \sin nt) \\ a_n=\frac{1}{\pi}\int_{-\pi}^{\pi} f(t)\cos nt\, dt, \quad b_n=\frac{1}{\pi}\int_{-\pi}^{\pi} f(t)\sin nt\, dt \end{cases}$$

$$f(t)\text{ even} \Rightarrow a_n=\frac{4}{T}\int_0^{T/2} f(t)\cos n\Omega t\, dt, \qquad b_n=0$$

$$f(t)\text{ odd} \Rightarrow a_n=0, \qquad b_n=\frac{4}{T}\int_0^{T/2} f(t)\sin n\Omega t\, dt$$

* Essentially following Jan Petersson: Fourieranalys, Chalmers University of Technology, Göteborg 1987.

Approximation in mean

$$s_n(t)=\frac{a_0}{2}+\sum_{k=1}^{n}(a_k\cos k\Omega t+b_k\sin k\Omega t)$$

$$Q_n=\frac{2}{T}\int_0^T[f(t)-s_n(t)]^2dt\quad(T=2\pi/\Omega)$$

Q_n minimal $\Leftrightarrow$ a_k, b_k are Fourier coefficients.

$$\min Q_n=\frac{2}{T}\int_0^T f^2(t)dt-\frac{a_0^2}{2}-\sum_{k=1}^{n}(a_k^2+b_k^2)$$

Amplitude – phase form

$\left(\Omega=\frac{2\pi}{T}, f(t) \text{ real}\right)$

$$f(t)=A_0+\sum_{n=1}^{\infty}A_n\cos(n\Omega t+\alpha_n)$$

$$A_0=\frac{a_0}{2},\ A_n=\sqrt{a_n^2+b_n^2},\qquad \alpha_n=\arg(a_n-ib_n)$$

$$\cos\alpha_n=a_n/A_n,\qquad \sin\alpha_n=-b_n/A_n,\qquad \tan\alpha_n=-b_n/a_n,\qquad a_n-ib_n=A_ne^{i\alpha_n}$$

Complex form ($\Omega=2\pi/T$)

Orthogonality

$$\int_0^T e^{ik\Omega t}e^{-in\Omega t}dt=T\delta_{kn}$$

$$f(t)=\sum_{n=-\infty}^{\infty}c_ne^{in\Omega t},\qquad c_n=\frac{1}{T}\int_a^{a+T}f(t)e^{-in\Omega t}\,dt$$

Relations between Fourier coefficients

$$\frac{a_0}{2}=c_0;\qquad a_n-ib_n=2c_n,\ a_n+ib_n=2c_{-n},\ a_n=c_n+c_{-n},\ b_n=i(c_n-c_{-n}),\qquad n\geq 1.$$

$$A_0=c_0;\ A_ne^{i\alpha_n}=2c_n,\ A_n=2|c_n|,\ \alpha_n=\arg c_n,\ n\geq 1$$

Parseval's identities

[Primed coefficients refer to $g(t)$]

$$\frac{1}{T}\int_a^{a+T}f(t)g(t)dt=\frac{1}{4}a_0a_0'+\frac{1}{2}\sum_{n=1}^{\infty}(a_na_n'+b_nb_n')=$$

$$=A_0A_0'+\frac{1}{2}\sum_{n=1}^{\infty}A_nA_n'\cos(\alpha_n-\alpha_n')$$

$$\frac{1}{T}\int_a^{a+T}f^2(t)dt=\frac{a_0^2}{4}+\frac{1}{2}\sum_{n=1}^{\infty}(a_n^2+b_n^2)=A_0^2+\frac{1}{2}\sum_{n=1}^{\infty}A_n^2$$

$$\frac{1}{T}\int_a^{a+T}|f(t)|^2dt=\sum_{n=-\infty}^{\infty}|c_n|^2\qquad \frac{1}{T}\int_a^{a+T}f(t)\,\overline{g(t)}\,dt=\sum_{-\infty}^{\infty}c_n\,\overline{c_n'}$$

Sine and cosine series

Orthogonality

$$\int_0^L \sin\frac{k\pi x}{L}\sin\frac{n\pi x}{L}\,dx=\frac{L}{2}\delta_{kn}\ (k,\ n>0)$$

$$\int_0^L \cos\frac{k\pi x}{L}\cos\frac{n\pi x}{L}\,dx=\begin{cases}0, & n\neq k\\ L, & n=k=0\\ L/2, & n=k>0\end{cases}$$

$f(x)$ given in the interval $(0, L)$:

$$f(x)=\sum_{n=1}^{\infty} b_n \sin\frac{n\pi x}{L},\qquad b_n=\frac{2}{L}\int_0^L f(x)\sin\frac{n\pi x}{L}\,dx$$

$$f(x)=\frac{a_0}{2}+\sum_{n=1}^{\infty} a_n \cos\frac{n\pi x}{L},\qquad a_n=\frac{2}{L}\int_0^L f(x)\cos\frac{n\pi x}{L}\,dx$$

Special Fourier series

$$f(t)=\frac{1}{2}a_0+\sum_{n=1}^{\infty}(a_n\cos n\Omega t+b_n\sin n\Omega t),\ \Omega=2\pi/T$$

Function $f(t)$	Fourier coefficients
(1) 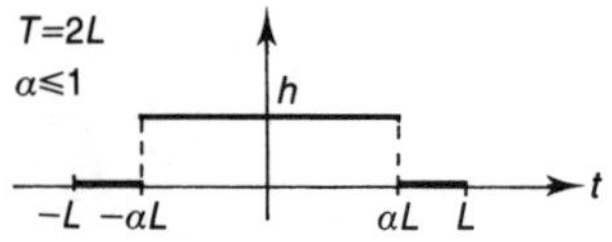	$a_0=2\alpha h;\ a_n=\dfrac{2h\sin n\pi\alpha}{n\pi},\ n\geqslant 1$ $b_n=0$
(2) 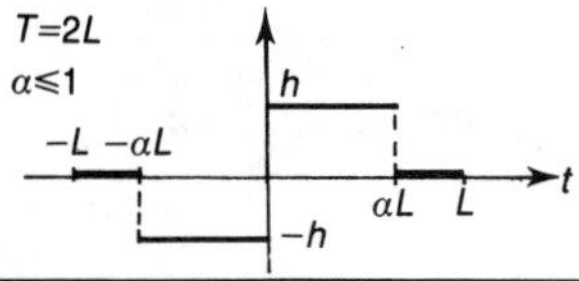	$a_n=0$ $b_n=\dfrac{2h(1-\cos n\pi\alpha)}{n\pi}$
(3) 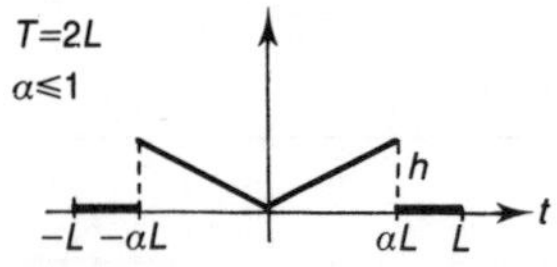	$a_0=\alpha h;\ a_n=\dfrac{2h}{n\pi}\sin n\pi\alpha-\dfrac{2h}{\alpha n^2\pi^2}(1-\cos n\pi\alpha)$ $b_n=0$
(4) 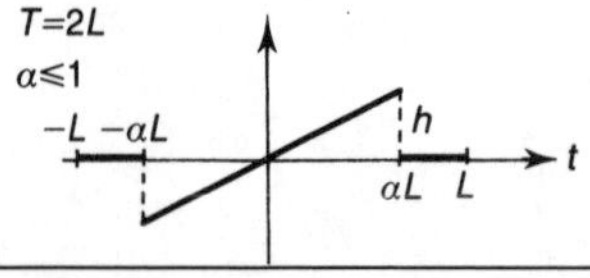	$a_n=0$ $b_n=-\dfrac{2h}{n\pi}\cos n\pi\alpha+\dfrac{2h}{\alpha n^2\pi^2}\sin n\pi\alpha$

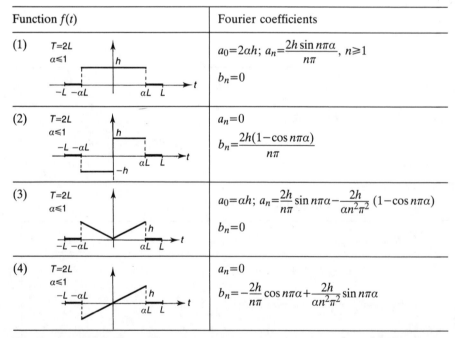

Fourier series for further *rectangular* and *triangular periodic* functions can be obtained by combining (1)–(4).

(5)	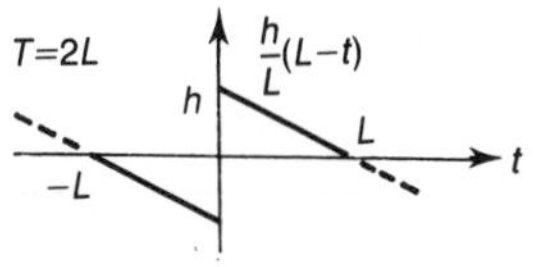	$a_n=0$ $b_n=\frac{2h}{\pi n}$
(6)	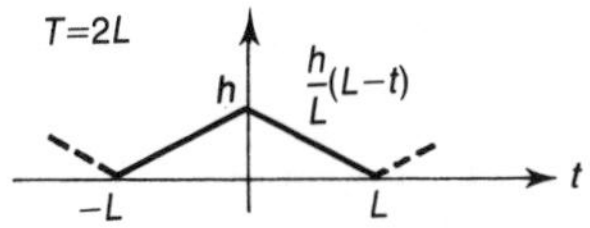	$a_0=h;\ a_n=\begin{cases}0, & n \text{ even}\\ \frac{4h}{\pi^2 n^2}, & n \text{ odd}\end{cases}$ $b_n=0$
(7)	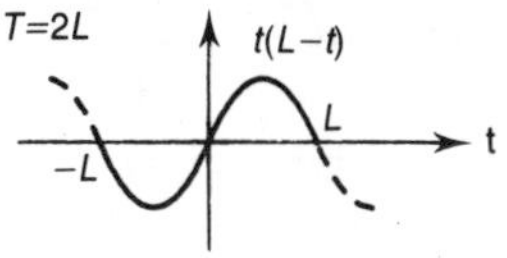	$a_n=0$ $b_n=\begin{cases}0, & n \text{ even}\\ \frac{8L^2}{\pi^3 n^3}, & n \text{ odd}\end{cases}$
(8)	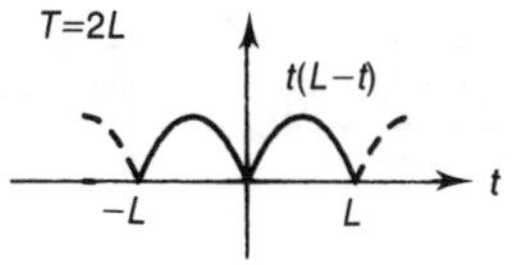	$a_0=\frac{L^2}{3};\ a_n=\begin{cases}0, & n \text{ odd}\\ -\frac{4L^2}{\pi^2 n^2}, & n \text{ even}\end{cases},\ n\geqslant 1$ $b_n=0$
(9)	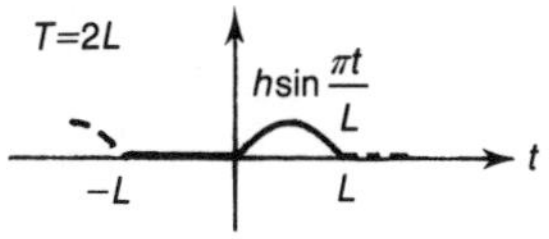	$a_n=\begin{cases}0, & n \text{ odd}\\ -\frac{2h}{\pi(n^2-1)}, & n \text{ even}\end{cases}$ $b_1=\frac{h}{2};\ b_n=0,\ n\geqslant 2$
(10)	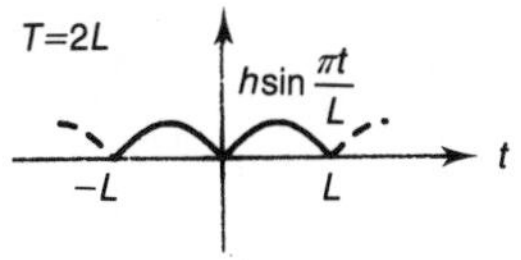	$a_n=\begin{cases}0, & n \text{ odd}\\ -\frac{4h}{\pi(n^2-1)}, & n \text{ even}\end{cases}$ $b_n=0$
(11)	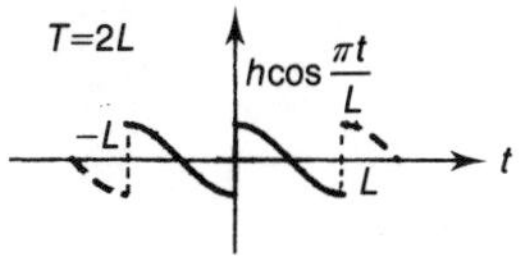	$a_0=0$ $b_n=\begin{cases}0, & n \text{ odd}\\ \frac{4nh}{\pi(n^2-1)}, & n \text{ even}\end{cases}$

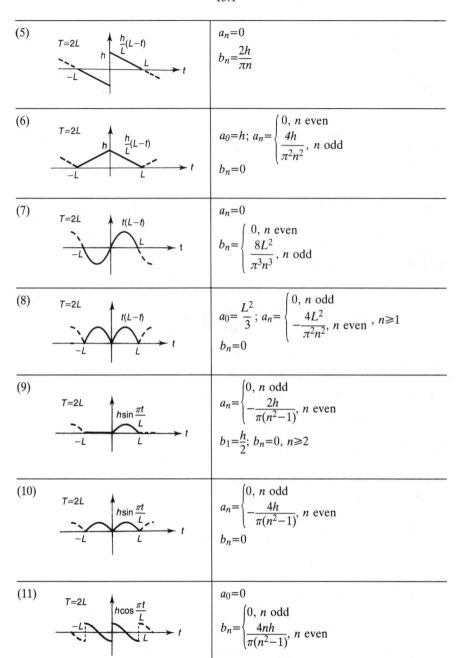

(12) $f(t)=t,\ -L<t<L$ $T=2L$	$a_n=0$ $b_n=\frac{2L}{n\pi}(-1)^{n+1}$
(13) $f(t)=t^2,\ -L<t<L$ $T=2L$	$a_0=\frac{2L^2}{3};\ a_n=(-1)^n\frac{4L^2}{n^2\pi^2},\ n\geq 1$ $b_n=0$
(14) $f(t)=t^3,\ -L<t<L$ $T=2L$	$a_n=0$ $b_n=(-1)^{n+1}\frac{2L^3}{n\pi}\left(1-\frac{6}{n^2\pi^2}\right)$
(15) $f(t)=\cos\alpha t,\ -\pi<t<\pi$ $T=2\pi,\ \alpha\neq$integer	$a_n=(-1)^{n+1}\frac{2\alpha\sin\alpha\pi}{\pi(n^2-\alpha^2)}$ $b_n=0$
(16) $f(t)=\sin\alpha t,\ -\pi<t<\pi$ $T=2\pi,\ \alpha\neq$integer	$a_n=0$ $b_n=(-1)^{n+1}\frac{2n\sin\alpha\pi}{\pi(n^2-\alpha^2)}$
(17) $f(t)=e^{-\alpha\|t\|}, -\pi<t<\pi$ T=2π	$a_n=\frac{2\alpha[1+(-1)^{n+1}e^{-\alpha\pi}]}{\pi(n^2+\alpha^2)}$ $b_n=0$

(18) $\sum_{n=1}^{\infty}\frac{\sin nt}{n}=\frac{\pi-t}{2},\ 0<t<2\pi$

(19) $\sum_{n=1}^{\infty}(-1)^{n+1}\frac{\sin nt}{n}=\frac{t}{2}, -\pi<t<\pi$

(20) $\sum_{n=1}^{\infty}\frac{\cos nt}{n}=-\ln\left(2\sin\frac{t}{2}\right),\ 0<t<2\pi$

(21) $\sum_{n=1}^{\infty}(-1)^{n+1}\frac{\cos nt}{n}=\ln\left(2\cos\frac{t}{2}\right),\ -\pi<t<\pi$

(22) $\sum_{n=1}^{\infty}\frac{\cos nt}{n^2}=\frac{\pi^2}{6}-\frac{\pi t}{2}+\frac{t^2}{4},\ 0<t<2\pi$

(23) $\sum_{n=1}^{\infty}\frac{\sin nt}{n^3}=\frac{\pi^2 t}{6}-\frac{\pi t^2}{4}+\frac{t^3}{12},\ 0<t<2\pi$

(24) $\sum_{n=0}^{\infty}\frac{\cos(2n+1)t}{2n+1}=-\frac{1}{2}\ln\left(\tan\frac{t}{2}\right),\ 0<t<\pi$

(25) $\sum_{n=0}^{\infty}\frac{\sin(2n+1)t}{2n+1}=\frac{\pi}{4},\ 0<t<\pi$

(26) $\sum_{n=0}^{\infty}(-1)^n\frac{\cos(2n+1)t}{2n+1}=\frac{\pi}{4}, -\frac{\pi}{2}<t<\frac{\pi}{2}$

(27) $\sum_{n=0}^{\infty}(-1)^n\frac{\sin(2n+1)t}{2n+1}=-\frac{1}{2}\ln\left[\tan\left(\frac{\pi}{4}-\frac{t}{2}\right)\right], -\frac{\pi}{2}<t<\frac{\pi}{2}$

(28) $\sum_{n=1}^{\infty}r^n\sin nt=\frac{r\sin t}{1-2r\cos t+r^2},\ |r|<1$

(29) $\sum_{n=0}^{\infty}r^n\cos nt=\frac{1-r\cos t}{1-2r\cos t+r^2},\ |r|<1$

13.2 Fourier Transforms

Assume that $f(t)$, $-\infty<t<\infty$, is piece-wise differentiable and absolutely integrable. (Fourier transforms of generalized functions, see sec. 12.9).

Fourier transform	$F(\omega)=\int_{-\infty}^{\infty} f(t)e^{-i\omega t}dt$
Inversion formula	$f(t)=\frac{1}{2\pi}\int_{-\infty}^{\infty} F(\omega)e^{i\omega t}d\omega$
(if $f(t)$ differentiable at t)	

Plancherel's (Parseval's) formulas

$$\int_{-\infty}^{\infty} f(t)\overline{g(t)}dt=\frac{1}{2\pi}\int_{-\infty}^{\infty} F(\omega)\overline{G(\omega)}d\omega$$

$$\int_{-\infty}^{\infty} |f(t)|^2dt=\frac{1}{2\pi}\int_{-\infty}^{\infty} |F(\omega)|^2d\omega$$

Cosine and sine transforms

$F_c(\beta)=\int_0^{\infty} f(x)\cos\beta x\,dx$	$f(x)=\frac{2}{\pi}\int_0^{\infty} F_c(\beta)\cos\beta x\,d\beta$
$F_s(\beta)=\int_0^{\infty} f(x)\sin\beta x\,dx$	$f(x)=\frac{2}{\pi}\int_0^{\infty} F_s(\beta)\sin\beta x\,d\beta$

An application of Fourier transformation, see example 5, sec. 10.9.

Properties and table of Fourier transforms

	$f(t)$		$F(\omega)$
F1.	$f(t)$		$\int_{-\infty}^{\infty} f(t)e^{-i\omega t}dt$
F2.	$\frac{1}{2\pi}\int_{-\infty}^{\infty} F(\omega)e^{i\omega t}d\omega$		$F(\omega)$
F3.	$af(t)+b\,g(t)$		$a\,F(\omega)+b\,G(\omega)$
F4.	$f(at)$	($a\neq 0$ real)	$\frac{1}{\lvert a\rvert}F\left(\frac{\omega}{a}\right)$
F5.	$f(-t)$		$F(-\omega)$

	$f(t)$		$F(\omega)$	
F6.	$\overline{f(t)}$		$\overline{F(-\omega)}$	
F7.	$f(t-T)$	(T real)	$e^{-i\omega T}F(\omega)$	
F8.	$e^{i\Omega t}f(t)$	(Ω real)	$F(\omega-\Omega)$	
F9.	$F(t)$		$2\pi f(-\omega)$	
F10.	$\left(\frac{d}{dt}\right)^n f(t)$		$(i\omega)^n F(\omega)$	
F11.	$(-it)^n f(t)$		$\left(\frac{d}{d\omega}\right)^n F(\omega)$	
F12.	$f*g(t)=\int_{-\infty}^{\infty} f(t-\tau)g(\tau)d\tau$		$F(\omega)\,G(\omega)$	
F13.	$f(t)\,g(t)$		$\frac{1}{2\pi}\,F*G(\omega)$	
F14.	$\delta(t)$	$[\delta(t-T)]$	1	$[e^{-i\omega T}]$
F15.	$\delta^{(n)}(t)$		$(i\omega)^n$	
F16.	$\theta(t)=\begin{cases}1, & t>0\\ 0, & t<0\end{cases}$		$\frac{1}{i\omega}+\pi\delta(\omega)$	
F17.	$\operatorname{sgn} t=\begin{cases}1, & t>0\\ -1, & t<0\end{cases}$		$\frac{2}{i\omega}$	
F18.	$\begin{cases}1, & \lvert t\rvert<a\\ 0, & \lvert t\rvert>a\end{cases}$		$\frac{2\sin a\omega}{\omega}$	
F19.	$\begin{cases}1, & 0<t<a\\ -1, & -a<t<0\\ 0, & \lvert t\rvert>a\end{cases}$		$\frac{4\sin^2 a\omega/2}{i\omega}$	
F20a.	1		$2\pi\,\delta(\omega)$	
F20b.	$e^{i\Omega t}$		$2\pi\,\delta(\omega-\Omega)$	
F21.	$\begin{cases}e^{i\Omega t}, & \lvert t\rvert<a\\ 0, & \lvert t\rvert>a\end{cases}$		$\frac{2\sin a(\Omega-\omega)}{\Omega-\omega}$	
F22.	$e^{-at}\theta(t)$	$(a>0)$	$\frac{1}{a+i\omega}$	
F23.	$e^{at}(1-\theta(t))$	$(a>0)$	$\frac{1}{a-i\omega}$	
F24.	$e^{-a\lvert t\rvert}$	$(a>0)$	$\frac{2a}{a^2+\omega^2}$	
F25.	$te^{-a\lvert t\rvert}$	$(a>0)$	$-\frac{4ia\omega}{(a^2+\omega^2)^2}$	

	$f(t)$		$F(\omega)$
F26.	$\|t\|e^{-a\|t\|}$	$(a>0)$	$\frac{2(a^2-\omega^2)}{(a^2+\omega^2)^2}$
F27.	$e^{-a\|t\|}\operatorname{sgn} t$	$(a>0)$	$-\frac{2i\omega}{a^2+\omega^2}$
F28.	$\frac{1}{\sqrt{4\pi a}}\, e^{-t^2/4a}$	$(a>0)$	$e^{-a\omega^2}$
F29.	$\|t\|^{-a}$	$(0<a<1)$	$2\Gamma(1-a)\sin\frac{\pi a}{2}\,\|\omega\|^{1-a}$
F30.	$\|t\|^{-a}\operatorname{sgn} t$	$(0<a<1)$	$-2i\operatorname{sgn}\omega\,\Gamma(1-a)\cos\frac{\pi a}{2}\,\|\omega\|^{1-a}$
F31.	$\frac{\sin\Omega t}{\pi t}$		$\theta(\omega+\Omega)-\theta(\omega-\Omega)$
F32a.	$\sin at^2$	$(a>0)$	$\sqrt{\frac{\pi}{a}}\cos\left(\frac{\omega^2}{4a}+\frac{\pi}{4}\right)$
F32b.	$\cos at^2$	$(a>0)$	$\sqrt{\frac{\pi}{a}}\cos\left(\frac{\omega^2}{4a}-\frac{\pi}{4}\right)$
F33.	$h_n(t)$		$i^{-n}\sqrt{2\pi}\, h_n(\omega)$
F34.	$\theta(t)\, l_n(t)$		$\left(i\omega-\frac{1}{2}\right)^n \Big/ \left(i\omega+\frac{1}{2}\right)^{n+1}$
F35.	$\begin{cases}(a^2-t^2)^{-1/2}, & \|t\|<a\\ 0, & \|t\|>a\end{cases}$		$\pi J_0(a\omega)$
F36.	$\frac{i}{2(n-1)!}(it)^{n-1}e^{ict}\operatorname{sgn} t$		$\frac{1}{(\omega-c)^n}$ (c real, $n=1, 2, \ldots$)
F37.	$\frac{1}{2a}e^{-a\|t\|}$		$\frac{1}{\omega^2+a^2}$ $(a>0)$
F38.	$\frac{i}{2}e^{-a\|t\|}\operatorname{sgn} t$		$\frac{\omega}{\omega^2+a^2}$ $(a>0)$
F39.	$\frac{1}{2a}e^{-a\|t\|+ict}(iak\operatorname{sgn} t+b+kc)$		$\frac{k\omega+b}{(\omega-c)^2+a^2}$ (c real, $a>0$)
F40.	$\frac{1}{4a^3}e^{-a\|t\|+ict}[ia^2kt+(b+kc)a\|t\|+b+kc]$		$\frac{k\omega+b}{[(\omega-c)^2+a^2]^2}$ (c real, $a>0$)
F41.	$\frac{\partial}{\partial a}f(t, a)$		$\frac{\partial}{\partial a}F(\omega, a)$

$$F^{(n)}(\omega) \subset g(t) \Rightarrow F(\omega) \subset \left(\frac{i}{t}\right)^n g(t)+C_0\delta(t)+C_1\delta'(t)+\ldots+\ C_{n-1}\delta^{(n-1)}(t)$$

[$\subset$ means: is Fourier transform of]

Table of Fourier cosine transform

	$f(x),\ x>0$		$F_c(\beta),\ \beta>0$
*F*36.	$\begin{cases} 1, & x<a \\ 0, & x>a \end{cases}$	$(a>0)$	$\dfrac{\sin a\beta}{\beta}$
*F*37.	e^{-ax}	$(a>0)$	$\dfrac{a}{a^2+\beta^2}$
*F*38.	e^{-ax^2}	$(a>0)$	$\dfrac{1}{2}\sqrt{\dfrac{\pi}{a}}\ e^{-\beta^2/4a}$
*F*39.	x^{a-1}	$(0<a<1)$	$\Gamma(a)\,\beta^{-a}\ \sin\dfrac{a\pi}{2}$
*F*40.	$\cos ax^2$		$\dfrac{1}{2}\sqrt{\dfrac{\pi}{a}}\cos\left(\dfrac{\beta^2}{4a}-\dfrac{\pi}{4}\right)$
*F*41.	$\sin ax^2$		$\dfrac{1}{2}\sqrt{\dfrac{\pi}{2}}\cos\left(\dfrac{\beta^2}{4a}+\dfrac{\pi}{4}\right)$

Table of Fourier sine transform

	$f(x),\ x>0$		$F_s(\beta),\ \beta>0$
*F*42	$\begin{cases} 1, & x<a \\ 0, & x>a \end{cases}$	$(a>0)$	$\dfrac{1-\cos a\beta}{\beta}$
*F*44	e^{-ax}		$\dfrac{\beta}{a^2+\beta^2}$
*F*44	xe^{-ax^2}	$(a>0)$	$\sqrt{\dfrac{\pi}{a}}\ \dfrac{\beta}{4a}\ e^{-\beta^2/4a}$
*F*45	x^{a-1}	$(0<a<1)$	$\Gamma(a)\,\beta^{-a}\sin\dfrac{a\pi}{2}$
*F*46	$\cos ax^2$		$\sqrt{\dfrac{\pi}{2a}}\left[\sin\dfrac{\beta^2}{4a}\,C\left(\dfrac{\beta}{\sqrt{2\pi a}}\right)-\cos\dfrac{\beta^2}{4a}\ S\left(\dfrac{\beta}{\sqrt{2\pi a}}\right)\right]$
*F*47	$\sin ax^2$		$\sqrt{\dfrac{\pi}{2a}}\left[\cos\dfrac{\beta^2}{4a}\,C\left(\dfrac{\beta}{\sqrt{2\pi a}}\right)+\sin\dfrac{\beta^2}{4a}\ S\left(\dfrac{\beta}{\sqrt{2\pi a}}\right)\right]$

Fourier transforms in higher dimensions

Two-dimensional Fourier transforms

Fourier transform $F(u, v)=\iint_{\mathbf{R}^2} f(x, y)e^{-i(ux+vy)}dxdy$

Inversion formula $f(x, y)=\dfrac{1}{(2\pi)^2}\iint_{\mathbf{R}^2} F(u, v)e^{i(ux+vy)}dudv$

Parseval's formulas $\iint f(x, y)\overline{g(x, y)}dxdy=\dfrac{1}{(2\pi)^2}\iint_{\mathbf{R}^2} F(u, v)\overline{G(u, v)}dudv$

$$\iint_{\mathbf{R}^2} |f(x, y)|^2dxdy=\frac{1}{(2\pi)^2}\iint_{\mathbf{R}^2} |F(u, v)|^2dudv$$

Properties and table of 2-dimensional Fourier transforms

	$f(x, y)$	$F(u, v)$
$F_2 1.$	$f(ax, by)$ (a, b real)	$\frac{1}{\|ab\|} F\left(\frac{u}{a}, \frac{v}{b}\right)$
$F_2 2.$	$f(x-a, y-b)$ (a, b real)	$e^{-i(au+bv)}F(u, v)$
$F_2 3.$	$e^{iax}e^{iby}f(x, y)$	$F(u-a, v-b)$
$F_2 4.$	$D_x^m D_y^n f(x, y)$	$(iu)^m(iv)^n F(u, v)$
$F_2 5.$	$(-ix)^m(-iy)^n f(x, y)$	$D_u^m D_v^n F(u, v)$
$F_2 6.$	$(f*g)(x, y)=\iint_{\mathbf{R}^2} f(\xi, \eta)g(x-\xi, y-\eta)d\xi d\eta$	$F(u, v)G(u, v)$
$F_2 7.$	$F(x, y)$	$(2\pi)^2 f(-u, -v)$
$F_2 8.$	$\delta(x-a, y-b)=\delta(x-a)\delta(y-b)$	$e^{-i(au+bv)}$
$F_2 9.$	$\frac{1}{4\pi\sqrt{ab}}\, e^{-x^2/4a-y^2/4b}$ ($a, b>0$)	$e^{-au^2-bv^2}$
$F_2 10.$	$\begin{cases}1, & \|x\|<a,\ \|y\|<b \quad \text{(rectangle)}\\ 0, & \text{otherwise}\end{cases}$	$\frac{4\sin au\,\sin bv}{uv}$
$F_2 11.$	$\begin{cases}1, & \|x\|<a \quad \text{(strip)}\\ 0, & \text{otherwise}\end{cases}$	$\frac{4\pi\sin au}{u}\,\delta(v)$
$F_2 12.$	$\begin{cases}1, & x^2+y^2<a^2 \quad \text{(circle)}\\ 0, & \text{otherwise}\end{cases}$	$\frac{2\pi a}{\varrho} J_1(a\varrho)$, $\varrho^2=u^2+v^2$

n-dimensional Fourier transforms

Notation: $f(\mathbf{x})=f(x_1, \ldots, x_n)$, $\mathbf{x}\cdot\mathbf{y}=\sum_{i=1}^{n} x_i y_i$

Fourier transform	$F(\mathbf{y})=\int_{\mathbf{R}^n} f(\mathbf{x})e^{-i\mathbf{x}\cdot\mathbf{y}}d\mathbf{x}$
Inversion formula	$f(\mathbf{x})=(2\pi)^{-n}\int_{\mathbf{R}^n} F(\mathbf{y})e^{i\mathbf{x}\cdot\mathbf{y}}d\mathbf{y}$
Parseval's formulas	$\int_{\mathbf{R}^n} f(\mathbf{x})\overline{g(\mathbf{x})}d\mathbf{x}=(2\pi)^{-n}\int_{\mathbf{R}^n} F(\mathbf{y})\overline{G(\mathbf{y})}d\mathbf{y}$
	$\int_{\mathbf{R}^n} \lvert f(\mathbf{x})\rvert^2 d\mathbf{x}=(2\pi)^{-n}\int_{\mathbf{R}^n} \lvert F(\mathbf{y})\rvert^2 d\mathbf{y}$

Properties and table of n (3)-dimensional Fourier transforms

	$f(\mathbf{x})$	$F(\mathbf{y})$
F_n1.	$f(a\mathbf{x})$ (a real)	$\lvert a\rvert^{-n}F(a^{-1}\mathbf{y})$
F_n2.	$f(\mathbf{x}-\mathbf{a})$	$e^{-i\mathbf{a}\cdot\mathbf{y}}F(\mathbf{y})$
F_n3.	$e^{i\mathbf{a}\cdot\mathbf{x}}f(\mathbf{x})$	$F(\mathbf{y}-\mathbf{a})$
F_n4.	$D^\alpha f(\mathbf{x})=D_1^{\alpha_1}\ldots D_n^{\alpha_n}f(\mathbf{x})$	$(i\mathbf{y})^\alpha F(\mathbf{y})=(iy_1)^{\alpha_1}\ldots(iy_n)^{\alpha_n}F(\mathbf{y})$
F_n5.	$(-i\mathbf{x})^\alpha f(\mathbf{x})$	$D^\alpha F(\mathbf{y})$
F_n6.	$(f*g)(\mathbf{x})=\int_{\mathbf{R}^n} f(\mathbf{x}-\mathbf{t})g(\mathbf{t})d\mathbf{t}$	$F(\mathbf{y})G(\mathbf{y})$
F_n7.	$F(\mathbf{x})$	$(2\pi)^n f(-\mathbf{y})$

Below, $n=3$, $\mathbf{x}=(x, y, z)$, $\mathbf{y}=(u, v, w)$

F_38.	$\delta(x-a,\ y-b,\ z-c)$	$e^{-i(au+bv+cw)}$
F_39.	$e^{-x^2/4a-y^2/4b-z^2/4c}$	$8\pi^{3/2}\sqrt{abc}\,e^{-au^2-bv^2-cw^2}$
F_310.	$\begin{cases}1, & \lvert x\rvert<a,\ \lvert y\rvert<b,\ \lvert z\rvert<c \quad \text{(box)}\\ 0, & \text{otherwise}\end{cases}$	$\dfrac{8\sin au\,\sin bv\,\sin vw}{uvw}$
F_311.	$\begin{cases}1, & \lvert x\rvert<a,\ \lvert y\rvert<b \quad \text{(bar)}\\ 0, & \text{otherwise}\end{cases}$	$\dfrac{8\pi\sin au\,\sin bv}{uv}\delta(w)$
F_312.	$\begin{cases}1, & \lvert x\rvert<a \quad \text{(slab)}\\ 0, & \text{otherwise}\end{cases}$	$\dfrac{8\pi^2\sin au}{u}\delta(v)\delta(w)$
F_313.	$\begin{cases}1, & \lvert x\rvert<a,\ y^2+z^2<b^2 \quad \text{(cylinder)}\\ 0, & \text{otherwise}\end{cases}$	$\dfrac{2\sin au}{u}\cdot\dfrac{2\pi b}{\varrho}J_1(b\varrho)$, $(\varrho^2=u^2+v^2)$
F_314.	$\begin{cases}1, & x^2+y^2+z^2<b^2 \quad \text{(ball)}\\ 0, & \text{otherwise}\end{cases}$	$\dfrac{4\pi}{s^3}(\sin as-as\cos as)$ $(s^2=u^2+v^2+w^2)$

13.3 Discrete Fourier Transforms

Periodic sequences

Let S^N donote the set of N-periodic complex-valued sequences $(x(n))_{n \in \mathbf{Z}}$, i.e.

$$x(n+N)=x(n),\ n \in \mathbf{Z}$$

Unit pulse at k *(mod N)*

$$d_k(n)=\begin{cases} 1, & n=k+mN,\ m \in \mathbf{Z} \\ 0, & \text{otherwise} \end{cases}$$

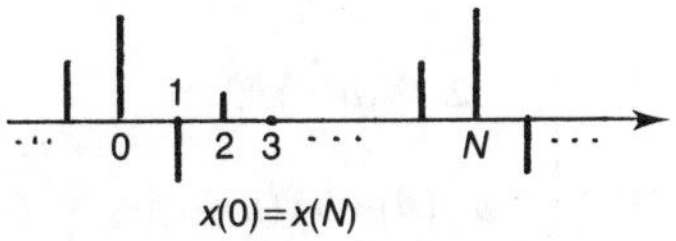

Discrete Fourier transforms. (DFT)

The *discrete Fourier transform* of $x \in S^N$:

$$X(\mu)=\frac{1}{N}\sum_{n=0}^{N-1} x(n)W^{-\mu n},\ \mu \in \mathbf{Z},\ W=e^{2i\pi/N}$$

The inversion formula:

$$x(n)=\sum_{\mu=0}^{N-1} X(\mu)W^{\mu n},\ n \in \mathbf{Z}$$

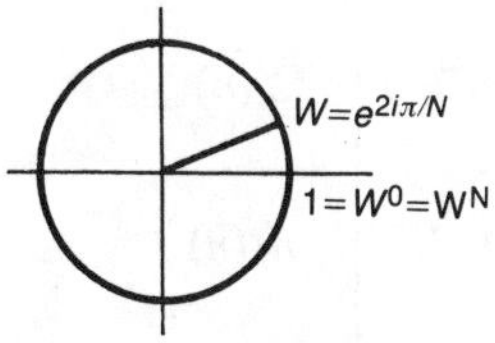

Parseval's formulas

$$\frac{1}{N}\sum_{n=0}^{N-1} x(n)\overline{y(n)}=\sum_{\mu=0}^{N-1} X(\mu)\overline{Y(\mu)}$$

$$\frac{1}{N}\sum_{n=0}^{N-1} |x(n)|^2=\sum_{\mu=0}^{N-1} |X(\mu)|^2$$

Fast Fourier transform (*FFT*)

FFT is an algorithm which reduces the number of operations for *DFT* if $N=2^m$. The number of operations using the definition $\approx N^2$ and using $FFT \approx N \cdot {}^2\log N$.

The idea of FFT:

$$X(\mu)=\sum_{n=0}^{2^m-1} x(n)W^{-\mu n}=\sum_{n\text{ even}}+\sum_{n\text{ odd}}=$$

$$=\sum_{k=0}^{2^{m-1}-1} x(2k)(W^2)^{-\mu k}+W^{-\mu}\sum_{k=0}^{2^{m-1}-1} x(2k+1)(W^2)^{-\mu k}$$

Since $W^2=\exp(-2\pi i/2^{m-1})$ these two sums are $DFT(N=2^{m-1})$. This halving procedure is then continued until a sum of $DFT(N=2)$ is obtained.

Properties and table of discrete Fourier transforms

	$x(n)$	$X(\mu)$
*DF*1.	$x(n)$	$\frac{1}{N}\sum_{n=0}^{N-1} x(n)\,W^{-\mu n}$
*DF*2.	$\sum_{\mu=0}^{N-1} X(\mu)\,W^{\mu n}$	$X(\mu)$
*DF*3.	$a\,x(n)+b\,y(n)$	$a\,X(\mu)+b\,Y(\mu)$
*DF*4.	$\frac{1}{N}x*y(n)=\frac{1}{N}\sum_{k=0}^{N-1} x(n-k)\,y(k)$	$X(\mu)\,Y(\mu)$
*DF*5.	$x(n-p)$	$W^{-\mu p}\,X(\mu)$
*DF*6.	$W^{\nu n}\,x(n)$	$X(\mu-\nu)$
*DF*7.	$X(n)$	$\frac{1}{N}\,x(-\mu)$
*DF*8.	$d_k(n)$	$\frac{1}{N}\,W^{-\mu k}$
*DF*9.	$d_0(n)$	$\frac{1}{N}$
*DF*10.	$W^{\nu n}$	$d_\nu(\mu)$
*DF*11.	1	$d_0(\mu)$

The sampling theorem

A continuous bandlimited signal is uniquely determined by its values at uniform sampling points if the sampling frequency is greater than twice the maximal frequency of the signal.

13.4 The z-transform

For sequences $(x(n))_{n=0}^{\infty}$ the z-transform is defined by

$$X(z)=\sum_{n=0}^{\infty} x(n)z^{-n}$$

Inversion formula. $x(n)=\frac{1}{2\pi i}\int_{|z|=r} X(z)z^{n-1}dz=\frac{1}{n!}\left(\frac{d}{dz^{-1}}\right)^n X(z)|_{z^{-1}=0}$ (r large enough)

In practice, $x(n)$ often is determined by series expansion or by using the table below.

Properties and table of z-transforms

Below $\theta(n)=\begin{cases}1, & n\geq 0\\ 0, & n<0\end{cases}$

	$x(n)$	$X(z)$
z1.	$x(n)$	$\sum_{n=0}^{\infty} x(n)z^{-n}$
z2.	$a\,x(n)+b\,y(n)$	$a\,X(z)+b\,Y(z)$
z3.	$x(n-k)\theta(n-k)=\begin{cases}0, & 0\leq n\leq k-1\\ x(n-k), & n\geq k\end{cases}$	$z^{-k}X(z)$
z4.	$x(n+k)$ $(k>0)$	$z^kX(z)-z^kx(0)-z^{k-1}x(1)-\ldots-$ $-z\,x(k-1)$
z5.	$a^{-n}x(n)$	$X(az)$
z6.	$nx(n)$	$-zX'(z)$
z7.	$x*y(n)=\sum_{k=0}^{n} x(n-k)\,y(k)$	$X(z)\,Y(z)$

	$x(n)$		$X(z)$	
z8.	$\delta_k(n)=\begin{cases}1, & n=k\\0, & n\neq k\end{cases}$		$\dfrac{1}{z^k}$	
z9.	$\begin{cases}0, & n=0\\a^{n-1}, & n\geqslant 1\end{cases}$	(a complex)	$\dfrac{1}{z-a}$	
z10.	$\begin{cases}0, & 0\leqslant n\leqslant k-1\\\binom{n-1}{k-1}a^{n-k}, & n\geqslant k\end{cases}$		$\dfrac{1}{(z-a)^k}$	
z11.	a^n		$\dfrac{z}{z-a}$	
z12.	$\binom{n}{k}a^{n-k}$	$[na^{n-1}]$	$\dfrac{z}{(z-a)^{k+1}}$	$\left[\dfrac{z}{(z-a)^2}\right]$
z13.	n^2a^n		$\dfrac{a(z+a)z}{(z-a)^3}$	
z14.	$\binom{n+k}{k}a^n$		$\left(\dfrac{z}{z-a}\right)^{k+1}$	
z15.	$a^{n-1}\sin\dfrac{n\pi}{2}$		$\dfrac{z}{z^2+a^2}$	
z16.	$\dfrac{1}{b}r^n\sin n\theta$ $\left(a,\ b>0,\ r=\sqrt{a^2+b^2},\ \theta=\arctan\dfrac{b}{a}\right)$		$\dfrac{z}{(z-a)^2+b^2}$	
z17.	$\dfrac{1}{b}r^n\sin n\theta$ $\left(a,\ b>0,\ r=\sqrt{a^2+b^2},\ \theta=\pi\text{-}\arctan\dfrac{b}{a}\right)$		$\dfrac{z}{(z+a)^2+b^2}$	
z18.	$r^n\cos n\theta$		$\dfrac{z(z-r\cos\theta)}{z^2-2rz\cos\theta+r^2}$	
z19.	$r^n\sin n\theta$		$\dfrac{rz\sin\theta}{z^2-2rz\cos\theta+r^2}$	
z20.	$a^n/n!$		$e^{a/z}$	

Recurrence (difference) equations

An Nth order *linear recurrence equation with constant coefficients* and N initial values:

$$\begin{cases}x(n+N)+a_{N-1}\,x(n+N-1)+\ldots+a_0x(n)=f(n),\ n=0,\ 1,\ 2,\ \ldots & (13.1)\\ x(0),\ x(1),\ \ldots,\ x(N-1)\ \text{given} & (13.2)\end{cases}$$

To find the solution, take z-transform of (13.1) and use $z4$ and (13.2). This gives $X(z)$, from which $x(n)$, $n=0$, 1, 2, ... are uninquely determined.

13.5 Laplace Transforms

Assume that $f(t)$ is piece-wise continuous, apart from finitely many impulses, and that $\int_{0^-}^{\infty} f(t)\, e^{-st} dt$ exists for Re $s \geqslant \alpha$.

Laplace transform

$$F(s) = \int_{0^-}^{\infty} e^{-st} f(t) dt = \lim_{\varepsilon \to 0^+} \int_{-\varepsilon}^{\infty} e^{-st} f(t) dt$$

Inversion formula

$$f(t) = \lim_{b \to \infty} \frac{1}{2\pi i} \int_{a-ib}^{a+ib} e^{st} F(s) ds, \ a \geqslant \alpha$$

(if $f(t)$ has no impulses and is differentiable at t).

Applications of the Laplace transform, see sec. 9.3, 10.9 and 13.6.

Functions with rational Laplace transform

Limit theorems

Assume that $f(t)$ is continuous and $F(s) = \frac{P(s)}{Q(s)}$ is rational. Then

$$\lim_{t \to 0+} f(t) = \lim_{s \to \infty} sF(s) \text{ if degree } P(s) < \text{degree } Q(s)$$

$$\lim_{t \to \infty} f(t) = \lim_{s \to 0} sF(s) \text{ if all poles (singularities) of } sF(s) \text{ lie in the half plane Re } s<0.$$

Rational transforms with simple poles

Let $P(s)$ and $Q(s)$ be polynomials with degree $P(s)<$degree $Q(s)$ and assume that the zeros s_k of $Q(s)$ are *simple* (real or complex). If

$$F(s) = \frac{P(s)}{Q(s)} = \frac{P(s)}{c(s-s_1)(s-s_2)\dots(s-s_n)} = \frac{A_1}{s-s_1} + \frac{A_2}{s-s_2} + \dots + \frac{A_n}{s-s_n}$$

then

$$f(t) = \frac{P(s_1)}{Q'(s_1)} e^{s_1 t} + \dots + \frac{P(s_n)}{Q'(s_n)} e^{s_n t} \qquad \left(\text{i.e. } A_k = \frac{P(s_k)}{Q'(s_k)} \right)$$

Properties and table of Laplace transforms

	$f(t)$	$F(s)$
$L1.$	$f(t)$	$\int_{0^-}^{\infty} e^{-st} f(t) dt$
$L2$	$a f(t) + b g(t)$	$a F(s) + b G(s)$
$L3.$	$f(at) \qquad (a>0)$	$\frac{1}{a} F\left(\frac{s}{a}\right)$
$L4.$	$f(t-T)\theta(t-T) = \begin{cases} f(t-T), & t>T \\ 0, & t<T \end{cases} \quad (T \geqslant 0)$	$e^{-Ts} F(s)$
$L5.$	$e^{-at} f(t)$	$F(s+a)$
$L6.$	$t^n f(t)$	$(-1)^n F^{(n)}(s)$
$L7.$	$f'(t)$	$s F(s) - f(0^-)$*
$L8.$	$f''(t)$	$s^2 F(s) - s f(0^-) - f'(0^-)$
$L9.$	$f^{(n)}(t)$	$s^n F(s) - \sum_{k=1}^{n} s^{n-k} f^{(k-1)}(0^-)$
$L10.$	$\int_{0^-}^{t} f(\tau) d\tau$	$\frac{1}{s} F(s)$
$L11.$	$\begin{cases} \int_{-\infty}^{\infty} f(\tau) g(t-\tau) d\tau \quad (f(t), g(t)=0, t<0) \\ \int_{0^-}^{t+} f(\tau) g(t-\tau) d\tau \end{cases}$	$F(s) G(s)$
$L12.$	$\frac{1}{t} f(t)$	$\int_{s}^{\infty} F(u) du$
$L13.$	$f(t+T) = f(t) \qquad$ (periodic)	$(1-e^{-Ts})^{-1} \int_{0}^{T} e^{-st} f(t) dt$
$L14.$	$\delta(t)$	1
$L14a.$	$\delta(t-T) \qquad (T \geqslant 0)$	e^{-Ts}
$L15.$	$\delta^{(n)}(t)$	s^n
$L16.$	$1, \theta(t)$	$\frac{1}{s}$
$L17.$	$t^n, n=0, 1, 2, \ldots$	$\frac{n!}{s^{n+1}}$
$L18.$	e^{at}	$\frac{1}{s-a}$

* In some texts this reads $s F(s) - f(0^+)$ if impulse functions are not considered.

	$f(t)$	$F(s)$
L19.	$t^n e^{at}$, $n=0, 1, 2, \ldots$	$\frac{n!}{(s-a)^{n+1}}$
L20.	$(1+at)e^{at}$	$\frac{s}{(s-a)^2}$
L21.	$\sin at$	$\frac{a}{s^2+a^2}$
L22.	$\cos at$	$\frac{s}{s^2+a^2}$
L23.	$\sinh at$	$\frac{a}{s^2-a^2}$
L24.	$\cosh at$	$\frac{s}{s^2-a^2}$
L25.	$\frac{e^{at}-e^{bt}}{a-b}$	$\frac{1}{(s-a)(s-b)}$
L26.	$\frac{ae^{at}-be^{bt}}{a-b}$	$\frac{s}{(s-a)(s-b)}$
L27.	$-\frac{(b-c)e^{at}+(c-a)e^{bt}+(a-b)e^{ct}}{(a-b)(b-c)(c-a)}$	$\frac{1}{(s-a)(s-b)(s-c)}$
L28.	$-\frac{a(b-c)e^{at}+b(c-a)e^{bt}+c(a-b)e^{ct}}{(a-b)(b-c)(c-a)}$	$\frac{s}{(s-a)(s-b)(s-c)}$
L29.	$-\frac{a^2(b-c)e^{at}+b^2(c-a)e^{bt}+c^2(a-b)e^{ct}}{(a-b)(b-c)(c-a)}$	$\frac{s^2}{(s-a)(s-b)(s-c)}$
L30.	$\frac{e^{at}-e^{bt}-(a-b)te^{bt}}{(a-b)^2}$	$\frac{1}{(s-a)(s-b)^2}$
L31.	$\frac{ae^{at}-ae^{bt}-b(a-b)te^{bt}}{(a-b)^2}$	$\frac{s}{(s-a)(s-b)^2}$
L32.	$\frac{a^2e^{at}-b(2a-b)e^{bt}-b^2(a-b)te^{bt}}{(a-b)^2}$	$\frac{s^2}{(s-a)(s-b)^2}$
L33.	$\frac{e^{at}-(a/b)\sin bt-\cos bt}{a^2+b^2}$	$\frac{1}{(s-a)(s^2+b^2)}$
L34.	$\frac{ae^{at}-a\cos bt+b\sin bt}{a^2+b^2}$	$\frac{s}{(s-a)(s^2+b^2)}$
L35.	$\frac{a^2e^{at}+ab\sin bt+b^2\cos bt}{a^2+b^2}$	$\frac{s^2}{(s-a)(s^2+b^2)}$
L36.	$\frac{at-\sin at}{a^3}$	$\frac{1}{s^2(s^2+a^2)}$

	$f(t)$		$F(s)$
$L37.$	$\frac{\sin at - at\cos at}{2a^3}$		$\frac{1}{(s^2+a^2)^2}$
$L38.$	$\frac{t\sin at}{2a}$		$\frac{s}{(s^2+a^2)^2}$
$L39.$	$\frac{1}{2a}(\sin at + at\cos at)$		$\frac{s^2}{(s^2+a^2)^2}$
$L40.$	$t\cos at$		$\frac{s^2-a^2}{(s^2+a^2)^2}$
$L41.$	$\frac{b\sin at - a\sin bt}{ab(b^2-a^2)}$		$\frac{1}{(s^2+a^2)(s^2+b^2)}$
$L42.$	$\frac{1}{b}e^{at}\sin bt$		$\frac{1}{(s-a)^2+b^2}$
$L43.$	$e^{at}\cos bt$		$\frac{s-a}{(s-a)^2+b^2}$
$L44.$	$e^{-at}-e^{at/2}\left(\cos\frac{at\sqrt{3}}{2}-\sqrt{3}\sin\frac{at\sqrt{3}}{2}\right)$		$\frac{3a^2}{s^3+a^3}$
$L45.$	$\sin at\cosh at - \cos at\sinh at$		$\frac{4a^3}{s^4+4a^4}$
$L46.$	t^α	$(\operatorname{Re}\alpha > -1)$	$\frac{\Gamma(\alpha+1)}{s^{\alpha+1}}$
$L47.$	$\frac{1}{\sqrt{t}}$		$\sqrt{\frac{\pi}{s}}$
$L48.$	$\sqrt{t}$		$\frac{1}{2s}\sqrt{\frac{\pi}{s}}$
$L49.$	$\ln t$		$-\frac{\gamma}{s}-\frac{\ln s}{s}$
$L50.$	$\frac{1}{\sqrt{\pi t}}e^{-a^2/4t}$	$a \geqslant 0$	$e^{-a\sqrt{s}}/\sqrt{s}$
$L51.$	$\frac{a}{2\sqrt{\pi t^3}}e^{-a^2/4t}$	$a>0$	$e^{-a\sqrt{s}}$
$L52.$	$1-\operatorname{erf}\left(\frac{a}{2\sqrt{t}}\right)$	$a>0$	$\frac{1}{s}e^{-a\sqrt{s}}$
$L53.$	$e^{-t^2/2a^2}$	$a>0$	$\sqrt{\frac{\pi}{2}}ae^{a^2s^2/2}\left(1-\operatorname{erf}\left(\frac{as}{\sqrt{2}}\right)\right)$
$L54.$	$\frac{\sin at}{t}$		$\arctan\frac{a}{s}$
$L55.$	$\sin a\sqrt{t}$		$\frac{\sqrt{\pi}}{2}ae^{-a^2/4s}/s^{3/2}$

	$f(t)$		$F(s)$
*L*56.	$\dfrac{\cos a\sqrt{t}}{\sqrt{t}}$		$\sqrt{\pi}\ e^{-a^2/4s}/\sqrt{s}$
*L*57.	$\sinh a\sqrt{t}$		$\dfrac{\sqrt{\pi}}{2}\ ae^{a^2/4s}/s^{3/2}$
*L*58.	$\dfrac{\cosh a\sqrt{t}}{\sqrt{t}}$		$\sqrt{\pi}\ e^{a^2/4s}/\sqrt{s}$
*L*59.	$\dfrac{\sin a\sqrt{t}}{t}$		$\pi\,\mathrm{erf}(a^2/4s)$
*L*60.	$J_0(at)$		$\dfrac{1}{\sqrt{s^2+a^2}}$
*L*61.	$J_n(at)$	$n=0, 1, \ldots$	$\dfrac{(\sqrt{s^2+a^2}-s)^n}{a^n\sqrt{s^2+a^2}}$
*L*62.	$tJ_0(at)$		$\dfrac{s}{(s^2+a^2)^{3/2}}$
*L*63.	$tJ_1(at)$		$\dfrac{a}{(s^2+a^2)^{3/2}}$
*L*64.	$tJ_n(at)$	$n=-1, 0, 1, \ldots$	$\dfrac{(\sqrt{s^2+a^2}-s)^n(s+n\sqrt{s^2+a^2})}{a^n(s^2+a^2)^{3/2}}$
*L*65.	$t^nJ_n(at)$	$n=1, 2, \ldots$	$\dfrac{a^n(2n-1)!!}{(s^2+a^2)^{(2n+1)/2}}$
*L*66.	$J_0(at)-atJ_1(at)$		$\dfrac{s^2}{(s^2+a^2)^{3/2}}$
*L*67.	$J_n(at)/t$	$n=1, 2, \ldots$	$\dfrac{(\sqrt{s^2+a^2}-s)^n}{na^n}$
*L*68.	$J_0(a\sqrt{t})$		$e^{-a^2/4s}/s$
*L*69.	$t^{n/2}J_n\ (a\sqrt{t})$	$n=0, 1, \ldots$	$\left(\dfrac{a}{2}\right)^n \dfrac{e^{-a^2/4s}}{s^{n+1}}$
*L*70.	$I_0(at)$		$\dfrac{1}{\sqrt{s^2-a^2}}$
*L*71.	$I_n(at)$	$n=0, 1, 2, \ldots$	$\dfrac{(s-\sqrt{s^2-a^2})^n}{a^n\sqrt{s^2-a^2}}$
*L*72.	$tI_0(at)$		$\dfrac{s}{(s^2-a^2)^{3/2}}$
*L*73.	$tI_1(at)$		$\dfrac{a}{(s^2-a^2)^{3/2}}$

	$f(t)$		$F(s)$
L74.	$I_0(at)+at\,I_1(at)$		$\dfrac{s^2}{(s^2-a^2)^{3/2}}$
L75.	$I_n(at)/t$	$n=1, 2, \ldots$	$\dfrac{(s-\sqrt{s^2-a^2})^n}{na^n}$
L76.	$I_0(a\sqrt{t})$		$e^{a^2/4s}/s$
L77.	$t^{n/2}I_n\,(a\sqrt{t})$	$n=0, 1, \ldots$	$\left(\dfrac{a}{2}\right)^n \dfrac{e^{a^2/4s}}{s^{n+1}}$
L78.	$P_0(\cos t)$		$1/s$
L79.	$P_1(\cos t)$		$\dfrac{s}{s^2+1}$
L80.	$P_{2n}(\cos t)$	$n=1, 2, \ldots$	$\dfrac{(s^2+1^2)(s^2+3^2)\ldots(s^2+(2n-1)^2)}{s(s^2+2^2)(s^2+4^2)\ldots(s^2+(2n)^2)}$
L81.	$P_{2n+1}(\cos t)$	$n=1, 2, \ldots$	$\dfrac{s(s^2+2^2)(s^2+4^2)\ldots(s^2+(2n)^2)}{(s^2+1^2)(s^2+3^2)\ldots(s^2+(2n+1)^2)}$
L82.	$L_n(t)$	$n=0, 1, \ldots$	$\dfrac{(s-1)^n}{s^{n+1}}$
L83.	$l_n(t)$		$\dfrac{(s-1/2)^n}{(s+1/2)^{n+1}}$
L84.	$H_{2n}(\sqrt{t})/\sqrt{t}$	$n=0, 1, \ldots$	$\dfrac{\sqrt{\pi}\,(2n)!\,(1-s)^n}{n!\,s^{n+1/2}}$
L85.	$H_{2n+1}(\sqrt{t})$	$n=0, 1, \ldots$	$\dfrac{\sqrt{\pi}\,(2n+1)!\,(1-s)^n}{n!\,s^{n+3/2}}$
L86.	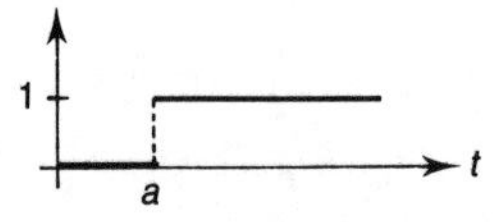 $f(t)=0,\ 0\leqslant t<a$ $f(t)=1,\ t\geqslant a$		e^{-as}/s
L87.	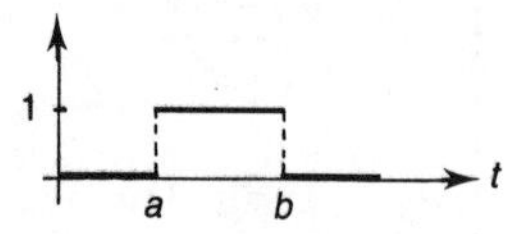 $f(t)=0,\ t<a,\ t>b$ $f(t)=1,\ a\leqslant t\leqslant b$		$(e^{-as}-e^{-bs})/s$

	$f(t)$	$F(s)$
L88.	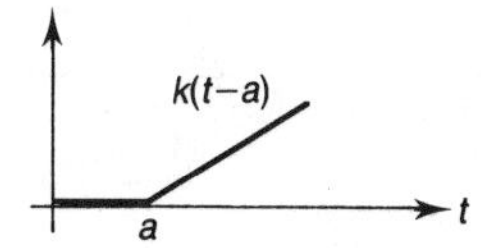 $f(t)=0,\ 0\leqslant t\leqslant a$ $f(t)=k(t-a),\ t\geqslant a$	ke^{-as}/s^2
L89.	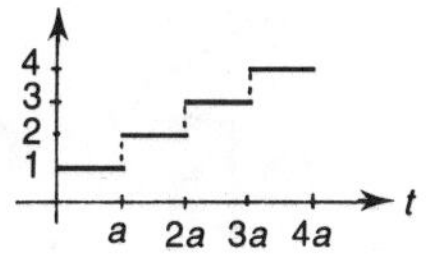	$\dfrac{1}{s(1-e^{-as})}=\dfrac{1+\coth(as/2)}{2s}$
L90.	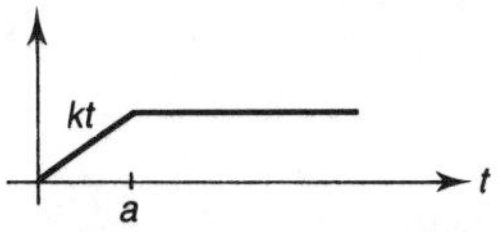 $f(t)=kt,\ 0\leqslant t<a$ $f(t)=ka,\ t\geqslant a$	$\dfrac{k(1-e^{-as})}{s^2}$
L91.	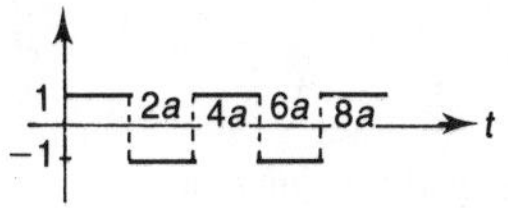	$\dfrac{\tanh(as)}{s}$
L92.	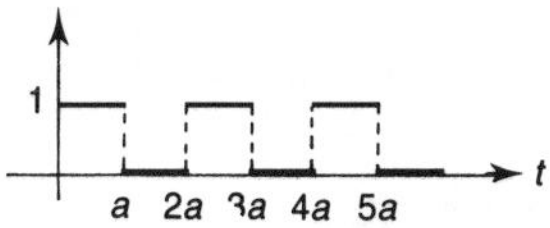	$\dfrac{1}{s(1+e^{-as})}$
L93.	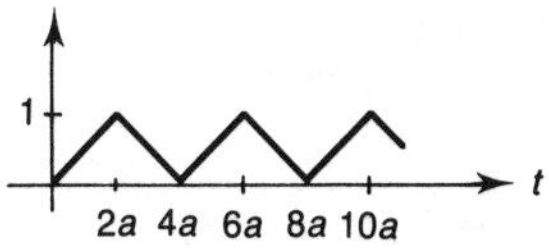	$\dfrac{\tanh(as)}{2as^2}$
L94.	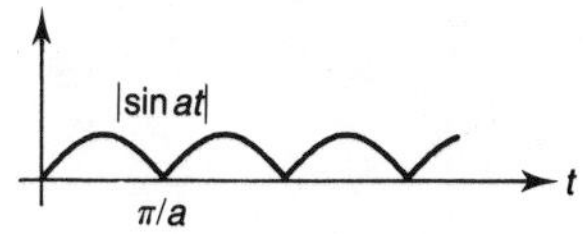	$\dfrac{a}{s^2+a^2}\coth\dfrac{\pi s}{2a}$
L95.	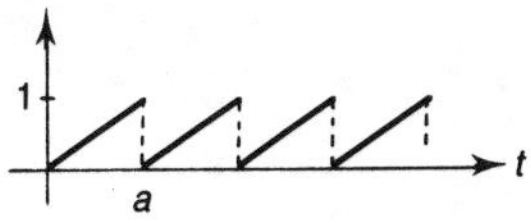	$\dfrac{1}{as^2}-\dfrac{e^{-as}}{s(1-e^{-as})}$

13.6 Dynamical Systems (Filters)

Filters

A *filter* (or a *dynamical system*) is characterized by an *operator* L, which uniquely transforms an *input signal* $x(t)$ to an *output signal* $y(t)=Lx(t)$). Filters are of two kinds: *Analog filters,* transforming continuous time signals, and *discrete filters,* transforming discrete time signals.

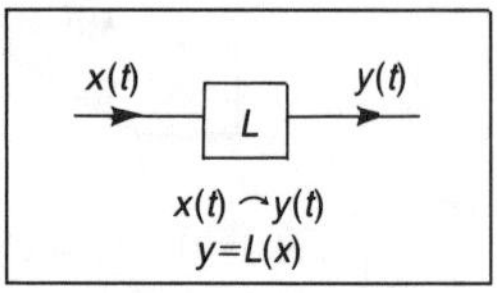

Composite filters

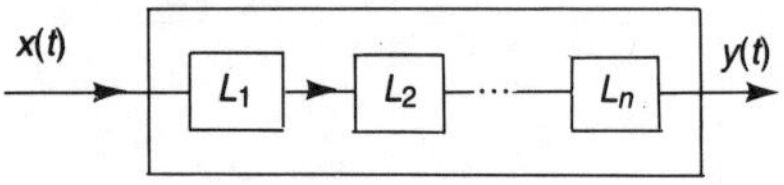

Filters in series

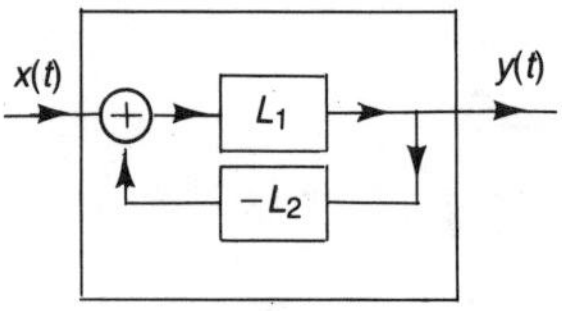

Feedback

Definitions

A filter L is

(*i*) *linear,* if $L(ax_1+bx_2)=aL(x_1)+bL(x_2)$, a and b constants
(*ii*) *time invariant,* if $x(t) \rightsquigarrow y(t) \Rightarrow x(t-T) \rightsquigarrow y(t-T)$, T real
(*iii*) *stable,* if there exists a constant K such that

$$|x(t)|\leq M \Rightarrow |y(t)|\leq KM, \text{ all } t.$$

If a filter L is linear and time invariant, then it is
(*iv*) *causal,* if $x(t)=0,\ t<0 \Rightarrow y(t)=0,\ t<0$

Note. In the following, all filters are assumed to be linear and time invariant.

Analog filters

$\hat{f}(\omega)$ is the Fourier transform, $F(s)$ the Laplace transform of $f(t)$.

Let $h(t)$ denote the *impulse response*, i.e. $\delta(t) \rightsquigarrow h(t)$.

Step response $J(t)$: $\theta(t) \rightsquigarrow J(t)=\int_{-\infty}^{t} h(\tau)d\tau;\ h(t)=J'(t)$

1. L is *stable* $\Leftrightarrow \int_{-\infty}^{\infty} |h(t)|dt<\infty$
2. L is *causal* $\Leftrightarrow h(t)=0$ for $t<0$

3. $x(t) \rightsquigarrow y(t)=h*x(t)=\int_{-\infty}^{\infty} h(\tau)x(t-\tau)d\tau$
4. $\hat{y}(\omega)=\hat{h}(\omega)\hat{x}(\omega)$
5. $e^{i\omega t} \rightsquigarrow \hat{h}(\omega)e^{i\omega t}$
6. $\cos\omega t \rightsquigarrow A(\omega)\cos[\omega t+\phi(\omega)]$
7. $\sin\omega t \rightsquigarrow A(\omega)\sin[\omega t+\phi(\omega)]$

where $A(\omega)=|\hat{h}(\omega)|$, $\phi(\omega)=\arg\hat{h}(\omega)$

The transfer function

Assume that L is causal. The *transfer function* $H(s)$ is the Laplace transform of the impulse response $h(t)$.

061

8. $Y(s)=H(s)X(s)$ if $x(t)=0,\ t<0$
9. Filters in series: $H(s)=H_1(s)H_2(s)\ \dots\ H_n(s)$
10. Feedback: $H(s)=\dfrac{H_1(s)}{1+H_1(s)H_2(s)}$

If L is causal and defined by a *state equation*

$$P(D)y(t)=Q(D)x(t),\ D=d/dt,$$

where $P(D)$ and $Q(D)$ are linear differential operators with constant coefficients, then

$$H(s)=Q(s)/P(s)$$

11. If the transfer function $H(s)$ is rational, then L is stable $\Leftrightarrow$ *All poles* of $H(s)$ lie in $\operatorname{Re} s<0$ and degree $Q(s)\leq$degree $P(s)$.

Discrete filters

$F(z)$ denotes the z-transform of $f(n)$. The input and output signals are denoted by $x(n)$ and $y(n)$, respectively and the unit pulse at k by $\delta_k(n)$.

Let $h(n)$ denote the response of the unit pulse at zero, i.e.

$$\delta_0(n) \rightharpoondown h(n)$$

1. L is *stable* $\Leftrightarrow \sum_{-\infty}^{\infty} |h(n)|<\infty$
2. L is *causal* $\Leftrightarrow h(n)=0,\ n<0$

Transfer function

Assume that L is causal.

The *transfer function* $H(z)$ is the z-transform of the unit pulse response $h(n)$.

3. $y(n)=h*x(n)=\sum_{k=-\infty}^{\infty} h(k)x(n-k)$ $[=\sum_{k=0}^{n} h(k)x(n-k)$ if L is causal and $x(n)=0,\ n<0]$
4. $Y(z)=H(z)X(z)$ if $x(n)=0,\ n<0$
5. Filters in series: $H(z)=H_1(z)H_2(z)\ \dots\ H_n(z)$
6. Feedback: $H(z)=\dfrac{H_1(z)}{1+H_1(z)H_2(z)}$

If L is causal and defined by a *recurrence equation*

$$\sum_{k=0}^{N} b_k y(n-k)=\sum_{k=0}^{M} a_k x(n-k)\ (a_k,\ b_k \text{ constants}) \text{ then}$$

$$H(z)=\sum_{k=0}^{M} a_k z^{-k}\Big/\sum_{k=0}^{N} b_k z^{-k}$$

7. If the transfer function $H(z)$ is rational, then L is stable $\Leftrightarrow$ *All poles* of $H(z)$ lie in $|z|<1$ and $M\leq N$.

Example

[$\oplus$=addition $\triangleright a$=multiplication by a $\boxed{z^{-k}}$=delay k steps]

From a *block diagram* to the corresponding *recurrence equation*:

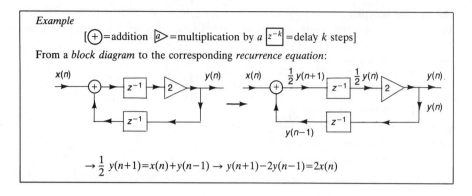

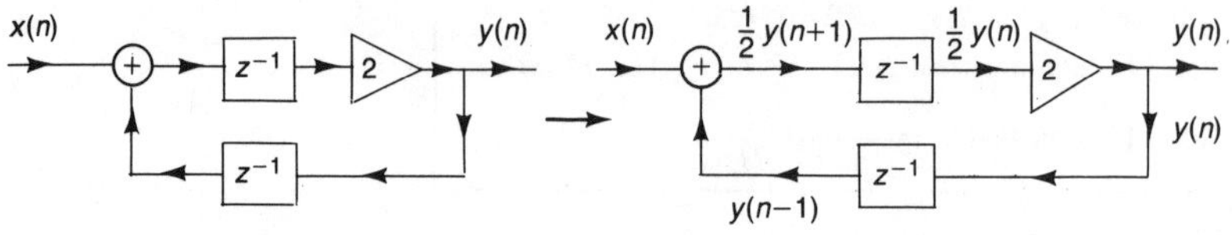

$$\rightarrow \frac{1}{2}\, y(n+1)=x(n)+y(n-1) \rightarrow y(n+1)-2y(n-1)=2x(n)$$

13.7 Hankel and Hilbert transforms

Hankel transforms

Hankel transform of order p	$F_p(y)=\int_0^\infty xf(x)J_p(xy)dx \quad \left(p>-\frac{1}{2}\right)$
Inversion formula	$f(x)=\int_0^\infty yF_p(y)J_p(xy)dy$
Parseval type formulas	$\int_0^\infty xf(x)\overline{g(x)}dx=\int_0^\infty yF_p(y)\overline{G_p(y)}dy$
	$\int_0^\infty x\lvert f(x)\rvert^2dx=\int_0^\infty y\lvert F_p(y)\rvert^2dy$

The Hankel transform of order zero is essentially the result of what one gets when making a polar substitution in the two-dimensional Fourier transform for functions with circular symmetry.

Properties of Hankel transforms of order p

	$f(x)$	$F_p(y)$
*Ha*1.	$f(ax) \quad (a>0)$	$\frac{1}{a^2} F_p\left(\frac{y}{a}\right)$
*Ha*2.	$x^{p-1}D\{x^{1-x}f(x)\}$	$-yF_{p-1}(y)$
*Ha*3.	$x^{-p-1}D\{x^{p+1}f(x)\}$	$yF_{p+1}(y)$

Properties and table of Hankel transforms of order zero

	$f(x)$	$F(y)=F_0(y)$
*Ha*4.	$f(ax) \quad (a>0)$	$\frac{1}{a^2} F\left(\frac{y}{a}\right)$
*Ha*5.	$F(x)$	$f(y)$
*Ha*6.	$xf'(x)$	$-2F(y)-yF'(y)$
*Ha*7.	$\delta(x-a) \quad (a>0)$	$aJ_0(ay)$
*Ha*8.	$\begin{cases} 1, & 0<x<a \\ 0, & x>a \end{cases}$	$\frac{a}{y} J_1(ay)$
*Ha*9.	$\frac{1}{x^a} \quad (0<a<2)$	$\dfrac{2^{1-a}\Gamma\left(1-\frac{a}{2}\right)}{y^{2-a}\Gamma\left(\frac{a}{2}\right)}$
*Ha*10.	$\frac{1}{x}$	$\frac{1}{y}$
*Ha*11.	$\frac{1}{x^2+a^2}$	$K_0(ay)$
*Ha*12.	$\frac{a}{(x^2+a^2)^2}$	$\frac{y}{2} K_1(ay)$
*Ha*13.	$\frac{1}{\sqrt{x^2+a^2}}$	$\frac{e^{-ay}}{y}$
*Ha*14.	$\frac{1}{(x^2+a^2)^{3/2}}$	$\frac{e^{-ay}}{a}$
*Ha*15.	e^{-ax^2}	$2ae^{-y^2/4a}$
*Ha*16.	$\frac{\cos ax}{x}$	$\frac{\theta(y-a)}{\sqrt{y^2-a^2}}$
*Ha*17.	$\frac{\sin ax}{x}$	$\frac{\theta(a-y)}{\sqrt{a^2-y^2}}$

Hilbert transforms

Hilbert transform $\quad F(y)=(\text{CPV})\frac{1}{\pi}\int_{-\infty}^{\infty}\frac{f(x)}{x-y}\,dx$ (Cauchy Principal Value)=

$$=\left[-\frac{1}{\pi\bullet}*f(\bullet)\right](y)$$

Inversion formula $\quad f(x)=-(\text{CPV})\frac{1}{\pi}\int_{-\infty}^{\infty}\frac{F(y)}{y-x}dy$

Parseval type formulas	$\int_{-\infty}^{\infty} f(x)\overline{g(x)}dx = \int_{-\infty}^{\infty} F(y)\overline{G(y)}dy$
	$\int_{-\infty}^{\infty} \lvert f(x)\rvert^2 dx = \int_{-\infty}^{\infty} \lvert F(y)\rvert^2 dy$

Properties and table of Hilbert transforms

	$f(x)$	$F(y)$
*Hi*1.	$F(x)$	$-f(y)$
*Hi*2.	$af(x)+bg(x)$	$aF(y)+bG(y)$
*Hi*3.	$f(x+a)$	$F(y+a)$
*Hi*4.	$f(ax) \quad (a>0)$	$F(ay)$
*Hi*5.	$f(-ax) \quad (a>0)$	$-F(-ay)$
*Hi*6.	$f'(x)$	$F'(y)$
*Hi*7.	$xf(x)$	$yf(y)+\frac{1}{\pi}\int_{-\infty}^{\infty} f(x)dx$
*Hi*8.	$(f*g)(x)= \int_{-\infty}^{\infty} f(t)g(x-t)dt$	$-(F*G)(y)$
*Hi*9.	$\begin{cases}1, & a<x<b\\ 0, & \text{otherwise}\end{cases}$	$\frac{1}{\pi}\ln\left\lvert\frac{b-y}{a-y}\right\rvert$
*Hi*10.	$\delta(x-a)$	$\frac{1}{\pi(a-y)}$
*Hi*11.	$\begin{cases}\frac{1}{x}, & a<x<\infty,\ a>0\\ 0, & \text{otherwise}\end{cases}$	$\frac{1}{\pi y}\ln\left\lvert\frac{a}{a-y}\right\rvert$
*Hi*12.	$\frac{1}{x+\alpha} \quad (\text{Im } \alpha>0)$	$\frac{i}{y+\alpha}$
*Hi*13.	$\frac{1}{x^2+\alpha^2} \quad (\text{Re } \alpha>0)$	$-\frac{y}{\alpha(y^2+\alpha^2)}$
*Hi*14.	$\frac{x}{x^2+\alpha^2} \quad (\text{Re } \alpha>0)$	$\frac{\alpha}{y^2+\alpha^2}$
*Hi*15.	$\sin ax \quad (a>0)$	$\cos ay$
*Hi*16.	$\cos ax \quad (a>0)$	$-\sin ay$
*Hi*17.	$\frac{\sin ax}{x} \quad (a>0)$	$\frac{\cos ay-1}{y}$

14 Complex Analysis

14.1 Functions of a Complex Variable

Complex numbers, see sec. 2.3.

Notation

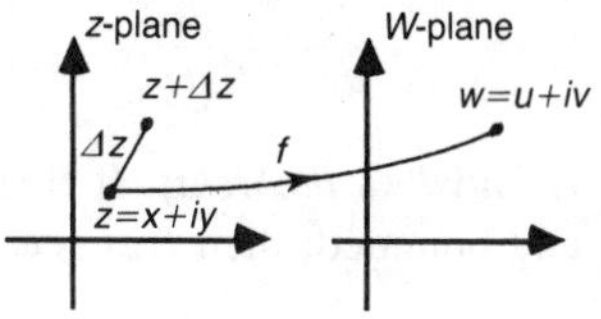

$$w=f(z)=f(x+iy)=u(x, y)+iv(x, y)$$

Differentiation

$f(z)$ is *differentiable* at z if

$$f'(z)=\lim_{\Delta z\to 0}\frac{f(z+\Delta z)-f(z)}{\Delta z} \text{ exists.}$$

Remark. $f'(z)=u_x'+iv_x'=v_y'-iu_y'$

Analytic functions

Definition. The function $f(z)$ is analytic in a domain Ω if $f(z)$ is differentiable at every point of Ω. [$f(z)$ is analytic at ∞ if $f(1/z)$ is analytic at 0.]

Remark. $|z|$ and $\bar{z}$ are not analytic functions.

Some properties of analytic functions

Assume that $f(z)$ is analytic in Ω with boundary C. Then in Ω,

1. Any order derivative of $f(z)$ exists and is an analytic function.
2. (*Cauchy–Riemann's equations:*)

$$\frac{\partial u}{\partial x}=\frac{\partial v}{\partial y} \qquad \frac{\partial u}{\partial y}=-\frac{\partial v}{\partial x}$$

The converse is true if the partial derivatives are continuous in Ω.

Remark. $f(z)=u(z, 0)+iv(z, 0)$; $f'(z)=u_x'(z, 0)+iv_x'(z, 0)=$
$=u_x'(z, 0)-iu_y'(z, 0)$ etc.; $f(z)=2u\left(\frac{z}{2}, -\frac{iz}{2}\right)+C=2iv\left(\frac{z}{2}, -\frac{iz}{2}\right)+C$
if $f(z)$ is analytic around zero.

3. $\Delta u = u''_{xx} + u''_{yy} = 0$, $\Delta v = 0$, i.e. u and v are (*conjugate*) *harmonic functions.*

4. $u(x, y) = C_1$, $v(x, y) = C_2$ represent two orthogonal families of curves.

5. l'Hospital's rule for limits is valid for a quotient of analytic functions.

6. (*Maximum – modulus principle.*)
$|f(z)| \leq M$ on C (C simple) $\Rightarrow |f(z)| < M$ in Ω (if $f(z)$ is not constant). [$|f(z)|$ attains its maximum (and minimum if $f(z) \neq 0$) on the boundary].

7. $f'(a) \neq 0 \Rightarrow w = f(z)$ has an analytic inverse function $z = f^{-1}(w)$ in a neighborhood of a and
$$\frac{dz}{dw} = 1 / \frac{dw}{dz}.$$

8. (*Liouville's theorem*). If $f(z)$ is analytic in the entire plane (i.e. an *entire function*) and bounded, then $f(z)$ is constant.

9. (*Schwarz' lemma*)
(*i*) $f(z)$ analytic for $|z| < 1$ (*ii*) $|f(z)| \leq 1$, $f(0) = 0 \Rightarrow$
$$|f(z)| \leq |z| \text{ (equality only if } f(z) = cz,\ |c| = 1)$$

Elementary functions

Single-valued functions

1. $z^n = (x + iy)^n$, n integer ($z \neq 0$ if $n < 0$)
2. $e^z = e^x e^{iy} = e^x(\cos y + i \sin y)$. Period$= 2\pi i$
3. $\cosh z = \frac{1}{2}(e^z + e^{-z})$, $\sinh z = \frac{1}{2}(e^z - e^{-z})$

$$\tanh z = \frac{\sinh z}{\cosh z} \quad \left(z \neq \left(k + \frac{1}{2}\right)\pi i\right), \quad \coth z = \frac{\cosh z}{\sinh z} \quad (z \neq k\pi i)$$

4. $\cos z = \frac{1}{2}(e^{iz} + e^{-iz})$, $\sin z = \frac{1}{2i}(e^{iz} - e^{-iz})$

$$\tan z = \frac{\sin z}{\cos z} \quad \left(z \neq \left(k + \frac{1}{2}\right)\pi\right), \quad \cot z = \frac{\cos z}{\sin z} \quad (z \neq k\pi)$$

$\cos iz = \cosh z$, $\sin iz = i \sinh z$ $\cosh iz = \cos z$, $\sinh iz = i \sin z$

(Formulas for real elementary functions (cf. chapt. 5) are valid also in the complex case.)

Multiple – valued functions

5. $\log z = \ln|z| + i \arg z = \ln r + i(\theta + 2n\pi)$
(infinitely-valued)

Principal branch:
$\text{Log}\, z = \ln r + i\theta,\ -\pi < \theta \leq \pi$

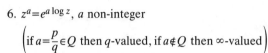

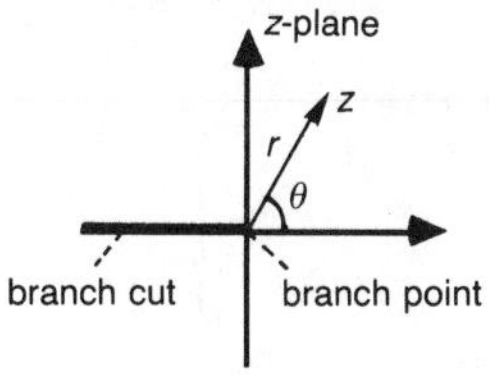

6. $z^a = e^{a \log z}$, a non-integer

$\left(\text{if } a = \frac{p}{q} \in Q \text{ then } q\text{-valued, if } a \notin Q \text{ then } \infty\text{-valued}\right)$

> *Example*
>
> 1. $\log 2i = \ln|2i| + i \arg 2i = \ln 2 + i\left(\frac{\pi}{2} + 2n\pi\right)$
>
> 2. $(2i)^i = e^{i \log 2i} = e^{-\left(\frac{\pi}{2}+2n\pi\right)+i\ln 2} = e^{-\pi/2-2n\pi}[\cos(\ln 2) + i\sin(\ln 2)]$

A survey of elementary functions

$$w = f(z) = f(x+iy) = u(x, y) + iv(x, y),\ r = |z| = \sqrt{x^2+y^2},\ \theta = \arg z$$

Function $w=f(z)$	Real part $u(x, y)$	Imaginary part $v(x, y)$	Zeros $(k=0, \pm1, \pm2, \ldots)$ m=order	Isolated singularities m=order	Inverse $z=f^{-1}(w)$
z	x	y	$0,\ m=1$	$\infty,\ m=1$ (pole)	w
z^2	x^2-y^2	$2xy$	$0,\ m=2$	$\infty,\ m=2$ (pole)	$w^{1/2}$
$1/z$	$\frac{x}{r^2}$	$-\frac{y}{r^2}$	$\infty,\ m=1$	$0,\ m=1$ (pole)	$1/w$
$1/z^2$	$\frac{x^2-y^2}{r^4}$	$-\frac{2xy}{r^4}$	$\infty,\ m=2$	$0,\ m=2$ (pole)	$w^{-1/2}$
$\sqrt{z}$	$\pm\left(\frac{x+r}{2}\right)^{\frac{1}{2}}$	$\pm\left(\frac{-x+r}{2}\right)^{\frac{1}{2}}$	0, branch point	$0,\ \infty$ branch points	w^2
e^z	$e^x \cos y$	$e^x \sin y$	–	∞ (ess. sing.)	$\log w$
$\cosh z$	$\cosh x \cos y$	$\sinh x \sin y$	$\left(k+\frac{1}{2}\right)\pi i,\ m=1$	∞ (ess. sing.)	$\log(w+\sqrt{w^2-1})$
$\sinh z$	$\sinh x \cos y$	$\cosh x \sin y$	$k\pi i,\ m=1$	∞ (ess. sing.)	$\log(w+\sqrt{w^2+1})$
$\tanh z$	$\frac{\sinh 2x}{\cosh 2x+\cos 2y}$	$\frac{\sin 2y}{\cosh 2x+\cos 2y}$	$k\pi i,\ m=1$	$\left(k+\frac{1}{2}\right)\pi i,\ m=1$ (poles) ∞, (ess. sing.)	$\frac{1}{2}\log\left(\frac{1+w}{1-w}\right)$
$\log z$	$\ln r$	$\theta+2n\pi$	1 (princ. branch), $m=1$	$0,\ \infty$ branch points	e^w
$\cos z$	$\cos x \cosh y$	$-\sin x \sinh y$	$\left(k+\frac{1}{2}\right)\pi,\ m=1$	∞ (ess. sing.)	$-i\log(w+\sqrt{w^2-1})$
$\sin z$	$\sin x \cosh y$	$\cos x \sinh y$	$k\pi,\ m=1$	∞ (ess. sing.)	$-i\log(iw+\sqrt{1-w^2})$
$\tan z$	$\frac{\sin 2x}{\cos 2x+\cosh 2y}$	$\frac{\sinh 2y}{\cos 2x+\cosh 2y}$	$k\pi,\ m=1$	$\left(k+\frac{1}{2}\right)\pi,\ m=1$ (poles) ∞, (ess. sing.)	$-\frac{i}{2}\log\left(\frac{1+iw}{1-iw}\right)$

14.2 Complex Integration

Basic properties

Definition

$$\int_C f(z)dz = \int_a^b f(z(t))z'(t)dt =$$
$$= \int_C (u+iv)(dx+i\,dy)$$

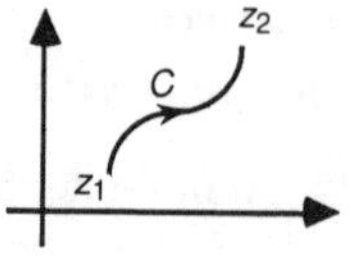

C: $z=z(t)$, $a \leq \vec{t} \leq b$

Properties

1. $\left|\int_C f(z)dz\right| \leq \int_C |f(z)| \cdot |dz| \leq M \cdot L$, if $|f(z)| \leq M$ on C, L=length of C.

2. If $f(z)$ is analytic in a domain containing C and $F(z)$ is a primitive function of $f(z)$, then
$$\int_C f(z)dz = F(z_2) - F(z_1)$$

3. (*Cauchy's theorem*)
$f(z)$ analytic on and inside a closed curve $C \Rightarrow \oint_C f(z)dz = 0$

4. (*Morera's theorem,* converse of Cauchy's theorem)
(*i*) $f(z)$ continuous in a region Ω
(*ii*) $\oint_C f(z)dz = 0$, every simple closed curve C in Ω
$\Rightarrow f(z)$ is analytic in Ω.

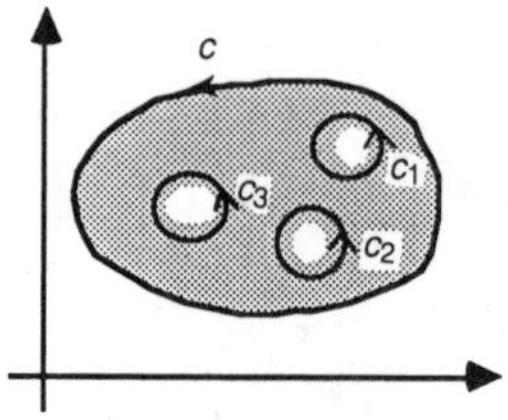

5. If $f(z)$ is analytic in a region with a finite number of "holes" (where $f(z)$ is not necessarily analytic), then
$$\int_C f(z)dz = \int_{C_1} f(z)dz + \int_{C_2} f(z)dz + \ldots$$

6. If $f(z)$ is analytic on and inside a simple closed curve C, and a is any point inside C, then
(*i*) (*Cauchy's integral formula*)

$$f(a) = \frac{1}{2\pi i} \oint_C \frac{f(z)}{z-a} dz$$
$$f^{(n)}(a) = \frac{n!}{2\pi i} \oint_C \frac{f(z)}{(z-a)^{n+1}} dz$$

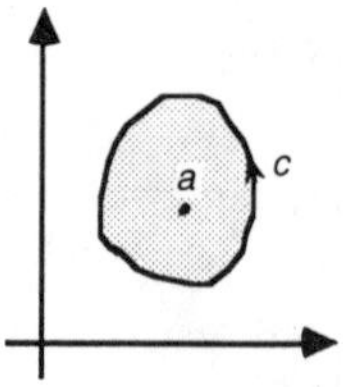

(*ii*) $|f^{(n)}(a)| \leq \dfrac{M \cdot n!}{R^n}$ if C is a circle with centre at a and radius$=R$, $|f(z)| \leq M$ on C.

Residues

$\operatorname*{Res}_{z=a} f(z)=c_{-1}$, i.e. the coefficient of $(z-a)^{-1}$ in the Laurent series expansion of $f(z)$ [cf. sec. 14.3].

The residue theorem

Assume that $f(z)$ is analytic on and inside C except at finitely many points $a_1, a_2, \ldots, a_n$. Then

$$\frac{1}{2\pi i}\oint_C f(z)dz=\sum_{k=1}^{n}\operatorname*{Res}_{z=a_k} f(z)$$

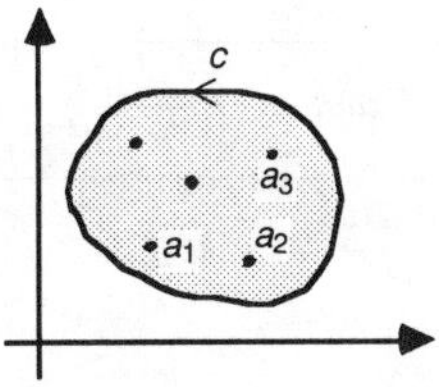

Calculation of residues

1. Determine c_{-1} in the Laurent series expansion.
2. Simple pole: $\operatorname*{Res}_{z=a} f(z)=\lim_{z\to a}(z-a)f(z)$. [l'Hospital's rule may be used].
   In particular, if $f(z)$, $g(z)$ analytic, $f(a)\neq 0$, $g(a)=0$, $g'(a)\neq 0$, then
   $$\operatorname*{Res}_{z=a}\frac{f(z)}{g(z)}=\frac{f(a)}{g'(a)}.$$
3. Pole of order m: $\operatorname*{Res}_{z=a} f(z)=\lim_{z\to a}\frac{1}{(m-1)!}\left(\frac{d}{dz}\right)^{m-1}\{(z-a)^m f(z)\}$.

Calculation of definite integrals

1. $\int_0^{2\pi} R(\sin\theta, \cos\theta)d\theta=[z=e^{i\theta}]=\oint_{|z|=1} R\left(\frac{z-z^{-1}}{2i}, \frac{z+z^{-1}}{2}\right)\frac{dz}{iz}$.
2. If $f(z)$ is analytic in the upper half-plane $\operatorname{Im} z\geq 0$ except for finitely points $a_1, \ldots, a_n$ above the real axis, and if $|zf(z)|\to 0$ as $z\to\infty$, then
   $$\int_{-\infty}^{\infty} f(x)dx=2\pi i\sum_{k=1}^{n}\operatorname*{Res}_{z=a_k} f(z)$$
3. If C_R: $z=Re^{i\theta}$, $0\leq\theta\leq\pi$ and if $|f(z)|\leq M\cdot R^{-k}$, (M, $k>0$ constants), then $\int_{C_R} f(z)e^{-\operatorname{Im} z}dz\to 0$ as $R\to\infty$.

Example. $I=\int_{-\infty}^{\infty}\frac{\cos x}{x^2+a^2}dx=\operatorname{Re}\int_{-\infty}^{\infty}\frac{e^{ix}}{x^2+a^2}dx$, $a>0$.

Set $f(z)=\frac{e^{iz}}{z^2+a^2}$: $\operatorname*{Res}_{z=ia} f(z)=e^{-a}\lim_{z\to ia}\frac{z-ia}{z^2+a^2}=$[l'Hospital's rule]$=$

$=e^{-a}\lim_{z\to ia}\frac{1}{2z}=\frac{e^{-a}}{2ia}$. Furthermore, $|zf(z)|=\frac{|z|e^{-y}}{|z^2+a^2|}\leq\frac{|z|}{|z^2+a^2|}\to 0$ as $z\to\infty$.

Thus, $I=\operatorname{Re}\left(2\pi i\cdot\frac{e^{-a}}{2ia}\right)=\frac{\pi e^{-a}}{a}$

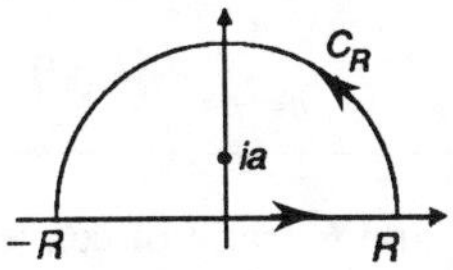

Calculation of sum of infinite series

Assume that $|f(z)| \leq \text{constant } |z|^{-a}$, $a>1$ as $z \to \infty$.

1. $\sum_{-\infty}^{\infty} f(n) = -$[sum of residues of $\pi f(z) \cot \pi z$ at all poles of $f(z)$].
2. $\sum_{-\infty}^{\infty} (-1)^n f(n) = -$[sum of residues of $\dfrac{\pi f(z)}{\sin \pi z}$ at all poles of $f(z)$].

Example. $\sum_{-\infty}^{\infty} \dfrac{1}{n^2+a^2} = \left(\operatorname*{Res}_{z=ia} \dfrac{\pi \cot \pi z}{z^2+a^2} + \operatorname*{Res}_{z=-ia} \dfrac{\pi \cot \pi z}{z^2+a^2} \right) = \dfrac{\pi}{a} \coth \pi a$

14.3 Power Series Expansions

Taylor series

If $f(z)$ is analytic in a neighborhood of $z=a$, then

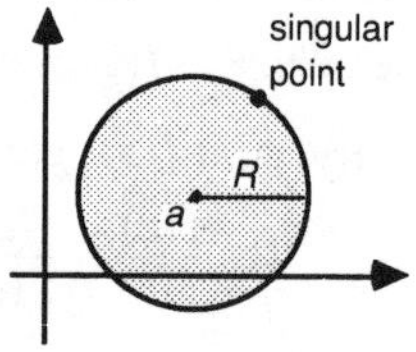

$$f(z) = \sum_{n=0}^{\infty} a_n (z-a)^n, \quad a_n = \frac{f^{(n)}(a)}{n!}$$

Radius of convergence R=distance to the nearest singular point, or

$$\frac{1}{R} = \limsup_{n\to\infty} \sqrt[n]{|a_n|} = \left[\lim_{n\to\infty} \sqrt[n]{|a_n|} = \lim_{n\to\infty} \left| \frac{a_{n+1}}{a_n} \right| \text{ if they exist} \right].$$

Example. Sought: Taylor series of $\text{Log}(2z-i)$ about $z=0$; $\text{Log}(2z-i) = \text{Log}[-i(1+2iz)] =$
$= \text{Log}(-i) + \text{Log}(1+2iz) = -i\pi/2 + 2iz - 1/2\,(2iz)^2 + \ldots$

Table of series expansions, see sec. 8.6.

Laurent series

If $f(z)$ is analytic in an annulus about $z=a$, then

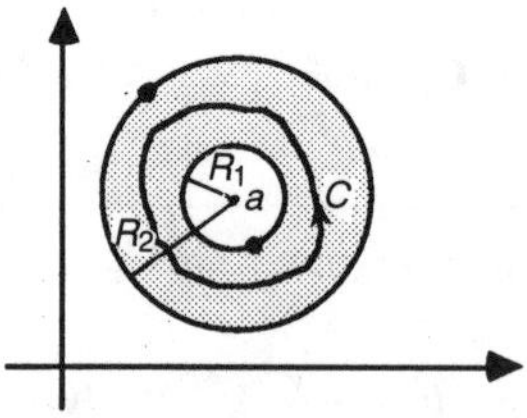

$$f(z) = \sum_{n=-\infty}^{\infty} c_n (z-a)^n, \quad c_n = \frac{1}{2\pi i} \oint_C \frac{f(z)}{(z-a)^{n+1}}\, dz$$

R_1 and R_2 radii of convergence:

$$\frac{1}{R_2} = \limsup_{n\to\infty} \sqrt[n]{|c_n|}; \qquad R_1 = \limsup_{n\to\infty} \sqrt[n]{|c_{-n}|}$$

$f(z)$ has singular points on the circles $|z-a| = R_i$, $i=1, 2$.

Example.

Sought: Laurent series expansion of $f(z)=\dfrac{2}{z^2-1}$. in the annulus $1<|z-2|<3$.

Solution. $f(z)=\dfrac{1}{z-1}-\dfrac{1}{z+1}=[z-2=w]=\dfrac{1}{w+1}-\dfrac{1}{w+3}=$

$$=\frac{1}{w\left(1+\dfrac{1}{w}\right)}-\frac{1}{3\left(1+\dfrac{w}{3}\right)}=\frac{1}{w}\left(1-\frac{1}{w}+\frac{1}{w^2}-\dots\right)-\frac{1}{3}\left(1-\frac{w}{3}+\frac{w^2}{9}-\dots\right)=$$

$$=\sum_{n=0}^{\infty}(-1)^n(z-2)^{-n-1}-\frac{1}{3}\sum_{n=0}^{\infty}\left(-\frac{1}{3}\right)^n(z-2)^n$$

14.4 Zeros and Singularities

Zeros

Assume that $f(z)$ is analytic (and $\not\equiv 0$) in a neighbourhood of $z=a$. The point a is a *zero* of order n if $f(z)=(z-a)^n g(z)$, where $g(z)$ is analytic and $g(a)\neq 0$.

Remark. a is a zero of order $n \Leftrightarrow$

$$f(a)=f'(a)=\dots=f^{(n-1)}(a)=0,\ f^{(n)}(a)\neq 0$$

Singularities

$z=a$ is a *singular point* of $f(z)$ if $f(z)$ fails to be analytic at a. It is isolated if there is a neighbourhood of a in which there are no more singular points.

Classification of isolated singularities.

The point $z=a$ is

(*i*) a *removable singularity* if $\lim_{z\to a} f(z)$ exists.

(*ii*) a *pole* of order n if $f(z)=(z-a)^{-n}g(z)$, where $g(z)$ is analytic, $g(a)\neq 0$. [The Laurent series expansion about a contains finitely many negative power terms.]

(*iii*) an *essential singularity* otherwise, in which case there are infinitely many negative power terms in the Laurent series expansion about a.

Furthermore, branch points of multiple-valued function are examples of non-isolated singular points.

1. (*Picard's theorem*)
The point $z=a$ is an essential singularity of $f(z) \Rightarrow$ Every neighborhood of a contains an infinte set of points z such that $f(z)=w$ for every complex number w (with the possible exception of a single value of w).

[*Example.* $f(z)=e^{1/z}$. Essential singularity at $z=0$, exceptional value $w=0$].

2. An isolated singular point $z=a$ is a pole $\Leftrightarrow \lim_{z\to a}|f(z)|=\infty$.

The argument principle

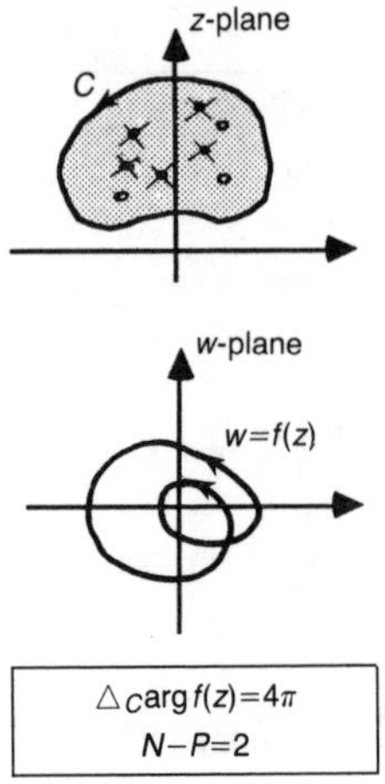

Assume that $f(z)$ is analytic inside and on a simple curve C except for a finite number of poles inside C, $f(z)\neq 0$ on C. Let N=number of zeros, P=number of poles inside C (including multiplicity). Then

$$N-P=\frac{1}{2\pi i}\oint_C \frac{f'(z)}{f(z)}dz=\frac{1}{2\pi}\triangle_C \arg f(z)$$

Rouché's theorem

Assume (*i*) $f(z)$, $g(z)$ analytic on and inside a simple closed curve C (*ii*) $|g(z)|<|f(z)|$ on C. Then $f(z)$ and $f(z)+g(z)$ have the same number of zeros inside C.

14.5 Conformal Mappings

Assume that $f(z)$ is analytic. The mapping $w=f(z)$ is *conformal* (i.e. preserves angles both in magnitude and sense) at z_0 if $f'(z_0)\neq 0$.

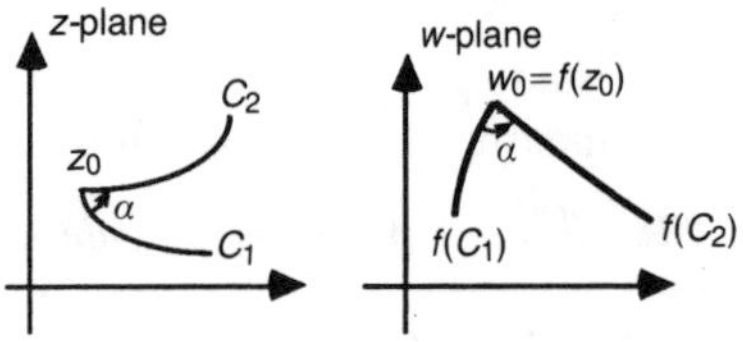

Remark. The Jacobian $\dfrac{\partial(u, v)}{\partial(x, y)}=|f'(z)|^2$.

Riemann's mapping theorem

Assume that Ω is a simply connected region with boundary C. Then there exists a mapping $w=f(z)$, analytic in Ω, which maps Ω one-to-one and conformally onto the unit disc and C onto the unit circle.

The bilinear (Möbius) transformation

The mapping $w=\dfrac{az+b}{cz+d}$ $(ad-bc\neq 0)$ maps

(*i*) circle → circle or straight line
(*ii*) straight line → circle or straight line

Invariance of cross ratio

$$\frac{(w-w_1)(w_2-w_3)}{(w-w_3)(w_2-w_1)}=\frac{(z-z_1)(z_2-z_3)}{(z-z_3)(z_2-z_1)}\quad [w_k=w(z_k)]$$

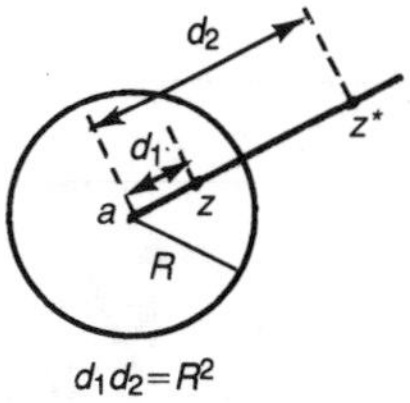

Inverse points

z and z^* are inverse points

(*i*) with respect to a circle if

$$(z^*-a)(\overline{z}-\overline{a})=R^2$$

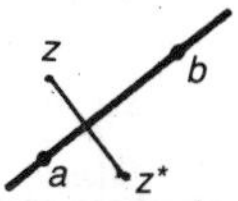

(*ii*) with respect to a line if

$$(\overline{b}-\overline{a})(z^*-a)=(b-a)(\overline{z}-\overline{a})$$

Invariance of inverse points

Pairs of inverse points are mapped to pairs of inverse points (with respect to corresponding circles or lines).

Preservation of harmonicity by conformal mappings

Assume that

(*i*) $h(u, v)$ is harmonic in w-plane.
(*ii*) $f(z)=u(x, y)+iv(x, y)$ is an analytic function mapping Ω conformally into Ω'.

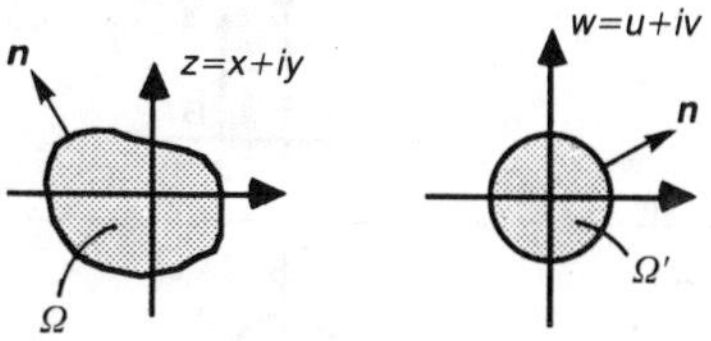

Then

$H(x, y)=h(u(x, y), v(x, y))$ is harmonic in Ω.

Remark. $\dfrac{\partial h}{\partial n}=0$ on $\partial\Omega' \Rightarrow \dfrac{\partial H}{\partial n}=0$ on $\partial\Omega$.

Cf. Poisson's integral formulas, sec. 10.9.

Special conformal mappings

Mappings onto the upper half plane

		Mapping
1.	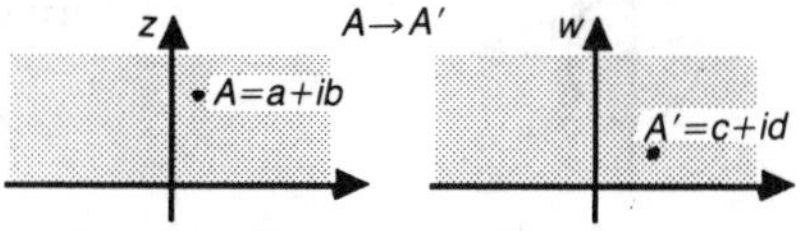	$w=\dfrac{d}{b}(z-a)+c$
2.	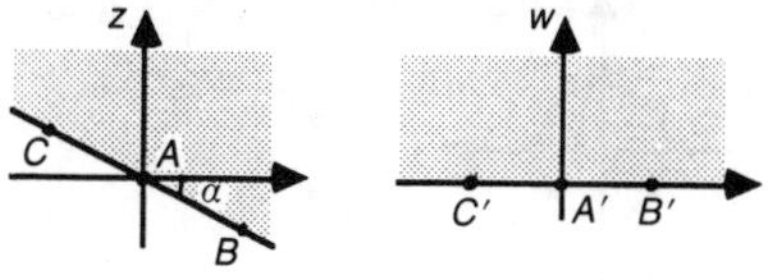	$w=e^{i\alpha}z$

		Mapping
3.		$w = z^{\pi/\alpha}$
4.	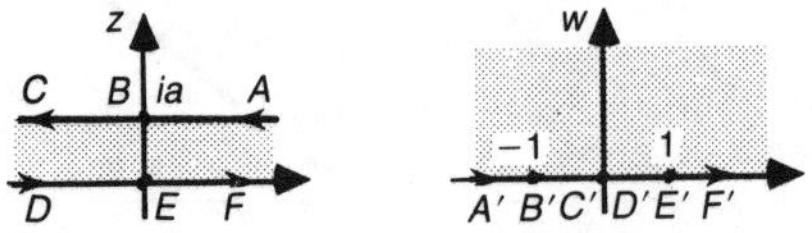	$w = e^{\pi z/a}$
5.	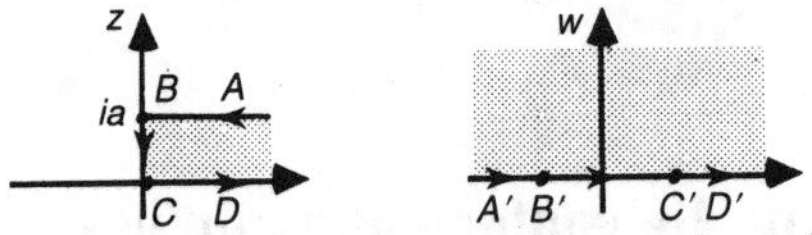	$w = \cosh \dfrac{\pi z}{a}$
6.	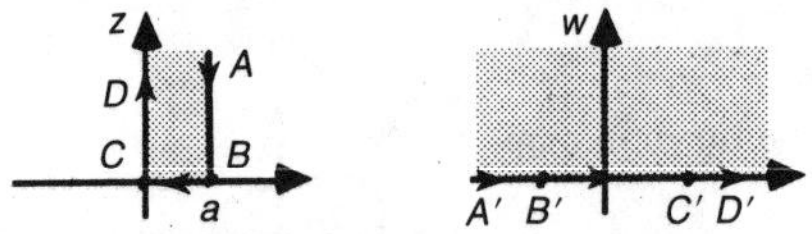	$w = \cos \dfrac{\pi z}{a}$
7.	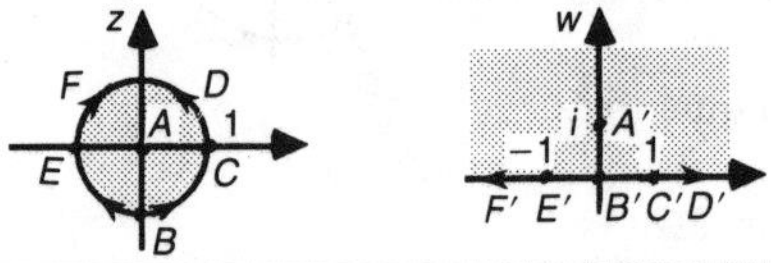	$w = \dfrac{1 - iz}{z - i}$
8.	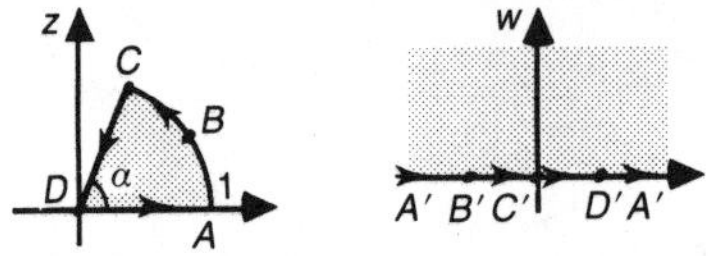	$w = \left(\dfrac{1 + z^{\pi/\alpha}}{1 - z^{\pi/\alpha}}\right)^2$

Mappings onto the unit circle

9.	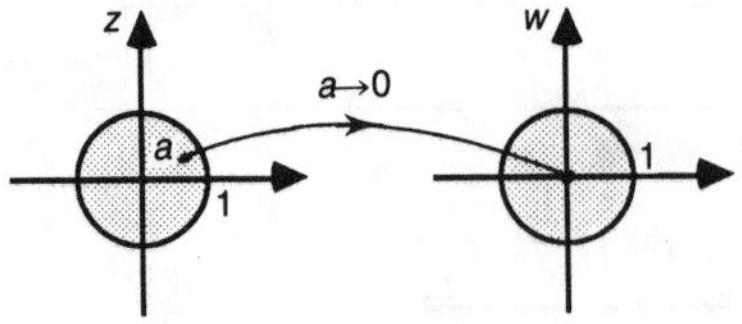	$w = e^{i\theta} \dfrac{z - a}{1 - \bar{a}z}$ (θ arbitrary)
10.	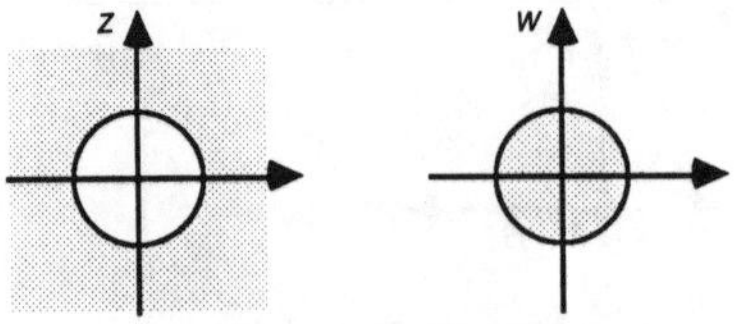	$w = \dfrac{1}{z}$

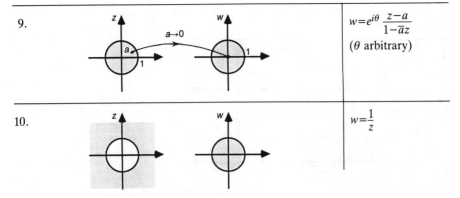

		Mapping
11.	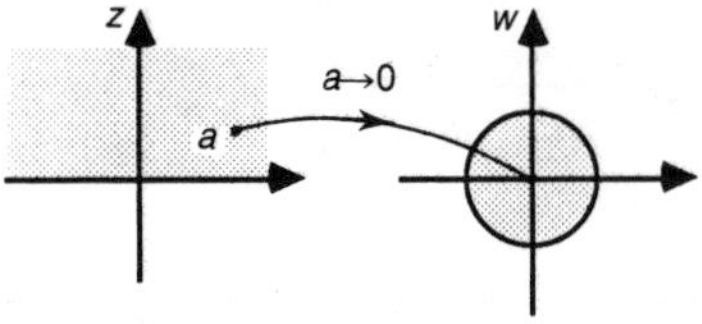	$w=\dfrac{z-a}{z-\overline{a}}$

Composite mappings

Example. Find a conformal mapping of the circle sector $0<\arg z<\pi/4$, $|z|<1$ onto the unit disc $|z|<1$.

Solution.

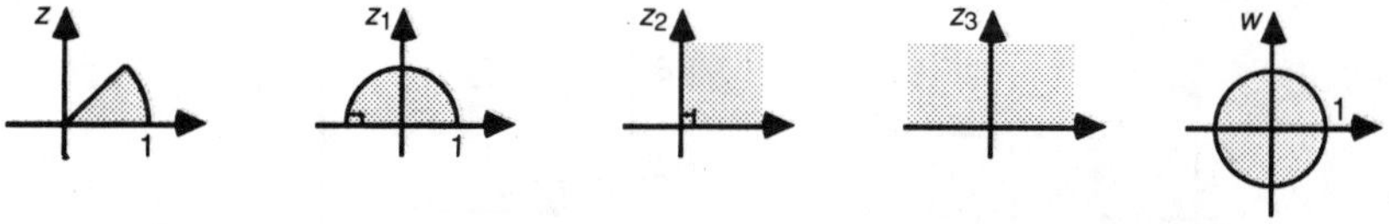

(*i*) $z_1=z^4$

(*ii*) $z_2=\dfrac{1+z_1}{1-z_1}$

(*iii*) $z_3=z_2{}^2$ or directly by 8: $z_3=\left(\dfrac{1+z^4}{1-z^4}\right)^2$

(*iv*) By 11: $w=\dfrac{z_3-i}{z_3+i}=\dfrac{(1+z^4)^2-i(1-z^4)^2}{(1+z^4)^2+i(1-z^4)^2}$

Miscellaneous mappings

12.	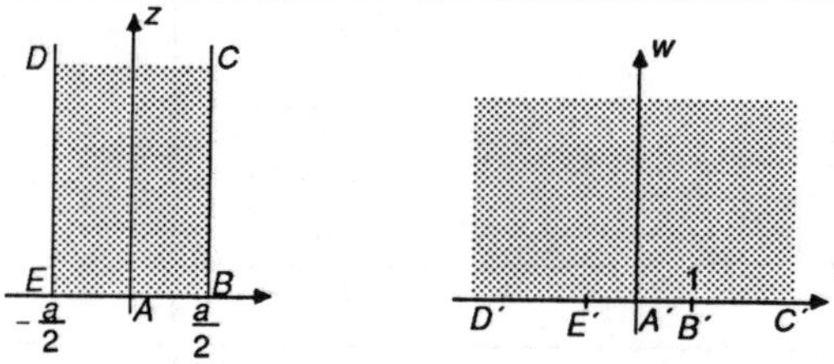	$w = \sin\dfrac{\pi z}{a}$
13.	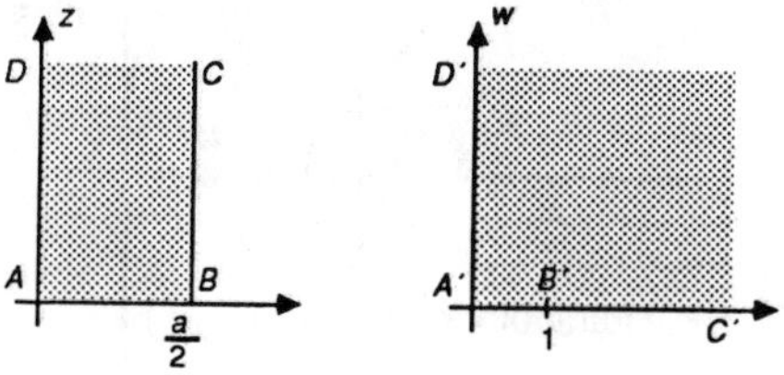	$w = \sin\dfrac{\pi z}{a}$

14.

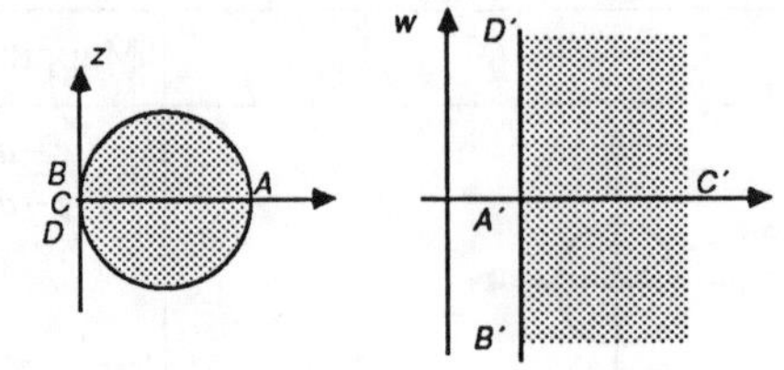

$w = \dfrac{1}{z}$

15.

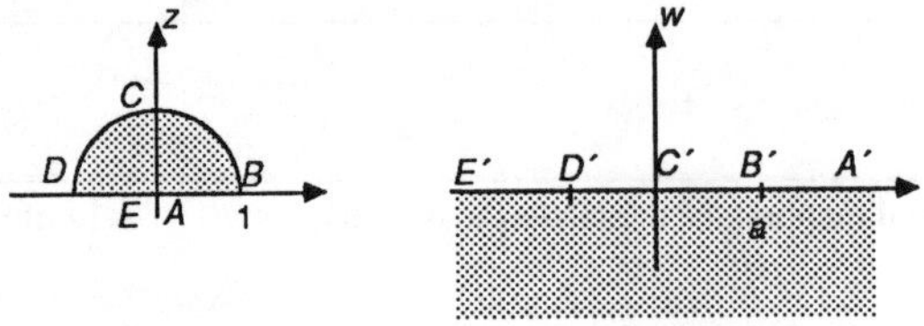

$w = \dfrac{a}{2}\left(z + \dfrac{1}{z}\right)$

16.

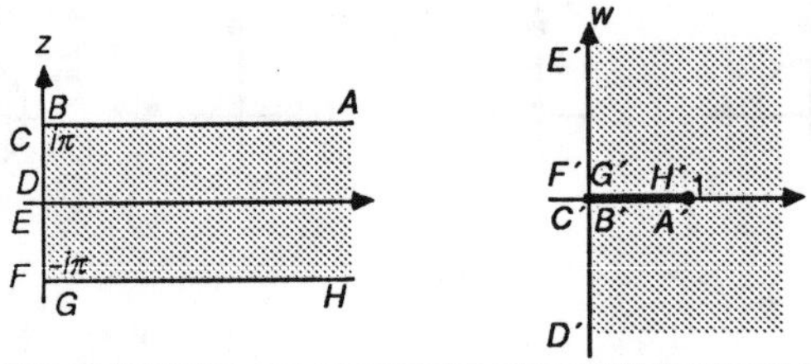

$w = \coth \dfrac{z}{2}$

$= \dfrac{e^z + 1}{e^z - 1}$

17.

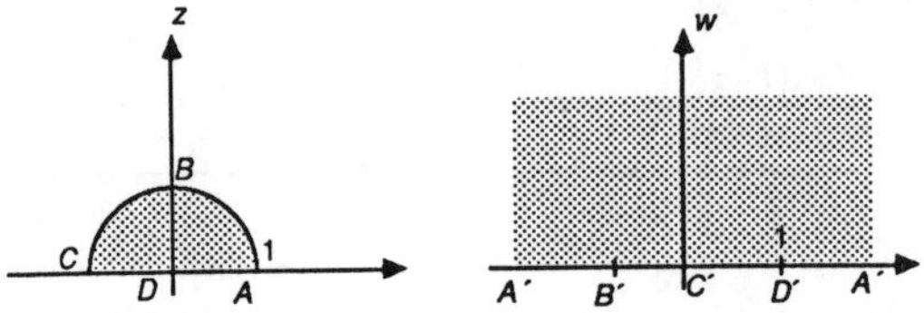

$w = \left(\dfrac{1 + z}{1 - z}\right)^2$

18.

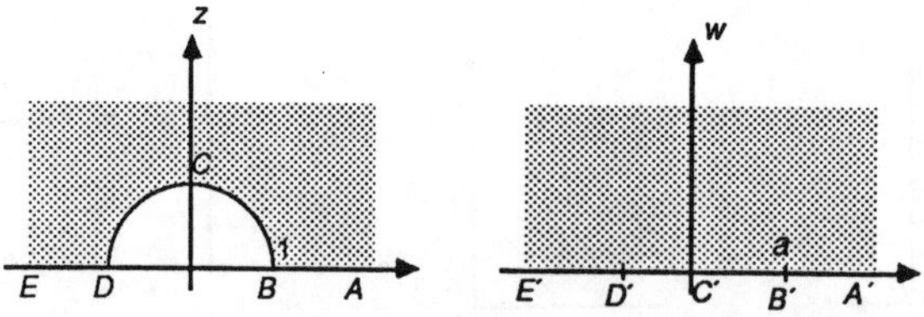

$w = \dfrac{a}{2}\left(z + \dfrac{1}{z}\right)$

19.

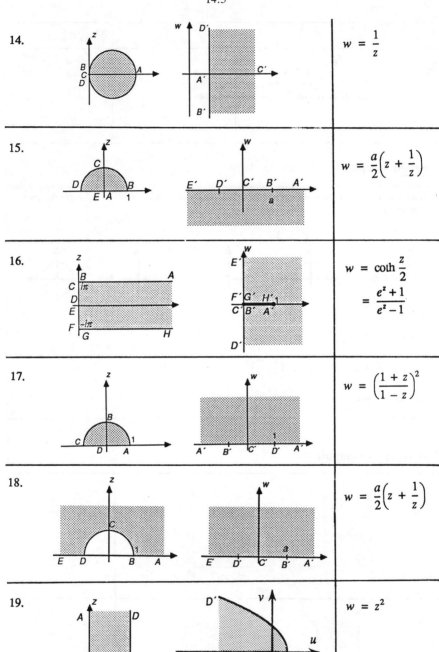

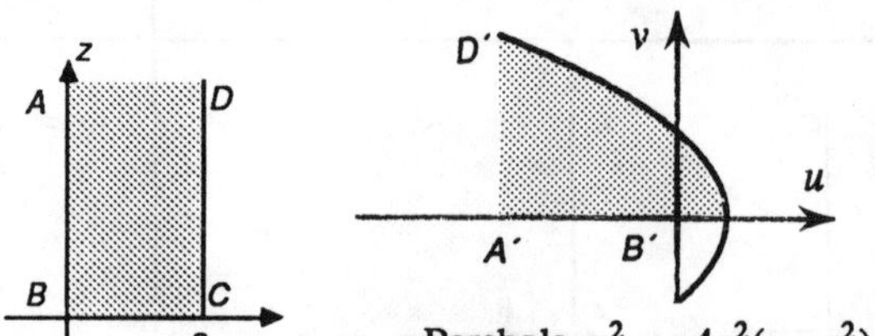

Parabola $v^2 = -4a^2(u - a^2)$

$w = z^2$

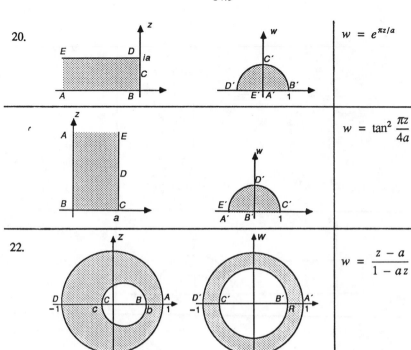

20.

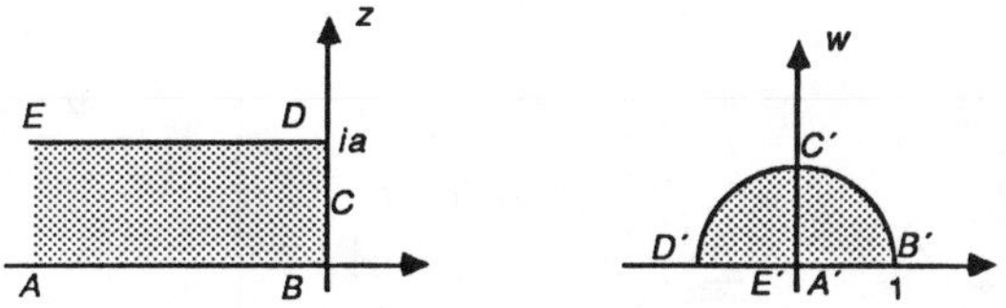

$$w = e^{\pi z/a}$$

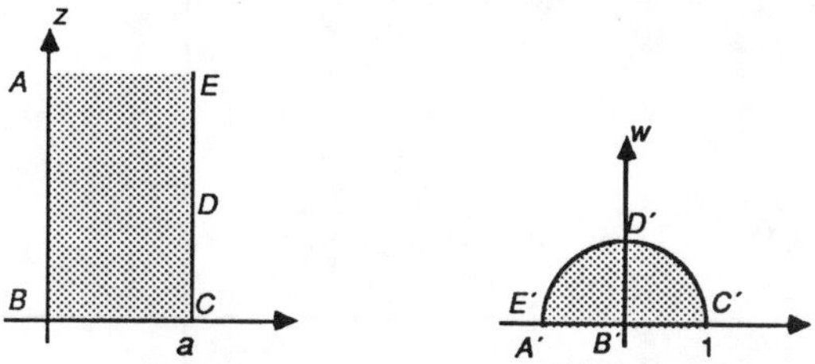

$$w = \tan^2 \frac{\pi z}{4a}$$

22.

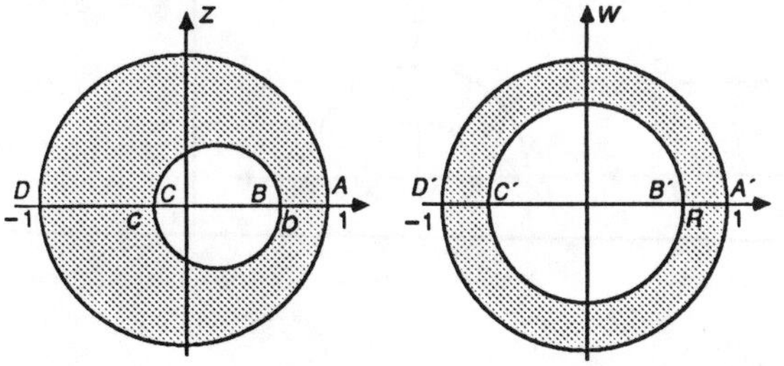

$$w = \frac{z - a}{1 - az}$$

$$a = \frac{1 + bc - \sqrt{(1-b^2)(1-c^2)}}{b+c} \qquad R = \frac{1 - bc - \sqrt{(1-b^2)(1-c^2)}}{b-c}$$

23.

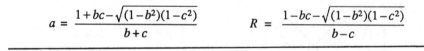

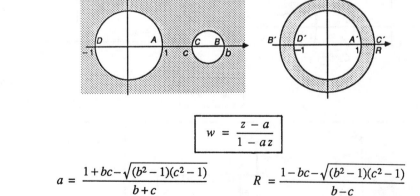

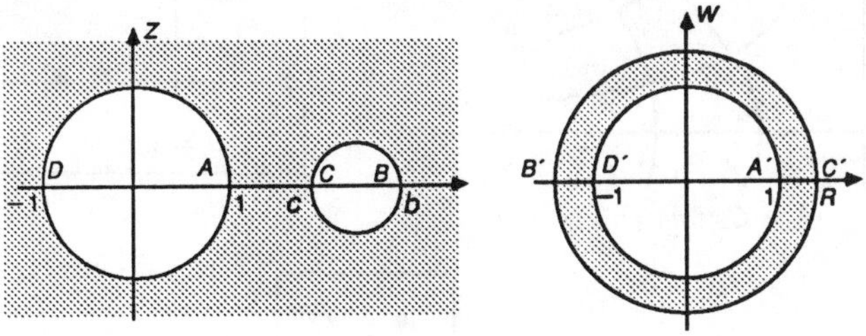

$$\boxed{w = \frac{z - a}{1 - az}}$$

$$a = \frac{1 + bc - \sqrt{(b^2-1)(c^2-1)}}{b+c} \qquad R = \frac{1 - bc - \sqrt{(b^2-1)(c^2-1)}}{b-c}$$

24.

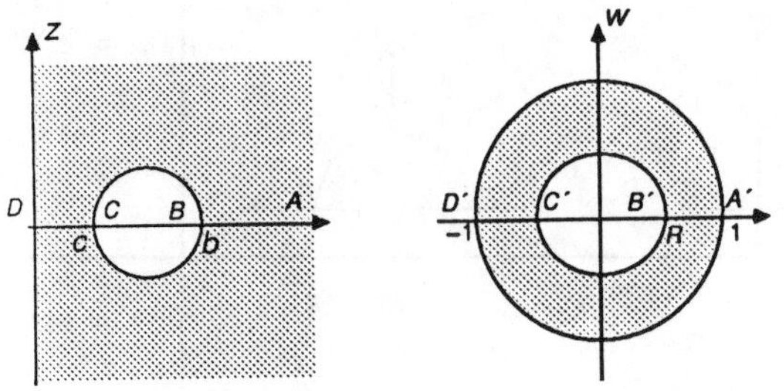

$a = \sqrt{bc}$

$$R = \frac{\sqrt{b}-\sqrt{c}}{\sqrt{b}+\sqrt{c}}$$

$$w = \frac{z-a}{z+a}$$

25.

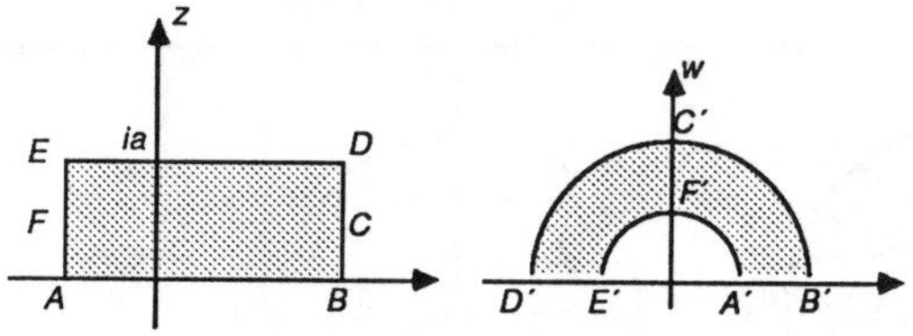

$w = e^{\pi z/a}$

26.

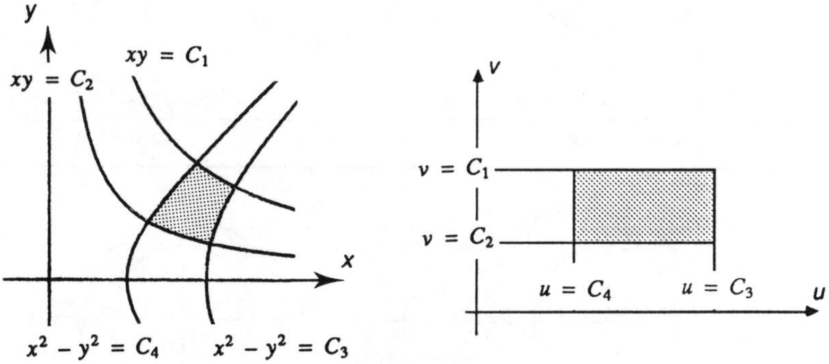

$w = z^2$

27.

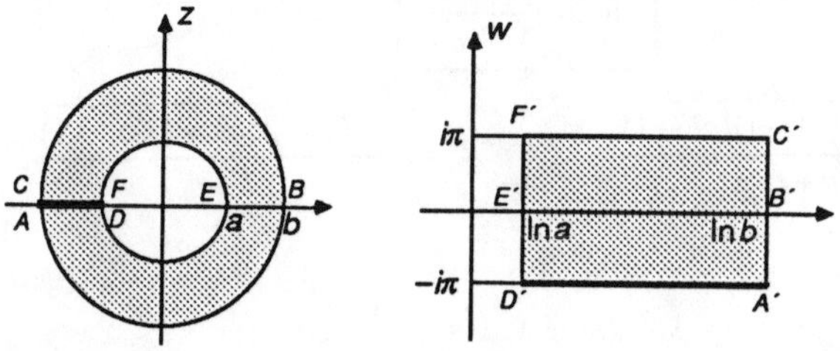

$w = \text{Log}\, z$

28.

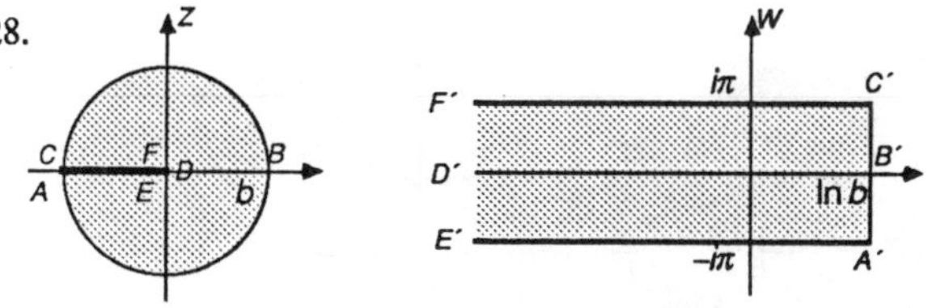

$w = \operatorname{Log} z$

29.

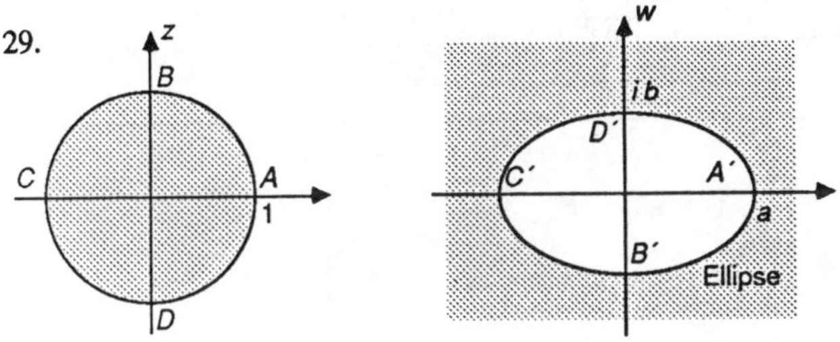

$w = pz + \dfrac{q}{z}$

$p = \dfrac{a-b}{2}$ $\qquad q = \dfrac{a+b}{2}$

30.

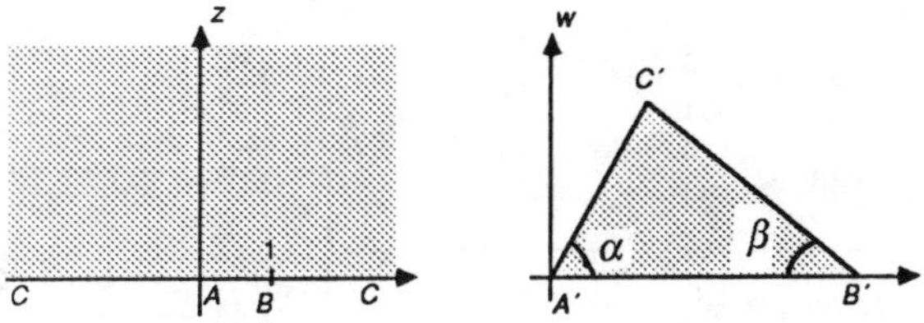

$$w = \int_0^z t^{\alpha/\pi - 1}(1-t)^{\beta/\pi - 1}\,dt$$

31.

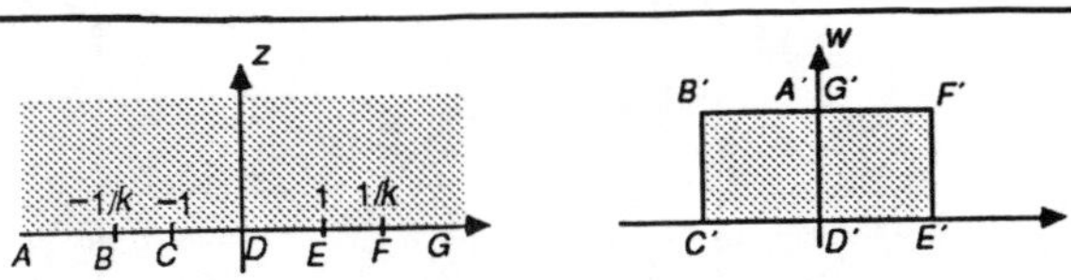

$$w = \int_0^z \frac{dt}{\sqrt{(1-t^2)(1-k^2t^2)}}\quad,\ 0 < k < 1$$

32.

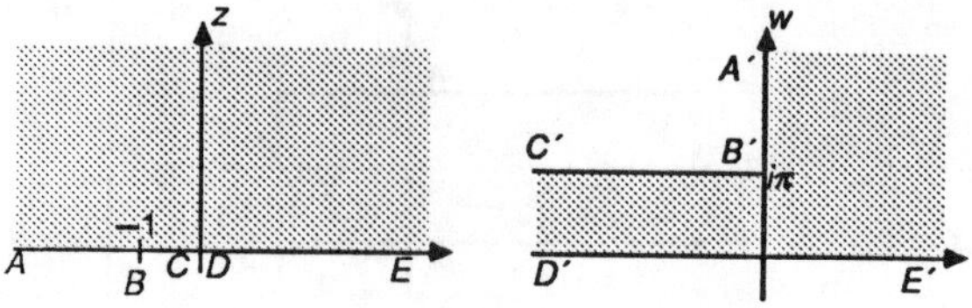

$$w = 2\sqrt{z+1} + \operatorname{Log}\frac{\sqrt{z+1}-1}{\sqrt{z+1}+1}$$

33.

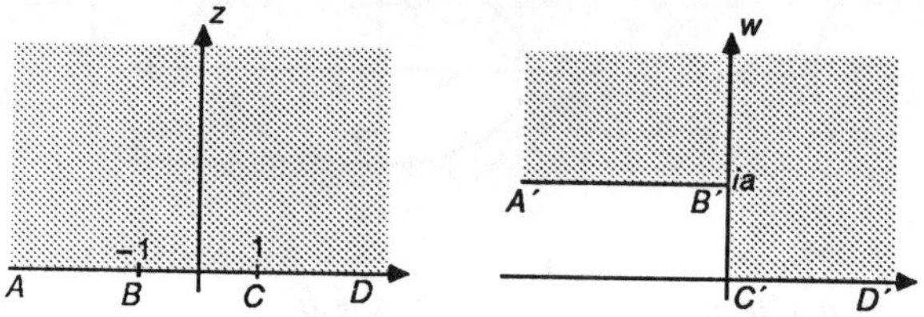

$$w = \frac{a}{\pi}\left(\sqrt{z^2-1} + \cosh^{-1}z\right)$$

34.

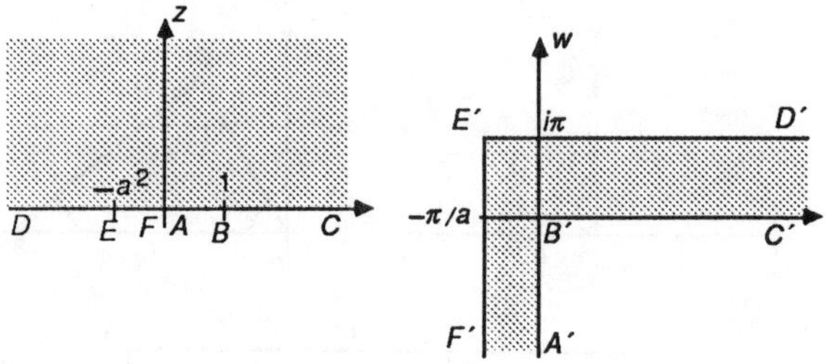

$$w = \frac{i}{a}\operatorname{Log}\frac{1+iat}{1-iat} + \operatorname{Log}\frac{1+t}{1-t} \quad , \quad t = \sqrt{\frac{z-1}{z+a^2}}$$

35.

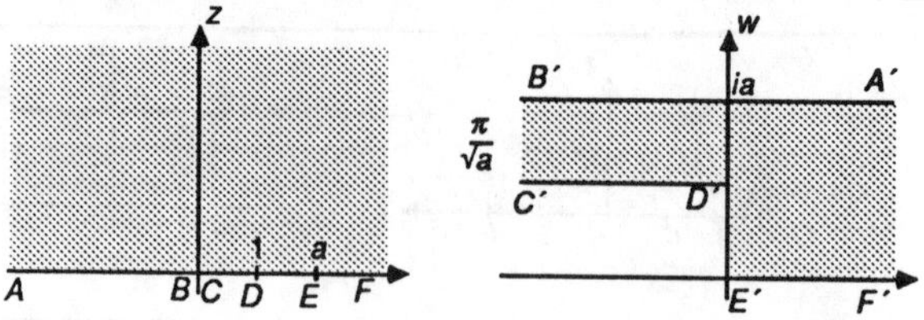

$$w = \cosh^{-1}\left(\frac{2z-a-1}{a-1}\right) - \frac{1}{\sqrt{a}}\cosh^{-1}\left[\frac{(a+1)z-2a}{(a-1)z}\right]$$

15 Optimization

(In this chapter all functions are assumed to be "smooth enough".)

15.1 Calculus of Variations

The calculus of variations treats the problem of finding extrema of *functionals,* i.e. real valued functions having *functions* as "independent variables". Below, *necessary* conditions (the *Euler-Lagrange equation* (15.1), the solutions of which are called *extremals*) are stated for some different kinds of variational problems. *Sufficient* conditions can be formulated (e.g. Weierstrass' theory on strong extrema). However, "common sense" may often be used to establish the sufficiency.

Problem 1 (fixed end points)

Find a function $y=y(x)$ that *minimizes*

$$I(y)=\int_a^b F(x, y, y')dx$$
$$y(a)=\alpha,\ y(b)=\beta$$

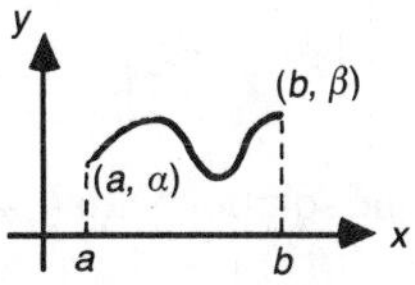

for a given function $F(x, y, y')$.

Necessary condition for solution:

(15.1) $$\frac{\partial F}{\partial y}-\frac{d}{dx}\left(\frac{\partial F}{\partial y'}\right)=0 \Leftrightarrow$$

$$F'_y-F''_{xy'}-y'\,F''_{yy'}-y''\,F''_{y'y'}=0$$

In particular, if $F=F(y, y')$ then (15.1) implies

(15.2) $$F-y'\,F'_{y'}=C \qquad (C \text{ constant})$$

Remark. The equation (15.1) is an ordinary differential equation of 2nd order. Combined with the boundary conditions $y(a)=\alpha$ and $y(b)=\beta$ the problem to be solved is a boundary-value problem.

Example.

Find that curve $y=y(x)$ connecting two points (a, α) and (b, β) which has the property that the area of the surface of revolution, arising as the curve rotates about the x-axis, is minimized.

The functional to be minimized is

$$I(y)=2\pi \int_a^b y\sqrt{1+y'^2}\,dx.$$

The equation (15.2) takes the form

$$y(1+y'^2)^{1/2}-yy'^2(1+y'^2)^{-1/2}=C$$

$$\Rightarrow\ 1+y'^2=c_1^2y^2 \text{ with solution}$$

$$y(x)=\frac{1}{c_1}\cosh c_1(x+c_2).$$

Thus, the extremals are *catenaries.*

Weierstrass-Erdmann corner condition

Assume that the system

$$\begin{cases} \left.\dfrac{\partial F}{\partial y'}\right|_{y'=k_1} = \left.\dfrac{\partial F}{\partial y'}\right|_{y'=k_2} \\ \left.\left(F-y'\dfrac{\partial F}{\partial y'}\right)\right|_{y'=k_1} = \left.\left(F-y'\dfrac{\partial F}{\partial y'}\right)\right|_{y'=k_2} \end{cases}$$

has at least one solution with $k_1 \neq k_2$. Then the curve $y=y(x)$ that minimizes $I(y)$ may have corners.

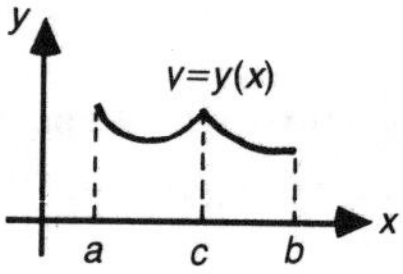

$k_1=y'(c-)$
$k_2=y'(c+)$

Example.

$I(y)=\int_0^2 y'^2(1-y')^2dx;\ y(0)=0,\ y(2)=1.$

In minimizing $I(y)$, note that $I(y)\geq 0$, all y. Furthermore, $I(y)>0$, if $y(x)$ has continuous derivative on $[0, 2]$, but $I(y_0)=0$ for $y_0(x)=\begin{cases} x, & 0<x<1 \\ 1, & 1<x<2 \end{cases}$, which is an extremal in this case.

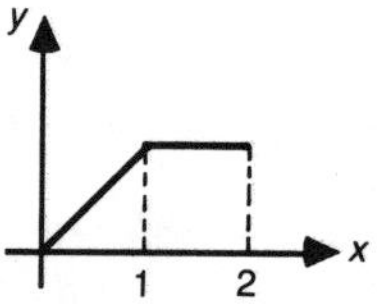

Problem 2 (several functions, fixed end points)

Find functions $y_i=y_i(x)$, $i=1, \ldots, n$, that minimize

$$I(y_1, \ldots, y_n)=\int_a^b F(x, y_1, \ldots, y_n, y_1', \ldots, y_n')dx$$

$$y_i(a)=\alpha_i,\ y_i(b)=\beta_i,\ i=1, \ldots, n$$

Necessary conditions for solution:

$$\frac{\partial F}{\partial y_i}-\frac{d}{dx}\left(\frac{\partial F}{\partial y_i'}\right)=0,\ i=1,\ \ldots,\ n$$

$$\Leftrightarrow F'_{y_i}-F''_{xy_i'}-\sum_{k=1}^{n} y_k'\, F''_{y_k y_i'}-\sum_{k=1}^{n} y_k''\, F''_{y_k' y_i'},\ i=1,\ \ldots,\ n$$

(System of n ODE of 2nd order.)

Problem 3 (Higher derivatives)

Find a function $y=y(x)$ that minimizes

$$I(y)=\int_a^b F(x,\ y,\ y',\ \ldots,\ y^{(n)})\,dx$$

$$y^{(k)}(a)=\alpha_k,\ \ y^{(k)}(b)=\beta_k,\ \ k=0,\ \ldots,\ n-1$$

Necessary condition for solution:

$$\frac{\partial F}{\partial y}-\frac{d}{dx}\left(\frac{\partial F}{\partial y'}\right)+\ldots+(-1)^n\,\frac{d^n}{dx^n}\left(\frac{\partial F}{\partial y^{(n)}}\right)=0$$

Problem 4 (Free boundary values)

Find a function $y=y(x)$ that minimizes

$$I(y)=\int_a^b F(x,\ y,\ y')\,dx$$

No conditions on $y(a)$ and (or) $y(b)$

Necessary conditions for solution:

Condition (15.1) and

$$\frac{\partial F}{\partial y'}\,(x,\ y(x),\ y'(x))=0 \text{ at } x=a \text{ (and } x=b).$$

(*natural boundary conditions*).

Problem 5 (Transversality)

Find a function $y=y(x)$ (and t) that minimizes

$$I(y)=\int_a^t F(x,\ y,\ y')dx$$

$y(a)=\alpha$, $(t,\ y(t))$ has to lie on the given curve $y=g(x)$.

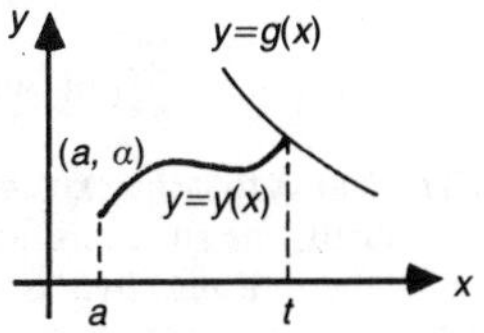

Necessary conditions for solution:

Condition (15.1) and

$$F+(g'-y')\frac{\partial F}{\partial y'}=0 \text{ at } x=t \text{ (\textit{transversality condition})}$$

(Analogous condition if the left end point lies on a given curve.)

Problem 6 (Side condition)

Find a function $y=y(x)$ that minimizes

$$I(y)=\int_a^b F(x, y, y')dx$$

$$y(a)=\alpha,\ y(b)=\beta \text{ and}$$

$$J(y)=\int_a^b G(x, y, y')dx=J_0 \quad \text{(constant)}$$

Necessary conditions for solution:

$y(x)$ has to satisfy the Euler-Lagrange equation for

$$\int_a^b (F+\lambda G)dx,\ \lambda=\text{constant},$$

[or for $\int_a^b G\,dx$ (degenerate case)].

The extremals may be written in the form

$$y=y(x, \lambda, c_1, c_2)$$

Determine λ, c_1 and c_2 from the boundary conditions and the side condition.

Example. (*The classical isoperimetric problem.*)
Find that curve $y=y(x)$ connecting the points A and B and having given length L which has the property that the area between the curve and the x-axis is maximal.

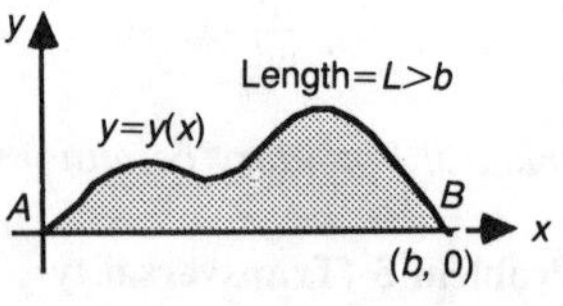

Set $I(y)=\int_0^b y\,dx$, $J(y)=\int_0^b (1+y'^2)^{1/2}dx=L$.

The Euler-Lagrange equation for the functional $\int_0^b [y+\lambda(1+y'^2)^{1/2}]dx$ is

$$1-\lambda\frac{d}{dx}[y'(1+y'^2)^{-1/2}]=0;\quad y''(1+y'^2)^{-3/2}=1/\lambda=\text{constant}.$$

This shows that the extremals have constant curvature and hence are parts of a circle. There is exactly one such through the given points having length L. This result can easily be used to show that *the circle is that curve of given length which encloses the greatest area.*

Variational formulation of boundary value problems

As an example, consider the boundary value problem

(15.3) $$\begin{cases} -[p(x)\,y'(x)]'+q(x)\,y(x)=h(x), & a<x<b \\ y(a)=y(b)=0. \end{cases}$$

In this case, the equation (15.3) is the Euler-Lagrange equation (15.1) of the functional

$$I(y)=\int_a^b (py'^2+qy^2-2\,hy)dx$$

Ritz' method

Often, the Euler-Lagrange differential equation cannot be solved exactly so that approximative methods have to be used. Consider *Problem* 1 above and assume $y(a)=y(b)=0$. [If this condition is not fullfilled, replace $y(x)$ by $y(x)-\alpha-(\beta-\alpha)(x-a)/(b-a)$]. Now, let $y_n=\alpha_1\varphi_1+\ldots+\alpha_n\varphi_n$, where $\varphi_1, \ldots, \varphi_n$ are linear independent functions on $[a, b]$ satisfying $\varphi_k(a)=\varphi_k(b)=0$, all k. Then, for y_n to be an approximate solution of the problem, the parameters α_k have to be chosen such that

(15.4) $$\boxed{\partial I(y_n)/\partial\alpha_k=0,\ k=1, \ldots, n}$$

Example.

$I(y)=\int_0^1 (y'^2-y^2-2xy)dx,\ y(0)=y(1)=0.$

Set $y_2(x)=\alpha_1\varphi_1(x)+\alpha_2\varphi_2(x),\ \varphi_1(x)=x(1-x),\ \varphi_2(x)=x^2(1-x).$

Determining α_1 and α_2 by (15.4) yields the approximate solution

$$y_2(x)=(71x-8x^2-63x^3)/369.$$

The exact solution, determined by (15.1) is

$$y(x)=\frac{\sin x}{\sin 1}-x.$$

The difference $y(x)-y_2(x)$ is of order 10^{-4}.

Problem 7 (Several variables)

Find a function $u=u(x, y)$ that minimizes

$$I(u)=\iint_D F(x, y, u, u_x, u_y)\ dxdy$$

u given on ∂D

Necessary condition for solution:

$$F_u-\frac{\partial}{\partial x}(F_{u_x})-\frac{\partial}{\partial y}(F_{u_y})=0$$

Control systems

Notation:

$\boldsymbol{x}=(x_1, \ldots, x_n)^t\in\boldsymbol{R}^n$

$\boldsymbol{u}=(u_1, \ldots, u_m)^t\in\boldsymbol{R}^m$ (control variable)

$\boldsymbol{f}=(f_1, \ldots, f_n)^t\in\boldsymbol{R}^n$

$\frac{\partial\boldsymbol{f}}{\partial\boldsymbol{x}}=\left(\frac{\partial f_i}{\partial x_j}\right)$, $n\times n-$matrix

Problem

Find $\min \int_{t_0}^{t_1} f_0(t, \boldsymbol{x}(t), \boldsymbol{u}(t))dt$ if

$$(15.5)\qquad \begin{cases} \dot{\boldsymbol{x}}(t)=\boldsymbol{f}(t, \boldsymbol{x}(t), \boldsymbol{u}(t)) \\ \boldsymbol{x}(t_0)=\boldsymbol{x}_0,\ \boldsymbol{x}(t_1)=\boldsymbol{x}_1 \\ \boldsymbol{u}(t)\in\Omega\subset\boldsymbol{R}^m \qquad \text{(admissible controls)} \\ (t_0 \text{ fixed, } t_1 \text{ free or fixed}) \end{cases}$$

Necessary conditions for extremum (Pontryagin's maximum principle):
Introduce the *Hamiltonian function*

$$H(t, \boldsymbol{x}, \boldsymbol{u}, \eta_0, \boldsymbol{\eta})=\eta_0 f_0(t, \boldsymbol{x}, \boldsymbol{u})+\boldsymbol{\eta}\boldsymbol{f}(t, \boldsymbol{x}, \boldsymbol{u}),$$

where η_0 is a scalar and $\boldsymbol{\eta}$ a row vector. If $\boldsymbol{u}^*$ is an optimal control on $[t_0, t_1^*]$ and $\boldsymbol{x}^*$ the corresponding solution of (15.5), then there is a constant $\eta_0\leq 0$ and a row vector function $\boldsymbol{\eta}(t)=(\eta_1(t), \ldots, \eta_n(t))$ with the following properties for $t\in[t_0, t_1^*]$:

(*i*) $\dot{\boldsymbol{\eta}}(t)=-\dfrac{\partial H}{\partial \boldsymbol{x}}=-\eta_0 \dfrac{\partial f_0}{\partial \boldsymbol{x}}(t, \boldsymbol{x}^*(t), \boldsymbol{u}^*(t))-\boldsymbol{\eta}(t)\dfrac{\partial \boldsymbol{f}}{\partial \boldsymbol{x}}(t, \boldsymbol{x}^*(t), \boldsymbol{u}^*(t))$

(*ii*) $(\eta_0, \boldsymbol{\eta}(t))\neq(0, \boldsymbol{0})$

(*iii*) $H(t, \boldsymbol{x}^*(t), \boldsymbol{u}^*(t), \eta_0, \boldsymbol{\eta}(t))= \max\limits_{\boldsymbol{u}\in\Omega} H(t, \boldsymbol{x}^*(t), \boldsymbol{u}, \eta_0, \boldsymbol{\eta}(t))$

(*iv*) If t_1 is free, then t_1^* also satisfies
$$H(t_1^*, \boldsymbol{x}^*(t_1^*), \boldsymbol{u}^*(t_1^*), \eta_0, \boldsymbol{\eta}(t_1^*))=0$$

(*v*) Furthermore, if $\boldsymbol{f}$ and f_0 are independent of t, then
$$H(\boldsymbol{x}^*(t), \boldsymbol{u}^*(t), \eta_0, \boldsymbol{\eta}(t))\equiv\text{constant}$$

Remarks.

1. In most applications $\eta_0\neq 0$, in which case one may take $\eta_0=-1$.
2. The condition $\boldsymbol{x}(t_1)=\boldsymbol{x}_1$ means that all components of the vector $\boldsymbol{x}(t_1)$ are fixed. If instead only some of the components are required to be fixed while the others are free, then the components of $\boldsymbol{\eta}(t_1^*)$ corresponding to the free components of $\boldsymbol{x}(t_1)$ are 0. (This is a transversality condition). In particular, if $\boldsymbol{x}(t_1)$ is free, then $\boldsymbol{\eta}(t_1^*)=\boldsymbol{0}$.

Special case

a. If $f_0\equiv 1$, t_1 free (minimum time problem), then there is $\eta(t)\neq 0$ such that for $t\in[t_0, t_1^*]$,

(*i*) $\dot{\boldsymbol{\eta}}(t)=-\boldsymbol{\eta}(t)\dfrac{\partial \boldsymbol{f}}{\partial x}(t, \boldsymbol{x}^*(t), \boldsymbol{u}^*(t))$

(*ii*) $\boldsymbol{\eta}(t)\boldsymbol{f}(t, \boldsymbol{x}^*(t), \boldsymbol{u}^*(t))=\max\limits_{\boldsymbol{u}\in\Omega} \boldsymbol{\eta}(t)\boldsymbol{f}(t, \boldsymbol{x}^*(t), \boldsymbol{u})$

b. In particular, if (15.5) is linear, i.e. $\dot{\boldsymbol{x}}=A(t)\boldsymbol{x}+B(t)\boldsymbol{u}$, then

(*i*) $\dot{\boldsymbol{\eta}}(t)=-\boldsymbol{\eta}(t)A(t)$

(*ii*) $\boldsymbol{\eta}(t)B(t)\boldsymbol{u}^*(t)= \max\limits_{\boldsymbol{u}\in\Omega} \boldsymbol{\eta}(t)B(t)\boldsymbol{u}$

15.2 Linear Optimization

Notation. $A=\begin{bmatrix} a_{11} \dots a_{1n} \\ \dots \\ a_{m1} \dots a_{mn} \end{bmatrix}$, $\boldsymbol{x}=\begin{bmatrix} x_1 \\ \dots \\ x_n \end{bmatrix}$, $\boldsymbol{x}^t=(x_1, \dots, x_n)$, $\boldsymbol{x}\leqslant\boldsymbol{y} \Leftrightarrow x_i\leqslant y_i$, all i

Linear optimization (linear programming)

Canonical form

(CLP) Find minimum of the *object function*

$$c_1x_1+\dots+c_nx_n$$

with the *constraints*

$$\begin{cases} a_{11}x_1+\dots+a_{1n}x_n=b_1 \\ \dots \\ a_{m1}x_1+\dots+a_{mn}x_n=b_m \\ x_i\geqslant 0, \text{ all } i \end{cases}$$

$\Leftrightarrow$ Find $\min \boldsymbol{c}^t\boldsymbol{x}$ if $\begin{cases} A\boldsymbol{x}=\boldsymbol{b} \\ \boldsymbol{x}\geqslant\boldsymbol{0} \end{cases}$

Standard form

(SLP) Find $\min(c_1x_1+\dots+c_nx_n)$

if $\begin{cases} a_{11}x_1+\dots+a_{1n}x_n\geqslant b_1 \\ \dots \\ a_{m1}x_1+\dots+a_{mn}x_n\geqslant b_m \\ x_i\geqslant 0, \text{ all } i \end{cases}$ $\Leftrightarrow$ Find $\min \boldsymbol{c}^t\boldsymbol{x}$ if $\begin{cases} A\boldsymbol{x}\geqslant\boldsymbol{b} \\ \boldsymbol{x}\geqslant\boldsymbol{0} \end{cases}$

An (SLP) can be transformed to the following (CLP) by introducing *slack variables* $s_1, \dots, s_m$ defined by $\boldsymbol{s}=A\boldsymbol{x}-\boldsymbol{b}$:

Find $\min(c_1x_1+\dots+c_nx_n)$

if $\begin{cases} a_{11}x_1+\dots+a_{1n}x_n-s_1=b_1 \\ \dots \\ a_{m1}x_1+\dots+a_{mn}x_n-s_m=b_m \\ x_i\geqslant 0, s_i\geqslant 0, \text{ all } i \end{cases}$ $\Leftrightarrow$ Find $\min \boldsymbol{c}^t\boldsymbol{x}$ if $\begin{cases} A\boldsymbol{x}-\boldsymbol{s}=\boldsymbol{b} \\ \boldsymbol{x}\geqslant\boldsymbol{0},\ \boldsymbol{s}\geqslant\boldsymbol{0} \end{cases}$

The dual problems

The *dual* problems (D) corresponding to the above *primal* (P) LP-problems are:

$(\text{CLP})_D$ Find $\max \boldsymbol{b}^t\boldsymbol{u}$ if $\begin{cases} A^t\boldsymbol{u}\leqslant\boldsymbol{c} \\ \text{No constraints on } \boldsymbol{u} \end{cases}$

$(\text{SLP})_D$ Find $\max \boldsymbol{b}^t\boldsymbol{u}$ if $\begin{cases} A^t\boldsymbol{u}\leqslant\boldsymbol{c} \\ \boldsymbol{u}\geqslant\boldsymbol{0} \end{cases}$

Optimality criterions

A *feasible solution* of an LP-problem is a vector $\boldsymbol{x}$ (or $\boldsymbol{u}$) that satisfies the constraints. An *optimal solution* $\hat{\boldsymbol{x}}$ (or $\hat{\boldsymbol{u}}$) is a feasible solution that minimizes (maximizes) the object function.

1. If $\boldsymbol{x}$ and $\boldsymbol{u}$ are feasible solutions of (P) and (D) respectively, then

   (*i*) $\boldsymbol{b}^t\boldsymbol{u} \leq \boldsymbol{c}^t\boldsymbol{x}$
   (*ii*) $\boldsymbol{b}^t\boldsymbol{u} = \boldsymbol{c}^t\boldsymbol{x} \Rightarrow \boldsymbol{x}$ and $\boldsymbol{u}$ are optimal.

2. (*Duality theorem.*)

   (*i*) If both (P) and (D) have feasible solutions, then there exist finite optimal solutions $\hat{\boldsymbol{x}}$ and $\hat{\boldsymbol{u}}$, and $\boldsymbol{c}^t\hat{\boldsymbol{x}} = \boldsymbol{b}^t\hat{\boldsymbol{u}}$.

   (*ii*) (P) [or (D)] has no feasible solution $\Leftrightarrow$ (D) [or (P)] has no finite optimal solution.

3. (*The complementarity theorem*)
   Assume that $\hat{\boldsymbol{x}}$ and $\hat{\boldsymbol{u}}$ are feasible. Then the vectors $\hat{\boldsymbol{x}}$ and $\hat{\boldsymbol{u}}$ are optimal solutions of (P) and (D) respectively if and only if (all i and j)

$$\begin{cases} \sum_{j=1}^{n} a_{ij}\hat{x}_j > b_i \Rightarrow \hat{u}_i = 0 \\ \sum_{i=1}^{m} a_{ij}\hat{u}_i < c_j \Rightarrow \hat{x}_j = 0 \end{cases} \quad \text{or equivalently} \quad \begin{cases} \hat{u}_i > 0 \Rightarrow \sum_{j=1}^{n} a_{ij}\hat{x}_j = b_i \\ \hat{x}_j > 0 \Rightarrow \sum_{i=1}^{m} a_{ij}\hat{u}_i = c_j \end{cases}$$

The Simplex method

Basic solutions

Assume that the system $A\boldsymbol{x}=\boldsymbol{b}$ has been transformed to echelon form by Gaussian elimination. A solution in which all free variables (i.e. variables not corresponding to pivots) are zero, is called a *basic solution*. [Referring to the example of sec. 4.3, $x=2$, $y=2$, $z=0$, $u=0$ (u free) is one basic solution of that system.] The number of basic solutions of $A\boldsymbol{x}=\boldsymbol{b}$ is at most $\binom{n}{m}$, where n=number of variables, m=rank of A. A *feasible basic solution* is a solution with all $x_i \geq 0$.

If an LP-problem in canonical form has an optimal solution, then at least one feasible basic solution is optimal.

The Simplex algorithm

Consider an LP-problem of canonical form,

$$\text{(LP 1)} \qquad \min(c_1x_1+\ldots+c_nx_n) = \min \boldsymbol{c}^t\boldsymbol{x}$$

$$\begin{cases} a_{11}x_1+\ldots+a_{1n}x_n = b_1 \\ \ldots \\ a_{m1}x_1+\ldots+a_{mn}x_n = b_m \\ x_i \geq 0,\ i=1, \ldots, n \end{cases} \Leftrightarrow \begin{cases} A\boldsymbol{x} = \boldsymbol{b} \\ \boldsymbol{x} \geq \boldsymbol{0} \end{cases}$$

and assume that rank $(A)=m$ and $\boldsymbol{b} \geq \boldsymbol{0}$ (if necessary, change the sign of some equations). The *simplex algorithm* is a method of finding an optimal basic solution of the problem.

To begin with, it may be difficult to find a feasible basic solution. Therefore, *artificial variables* $y_1, \ldots, y_m$ are introduced as follows: The original problem is replaced by the new LP-problem

(LP 2) $\min(c_1x_1+\ldots+c_nx_n+y_1+\ldots+y_m)$

$$\begin{cases} a_{11}x_1+\ldots+a_{1n}x_n+y_1=b_1 \\ \ldots \\ a_{m1}x_1+\ldots+a_{mn}x_n+y_m=b_m \\ x_i\geqslant 0,\ y_k\geqslant 0,\ i=1,\ \ldots,\ n,\ k=1,\ \ldots,\ m \end{cases} \Leftrightarrow \begin{cases} \boldsymbol{Ax+y=b} \\ \boldsymbol{x\geqslant 0},\ \boldsymbol{y\geqslant 0} \end{cases}$$

If (LP 2) admits an optimal basic solution with $\boldsymbol{y=0}$, then the corresponding vector $\boldsymbol{x}$ is an optimal basic solution of (LP 1). Note that $x_i=0$, $i=1$, ..., n and $y_k=b_k\geqslant 0$, $k=1$, ..., m, is a feasible basic solution of (LP 2). The simplex algorithm now consists of two phases. (In order to get all $y_i=0$, phase 1 obtains a solution of the problem $\min(y_1+\ldots+y_m)$ if $\boldsymbol{Ax+y=b}$, $\boldsymbol{x\geqslant 0}$, $\boldsymbol{y\geqslant 0}$):

Phase 1 (Finding a feasible basic solution of (LP 1))

Initial tableau

	x_1 ... x_j ... x_n	y_1 ... y_m	$\boldsymbol{b}$
	a_{11} ... a_{1n}	1 0 ... 0	b_1
	a_{21} ... a_{2n}	0 1 ... 0	b_2
i	... (a_{ij}) ...	...	...
	a_{m1} ... a_{mn}	0 ... 01	b_m
	c_1 ... c_n	0 ... 0	0
	d_1 ... d_j ... d_n	0 ... 0	β
	↑	* * ... *	

Box scheme

x_1 ... x_n	y_1 ... y_m	$\boldsymbol{b}$
(1)	(2)	(3)
(4)	(5)	(6)
(7)	(8)	(9)

$$d_j=-\sum_{i=1}^{m} a_{ij} \quad \beta=-\sum_{i=1}^{m} b_i$$

The stars are below the basic variables (unit vectors). In box (2) is an identity matrix.

Change of basic variables. (Cf. the example below)

1. Look for the smallest number in box (7), say d_j. Mark with an arrow. This means that x_j will enter the basic solution.
2. Look for that number in box (1) below x_j, say a_{ij}, which is positive and for which the quotient of the $\boldsymbol{b}$-element of the ith row and a_{ij} (i.e. b_i/a_{ij}) is minimal. Mark with a ring.
3. Divide each element of the ith row by a_{ij}.
4. Pivot (Gauss eliminate) so that all other numbers of the jth column in the boxes (1), (4), (7) will be 0. Move the star from below y_i to below x_j.
5. Repeat the above procedure until all the basic variables are components of the $\boldsymbol{x}$-vector (i.e. all stars below box (7). Check that the number in box (9) is zero.

Example (Phase 1)

$$\min(x_1+x_2+x_3)$$
$$\begin{cases} x_1-x_2+x_3=1 \\ -x_1+2x_2+2x_3=3 \\ x_1,\ x_2,\ x_3\geqslant 0 \end{cases}$$

Tableau 1.

x_1	x_2	x_3	y_1	y_2	b
1	−1	(1)	1	0	1
−1	2	2	0	1	3
1	1	1	0	0	0
0	−1	−3	0	0	−4
		↑	*	*	

(−2) (−) (3) $\left(\text{Note, } \frac{1}{1}<\frac{3}{2}\right)$

Tableau 2.

x_1	x_2	x_3	y_1	y_2	b
1	−1	1	1	0	1
−3	(4)	0	−2	1	1
0	2	0	−1	0	−1
3	−4	0	3	0	−1
	↑	*		*	

(1/4) (+) (−2) (4)

Tableau 3.

x_1	x_2	x_3	y_1	y_2	b
1/4	0	1	1/2	1/4	5/4
−3/4	1	0	−1/2	1/4	1/4
3/2	0	0	0	−1/2	−3/2
0	0	0	1	1	0
	*	*			

Since there is no negative number in box (7), Phase 1 is now finished. A feasible basic solution of (LP 1) is $x_1=0$, $x_2=1/4$, $x_3=5/4$ (look at box (3)). This solution is in fact *optimal* because there is no negative number in box (4). Also the *dual solution* may be obtained by tableau 3. Changing the signs in box (5), the dual solution is $u_1=0$, $u_2=1/2$. Also changing the sign in box (6), this is the value of the object function.

Phase 2 (Finding an optimal basic solution of (LP 1))

(If there is a negative number in box (4)):

By phase 1 a feasible basic solution of the original problem is obtained. But is it optimal? The *optimum criterion* is that alla numbers in box (4), the so-called *reduced prices,* are $\geqslant 0$. If this is not the case, continue the procedure above until all elements in box (4) are $\geqslant 0$.

Remark. In phase 2 the boxes (2), (5), (7), (8), (9) may be cancelled if only the solution of the primal problem is required.

The linear transportation problem

Goods are transported from m *depots* ($i=1, \dots, m$) to n *consumers* ($j=1, \dots, n$). Assume that

(i) the transportation cost of one unit from depot i to consumer j is c_{ij},
(ii) b_i=quantity of goods available at depot i
a_j=demand of consumer j
(iii) $\sum_{i=1}^{m} b_i = \sum_{j=1}^{n} a_j$

The transportation problem

Find $\min \sum_{i,j} c_{ij} x_{ij}$ if

$$\begin{cases} \sum_{i=1}^{m} x_{ij} = a_j, \ j=1, \dots, n \\ \sum_{j=1}^{n} x_{ij} = b_i, \ i=1, \dots, m \\ x_{ij} \geq 0 \end{cases}$$

For the solution the simplex algorithm may be used or, alternatively, the more efficient *transportation algorithm.*

15.3 Non-linear Optimization

(Cf. methods in sec. 10.5)

Problem

(NLP) Find $\min f(x_1, \dots, x_n)$ if $\Leftrightarrow$ Find $\min f(\boldsymbol{x})$ if $\boldsymbol{g}(x) \leq \boldsymbol{0}$

$$D: \begin{cases} g_1(x_1, \dots, x_n) \leq 0 \\ \dots \\ g_m(x_1, \dots, x_n) \leq 0 \end{cases} \qquad (15.6)$$

Remark. A constraint of the form $g_i=0$ may be replaced by $g_i \leq 0$ and $-g_i \leq 0$.

Optimum conditions

The *Lagrange function* $L(\boldsymbol{x}, \boldsymbol{u})$ is defined by

$$L(\boldsymbol{x}, \boldsymbol{u}) = f(\boldsymbol{x}) + \boldsymbol{u}^t \boldsymbol{g}(\boldsymbol{x}) = f(x_1, \dots, x_n) + \sum_{i=1}^{m} u_i g_i(x_1, \dots, x_n)$$

(The u_i are called *Lagrange multipliers* or *dual variables.*)

Necessary condition for optimum (the Kuhn-Tucker theorem)

Assuming a *convex** (admissible) domain:

If $\hat{\boldsymbol{x}}$ is optimal of (NLP), then there exists a vector $\hat{\boldsymbol{u}} \geq \boldsymbol{0}$ such that

$$\begin{cases} \dfrac{\partial L}{\partial x_i} = 0 \text{ at } (\hat{\boldsymbol{x}}, \hat{\boldsymbol{u}}), \ i=1, \dots, n \\ \hat{u}_i g_i(\hat{\boldsymbol{x}}) = 0, \ i=1, \dots, m \end{cases} \qquad (15.7)$$

Vectors $\boldsymbol{x}$ satisfying (15.6) and (15.7) are called KT-points.

* The result is valid also for more general domains satisfying "constraint qualifications", i.e. all "normal" domains.

Convex functions

A function f of n variables is *convex* if

$$f(\lambda \boldsymbol{x}+(1-\lambda)\boldsymbol{y}) \leqslant \lambda f(\boldsymbol{x})+(1-\lambda)f(\boldsymbol{y}), \quad 0 \leqslant \lambda \leqslant 1$$

The function $f(\boldsymbol{x})$ is convex in a domain $D \Leftrightarrow$ the *Hessian matrix* $\left(\dfrac{\partial f}{\partial x_i \partial x_j}\right)$

is positive semi-definite (cf. sec. 4.6) in D.

If $f, g_1, \ldots, g_m$ are convex, then every *KT*-point is optimal.

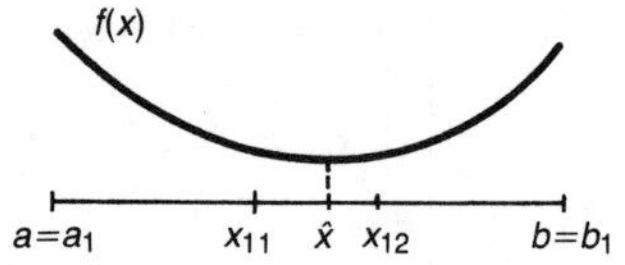

Algorithms for finding extrema

The golden ratio method (one variable)

Algorithm for finding the minimum point $\hat{x}$ of a (convex) function $f(x)$ in the interval $[a, b]$. Below, a sequence of intervals $[a_n, b_n]$ is constructed recursively so that they form a nest of intervals contracting to the minimum point $\hat{x}$:

Let $r=(\sqrt{5}-1)/2$.

Step 1. Set $a_1=a,\ b_1=b,\ x_{11}=a_1+(1-r)(b_1-a_1),\ x_{12}=a_1+r(b_1-a_1)$

Step n.

Set	If $f(x_{n2})>f(x_{n1})$	If $f(x_{n2})\leqslant f(x_{n1})$
$a_{n+1}=$	a_n	x_{n1}
$b_{n+1}=$	x_{n2}	b_n
$x_{n+1,1}=$	$a_n+(1-r)(x_{n2}-a_n)$	x_{n2}
$x_{n+1,2}=$	x_{n1}	$b_n-(1-r)(b_n-x_{n1})$

Gradient method (several variables)

Iterative method of finding the optimum point $\hat{\boldsymbol{x}}$.

1. of (NLP) *without constraints*:

> 1. Starting point $\boldsymbol{x}_0$.
> 2. $\boldsymbol{x}_{k+1}=\boldsymbol{x}_k+t\,\boldsymbol{v}_k$, t realizing $\min f(\boldsymbol{x}_k+t\,\boldsymbol{v}_k)$, where the direction $\boldsymbol{v}_k$ is chosen as
>
>   (*a*) $\boldsymbol{v}_k=-\text{grad}\, f(\boldsymbol{x}_k)$ or
>
>   (*b*) $\boldsymbol{v}_k=Hf(\boldsymbol{x}_k)^{-1}$ grad $f(\boldsymbol{x}_k)$, where Hf is the Hessian matrix of f.
>
> (To find the minimum, use e.g. the golden ratio algorithm.)

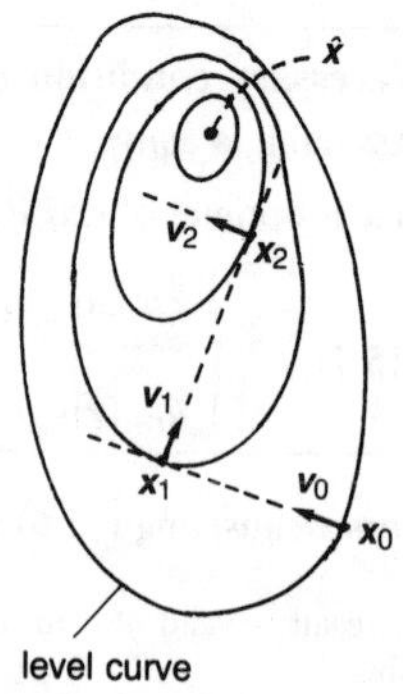

Note that if f is quadratic, i.e. $f(\boldsymbol{x})=\boldsymbol{x}^t A\boldsymbol{x}+\boldsymbol{b}^t\boldsymbol{x}+c$, then grad $f(\boldsymbol{x})=2A\boldsymbol{x}+\boldsymbol{b}$ (A symmetric).

Other *descent methods* may be used, e.g. cyclic search in the coordinate directions.

2. of (NLP) *with constraints*:
Transform problem (15.6) to an (NLP) without constraints, using a *penalty function* $R(\boldsymbol{x})$:

If $R(\boldsymbol{x})=\begin{cases} 0, & \boldsymbol{x}\in D \\ \infty, & \boldsymbol{x}\notin D \end{cases}$, then (15.6) is equivalent to:

Find $\min(f(\boldsymbol{x})+R(\boldsymbol{x}))$ without constraints. In practice $R(\boldsymbol{x})$ is approximated by $R_\lambda(\boldsymbol{x})[R_\lambda(\boldsymbol{x}) \to R(\boldsymbol{x})$ as $\lambda \to \infty]$, e.g. $R_\lambda(\boldsymbol{x})=\sum_i \exp(\lambda_i g_i(\boldsymbol{x}))$.

(Key-word: SUMT = Sequential Unconstrained Minimization Techniques)

15.4 Dynamic Optimization

$t=N, N-1, \ldots, 1, 0$ *recursive variable*, $x(t)$=*state variable*, $u(t)$=*decision variable* (x and u may be vectors).

Discrete deterministic dynamic programming

Problem

Find $\min_{x,u}[f_0(x(0))+\sum_{t=1}^{N} f_t(x(t), u(t))]$

if $\begin{cases} x(t)\in X(t), \ 0\le t\le N \\ u(t)\in U(t, x(t)), \ 1\le t\le N \\ x(t-1)=T_t(x(t), u(t)), \ 1\le t\le N \\ (x(N) \text{ given}) \end{cases}$

A *policy* s_N consists of values of $x(N)$ and $u(N), \ldots, u(1)$ with $x(t)\in X(t)$ and $u(t)\in U(t, x(t))$, $t=1, \ldots, n$. If the value of the above object function (obtained by s_N) is denoted by $f(s_N)$, then an *optimal policy* $\hat{s}_N$ satisfies $f(\hat{s}_N)\le f(s_N)$ for all admissible s_N.

Bellman's optimality principle

An optimal policy $\hat{s}_N$ has the property that $\hat{s}_k$ is an optimal policy from level $t=k$ to $t=0$ with initial value $x(k)$.

Algorithm for dynamic programming

1. Let $\hat{f}_0(x(0))=f_0(x(0))$ for each $x(0)\in X(0)$
2. Recursively for increasing $t=1, \ldots, N$, find for each $x(t)\in X(t)$,
$$\hat{f}_t(x(t))=\min_u[f_t(x(t), u(t))+\hat{f}_{t-1}(x(t-1))],$$
where $x(t-1)=T_t(x(t), u(t))$ and $u(t)\in U(t, x(t))$.
3. For each t and $x(t)$, store one value of $u(t)$, denoted $\hat{u}_t(x(t))$, which realizes $\hat{f}_t(x(t))$.
4. Set (if $x(N)$ is not given) $\hat{f}_N=\min_{x(N)\in X(N)} \hat{f}_N(x(N))$, realized by $\hat{x}(N)$.

Then $\hat{f}=\hat{f}_N$ is the minimal value of the object function and an optimal policy $\hat{s}_N$ is defined by
$$x(N)=\hat{x}(N),\ u(t)=\hat{u}_t(x(t)),\ x(t-1)=T_t(x(t), u(t)),\ t=N, N-1, \ldots, 1.$$

Stochastic dynamic programming

Given:

(*i*) *Object function*: $w_0(x(0))+\sum_{t=1}^{N} w_t(x(t), u(t))$

with state variable $x(t)$ and decision variable $u(t)=u(t, x(t))$

(*ii*) *probabilities*: For each $u(t)$, $p_{jk}(t, u(t, j))=P[x(t-1)=k|x(t)=j$, decision $u(t, j)]$

For a *strategy* z, let $f(z)$ be the value of the object function and let $Ef(z)$ be the *expected value of the object function.*

Problem

Find an optimal strategy $\hat{z}$ that minimizes the expected value of the object function, i.e.

$$Ef(\hat{z})\leq Ef(z)$$

for all strategies z.

Algorithm for stochastic dynamic programming

1. For each state value $x(1)=j$, find

$$\hat{f}_1(j)=\min_{u(1,j)}[w_1(j, u(1, j))+\sum_{k=1}^{J} p_{jk}(1, u(1, j))w_0(k)]$$

2. Recursively for $t=1, \ldots, N$, find for each j,

$$\hat{f}_t(j)=\min_{u(t,j)}[w_t(j, u(t, j))+\sum_{k=1}^{J} p_{jk}(t, u(t, j))\hat{f}_{t-1}(k)]$$

3. The minimum realization values of $u(t, j)$ are denoted $\hat{u}_t(j)$ and they constitute the optimal strategy $\hat{z}$.

16 Numerical Analysis and Programming

16.1 Approximations and Errors

Errors

Errors in numerical computations are due to

a) errors in *input data,*
b) *round-off errors,* due to the use of a finite number of digits,
c) *truncation errors,* due to approximations,
d) *mistakes* in numerical computations.

Let a_0 be an approximation to a.

The absolute error ε_a of a_0 is $\varepsilon_a = a_0 - a$.

The relative error of a_0 is $\dfrac{\varepsilon_a}{a} \approx \dfrac{\varepsilon_a}{a_0}$.

$\varepsilon_{a+b} = \varepsilon_a + \varepsilon_b$	$\varepsilon_{a-b} = \varepsilon_a - \varepsilon_b$
$\dfrac{\varepsilon_{ab}}{ab} \approx \dfrac{\varepsilon_a}{a} + \dfrac{\varepsilon_b}{b}$	$\dfrac{\varepsilon_{a/b}}{a/b} \approx \dfrac{\varepsilon_a}{a} - \dfrac{\varepsilon_b}{b}$
$\varepsilon_{f(a)} \approx f'(a_0)\,\varepsilon_a$	$\varepsilon_{f(a,\,b)} \approx \varepsilon_a\, f_a'(a_0, b_0) + \varepsilon_b\, f_b'(a_0, b_0)$

If $|\varepsilon_a| \leq \delta_a$ and $|\varepsilon_b| \leq \delta_b$ then

$$|\varepsilon_{a+b}| \leq \delta_a + \delta_b \qquad |\varepsilon_{a-b}| \leq \delta_a + \delta_b$$

$$\left|\frac{\varepsilon_{ab}}{ab}\right| \lesssim \frac{\delta_a}{|a|} + \frac{\delta_b}{|b|} \qquad \left|\frac{\varepsilon_{a/b}}{a/b}\right| \lesssim \frac{\delta_a}{|a|} + \frac{\delta_b}{|b|}$$

More general let $x_1^0, \ldots, x_n^0$ be approximations to $x_1, x_2, \ldots x_n$ and let $f = f(x_1, \ldots, x_n)$. Then with $\varepsilon_i = x_i^0 - x_i$

$$\varepsilon_f \approx \varepsilon_1 f'_{x_1}(x_1^0, \ldots, x_n^0) + \ldots + \varepsilon_n f'_{x_n}(x_1^0, \ldots, x_n^0)$$

In the special case $f = x_1^{k_1} \ldots x_n^{k_n}$

$$r_f \approx k_1 r_1 + \ldots + k_n r_n,$$

where $r_f, r_1, \ldots, r_n$ are the relative errors of $f, x_1, \ldots, x_n$.

Convergence of order k

Let $\{a_n\}_1^\infty$ be a sequence of approximations to a with

$$\lim_{n\to\infty} a_n = a$$

The convergence is *of order k* if

$$|a_{n+1}-a|<A|a_n-a|^k,\ n>N,\ A \text{ constant.}$$

For $k=1$ the convergence is *linear,* for $k=2$ it is *quadratic.*

16.2 Numerical Solution of Equations

(Algorithms for finding extrema of functions, see sec. 15.3)

Solving an equation by bisection

Let the equation $F(X)=0$ have a root in the interval $A<X<B$. A Basic computer-program according to the following flow charts determines this root with an error $\pm E$. It is assumed that $F(A)$ and $F(B)$ have different signs.

A program may be tested using the equation $x^3-2x-5=0$, which has the root 2.094551482 in the interval $1<x<3$.

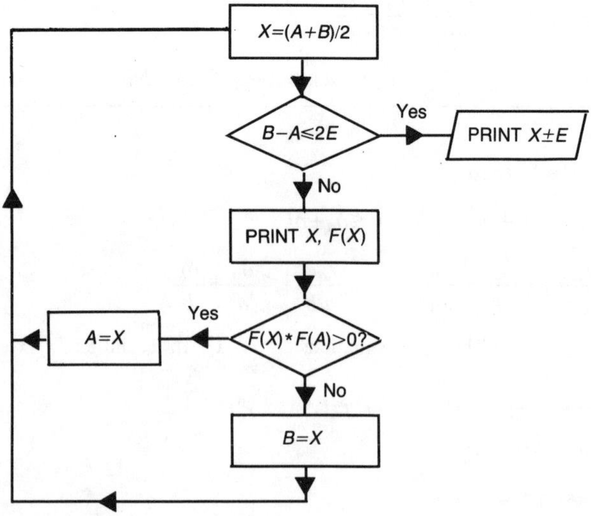

Iteration

$f(x)=0 \Leftrightarrow x=\phi(x)$ $\qquad$ $x_{n+1}=\phi(x_n)$

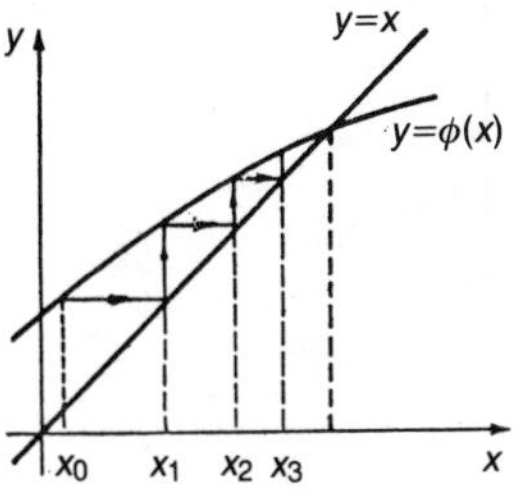

Convergence

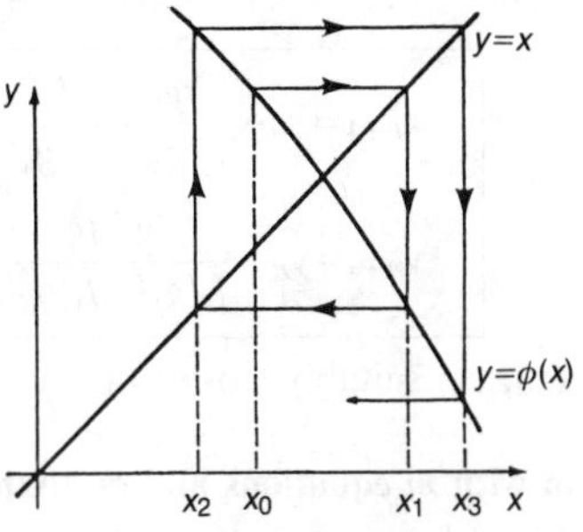

Divergence

The iteration process converges if $|\phi'(x)| \leqslant k < 1$. If $|\phi'(x)| \leqslant k$ in a neighbourhood of the root a, then

$$|x_{n+1}-a| \leqslant \frac{k}{1-k}\,|x_{n+1}-x_n|$$

The secant method (Regula falsi)

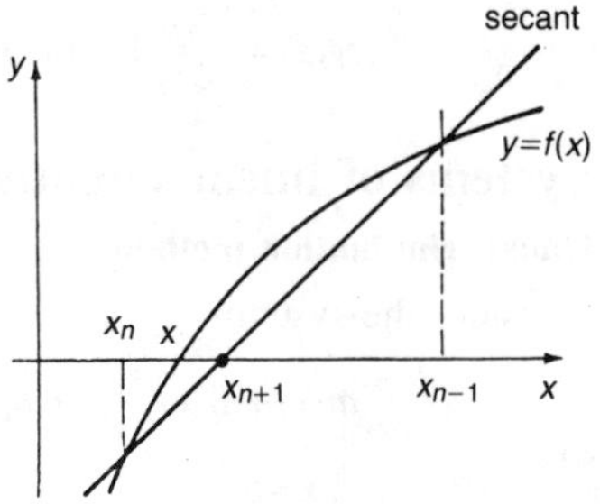

$$x_{n+1}=x_n-f(x_n)\cdot\frac{x_n-x_{n-1}}{f(x_n)-f(x_{n-1})}$$

The convergence is of order approximately 1.6. Choose x_0 and x_1 close to x, not nessessarily on the same side.

The Newton-Raphson method

1. One single equation

$f(x)=0$

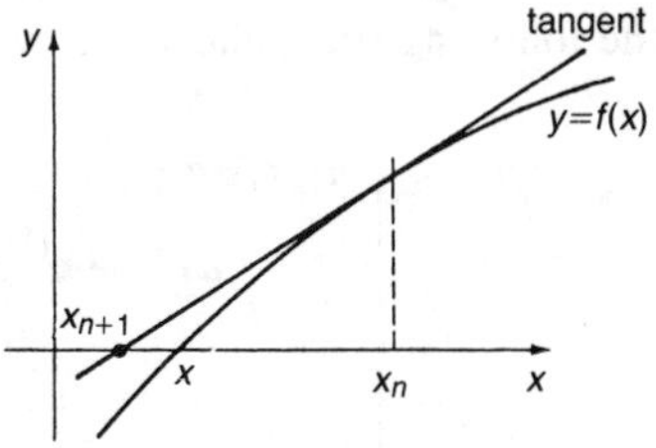

$$x_{n+1}=x_n-\frac{f(x_n)}{f'(x_n)}$$

Choose x_0 suitably close to x.

The convergence is at least quadratic for a simple root if $f''(x)$ is continuous.

2. System with two equations and two unknowns

$$\begin{cases} f(x, y)=0 \\ g(x, y)=0 \end{cases}$$

Iteration $(x_n, y_n) \rightarrow (x_{n+1}, y_{n+1})$ given by

$$x_{n+1}=x_n-\frac{fg_y'-f_y'g}{f_x'g_y'-f_y'g_x'}(x_n, y_n)$$

$$y_{n+1}=y_n-\frac{f_x'g-fg_x'}{f_x'g_y'-f_y'g_x'}(x_n, y_n)$$

Choose (x_0, y_0) suitably close to (x, y).

3. System with *m* equations and *m* unknowns

$f_k(x_1, \ldots, x_m)=0$, $k=1, \ldots, m$ or
$\boldsymbol{f}(\boldsymbol{x})=\boldsymbol{0}$

Iteration $\boldsymbol{x}^{(n)} \rightarrow \boldsymbol{x}^{(n+1)}$ by solving the linear system

$$M(\boldsymbol{x}^{(n)})(\boldsymbol{x}^{(n+1)}-\boldsymbol{x}^{(n)})+\boldsymbol{f}(\boldsymbol{x}^{(n)})=\boldsymbol{0}$$

$\boldsymbol{x}^{(n+1)}=\boldsymbol{x}^{(n)}-M(\boldsymbol{x}^{(n)})^{-1}\,\boldsymbol{f}(\boldsymbol{x}^{(n)})$, where

$M(\boldsymbol{x})=\left(\dfrac{\partial f_i}{\partial x_j}\right)$ is the functional matrix [cf. sec. 10.6].

Systems of linear equations

Gauss' elimination method

Consider the system

$$a_{i1}x_1+a_{i2}x_2+\ldots+a_{in}x_n=b_i,\ i=1, 2, \ldots, n$$

or

$$A\boldsymbol{x}=\boldsymbol{b}$$

with

$$A=\begin{bmatrix} a_{11} & \ldots & a_{1n} \\ & \ldots\ldots\ldots\ldots & \\ a_{n1} & \ldots & a_{nn} \end{bmatrix},\quad \boldsymbol{x}=\begin{bmatrix} x_1 \\ \vdots \\ x_n \end{bmatrix},\quad \boldsymbol{b}=\begin{bmatrix} b_1 \\ \vdots \\ b_n \end{bmatrix}$$

The system is assumed to be *non-singular,* i.e. $\det A \neq 0$.

When using the Gauss elimination method (see also sec. 4.3) the system is brought to the following triangular form by successive elimination

$$\begin{cases} a_{11}x_1+a_{12}x_2+a_{13}x_3+\ldots+a_{1n}x_n=b_1 \\ \quad a_{22}^{(1)}x_2+a_{23}^{(1)}x_3+\ldots+a_{2n}^{(1)}x_n=b_2^{(1)} \\ \quad\quad a_{33}^{(2)}x_3+\ldots+a_{3n}^{(2)}x_n=b_3^{(2)} \\ \quad\quad\quad \ldots\ldots\ldots\ldots\ldots\ldots \\ \quad\quad\quad\quad a_{n,n}^{(n-1)}x_n=b_n^{(n-1)} \end{cases}$$

This system is then solved by *back substitution,* computing x_n from the *nth* equation, x_{n-1} from the $(n-1)$st equation and so on in that order. The numbers a_{11}, $a_{22}^{(1)}$, $a_{33}^{(2)}$... are called *pivot elements.*

Partial pivoting: At the *rth* elimination step use that equation of the $(n-r)$ last ones for the elimination where x_r has the largest coefficient in absolute value.

Complete pivoting: At the *rth* elimination step the largest element $|a_{ij}^{(r)}|$, $i, j \geqslant r+1$ is determined. If this is $|a_{pq}^{(r)}|$, then the *pth* equation is used to eliminate x_q from the remaining $n-r-1$ equations.

Scaling: Multiply equations by numbers $\neq 0$ and/ore make substitutions $x_i = c_i y_i$, $c_i \neq 0$.

For an example of Gauss-elimination see below under *LR*-decomposition.

LR-decomposition

Decompose the coefficient matrix A such that

$$A = LR,$$

where L is a *unit left triangular* matrix ($l_{ij}=0$ for $i<j$ and $l_{ii}=1$) and R is a *right triangular matrix* ($r_{ij}=0$ for $i>j$). Then

$$A\boldsymbol{x}=\boldsymbol{b} \Leftrightarrow R\boldsymbol{x}=z,\ Lz=\boldsymbol{b}$$

First solve $Lz=\boldsymbol{b}$ by forward substitution and then $R\boldsymbol{x}=\boldsymbol{b}$ by back substitution. The matrix A can be uniquely factorized, $A=LR$, if $\det A_k \neq 0$, $k=1, 2, \ldots, n-1$, where A_k is obtained from matrix A by deleting rows and columns $k+1$, $k+2$, ..., n.

The actual decomposition can be performed using Gauss elimination.

Example

$$\begin{cases} 2x+\ 4y+\ 6z=0 \\ 4x+11y+17z=-2 \\ 6x+24y+39z=-9 \end{cases}$$

We only write the coefficients when we solve this system by Gauss-elimination. We write the factors used to multiply equations to the right.

$$\left[\begin{array}{ccc|c} 2 & 4 & 6 & 0 \\ 4 & 11 & 17 & -2 \\ 6 & 24 & 39 & -9 \end{array}\right] \begin{array}{c} (-2)\ (-3) \\ \\ \\ \end{array} \left[\begin{array}{ccc|c} 2 & 4 & 6 & 0 \\ 0 & 3 & 5 & -2 \\ 0 & 12 & 21 & -9 \end{array}\right] \begin{array}{c} \\ (-4) \\ \\ \end{array} \left[\begin{array}{ccc|c} 2 & 4 & 6 & 0 \\ 0 & 3 & 5 & -2 \\ 0 & 0 & 1 & -1 \end{array}\right]$$

The system is equivalent to $\begin{cases} 2x+4y+6z=\ 0 \\ \qquad 3y+5z=-2 \\ \qquad\qquad z=-1 \end{cases}$ or $\begin{array}{l} x=1 \\ y=1 \\ z=-1 \end{array}$

From the solution above we get the LR-decomposition

$$\begin{bmatrix} 2 & 4 & 6 \\ 4 & 11 & 17 \\ 6 & 24 & 39 \end{bmatrix} = \begin{bmatrix} 1 & 0 & 0 \\ 2 & 1 & 0 \\ 3 & 4 & 1 \end{bmatrix} \begin{bmatrix} 2 & 4 & 6 \\ 0 & 3 & 5 \\ 0 & 0 & 1 \end{bmatrix}$$

Iterative methods

Write $A=D-L-R$, where D is a diagonal matrix and L and R are left and right triangular matrises. Start with an arbitrary initial vector $\boldsymbol{x}^{(0)}$.

The Jacobi method

$$\boldsymbol{x}^{(k)}=D^{-1}(L+R)\boldsymbol{x}^{(k-1)}+D^{-1}\boldsymbol{b}$$

The Gauss-Seidel method

$$\boldsymbol{x}^{(k)}=(D-L)^{-1}R\boldsymbol{x}^{(k-1)}+(D-L)^{-1}\boldsymbol{b}$$

Example

$$\begin{cases} 10x_1+x_2+7x_3=33 \\ x_1+10x_2+6x_3=39 \\ 3x_1+5x_2+10x_3=43 \end{cases}$$

In this case the Jacobi method will use the following recursive formulas.

$$x_1^{(k)}=-0.1x_2^{(k-1)}-0.7x_3^{(k-1)}+3.3$$
$$x_2^{(k)}=-0.1x_2^{(k-1)}-0.6x_3^{(k-1)}+3.9$$
$$x_3^{(k)}=-0.3x_1^{(k-1)}-0.5x_2^{(k-1)}+4.3$$

While the Gauss-Seidel method will make use of the following recursive formulas.

$$x_1^{(k)}=-0.1x_2^{(k-1)}-0.7x_3^{(k-1)}+3.3$$
$$x_2^{(k)}=-0.1x_1^{(k)}\quad-0.6x_3^{(k-1)}+3.9$$
$$x_3^{(k)}=-0.3x_1^{(k)}\quad-0.5x_2^{(k)}+4.3$$

A simple sufficient condition for the convergence of the Jacobi and Gauss-Seidel methods is that the absolute values of each diagonal element in A is larger than the sum of the absolute values of the other elements in the same row of A.

Iterative method to find eigenvalues and eigenvectors (Power method)

Assume that $|\lambda_1|>|\lambda_2|\geq\ldots$ where λ_i are the eigenvalues of a square matrix A. Starting with a column vector $\boldsymbol{v}_1$ (almost arbitrarily chosen), then $\boldsymbol{v}_{k+1}=A\boldsymbol{v}_k/|A\boldsymbol{v}_k|$ (if k is big enough) can be used as an approximation of an eigenvector of length 1 corresponding to λ_1, and $\lambda_1\approx\boldsymbol{v}_k{}^tA\boldsymbol{v}_k$.

Note. (*i*) The eigenvalue with the *smallest* absolute value and the corresponding eigenvector may be calculated as the eigenvalue with the *greatest* absolute value and the corresponding eigenvector, respectively, of A^{-1}.

(*ii*) If A is symmetric, then the corresponding eigenvector of the second eigenvalue may be calculated similarly by starting with a vector $\boldsymbol{u}_1$ which is orthogonal to the first eigenvector etc.

Perturbation analysis

Vector norms and matrix norms

Let $\mathbf{x}=(x_1, \ldots, x_n)^t$ and $A=(a_{ij})$ be of order $n\times n$. In any *vector norm* $\|x\|$ the corresponding *matrix norm* is

$$\|A\|=\sup_{\mathbf{x}\neq\mathbf{0}}\frac{\|A\mathbf{x}\|}{\|\mathbf{x}\|}=\sup_{\|\mathbf{x}\|=1}\|A\mathbf{x}\|$$

$$\|A\mathbf{x}\|\leq\|A\|\cdot\|\mathbf{x}\| \qquad \|A+B\|\leq\|A\|+\|B\| \qquad \|AB\|\leq\|A\|\cdot\|B\|$$

The *condition number* of A is $\varkappa(A)=\|A\|\cdot\|A^{-1}\|$

Special norms

l_p-norm $\|\mathbf{x}\|_p=\left(\sum_{i=1}^{n}|x_i|^p\right)^{1/p}$. In particular,

Euclidean norm $\|\mathbf{x}\|_2=\left(\sum_{i=1}^{n}x_i^2\right)^{1/2}$. $\|A\|_2=\sqrt{\text{greatest eigenvalue of } A'A}$

Maximum norm $\|\mathbf{x}\|_\infty=\max_{1\leq i\leq n}|x_i|$. $\|A\|_\infty=\max_{1\leq i\leq n}\left(\sum_{j=1}^{n}|a_{ij}|\right)$

Perturbation

Let A be a non-singular matrix.

1. Perturbation of the right hand side.

$A\mathbf{x}=\mathbf{b}$ (exact system)

$A(\mathbf{x}+\delta\mathbf{x})=\mathbf{b}+\delta\mathbf{b}$ (perturbed system)

Error estimate:

$$\frac{\|\delta\mathbf{x}\|}{\|\mathbf{x}\|}\leq\varkappa(A)\frac{\|\delta\mathbf{b}\|}{\|\mathbf{b}\|}$$

2. Perturbation of the coefficient matrix and of the right hand side.

$A\mathbf{x}=\mathbf{b}$ (exact system)

$(A+\delta A)\mathbf{y}=\mathbf{b}+\delta\mathbf{b}$ (perturbed system)

Error estimate: If $\|\delta A\|\cdot\|A^{-1}\|=r<1$ then

$$\frac{\|\mathbf{y}-\mathbf{x}\|}{\|\mathbf{x}\|}\leq\frac{1}{1-r}\cdot\varkappa(A)\left(\frac{\|\delta A\|}{\|A\|}+\frac{\|\delta\mathbf{b}\|}{\|\mathbf{b}\|}\right)$$

16.3 Interpolation

The interpolation problem is to find a polynomial $P(x)$ of degree at most n such that $P(x_i)=y_i$, $i=0, 1, \ldots, n$.

Often $y_i=f(x_i)$ for a given function $f(x)$. There exists exactly one such interpolation polynomial.

Lagrange's interpolation formula

General form

$$P(x)=\sum_{k=0}^{n} f(x_k)\, L_k(x)$$

$$L_k(x)=\frac{(x-x_0)(x-x_1)\ldots(x-x_{k-1})(x-x_{k+1})\ldots(x-x_n)}{(x_k-x_0)(x_k-x_1)\ldots(x_k-x_{k-1})(x_k-x_{k+1})\ldots(x_k-x_n)}$$

$$L_k(x_k)=1,\ L_k(x_i)=0 \text{ for } i\neq k$$

Remainder term $R(x)=f(x)-P(x)=(x-x_0)(x-x_1)\ldots(x-x_n)f^{(n+1)}(\xi)/(n+1)!$ where $\xi\in(a, b)$ if $x, x_1, x_2, \ldots, x_n\in[a, b]$.

The case $n=2$ (three point case)

$$P(x)=f(x_0)\,\frac{(x-x_1)(x-x_2)}{(x_0-x_1)(x_0-x_2)}+f(x_1)\,\frac{(x-x_0)(x-x_2)}{(x_1-x_0)(x_1-x_2)}+f(x_2)\,\frac{(x-x_0)(x-x_1)}{(x_2-x_0)(x_2-x_1)}$$

Remainder term $R=(x-x_0)(x-x_1)(x-x_2)f^{(3)}(\xi)/3!$

The case $n=2$ with three equally spaced points

$$P(x_0+th)=\frac{t(t-1)}{2}f(x_0-h)+(1-t^2)f(x_0)+\frac{t(t+1)}{2}f(x_0+h)$$

$$|R|\leq 0.065\,h^3\,|f^{(3)}(\xi)|\approx 0.065\,\Delta^3 \qquad \text{if } |t|\leq 1$$

The case $n=3$ with four equally spaced points

$$P(x_0+th)=-\frac{t(t-1)(t-2)}{6}f(x_0-h)+\frac{(t^2-1)(t-2)}{2}f(x_0)-$$

$$-\frac{t(t+1)(t-2)}{2}f(x_0+h)+\frac{t(t^2-1)}{6}f(x_0+2h)$$

$$|R|\leq 0.024h^4|f^{(4)}(\xi)|\approx 0.024\,\Delta^4 \qquad \text{for } 0<t<1$$

$$|R|\leq 0.042h^4|f^{(4)}(\xi)|\approx 0.042\,\Delta^4 \qquad \text{for } -1<t<0,\ 1<t<2$$

The case $n=4$ with five equally spaced points

$$P(x_0+th)=\frac{(t^2-1)t(t-2)}{24}f(x_0-2h)-\frac{(t-1)t(t^2-4)}{6}f(x_0-h)+$$
$$+\frac{(t^2-1)(t^2-4)}{4}f(x_0)-\frac{(t+1)t(t^2-4)}{6}f(x_0+h)+\frac{(t^2-1)t(t+2)}{24}f(x_0+2h)$$

$|R|\lesssim 0.012h^5|f^{(5)}(\xi)|\approx 0.012\Delta^5$ for $|t|<1$

$|R|\lesssim 0.031h^5|f^{(5)}(\xi)|\approx 0.031\Delta^5$ for $|<|t|<2$

Horner's scheme

Writing a polynomial $P(x)$ in *nested form*

$$P(x)=a_nx^n+a_{n-1}x^{n-1}+a_{n-2}x^{n-2}+\ldots+a_1x+a_0=$$
$$=(\ldots(a_nx+a_{n-1})x+a_{n-2})x+\ldots)x+a_0$$

the value $P(x)=b_0$ can be calculated using the algorithm

$$b_n=a_n$$
$$b_k=b_{k+1}x+a_k,\ k=n-1,\ \ldots,\ 0$$

Newton's interpolation formula

$$P(x)=A_0+A_1(x-x_0)+A_2(x-x_0)(x-x_1)+A_3(x-x_0)(x-x_1)(x-x_2)+$$
$$+\ldots+A_n(x-x_0)(x-x_1)\ \ldots\ (x-x_{n-1})$$

$$P(x)=f(x_0)+\Delta_1(x_0,\ x_1)(x-x_0)+\Delta_2(x_0,\ x_1,\ x_2)(x-x_0)(x-x_1)+$$
$$+\Delta_3(x_0,\ x_1,\ x_2,\ x_3)(x-x_0)(x-x_1)(x-x_2)+\ldots+$$
$$+\Delta_n(x_0,\ x_1,\ \ldots,\ x_n)(x-x_0)(x-x_1)\ldots(x-x_{n-1})$$

The *divided differences* $\Delta_k(x_0,\ x_1,\ \ldots,\ x_k)$ are defined as follows

$$\Delta_1(x_0,\ x_1)=\frac{f(x_1)-f(x_0)}{x_1-x_0}$$

$$\Delta_2(x_0,x_1,x_2)=\frac{\Delta_1(x_1,x_2)-\Delta_1(x_0,x_1)}{x_2-x_0}$$

....

$$\Delta_k(x_0,\ x_1,\ \ldots,\ x_k)=\frac{\Delta_{k-1}(x_1,\ x_2,\ \ldots,\ x_k)-\Delta_{k-1}(x_0,\ x_1,\ \ldots,\ x_{k-1})}{x_k-x_0}$$

Remainder term$=(x-x_0)(x-x_1)\ \ldots\ (x-x_n)f^{(n+1)}(\xi)/(n+1)!$,
where $\xi\in(a,\ b)$ if $x,\ x_1,\ x_2,\ \ldots,\ x_n\in[a,\ b]$.

Stirling numbers

Factorial polynomials

The *factorial power* $x^{(n)}$ is defined by

$$x^{(n)}=x(x-1)(x-2)\ \dots\ (x-n+1),\ n=1,2,3,\dots$$
$$x^{(0)}=1$$
$$x^{(-n)}=1/(x+n)^{(n)},\ n=1,2,3,\dots$$

With $\Delta x^{(n)}=(x+1)^{(n)}-x^{(n)}$ it follows for all integers n,

$$\Delta x^{(n)}=n\,x^{(n-1)}$$
$$\Delta^k x^{(n)}=n(n-1)\ \dots\ (n-k+1)x^{(n-k)}=n^{(k)}x^{(n-k)}$$

In particular $\Delta^n x^{(n)}=n!$

Stirling numbers $\alpha_k^{(n)}$ of the first kind

$$x^{(n)}=\sum_{k=1}^{n}\alpha_k^{(n)}x^k$$

Recursion: $\alpha_k^{(n)}=\alpha_{k-1}^{(n-1)}-(n-1)\alpha_k^{(n-1)},\ \alpha_0^{(n)}=0$

E.g. $x^{(5)}=x^5-10x^4+35x^3-50x^2+24x$

n \ k	1	2	3	4	5	6	7	8	9	10
1	1									
2	−1	1								
3	2	−3	1							
4	−6	11	−6	1						
5	24	−50	35	−10	1					
6	−120	274	−225	85	−15	1				
7	720	−1764	1624	−735	175	−21	1			
8	−5040	13068	−13132	6769	−1960	322	−28	1		
9	40320	−109584	118124	−67284	22449	−4536	546	−36	1	
10	−362880	1026576	−1172700	723680	−269325	63273	−9450	870	−45	1

Stirling numbers $\beta_k^{(n)}$ of the second kind

$$x^n=\sum_{k=1}^{n}\beta_k^{(n)}x^{(k)}$$

Recursion: $\beta_k^{(n)}=\beta_{k-1}^{(n-1)}+k\beta_k^{(n-1)},\ \beta_0^{(n)}=0$

E.g. $x^5=x^{(5)}+10x^{(4)}+25x^{(3)}+15x^{(2)}+x^{(1)}$

n \ k	1	2	3	4	5	6	7	8	9	10
1	1									
2	1	1								
3	1	3	1							
4	1	7	6	1						
5	1	15	25	10	1					
6	1	31	90	65	15	1				
7	1	63	301	350	140	21	1			
8	1	127	966	1701	1050	266	28	1		
9	1	255	3025	7770	6951	2646	462	36	1	
10	1	511	9330	34105	42525	22827	5880	750	45	1

Finite differences

$$y=f(x),\ y_k=f(x_k),\ x_k=x_0+kh,\ k=0,\ \pm1,\ \pm2,\ \pm3,\ \ldots$$

$\Delta y_k=y_{k+1}-y_k$

$\Delta^2 y_k=\Delta y_{k+1}-\Delta y_k=y_{k+2}-2y_{k+1}+y_k$

....

$\Delta^r y_k=\Delta^{r-1}y_{k+1}-\Delta^{r-1}y_k=\sum\limits_{i=0}^{r}(-1)^i\binom{r}{i}y_{k+r-i}$

$$\begin{array}{llllll}
x_{-2} & y_{-2} & & & & \\
 & & \Delta y_{-1} & & & \\
x_{-1} & y_{-1} & & \Delta^2 y_{-1} & & \\
 & & \Delta y_0 & & \Delta^3 y_{-1} & \\
x_0 & y_0 & & \Delta^2 y_0 & & \Delta^4 y_{-1} \\
 & & \Delta y_1 & & \Delta^3 y_0 & \\
x_1 & y_1 & & \Delta^2 y_1 & & \\
 & & \Delta y_2 & & & \\
x_2 & y_2 & & & &
\end{array}$$

$\Delta(yz)_k=y_k\Delta z_k+z_{k+1}\Delta y_k$

$\Delta\left(\frac{y}{z}\right)_k=\frac{z_k\Delta y_k-y_k\Delta z_k}{z_k\,z_{k+1}}$

$\Delta^r(yz)_k=\sum\limits_{i=0}^{r}\binom{r}{i}\Delta^{r-i}y_{k+i}\Delta^i z_k$ (*the discrete Leibniz formula*)

$\sum\limits_{k=0}^{n-1}y_k\,\Delta z_k=y_n z_n-y_0 z_0-\sum\limits_{k=0}^{n-1}\Delta y_k\cdot z_{k+1}$ (summation by parts)

Newton's interpolation formula

$$f(x_0+th)\approx y_0+t\Delta y_0+\binom{t}{2}\Delta^2 y_0+\ldots+\binom{t}{n}\Delta^n y_0$$

Remainder term $=h^{n+1}\binom{t}{n+1}f^{(n+1)}(\xi)\approx\binom{t}{n+1}\Delta^{n+1}y_0$

Some linear operators

E: $Ey(x)=y(x+h)$	(Displacement operator)
Δ: $\Delta y(x)=y(x+h)-y(x)$	(Forward difference operator)
∇: $\nabla y(x)=y(x)-y(x-h)$	(Backward difference operator)
δ: $\delta y(x)=y(x+h/2)-y(x-h/2)$	(Central difference operator)
μ: $\mu y(x)=[y(x+h/2)+y(x-h/2)]/2$	(Averaging operator)
D: $Dy(x)=y'(x)$	(Differentiation operator)

Relations between operators

	E	Δ	∇	δ	D
E	–	$1+\Delta$	$(1-\nabla)^{-1}$	$1+\delta^2/2+\delta\sqrt{1+\delta^2/4}$	e^{hD}
Δ	$E-1$	–	$(1-\nabla)^{-1}-1$	$\delta^2/2+\delta\sqrt{1+\delta^2/4}$	e^{hD-1}
∇	$1-E^{-1}$	$1-(1+\Delta)^{-1}$	–	$-\delta^2/2+\delta\sqrt{1+\delta^2/4}$	$1-e^{-hD}$
δ	$E^{1/2}-E^{-1/2}$	$\Delta(1+\Delta)^{-1/2}$	$\nabla(1-\nabla)^{-1/2}$	–	$2\sinh(hD/2)$
μ	$[E^{1/2}+E^{-1/2}]/2$	$(1+\Delta/2)(1+\Delta)^{-1/2}$	$(1-\nabla/2)(1-\nabla)^{-1/2}$	$\sqrt{1+\delta^2.4}$	$\cosh(hD/2)$
D	$h^{-1}\ln E$	$h^{-1}\ln(1+\Delta)$	$-h^{-1}\ln(1--\nabla)$	$2h^{-1}\sinh^{-1}(\delta/2)$	–

E.g. $\Delta=(1-\nabla)^{-1}-1$

Inverse interpolation

By inverse interpolation is meant that t is to be calculated such that

$$f(x_0+th)=y_t$$

for given y_t.

Linear inverse interpolation

$$t\approx\frac{y_t-y_0}{y_1-y_0}$$

Reversion of series

Write an interpolation as a power series

$$y_t\approx a_0+a_1t+a_2t^2+\ldots$$

Then with $u=(y_t-a_0)/a_1$

$$t\approx u+c_2u^2+c_3u^3+\ldots$$

where

$$c_2=-a_2/a_1$$
$$c_3=-a_3/a_1+2(a_2/a_1)^2$$
$$c_4=-a_4/a_1+5a_2a_3/a_1^2-5(a_2/a_1)^3$$
$$c_5=-a_5/a_1+6a_2a_4/a_1^2+3a_3^2/a_1^2-21a_2^2a_3/a_1^3+14(a_2/a_1)^4$$

Bisection

Also bisection can be used se sec. 16.2.

Richardson extrapolation

Assume the following holds when y is approximated by $y(h)$

$$y=y(h)+kh^p$$

If y is approximated by $y(h)$ and $y(2h)$ then

$$y\approx y(h)+\frac{y(h)-y(2h)}{2^p-1}$$

$$p=2 \text{ gives } y\approx y(h)+\frac{y(h)-y(2h)}{3}=\frac{4y(h)-y(2h)}{3} \quad \text{"}(2^2-1)\text{-rule"}$$

$$p=4 \text{ gives } y\approx y(h)+\frac{y(h)-y(2h)}{15}=\frac{16y(h)-y(2h)}{15} \quad \text{"}(2^4-1)\text{-rule"}$$

If more generally y is approximated by $y(h)$ and $y(qh)$, $q>1$,

$$y\approx y(h)+\frac{y(h)-y(qh)}{q^p-1}$$

Richardson extrapolation can be used e.g. in connection with numerical integration.

Curve fitting

Given N points $(x_1, y_1), \ldots, (x_N, y_N)$, $x_i \neq x_j$, $i \neq j$, find a curve $y=f(x)$ of given type that is of best fit in the sense of the least square method, i.e. such that $Q=\sum_{k=1}^{N}(f(x_k)-y_k)^2$ is minimized. (See also sec. 4.3).

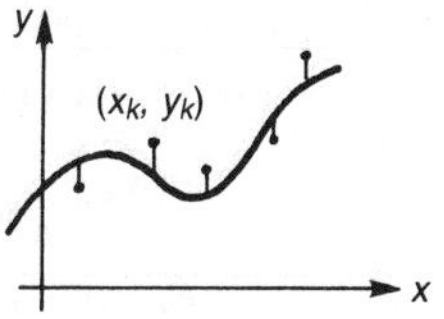

(A) General case

Function $y=f(x;\, a_1, \ldots, a_n)$ depending on n parameters a_i $(n \leq N)$. Set $Q=\sum_{k=1}^{N}[f(x_k;\, a_1, \ldots, a_n)-y_k]^2$. The constants a_i may be found from the equations

$$\boxed{\partial Q/\partial a_i=0,\ i=1, \ldots, n} \qquad \text{(the \textit{normal equations})}$$

In particular, if the desired function is to be a linear combination of some "simple" functions $\varphi_1(x), \ldots, \varphi_n(x)$, i.e.

$$y=f(x)=c_1\varphi_1(x)+\ldots+c_n\varphi_n(x),$$

then c_k has to satisfy

$$\begin{cases} c_1(\varphi_1, \varphi_1)+c_2(\varphi_1, \varphi_2)+\ldots+c_n(\varphi_1, \varphi_n)=(f, \varphi_1) \\ c_1(\varphi_2, \varphi_1)+c_2(\varphi_2, \varphi_2)+\ldots+c_n(\varphi_2, \varphi_n)=(f, \varphi_2) \\ \cdots \\ c_1(\varphi_n, \varphi_1)+c_2(\varphi_n, \varphi_2)+\ldots+c_n(\varphi_n, \varphi_n)=(f, \varphi_n) \end{cases}$$

where the vectors $\boldsymbol{\varphi}_k$ and $\boldsymbol{f}$ are defined by $\boldsymbol{\varphi}_k{}^t=(\varphi_k(x_1), \ldots, \varphi_k(x_N))$ and $\boldsymbol{f}^t=(y_1, \ldots, y_N)$, respectively, and $(\boldsymbol{u}, \boldsymbol{v})$ is the scalar product $\boldsymbol{u}^t\boldsymbol{v}$.

(B) Fitting of polynomial $y=\sum_{k=0}^{n} a_k x^k$.

The coefficients a_k are determined from the linear system

$$\boxed{a_0\sum_{k=1}^{N}x_k^i+a_1\sum_{k=1}^{N}x_k^{1+i}+\ldots+a_n\sum_{k=1}^{N}x_k^{n+i}=\sum_{k=1}^{N}x_k^i y_k, \qquad i=0, \ldots, n}$$

(C) Fitting of straight line $y=ax+b$.
Set $D=N\Sigma x_k^2-(\Sigma x_k)^2$.

$$\boxed{\begin{aligned} a&=(N\Sigma x_k y_k-\Sigma x_k\Sigma y_k)/D \\ b&=(\Sigma y_k-a\Sigma x_k)/N \end{aligned}}$$

Least square method with respect to log *y*

(D) *Power function* $y=ax^b$. This equation is, by taking logarithms, transformed to $\log y=\log a+b\log x$. Use (C) to find $\log a$ and b. (In a diagram with logarithmic scales on both axis $y=ax^b$ becomes a straight line).

(E) *Exponential function* $y=ab^x$. This equation is, by taking logarithms, transformed to $\log y=\log a+x\log b$. Use (C) to find $\log a$ and $\log b$. (In a diagram with logarithmic scale on the y-axis and uniform scale on the x-axis $y=ab^x$ becomes a straight line.)

Bezier curves

Let $\boldsymbol{r}_0, \boldsymbol{r}_1, \ldots, \boldsymbol{r}_n$ be position vectors of $n+1$ *control points* $P_0, P_1, \ldots, P_n$.

A *Bezier curve* is then given by

$$\boldsymbol{r}(t)=\sum_{k=0}^{n} B_{n,k}(t)\boldsymbol{r}_k$$

where $B_{n,k}(t)=\binom{n}{k}t^k(1-t)^{n-k}$ are *Bernstein polynomials*. Then

$$\boldsymbol{r}(0)=\boldsymbol{r}_0,\ \boldsymbol{r}(1)=\boldsymbol{r}_n,\ \boldsymbol{r}'(0)=n(\boldsymbol{r}_1-\boldsymbol{r}_0),\ \boldsymbol{r}'(1)=n(\boldsymbol{r}_n-\boldsymbol{r}_{n-1}).$$

A Bezier curve lies in the convex hull of its control points.

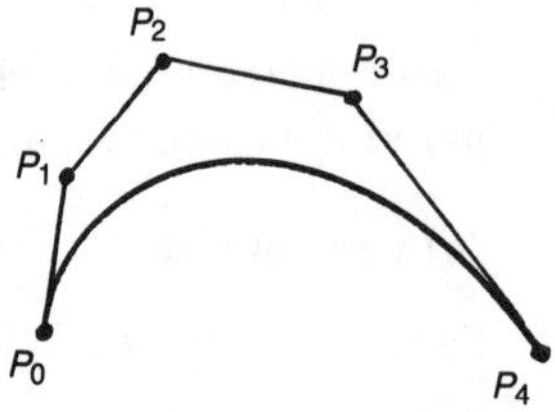

16.4 Numerical Integration and Differentiation

Numerical integration

Midpoint rule

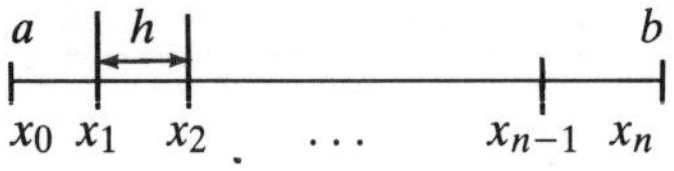

R_T=truncation error

$$\int_a^b f(x)dx=h\left[f\left(\frac{x_0+x_1}{2}\right)+f\left(\frac{x_1+x_2}{2}\right)+\ldots+f\left(\frac{x_{n-1}+x_n}{2}\right)\right]+R_T=M_n+R_T$$

$$h=\frac{b-a}{n}\qquad x_k=a+\frac{k}{n}(b-a)\qquad R_T=\frac{(b-a)h^2}{24}f''(\xi),\ a<\xi<b$$

Trapezoidal rule

$$\int_a^b f(x)dx=\frac{h}{2}[f(x_0)+2f(x_1)+2f(x_2)+\ldots+2f(x_{n-1})+f(x_n)]+R_T=T_n+R_T$$

$$h=\frac{b-a}{n}\qquad x_k=a+\frac{k}{n}(b-a)\qquad R_T=-\frac{(b-a)h^2}{12}f''(\xi),\ a<\xi<b$$

Simpson's rule

$$\int_a^b f(x)dx=\frac{h}{3}[f(x_0)+4f(x_1)+2f(x_2)+\ldots+2f(x_{2n-2})+4f(x_{2n-1})+f(x_{2n})]+R_T=S_n+R_T$$

$$h=\frac{b-a}{2n}\qquad x_k=a+\frac{k}{2n}(b-a)\qquad R_T=-\frac{(b-a)h^4}{180}f^{(4)}(\xi),\ a<\xi<b$$

Remarks

(a) Simpson's rule is exact for polynomials of at most third degree.

(b) $S_n = \frac{1}{3}\,(T_n + 2M_n) \qquad T_{n+1} = \frac{1}{2}\,(T_n + M_n)$

The flow chart in the following figure describes a computer program where these formulas are used to find the Simpson approximation to $\int_a^b f(x)dx$ by interval halving such that the relative error is at most E.

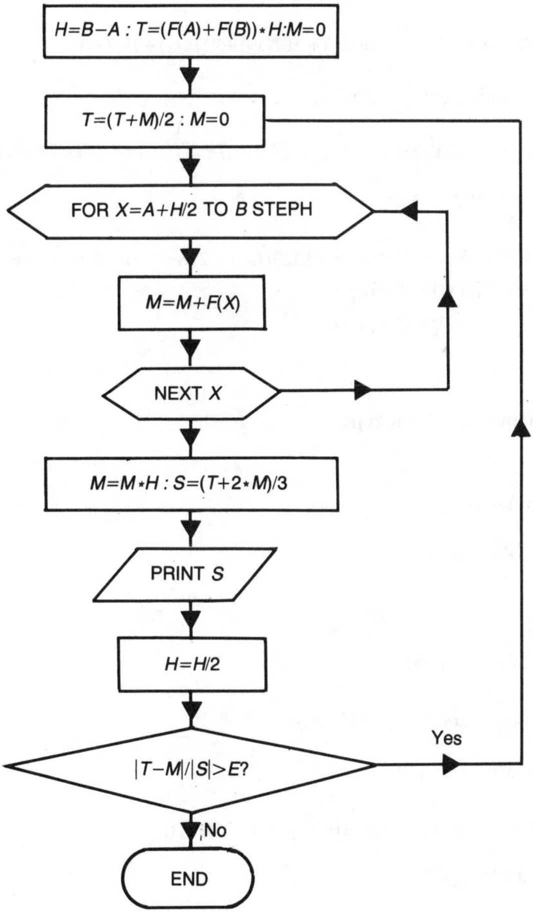
H=B−A : T=(F(A)+F(B))∗H:M=0
T=(T+M)/2 : M=0
FOR X=A+H/2 TO B STEPH
M=M+F(X)
NEXT X
M=M∗H : S=(T+2∗M)/3
PRINT S
H=H/2
|T−M|/|S|>E?
Yes
No
END

Newton-Cotes formulas

Newton-Cotes formulas (Closed type)

$x_k = x_0 + kh,\ k = 1, 2, \ldots$

$$\int_{x_0}^{x_3} f(x)dx \approx \frac{3h}{8}\ (f(x_0)+3f(x_1)+3f(x_2)+f(x_3))$$

Remainder term $= -3f^{(4)}(\xi)h^5/80$

$$\int_{x_0}^{x_4} f(x)dx \approx \frac{2h}{45}\ (7f(x_0)+32f(x_1)+12f(x_2)+32f(x_3)+7f(x_4))$$

Remainder term $= -8f^{(6)}(\xi)h^7/945$

$$\int_{x_0}^{x_5} f(x)dx \approx \frac{5h}{288}\ (19f(x_0)+75f(x_1)+50f(x_2)+50f(x_3)+75f(x_4)+19f(x_5))$$

Remainder term $= -275f^{(6)}(\xi)h^7/12096$

$$\int_{x_0}^{x_6} f(x)dx \approx \frac{h}{140}\ (41f(x_0)+216f(x_1)+27f(x_2)+272f(x_3)+27f(x_4)+216f(x_5)+41f(x_6))$$

Remainder term $= -9f^{(8)}(\xi)h^9/1400$

$$\int_{x_0}^{x_7} f(x)dx \approx \frac{7h}{17280}\ (751f(x_0)+3577f(x_1)+ +1323f(x_2)+2989f(x_3)+2989f(x_4)+1323f(x_5)+ +3577f(x_6)+751f(x_7))$$

Remainder term $= -8183f^{(8)}(\xi)h^9/518400$

Newton-Cotes formulas (Open type)

$x_k = x_0 + kh,\ k = 1, 2, \ldots$

$$\int_{x_0}^{x_3} f(x)dx \approx \frac{3h}{2}\ (f(x_1)+f(x_2))$$

Remainder term $= 3f^{(2)}(\xi)h^3/4$

$$\int_{x_0}^{x_4} f(x)dx \approx \frac{4h}{3}\ (2f(x_1)-f(x_2)+2f(x_3))$$

Remainder term $= 28f^{(4)}(\xi)h^5/90$

$$\int_{x_0}^{x_5} f(x)dx \approx \frac{5h}{24}\ (11f(x_1)+f(x_2)+f(x_3)+11f(x_4))$$

Remainder term $= 95f^{(4)}(\xi)h^5/144$

$$\int_{x_0}^{x_6} f(x)dx \approx \frac{6h}{20}\ (11f(x_1)-14f(x_2)+26f(x_3)-14f(x_4)+11f(x_5))$$

Remainder term $= 41f^{(6)}(\xi)h^7/140$

$$\int_{x_0}^{x_7} f(x)dx \approx \frac{7h}{1440}\ (611f(x_1)-453f(x_2)+562f(x_3)+562f(x_4)-453f(x_5)+611f(x_6))$$

Remainder term $= 5257f^{(6)}(\xi)h^7/8640$

Gauss' quadrature formula

$$\int_{-1}^{1} f(x)dx \approx \sum_{i=1}^{n} a_i f(x_i)$$

$$\text{Truncation error} = \frac{2^{2n+1}(n!)^4}{(2n+1)((2n)!)^3} f^{(2n)}(\xi)$$

The weights a_i are given by

$$a_i = \frac{2}{(1-x_i^2)(P_n'(x_i))^2},$$

where the abscissas x_i are the n zeros of the Legendre polynomial P_n.

For the integral $\int_a^b f(x)dx$ make the substitution $x = \frac{b-a}{2}t + \frac{a+b}{2}$.

Abscissas x_i and weights for some n are given in the following table

n	$\pm x_i$	a_i	n	$\pm x_i$	a_i
2	0.57735 02692	1	6	0.23861 91861	0.46791 39346
				0.66120 93865	0.36076 15730
3	0	0.88888 88889		0.93246 95142	0.17132 44924
	0.77459 66692	0.55555 55556			
			8	0.18343 46425	0.36268 37834
4	0.33998 10436	0.65214 51549		0.52553 24099	0.31370 66459
	0.86113 63116	0.34785 48451		0.79666 64774	0.22238 10345
				0.96028 98565	0.10122 85363
5	0	0.56888 88889			
	0.53846 93101	0.47862 86705	10	0.14887 43390	0.29552 42247
	0.90617 98459	0.23692 68851		0.43339 53941	0.26926 67193
				0.67940 95683	0.21908 63625
				0.86506 33667	0.14945 13492
				0.97390 65285	0.06667 13443

Romberg integration

This method to approximate $\int_a^b f(x)dx$ uses Richardson extrapolation, see sec. 16.3, starting with the trapezoidal rule

$$T(h) = \frac{h}{2}(f(x_0)+2f(x_1)+\ldots+2f(x_{n-1})+f(x_n)),\ h = \frac{b-a}{n}$$

with truncation error $\approx kh^2$

$$T_1(h) = \frac{4T(h)-T(2h)}{3} \qquad T_2(h) = \frac{16T_1(h)-T_1(2h)}{15}$$

The calculations can be organized according to the following scheme

$T(8h)$				
$T(4h)$	$T_1(4h)$			
$T(2h)$	$T_1(2h)$	$T_2(2h)$		
$T(h)$	$T_1(h)$	$T_2(h)$	$T_3(h)$	
$T(h/2)$	$T_1(h/2)$	$T_2(h/2)$	$T_3(h/2)$	$T_4(h/2)$
⋮	⋮	⋮	⋮	⋮
	$\longrightarrow$ (2^2-1)-rule	$\longrightarrow$ (2^4-1)-rule	$\longrightarrow$ (2^6-1)-rule	$\longrightarrow$ (2^8-1)-rule

Improper integrals

Given an improper integral or an integral with a non-regular integrand (i.e. with unbounded derivatives). Before using some numerical method (e.g. Simpson's formula), the integral may be transformed by substitution, integration by parts or series expansion etc.

Bounded interval

If origin is the singular point, try the substitution $x=t^\alpha$ with α large enough.

Example. The substitution $x=t^2$ transforms the integrals

$$\int_0^1 \tan\sqrt{x}\,dx \text{ and } \int_0^1 \frac{\cos x}{\sqrt{x}}\,dx$$

to the regular forms $\int_0^1 2t\tan t\,dt$ and $\int_0^1 2\cos t^2 dt$, respectively.

Unbounded interval

Suitable substitutions: $t=\arctan x$, $x=t^{-\alpha}$, $e^{-ax}=t$ etc.

Example. The substitution $x=t^{-\alpha}$ transforms the integral

$$\int_1^\infty (1+x^6)^{-1/2}\,dx \text{ to } \alpha\int_0^1 t^{2\alpha-1}(1+t^{6\alpha})^{-1/2}\,dt$$

with a regular integrand if $\alpha=1/2$ (or $\alpha=n/2$, $n=2, 3, 4, \ldots$).

Double integrals

The integral $\int_0^1\int_0^1 f(x, y)\,dxdy$ may be calculated by successive use of some of the above rules of integration. Thus, using a one-dimensional approximation

$$\int_0^1 f(x)dt \approx \sum_{i=1}^n a_i f(x_i), \text{ then } \quad \boxed{\int_0^1\int_0^1 f(x, y)dxdy \approx \sum_{i,\,j=1}^n a_i a_j f(x_i, y_j)}$$

becomes a two-dimensional approximation.

Also, cf. the Monte Carlo method below.

Numerical differentiation

(16.1)

Derivative	Approximation	Truncation error
(a) $f'(x)$	$\frac{1}{2h}(f(x+h)-f(x-h))$	$0(h^2)$
(b) $f'(x)$	$\frac{1}{12h}(-f(x+2h)+8f(x+h)-8f(x-h)+f(x-2h))$	$0(h^4)$
(c) $f''(x)$	$\frac{1}{h^2}(f(x+h)-2f(x)+f(x-h))$	$0(h^2)$
(d) $f''(x)$	$\frac{1}{12h^2}(-f(x+2h)+16f(x+h)-30f(x)+16f(x-h)-f(x-2h))$	$0(h^4)$
(e) $f'''(x)$	$\frac{1}{2h^3}(f(x+2h)-2f(x+h)+2f(x-h)-f(x-2h))$	$0(h^2)$

These approximations can be improved by the use of Richardson extrapolation, see sec. 16.3.

General differentiation formula

$$y^{(n)}(x)=h^{-n}\left(\alpha_n^{(n)}\Delta^n y(x)+\frac{\alpha_n^{(n+1)}}{n+1}\Delta^{n+1}y(x)+\frac{\alpha_n^{(n+2)}}{(n+1)(n+2)}\Delta^{n+2}y(x)+\dots\right)=$$

$$=h^{-n}\left(\alpha_n^{(n)}\nabla^n y(x)-\frac{\alpha_n^{(n+1)}}{n+1}\nabla^{n+1}y(x)+\frac{\alpha_n^{(n+2)}}{(n+1)(n+2)}\nabla^{n+2}y(x)-\dots\right),$$

where $\alpha_n^{(k)}$ are Stirling numbers of the first kind.

In particular

1. $$y'(x)=\frac{1}{h}\left(\Delta-\frac{1}{2}\Delta^2+\frac{1}{3}\Delta^3-\frac{1}{4}\Delta^4+\dots\right)y(x)=\frac{1}{h}\left(\nabla+\frac{1}{2}\nabla^2+\frac{1}{3}\nabla^3+\frac{1}{4}\nabla^4+\dots\right)y(x)=$$

$$=\frac{1}{h}\left(\mu\delta-\frac{1}{6}\mu\delta^3+\frac{1}{30}\mu\delta^5-\frac{1}{140}\mu\delta^7+\frac{1}{630}\mu\delta^9-\dots\right)y(x)=$$

$$=\frac{\mu}{h}\sum_{k=0}^{\infty}(-1)^k\cdot\frac{(k!)^2}{(2k+1)!}\delta^{2k+1}y(x)$$

2. $$y''(x)=\frac{1}{h^2}\left(\Delta^2-\Delta^3+\frac{11}{12}\Delta^4-\frac{5}{6}\Delta^5+\frac{137}{180}\Delta^6-\frac{7}{10}\Delta^7+\dots\right)y(x)=$$

$$=\frac{1}{h^2}\left(\nabla^2+\nabla^3+\frac{11}{12}\nabla^4+\frac{5}{6}\nabla^5+\frac{137}{180}\nabla^6+\frac{7}{10}\nabla^7+\dots\right)y(x)=$$

$$=\frac{1}{h^2}\left(\delta^2-\frac{1}{12}\delta^4+\frac{1}{90}\delta^6-\frac{1}{560}\delta^8+\frac{1}{3150}\delta^{10}+\dots\right)y(x)=$$

$$=\frac{1}{h^2}\sum_{k=0}^{\infty}2(-1)^k\cdot\frac{(k!)^2}{(2k+2)!}\delta^{2k+2}y(x)$$

Using *operator multiplication* foımulas for higher order derivatives can be derived from the above formulas. [In that case the identity $\mu^2=1+\delta^2/4$ may be used in order to avoid higher powers than one of μ in the formulas.] E.g.

$$y^{(4)}=D^4y=D^2D^2y=\left[\frac{1}{h^2}\left(\delta^2-\frac{1}{12}\delta^4+\frac{1}{90}\delta^6-\frac{1}{560}\delta^8+\frac{1}{3150}\delta^{10}+\ldots\right)\right]^2 y=$$

$$=\frac{1}{h^4}\left(\delta^4-\frac{1}{6}\delta^6+\frac{7}{240}\delta^8-\frac{41}{7560}\delta^{10}+\ldots\right)y$$

Formulas for numerical differentiation

(For calculation of derivatives at non-tabular points the table below may be combined with interpolation.)

Notation: $y_k = y(x+kh)$		$y^{(k)}= y^{(k)}(\xi)$, ξ in the relevant interval | Truncation error | =
$y_0' =$	$(y_1-y_0)/h$	$hy''/2$
	$(y_1-y_{-1})/2h$	$h^2y^{(3)}/6$
	$(-y_2+4y_1-3y_0)/2h = (3y_0-4y_{-1}+y_{-2})/2h$	$h^2y^{(3)}/3$
	$(2y_3-9y_2+18y_1-11y_0)/6h = (11y_0-18y_{-1}+9y_{-2}-2y_{-3})/6h$	$h^3y^{(4)}/12$
	$(-3y_4+16y_3-36y_2+48y_1-25y_0)/12h$	$h^4y^{(5)}/5$
	$(25y_0-48y_{-1}+36y_{-2}-16y_{-3}+3y_{-4})/12h$	$h^4y^{(5)}/5$
	$(-y_2+8y_1-8y_{-1}+y_{-2})/12h$	$h^4y^{(5)}/30$
	$(12y_5-75y_4+200y_3-300y_2+300y_1-137y_0)/60h$	$h^5y^{(6)}/6$
	$(137y_0-300y_{-1}+300y_{-2}-200y_{-3}+75y_{-4}-12y_{-5})/60h$	$h^5y^{(6)}/6$
	$(y_3-9y_2+45y_1-45y_{-1}+9y_{-2}-y_{-3})/60h$	$h^6y^{(7)}/140$
$y_0'' =$	$(y_2-2y_1+y_0)/h^2 = (y_0-2y_{-1}+y_{-2})/h^2$	$hy^{(3)}$
	$(y_1-2y_0+y_{-1})/h^2$	$h^2y^{(4)}/12$
	$(-y_3+4y_2-5y_1+2y_0)/h^2 = (2y_0-5y_{-1}+4y_{-2}-y_{-3})/h^2$	$11h^2y^{(4)}/12$
	$(11y_4-56y_3+114y_2-104y_1+35y_0)/12h^2$	$5h^3y^{(5)}/6$
	$(35y_0-104y_{-1}+114y_{-2}-56y_{-3}+11y_{-4})/12h^2$	$5h^3y^{(5)}/6$
	$(-y_2+16y_1-30y_0+16y_{-1}-y_{-2})/12h^2$	$h^4y^{(5)}/90$
	$(-10y_5+61y_4-156y_3+214y_2-154y_1+45y_0)/12h^2$	$137h^4y^{(5)}/180$
	$(45y_0-154y_{-1}+214y_{-2}-156y_{-3}+61y_{-4}-10y_{-5})/12h^2$	$137h^4y^{(5)}/180$
	$(2y_3-27y_2+270y_1-490y_0+270y_{-1}-27y_{-2}+2y_{-3})/180h^2$	$h^6y^{(8)}/560$
$y_0^{(3)} =$	$(y_3-3y_2+3y_1-y_0)/h^3 = (y_0-3y_{-1}+3y_{-2}-y_{-3})/h^3$	$3hy^{(4)}/2$
	$(-3y_4+14y_3-24y_2+18y_1-5y_0)/2h^3$	$7h^2y^{(5)}/4$
	$(5y_0-18y_{-1}+24y_{-2}-14y_{-3}+3y_{-4})/2h^3$	$7h^2y^{(5)}/4$
	$(y_2-2y_1+2y_{-1}-y_{-2})/2h^3$	$h^2y^{(5)}/4$
	$(-y_3+8y_2-13y_1+13y_{-1}-8y_{-2}+y_{-3})/8h^3$	$7h^4y^{(7)}/120$
$y_0^{(4)} =$	$(y_4-4y_3+6y_2-4y_1+y_0)/h^4$	$2hy^{(5)}$
	$(y_0-4y_{-1}+6y_{-2}-4y_{-3}+y_{-4})/h^4$	$2hy^{(5)}$
	$(y_2-4y_1+6y_0-4y_{-1}+y_{-2})/h^4$	$h^2y^{(5)}/6$

In the following table with approximations to partial derivatives the following notation is used $u_{0,0}=u(x, y)$ $u_{m,n}=u(x+mh, y+nh)$

Thus $\left(\frac{\partial u}{\partial x}\right)_{0,0}$ is the partial derivative $\frac{\partial u}{\partial x}$ in the point (x, y)

(16.2)

Partial derivative	Approximation	Truncation error	Configuration
(a) $\left(\frac{\partial u}{\partial x}\right)_{0,0}$	$\frac{1}{2h}(u_{1,0}-u_{-1,0})$	$0(h^2)$	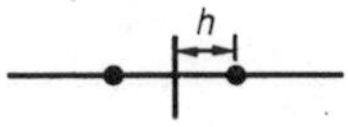
(b) $\left(\frac{\partial^2 u}{\partial x^2}\right)_{0,0}$	$\frac{1}{h^2}(u_{1,0}-2u_{0,0}+u_{-1,0})$	$0(h^2)$	
(c) $\left(\frac{\partial^2 u}{\partial x^2}\right)_{0,0}$	$\frac{1}{12h^2}(-u_{2,0}+16u_{1,0}-30u_{0,0}+16u_{-1,0}-u_{-2,0})$	$0(h^4)$	
(d) $\left(\frac{\partial^3 u}{\partial x^3}\right)_{0,0}$	$\frac{1}{2h^3}(u_{2,0}-2u_{1,0}+2u_{-1,0}-u_{-2,0})$	$0(h^2)$	
(e) $\left(\frac{\partial^4 u}{\partial x^4}\right)_{0,0}$	$\frac{1}{h^4}(u_{2,0}-4u_{1,0}+6u_{0,0}-4u_{-1,0}+u_{-2,0})$	$0(h^2)$	
(f) $\left(\frac{\partial^2 u}{\partial x \partial y}\right)_{0,0}$	$\frac{1}{4h^2}(u_{1,1}-u_{1,-1}-u_{-1,1}+u_{-1,-1})$	$0(h^2)$	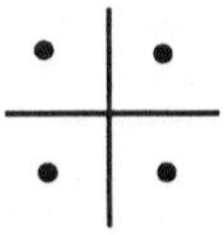

(g) $\Delta u=\left(\frac{\partial^2 u}{\partial x^2}+\frac{\partial^2 u}{\partial y^2}\right)_{0,0}=\frac{1}{h^2}(u_{1,0}+u_{0,1}+u_{-1,0}+u_{0,-1}-4u_{0,0})+0(h^2)$

(h) $\Delta^2 u=\left(\frac{\partial^4 u}{\partial x^4}+2\frac{\partial^4 u}{\partial x^2 \partial y^2}+\frac{\partial^4 u}{\partial y^4}\right)_{0,0}=$

$=\frac{1}{h^4}[u_{2,0}+u_{0,2}+u_{-2,0}+u_{0,-2}-8(u_{1,0}+u_{0,1}+u_{-1,0}+u_{0,-1})+$

$+2(u_{1,1}+u_{1,-1}+u_{-1,1}+u_{-1,-1})+20u_{0,0}]+0(h^2)$

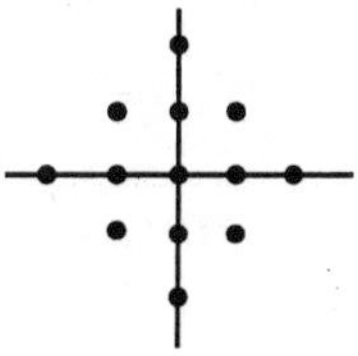

Monte Carlo methods

A Monte Carlo method is a method to solve a mathematical problem by an experiment with random numbers. Such methods can be used to estimate areas, volymes, multiple integrals and solutions to partical differential equations. E.g.: Let X and Y be independent random variables uniformly distributed in (a, b) and (c, d). Suppose that M out of N generated points (X, Y) belong to the subregion D of the rectangle R. Then $(M/N)\,(b-a)\,(d-c)$ for large N is an estimate of the area of D.

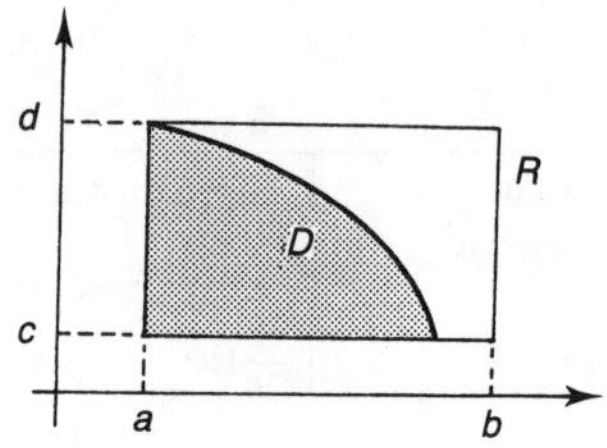

16.5 Numerical Solutions of Differential Equations

Ordinary differential equations. Initial value problems

The methods below may be combined with Richardson extrapolation.

Problem. Find $y=y(x)$ such that

$$\begin{cases} y'=f(x, y) \\ y(x_0)=y_0 \end{cases}$$

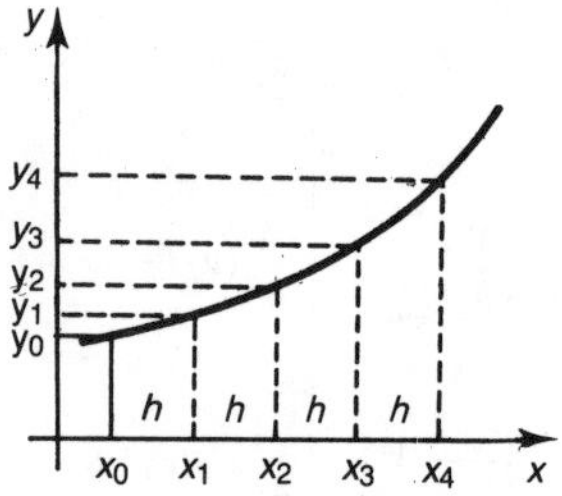

$$x_n=x_0+nh$$

Euler's method

$$y_{n+1}=y_n+hf(x_n, y_n)$$

Global truncation error$=O(h)$.

System in vector form: $\begin{cases} \mathbf{y}'=\mathbf{f}(x, \mathbf{y}) \\ \mathbf{y}(x_0)=\mathbf{y}_0 \end{cases}$ $\quad \mathbf{y}=(y_1, \ldots, y_m)$, $\mathbf{f}=(f_1, \ldots, f_m)$

$$\mathbf{y}_{n+1}=\mathbf{y}_n+h\mathbf{f}(x_n, \mathbf{y}_n)$$

Midpoint method

$$y_{n+1}=y_n+hf\left(x_n+\frac{h}{2}, y_n+\frac{hk_1}{2}\right)$$

$$k_1=f(x_n, y_n)$$

Global truncation error$=0(h^2)$

Heun's method

$$y_{n+1}=y_n+\frac{h}{2}(k_1+k_2)$$

$$k_1=f(x_n, y_n)$$

$$k_2=f(x_n+h, y_n+hk_1)$$

Global truncation error$=0(h^2)$

The Runge-Kutta methods

1. First order equation. Problem: $y'=f(x, y)$, $y(x_0)=y_0$

Iteration: $(x_{n+1}=x_n+h)$

$$y_{n+1}=y_n+\frac{1}{6}(k_1+2k_2+2k_3+k_4)$$

$$k_1=hf(x_n, y_n)$$

$$k_2=hf\left(x_n+\frac{h}{2}, y_n+\frac{k_1}{2}\right)$$

$$k_3=hf\left(x_n+\frac{h}{2}, y_n+\frac{k_2}{2}\right)$$

$$k_4=hf(x_n+h, y_n+k_3)$$

Global truncation error$=0(h^4)$.

2. System of first order equations

Problem: $\begin{cases} y'=f(x, y, z, \ldots) & y(x_0)=y_0 \\ z'=g(x, y, z, \ldots) & z(x_0)=z_0 \\ \ldots & \ldots \end{cases}$

Iteration: $(x_{n+1}=x_n+h)$

$$y_{n+1}=y_n+\frac{1}{6}(k_1+2k_2+2k_3+k_4)$$
$$z_{n+1}=z_n+\frac{1}{6}(m_1+2m_2+2m_3+m_4)$$
$$\ldots$$

$$k_1=hf(x_n, y_n, z_n, \ldots)$$
$$m_1=hg(x_n, y_n, z_n, \ldots)$$
$$\ldots$$
$$k_2=hf\left(x_n+\frac{h}{2}, y_n+\frac{k_1}{2}, z_n+\frac{m_1}{2}, \ldots\right)$$
$$m_2=hg\left(x_n+\frac{h}{2}, y_n+\frac{k_1}{2}, z_n+\frac{m_1}{2}, \ldots\right)$$
$$\ldots$$
$$k_3=hf\left(x_n+\frac{h}{2}, y_n+\frac{k_2}{2}, z_n+\frac{m_2}{2}, \ldots\right)$$
$$m_3=hg\left(x_n+\frac{h}{2}, y_n+\frac{k_2}{2}, z_n+\frac{m_2}{2}, \ldots\right)$$
$$\ldots$$

$$k_4=hf(x_n+h,\ y_n+k_3,\ z_n+m_3,\ \ldots)$$
$$m_4=hg(x_n+h,\ y_n+k_3,\ z_n+m_3,\ \ldots)$$
$$\ldots$$

3. Second order equation

Problem:

$y''=f(x, y, y'),\ y(x_0)=y_0,\ y'(x_0)=y_0'$

Iteration: $(x_{n+1}=x_n+h)$

$$y_{n+1}=y_n+hy_n'+\frac{h}{6}(k_1+k_2+k_3)$$
$$y_{n+1}'=y_n'+\frac{1}{6}(k_1+2k_2+2k_3+k_4)$$

$$k_1=hf(x_n,\ y_n,\ y_n')$$
$$k_2=hf\left(x_n+\frac{h}{2},\ y_n+y_n'\frac{h}{2},\ y_n'+\frac{k_1}{2}\right)$$
$$k_3=hf\left(x_n+\frac{h}{2},\ y_n+y_n'\frac{h}{2}+\frac{k_1h}{4},\ y_n'+\frac{k_2}{2}\right)$$
$$k_4=hf\left(x_n+h,\ y_n+y_n'h+\frac{k_2h}{2},\ y_n'+k_3\right)$$

4. n'th order equation

Problem: $\begin{cases} y^{(n)}=f(x,\ y,\ y',\ \ldots,\ y^{(n-1)}) \\ y(x_0)=y_0,\ y'(x_0)=y_0',\ \ldots,y^{(n-1)}(x_0)=y_0^{(n-1)} \end{cases}$

The problem may be transformed to the case 2 by the substitutions $y_1=y,\ y_2=y',\ \ldots,\ y_n=y^{(n-1)}$.

Taylor series method

Problem:

(16.3) $\begin{cases} y'=f(x,\ y) \\ y(x_0)=y_0 \end{cases}$

By (16.3), $y'(x_0)=f(x_0,\ y_0)$.

Differentiating (16.3) yields

$$y''=f_x'(x,\ y)+y'\cdot f_y'(x,\ y) \Rightarrow y''(x_0)=f_x'(x_0,\ y_0)+y'(x_0)f_y'(x_0,\ y_0)$$

etc.

Then use Taylor's formula [cf. sec. 8.5].

Remark. Sometimes, in linear differential equations, the power series substitution $y(x)=\sum_{n=0}^{\infty} a_nx^n$ (identifying coefficients), is useful.

Boundary value problems

Two-point boundary problem.

$$\begin{cases} y''=f(x, y, y'), \; a<x<b \\ y(a)=\alpha \\ y(b)=\beta \end{cases}$$

Shooting method

With a suitable start solution at one endpoint the value at the other endpoint may be calculated by any of the above methods (e.g. the Runge-Kutta method). Varying the start solution the boundary conditions will be satisfied with desired accuracy. Assuming that $y'(a)=\gamma_1$ and $y'(a)=\gamma_2$ give ending values β_1 and β_2 respectively, then try

$$y'(a)=\gamma_3=\gamma_2+\frac{\gamma_1-\gamma_2}{\beta_1-\beta_2}(\beta-\beta_2) \text{ etc.}$$

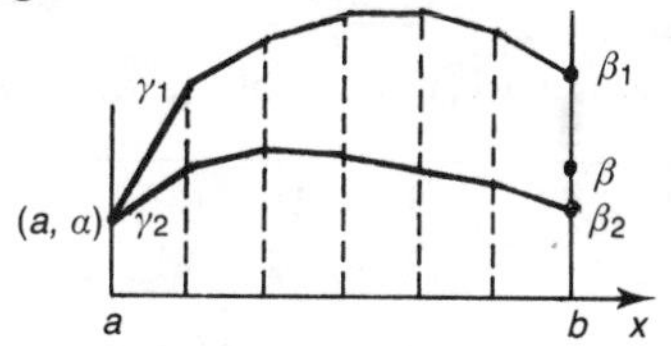

Finite difference method

Derivatives are approximated by finite difference quotients.

Example (one dimension)

$$y''+a^2y=0; \qquad y(0)=0, \; y(1)=1$$

Dividing the interval [0, 1] into n subintervals of length $h=1/n$ and using (16.1 c), the corresponding difference equations will be $[y_k \approx y(kh)]$

$$\begin{cases} y_{k+1}-2y_k+y_{k-1}+a^2h^2y_k=0, \; k=1, \ldots, n-1 \\ y_0=0, \; y_n=1 \end{cases}$$

The finite element method (Ritz-Galerkin method)

Model problem

(16.4) $$\begin{cases} -u''(x)+q(x)u(x)=f(x), \; 0<x<1 \quad (q(x)\geq 0) \\ u(0)=u(1)=0 \end{cases}$$

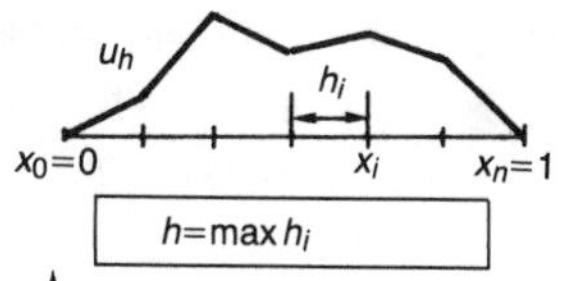

Let V denote the space consisting of all continuous and piecewise differentiable functions $v(x)$ with $v(0)=v(1)=0$. Using the notation $(u, v)=\int_0^1 u(x)\,v(x)dx$, the *variational formulation* of (16.4) is: Find $u \in V$ such that

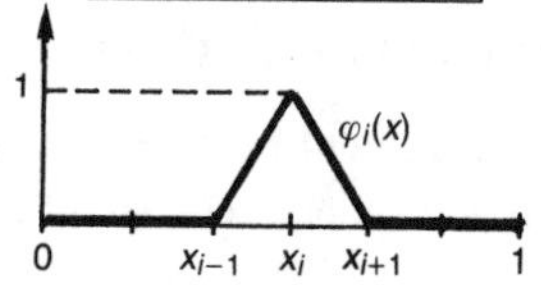

(16.5) $$(u', v')+(qu, v)=(f, v), \text{ all } v \in V$$

Now, given a subdivision $0=x_0<x_1<\ldots<x_n=1$ of [0, 1], let V_h denote the finite dimensional space consisting of all continuous and piecewise linear functions $v(x)$, subordinate to the subdivision, with $v(0)=v(1)=0$ (cf. the figure). [In greater generality: V_h consists of all piecewise polynomials $v(x)$ of degree $\leq k$ with $v(0)=v(1)=0$ and such that the derivatives of $v(x)$ of order $\leq k-1$ are continuous, so-called *splines*.]

The space V_h has basis functions $\varphi_i(x)$, $i=1, \dots, n$ (cf. the figure) so that every $u_h \in V_h$ can be written

$$u_h = \sum_{i=1}^{n} \alpha_i \varphi_i(x), \ \alpha_i \text{ constants}$$

The corresponding *discrete analog* of (16.5) is:

Find $u_h \in V_h$ such that

$$\boxed{(u_h', v_h') + (q u_h, v_h) = (f, v_h), \text{ all } v_h \in V_h}$$

or, equivalently:

Find the constants α_i such that

$$\boxed{\sum_{i=1}^{n} \alpha_i(\varphi_i', \varphi_j') + \sum_{i=1}^{n} \alpha_i(q\varphi_i, \varphi_j) = (f, \varphi_j), \ j=1, \dots, n} \tag{16.6}$$

This is a linear system of $\alpha_1, \dots, \alpha_n$ with the coefficients

$$(\varphi_i', \varphi_i') = \frac{1}{h_i} + \frac{1}{h_{i+1}}$$

$$(\varphi_i', \varphi_{i-1}') = (\varphi_{i-1}', \varphi_i') = -\frac{1}{h_i}$$

$$(\varphi_i', \varphi_j') = 0 \text{ if } |i-j| > 1$$

$$(q\varphi_i, \varphi_i) \begin{cases} \approx \frac{1}{6} h_i q(x_{i-1}) + \frac{1}{6}(h_i + h_{i+1})\, q(x_i) + \frac{1}{6}\, h_{i+1} q(x_{i+1}) \\ = \frac{1}{3}\, q_0(h_i + h_{i+1}) \text{ if } q(x) \equiv q_0 \text{ constant} \end{cases}$$

$$(q\varphi_i, \varphi_{i-1}) = (q\varphi_{i-1}, \varphi_i) \begin{cases} \approx \frac{1}{12}\, h_i[q(x_{i-1}) + q(x_i)] \\ = \frac{1}{6}\, q_0 h_i \text{ if } q(x) \equiv q_0 \end{cases}$$

$$(q\varphi_i, \varphi_j) = 0 \text{ if } |i-j| > 1$$

The system (16.6) always admits a unique solution, because the coefficient matrix is positive definite. If q, f are smooth, then u_h converges to u in the norm

$\|v\| = (v, v)^{1/2}$ with the rate $O(h^2)$, i.e.

$\|u - u_h\| \leq Ch^2$, $h = \max h_i$

Integral equations

Replacement by Riemann sum

As an example, consider a Fredholm equation of second kind (cf. sec. 9.4)

$$y(x) - \int_a^b K(x, t)\, y(t)\, dt = h(x) \tag{16.7}$$

Setting $x_j = a + j(b-a)/n$ and $y_j = y(x_j)$, $j=1, \dots, n$, then (16.7) can be approximated by the system of linear equations

$$y_i - \sum_{j=1}^{n} K(x_i, x_j)\, y_j (b-a)/n = h(x_i), \ i=1, \dots, n$$

Writing this system $(E-A)Y=H$, then if $\det(E-A)\neq 0$, $Y=(E-A)^{-1}H$ which converges to $y(x)$ as $n\to\infty$, under suitable assumptions.

Remark. Instead of a Riemann sum, other integral approximations may be used.

Partial differential equations

Some illustrating examples.

Example 1. The finite element method

The Dirichlet problem in the plane.

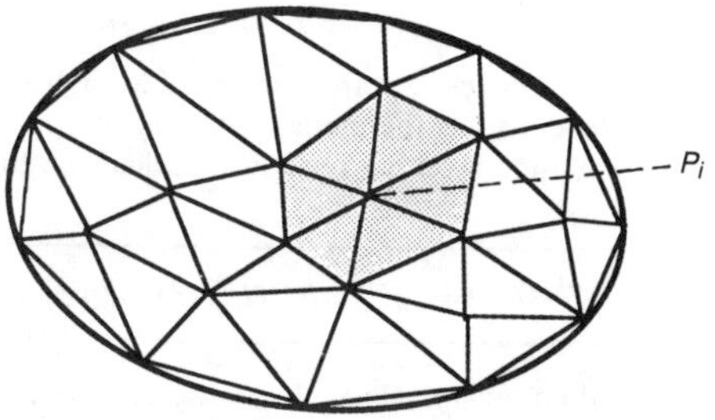

$$(16.8)\qquad \begin{cases} -\Delta u=f \text{ in } \Omega \\ \quad u=0 \text{ on the boundary } \partial\Omega \end{cases}$$

In analogy with the one-dimensional case above, *basis functions* $\varphi_i(x, y)$ of the space V_h consisting of all continuous piece-wise linear functions in Ω_h which are zero on $\partial\Omega_h$, are defined in the following way: Assume that there are N interior *node points* P_i of the triangulation. Then $\varphi_i(x, y)$ is chosen such that $\varphi_i(P_i)=1$ and $\varphi_i(P_j)=0, j\neq i$. An approximate solution $u_h=\sum_{i=1}^{N}\alpha_i\varphi_i$ of (16.8) is then obtained as the unique solution of the linear system

Ω=original domain
Ω_h=triangulated domain
$\varphi_i\equiv 0$ outside the shadowed region.
h=max diam T_i ($T_i\in\Omega_h$)

$$\sum_{i=1}^{N}\alpha_i\iint_{\Omega}(\text{grad }\varphi_i)(\text{grad }\varphi_j)dxdy=\iint_{\Omega}f(x, y)\,\varphi_j(x, y)dxdy,\ j=1, \dots, N$$

Rate of convergence$=O(h^2)$ (smooth data).

Example 2. Finite difference method

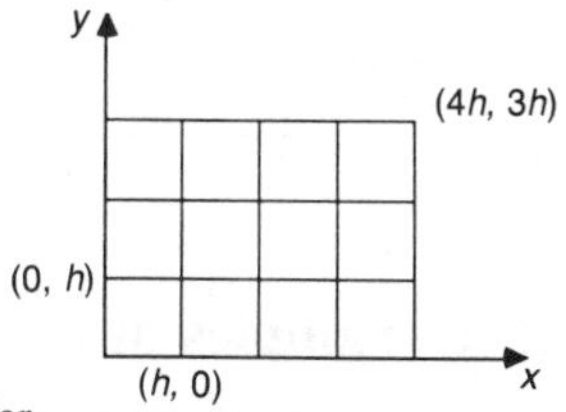

The Dirichlet problem in a rectangular domain.

$$\begin{cases} -\Delta u=f \text{ in } \Omega \\ \quad u=g \text{ on } \partial\Omega \end{cases}$$

Using the 5-points-formula (16.2 g), the corresponding difference scheme will be

$U_{i,j}=U(ih, jh)$

$$\begin{cases} -h^2\Delta_h U_{i,j}\equiv 4U_{i,j}-U_{i+1,j}-U_{i-1,j}-U_{i,j+1}-U_{i,j-1}=h^2 f(ih, jh),\ 1\leq i,\ j\leq 3 \\ \qquad U_{i,j}=g(ih, jh) \text{ on } \partial\Omega \end{cases}$$

$\Leftrightarrow$

$$\begin{bmatrix} 4 & -1 & 0 & -1 & 0 & 0 \\ -1 & 4 & -1 & 0 & -1 & 0 \\ 0 & -1 & 4 & 0 & 0 & -1 \\ -1 & 0 & 0 & 4 & -1 & 0 \\ 0 & -1 & 0 & -1 & 4 & -1 \\ 0 & 0 & -1 & 0 & -1 & 4 \end{bmatrix}\begin{bmatrix} U_{11} \\ U_{21} \\ U_{31} \\ U_{12} \\ U_{22} \\ U_{32} \end{bmatrix}=\begin{bmatrix} h^2f_{11}+U_{01}+U_{10} \\ h^2f_{21}+U_{20} \\ h^2f_{31}+U_{41}+U_{30} \\ h^2f_{12}+U_{02}+U_{13} \\ h^2f_{22}+U_{23} \\ h^2f_{32}+U_{42}+U_{33} \end{bmatrix}$$

Note that the coefficient matrix is symmetric and sparsely filled, i.e. it contains many zeros.

Remark. If Ω is not rectangular, (16.2 g) has to be modified near the boundary.

Example 3. Finite difference method

Heat equation [$u=u(x, y, t)$, Ω: $0<x<1$, $0<y<1$]

$$\begin{cases} u_t'=\Delta u, \ (x, y)\in\Omega, \ 0<t<T & \\ u(x, y, t)=0, \ (x, y)\in\partial\Omega, \ 0<t<T & \text{(boundary condition)} \\ u(x, y, 0)=g(x, y), \ (x, y)\in\Omega & \text{(initial condition)} \end{cases}$$

The space variables x, y are discretized as in example 2 above. The time interval is divided into parts of length τ. At the mesh points $(ih, jh, n\tau)$ the approximate solution $U=U^n_{i,j}=U(ih, jh, n\tau)$ can, at any time level, be recursively calculated by the difference equation

$$\frac{1}{\tau}(U^{n+1}_{i,j}-U^n_{i,j})=\Delta_h U^n_{i,j}\equiv h^{-2}(U^n_{i+1,j}+U^n_{i-1,j}+U^n_{i,j+1}+U^n_{i,j-1}-4U^n_{i,j})$$

Initial and boundary values of the discrete problem are obtained in an obvious manner from the corresponding values of the original continuous problem.

Convergence $U\to u$ as h, $\tau\to 0$ if $\tau/h^2=\text{constant}\leq 1/4$.

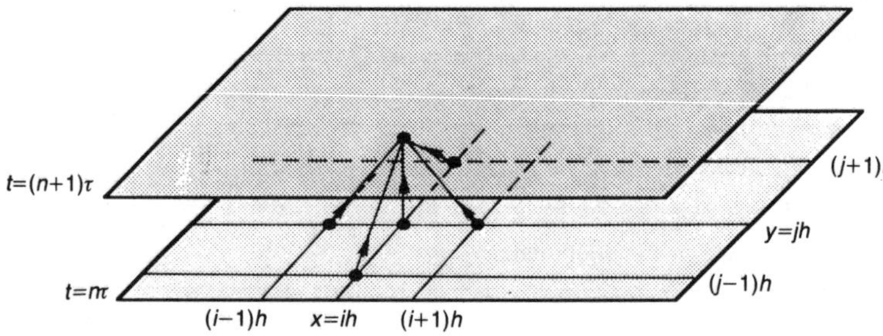

Further examples of the finite difference method

Notation.

$U^n_i = U(ih, nk) \qquad f^n = f(nk) \qquad f_i = f(ih)$

$\delta_x U^n_i = U^n_{i+1} - U^n_i \qquad \bar{\delta}_x U^n_i = U^n_i - U^n_{i-1}$

$\delta_x \bar{\delta}_x U^n_i = U^n_{i+1} - 2U^n_i + U^n_{i-1}$

Finite difference methods for the equation

$$\frac{\partial u}{\partial t} = a^2\frac{\partial^2 u}{\partial x^2} \qquad (a=\text{constant})$$

1. Forward difference scheme (*explicit*), $q = \dfrac{ka^2}{h^2}$ = constant

$$\frac{1}{k}\delta_t U_i^n = \frac{a^2}{h^2}\delta_x \bar{\delta}_x U_i^n \quad \Rightarrow$$
$$U_i^{n+1} = U_i^n + q(U_{i+1}^n - 2U_i^n + U_{i-1}^n)$$

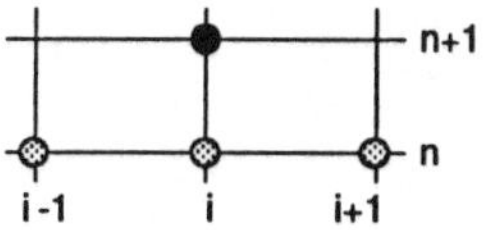

Local truncation error = $O(k+h^2)$. *Stable* if $q \le \dfrac{1}{2}$.

Special cases :

(*i*) $q = \dfrac{1}{2}$: $U_i^{n+1} = \dfrac{1}{2}(U_{i+1}^n + U_{i-1}^n)$

(*ii*) $q = \dfrac{1}{6}$: *Local truncation error* = $O(k^2 + h^4)$.

2. Backward difference scheme (*implicit*), $q = \dfrac{ka^2}{h^2}$ = constant

$$\frac{1}{k}\delta_t U_i^n = \frac{a^2}{h^2}\delta_x \bar{\delta}_x U_i^{n+1} \quad \Rightarrow$$
$$-qU_{i+1}^{n+1} + (1+2q)U_i^{n+1} - qU_{i-1}^{n+1} = U_i^n$$

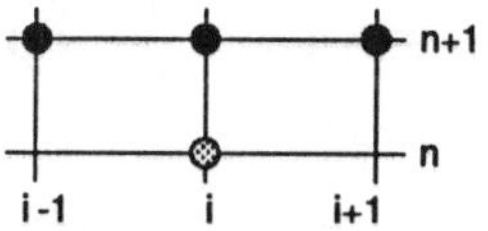

Local truncation error = $O(k+h^2)$. *Always stable.*

3. Crank - Nicolson scheme (*implicit*), $q = \dfrac{ka^2}{h^2}$ = constant

$$\frac{1}{k}\delta_t U_i^n = \frac{a^2}{2h^2}\delta_x \bar{\delta}_x (U_i^n + U_i^{n+1}) \quad \Rightarrow$$
$$-qU_{i+1}^{n+1} + 2(1+2q)U_i^{n+1} - qU_{i-1}^{n+1} =$$
$$= qU_{i+1}^n + 2(1-2q)U_i^n + qU_{i-1}^n$$

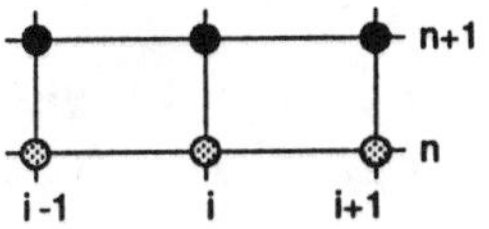

Local truncation error = $O(k^2+h^2)$. *Always stable.*

Initial - Boundary - Value Problem.

(*PDE*) $u_t = a^2 u_{xx}, \quad 0 < x < L,\ t > 0$

(*BC*1,2) $u(0,t) = f(t),\ u(L,t) = g(t), \quad t > 0$

(*IC*) $u(x,0) = p(x), \quad 0 < x < L$

Finite difference approximation
($h = L/M$, M = number of subintervals) :

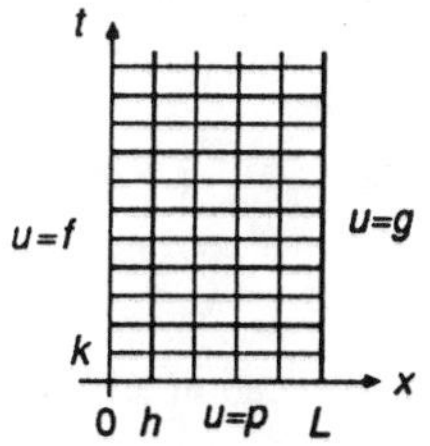

(*PFD*) Any of the schemes above.

(*BC*1,2) $U_0^n = f^n = f(nk) \quad U_M^n = g^n$

(*IC*) $U_i^0 = p_i = p(ih)$

A boundary condition of the form $u_x(0,t) = f(t)$ may be approximated by

$$\frac{1}{h}(U_1^n - U_0^n) = f^n \text{ with truncation error } = O(h) \text{ or}$$

$$\frac{1}{2h}(-U_2^n + 4U_1^n - 3U_0^n) = f^n \text{ with truncation error } = O(h^2)$$

etc. (cf. table in sec. 16.4)

Finite difference methods for the equation

$$\frac{\partial^2 u}{\partial t^2} = a^2 \frac{\partial^2 u}{\partial x^2} \qquad (a = \text{constant})$$

1. *Explicit* scheme. $\quad q = \dfrac{a^2k^2}{h^2} = \text{constant}$

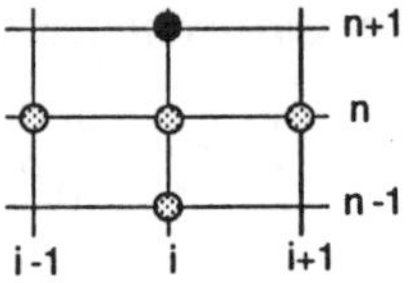

$$\frac{1}{k^2}\delta_t \bar{\delta}_t U_i^n = \frac{a^2}{h^2}\delta_x \bar{\delta}_x U_i^n \quad \Rightarrow$$

$$U_i^{n+1} = qU_{i+1}^n + 2(1-q)U_i^n + qU_{i-1}^n - U_i^{n-1}$$

Local truncation error $= O(k^2 + h^2)$. *Stable* if $q \le 1$.

2. *Implicit* scheme. $\quad q = \dfrac{a^2k^2}{h^2} = \text{constant}$

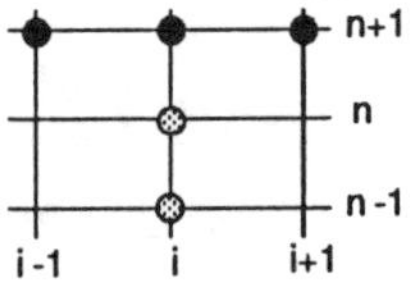

$$\frac{1}{k^2}\delta_t \bar{\delta}_t U_i^n = \frac{a^2}{h^2}\delta_x \bar{\delta}_x U_i^{n+1} \quad \Rightarrow$$

$$-qU_{i+1}^{n+1} + (1+2q)U_i^{n+1} - qU_{i-1}^{n+1} = 2U_i^n - U_i^{n-1}$$

Local truncation error $= O(k^2 + h^2)$. *Always stable.*

Initial - Value Problem.

(PDE)	$u_{tt} = a^2 u_{xx}$,	$t > 0$
(IC1)	$u(x,0) = f(x)$	
(IC2)	$u_t(x,0) = g(x)$	

Finite difference approximation :
The initial conditions may be approximated by

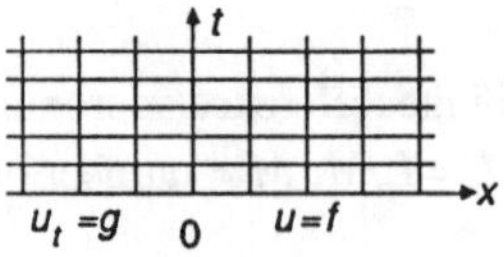

(IC1) $\quad U_i^0 = f_i \quad$ (exact)

(IC2) $\quad \partial_t U_i^0 = g_i \Rightarrow U_i^1 = U_i^0 + kg_i$

with truncation error $= O(k)$ or

$$U_i^1 = U_i^0 + kg_i + \frac{1}{2}qk(f_{i+1} - 2f_i + f_{i-1})$$

with truncation error $= O(kh^2 + k^2)$

16.6 Numerical summation

Euler-MacLaurin summation formula

Let $x_k = x_0 + hk$. Then

$$f(x_0)+f(x_1)+\dots+f(x_n)=\frac{1}{h}\int_{x_0}^{x_n} f(x)dx+\frac{1}{2}[f(x_0)+f(x_n)]+$$

$$+\sum_{k=1}^{m}\frac{h^{2k-1}B_{2k}}{(2k)!}[f^{(2k-1)}(x_n)-f^{(2k-1)}(x_0)]+\frac{nB_{2m+2}h^{2m+2}}{(2m+2)!}f^{(2m+2)}(\xi),\ x_0<\xi<x_n,$$

where B_j are the Bernoulli numbers (see sec. 12.3)

Remark. The formula may also be used for numerical integration.

In particular,

$$\sum_{k=m}^{n} f(k)=f(m)+f(m+1)+\dots+f(n)=\int_m^n f(x)dx+\frac{1}{2}[f(m)+f(n)]+\frac{1}{12}[f'(n)-f'(m)]-$$

$$-\frac{1}{720}[f^{(3)}(n)-f^{(3)}(m)]+\frac{1}{30240}[f^{(5)}(n)-f^{(5)}(m)]-\frac{1}{1209600}[f^{(7)}(n)-f^{(7)}(m)]+$$

$$+\dots+\frac{B_{2m}}{(2m)!}[f^{(2m-1)}(n)-f^{(2m-1)}(m)]+\dots$$

Numerical calculation of sums of infinite series

Notation:

$$s=\sum_{n=1}^{\infty} a_n,\qquad s_N=\sum_{n=1}^{N} a_n,\qquad R_N=\sum_{n=N+1}^{\infty} a_n$$

Improvement of convergence

1. Kummer's transformation

Given two convergent series

$$s=\sum_{n=1}^{\infty} a_n \text{ and } s_1=\sum_{n=1}^{\infty} b_n \text{ with } \frac{a_n}{b_n}\to c\neq 0 \text{ as } n\to\infty.$$

Then

$$s=\sum_{n=1}^{\infty} a_n=cs_1+\sum_{n=1}^{\infty}\left(1-c\frac{b_n}{a_n}\right)a_n.$$

If s_1 is known this transformation may improve the convergence.

$$\left[\text{E.g. } \sum_{n=1}^{\infty}\frac{1}{n^2+1}=\sum_{n=1}^{\infty}\frac{1}{n^2}+\sum_{n=1}^{\infty}\left(\frac{1}{n^2+1}-\frac{1}{n^2}\right)=\frac{\pi^2}{6}-\sum_{n=1}^{\infty}\frac{1}{n^2(n^2+1)}\right]$$

2. Euler's transformation

If the series to the left is convergent then

$$\sum_{n=0}^{\infty} a_n x^n=\frac{1}{1-x}\sum_{n=0}^{\infty}\left(\frac{x}{1-x}\right)^n \Delta^n a_0 \text{ where } \Delta^n a_0=\sum_{k=0}^{n}(-1)^k\binom{n}{k}a_{n-k}$$

Series with positive terms

Assume that $f(x)\geq 0$ and set $a_n=f(n)$.

Rapidly convergent series

3. Estimate by an integral:

If $f(x)$ is *decreasing* for $x>N$ then

$$\int_{N+1}^{\infty} f(x)dx \leq R_N \leq \int_{N}^{\infty} f(x)dx$$

Slowly convergent series

4. Lagrange interpolation:

Consider s_N as $P\left(\frac{1}{N}\right)$ and extrapolate according to Lagrange's interpolation formula (see sec. 16.3) towards $\frac{1}{N}=0$.

> *Example*
>
> $$s=\sum_1^{\infty}\frac{1}{n^2} \qquad s_N=P\left(\frac{1}{N}\right)=\sum_1^{N}\frac{1}{n^2}$$
>
> $P(0.1)=s_{10}=1.549767731$
> $P(0.05)=s_{20}=1.596163244$
> $P(0.04)=s_{25}=1.605723404$
>
> By Lagrange interpolation (extrapolation),
>
> $$s=P(0)=\frac{2}{3}P(0.1)-8P(0.05)+\frac{25}{3}P(0.04)=1.644901$$
>
> Correct value $=s=\frac{\pi^2}{6}=1.644934$

5. Euler-Maclurin's formula:

$$R_{N-1}=\sum_{n=N}^{\infty} f(n)=\int_N^{\infty} f(x)dx+\frac{1}{2}f(N)-\frac{1}{12}f'(N)+\frac{1}{720}f^{(3)}(N)-$$
$$-\frac{1}{30240}f^{(5)}(N)+\frac{1}{1209600}f^{(7)}(N)-\ldots-\frac{B_{2m}}{(2m)!}f^{(2m-1)}(N)-\ldots$$

If the first neglected term to the right is decreasing for $x\geq N$ then the error is less than twice the modulus of that term.

> *Example*
>
> As in the previous example, let $s=\sum_1^{\infty}\frac{1}{n^2}$.
>
> $s_9=1.53976773$
>
> $f(x)=x^{-2}$, $f^{(k)}(x)=(-1)^k(k+1)!x^{-k-2}$. Thus,
>
> $$R_9\approx\int_{10}^{\infty}\frac{dx}{x^2}+\frac{1}{2}\cdot 10^{-2}+\frac{1}{12}\cdot 2\cdot 10^{-3}-\frac{1}{720}\cdot 4!\cdot 10^{-5}=0.10516633 \text{ and}$$
>
> $s=1.64493406$
>
> Correct value $=\frac{\pi^2}{6}=1.644934066\ \ldots$

Alternating series

Remark. An alternating series $a_0-a_1+a_2-a_3+\ldots=\sum_{n=0}^{\infty}(-1)^n a_n$, $a_n>0$ may be transformed to a positive series by grouping the terms as $(a_0-a_1)+(a_2-a_3)+\ldots$ if $\{a_n\}$ is decreasing.

Rapidly convergent alternating series

6. If $a_n \downarrow 0$ as $n\to\infty$ then $|R_N|\leq a_{N+1}$.

Slowly convergent alternating series

7. Repeated means:

Example: It is known that

$$\sum_{n=1}^{\infty}\frac{(-1)^{n-1}}{2n-1}=1-\frac{1}{3}+\frac{1}{5}-\frac{1}{7}+\ldots=\frac{\pi}{4}=0.78539816\ldots$$

In the scheme below, M_j is the mean of the "northwestern" and the "southwestern" neighbours.

N	S_N	M_1	M_2	M_3	M_4	M_5
6	0.744 012					
		0.782 473				
7	0.820 935		0.785 037			
		0.787 601		0.785 339		
8	0.754 268		0.785 641		0.785 386	
		0.783 680		0.785 434		0.785 396
9	0.813 091		0.785 228		0.785 405	
		0.786 776		0.785 375		
10	0.760 460		0.785 523			
		0.784 269				
11	0.808 079					

8. Euler-Maclurin's formula

$$R_{N-1}=f(N)-f(N+1)+f(N+2)-f(N+3)+\ldots=$$

$$=\frac{1}{2}\left[f(N)-\frac{1}{2}f'(N)+\frac{1}{24}f^{(3)}(N)-\frac{1}{240}f^{(5)}(N)+\frac{17}{40320}f^{(7)}(N)-\right.$$

$$\left.-\frac{31}{725760}f^{(9)}(N)+\ldots-\frac{2^{2m}-1}{(2m)!}B_{2m}f^{(2m-1)}(N)+\ldots\right]$$

Example. See the previous example. Let

$f(x)=(2x-1)^{-1}$, $f^{(k)}(x)=(-1)^k k!2^k(2x-1)^{-k-1}$ and

$s=1-\frac{1}{3}+\frac{1}{5}-\frac{1}{7}+\ldots=\sum_{n=1}^{\infty}(-1)^{n-1}f(n)$. Then

$$s_{10}=1-\frac{1}{3}+\ldots-\frac{1}{19}=0.7604599$$

$$R_{10}=\frac{1}{21}-\frac{1}{23}+\ldots\approx\frac{1}{2}\left(\frac{1}{21}+\frac{1}{2}\cdot\frac{2!}{21^2}-\frac{1}{24}\cdot\frac{3!2^3}{21^4}+\frac{1}{240}\cdot\frac{5!2^5}{21^6}\right)=0.0249383$$

Thus, $s=s_{10}+R_{10}=0.785398$. Correct value$=\frac{\pi}{4}=0.785398163\ldots$

16.7 Programming

Symbols for flow charts

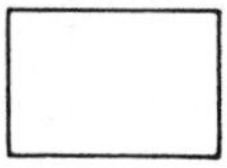

General symbol for process e.g. assigning values to variables.

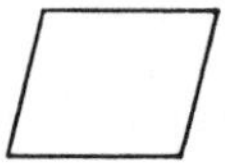

Input and output.

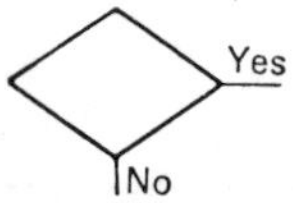

Decision with two alternatives.

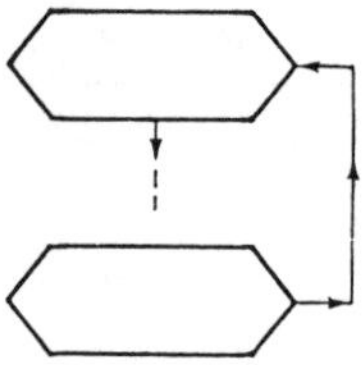

Loop in program.

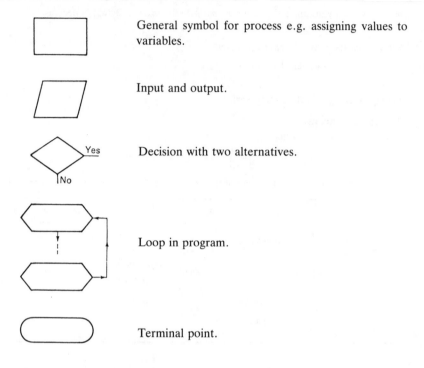

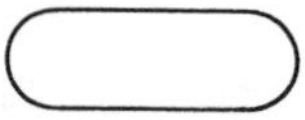

Terminal point.

BASIC

Instruction	**Purpose**
DATA	Assigns values to variables in READ statement.
DEF FN...(...)	Defines a function.
DIM	Reserves computer memory for range of variables.
END	Tells the computer to cease the execution of a program. On some computers also all variables are cleared.
FOR ... TO ...	Executes part of program for all integer values of variable from given starting value to given end value.
FOR ... TO ... STEP ...	Gives step length in connection with the instruction FOR ... TO ...
GOSUB	Jump to subroutine.

GOTO	Jump to specified line number.
IF ... THEN ...	If the condition after IF is satisfied, the computer follows the instruction after THEN. If not the computer follows the instruction in the next program line.
IF ... THEN ... ELSE ...	If the condition after IF is staisfied, the computer follows the instruction after THEN. If not the computer follows the instruction after ELSE.
INPUT	The computer prints a question mark and waits for values of one or more variables.
LET	Makes it possible to change the value of a variable.
NEXT	Used in connection with the instruction FOR ... TO ... to indicate point of looping back to the instruction after the FOR ... TO ... instruction.
ON ... GOSUB ...	Gives jump to subroutine depending upon the value of indicated variable.
ON ... GOTO ...	Gives jump to program line depending upon value of indicated variable.
ON ... RESTORE	Gives RESTORE to item in DATA statement depending upon value of variable.
PRINT	Gives transcription on the screen.
PRINT X;	Gives dense transcription.
PRINT X,	Gives transcription in columns.
PRINT 'X'	Expression between quotation marks will be printed.
PRINT TAB(N)	Gives transcription in position N on line.
RANDOMIZE	Gives random start to the random number generator, which is initiated by the instruction RND.
READ	Gives values to one or several variables according to DATA statement.
REM	Gives a possibility to write comments in a program.
RESTORE	Sets the READ pointer back to the first item in the DATA statement.
RETURN	A subroutine must be finished with a RETURN statement. Will return operation to the line following the one which called the subroutine.

Arithmetic operations

A+B	Addition	
A−B	Subtraction	
A∗B	Multiplication	
A / B	Division	
A ↑ B, A∗∗ B, AüB	Powers A^B	

Priority order

1. Functions, powers, minus sign
2. Multiplication, division
3. Addition, subtraction

Equalities and inequalities

=	equals
<	less than
<=	less than or equal to
>	greater than
>=	greater than or equal to
<>	not equal to

Mathematical functions

SIN(X), COS(X), TAN(X)	Trigonometric functions with X in radians.
ASN(X), ACS(X), ATN(X)	$\arcsin X$, $\arccos X$, $\arctan X$
LOG(X)	$\ln X$ or $\log_e X$
LGT(X)	$\log_{10} X$ or lg X
EXP(X)	e^x or exp(X)
SQR(X)	$\sqrt{X}$
INT(X)	The integer part of X, the greatest integer, which is less than or equal to X.
ABS(X)	Absolute value $\|X\|$ of X
SGN(X)	Signature function, that is the function gives −1 if X is negative, 0 if X is 0 and 1 if X is positive.
PI	Gives decimal approximation of π.
RND	Gives random number between 0 and 1.

Useful program instructions

1 Calculation of sum

```
20 LET S=S+X
```

The variable S gives the sum of the values of the variable X.

2 Calculation of sum of squares

```
30 LET S=S+X*X
```

The variable S gives the sum of the squares of the values of the variable X.

3 Rounding off to nearest integer

```
30 LET X=INT(X+ .5)
```

4 Rounding off to one decimal place

```
30 LET X=INT(10*X+ .5)/10
```

5 Rounding off to two decimal places

```
30 LET X=INT(100*X+ .5)/100
```

6 Test of divisibility

```
40 IF X/Y=INT(X/Y) THEN ...
```

X is divisible by Y, if the condition is satisfied.

7 Is X odd or even?

```
40 IF X/2=INT(X/2) THEN ...
```

X is an even number, if the condition is satisfied.

8 Remainder by division

```
30 LET R=X-INT(X/Y)*Y
```

The variable R is the remainder, when integer X is divided by the integer Y.

9 To truncate digits one by one from right to left

```
40 LET Z=10*(X/10-INT(X/10)): PRINT Z
50 LET X=INT(X/10)
60 IF X<>0 THEN 40
```

10 Frequencies

```
50 LET F(X)=F(X)+1
```

The variable F(X) gives the number of times the variable X has taken different values.

11 Setting flags

```
60 LET F(X)=1
```

If the variable X has been assigned a certain value then the flag F(X) is 1. If not F(X)=0.

12 All combinations of variable values

```
 40 FOR X=1 TO N
 50 FOR Y=1 TO M
.........................
100 NEXT Y
110 NEXT X
```

13 Interest

```
50 LET X=K*R ↑ N
```

X is the value of the amount K after N years with compound interest with interest factor R. When the interest is 5 percent, the interest factor is 1.05.

14 Present value

```
60 LET X=K/R ↑ N
```

X is the present value of the amount K to be paid in N years. The interest factor is R.

17 Probability Theory

17.1 Basic Probability Theory

Outcome sets (sample spaces) and events

Outcome set Ω=set of all possible outcomes. *Events* are subsets of the outcome set Ω.

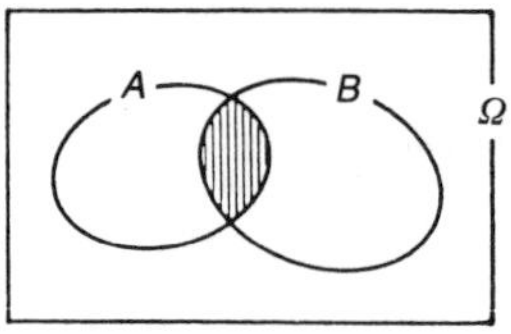

A and *B*
A and *B* occur
$A \cap B$

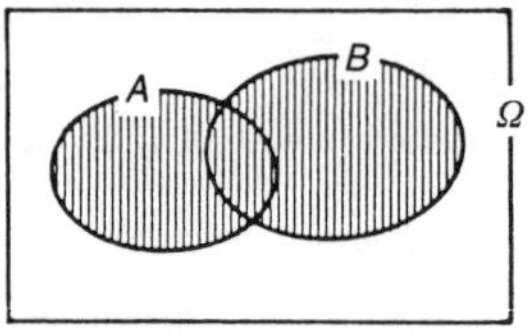

A or *B*
At least one of *A* and *B* occurs
$A \cup B$

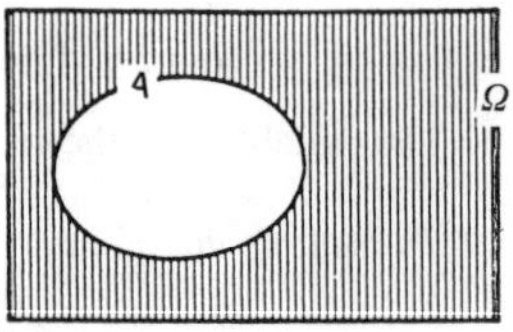

Not *A*
A does not occur
$\complement A$ or A^c

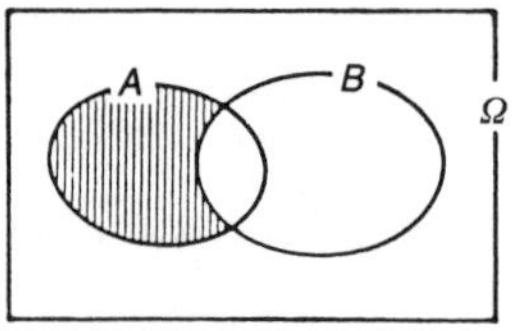

A but not *B*
A occurs and *B* does not occur
$A \setminus B$

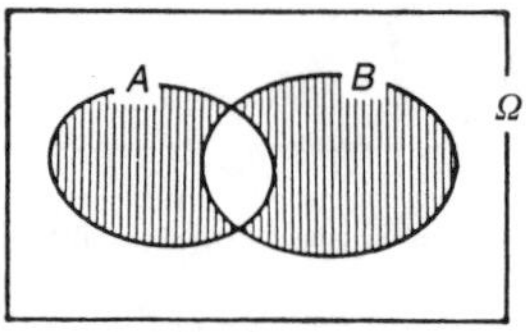

Exactly one of *A* and *B* occurs
$A \triangle B$

Definition of probability measure

1) $0 \leq P(A) \leq 1$ for each event *A*
2) $P(\Omega)=1$
3) $A \cap B=\varnothing \Rightarrow P(A \cup B)=P(A)+P(B)$

4) $A_i \cap A_j = \emptyset,\ i \neq j \Rightarrow P(\bigcup_{1}^{\infty} A_i) = \sum_{i=1}^{\infty} P(A_i)$

Property 4 is equivalent to $A_i \supset A_{i+1},\ \bigcap_{i=1}^{\infty} A_i = \emptyset \Rightarrow \lim_{n\to\infty} P(A_n) = 0$

Probability laws

$$P(\complement A) = 1 - P(A)$$
$$P(A \cup B) = P(A) + P(B) - P(A \cap B)$$
$$P(A \cup B \cup C) = P(A) + P(B) + P(C) - \\ - P(A \cap B) - P(A \cap C) - P(B \cap C) + P(A \cap B \cap C)$$
$$P(B \setminus A) = P(B) - P(A \cap B)$$
$$P(A \triangle B) = P(A) + P(B) - 2P(A \cap B)$$

$A_1, A_2, \ldots, A_n$ are events and

$$S_1 = \sum_i P(A_i),\ S_2 = \sum\sum_{i<j} P(A_i \cap A_j),\ S_3 = \sum\sum\sum_{i<j<k} P(A_i \cap A_j \cap A_k),\ \ldots$$

P(at least one of the events occur)=

$$= P(\bigcup_{i=1}^{n} A_i) = S_1 - S_2 + S_3 - S_4 + \ldots \pm S_n$$

P(at least k of the events occur)=

$$= S_k - \binom{k}{k-1} S_{k+1} + \binom{k+1}{k-1} S_{k+2} - \binom{k+2}{k-1} S_{k+3} + \ldots \pm \binom{n-1}{k-1} S_n$$

P(exactly k of the events occur)=

$$= S_k - \binom{k+1}{k} S_{k+1} + \binom{k+2}{k} S_{k+2} - \binom{k+3}{k} S_{k+3} + \ldots \pm \binom{n}{k} S_n$$

$$P(\bigcup_{i=1}^{n} A_i) \leq \sum_{i=1}^{n} P(A_i) \qquad \textit{(Boole's inequality)}$$

$$P(\bigcap_{i=1}^{n} A_i) \geq 1 - \sum_{i=1}^{n} (1 - P(A_i)) \qquad \textit{(Bonferroni's inequality)}$$

For a uniform probability distribution

$$P(A) = \frac{\text{Number of outcomes in } A}{\text{Total number of outcomes}}$$

Combinatorics

Multiplication principle

A multistage experiment is performed in k stages in a given order. The number of outcomes in the stages are $n_1, n_2, \ldots, n_k$. Then the total number of outcomes of the multistage experiment is

$$n_1 \cdot n_2 \cdot \ldots \cdot n_k$$

Combinatorial formulas

Set Ω has n elements

The number of *ordered k-tuples* (variations) of Ω is

$$(n)_k = n(n-1) \ldots (n-k+1)$$

The number of *permutations* of Ω is

$$n! = n(n-1) \ldots 2 \cdot 1$$

The number of *subsets* of Ω with k elements is

$$\binom{n}{k} = \frac{(n)_k}{k!} = \frac{n(n-1) \cdot \ldots \cdot (n-k+1)}{k!} = \frac{n!}{k!(n-k)!}$$

The number of ways Ω can be partioned in r subsets with $k_1, k_2, \ldots, k_r$ elements $(n = \sum_{i=1}^{r} k_i)$ is

$$\frac{n!}{k_1!\, k_2! \ldots k_r!}$$

Sampling

Number of ways of taking k elements from a population of n elements		
	Without replacement	With replacement
With regard to order	$(n)_k$	n^k
Without regard to order	$\binom{n}{k}$	$\binom{n+k-1}{k}$

Conditional probability and independent events

$$P(A \mid B) = \frac{P(A \cap B)}{P(B)} \qquad P(B \mid A) = \frac{P(A \cap B)}{P(A)}$$

$$P(A \cap B) = P(A) \cdot P(B \mid A)$$

$$P(A \cap B \cap C) = P(A) \cdot P(B \mid A) \cdot P(C \mid A \cap B)$$

$$P(A_1 \cap \ldots \cap A_n) =$$
$$= P(A_1)\, P(A_2 \mid A_1)\; P(A_3 \mid A_1 \cap A_2) \cdot \ldots \cdot P(A_n \mid A_1 \cap \ldots \cap A_{n-1})$$

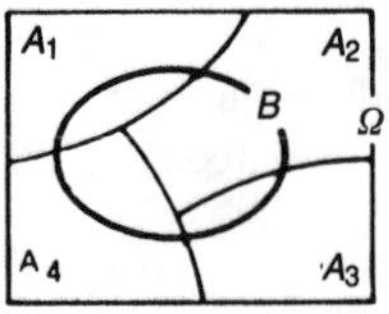

$$P(A_i \mid B) = \frac{P(A_i)\, P(B \mid A_i)}{\sum_{i=1}^{n} P(A_i)\, P(B \mid A_i)}$$ *(Bayes' formula)*

$$P(B) = \sum_{i=1}^{n} P(A_i)\, P(B \mid A_i)$$ *(Total probability formula)*

For *independent* events the following holds

$P(A \cap B) = P(A)\, P(B)$

$P(B \mid A) = P(B), \quad P(A \mid B) = P(A)$

$P(A \cap B \cap C) = P(A)\, P(B)\, P(C)$

$P(A_1 \cap A_2 \cap \ldots \cap A_n) = P(A_1)\, P(A_2) \cdot \ldots \cdot P(A_n)$

Calculation of probabilities with the aid of tree diagrams

Multistage random experiments can be portrayed by tree diagrams.

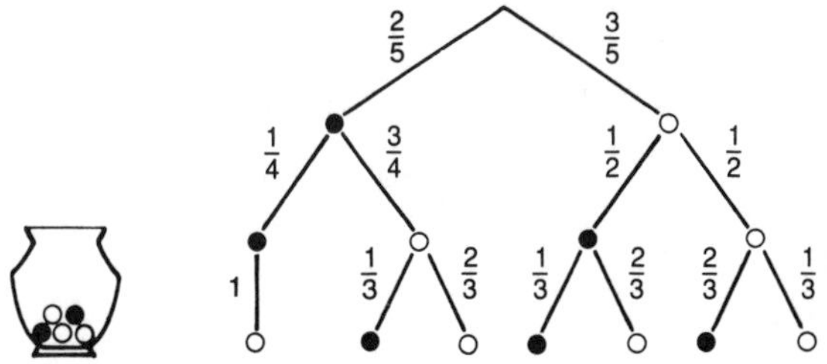

Sum rule

The probability of an event is the sum of the probabilities for the outcomes of the event.

Product rule

The probability for a path is the product of the probabilities along the path.

Random (stochastic) variables

Basic definitions

Discrete random variable X

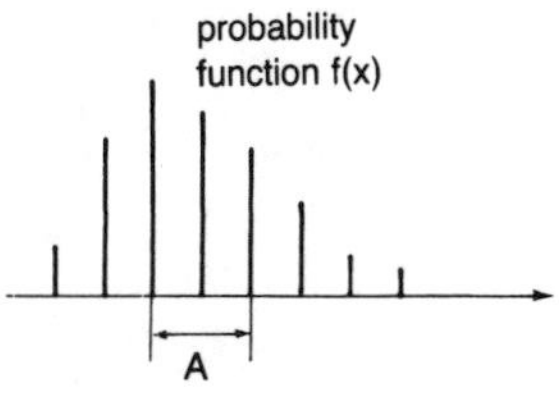

$P(X \in A) = \sum_{x \in A} f(x)$

Continuous random variable X

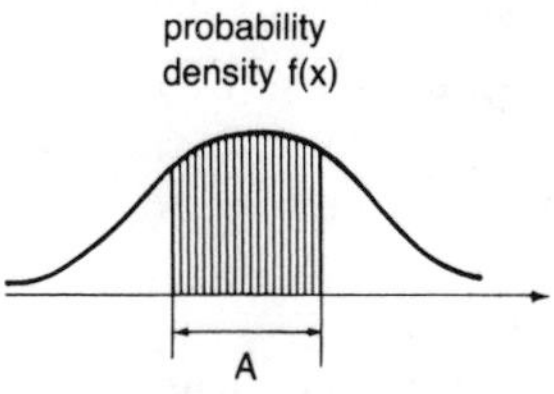

$P(X \in A) = \int_A f(x)\,dx$

Expectation $E[X]$ or $\mu=$

$$\sum_{x\in\Omega} xf(x)$$

Expectation $E[X]$ or $\mu=$

$$\int_{\Omega} xf(x)dx$$

Distribution function F

$$F(x)=P(X\leqslant x)=\sum_{t\leqslant x} f(t)$$

Distribution function F

$$F(x)=\int_{-\infty}^{x} f(t)dt$$

Variance $Var[X]$ or σ^2

$$\sigma^2=E[(X-\mu)^2]=\sum_{x\in\Omega}(x-\mu)^2 f(x)$$

Variance $Var[X]$ or σ^2

$$\sigma^2=E[(X-\mu)^2]=\int_{\Omega}(x-\mu)^2 f(x)dx$$

Standard deviation$=\sigma$

Entropy $H(X)$

$$H(X)=E[{}^2\log(1/f(X))]=$$
$$=-\sum_{x\in\Omega} f(x)({}^2\log f(x))$$

Entropy $H(X)$

$$H(X)=E[{}^2\log(1/f(X))]=$$
$$=-\int_{\Omega}{}^2\log f(x)f(x)dx$$

Expectations

$E[aX]=a\,E[X]$

$E[X+Y]=E[X]+E[Y]$

$E[g(X)]=\sum_{x\in\Omega} g(x)\,f_X(x)$ (discrete random variable)

$E[g(X)]=\int_{\Omega} g(x)\,f_X(x)dx$ (continuous random variable)

Variances

$\mathrm{Var}[aX]=a^2\,\mathrm{Var}[X]$

$\mathrm{Var}[X]=E[X^2]-(E[X])^2$ *(Steiner's theorem)*

$\mathrm{Var}[X+Y]=\mathrm{Var}[X]+\mathrm{Var}[Y]+2\mathrm{Cov}[X,\ Y]$

Chebyshev inequality

$P(|X|\geqslant a)\leqslant E[X^2]/a^2$

$P(|X-\mu|\geqslant a)\leqslant \mathrm{Var}[X]/a^2$

$P(|X-\mu|\geqslant k\sigma)\leqslant 1/k^2$

Jensen inequality

If $f(x)$ is a convex function, then

$$E[f(X)]\geqslant f(E[X]).$$

Moments

The *kth central moment* μ_k is defined as

$$\mu_k = E[(X-\mu)^k]$$

The *skewness* γ_1 and *curtosis* γ_2 are defined as

$$\gamma_1 = \mu_3/\sigma^3 \qquad \gamma_2 = (\mu_4/\sigma^4) - 3$$

For the $N(\mu, \sigma)$ normal distribution

$$\mu_{2k+1} = 0 \qquad \mu_{2k} = \sigma^{2k}(2k-1)!! \qquad \gamma_1 = \gamma_2 = 0$$

Convergence

Convergence in probability

$$\underset{n\to\infty}{\text{limp}}\, X_n = X \Leftrightarrow \lim_{n\to\infty} P(|X_n - X| > \varepsilon) = 0 \text{ for each } \varepsilon > 0$$

Convergence almost surely

$$\underset{n\to\infty}{\text{plim}}\, X_n = X \Leftrightarrow P(\lim_{n\to\infty} X_n = X) = 1$$

Convergence in distribution

$$X_n \to X \Leftrightarrow \lim_{n\to\infty} P(X_n \leqslant x) = P(X \leqslant x) \text{ for each } x \text{ such that } P(X \leqslant x) \text{ is continuous in } x$$

Convergence in mean

$$\underset{n\to\infty}{\text{l.i.m.}}\, X_n = X \Leftrightarrow \lim_{n\to\infty} E[|X_n - X|^2] = 0$$

$$\left.\begin{array}{l} \underset{n\to\infty}{\text{p lim}}\, X_n = X \\ \text{l.i.m.}\, X_n = X \end{array}\right\} \Rightarrow \underset{n\to\infty}{\text{lim p}}\, X_n = X \Rightarrow X_n \to X \text{ in distribution}$$

Twodimensional (bivariate) random variable

Discrete variable (X, Y)

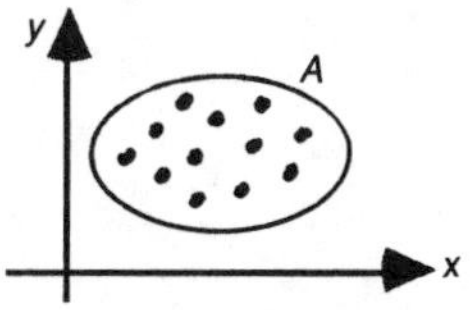

$$P((X, Y) \in A) = \sum_{(x,y)\in A}\sum f(x, y)$$

f=twodimensional probability function

Distribution function F

$$F(x, y) = P(X \leqslant x, Y \leqslant y) = \sum_{u \leqslant x,}\sum_{v \leqslant y} f(u, v)$$

Continuous variable (X, Y)

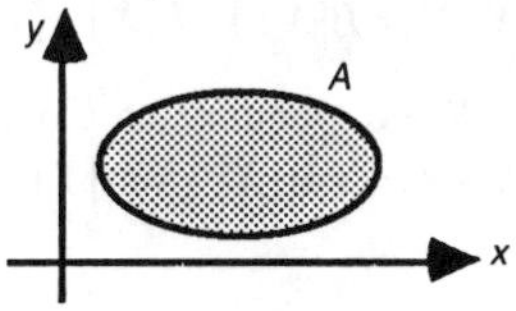

$$P((X, Y) \in A) = \iint_A f(x, y)\,dx\,dy$$

f=twodimensional probability density

Distribution function F

$$F(x, y) = P(X \leqslant x, Y \leqslant y) = \int_{-\infty}^{x}\int_{-\infty}^{y} f(u, v)\,du\,dv$$

Marginal distributions

$$f_X(x)=\sum_y f(x, y)$$

$$f_Y(y)=\sum_x f(x, y)$$

Expectation $E[g(X, Y)]$

$$E[g(X, Y)]=\sum_x \sum_y g(x, y) f(x, y)$$

$$(E[XY])^2 \leqslant E[X^2]\, E[Y^2]$$

Marginal distributions

$$f_X(x)=\int_{-\infty}^{\infty} f(x, y)dy$$

$$f_Y(y)=\int_{-\infty}^{\infty} f(x, y)dx$$

Expectation $E[g(X, Y)]$

$$E[g(X, Y)]=\int_{-\infty}^{\infty}\int_{-\infty}^{\infty} g(x, y)\, f(x, y)\, dxdy$$

(Schwarz' inequality)

For *independent* random variables X and Y

$$f(x, y)=f_X(x)\, f_Y(y)$$
$$F(x, y)=F_X(x)\, F_Y(y)$$
$$E[XY]=E[X]\, E[Y]$$
$$\mathrm{Var}[X+Y]=\mathrm{Var}[X]+\mathrm{Var}[Y]$$

For independent continuous random variables X and Y

$$f_{X+Y}(x)=\int_{-\infty}^{\infty} f_X(t)\, f_Y(x-t)dt$$

For independent non-negative random variables X and Y

$$f_{X+Y}(x)=\int_0^x f_X(t)\, f_Y(x-t)dt$$

Covariance $\mathrm{Cov}[X, Y]$

$$\mathrm{Cov}[X, Y]=E[(X-\mu_1)(Y-\mu_2)],\quad E[X]=\mu_1,\ E[Y]=\mu_2$$

$$\mathrm{Cov}[X, X]=\mathrm{Var}[X]$$
$$\mathrm{Cov}[X, Y]=E[XY]-E[X]\, E[Y]=\mathrm{Cov}[Y, X]$$

X and Y independent $\Rightarrow \mathrm{Cov}[X, Y]=0$

$$\mathrm{Var}[X+Y]=\mathrm{Var}[X]+\mathrm{Var}[Y]+2\,\mathrm{Cov}[X, Y]$$

Correlation coefficient ϱ

$$\varrho=\frac{\mathrm{Cov}[X, Y]}{\sqrt{\mathrm{Var}[X]\ \mathrm{Var}[Y]}} \qquad -1\leqslant\varrho\leqslant 1$$

X and Y independent $\Rightarrow \varrho=0$

$$\mathrm{Var}[X+Y]=\mathrm{Var}[X]+\mathrm{Var}[Y]+2\,\varrho\sqrt{\mathrm{Var}(X)\,\mathrm{Var}(Y)}$$

For random variables $X_1, X_2, \ldots, X_n$

$$\mathrm{Var}[\sum_{i=1}^{n} X_i] = \sum_{i=1}^{n}\sum_{j=1}^{n} \mathrm{Cov}[X_i, X_j] = \sum_{i=1}^{n} \mathrm{Var}[X_i] + \sum\sum_{i \neq j} \mathrm{Cov}[X_i, X_j] =$$

$$= \sum_{i=1}^{n} \mathrm{Var}[X_i] + 2 \sum\sum_{i<j} \mathrm{Cov}[X_i, X_j]$$

$$X_1, X_2, \ldots, X_n \text{ independent} \Rightarrow \mathrm{Var}[\sum_{i=1}^{n} X_i] = \sum_{i=1}^{n} \mathrm{Var}[X_i]$$

Conditional distributions $f(x|y)$ and $f(y|x)$

$f(y|x) = f(x, y)/f_X(x)$ $\qquad$ $f(x|y) = f(x, y)/f_Y(y)$

$f_X(x) = \sum_y f_Y(y) f(x|y)$

$f_Y(y) = \sum_x f_X(x) f(y|x)$

(Discrete random variables)

$f_X(x) = \int_{-\infty}^{\infty} f_Y(y) f(x|y) dy$

$f_Y(y) = \int_{-\infty}^{\infty} f_X(x) f(y|x) dx$

(Continuous random variables)

For independent random variables X and Y

$$f(y|x) = f_Y(y) \qquad f(x|y) = f_X(x) \qquad f(x, y) = f_X(x) f_Y(y)$$

$$f(x|y) = \frac{f_X(x) f(y|x)}{\sum_x f_X(x) f(y|x)} \quad \text{(\textit{Bayes' rule}, discrete case)}$$

$$f(x|y) = \frac{f_X(x) f(y|x)}{\int_{-\infty}^{\infty} f_X(x) f(y|x) dx} \quad \text{(\textit{Bayes' rule}, continuous case)}$$

Conditional expectations

$E[X|Y] = \sum_x x f(x|y)$ $\qquad$ $E[Y|X] = \sum_y y f(y|x)$

$E[X|Y] = \int_{-\infty}^{\infty} x f(x|y) dx$ $\qquad$ $E[Y|X] = \int_{-\infty}^{\infty} y f(y|x) dy$

$E[X] = \sum_y E[X|y] f_Y(y)$ $\qquad$ $E[Y] = \sum_x E[Y|x] f_X(x)$

$E[X] = \int_{-\infty}^{\infty} E[X|y] f_Y(y) dy$ $\qquad$ $E[Y] = \int_{-\infty}^{\infty} E[Y|x] f_X(x) dx$

$E[X] = E[E[X|Y]]$ $\qquad$ $E[Y] = E[E[Y|X]]$

$$\mathrm{Var}[X] = E[\mathrm{Var}[X|Y]] + \mathrm{Var}[E[X|Y]]$$
$$\mathrm{Var}[Y] = E[\mathrm{Var}[Y|X]] + \mathrm{Var}[E[Y|X]]$$

X has *linear regression* with regard to Y if $E[X|y]$ is a linear function of y, in which case

$$E[X|y]=\mu_1+\varrho\,\frac{\sigma_1}{\sigma_2}\,(y-\mu_2)$$

where $\mu_1=E[X]$, $\mu_2=E[Y]$, $\sigma_1{}^2=\mathrm{Var}[X]$, $\sigma_2{}^2=\mathrm{Var}[Y]$ and ϱ is the correlation coefficient.

If Y has linear regression with regard to X then

$$E[Y|x]=\mu_2+\varrho\frac{\sigma_2}{\sigma_1}\,(x-\mu_1)$$

17.2 Probability Distributions

Discrete probability distributions (random variables)

Distribution	**$P(X=x)$**	**Expectation μ**	**Variance σ^2**	**Example of application**
Binomial $B(n, p)$	$\binom{n}{x}p^x(1-p)^{n-x}$ $x=0, 1, \ldots, n$	np	$np(1-p)$	The frequency in n independent trials has a binomial distribution. Probability in each trial$=p$
Geometric $G(p)$	$(1-p)^{x-1}p$ $x=1, 2, 3, \ldots$	$\frac{1}{p}$	$\frac{1-p}{p^2}$	The number of required trials until an event with probability p occurs has a geometric distribution
Poisson $P(\lambda)$	$e^{-\lambda}\lambda^x/x!$ $x=0, 1, 2, \ldots$	λ	λ	Distribution of number of points in random point process under certain simple assumptions. Approximation to the binomial distribution when n is large and p is small, $\lambda=np$.
Hypergeometric $H(N, n, p)$	$\frac{\binom{Np}{x}\binom{N-Np}{n-x}}{\binom{N}{n}}$	np	$np(1-p)\,\frac{N-n}{N-1}$	This distribution is used in connection with sampling without replacement from a finite population with elements of two different kinds
Negative binomial or Pascal $NB(r, p)$	$\binom{x-1}{r-1}p^r(1-p)^{x-r}$ $x=r, r+1, \ldots$	$\frac{r}{p}$	$\frac{r(1-p)}{p^2}$	The number of required trials until an event with probability p occurs for the rth time has a negative binomial distribution

Continuous probability distributions (random variables)

Distribution	$f(x)$	Expectation μ	Variance σ^2	Example of application
Uniform $U(a, b)$	$\frac{1}{b-a}$ $a \leqslant x \leqslant b$	$\frac{a+b}{2}$	$\frac{(b-a)^2}{12}$	Certain waiting times Rounding off errors
Exponential $E(\lambda)$	$\lambda e^{-\lambda x}$ $x \geqslant 0$	$1/\lambda$	$1/\lambda^2$	Distribution of length of life when no aging
Normed normal distribution $N(0, 1)$	$\varphi(x)=\frac{1}{\sqrt{2\pi}} e^{-x^2/2}$	0	1	If X has a general normal distribution, then $(X-\mu)/\sigma$ has a normed normal distribution
General normal distribution $N(\mu, \sigma)$	$\frac{1}{\sigma}\varphi\left(\frac{x-\mu}{\sigma}\right)$	μ	σ^2	Under general conditions, the sum of a large number of random variables is approximately normally distributed *(the central limit theorem)*
Gamma $\Gamma(n, \lambda)$	$\frac{\lambda^n}{\Gamma(n)} x^{n-1} e^{-\lambda x}$	$\frac{n}{\lambda}$	$\frac{n}{\lambda^2}$	Distribution of the sum of n independent random variables with an exponential distribution with parameter λ
χ^2 $\chi^2(r)$	$\frac{1}{2^{r/2}\Gamma(\frac{r}{2})} x^{\frac{r}{2}-1} e^{-\frac{x}{2}}$ $x \geqslant 0$ The parameter r is called the "number of degrees of freedom"	r	$2r$	Distribution of $u_1^2+u_2^2+\ldots+u_r^2$, where $u_1, u_2, \ldots, u_r$ are independent and have a normed normal distribution
t $t(r)$	$\frac{a_r}{b_r}\left(1+\frac{x^2}{r}\right)^{-\frac{r+1}{2}}.$ $a_r=\Gamma\left(\frac{r+1}{2}\right)$ $b_r=\sqrt{r\pi}\,\Gamma\left(\frac{r}{2}\right)$	0, $r>1$	$\frac{r}{r-2}$, $r>2$	Distribution of $u/\sqrt{X/r}$ where u and X are independent, u has a normed normal distribution and X a χ^2-distribution with r degrees of freedom
F $F(r_1, r_2)$	$\frac{a_r x^{(r_1/r_2)-1}}{b_r(r_2+r_1 x)^{\frac{r_1+r_2}{2}}}$ $x \geqslant 0$ $a_r=\Gamma(\frac{r_1+r_2}{2})\, r_1^{r_1/2} r_2^{r_2/2}$ $b_r=\Gamma(\frac{r_1}{2})\,\Gamma(\frac{r_2}{2})$	$\frac{r_2}{r_2-2}$ $r_2>2$	$\frac{2r_2^2(r_1+r_2-2)}{r_1(r_2-2)^2(r_2-4)}$ $r_2>4$	Distribution of $(X_1/r_1)/(X_2/r_2)$ where X_1 and X_2 are independent and have χ^2-distributions with r_1 and r_2 degrees of freedom

Beta $\beta(p, q)$	$a_{p,q}\,x^{p-1}(1-x)^{q-1}$ $0 \leqslant x \leqslant 1$ $a_{p,q}=\dfrac{\Gamma(p+q)}{\Gamma(p)\,\Gamma(q)}$ $p>0,\ q>0$	$\dfrac{p}{p+q}$	$\dfrac{pq}{(p+q)^2(p+q+1)}$	Useful as apriori distribution for unknown probability in Bayesian models and in PERT-analysis
Weibull $W(\lambda, \beta)$	$\lambda^\beta \beta x^{\beta-1} e^{-(\lambda x)^\beta}$ $x \geqslant 0$ $F(x)=1-e^{-(\lambda x)^\beta}$	$\dfrac{1}{\lambda}\Gamma\left(1+\dfrac{1}{\beta}\right)$	$\dfrac{1}{\lambda^2}(A-B)$ $A=\Gamma\left(1+\dfrac{2}{\beta}\right)$ $B=\Gamma^2\left(1+\dfrac{1}{\beta}\right)$	Useful as length of life distribution in reliability theory
Rayleigh $R(\sigma)$	$\dfrac{x}{\sigma^2}\,e^{-x^2/2\sigma^2}$ $x \geqslant 0$	$\sigma\sqrt{\dfrac{\pi}{2}}$	$2\sigma^2\left(1-\dfrac{\pi}{4}\right)$	Useful in communications systems and in reliability theory
Cauchy $C(a)$	$\dfrac{a}{\pi(a^2+x^2)}$	Does not exist	Does not exist	If angle φ has the $U(-\pi/2, \pi/2)$ distribution, then $a \tan\varphi$ has the $C(a)$ distribution 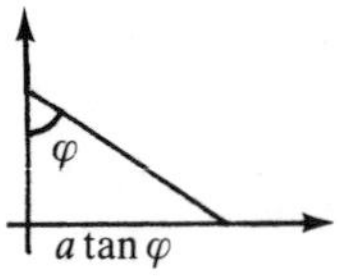

The bivariate normal distribution

The two-dimensional random variable (X, Y) is $N(\mu_1, \mu_2, \sigma_1, \sigma_2, \varrho)$ if its density function is

$$f(x, y)=\frac{1}{2\pi\sigma_1\sigma_2\sqrt{1-\varrho^2}}\cdot$$

$$\cdot\exp\left(-\frac{1}{2(1-\varrho^2)}\left(\left(\frac{x-\mu_1}{\sigma_1}\right)^2-2\varrho\left(\frac{x-\mu_1}{\sigma_1}\right)\left(\frac{y-\mu_2}{\sigma_2}\right)+\left(\frac{y-\mu_2}{\sigma_2}\right)^2\right)\right)$$

The conditional distribution of Y given $X=x$ is normal $N(\mu, \sigma)$ with

$$\mu=\mu_2+\varrho\,\frac{\sigma_2}{\sigma_1}\,(x-\mu_1) \text{ and } \sigma^2=\sigma_2{}^2(1-\varrho^2)$$

(X, Y) is $N(\mu_1, \mu_2, \sigma_1, \sigma_2, \varrho) \Rightarrow ((X-\mu_1)/\sigma_1, (Y-\mu_2)/\sigma_2)$ is $N(0, 0, 1, 1, \varrho)$

If (X, Y) is $N(0, 0, 1, 1, \varrho)$ then

$$P(XY>0)=\frac{1}{2}+\frac{1}{\pi}\arcsin\varrho \qquad P(XY<0)=\frac{1}{2}-\frac{1}{\pi}\arcsin\varrho$$

Let Z_1 and Z_2 be independent and $N(0, 1)$ and

$$X=\mu_1+\sigma_1 Z_1 \text{ and } Y=\mu_2+\sigma_2(\varrho Z_1+\sqrt{1-\varrho^2}\,Z_2)$$

Then (X, Y) is $N(\mu_1, \mu_2, \sigma_1, \sigma_2, \varrho)$

The density function of the bivariate normal distribution can be written

$$\frac{1}{2\pi\sqrt{\det\Sigma}}\exp\left[-\frac{1}{2}(\boldsymbol{x}-\boldsymbol{\mu})\,\Sigma^{-1}\,(\boldsymbol{x}-\boldsymbol{\mu})^t\right]$$

where $\mathbf{x}-\boldsymbol{\mu}=(x_1-\mu_1,\ x_2-\mu_2)$ and Σ is the covariance matrix

$$\Sigma=\begin{bmatrix}\sigma_1^2 & \varrho\sigma_1\sigma_2\\ \varrho\sigma_1\sigma_2 & \sigma_2^2\end{bmatrix}$$

For the k-dimensional normal distribution $N(\boldsymbol{\mu}, \Sigma)$ the density function is

$$\frac{1}{(2\pi)^{k/2}(\det\Sigma)^{1/2}}\exp\left(-\frac{1}{2}(\boldsymbol{x}-\boldsymbol{\mu})\,\Sigma^{-1}\,(\boldsymbol{x}-\boldsymbol{\mu})^t\right)$$

where $\boldsymbol{\mu}=(\mu_1, \mu_2, \ldots, \mu_k)$ is the vector of expectations and Σ is the covariance matrix.

Addition theorems for some distributions

X_1 and X_2 are independent random variables.

X_1	X_2	X_1+X_2
$B(n_1, p)$	$B(n_2, p)$	$B(n_1+n_2, p)$
$P(\lambda_1)$	$P(\lambda_2)$	$P(\lambda_1+\lambda_2)$
$NB(r_1, p)$	$NB(r_2, p)$	$NB(r_1+r_2, p)$
$E(\lambda)$	$E(\lambda)$	$\Gamma(2, \lambda)$
$\Gamma(n_1, \lambda)$	$\Gamma(n_2, \lambda)$	$\Gamma(n_1+n_2, \lambda)$
$N(\mu_1, \sigma_1)$	$N(\mu_2, \sigma_2)$	$N(\mu_1+\mu_2, \sqrt{\sigma_1^2+\sigma_2^2})$
$\chi^2(r_1)$	$\chi^2(r_2)$	$\chi^2(r_1+r_2)$
$C(a)$	$C(a)$	$C(2a)$

Relations between distributions

X_1 and X_2 are independent random variables and Y is a function of X_1 or of X_1 and X_2.

X_1	X_2	Y	Distribution of Y
$E(\lambda)$	–	λX_1	$E(1)$
$E(\lambda)$	$E(\lambda)$	$\lambda(X_1+X_2)$	$\Gamma(2, 1)$
$E(\lambda)$	$E(\lambda)$	$2\lambda(X_1+X_2)$	$\chi^2(4)$
$\Gamma(r, \lambda)$	–	λX_1	$\Gamma(r, 1)$
$\Gamma(r_1, \lambda)$	$\Gamma(r_2, \lambda)$	$2\lambda(X_1+X_2)$	$\chi^2(2r_1+2r_2)$
$E(\lambda_1)$	$E(\lambda_2)$	$\text{Min}(X_1, X_2)$	$E(\lambda_1+\lambda_2)$
$N(0, 1)$	$\chi^2(r)$	$X_1/\sqrt{X_2/r}$	$t(r)$
$t(r)$	–	X_1^2	$F(1, r)$
$\chi^2(r_1)$	$\chi^2(r_2)$	$(X_1/r_1)/(X_1/r_2)$	$F(r_1, r_2)$
$F(r_1, r_2)$	–	$1/\left(1+\frac{r_1}{r_2}X_1\right)$	$\beta(r_2/2, r_1/2)$
$\Gamma(r_1, \lambda)$	$\Gamma(r_2, \lambda)$	$X_1/(X_1+X_2)$	$\beta(r_1, r_2)$
$\chi^2(r_1)$	$\chi^2(r_2)$, $r_2<r_1$	X_1-X_2	$\chi^2(r_1-r_2)$
$N(0, \sigma)$	$N(0, \sigma)$	X_1/X_2	$C(1)$

Transforms of probability distributions (random variables)

Definitions

Transform	Definition	$E[X]$	$\text{Var}[X]$
Probability generating function or geometric transform (integer-valued random variables only)	$\psi(s)=E[s^X]$	$\psi'(1)$	$\psi''(1)+\psi'(1)-(\psi'(1))^2$
Moment generating function	$\psi(s)=E[e^{sX}]$	$\psi'(0)$	$\psi''(0)-(\psi'(0))^2$
Characteristic function	$\psi(s)=E[e^{isX}]$	$-i\psi'(0)$	$-\psi''(0)+(\psi'(0))^2$
Laplace transform (non-negative random variables only)	$\psi(s)=E[e^{-sX}]$	$-\psi'(0)$	$\psi''(0)-\psi'(0)^2$

In each case the transform of X_1+X_2 is the product of the transforms of X_1 and X_2, if X_1 and X_2 are independent.

$$\psi_{X_1+X_2}(s)=\psi_{X_1}(s)\,\psi_{X_2}(s)$$

Transforms for specific distributions

Distribution	Probability generating function	Moment generating function	Characteristic function	Laplace transform
Binomial $B(n, p)$	$(1-p+ps)^n$	$(1-p+pe^s)^n$	$(1-p+pe^{is})^n$	–
Geometric $G(p)$	$\frac{ps}{1-(1-p)s}$	$\frac{pe^s}{1-(1-p)e^s}$	$\frac{pe^{is}}{1-(1-p)e^{is}}$	–
Poisson $P(\lambda)$	$e^{\lambda(s-1)}$	$e^{\lambda(e^s-1)}$	$e^{\lambda(e^{is}-1)}$	–
Negative binomial $NB(r, p)$	$\left(\frac{ps}{1-(1-p)s}\right)^r$	$\left(\frac{pe^s}{1-(1-p)e^s}\right)^r$	$\left(\frac{pe^{is}}{1-(1-p)e^{is}}\right)^r$	–
Uniform $U(a, b)$	–	$\frac{e^{bs}-e^{as}}{s(b-a)}$	$\frac{e^{ibs}-e^{ias}}{is(b-a)}$	–
Exponential $E(\lambda)$	–	$\frac{\lambda}{\lambda-s}$	$\frac{\lambda}{\lambda-is}$	$\frac{\lambda}{\lambda+s}$
Normed normal distribution $N(0, 1)$	–	$e^{s^2/2}$	$e^{-s^2/2}$	–
General normal distribution $N(\mu, \sigma)$	–	$e^{\mu s+\frac{1}{2}\sigma^2 s^2}$	$e^{i\mu s-\frac{1}{2}\sigma^2 s^2}$	–
Gamma $\Gamma(r, \lambda)$		$\left(\frac{\lambda}{\lambda-s}\right)^r$	$\left(\frac{\lambda}{\lambda-is}\right)^r$	$\left(\frac{\lambda}{s+\lambda}\right)^r$
χ^2-distribution $\chi^2(r)$	–	$(1-2s)^{-r/2}$	$(1-2is)^{-r/2}$	$(1+2s)^{-r/2}$

17.3 Stochastic Processes

Markov chains

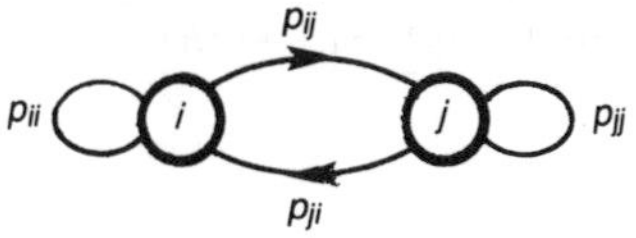

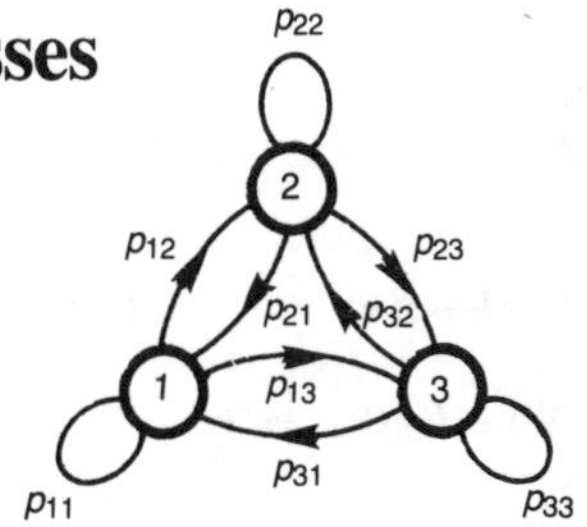

$$P=\begin{bmatrix} p_{11} & p_{12} & p_{13} \\ p_{21} & p_{22} & p_{23} \\ p_{31} & p_{32} & p_{33} \end{bmatrix}$$

$P(X_n=i \mid X_0, X_1, \ldots, X_{n-1})=P(X_n=i \mid X_{n-1})$

$p_{ij}=P(X_{n+1}=j \mid X_n=i) \qquad p_{ij}(n)=P(X_{m+n}=j \mid X_m=i)$

$P=[p_{ij}]=$transition matrix

$P_n=[p_{ij}(n)]=n$-step transition matrix

$P_{m+n}=P_mP_n,\ P_n=P^n$ *(Chapman–Kolmogorov equations)*

The Poisson process

$P(X(t+h)=i+1|X(t)=i) = \lambda h+o(h)$ λ=intensity of the process

$X(t)$ is $P(\lambda t)$

$P(X(t) = i)=e^{-\lambda t}\frac{(\lambda t)^i}{i!}$, $i=0, 1, 2, \ldots$ $E[X(t)]=\mathrm{Var}[X(t)]=\lambda t$

$P(X(h) = 1)=\lambda h+o(h)$ $P(X(h)=0)=1-\lambda h+o(h)$

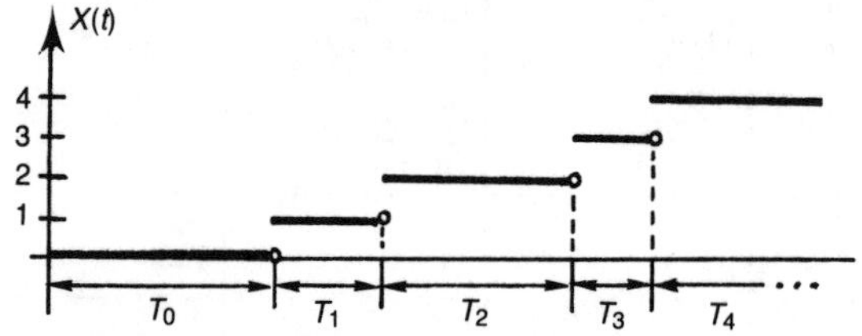

The lengths of times T_0, T_1, T_2, ... in states 0, 1, 2, ... are independent and $E(\lambda)$,

$$P(T_i>t)=e^{-\lambda t},\ t>0.$$

Birth and death processes

Birth and death process with two states

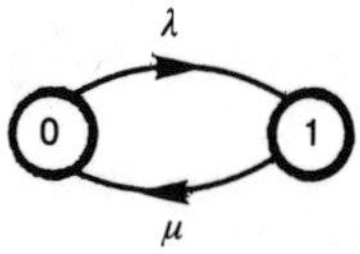

$P(X(t+h)=1|X(t)=0)=\lambda h+o(h)$ $P(X(t+h)=0|X(t)=1)=\mu h+o(h)$

$$P_0(t)=\frac{\mu}{\lambda+\mu}(1-e^{-(\lambda+\mu)t})+P_0(0)e^{-(\lambda+\mu)t}$$

$$P_1(t)=\frac{\lambda}{\lambda+\mu}(1-e^{-(\lambda+\mu)t})+P_1(0)e^{-(\lambda+\mu)t}$$

$\pi_0=\lim_{t\to\infty} P_0(t)=\frac{\mu}{\lambda+\mu}$ $\pi_1=\lim_{t\to\infty} P_1(t)=\frac{\lambda}{\lambda+\mu}$

General birth and death process

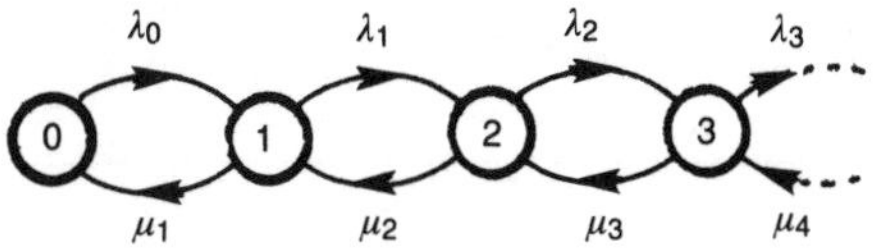

$P(X(t+h)=i+1|X(t)=i)=\lambda_i h+o(h)$ $\qquad$ $P(X(t+h)=i-1|X(t)=i)=\mu_i h+o(h)$

$\lambda_0, \lambda_1, \lambda_2, \ldots$ birth intensities of the process

$\mu_1, \mu_2, \mu_3, \ldots$ death intensities of the process

π_i=stationary (and asymptotic for $t\to\infty$) probability of state i

$$\pi_i=\frac{\lambda_{i-1}}{\mu_i}\pi_{i-1}=\frac{\lambda_0\lambda_1\ldots\lambda_{i-1}}{\mu_1\mu_2\ldots\mu_i}\pi_0, \quad \sum_i \pi_i=1$$

The lengths of time $T_0, T_1, T_2, \ldots$ in states $0, 1, 2, \ldots$ are independent and T_i is $E[\lambda_i+\mu_i]$, $\mu_0=0$.

Stationary processes

Basic definitions

Expectation function $\mu(t)=E[X(t)]$

Variance function $\sigma^2(t)=\mathrm{Var}[X(t)]$

Covariance kernel $K(s, t)=\mathrm{Cov}[X(s), X(t)]$

A stochastic process $\{X(t)\}$ is (strictly) *stationary* if $\{X(t_1), X(t_2), \ldots, X(t_n)\}$ and $\{X(t_1+h), X(t_2+h), \ldots, X(t_n+h)\}$ have the same probability distribution for all $t_1, t_2, \ldots, t_n$ and $h>0$.

A stochastic process $\{X(t)\}$ is *weakly* (second order) *stationary*, if $\mathrm{Cov}[X(s), X(s+t)]$ is independent of s and thus a function of t and $E[X(s)]$ and $E[X^2(s)]$ are independent of s.

Weakly stationary processes

Autocorrelation function $R_X(t)=\mathrm{Cov}[X(s), X(s+t)]$

$$E[X(t)]=0 \Rightarrow R_X(t)=E[X(s)X(s+t)]$$

Power spectral density $\varphi_X(\omega)=\int_{-\infty}^{\infty} R_X(s)e^{-i\omega s}ds=2\int_0^{\infty} R_X(s)\cos\omega s\, ds$

$R(-t)=R(t)$	$\|R(t)\|\leq R(0)$
$\varphi(-\omega)=\varphi(\omega)$	$\varphi(\omega)\geq 0$
$R_X(t)=\frac{1}{2\pi}\int_{-\infty}^{\infty}\varphi_X(\omega)e^{i\omega t}d\omega$	

Cross-correlation function $R_{X,Y}(t)=\mathrm{Cov}[X(s), Y(s+t)]$

Cross-spectral density $\varphi_{X,Y}(t)=\int_{-\infty}^{\infty} R_{X,Y}(s)e^{-i\omega s}ds$

$$R_{X,Y}(t)=\frac{1}{2\pi}\int_{-\infty}^{\infty} \varphi_{X,Y}(\omega)e^{i\omega t}d\omega$$

$X(t)=\alpha U(t) \Rightarrow$

$R_X(t)=\alpha^2 R_U(t)$	$R_{X,Y}(t)=\alpha R_{U,Y}(t)$	$R_{Y,X}(t)=\alpha R_{Y,U}(t)$
$\varphi_X(\omega)=\alpha^2 \varphi_U(\omega)$	$\varphi_{X,Y}(\omega)=\alpha \varphi_{U,Y}(\omega)$	$\varphi_{Y,X}(\omega)=\alpha \varphi_{Y,U}(\omega)$

$X(t)=U(t)+V(t) \Rightarrow$

$$R_X(t)=R_U(t)+R_V(t)+R_{U,V}(t)+R_{V,U}(t)$$

$$\varphi_X(\omega)=\varphi_U(\omega)+\varphi_V(\omega)+\varphi_{U,V}(\omega)+\varphi_{V,U}(\omega)$$

$$R_{X,Y}(t)=R_{U,Y}(t)+R_{V,Y}(t)$$

$$\varphi_{X,Y}(\omega)=\varphi_{U,Y}(\omega)+\varphi_{V,Y}(\omega)$$

Ergodic processes

$$R_X(t)=\lim_{T\to\infty}\frac{1}{2T}\int_{-T}^{T} X(u)\,X(u+t)\,du$$

$$R_{X,Y}(t)=\lim_{T\to\infty}\int_{-T}^{T} X(u)\,Y(u+t)\,du$$

Linear filters

$$L(a_1 X_1(t)+a_2 X_2(t))=a_1 L(X_1(t))+a_2 L(X_2(t))$$

$$Y(t)=\int_{-\infty}^{\infty} h(\tau)\,X(t-\tau)d\tau$$

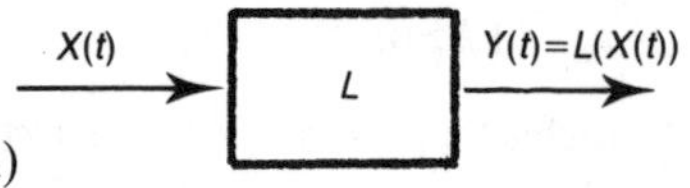

h=*impulse response* (weighting function)

Frequency-response function $G(i\omega)=\int_0^{\infty} e^{-i\omega t}h(t)dt$

$$E[Y]=E[X]\int_0^{\infty} h(\tau)d\tau$$

$$R_Y(t)=\int_0^{\infty}\int_0^{\infty} R_X(t+u-v)h(u)h(v)dudv$$

$$\varphi_Y(\omega)=|G(i\omega)|^2 S_X(\omega)$$

MA-, AR- and ARMA-processes

Here $\{e_t\}$ is a *white-noise process.*

$$E[e_t]=0,\ E[e_t^2]=\sigma^2,\ E[e_se_t]=0,\ s\neq t.$$

Moving Average Process, MA(q)

$$X_t=e_t+b_1e_{t-1}+\ldots+b_qe_{t-q}$$

Autoregressive Process AR(p)

$$X_t = e_t - a_1 X_{t-1} - \ldots - a_p X_{t-p}$$

Autoregressive-Moving Average Process ARMA(p, q)

$$X_t = e_t + b_1 e_{t-1} + \ldots + b_q e_{t-q} - a_1 X_{t-1} - \ldots - a_p X_{t-p}$$

Random sine signal process

$X(t) = A \sin(\omega t + \alpha)$ where α is $U(0, 2\pi)$

$$E[X(t)] = 0$$

$$R(t) = \frac{A^2}{2} \cos \omega t$$

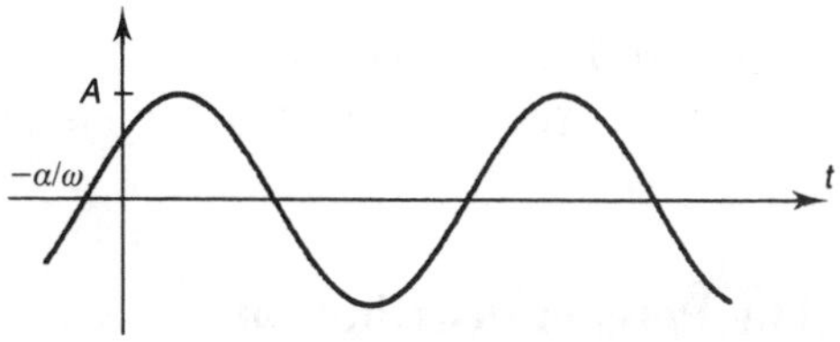

Random telegraph signal process

$X(t) = (-1)^{\alpha + Y(t)}$, $t \geq 0$ where $P(\alpha=0) = P(\alpha=1) = 0.5$ and $Y(t)$ is a Poisson process with intensity λ.

$$E[X(t)] = 0$$

$$R(t) = e^{-2\lambda|t|}$$

$$\varphi(\omega) = \frac{4\lambda}{4\lambda^2 + \omega^2}$$

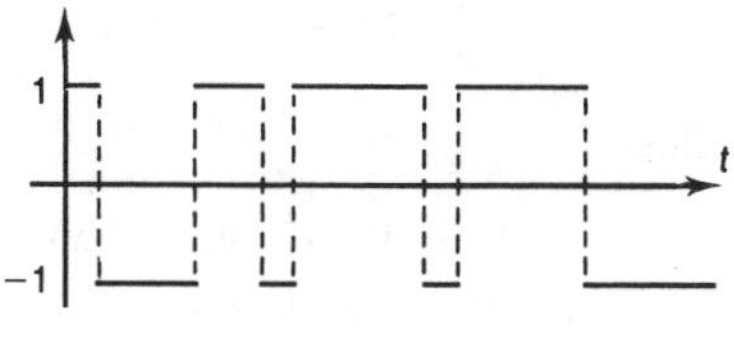

Filtered Poisson process (shot noise)

$X(t) = \sum_{t_k < t} g(t - t_k, Y_k)$, where $\{t_k\}_{-\infty}^{\infty}$ are time epochs of a Poisson process with intensity λ, Y_k are independent random variables independent of the Poisson process and $g(t-s, Y) = 0$ for $s > t$

$$E[X(t)] = \lambda \int_0^{\infty} E[g(u, Y)]du$$

$$R(t) = \lambda \int_0^{\infty} E[g(t+u, Y) g(u, Y)]du$$

17.4 Algorithms for Calculation of Probability Distributions

The binomial distribution

The probability function $f(x)$ and the distribution function $F(x)$,

$$f(x) = \binom{n}{x} p^x (1-p)^{n-x} \qquad \text{and} \qquad F(x) = \sum_{k=0}^{x} \binom{n}{k} p^k (1-p)^{n-k}$$

can be calculated for given values of n, p and x according to the formulas

$$f(x)=\frac{p}{1-p}\cdot\frac{n-x+1}{x}\cdot f(x-1),\qquad f(0)=(1-p)^n$$

Test values

$n=10$, $p=0.5$ and $x=5$ gives

$f(x)=0.24609375$ and $F(x)=0.623046875$

$n=20$, $p=0.7$ and $x=16$ gives

$f(x)=0.130420974$ and $F(x)=0.8929131955$

The Poisson distribution

The probablity function $f(x)=e^{-\lambda}\lambda^x/x!$ and the distribution function $F(x)=\sum_{k=0}^{x} e^{-\lambda}\lambda^k/k!$ can be calculated for given values of λ and x according to the formulas

$$f(x)=\frac{\lambda}{x}\cdot f(x-1),\ f(0)=e^{-\lambda}.$$

Test values

$\lambda=4$ and $x=5$ gives $f(x)=0.1562934518$ and $F(x)=0.785130387$

$\lambda=10$ and $x=10$ gives $f(x)=0.1251100357$ and $F(x)=0.5830397502$

The hypergeometric distribution

The binomial coefficient $\binom{n}{x}$ for given values of n and x can be calculated according to the formulas

$$\binom{n}{x}=\frac{n-x+1}{x}\cdot\binom{n}{x-1},\ \binom{n}{0}=1$$

This can be used to calculate

$$f(x)=\binom{Np}{x}\binom{N-Np}{n-x}\bigg/\binom{N}{n}\qquad\text{and}\qquad F(x)=\sum_{k=0}^{x} f(k)$$

Test values

$Np=200$, $N-Np=300$, $n=50$ and $x=30$ gives $f(x)=0.0013276703$

$Np=50$, $N-Np=450$, $n=40$ and $x=6$ gives $f(x)=0.1080810796$

Calculation of the normal, χ^2, t- and F-distributions

Programs for the calculation of the distribution functions for the normal, t-, χ^2- and F-distributions can be based on the following approximations.

Normal $N(0, 1)$ distribution

$\Phi(x) \approx 1/(1+e^{-p(x)})$, $p(x)=x(1.5976+0.070566x^2)$, $x \geq 0$

χ^2-distribution $\chi^2(r)$

$F(x) \approx H(A/B)$, $A=(x/r)^{1/3}+(2/9r)-1$, $B=\sqrt{2/9r}$

$H(x)=1/(1+e^{-p(x)})$, $p(x)=x(1.5976+0.070566x^2)$, $x \geq 0$

t-distribution $t(r)$

$G(x)=H(A/\sqrt{B})$, $A=x^{2/3}+2/9-2x^{2/3}/(9r)-1$, $B=2(1+x^{4/3}/r)/9$, $F(x) \approx (1+G(x))/2$

For $H(x)$, see above.

F-distribution $F(r_1, r_2)$

$$F(x) \approx \begin{cases} H(z) \text{ if } r_2 \geq 4 \\ H(z+0.08\, z^5/r_2^3) \text{ if } r_2 < 4 \end{cases}$$

$z=A/\sqrt{B}$, $A=x^{1/3}+2/(9r_1)-2x^{1/3}/(9r_2)-1$, $B=2(x^{2/3}/r_2+1/r_1)/9$

For $H(x)$, see above.

Programs for the calculation of x for given $F(x)$ can be written with the use of the above approximations and inverse interpolation e.g. by bisection.

17.5 Simulation

Random number generators

General methods

$x_{n+1}=ax_n(\bmod m)$ (multiplicative congruential generator)

$x_{n+1}=ax_n+b(\bmod m)$ (mixed congruential generator)

($x \bmod y=x-y\ \mathrm{INT}(x/y)$)

Some specific congruential generators

a	b	m	Name or source
23	0	10^8+1	Lehmer
2^7+1	1	2^{35}	Rotenberg
7^5	0	$2^{31}-1$	GGL
131	0	2^{35}	Neave
16333	25887	2^{15}	Oakenfull
3432	6789	9973	Oakenfull
171	0	30269	Wichmann-Hill

The following is a useful random number generator for modest simulations on micro and pocket computers.

$$x_{n+1} = \mathrm{FRAC}(147 x_n)$$

Instead of 147 the factors 83, 117, 123, 133, 163, 173, 187 and 197 can be used.

Random numbers in BASIC

In BASIC the command "RND" will produce a random number between 0 and 1. The command "RANDOMIZE" will give the generator a random start.

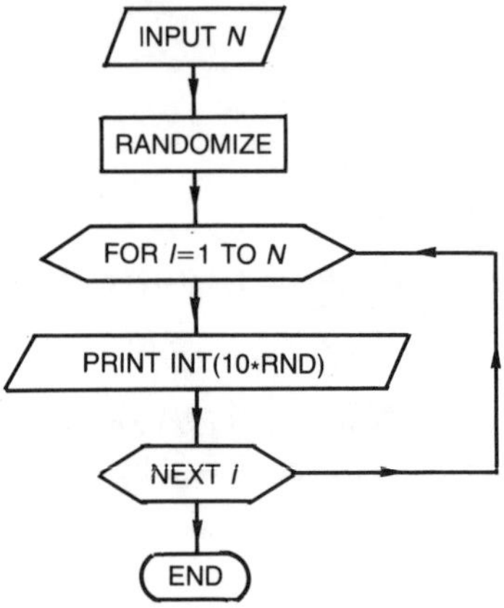

The BASIC program above will produce N random digits.

Simulation of specific distributions

In the following U, U_1, ..., U_n denote independent random numbers between 0 and 1.

Distribution		Simulation formula
Symmetric two point distribution		INT($2U$) or INT(U+0.5)
Two point distribution	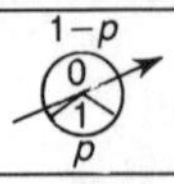	INT($U+P$)
Two point distribution (−1, 1)		2 INT($U+P$)−1
Three point distribution (0, 1, 2)	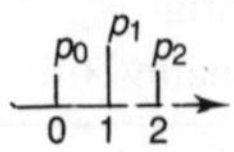	INT($U+P_2$)+INT($U+P_1+P_2$)

Uniform distribution on $\{0, 1, \ldots, N-1\}$	INT(NU)
Uniform distribution on $\{1, 2, \ldots, N\}$	INT(NU)+1
Binomial distribution $B(n, p)$	$\sum_{i=1}^{n}$ INT(U_i+p)
Exponential distribution $E(\lambda)$	$-\frac{1}{\lambda} \ln U$
Gamma distribution $\Gamma(n, \lambda)$	$-\frac{1}{\lambda} \sum_{i=1}^{n} \ln U_i$
Weibull distribution $W(\lambda, \beta)$	$\frac{1}{\lambda} (-\ln U)^{1/\beta}$

Simulation of the Poisson process and the Poisson distribution

The following flow chart describes a computer program in BASIC for simulation of the Poisson process. It uses the fact that in a Poisson process with intensity c the lengths of the time intervals are $E(c)$-variables.

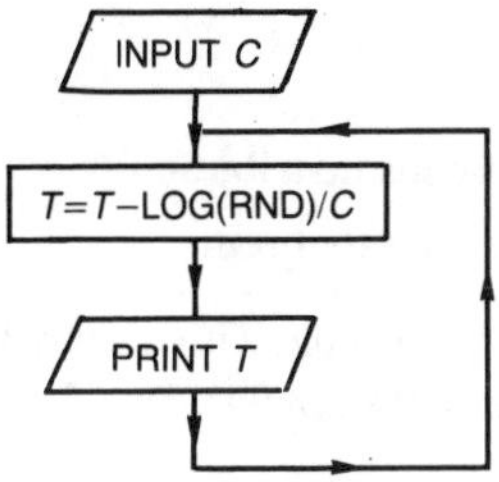

The following flow chart describes a computer program in BASIC for simulation of a Poisson distribution. It uses the fact that in a Poisson process with intensity c the number of points in a unit length interval has a Poisson distribution with parameter c.

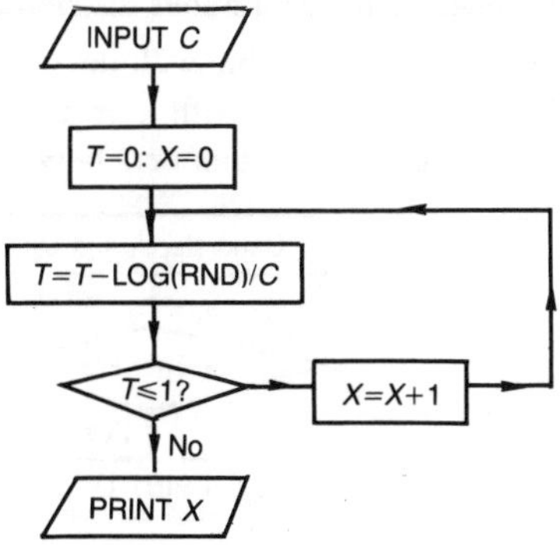

Simulation of the normal distribution $N(\mu, \sigma)$

Use of the central limit theorem

Calculate the sum $T=\sum_{i=1}^{12} U_i$ of twelve random numbers between 0 and 1 and calculate $(T-6)\sigma+\mu$.

The Box-Müller method

This method uses the following result. Calculate X_1 and X_2 according to

$$X_1=\sqrt{-2\ln U_2}\,\cos(2\pi U_1)$$

$$X_2=\sqrt{-2\ln U_2}\,\sin(2\pi U_1)$$

If U_1 and U_2 are independent random numbers between 0 and 1, then X_1 and X_2 are independent and $N(0, 1)$.

General methods for simulation of probability distributions

The "table-look-up" method for discrete distributions

Let X be such that $P(X=k)=p_k$, $k=0, 1, 2, \ldots$ Let U be a random number beetween 0 and 1. The following algorithm will give an observation on X.

If $0\leq U\leq p_0$ set $X=0$.

If $\sum_{i=0}^{k-1} p_i\leq U< \sum_{i=0}^{k} p_i$ set $X=k$.

The inverse method for continuous distribution

This method applies the following theorem.

Let X be a continuous random variable with strictly increasing distribution function F and let U be a random number between 0 and 1. Then $F^{-1}(U)$ has distribution function F.

The inverse method can be used when $F^{-1}(U)$ can be obtained explicitly or can be conveniently approximated.

Acceptance-rejection method for continuous distributions

Let X be a continuous random variable with density $f(x)$, $0\leq x\leq 1$, such that $f(x)\leq M$. Observations on X can be simulated with a program according to the following flow chart, where U_1 and U_2 are random numbers between 0 and 1.

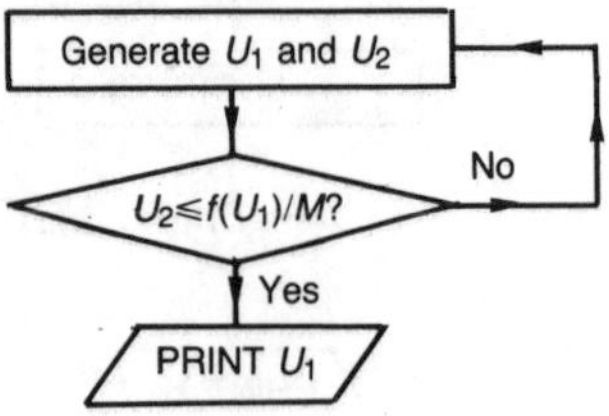

Variance reduction

Use of antithetic variable

Let the aim of a simulation be to estimate the expectation $E[X]$ of a random variable X. An *antithetic variable* is a random variable Y with the same distribution as X and negatively correlated to X. If ϱ is the correlation coefficient and $\sigma^2=\mathrm{Var}[X]=\mathrm{Var}[Y]$, then

$$\mathrm{Var}\left[\frac{X+Y}{2}\right]=\frac{\sigma^2}{2}(1+\varrho)\leqslant\sigma^2$$

The use of simulated observations of $(X+Y)/2$ will thus reduce the variance and especially so when ϱ is negative.

Example: Let U be a random number between 0 and 1 and let $X=-\lambda^{-1}\ln U$ and $Y=-\lambda^{-1}\ln(1-U)$. Then both X and Y are $E(\lambda)$-variables and it can be shown that $\varrho=1-\pi^2/6$.

Use of control variable

Let Y be a random variable with *known* expectation $E[Y]=\Theta$. Then simulated observations of $Z=X-(Y-\Theta)$ can be used to estimate $E[X]$, because $E[Z]=E[X]$. From

$$\mathrm{Var}[Z]=\mathrm{Var}[X]+\mathrm{Var}[Y]-2\,\mathrm{Cov}[X,\ Y]$$

follows that it is especially useful if X and the *control variable* Y have positive correlation.

17.6 Queueing Systems

*M/M/*1, *M/M/n* and *M/G/*1, Basic formulas

The formulas below hold for stationary conditions.

$X=$ number of customers in the system
$Y=$ number of customers in the queue
$W=$ waiting (queueing) time
$U=$ total time in system (system time)
$Z=$ length of busy period
$B=$ service time
$\lambda=$ intensity of arrival Poisson process
$\mu=$ intensity of exponential service time

	$M/M/1$	$M/M/n$	$M/G/1$
Traffic intensity ϱ	$\varrho=\lambda/\mu$ $0<\varrho<1$	$\varrho=\lambda/(n\mu)$ $0<\varrho<1$	$\varrho=\lambda E[B]$ $0<\varrho<1$
$P(X=k)=\pi_k$ $k=0, 1, 2, \ldots$	$(1-\varrho)\varrho^k$	$\pi_0=(A+B)^{-1}$ $A=\sum_{k=0}^{n-1}(n\varrho)^k/k!$ $B=(n\varrho)^n/((1-\varrho)n!)$ $\pi_k=(n\varrho)^k\pi_0/k!,\ k\leq n$ $\pi_k=\varrho^{k-n}\pi_n,\ k\geq n$	$E[s^X]$ $=\dfrac{(1-\varrho)(1-s)f^*(\lambda(1-s))}{f^*(\lambda(1-s))-s}$ $f^*(s)=E[e^{-sB}]$
$E[X]$	$\varrho/(1-\varrho)$	$\varrho(n+\pi/(1-\varrho)),\ \pi=\sum_{k=n}^{\infty}\pi_k$ $\pi=B/(A+B)$, A, B see above	$\varrho+\varrho^2E[B^2]/(2(1-\varrho)E^2[B])$ $E[B^2]=\mathrm{Var}[B]+E^2[B]$
$E[Y]$	$\varrho^2/(1-\varrho)$	$\varrho\pi/(1-\varrho),\ \pi=B/(A+B)$	$\varrho^2E[B^2]/(2(1-\varrho)E^2[B])$
$E[W]$	$\varrho/(\mu(1-\varrho))$	$\pi/(\mu n(1-\varrho))$	$\varrho^2E[B^2]/(2(1-\varrho)E[B])=$ $=\lambda E[B^2]/(2(1-\varrho))$
$E[U]$	$1/(\mu(1-\varrho))$	$(1+\pi/(n(1-\varrho)))/\mu$	$E[W]+E[B]$
$E[Z]$	$1/(\mu(1-\varrho))$	-	$E[B]/(1-\varrho)$

Additional formulas for $M/M/1$ and $M/G/1$

M/M/1

$E[s^X]=(1-\varrho)/(1-\varrho s)$

$E[s^Y]=(1-\varrho)(1+\varrho-\varrho s)/(1-\varrho s)$

$P(W\leq x)=1-\varrho e^{-\mu(1-\varrho)x}$

$P(U\leq x)=1-e^{-\mu(1-\varrho)x}$

$\mathrm{Var}[X]=\varrho/(1-\varrho)^2$

$\mathrm{Var}[Y]=\varrho^2(1+\varrho-\varrho^2)/(1-\varrho)^2$

$\mathrm{Var}[W]=\varrho(2-\varrho)/(\mu(1-\varrho))^2$

$\mathrm{Var}[U]=1/(\mu^2(1-\varrho)^2)$

M/G/1

$$E[e^{-sW}]=\frac{(1-\varrho)s}{s-\lambda+\lambda f^*(s)}$$

$$E[e^{-sU}]=\frac{(1-\varrho)sf^*(s)}{s-\lambda+\lambda f^*(s)}$$

$f_Z^*(s)=E[e^{-sZ}]=f^*(s+\lambda-\lambda f_Z^*(s))$

$f^*(s)=E[e^{-sB}]$

Little's formula

For a steady state system

$$L=\lambda W$$

$\lambda=$ intensity of arrival process
$L=$ expected number of customers in the system
$W=$ expected time of customer in system

Erlang's loss formula

A queueing system has c servers and no waiting line. The traffic intensity is ϱ. Then Erlang's loss formula gives the probability $\pi(c, \varrho)$ that all servers are busy under stationary conditions.

$$\pi(c, \varrho)=(\varrho^c/c!)/\sum_{i=0}^{c}(\varrho^i/i!)$$

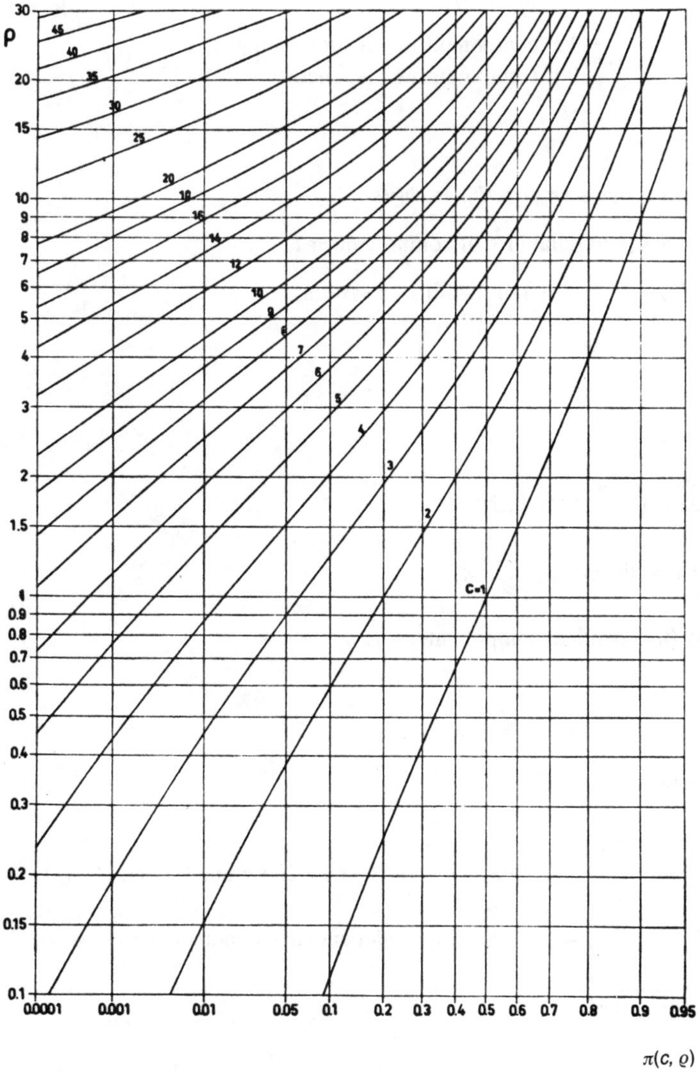

17.7 Reliability

Coherent systems

System functions and reliability

System function $\Phi(\boldsymbol{x})=\begin{cases}1 \text{ if system is functioning for vector } \boldsymbol{x}\\ 0 \text{ if system is failed for vector } \boldsymbol{x}\end{cases}$

$\boldsymbol{x}=(x_1, x_2, \ldots, x_n)$, where

$x_i=\begin{cases}1 \text{ if component nr } i \text{ is functioning}\\ 0 \text{ if component nr } i \text{ is failed}\end{cases}$

For *coherent systems* $\boldsymbol{x}<\boldsymbol{y} \Rightarrow \Phi(\boldsymbol{x})\leq\Phi(\boldsymbol{y})$.

$p_i=P(x_i=1)=E[x_i]=$reliability of component nr i

$E[\Phi(\boldsymbol{x})]=h(p_1, p_2, \ldots, p_n)=$reliability of system

Series system

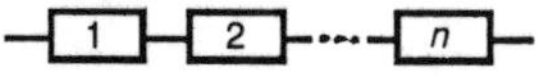

$\Phi(\boldsymbol{x})=\prod_{i=1}^{n} x_i=\min(x_1, x_2, \ldots, x_n)$

Reliability (independent components)$=\prod_{i=1}^{n} p_i$

Parallel system

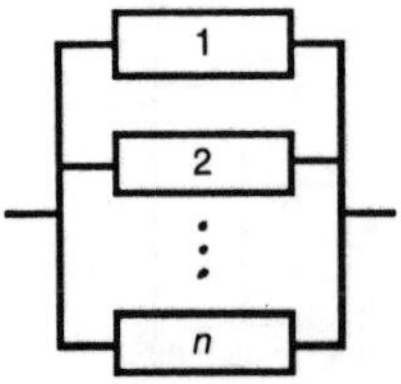

$\Phi(\boldsymbol{x})=1-\prod_{i=1}^{n}(1-x_i)=\coprod_{i=1}^{n} x_i=\max(x_1, x_2, \ldots, x_n)$

Reliability (independent components)$=\coprod_{i=1}^{n} p_i=1-\prod_{i=1}^{n}(1-p_i)$

k-out-of-n system

The system is functioning if at least k components are functioning.

$$\Phi(\boldsymbol{x})=\begin{cases}1 \text{ if } \sum_{i=1}^{n} x_i \geqslant k\\ 0 \text{ if } \sum_{i=1}^{n} x_i < k\end{cases}$$

Reliability (independent components with the same reliability p)=

$$=\sum_{i=k}^{n}\binom{n}{i}p^i(1-p)^{n-i}$$

Specific coherent systems

System	System function	Reliability (independent components)
	x_1x_2	p_1p_2
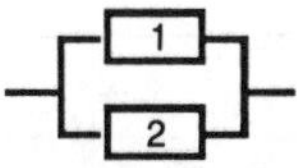	$x_1 \amalg x_2=$ $=1-(1-x_1)(1-x_2)$	$p_1 \amalg p_2=$ $=1-(1-p_1)(1-p_2)$
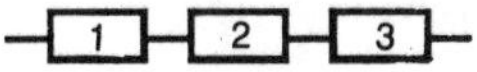	$x_1x_2x_3$	$p_1p_2p_3$
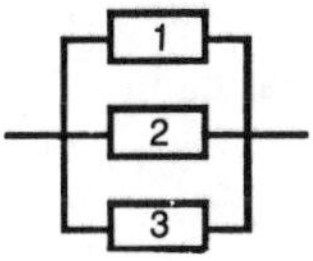	$x_1 \amalg x_2 \amalg c_3=$ $=1-(1-x_1)(1-x_2)(1-x_3)$	$p_1 \amalg p_2 \quad p_3=$ $=1-(1-p_1)(1-p_2)(1-p_3)$
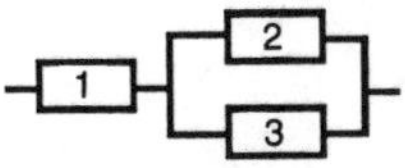	$x_1(x_2 \amalg x_3)=$ $=x_1(x_2+x_3-x_2x_3)$	$p_1(p_2+p_3-p_2p_3)$
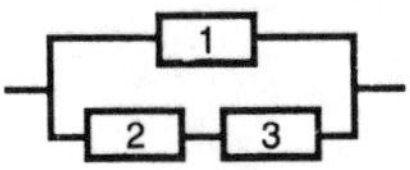	$x_1 \amalg (x_2x_3)=$ $=x_1+x_2x_3-x_1x_2x_3$	$p_1+p_2p_3-p_1p_2p_3$
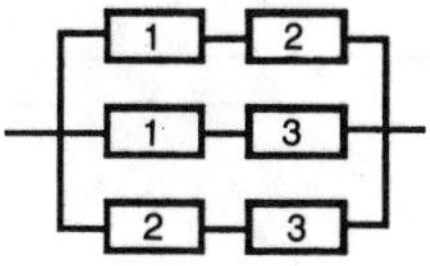	$(x_1x_2) \amalg (x_1x_3) \amalg (x_2x_3)=$ $=1-(1-x_1x_2)(1-x_1x_3)(1-x_2x_3)=$ $=x_1x_2+x_1x_3+x_2x_3-2x_1x_2x_3$	$p_1p_2+p_1p_3+p_2p_3-2p_1p_2p_3$

Life distributions

Basic definitions

X=length of life for component or system of components

Distribution function	$F(x)=P(X\leqslant x)$
Survival probability (reliability)	$G(x)=P(X>x)=1-F(x)$
Probability density	$f(x)=F'(x)=-G'(x)$
Failure (hazard) rate	$r(x)=f(x)/G(x)=F'(x)/(1-F(x))$
Expected lifetime	$\mu=\int_0^\infty xf(x)dx=\int_0^\infty G(x)dx$
Variance	$\sigma^2=\int_0^\infty (x-\mu)^2 f(x)dx$

$$P(X\leqslant t+h|X>t)=r(t)h+o(h)$$

$$G(x)=1-F(x)=e^{-\int_0^x r(t)dt}$$

Properties of life distributions

Property	Notation	Definition
Increasing failure rate	IFR	$r(t)$ is increasing in t
Decreasing failure rate	DFR	$r(t)$ is decreasing in t
Increasing failure rate average	IFRA	$\int_0^t r(x)dx/t$ is increasing in t
Decreasing failure rate average	DFRA	$\int_0^t r(x)dx/t$ is decreasing in t
New better than used	NBU	$G(x+y)\leqslant G(x)G(y)$
New worse than used	NWU	$G(x+y)\geqslant G(x)G(y)$
New better than used in expectation	NBUE	$\int_t^\infty G(x)dx\leqslant \mu G(t)$
New worse than used in expectation	NWUE	$\int_t^\infty G(x)dx\geqslant \mu G(t)$
Harmonic new better than used in expectation	HNBUE	$\int_t^\infty G(x)dx\leqslant \mu e^{-t/\mu}$
Harmonic new worse than used in expectation	HNWUE	$\int_t^\infty G(x)dx\geqslant \mu e^{-t/\mu}$

IFR $\Rightarrow$ IFRA $\Rightarrow$ NBU $\Rightarrow$ NBUE $\Rightarrow$ HNBUE

DFR $\Rightarrow$ DFRA $\Rightarrow$ NWU $\Rightarrow$ NWUE $\Rightarrow$ HNWUE

Specific life distributions

Name	f, G, r, μ, σ^2	Properties
Exponential	$f(x)=\lambda e^{-\lambda x},\ x\geqslant 0$ $G(x)=e^{-\lambda x},\ x\geqslant 0$ $r(x)=\lambda$ $\mu=1/\lambda$ $\sigma^2=1/\lambda^2$	Constant failure rate
Weibull	$f(x)=\beta\lambda^\beta x^{\beta-1}e^{-(\lambda x)^\beta}$ $G(x)=e^{-(\lambda x)^\beta}$ $r(x)=\beta\lambda^\beta x^{\beta-1}$ $\mu=\lambda^{-1}\Gamma(1+1/\beta)$ $\sigma^2=\lambda^{-2}(\Gamma(1+2/\beta)-\Gamma^2(1+1/\beta))$	IFR for $\beta\geqslant 1$ DFR for $\beta\leqslant 1$
Lognormal	$f(x)=\dfrac{1}{\beta x\sqrt{2\pi}}e^{-(\ln x-\alpha)^2/2\beta^2}$ $\mu=e^{\alpha+\beta^2/2}$ $\sigma^2=e^{2\alpha+\beta^2}(e^{\beta^2}-1)$	
Gamma	$f(x)=\dfrac{\lambda^n}{\Gamma(n)}x^{n-1}e^{-\lambda x},\ x\geqslant 0$ For positive integers n $G(x)=\sum_{i=0}^{n-1}\dfrac{(\lambda x)^i}{i!}e^{-\lambda x}$ $\mu=n/\lambda$ $\sigma^2=n/\lambda^2$	IFR for $n\geqslant 1$ DFR for $0<n\leqslant 1$
Uniform	$f(x)=1/(b-a),\ a\leqslant x\leqslant b$ $G(x)=(b-x)/(b-a)$ $r(x)=1/(b-x)$ $\mu=(a+b)/2$ $\sigma^2=(b-a)^2/12$	IFR

Life distribution for systems

General component life distributions

$F(x)=1-(1-F_1(x))(1-F_2(x))$
$G(x)=G_1(x)\,G_2(x)$

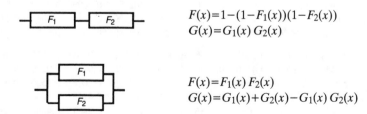

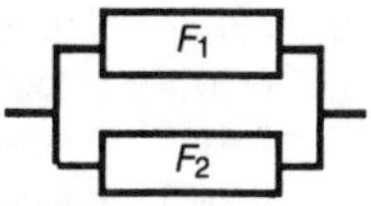

$F(x)=F_1(x)\,F_2(x)$
$G(x)=G_1(x)+G_2(x)-G_1(x)\,G_2(x)$

17.7

Exponential component life distributions

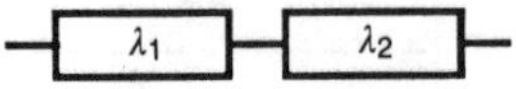

$$F(x)=1-e^{-(\lambda_1+\lambda_2)x}$$
$$G(x)=e^{-(\lambda_1+\lambda_2)x}$$
$$\mu=1/(\lambda_1+\lambda_2)$$

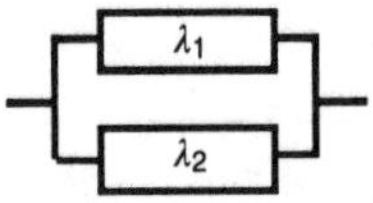

$$F(x)=(1-e^{-\lambda_1 x})(1-e^{-\lambda_2 x})$$
$$G(x)=e^{-\lambda_1 x}+e^{-\lambda_2 x}-e^{-(\lambda_1+\lambda_2)x}$$
$$\mu=\frac{1}{\lambda_1}+\frac{1}{\lambda_2}-\frac{1}{\lambda_1+\lambda_2}$$

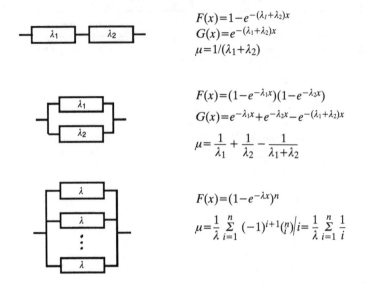

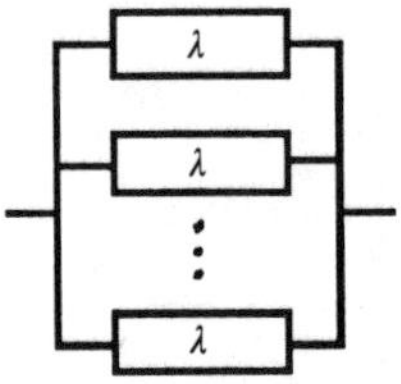

$$F(x)=(1-e^{-\lambda x})^n$$
$$\mu=\frac{1}{\lambda}\sum_{i=1}^{n}(-1)^{i+1}\binom{n}{i}\Big/ i=\frac{1}{\lambda}\sum_{i=1}^{n}\frac{1}{i}$$

17.8 Tables

The binomial distribution

Probability function

The table gives $f(x)=\binom{n}{x}p^x(1-p)^{n-x}$, $x=0, 1, \ldots, n$ for different n, p and x. Concerning calculation of $f(x)$ with programmable machine, see section 17.4.

n	*x*	*p*=0.1	0.2	0.3	0.4	0.5	0.6	0.7	0.8	0.9
2	**0**	0.8100	0.6400	0.4900	0.3600	0.2500	0.1600	0.0900	0.0400	0.0100
	1	0.1800	0.3200	0.4200	0.4800	0.5000	0.4800	0.4200	0.3200	0.1800
	2	0.0100	0.0400	0.0900	0.1600	0.2500	0.3600	0.4900	0.6400	0.8100
3	**0**	0.7290	0.5120	0.3430	0.2160	0.1250	0.0640	0.0270	0.0080	0.0010
	1	0.2430	0.3840	0.4410	0.4320	0.3750	0.2880	0.1890	0.0960	0.0270
	2	0.0270	0.0960	0.1890	0.2880	0.3750	0.4320	0.4410	0.3840	0.2430
	3	0.0010	0.0080	0.0270	0.0640	0.1250	0.2160	0.3430	0.5120	0.7290
4	**0**	0.6561	0.4096	0.2401	0.1296	0.0625	0.0256	0.0081	0.0016	0.0001
	1	0.2916	0.4096	0.4116	0.3456	0.2500	0.1536	0.0756	0.0256	0.0036
	2	0.0486	0.1536	0.2646	0.3456	0.3750	0.3456	0.2646	0.1536	0.0486
	3	0.0036	0.0256	0.0756	0.1536	0.2500	0.3456	0.4116	0.4096	0.2916
	4	0.0001	0.0016	0.0081	0.0256	0.0625	0.1296	0.2401	0.4096	0.6561

n	x	p=0.1	0.2	0.3	0.4	0.5	0.6	0.7	0.8	0.9
5	**0**	0.5905	0.3277	0.1681	0.0778	0.0312	0.0102	0.0024	0.0003	0.0000
	1	0.3280	0.4096	0.3602	0.2592	0.1562	0.0768	0.0284	0.0064	0.0004
	2	0.0729	0.2048	0.3087	0.3456	0.3125	0.2304	0.1323	0.0512	0.0081
	3	0.0081	0.0512	0.1323	0.2304	0.3125	0.3456	0.3087	0.2048	0.0729
	4	0.0004	0.0064	0.0284	0.0768	0.1562	0.2592	0.3602	0.4096	0.3280
	5	0.0000	0.0003	0.0024	0.0102	0.0312	0.0778	0.1681	0.3277	0.5905
6	**0**	0.5314	0.2621	0.1176	0.0467	0.0156	0.0041	0.0007	0.0001	0.0000
	1	0.3543	0.3932	0.3025	0.1866	0.0938	0.0369	0.0102	0.0015	0.0001
	2	0.0984	0.2458	0.3241	0.3110	0.2344	0.1382	0.0595	0.0154	0.0012
	3	0.0146	0.0819	0.1852	0.2765	0.3125	0.2765	0.1852	0.0819	0.0146
	4	0.0012	0.0154	0.0595	0.1382	0.2344	0.3110	0.3241	0.2458	0.0984
	5	0.0001	0.0015	0.0102	0.0369	0.0938	0.1866	0.3025	0.3932	0.3543
	6	0.0000	0.0001	0.0007	0.0041	0.0156	0.0467	0.1176	0.2621	0.5314
7	**0**	0.4783	0.2097	0.0824	0.0280	0.0078	0.0016	0.0002	0.0000	0.0000
	1	0.3720	0.3670	0.2471	0.1306	0.0547	0.0172	0.0036	0.0004	0.0000
	2	0.1240	0.2753	0.3177	0.2613	0.1641	0.0774	0.0250	0.0043	0.0002
	3	0.0230	0.1147	0.2269	0.2903	0.2734	0.1935	0.0972	0.0287	0.0026
	4	0.0026	0.0287	0.0972	0.1935	0.2734	0.2903	0.2269	0.1147	0.0230
	5	0.0002	0.0043	0.0250	0.0774	0.1641	0.2613	0.3177	0.2753	0.1240
	6	0.0000	0.0004	0.0036	0.0172	0.0547	0.1306	0.2471	0.3670	0.3720
	7	0.0000	0.0000	0.0002	0.0016	0.0078	0.0280	0.0824	0.2097	0.4783
8	**0**	0.4305	0.1678	0.0576	0.0168	0.0039	0.0007	0.0001	0.0000	0.0000
	1	0.3826	0.3355	0.1977	0.0896	0.0312	0.0079	0.0012	0.0001	0.0000
	2	0.1488	0.2936	0.2965	0.2090	0.1094	0.0413	0.0100	0.0011	0.0000
	3	0.0331	0.1468	0.2541	0.2787	0.2188	0.1239	0.0467	0.0092	0.0004
	4	0.0046	0.0459	0.1361	0.2322	0.2734	0.2322	0.1361	0.0459	0.0046
	5	0.0004	0.0092	0.0467	0.1239	0.2188	0.2787	0.2541	0.1468	0.0331
	6	0.0000	0.0011	0.0100	0.0413	0.1094	0.2090	0.2965	0.2936	0.1488
	7	0.0000	0.0001	0.0012	0.0079	0.0312	0.0896	0.1977	0.3355	0.3826
	8	0.0000	0.0000	0.0001	0.0007	0.0039	0.0168	0.0576	0.1678	0.4305
9	**0**	0.3874	0.1342	0.0404	0.0101	0.0020	0.0003	0.0000	0.0000	0.0000
	1	0.3874	0.3020	0.1556	0.0605	0.0176	0.0035	0.0004	0.0000	0.0000
	2	0.1722	0.3020	0.2668	0.1612	0.0703	0.0212	0.0039	0.0003	0.0000
	3	0.0446	0.1762	0.2668	0.2508	0.1641	0.0743	0.0210	0.0028	0.0001
	4	0.0074	0.0661	0.1715	0.2508	0.2461	0.1672	0.0735	0.0165	0.0008
	5	0.0008	0.0165	0.0735	0.1672	0.2461	0.2508	0.1715	0.0661	0.0074
	6	0.0001	0.0028	0.0210	0.0743	0.1641	0.2508	0.2668	0.1762	0.0446
	7	0.0000	0.0003	0.0039	0.0212	0.0703	0.1612	0.2668	0.3020	0.1722
	8	0.0000	0.0000	0.0004	0.0035	0.0176	0.0605	0.1556	0.3020	0.3874
	9	0.0000	0.0000	0.0000	0.0003	0.0020	0.0101	0.0404	0.1342	0.3874
10	**0**	0.3487	0.1074	0.0282	0.0060	0.0010	0.0001	0.0000	0.0000	0.0000
	1	0.3874	0.2684	0.1211	0.0403	0.0098	0.0016	0.0001	0.0000	0.0000
	2	0.1937	0.3020	0.2335	0.1209	0.0439	0.0106	0.0014	0.0001	0.0000
	3	0.0574	0.2013	0.2668	0.2150	0.1172	0.0425	0.0090	0.0008	0.0000

n	x	p=0.1	0.2	0.3	0.4	0.5	0.6	0.7	0.8	0.9
	4	0.0112	0.0881	0.2001	0.2508	0.2051	0.1115	0.0368	0.0055	0.0001
	5	0.0015	0.0264	0.1029	0.2007	0.2461	0.2007	0.1029	0.0264	0.0015
	6	0.0001	0.0055	0.0368	0.1115	0.2051	0.2508	0.2001	0.0881	0.0112
	7	0.0000	0.0008	0.0090	0.0425	0.1172	0.2150	0.2668	0.2013	0.0574
	8	0.0000	0.0001	0.0014	0.0106	0.0439	0.1209	0.2335	0.3020	0.1937
	9	0.0000	0.0000	0.0001	0.0016	0.0098	0.0403	0.1211	0.2684	0.3874
	10	0.0000	0.0000	0.0000	0.0001	0.0010	0.0060	0.0282	0.1074	0.3487
12	**0**	0.2824	0.0687	0.0138	0.0022	0.0002	0.0000	0.0000	0.0000	0.0000
	1	0.3766	0.2062	0.0712	0.0174	0.0029	0.0003	0.0000	0.0000	0.0000
	2	0.2301	0.2835	0.1678	0.0639	0.0161	0.0025	0.0002	0.0000	0.0000
	3	0.0852	0.2362	0.2397	0.1419	0.0537	0.0125	0.0015	0.0001	0.0000
	4	0.0213	0.1329	0.2311	0.2128	0.1208	0.0420	0.0078	0.0005	0.0000
	5	0.0038	0.0532	0.1585	0.2270	0.1934	0.1009	0.0291	0.0033	0.0000
	6	0.0005	0.0155	0.0792	0.1766	0.2256	0.1766	0.0792	0.0155	0.0005
	7	0.0000	0.0033	0.0291	0.1009	0.1934	0.2270	0.1585	0.0532	0.0038
	8	0.0000	0.0005	0.0078	0.0420	0.1208	0.2128	0.2311	0.1329	0.0213
	9	0.0000	0.0001	0.0015	0.0125	0.0537	0.1419	0.2397	0.2362	0.0852
	10	0.0000	0.0000	0.0002	0.0025	0.0161	0.0639	0.1678	0.2835	0.2301
	11	0.0000	0.0000	0.0000	0.0003	0.0029	0.0174	0.0712	0.2062	0.3766
	12	0.0000	0.0000	0.0000	0.0000	0.0002	0.0022	0.0138	0.0687	0.2824
20	**0**	0.1216	0.0115	0.0008	0.0000	0.0000	0.0000	0.0000	0.0000	0.0000
	1	0.2702	0.0576	0.0068	0.0005	0.0000	0.0000	0.0000	0.0000	0.0000
	2	0.2852	0.1369	0.0278	0.0031	0.0002	0.0000	0.0000	0.0000	0.0000
	3	0.1901	0.2054	0.0716	0.0123	0.0011	0.0000	0.0000	0.0000	0.0000
	4	0.0898	0.2182	0.1304	0.0350	0.0046	0.0003	0.0000	0.0000	0.0000
	5	0.0319	0.1746	0.1789	0.0746	0.0148	0.0013	0.0000	0.0000	0.0000
	6	0.0089	0.1091	0.1916	0.1244	0.0370	0.0049	0.0002	0.0000	0.0000
	7	0.0020	0.0545	0.1643	0.1659	0.0739	0.0146	0.0010	0.0000	0.0000
	8	0.0004	0.0222	0.1144	0.1797	0.1201	0.0355	0.0039	0.0001	0.0000
	9	0.0001	0.0074	0.0654	0.1597	0.1602	0.0710	0.0120	0.0005	0.0000
	10	0.0000	0.0020	0.0308	0.1171	0.1762	0.1171	0.0308	0.0020	0.0000
	11	0.0000	0.0005	0.0120	0.0710	0.1602	0.1597	0.0654	0.0074	0.0001
	12	0.0000	0.0001	0.0039	0.0355	0.1201	0.1797	0.1144	0.0222	0.0004
	13	0.0000	0.0000	0.0010	0.0146	0.0739	0.1659	0.1643	0.0545	0.0020
	14	0.0000	0.0000	0.0002	0.0049	0.0370	0.1244	0.1916	0.1091	0.0089
	15	0.0000	0.0000	0.0000	0.0013	0.0148	0.0746	0.1789	0.1746	0.0319
	16	0.0000	0.0000	0.0000	0.0003	0.0046	0.0350	0.1304	0.2182	0.0898
	17	0.0000	0.0000	0.0000	0.0000	0.0011	0.0123	0.0716	0.2054	0.1901
	18	0.0000	0.0000	0.0000	0.0000	0.0002	0.0031	0.0278	0.1369	0.2852
	19	0.0000	0.0000	0.0000	0.0000	0.0000	0.0005	0.0068	0.0576	0.2702
	20	0.0000	0.0000	0.0000	0.0000	0.0000	0.0000	0.0008	0.0115	0.1216

Binomial distribution, distribution function

The table gives the distribution function $F(x) = \sum_{k=0}^{x} \binom{n}{k} p^k (1-p)^{n-k}$ for different values of n and p.

Concerning calculation of $F(x)$ with programmable machine, see section 17.4.

n	x	p=0.10	0.20	0.30	0.40	0.50	0.60	0.70	0.80	0.90
2	**0**	0.8100	0.6400	0.4900	0.3600	0.2500	0.1600	0.0900	0.0400	0.0100
	1	0.9900	0.9600	0.9100	0.8400	0.7500	0.6400	0.5100	0.3600	0.1900
3	**0**	0.7290	0.5120	0.3430	0.2160	0.1250	0.0640	0.0270	0.0080	0.0010
	1	0.9720	0.8960	0.7840	0.6480	0.5000	0.3520	0.2160	0.1040	0.0280
	2	0.9990	0.9920	0.9730	0.9360	0.8750	0.7840	0.6570	0.4880	0.2710
4	**0**	0.6561	0.4096	0.2401	0.1296	0.0625	0.0256	0.0081	0.0016	0.0001
	1	0.9477	0.8192	0.6517	0.4752	0.3125	0.1792	0.0837	0.0272	0.0037
	2	0.9963	0.9728	0.9163	0.8208	0.6875	0.5248	0.3483	0.1808	0.0523
	3	0.9999	0.9984	0.9919	0.9744	0.9375	0.8704	0.7599	0.5904	0.3439
5	**0**	0.5905	0.3277	0.1681	0.0778	0.0313	0.0102	0.0024	0.0003	0.0000
	1	0.9185	0.7373	0.5282	0.3370	0.1875	0.0870	0.0308	0.0067	0.0005
	2	0.9914	0.9421	0.8369	0.6826	0.5000	0.3174	0.1631	0.0579	0.0086
	3	0.9995	0.9933	0.9692	0.9130	0.8125	0.6630	0.4718	0.2627	0.0815
	4	1.0000	0.9997	0.9976	0.9898	0.9688	0.9222	0.8319	0.6723	0.4095
6	**0**	0.5314	0.2621	0.1176	0.0467	0.0156	0.0041	0.0007	0.0001	0.0000
	1	0.8857	0.6554	0.4202	0.2333	0.1094	0.0410	0.0109	0.0016	0.0001
	2	0.9841	0.9011	0.7443	0.5443	0.3438	0.1792	0.0705	0.0170	0.0013
	3	0.9987	0.9830	0.9295	0.8208	0.6563	0.4557	0.2557	0.0989	0.0159
	4	0.9999	0.9984	0.9891	0.9590	0.8906	0.7667	0.5798	0.3446	0.1143
	5	1.0000	0.9999	0.9993	0.9959	0.9844	0.9533	0.8824	0.7379	0.4686
7	**0**	0.4783	0.2097	0.0824	0.0280	0.0078	0.0016	0.0002	0.0000	0.0000
	1	0.8503	0.5767	0.3294	0.1586	0.0625	0.0188	0.0038	0.0004	0.0000
	2	0.9743	0.8520	0.6471	0.4199	0.2266	0.0963	0.0288	0.0047	0.0002
	3	0.9973	0.9667	0.8740	0.7102	0.5000	0.2898	0.1260	0.0333	0.0027
	4	0.9998	0.9953	0.9712	0.9037	0.7734	0.5801	0.3529	0.1480	0.0257
	5	1.0000	0.9996	0.9962	0.9812	0.9375	0.8414	0.6706	0.4233	0.1497
	6	1.0000	0.0000	0.9998	0.9984	0.9922	0.9720	0.9176	0.7903	0.5217
8	**0**	0.4305	0.1678	0.0576	0.0168	0.0039	0.0007	0.0001	0.0000	0.0000
	1	0.8131	0.5033	0.2553	0.1064	0.0352	0.0085	0.0013	0.0001	0.0000
	2	0.9619	0.7969	0.5518	0.3154	0.1445	0.0498	0.0113	0.0012	0.0000
	3	0.9950	0.9437	0.8059	0.5941	0.3633	0.1737	0.0580	0.0104	0.0004
	4	0.9996	0.9896	0.9420	0.8263	0.6367	0.4059	0.1941	0.0563	0.0050
	5	1.0000	0.9988	0.9887	0.9502	0.8555	0.6846	0.4482	0.2031	0.0381
	6	1.0000	0.9999	0.9987	0.9915	0.9648	0.8936	0.7447	0.4967	0.1869
	7	1.0000	1.0000	0.9999	0.9993	0.9961	0.9832	0.9424	0.8322	0.5695
9	**0**	0.3874	0.1342	0.0404	0.0101	0.0020	0.0003	0.0000	0.0000	0.0000
	1	0.7748	0.4362	0.1960	0.0705	0.0195	0.0038	0.0004	0.0000	0.0000

n	x	p=0.10	0.20	0.30	0.40	0.50	0.60	0.70	0.80	0.90
	2	0.9470	0.7382	0.4628	0.2318	0.0898	0.0250	0.0043	0.0003	0.0000
	3	0.9917	0.9144	0.7297	0.4826	0.2539	0.0994	0.0253	0.0031	0.0001
	4	0.9991	0.9804	0.9012	0.7334	0.5000	0.2666	0.0988	0.0196	0.0009
	5	0.9999	0.9969	0.9747	0.9006	0.7461	0.5174	0.2703	0.0856	0.0083
	6	1.0000	0.9997	0.9957	0.9750	0.9102	0.7682	0.5372	0.2618	0.0530
	7	1.0000	1.0000	0.9996	0.9962	0.9805	0.9295	0.8040	0.5638	0.2252
	8	1.0000	1.0000	1.0000	0.9997	0.9980	0.9899	0.9596	0.8658	0.6126
10	0	0.3487	0.1074	0.0282	0.0060	0.0010	0.0001	0.0000	0.0000	0.0000
	1	0.7361	0.3758	0.1493	0.0464	0.0107	0.0017	0.0001	0.0000	0.0000
	2	0.9298	0.6778	0.3828	0.1673	0.0547	0.0123	0.0016	0.0001	0.0000
	3	0.9872	0.8791	0.6496	0.3823	0.1719	0.0548	0.0106	0.0009	0.0000
	4	0.9984	0.9672	0.8497	0.6331	0.3770	0.1662	0.0473	0.0064	0.0001
	5	0.9999	0.9936	0.9527	0.8338	0.6230	0.3669	0.1503	0.0328	0.0016
	6	1.0000	0.9991	0.9894	0.9452	0.8281	0.6177	0.3504	0.1209	0.0128
	7	1.0000	0.9999	0.9984	0.9877	0.9453	0.8327	0.6172	0.3222	0.0702
	8	1.0000	1.0000	0.9999	0.9983	0.9893	0.9536	0.8507	0.6242	0.2639
	9	1.0000	1.0000	1.0000	0.9999	0.9990	0.9940	0.9718	0.8926	0.6513
12	0	0.2824	0.0687	0.0138	0.0022	0.0002	0.0000	0.0000	0.0000	0.0000
	1	0.6590	0.2749	0.0850	0.0196	0.0032	0.0003	0.0000	0.0000	0.0000
	2	0.8891	0.5583	0.2528	0.0834	0.0193	0.0028	0.0002	0.0000	0.0000
	3	0.9744	0.7946	0.4925	0.2253	0.0730	0.0153	0.0017	0.0001	0.0000
	4	0.9957	0.9274	0.7237	0.4382	0.1938	0.0573	0.0095	0.0006	0.0000
	5	0.9995	0.9806	0.8822	0.6652	0.3872	0.1582	0.0386	0.0039	0.0001
	6	0.9999	0.9961	0.9614	0.8418	0.6128	0.3348	0.1178	0.0194	0.0005
	7	1.0000	0.9994	0.9905	0.9427	0.8062	0.5618	0.2763	0.0726	0.0043
	8	1.0000	0.9999	0.9983	0.9847	0.9270	0.7747	0.5075	0.2054	0.0256
	9	1.0000	1.0000	0.9998	0.9972	0.9807	0.9166	0.7472	0.4417	0.1109
	10	1.0000	1.0000	1.0000	0.9997	0.9968	0.9804	0.9150	0.7251	0.3410
	11	1.0000	1.0000	1.0000	1.0000	0.9998	0.9978	0.9862	0.9313	0.7176
20	0	0.1216	0.0115	0.0008	0.0000	0.0000	0.0000	0.0000	0.0000	0.0000
	1	0.3917	0.0692	0.0076	0.0005	0.0000	0.0000	0.0000	0.0000	0.0000
	2	0.6769	0.2061	0.0355	0.0036	0.0002	0.0000	0.0000	0.0000	0.0000
	3	0.8670	0.4114	0.1071	0.0160	0.0013	0.0000	0.0000	0.0000	0.0000
	4	0.9568	0.6296	0.2375	0.0510	0.0059	0.0003	0.0000	0.0000	0.0000
	5	0.9887	0.8042	0.4164	0.1256	0.0207	0.0016	0.0000	0.0000	0.0000
	6	0.9976	0.9133	0.6080	0.2500	0.0577	0.0065	0.0003	0.0000	0.0000
	7	0.9996	0.9679	0.7723	0.4159	0.1316	0.0210	0.0013	0.0000	0.0000
	8	0.9999	0.9900	0.8867	0.5956	0.2517	0.0565	0.0051	0.0001	0.0000
	9	1.0000	0.9974	0.9520	0.7553	0.4119	0.1275	0.0171	0.0006	0.0000
	10	1.0000	0.9994	0.9829	0.8725	0.5881	0.2447	0.0480	0.0026	0.0000
	11	1.0000	0.9999	0.9949	0.9435	0.7483	0.4044	0.1133	0.0100	0.0001
	12	1.0000	1.0000	0.9987	0.9790	0.8684	0.5841	0.2277	0.0321	0.0004
	13	1.0000	1.0000	0.9997	0.9935	0.9423	0.7500	0.3920	0.0867	0.0024
	14	1.0000	1.0000	1.0000	0.9984	0.9793	0.8744	0.5836	0.1958	0.0113
	15	1.0000	1.0000	1.0000	0.9997	0.9941	0.9490	0.7625	0.3704	0.0432

n	x	p=0.10	0.20	0.30	0.40	0.50	0.60	0.70	0.80	0.90
	16	1.0000	1.0000	1.0000	1.0000	0.9987	0.9840	0.8929	0.5886	0.1330
	17	1.0000	1.0000	1.0000	1.0000	0.9998	0.9964	0.9645	0.7939	0.3231
	18	1.0000	1.0000	1.0000	1.0000	1.0000	0.9995	0.9924	0.9308	0.6083
	19	1.0000	1.0000	1.0000	1.0000	1.0000	1.0000	0.9992	0.9885	0.8784

The Poisson distribution

Probability function

The table gives $f(x)=e^{-\lambda} \lambda^x/x!$ for different values of λ and x.

Concerning calculation of $f(x)$ with programmable machine, see section 17.4.

x	$\lambda=0.1$	$\lambda=0.2$	$\lambda=0.3$	$\lambda=0.4$	$\lambda=0.5$	$\lambda=0.6$
0	0.9048	0.8187	0.7408	0.6703	0.6065	0.5488
1	0.0905	0.1637	0.2222	0.2681	0.3033	0.3293
2	0.0045	0.0164	0.0333	0.0536	0.0758	0.0988
3	0.0002	0.0011	0.0033	0.0072	0.0126	0.0198
4		0.0001	0.0002	0.0007	0.0016	0.0030
5				0.0001	0.0002	0.0004

x	$\lambda=0.7$	$\lambda=0.8$	$\lambda=0.9$	$\lambda=1.0$	$\lambda=1.1$	$\lambda=1.2$
0	0.4966	0.4493	0.4066	0.3679	0.3329	0.3012
1	0.3476	0.3595	0.3659	0.3679	0.3662	0.3614
2	0.1217	0.1438	0.1647	0.1839	0.2014	0.2169
3	0.0284	0.0383	0.0494	0.0613	0.0738	0.0867
4	0.0050	0.0077	0.0111	0.0153	0.0203	0.0260
5	0.0007	0.0012	0.0020	0.0031	0.0045	0.0062
6	0.0001	0.0002	0.0003	0.0005	0.0008	0.0012
7				0.0001	0.0001	0.0002

x	$\lambda=1.3$	$\lambda=1.4$	$\lambda=1.5$	$\lambda=1.6$	$\lambda=1.7$	$\lambda=1.8$
0	0.2725	0.2466	0.2231	0.2019	0.1827	0.1653
1	0.3543	0.3452	0.3347	0.3230	0.3106	0.2975
2	0.2303	0.2417	0.2510	0.2584	0.2640	0.2678
3	0.0998	0.1128	0.1255	0.1378	0.1496	0.1607
4	0.0324	0.0395	0.0471	0.0551	0.0636	0.0723
5	0.0084	0.0111	0.0141	0.0176	0.0216	0.0260
6	0.0018	0.0026	0.0035	0.0047	0.0061	0.0078
7	0.0003	0.0005	0.0008	0.0011	0.0015	0.0020
8	0.0001	0.0001	0.0001	0.0002	0.0003	0.0005
9					0.0001	0.0001

x	λ=1.9	λ=2.0	λ=2.5	λ=3.0	λ=3.5	λ=4
0	0.1496	0.1353	0.0821	0.0498	0.0302	0.0183
1	0.2842	0.2707	0.2052	0.1494	0.1057	0.0733
2	0.2700	0.2707	0.2565	0.2240	0.1850	0.1465
3	0.1710	0.1804	0.2138	0.2240	0.2158	0.1954
4	0.0812	0.0902	0.1336	0.1680	0.1888	0.1954
5	0.0309	0.0361	0.0668	0.1008	0.1322	0.1563
6	0.0098	0.0120	0.0278	0.0504	0.0771	0.1042
7	0.0027	0.0034	0.0099	0.0216	0.0385	0.0595
8	0.0006	0.0009	0.0031	0.0081	0.0169	0.0298
9	0.0001	0.0002	0.0009	0.0027	0.0066	0.0132
10			0.0002	0.0008	0.0023	0.0053
11				0.0002	0.0007	0.0019
12				0.0001	0.0002	0.0006
13					0.0001	0.0002
14						0.0001

x	λ=5	λ=6	λ=7	λ=8	λ=9	λ=10
0	0.0067	0.0025	0.0009	0.0003	0.0001	0.0000
1	0.0337	0.0149	0.0064	0.0027	0.0011	0.0005
2	0.0842	0.0446	0.0223	0.0107	0.0050	0.0023
3	0.1404	0.0892	0.0521	0.0286	0.0150	0.0076
4	0.1755	0.1339	0.0912	0.0573	0.0337	0.0189
5	0.1755	0.1606	0.1277	0.0916	0.0607	0.0378
6	0.1462	0.1606	0.1490	0.1221	0.0911	0.0631
7	0.1044	0.1377	0.1490	0.1396	0.1171	0.0901
8	0.0653	0.1033	0.1304	0.1396	0.1318	0.1126
9	0.0363	0.0688	0.1014	0.1241	0.1318	0.1251
10	0.0181	0.0413	0.0710	0.0993	0.1186	0.1251
11	0.0082	0.0225	0.0452	0.0722	0.0970	0.1137
12	0.0034	0.0113	0.0264	0.0481	0.0728	0.0948
13	0.0013	0.0052	0.0142	0.0296	0.0504	0.0729
14	0.0005	0.0022	0.0071	0.0169	0.0324	0.0521
15	0.0002	0.0009	0.0033	0.0090	0.0194	0.0347
16		0.0003	0.0014	0.0045	0.0109	0.0217
17		0.0001	0.0006	0.0021	0.0058	0.0128
18			0.0002	0.0009	0.0029	0.0071
19			0.0001	0.0004	0.0014	0.0037
20				0.0002	0.0006	0.0019
21				0.0001	0.0003	0.0009
22					0.0001	0.0004
23						0.0002
24						0.0001

Poisson distribution, distribution function

The table gives the distribution function $F(x)=\sum_{k=0}^{x} e^{-\lambda} \lambda^k/k!$ for different values of λ and x.

Concerning calculation of $F(x)$ with programmable machine, see section 17.4.

x	λ=0.1	λ=0.2	λ=0.3	λ=0.4	λ=0.5	λ=0.6
0	0.9048	0.8187	0.7408	0.6703	0.6065	0.5488
1	0.9953	0.9825	0.9631	0.9384	0.9098	0.8781
2	0.9998	0.9989	0.9964	0.9921	0.9856	0.9769
3	1.0000	0.9999	0.9997	0.9992	0.9982	0.9966
4		1.0000	1.0000	0.9999	0.9998	0.9996
5				1.0000	1.0000	1.0000

x	λ=0.7	λ=0.8	λ=0.9	λ=1.0	λ=1.1	λ=1.2
0	0.4966	0.4493	0.4066	0.3679	0.3329	0.3012
1	0.8442	0.8088	0.7725	0.7358	0.6990	0.6626
2	0.9659	0.9526	0.9371	0.9197	0.9004	0.8795
3	0.9942	0.9909	0.9865	0.9810	0.9743	0.9662
4	0.9992	0.9986	0.9977	0.9963	0.9946	0.9923
5	0.9999	0.9998	0.9997	0.9994	0.9990	0.9985
6	1.0000	1.0000	1.0000	0.9999	0.9999	0.9997
7				1.0000	1.0000	1.0000

x	λ=1.3	λ=1.4	λ=1.5	λ=1.6	λ=1.7	λ=1.8
0	0.2725	0.2466	0.2231	0.2019	0.1827	0.1653
1	0.6268	0.5918	0.5578	0.5249	0.4932	0.4628
2	0.8571	0.8335	0.8088	0.7834	0.7572	0.7306
3	0.9569	0.9463	0.9344	0.9212	0.9068	0.8913
4	0.9893	0.9857	0.9814	0.9763	0.9704	0.9636
5	0.9978	0.9968	0.9955	0.9940	0.9920	0.9896
6	0.9996	0.9994	0.9991	0.9987	0.9981	0.9974
7	0.9999	0.9999	0.9998	0.9997	0.9996	0.9994
8	1.0000	1.0000	1.0000	1.0000	0.9999	0.9999
9					1.0000	1.0000

x	$\lambda=1.9$	$\lambda=2.0$	$\lambda=2.5$	$\lambda=3.0$	$\lambda=3.5$	$\lambda=4.0$
0	0.1496	0.1353	0.0821	0.0498	0.0302	0.0183
1	0.4338	0.4060	0.2873	0.1991	0.1359	0.0916
2	0.7037	0.6767	0.5438	0.4232	0.3208	0.2381
3	0.8747	0.8571	0.7576	0.6472	0.5366	0.4335
4	0.9559	0.9473	0.8912	0.8153	0.7254	0.6288
5	0.9868	0.9834	0.9580	0.9161	0.8576	0.7851
6	0.9966	0.9955	0.9858	0.9665	0.9347	0.8893
7	0.9992	0.9989	0.9958	0.9881	0.9733	0.9489
8	0.9998	0.9998	0.9989	0.9962	0.9901	0.9786
9	1.0000	1.0000	0.9997	0.9989	0.9967	0.9919
10			0.9999	0.9997	0.9990	0.9972
11			1.0000	0.9999	0.9997	0.9991
12				1.000	0.9999	0.9997
13					1.0000	0.9999
14						1.0000

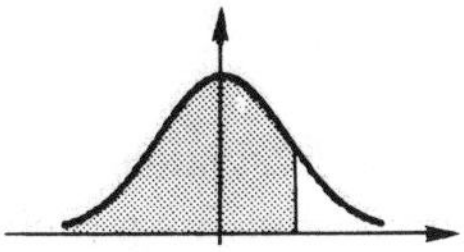

Normal distribution

Distribution function

The table gives $\Phi(x)=\frac{1}{\sqrt{2\pi}}\int_{-\infty}^{x} e^{-t^2/2}\,dt$. For $x<0$ values of $\Phi(x)$ can be obtained from $\Phi(-x)=1-\Phi(x)$.

x	**0**	**1**	**2**	**3**	**4**	**5**	**6**	**7**	**8**	**9**
0.0	0.5000	0.5040	0.5080	0.5120	0.5160	0.5199	0.5239	0.5279	0.5319	0.5359
0.1	0.5398	0.5438	0.5478	0.5517	0.5557	0.5596	0.5636	0.5675	0.5714	0.5753
0.2	0.5793	0.5832	0.5871	0.5910	0.5948	0.5987	0.6026	0.6064	0.6103	0.6141
0.3	0.6179	0.6217	0.6255	0.6293	0.6331	0.6368	0.6406	0.6443	0.6480	0.6517
0.4	0.6554	0.6591	0.6628	0.6664	0.6700	0.6736	0.6772	0.6808	0.6844	0.6879
0.5	0.6915	0.6950	0.6985	0.7019	0.7054	0.7088	0.7123	0.7157	0.7190	0.7224
0.6	0.7257	0.7291	0.7324	0.7357	0.7389	0.7422	0.7454	0.7486	0.7517	0.7549
0.7	0.7580	0.7611	0.7642	0.7673	0.7703	0.7734	0.7764	0.7794	0.7823	0.7852
0.8	0.7881	0.7910	0.7939	0.7967	0.7995	0.8023	0.8051	0.8078	0.8106	0.8133
0.9	0.8159	0.8186	0.8212	0.8238	0.8264	0.8289	0.8315	0.8340	0.8365	0.8389
1.0	0.8413	0.8438	0.8461	0.8485	0.8508	0.8531	0.8554	0.8577	0.8599	0.8621
1.1	0.8643	0.8665	0.8686	0.8708	0.8729	0.8749	0.8770	0.8790	0.8810	0.8830
1.2	0.8849	0.8869	0.8888	0.8907	0.8925	0.8944	0.8962	0.8980	0.8997	0.9015
1.3	0.9032	0.9049	0.9066	0.9082	0.9099	0.9115	0.9131	0.9147	0.9162	0.9177
1.4	0.9192	0.9207	0.9222	0.9236	0.9251	0.9265	0.9279	0.9292	0.9306	0.9319
1.5	0.9332	0.9345	0.9357	0.9370	0.9382	0.9394	0.9406	0.9418	0.9429	0.9441
1.6	0.9452	0.9463	0.9474	0.9484	0.9495	0.9505	0.9515	0.9525	0.9535	0.9545
1.7	0.9554	0.9564	0.9573	0.9582	0.9591	0.9599	0.9608	0.9616	0.9625	0.9633
1.8	0.9641	0.9649	0.9656	0.9664	0.9671	0.9678	0.9686	0.9693	0.9699	0.9706
1.9	0.9713	0.9719	0.9726	0.9732	0.9738	0.9744	0.9750	0.9756	0.9761	0.9767
2.0	0.9772	0.9778	0.9783	0.9788	0.9793	0.9798	0.9803	0.9808	0.9812	0.9817
2.1	0.9821	0.9826	0.9830	0.9834	0.9838	0.9842	0.9846	0.9850	0.9854	0.9857
2.2	0.9861	0.9864	0.9868	0.9871	0.9875	0.9878	0.9881	0.9884	0.9887	0.9890
2.3	0.9893	0.9896	0.9898	0.9901	0.9904	0.9906	0.9909	0.9911	0.9913	0.9916
2.4	0.9918	0.9920	0.9922	0.9925	0.9927	0.9929	0.9931	0.9932	0.9934	0.9936
2.5	0.9938	0.9940	0.9941	0.9943	0.9945	0.9946	0.9948	0.9949	0.9951	0.9952
2.6	0.9953	0.9955	0.9956	0.9957	0.9959	0.9960	0.9961	0.9962	0.9963	0.9964
2.7	0.9965	0.9966	0.9967	0.9968	0.9969	0.9970	0.9971	0.9972	0.9973	0.9974
2.8	0.9974	0.9975	0.9976	0.9977	0.9977	0.9978	0.9979	0.9979	0.9980	0.9981
2.9	0.9981	0.9982	0.9982	0.9983	0.9984	0.9984	0.9985	0.9985	0.9986	0.9986
3.0	0.9987	0.9987	0.9987	0.9988	0.9988	0.9989	0.9989	0.9989	0.9990	0.9990
3.1	$0.9^3 03$	$0.9^3 06$	$0.9^3 10$	$0.9^3 13$	$0.9^3 16$	$0.9^3 18$	$0.9^3 21$	$0.9^3 24$	$0.9^3 26$	$0.9^3 29$
3.2	$0.9^3 31$	$0.9^3 34$	$0.9^3 36$	$0.9^3 38$	$0.9^3 40$	$0.9^3 42$	$0.9^3 44$	$0.9^3 46$	$0.9^3 48$	$0.9^3 50$
3.3	$0.9^3 52$	$0.9^3 53$	$0.9^3 55$	$0.9^3 57$	$0.9^3 58$	$0.9^3 60$	$0.9^3 61$	$0.9^3 62$	$0.9^3 64$	$0.9^3 65$
3.4	$0.9^3 66$	$0.9^3 68$	$0.9^3 69$	$0.9^3 70$	$0.9^3 71$	$0.9^3 72$	$0.9^3 73$	$0.9^3 74$	$0.9^3 75$	$0.9^3 76$

For large values of x the following approximation can be used.

$$\frac{1}{\sqrt{2\pi}}\cdot e^{-x^2/2}\cdot\left(\frac{1}{x}-\frac{1}{x^3}\right)<1-\Phi(x)<\frac{1}{\sqrt{2\pi}}\cdot e^{-x^2/2}\cdot\frac{1}{x}$$

Normal distribution: x and P for given value of $\Phi(x)$

The table gives x for given values of the distribution function

$$\Phi(x)=\frac{1}{\sqrt{2\pi}}\int_{-\infty}^{x} e^{-t^2/2}\,dt$$

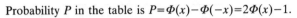

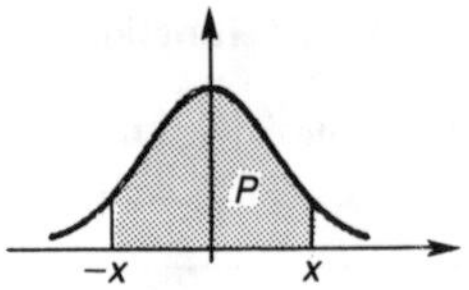

Probability P in the table is $P=\Phi(x)-\Phi(-x)=2\Phi(x)-1$.

$\Phi(x)$	x	P
0.50	0.0000	
0.51	0.0251	0.02
0.52	0.0502	0.04
0.53	0.0753	0.06
0.54	0.1004	0.08
0.55	0.1257	0.10
0.56	0.1510	0.12
0.57	0.1764	0.14
0.58	0.2019	0.16
0.59	0.2275	0.18
0.60	0.2533	0.20
0.61	0.2793	0.22
0.62	0.3055	0.24
0.63	0.3319	0.26
0.64	0.3585	0.28
0.65	0.3853	0.30
0.66	0.4125	0.32
0.67	0.4399	0.34
0.68	0.4677	0.36
0.69	0.4959	0.38
0.70	0.5244	0.40
0.71	0.5534	0.42
0.72	0.5828	0.44
0.73	0.6128	0.46
0.74	0.6433	0.48
0.75	0.6745	0.50
0.76	0.7063	0.52
0.77	0.7388	0.54
0.78	0.7722	0.56
0.79	0.8064	0.58
0.80	0.8416	0.60
0.81	0.8779	0.62
0.82	0.9154	0.64
0.83	0.9542	0.66
0.84	0.9945	0.68

$\Phi(x)$	x	P
0.85	1.0364	0.70
0.86	1.0803	0.72
0,87	1.1264	0.74
0.88	1.1750	0.76
0.89	1.2265	0.78
0.90	1.2816	0.80
0.91	1.3408	0.82
0.92	1.4051	0.84
0.93	1.4758	0.86
0.94	1.5548	0.88
0.950	1.6449	0.90
0.955	1.6954	0.91
0.960	1.7507	0.92
0.965	1.8119	0.93
0.970	1.8808	0.94
0.975	1.9600	0.95
0.980	2.0537	0.96
0.985	2.1701	0.97
0.990	2.3263	0.980
0.991	2.3656	0.982
0.992	2.4089	0.984
0.993	2.4573	0.986
0.994	2.5121	0.988
0.995	2.5758	0.990
0.996	2.6521	0.992
0.997	2.7478	0.994
0.998	2.8782	0.996
0.999	3.0902	0.998
0.9992	3.1559	0.9984
0.9994	3.2389	0.9988
0.9995	3.2905	0.9990
0.9996	3.3528	0.9992
0.9998	3.5401	0.9996
0.9999	3.7190	0.9998
0.99995	3.8906	0.9999

The χ^2-distribution

The table gives x for given values of the distribution function $F(x)$ for a χ^2-distribution with r degrees of freedom.

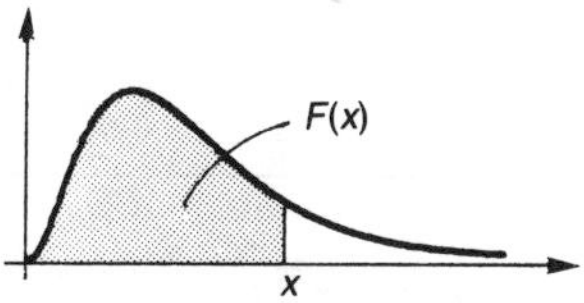

$F(x)$=	0.0005	0.001	0.005	0.010	0.025	0.05	0.10	0.25	0.50
r= 1	$0.0^{6}39$	$0.0^{5}16$	$0.0^{4}39$	$0.0^{3}16$	$0.0^{3}98$	0.0039	0.0158	0.1015	0.4549
2	0.0010	0.0020	0.0100	0.0201	0.0506	0.1026	0.2107	0.5754	1.386
3	0.0153	0.0243	0.0717	0.1148	0.2158	0.3518	0.5844	1.213	2.366
4	0.0639	0.0908	0.2070	0.2971	0.4844	0.7107	1.064	1.923	3.357
5	0.158	0.210	0.412	0.554	0.831	1.145	1.610	2.675	4.351
6	0.299	0.381	0.676	0.872	1.237	1.635	2.204	3.455	5.348
7	0.485	0.598	0.989	1.239	1.690	2.167	2.833	4.255	6.346
8	0.710	0.857	1.344	1.646	2.180	2.733	3.490	5.071	7.344
9	0.972	1.153	1.735	2.088	2.700	3.325	4.168	5.899	8.343
10	1.265	1.479	2.156	2.558	3.247	3.940	4.865	6.737	9.342
11	1.587	1.834	2.603	3.053	3.816	4.575	5.578	7.584	10.34
12	1.934	2.214	3.074	3.571	4.404	5.226	6.304	8.438	11.34
13	2.305	2.617	3.565	4.107	5.009	5.892	7.042	9.299	12.34
14	2.697	3.041	4.075	4.660	5.629	6.571	7.790	10.17	13.34
15	3.108	3.483	4.601	5.229	6.262	7.261	8.547	11.04	14.34
16	3.536	3.942	5.142	5.812	6.908	7.962	9.312	11.91	15.34
17	3.980	4.416	5.697	6.408	7.564	8.672	10.09	12.79	16.34
18	4.439	4.905	6.265	7.015	8.231	9.390	10.86	13.68	17.34
19	4.912	5.407	6.844	7.633	8.907	10.12	11.65	14.56	18.34
20	5.398	5.921	7.434	8.260	9.591	10.85	12.44	15.45	19.34
21	5.896	6.447	8.034	8.897	10.28	11.59	13.24	16.34	20.34
22	6.405	6.983	8.643	9.542	10.98	12.34	14.04	17.24	21.34
23	6.924	7.529	9.260	10.20	11.69	13.09	14.85	18.14	22.34
24	7.453	8.085	9.886	10.86	12.40	13.85	15.66	19.04	23.34
25	7.991	8.649	10.52	11.52	13.12	14.61	16.47	19.94	24.34
26	8.538	9.222	11.16	12.20	13.84	15.38	17.29	20.84	25.34
27	9.093	9.803	11.81	12.88	14.57	16.15	18.11	21.75	26.34
28	9.656	10.39	12.46	13.56	15.31	16.93	18.94	22.66	27.34
29	10.23	10.99	13.12	14.26	16.05	17.71	19.77	23.57	28.34
30	10.80	11.59	13.79	14.95	16.79	18.49	20.60	24.48	29.34
34	13.18	14.06	16.50	17.79	19.81	21.66	23.95	28.14	33.34
39	16.27	17.26	20.00	21.43	23.65	25.70	28.20	32.74	38.34
44	19.48	20.58	23.58	25.15	27.58	29.79	32.49	37.36	43.34
49	22.79	23.98	27.25	28.94	31.56	33.93	36.82	42.01	48.34
59	29.64	31.02	34.77	36.70	39.66	42.34	45.58	51.36	58.34
69	36.74	38.30	42.49	44.64	47.92	50.88	54.44	60.76	68.33
79	44.05	45.76	50.38	52.72	56.31	59.52	63.38	70.20	78.33
89	51.52	53.39	58.39	60.93	64.79	68.25	72.39	79.68	88.33
99	59.13	61.14	66.51	69.23	73.36	77.05	81.45	89.18	98.33

Ex. $0.0^3\,16 = 0.00016$

$F(x)=$	0.75	0.90	0.95	0.975	0.990	0.995	0.999	0.9995
$r=$ 1	1.323	2.706	3.841	5.024	6.635	7.879	10.83	12.12
2	2.773	4.605	5.991	7.378	9.210	10.60	13.82	15.20
3	4.108	6.251	7.815	9.348	11.34	12.84	16.27	17.73
4	5.385	7.779	9.488	11.14	13.28	14.86	18.47	20.00
5	6.626	9.236	11.07	12.83	15.09	16.75	20.52	22.10
6	7.841	10.64	12.59	14.45	16.81	18.55	22.46	24.10
7	9.037	12.02	14.07	16.01	18.48	20.28	24.32	26.02
8	10.22	13.36	15.51	17.53	20.09	21.96	26.12	27.87
9	11.39	14.68	16.92	19.02	21.67	23.59	27.88	29.67
10	12.55	15.99	18.31	20.48	23.21	25.19	29.59	31.42
11	13.70	17.28	19.68	21.92	24.72	26.76	31.26	33.14
12	14.85	18.55	21.03	23.34	26.22	28.30	32.91	34.82
13	15.98	19.81	22.36	24.74	27.69	29.82	34.53	36.48
14	17.12	21.06	23.68	26.12	29.14	31.32	36.12	38.11
15	18.25	22.31	25.00	27.49	30.58	32.80	37.70	39.72
16	19.37	23.54	26.30	28.85	32.00	34.27	39.25	41.31
17	20.49	24.77	27.59	30.19	33.41	35.72	40.79	42.88
18	21.60	25.99	28.87	31.53	34.81	37.16	42.31	44.43
19	22.72	27.20	30.14	32.85	36.19	38.58	43.82	45.97
20	23.83	28.41	31.41	34.17	37.57	40.00	45.32	47.50
21	24.93	29.62	32.67	35.48	38.93	41.40	46.80	49.01
22	26.04	30.81	33.92	36.78	40.29	42.80	48.27	50.51
23	27.14	32.01	35.17	38.08	41.64	44.18	49.73	52.00
24	28.24	33.20	36.42	39.36	42.98	45.56	51.18	53.48
25	29.34	34.38	37.65	40.65	44.31	46.93	52.62	54.95
26	30.43	35.56	38.89	41.92	45.64	48.29	54.05	56.41
27	31.53	36.74	40.11	43.19	49.96	49.64	55.48	57.86
28	32.62	37.92	41.34	44.46	48.28	50.99	56.89	59.30
29	33.71	39.09	42.56	45.72	49.59	52.34	58.30	60.73
30	34.80	40.26	43.77	46.98	50.89	53.67	59.70	62.16
34	39.14	44.90	48.60	51.97	56.06	58.96	65.25	67.80
39	44.54	50.66	54.57	58.12	62.43	65.48	72.06	74.72
44	49.91	56.37	60.48	64.20	68.71	71.89	78.75	81.53
49	55.26	62.04	66.34	70.22	74.92	78.23	85.35	88.23
59	65.92	73.28	77.93	82.12	87.17	90.72	98.32	101.4
69	76.52	84.42	89.39	93.86	99.23	103.0	111.1	114.3
79	87.08	95.48	100.7	105.5	111.1	115.1	123.6	127.0
89	97.60	106.5	112.0	117.0	122.9	127.1	136.0	139.5
99	108.1	117.4	123.2	128.4	134.6	139.0	148.2	151.9

For $r>30$, the variable $\sqrt{2X}$ is approximately $N(\sqrt{2r-1}, 1)$.

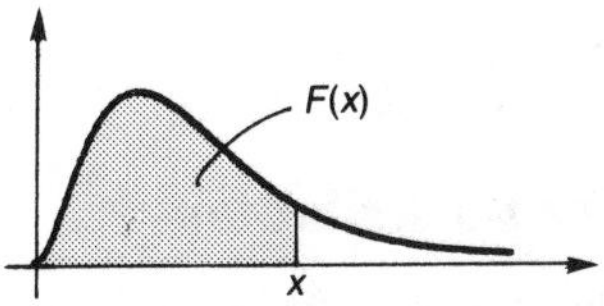

The gamma distribution

The table gives x for given values of the distribution function $F(x)$ for a gamma distribution with parameters r and $\lambda=1$.

$F(x)$	0.005	0.025	0.05	0.95	0.975	0.995	$F(x)$
$r=$ 1	0.005	0.025	0.051	3.00	3.69	5.30	$r=$ 1
2	0.103	0.242	0.355	4.74	5.57	7.43	2
3	0.338	0.619	0.818	6.30	7.22	9.27	3
4	0.672	1.09	1.37	7.75	8.77	11.0	4
5	1.08	1.62	1.97	9.15	10.2	12.6	5
6	1.54	2.20	2.61	10.5	11.7	14.1	6
7	2.04	2.81	3.29	11.8	13.1	15.7	7
8	2.57	3.45	3.98	13.1	14.4	17.1	8
9	3.13	4.12	4.70	14.4	15.8	18.6	9
10	3.72	4.80	5.43	15.7	17.1	20.0	10
11	4.32	5.49	6.17	17.0	18.4	21.4	11
12	4.94	6.20	6.92	18.2	19.7	22.8	12
13	5.58	6.92	7.69	19.4	21.0	24.1	13
14	6.23	7.65	8.46	20.7	22.2	25.5	14
15	6.89	8.40	9.25	21.9	23.5	26.8	15
16	7.57	9.14	10.0	23.1	24.7	28.2	16
17	8.25	9.90	10.8	24.3	26.0	29.5	17
18	8.94	10.7	11.6	25.5	27.2	30.8	18
19	9.64	11.4	12.4	26.7	28.4	32.1	19
20	10.4	12.2	13.3	27.9	29.7	33.4	20
21	11.1	13.0	14.1	29.1	30.9	34.7	21
22	11.8	13.8	14.9	30.2	32.1	35.9	22
23	12.5	14.6	15.7	31.4	33.3	37.2	23
24	13.3	15.4	16.5	32.6	34.5	38.5	24
25	14.0	16.2	17.4	33.8	35.7	39.7	25
26	14.7	17.0	18.2	34.9	36.9	41.0	16
27	15.5	17.8	19.1	36.1	38.1	42.3	27
28	16.2	18.6	19.9	37.2	39.3	43.5	28
29	17.0	19.4	20.7	38.4	40.5	44.7	29
30	17.8	20.2	21.6	39.5	41.6	46.0	30
31	18.5	21.1	22.4	40.7	42.8	47.2	31
32	19.3	21.9	23.3	41.8	44.0	48.4	32
33	20.1	22.7	24.2	43.0	45.2	49.7	33
34	20.9	23.5	25.0	44.1	46.3	50.9	34

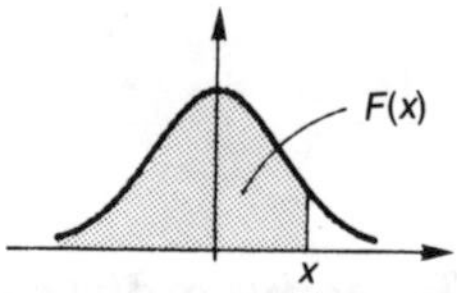

The *t*-distribution

The table gives x for given values of the distribution function $F(x)$ for a t-distribution with r degrees of freedom. For $x<0$ values of $F(x)$ can be obtained from $F(-x)=1-F(x)$.

$F(x)=$	0.75	0.90	0.95	0.975	0.990	0.995	0.9975	0.9995
$r=$ 1	1.0000	3.078	6.314	12.71	31.82	63.66	127.3	636.6
2	0.8165	1.886	2.920	4.303	6.965	9.925	14.09	31.60
3	0.7649	1.638	2.353	3.182	4.541	5.841	7.453	12.92
4	0.7407	1.533	2.132	2.776	3.747	4.604	5.598	8.610
5	0.7267	1.476	2.015	2.571	3.365	4.032	4.773	6.869
6	0.7176	1.440	1.943	2.447	3.143	3.707	4.317	5.959
7	0.7111	1.415	1.895	2.365	2.998	3.499	4.029	5.408
8	0.7064	1.397	1.860	2.306	2.896	3.355	3.832	5.041
9	0.7027	1.383	1.833	2.262	2.821	3.250	3.690	4.781
10	0.6998	1.372	1.812	2.228	2.764	3.169	3.581	4.587
11	0.6974	1.363	1.796	2.201	2.718	3.106	3.497	4.437
12	0.6955	1.356	1.782	2.179	2.681	3.055	3.428	4.318
13	0.6938	1.350	1.771	2.160	2.650	3.012	3.372	4.221
14	0.6924	1.345	1.761	2.145	2.624	2.977	3.326	4.140
15	0.6912	1.341	1.753	2.131	2.602	2.947	3.286	4.073
16	0.6901	1.337	1.746	2.120	2.583	2.921	3.252	4.015
17	0.6892	1.333	1.740	2.110	2.567	2.898	3.222	3.965
18	0.6884	1.330	1.734	2.101	2.552	2.878	3.197	3.922
19	0.6876	1.328	1.729	2.093	2.539	2.861	3.174	3.883
20	0.6870	1.325	1.725	2.086	2.528	2.845	3.153	3.850
21	0.6864	1.323	1.721	2.080	2.518	2.831	3.135	3.819
22	0.6858	1.321	1.717	2.074	2.508	2.819	3.119	3.792
23	0.6853	1.319	1.714	2.069	2.500	2.807	3.104	3.767
24	0.6848	1.318	1.711	2.064	2.492	2.797	3.090	3.745
25	0.6844	1.316	1.708	2.060	2.485	2.787	3.078	3.725
26	0.6840	1.315	1.706	2.056	2.479	2.779	3.069	3.707
27	0.6837	1.314	1.703	2.052	2.473	2.771	3.056	3.690
28	0.6834	1.313	1.701	2.048	2.467	2.763	3.047	3.674
29	0.6830	1.311	1.699	2.045	2.462	2.756	3.038	3.659
30	0.6828	1.310	1.697	2.042	2.457	2.750	3.030	3.646
34	0.6818	1.307	1.691	2.032	2.441	2.728	3.002	3.601
39	0.6808	1.304	1.685	2.023	2.426	2.708	2.976	3.559
44	0.6801	1.301	1.680	2.015	2.414	2.692	2.956	3.526
49	0.6795	1.299	1.677	2.010	2.405	2.680	2.940	3.501
59	0.6787	1.296	1.671	2.001	2.391	2.662	2.916	3.464
69	0.6781	1.294	1.667	1.995	2.382	2.649	2.900	3.438
79	0.6776	1.292	1.664	1.990	2.374	2.640	2.888	3.418
89	0.6773	1.291	1.662	1.987	2.369	2.632	2.879	3.404
99	0.6770	1.290	1.660	1.984	2.365	2.626	2.871	3.392
∞	0.6745	1.282	1.645	1.960	2.326	2.576	2.807	3.291

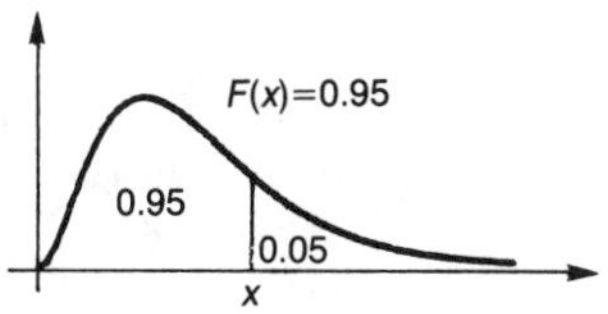

The F-distribution

The table gives x such that $F(x)=0.95$ for an F-distribution with r_1 degrees of freedom for the numerator and r_2 degrees of freedom for the denominator.

r_2/r_1	1	2	3	4	5	6	7	8	9	10
1	161.45	199.50	215.71	224.58	230.16	233.99	236.77	238.88	240.54	241.88
2	18.51	19.00	19.16	19.25	19.30	19.33	19.35	19.37	19.38	19.40
3	10.13	9.55	9.28	9.12	9.01	8.94	8.89	8.85	8.81	8.79
4	7.71	6.94	6.59	6.39	6.26	6.16	6.09	6.04	6.00	5.97
5	6.61	5.79	5.41	5.19	5.05	4.95	4.88	4.82	4.77	4.73
6	5.99	5.14	4.76	4.53	4.39	4.28	4.21	4.15	4.10	4.06
7	5.59	4.74	4.35	4.12	3.97	3.87	3.79	3.73	3.68	3.64
8	5.32	4.46	4.07	3.84	3.69	3.58	3.50	3.44	3.39	3.35
9	5.12	4.26	3.86	3.63	3.48	3.37	3.29	3.23	3.18	3.14
10	4.96	4.10	3.71	3.48	3.33	3.22	3.14	3.07	3.02	2.98
11	4.84	3.98	3.59	3.36	3.20	3.09	3.01	2.95	2.90	2.85
12	4.75	3.89	3.49	3.26	3.11	3.00	2.91	2.85	2.80	2.75
13	4.67	3.81	3.41	3.18	3.03	2.92	2.83	2.77	2.71	2.67
14	4.60	3.74	3.34	3.11	2.96	2.85	2.76	2.70	2.65	2.60
15	4.54	3.68	3.29	3.06	2.90	2.79	2.71	2.64	2.59	2.54
16	4.49	3.63	3.24	3.01	2.85	2.74	2.66	2.59	2.54	2.49
17	4.45	3.59	3.20	2.96	2.81	2.70	2.61	2.55	2.49	2.45
18	4.41	3.55	3.16	2.93	2.77	2.66	2.58	2.51	2.46	2.41
19	4.38	3.52	3.13	2.90	2.74	2.63	2.54	2.48	2.42	2.38
20	4.35	3.49	3.10	2.87	2.71	2.60	2.51	2.45	2.39	2.35
21	4.32	3.47	3.07	2.84	2.68	2.57	2.49	2.42	2.37	2.32
22	4.30	3.44	3.05	2.82	2.66	2.55	2.46	2.40	2.34	2.30
23	4.28	3.42	3.03	2.80	2.64	2.53	2.44	2.37	2.32	2.27
24	4.26	3.40	3.01	2.78	2.62	2.51	2.42	2.36	2.30	2.25
25	4.24	3.39	2.99	2.76	2.60	2.49	2.40	2.34	2.28	2.24
26	4.23	3.37	2.98	2.74	2.59	2.47	2.39	2.32	2.27	2.22
27	4.21	3.35	2.96	2.73	2.57	2.46	2.37	2.31	2.25	2.20
28	4.20	3.34	2.95	2.71	2.56	2.45	2.36	2.29	2.24	2.19
29	4.18	3.33	2.93	2.70	2.55	2.43	2.35	2.28	2.22	2.18
30	4.17	3.32	2.92	2.69	2.53	2.42	2.33	2.27	2.21	2.16
35	4.12	3.27	2.87	2.64	2.49	2.37	2.29	2.22	2.16	2.11
40	4.08	3.23	2.84	2.61	2.45	2.34	2.25	2.18	2.12	2.08
50	4.03	3.18	2.79	2.56	2.40	2.29	2.20	2.13	2.07	2.03
60	4.00	3.15	2.76	2.53	2.37	2.25	2.17	2.10	2.04	1.99
80	3.96	3.11	2.72	2.49	2.33	2.21	2.13	2.06	2.00	1.95
100	3.94	3.09	2.70	2.46	2.31	2.19	2.10	2.03	1.97	1.93

r_2/r_1	11	12	15	20	25	30	40	50	100
1	242.98	243.91	245.96	248.01	249.26	250.08	251.15	251.77	253.01
2	19.40	19.41	19.43	19.45	19.46	19.46	19.47	19.48	19.49
3	8.76	8.74	8.70	8.66	8.63	8.62	8.59	8.58	8.55
4	5.94	5.91	5.86	5.80	5.77	5.74	5.72	5.70	5.66
5	4.70	4.68	4.62	4.56	4.52	4.50	4.46	4.44	4.41
6	4.03	4.00	3.94	3.87	3.84	3.81	3.77	3.75	3.71
7	3.60	3.57	3.51	3.44	3.40	3.38	3.34	3.32	3.27
8	3.31	3.28	3.22	3.15	3.11	3.08	3.04	3.02	2.97
9	3.10	3.07	3.01	2.94	2.89	2.86	2.83	2.80	2.76
10	2.94	2.91	2.85	2.77	2.73	2.70	2.66	2.64	2.59
11	2.82	2.79	2.72	2.65	2.60	2.57	2.53	2.51	2.46
12	2.72	2.69	2.62	2.54	2.50	2.47	2.43	2.40	2.35
13	2.63	2.60	2.53	2.46	2.41	2.38	2.34	2.31	2.26
14	2.57	2.53	2.46	2.39	2.34	2.31	2.27	2.24	2.19
15	2.51	2.48	2.40	2.33	2.28	2.25	2.20	2.18	2.12
16	2.46	2.42	2.35	2.28	2.23	2.19	2.15	2.12	2.07
17	2.41	2.38	2.31	2.23	2.18	2.15	2.10	2.08	2.02
18	2.37	2.34	2.27	2.19	2.14	2.11	2.06	2.04	1.98
19	2.34	2.31	2.23	2.16	2.11	2.07	2.03	2.00	1.94
20	2.31	2.28	2.20	2.12	2.07	2.04	1.99	1.97	1.91
21	2.28	2.25	2.18	2.10	2.05	2.01	1.96	1.94	1.88
22	2.26	2.23	2.15	2.07	2.02	1.98	1.94	1.91	1.85
23	2.24	2.20	2.13	2.05	2.00	1.96	1.91	1.88	1.82
24	2.22	2.18	2.11	2.03	1.97	1.94	1.89	1.86	1.80
25	2.20	2.16	2.09	2.01	1.96	1.92	1.87	1.84	1.78
26	2.18	2.15	2.07	1.99	1.94	1.90	1.85	1.82	1.76
27	2.17	2.13	2.06	1.97	1.92	1.88	1.84	1.81	1.74
28	2.15	2.12	2.04	1.96	1.91	1.87	1.82	1.79	1.73
29	2.14	2.10	2.03	1.94	1.89	1.85	1.81	1.77	1.71
30	2.13	2.09	2.01	1.93	1.88	1.84	1.79	1.76	1.70
35	2.07	2.04	1.96	1.88	1.82	1.79	1.74	1.70	1.63
40	2.04	2.00	1.92	1.84	1.78	1.74	1.69	1.66	1.59
50	1.99	1.95	1.87	1.78	1.73	1.69	1.63	1.60	1.52
60	1.95	1.92	1.84	1.75	1.69	1.65	1.59	1.56	1.48
80	1.91	1.88	1.79	1.70	1.64	1.60	1.54	1.51	1.43
100	1.89	1.85	1.77	1.68	1.62	1.57	1.52	1.48	1.39

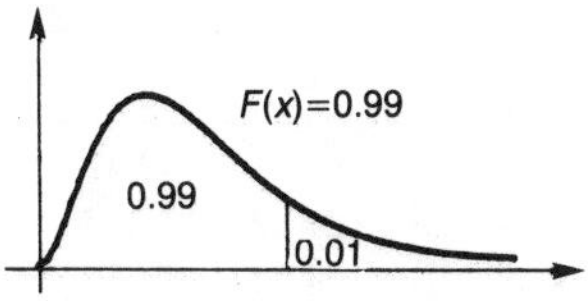

The F-distribution

The table gives x such that $F(x)=0.99$ for an F-distribution with r_1 degrees of freedom for the numerator and r_2 degrees of freedom for the denominator.

r_2/r_1	1	2	3	4	5	6	7	8	9	10
2	98.50	99.00	99.17	99.25	99.30	99.33	99.36	99.37	99.39	99.40
3	34.12	30.82	29.46	28.71	28.24	27.91	27.67	27.50	27.34	27.22
4	21.20	18.00	16.69	15.98	15.52	15.21	14.98	14.80	14.66	14.55
5	16.26	13.27	12.06	11.39	10.97	10.67	10.46	10.29	10.16	10.05
6	13.75	10.92	9.78	9.15	8.75	8.47	8.26	8.10	7.98	7.87
7	12.25	9.55	8.45	7.85	7.46	7.19	6.99	6.84	6.72	6.62
8	11.26	8.65	7.59	7.01	6.63	6.37	6.18	6.03	5.91	5.81
9	10.56	8.02	6.99	6.42	6.06	5.80	5.61	5.47	5.35	5.26
10	10.04	7.56	6.55	5.99	5.64	5.39	5.20	5.06	4.94	4.85
11	9.65	7.21	6.22	5.67	5.32	5.07	4.89	4.74	4.63	4.54
12	9.33	6.93	5.95	5.41	5.06	4.82	4.64	4.50	4.39	4.30
13	9.07	6.70	5.74	5.21	4.86	4.62	4.44	4.30	4.19	4.10
14	8.86	6.51	5.56	5.04	4.69	4.46	4.28	4.14	4.03	3.94
15	8.68	6.36	5.42	4.89	4.56	4.32	4.14	4.00	3.89	3.80
16	8.53	6.23	5.29	4.77	4.44	4.20	4.03	3.89	3.78	3.69
17	8.40	6.11	5.18	4.67	4.34	4.10	3.93	3.79	3.68	3.59
18	8.29	6.01	5.09	4.58	4.25	4.01	3.84	3.71	3.60	3.51
19	8.18	5.93	5.01	4.50	4.17	3.94	3.77	3.63	3.52	3.43
20	8.10	5.85	4.94	4.43	4.10	3.87	3.70	3.56	3.46	3.37
21	8.02	5.78	4.87	4.37	4.04	3.81	3.64	3.51	3.40	3.31
22	7.95	5.72	4.82	4.31	3.99	3.76	3.59	3.45	3.35	3.26
23	7.88	5.66	4.76	4.26	3.94	3.71	3.54	3.41	3.30	3.21
24	7.82	5.61	4.72	4.22	3.90	3.67	3.50	3.36	3.26	3.17
25	7.77	5.57	4.68	4.18	3.85	3.63	3.46	3.32	3.22	3.13
26	7.72	5.53	4.64	4.14	3.82	3.59	3.42	3.29	3.18	3.09
27	7.68	5.49	4.60	4.11	3.78	3.56	3.39	3.26	3.15	3.06
28	7.64	5.45	4.57	4.07	3.75	3.53	3.36	3.23	3.12	3.03
29	7.60	5.42	4.54	4.04	3.73	3.50	3.33	3.20	3.09	3.00
30	7.56	5.39	4.51	4.02	3.70	3.47	3.30	3.17	3.07	2.98
35	7.42	5.27	4.40	3.91	3.59	3.37	3.20	3.07	2.96	2.88
40	7.31	5.18	4.31	3.83	3.51	3.29	3.12	2.99	2.89	2.80
50	7.17	5.06	4.20	3.72	3.41	3.19	3.02	2.89	2.78	2.70
60	7.08	4.98	4.13	3.65	3.34	3.12	2.95	2.82	2.72	2.63
80	6.96	4.88	4.04	3.56	3.26	3.04	2.87	2.74	2.64	2.55
100	6.90	4.82	3.98	3.51	3.21	2.99	2.82	2.69	2.59	2.50

r_2/r_1	**11**	**12**	**15**	**20**	**25**	**30**	**40**	**50**	**100**
2	99.41	99.42	99.43	99.45	99.46	99.46	99.47	99.48	99.49
3	27.12	27.03	26.85	26.67	26.58	26.50	26.41	26.35	26.24
4	14.45	14.37	14.19	14.02	13.91	13.84	13.75	13.69	13.58
5	9.96	9.89	9.72	9.55	9.45	9.38	9.30	9.24	9.13
6	7.79	7.72	7.56	7.40	7.29	7.23	7.15	7.09	6.99
7	6.54	6.47	6.31	6.16	6.06	5.99	5.91	5.86	5.75
8	5.73	5.67	5.52	5.36	5.26	5.20	5.12	5.07	4.96
9	5.18	5.11	4.96	4.81	4.71	4.65	4.57	4.52	4.41
10	4.77	4.71	4.56	4.41	4.31	4.25	4.17	4.12	4.01
11	4.46	4.40	4.25	4.10	4.00	3.94	3.86	3.81	3.71
12	4.22	4.16	4.01	3.86	3.76	3.70	3.62	3.57	3.47
13	4.02	3.96	3.82	3.66	3.57	3.51	3.43	3.38	3.27
14	3.86	3.80	3.66	3.51	3.41	3.35	3.27	3.22	3.11
15	3.73	3.67	3.52	3.37	3.28	3.21	3.13	3.08	2.98
16	3.62	3.55	3.41	3.26	3.16	3.10	3.02	2.97	2.86
17	3.52	3.46	3.31	3.16	3.07	3.00	2.92	2.87	2.76
18	3.43	3.37	3.23	3.08	2.98	2.92	2.84	2.78	2.68
19	3.36	3.30	3.15	3.00	2.91	2.84	2.76	2.71	2.60
20	3.29	3.23	3.09	2.94	2.84	2.78	2.69	2.64	2.54
21	3.24	3.17	3.03	2.88	2.78	2.72	2.64	2.58	2.48
22	3.18	3.12	2.98	2.83	2.73	2.67	2.58	2.53	2.42
23	3.14	3.07	2.93	2.78	2.69	2.62	2.54	2.48	2.37
24	3.09	3.03	2.89	2.74	2.64	2.58	2.49	2.44	2.33
25	3.06	2.99	2.85	2.70	2.60	2.54	2.45	2.40	2.29
26	3.02	2.96	2.81	2.66	2.57	2.50	2.42	2.36	2.25
27	2.99	2.93	2.78	2.63	2.54	2.47	2.38	2.33	2.22
28	2.96	2.90	2.75	2.60	2.51	2.44	2.35	2.30	2.19
29	2.93	2.87	2.73	2.57	2.48	2.41	2.33	2.27	2.16
30	2.91	2.84	2.70	2.55	2.45	2.39	2.30	2.24	2.13
35	2.80	2.74	2.60	2.44	2.35	2.28	2.19	2.14	2.02
40	2.73	2.66	2.52	2.37	2.27	2.20	2.11	2.06	1.94
50	2.62	2.56	2.42	2.27	2.17	2.10	2.01	1.95	1.82
60	2.56	2.50	2.35	2.20	2.10	2.03	1.94	1.88	1.75
80	2.48	2.42	2.27	2.12	2.01	1.94	1.85	1.79	1.65
100	2.43	2.37	2.22	2.07	1.97	1.89	1.80	1.74	1.60

Random digits

Concerning generation of random numbers with programmable machines see section 17.5.

44955	16384	62827	82305	32836	96761	11602	81743	04141	47108
17932	78415	89813	17856	00680	71694	52288	75979	33302	99361
41763	11665	63153	43438	46603	03827	29956	00038	75401	94972
24368	09593	27757	44838	12770	91420	93676	66719	90221	16232
15642	24041	12815	18518	06378	99162	40329	24883	46760	26236
85537	15524	99132	95641	43956	98043	60034	02098	30631	12463
15677	42470	26268	40123	29130	50944	39644	13782	03367	77646
48595	88058	73988	87135	22800	20225	53898	45156	63801	34295
41738	27261	14091	40545	09782	97321	28817	81141	37045	11829
54523	09552	56660	53594	56115	56811	60488	23350	44662	77605
89334	90573	07140	59493	51322	97035	79963	62688	01059	37140
58339	58474	48617	34156	08020	37190	55787	46350	86923	42659
54199	58469	07812	10144	12042	02875	65886	32141	77782	81310
19148	30559	59869	13381	30812	42690	11672	62036	51495	38737
48331	65457	13151	59708	88927	51889	98772	73912	16399	37448
25293	52004	49064	12356	75433	73997	53983	52831	12185	76572
63951	93582	82641	51223	43848	93627	92107	17974	15294	94484
29565	62944	74131	26636	26962	21246	34327	05938	79038	97533
01089	21886	15310	67429	63405	63559	34930	68284	60604	48349
60220	63072	26778	59404	04745	44621	38544	85741	83060	96768
79683	54745	94840	86867	07609	58465	52296	32327	63997	53752
53064	18997	08430	77163	92571	80804	65540	16726	72245	94150
33819	07200	74681	57676	93974	17337	91193	82123	24452	78148
64553	23559	80327	45480	24850	41763	13819	70349	07650	57147
32597	64944	71337	48485	19982	30264	91456	37063	39605	54095
17544	50752	91544	93192	58536	84910	03137	50084	05482	67794
24026	54944	37891	13879	67888	88580	60992	91701	39938	49102
87362	32581	05670	90871	59193	71763	00730	43520	69073	30795
41673	16726	62427	18765	41364	87630	12355	95964	24665	96386
97223	50516	94212	70881	45125	59221	91447	28360	03518	40692
04146	49156	14321	30145	57476	26316	57831	21491	50325	79647
23432	90904	87099	30489	97607	11283	99215	47428	72654	58559
74381	28845	29786	66906	26377	96663	42434	83312	05480	72825
72999	12066	87644	29770	65753	64923	93435	03391	44963	76260
05670	41529	91943	47655	48027	24013	48716	79298	70093	13525
24179	78159	53752	08593	28764	08332	58345	83802	24289	27143
89836	35105	97261	96261	50601	14638	97187	20524	59107	53331
83810	31299	20328	24967	37923	25802	91158	79410	49566	63902
25069	64048	17067	73386	99206	77203	97801	49056	76395	19221
69768	65339	24077	45499	17472	09554	16845	75439	23694	10906

Normally distributed random numbers

The table gives observations on a random variable with an $N(0, 1)$ distribution. Concerning generation of such observations with a programmable machine see section 17.5.

−2.208	0.926	−0.518	−0.904	1.532	1.070	−0.993	−0.106	−0.733	−1.058
−0.302	−0.092	−0.696	0.373	1.174	−1.504	0.190	−0.111	−0.328	−1.075
−0.600	−1.241	0.916	−0.317	−0.711	−2.028	−0.119	0.218	−1.825	−0.241
−0.042	0.989	−0.092	0.631	−0.495	1.065	0.142	−0.444	0.210	−0.187
−0.284	−0.548	0.774	1.780	0.677	0.231	0.203	−1.221	1.657	0.847
1.170	0.386	−2.184	1.067	−0.873	−0.437	0.531	−2.506	−0.302	−0.601
−1.194	0.026	0.127	−0.979	0.025	1.009	1.659	−1.328	−0.227	−1.518
−0.169	0.136	−1.323	0.851	1.272	1.010	−0.929	0.451	1.025	0.368
−0.880	1.077	0.369	−0.576	0.262	0.266	−0.561	0.412	0.917	1.067
−1.645	1.687	0.412	−0.992	−0.965	−0.388	0.034	1.140	0.122	−0.258
−1.606	0.782	−2.341	0.739	−2.167	0.261	−0.657	−1.103	−0.036	−0.378
−0.950	−0.520	0.354	0.559	−0.178	−1.262	1.459	−0.900	0.980	1.052
−0.945	−0.077	−0.225	−3.098	−1.051	−0.275	−0.481	0.003	1.031	0.265
0.212	1.075	0.663	−0.797	−2.015	1.241	−1.243	1.634	0.837	−2.757
−0.620	0.978	0.140	2.095	−0.579	0.923	0.138	−0.621	−0.331	−0.690
2.446	0.437	−0.220	0.908	0.393	0.062	−0.494	0.107	−0.491	0.540
−1.610	0.350	−0.964	0.642	−0.666	−0.411	0.190	−0.143	0.377	0.270
−1.266	−0.307	−0.588	−0.991	0.780	−0.259	−0.511	0.339	0.144	0.361
−0.298	1.494	1.656	0.436	2.325	−0.544	0.813	−0.573	0.205	0.565
0.905	−0.248	0.515	0.732	−1.513	0.549	0.172	−0.449	0.030	0.902
1.003	−1.715	0.954	−2.255	−2.806	−1.358	0.637	0.763	−0.554	−0.821
−0.630	0.796	0.767	−0.567	−0.657	−1.485	−0.083	0.415	1.844	0.089
0.487	0.503	−0.132	−0.895	−1.560	0.303	−1.392	2.431	1.030	0.997
−0.780	0.885	0.151	−1.715	0.458	1.454	−1.445	−0.126	1.374	0.959
−0.454	0.854	−1.495	0.244	−2.014	−0.142	0.064	−0.428	1.229	2.013
0.189	0.387	0.129	1.173	0.614	1.406	0.171	0.258	−0.482	−0.021
1.385	1.333	0.479	0.553	−1.136	0.020	2.774	0.253	−0.389	−0.056
−0.651	0.014	−0.325	1.009	−1.064	1.891	−0.466	0.944	0.610	2.120
−0.480	−0.034	0.552	−0.204	0.645	1.104	−0.979	−0.081	0.130	0.646
−0.300	−0.632	−0.154	−0.872	−0.090	1.186	0.382	−0.457	0.456	−1.197
0.374	0.084	−0.575	0.683	0.350	−0.078	−0.958	−0.787	0.644	−0.466
−0.349	−1.932	−0.681	0.438	1.214	0.795	1.742	0.603	−2.538	−1.243
−0.321	0.747	−1.026	1.451	0.383	2.195	−0.646	−1.146	0.672	−0.761
−0.602	−1.920	−0.381	0.008	1.342	1.701	−0.370	0.444	0.011	−0.966
−0.945	1.450	−0.701	−0.938	−0.643	−0.410	0.825	−0.864	0.133	−1.295

18 Statistics

18.1 Descriptive Statistics

Diagrams

Bar diagram

Observation	1	2	3	4	5	6	7	8
Frequency	3	8	7	9	6	4	2	1

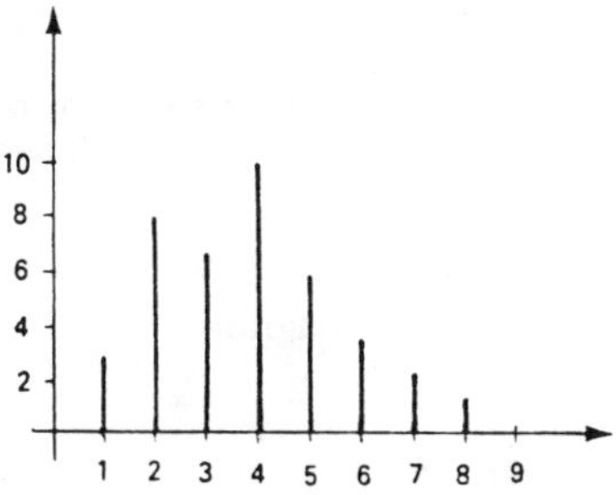

Histogram

Class		Frequency	Relative frequency
8– 9		1	0.02
9–10		1	0.02
10–11		5	0.10
11–12		8	0.16
12–13		9	0.18
13–14		9	0.18
14–15		7	0.14
15–16		4	0.08
16–17		4	0.08
17–18		2	0.04

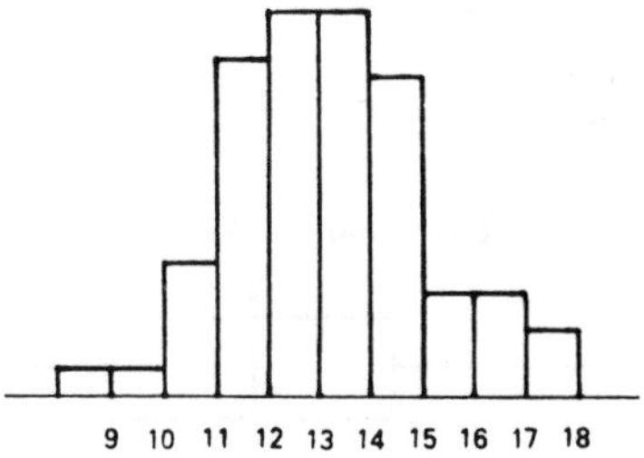

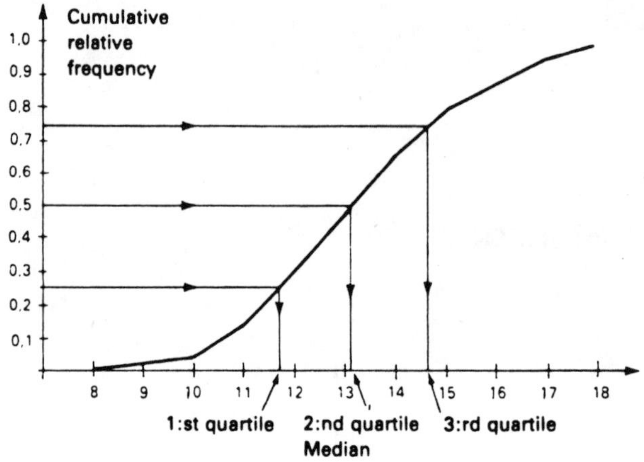

Stem and leaf diagram

8	8										(1)
9	9										(1)
10	3	3	5	9							(4)
11	0	2	2	3	4	4	4	6	8		(9)
12	3	3	3	3	5	6	9				(7)
13	0	0	1	2	2	4	4	4	7	7	(10)
14	0	2	3	4	5	5	5	5			(8)
15	1	6	6	9							(4)
16	1	3	5								(3)
17	0	3	7								(3)
											(50)

Box plot

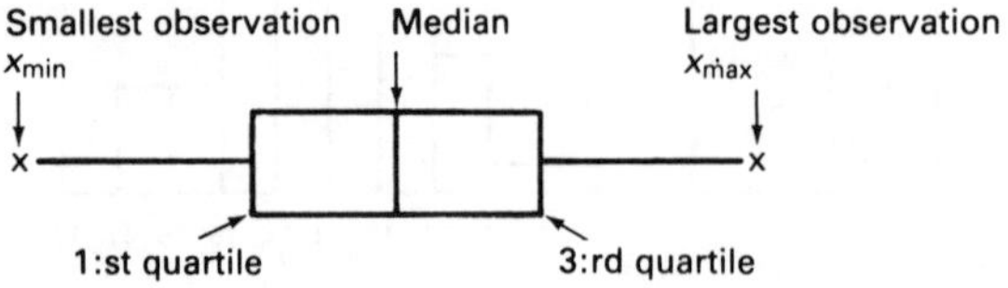

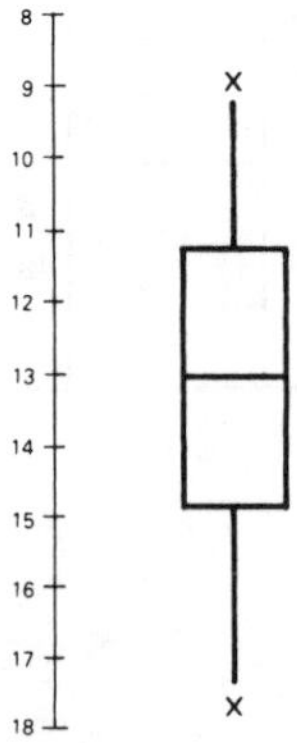

POPULATION – DISTRIBUTED ACCORDING TO SIZE OF HOUSEHOLD

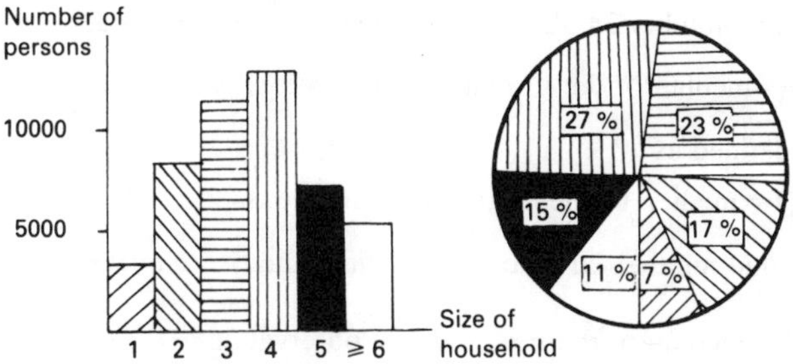

Household with ... persons	Number of persons
1	3 292
2	8 310
3	11 304
4	12 932
5	7 035
≥ 6	5 150
	48 023

India 1951

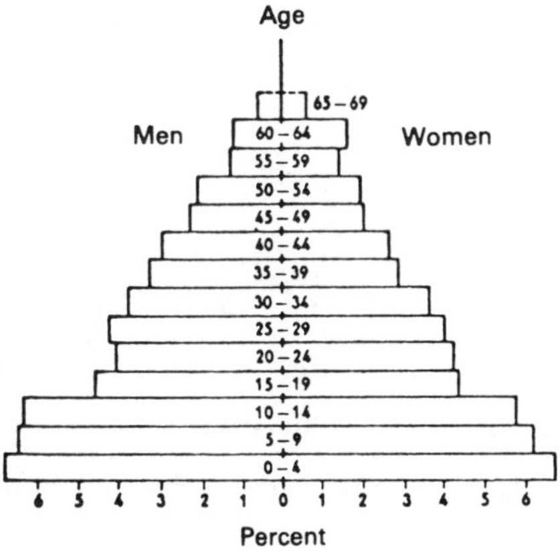

Sweden 1950

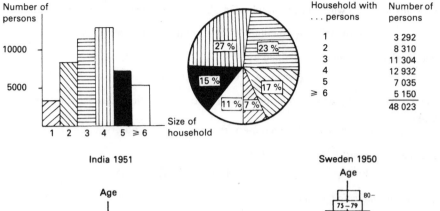

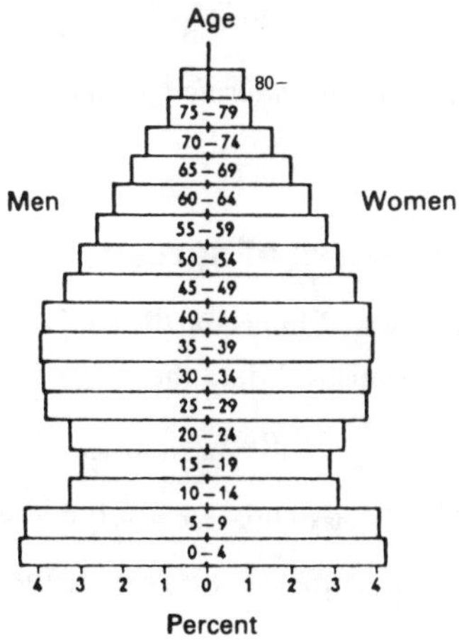

Measures of location

Mean and median

For statistical data with observations $x_1, x_2, \ldots, x_n$ the *mean (average)* $\bar{x}$ is

$$\bar{x}=\frac{x_1+x_2+\ldots+x_n}{n}=\frac{1}{n}\sum_{i=1}^{n} x_i.$$

For statistical data, where the observations $x_1, x_2, \ldots, x_n$ occur with the frequencies $f_1, f_2, \ldots, f_n$ the mean (average) $\bar{x}$ is $(N=f_1+f_2+\ldots+f_n)$

$$\bar{x}=\frac{f_1x_1+f_2x_2+\ldots+f_nx_n}{N}=\frac{1}{N}\sum_{i=1}^{n} f_ix_i.$$

For statistical data with an *odd* number of observations the *median* Q_2 is defined as

median Q_2=middle value.

For statistical data with an *even* number of observations the *median* is defined as

median Q_2=mean of the two middle values.

Quartiles

For statistical data with an even number $2n$ observations the *first quartile* Q_1 is defined as

first quartile Q_1=median of the n smallest observations

and the *third quartile* Q_3 is defined as

third quartile Q_3=median of the n largest observations.

For stastistical data with an odd number $2n+1$ of observations the *first quartile* Q_1 is defined as

first quartile Q_1=median of the $n+1$ smallest observations

and the third quartile Q_3 is defined as

third quartile Q_3=median of the $n+1$ largest observations.

The *second quartile* Q_2=median.

Measures of spread

Range and interquartile range

For statistical data the *range* R is defined by

$$R=x_{\max}-x_{\min},$$

where $x_{\max}$ and $x_{\min}$ are the largest and the smallest values in the data.

The *interquartile range* is defined as Q_3-Q_1, where Q_1 and Q_3 are the first and third quartiles.

Variance and standard deviation

For statistical data with observations $x_1, x_2, \ldots, x_n$ the *variance* s^2 and the *standard deviation* s are defined by

$$s^2=\frac{1}{n-1}\sum_{i=1}^{n}(x_i-\bar{x})^2, \qquad s=\sqrt{s^2}.$$

For statistical data, where the observations $x_1, x_2, \ldots, x_n$ occur with frequencies $f_1, f_2, \ldots, f_n$, the following holds ($N=f_1+f_2+\ldots f_n$)

$$s^2=\frac{1}{N-1}\sum_{i=1}^{n}f_i(x_i-\bar{x})^2, \qquad s=\sqrt{s^2}.$$

Sometimes the variance s^2 is defined with the numerator n instead of $n-1$. The following formula holds for the variance

$$s^2=\frac{1}{N-1}(K-S^2/N),$$

where N is the total number of observations, S is the sum of the observations and K is the sum of the squares of the observations.

Calculation of mean, variance and standard deviation with calculator and computer

Calculator

The following flow chart describes how to find mean, standard deviation and variance for statistical data $x_1, x_2, \ldots, x_n$ with the aid of a calculator with keys for such statistical data analysis. Instead of "$\Sigma+$" other notations like "x_D" are used. Check in the manual of your calculator if it is programmed to calculate the variance with the denominator n or $n-1$. Usually, calculators are preprogrammed to handle the case when observations $x_1, x_2, \ldots, x_n$ occur with frequencies $f_1, f_2, \ldots, f_n$.

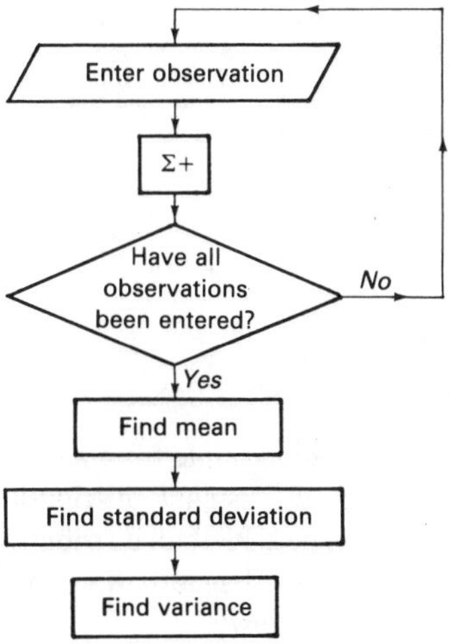

Central moments

For statistical data $x_1, x_2, \ldots, x_n$ the *central moments* m_k are defined as

$$m_k=\frac{1}{n}\sum_{i=1}^{n}(x_i-\bar{x})^k$$

With $h_k=\sum_{i=1}^{n} x_i{}^k/n$,

$$m_2=\frac{n-1}{n}s^2=h_2-h_1{}^2$$

$$m_3=h_3-3h_1h_2+2h_1{}^3$$

$$m_4=h_4-4h_1h_3+6h_1{}^2h_2-3h_1{}^4$$

The *skewness* g_1 and *curtosis* g_2 are defined as

$$g_1=m_3/m_2^{3/2} \qquad g_2=(m_4/m_2^2)-3$$

Weighted aggregate index

P_{0i}= base period prices
P_{1i}= current period prices
Q_{0i}= quantities in the base period
Q_{1i}= current period prices

$$\text{Laspeyres index} = \frac{\sum_i P_{1i} Q_{0i}}{\sum_i P_{0i} Q_{0i}}$$

$$\text{Paasche index} = \frac{\sum_i P_{1i} Q_{1i}}{\sum_i P_{0i} Q_{1i}}$$

Chain index $P_{0t} = P_{0,\, t-1}\, P_{t-1,\, t}$

Median ranks

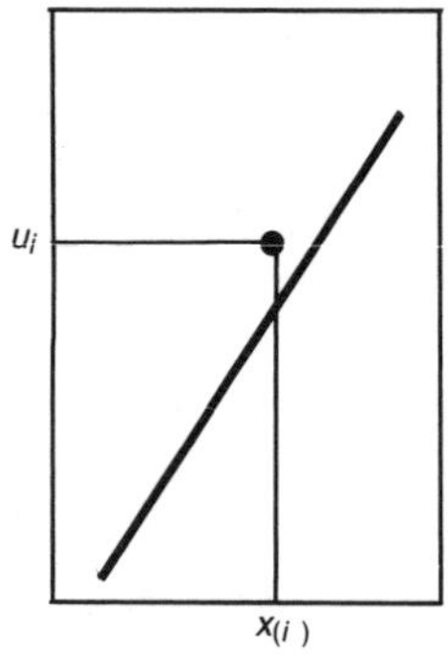

Let $x_{(1)} \leqslant x_{(2)} \leqslant \ldots \leqslant x_{(n)}$ be an ordered sample from a distribution with distribution function F. Using F-distribution paper (e.g. normal distribution paper, Weibull distribution paper) the median ranks u_i are plotted against $x_{(i)}$, where u_i in percent is found in the table below.

i	Sample size n									
	1	2	3	4	5	6	7	8	9	10
1	50.0	29.3	20.6	15.9	12.9	10.9	9.43	8.30	7.41	6.70
2		70.7	50.0	38.6	31.5	26.5	22.9	20.2	18.1	16.3
3			79.4	61.4	50.0	42.2	36.5	32.1	28.7	25.9
4				84.1	68.5	57.8	50.0	44.0	39.4	35.6
5					87.1	73.5	63.5	56.0	50.0	45.2
6						89.1	77.1	67.9	60.7	54.8
7							90.6	79.8	71.3	64.4
8								91.7	82.0	74.1
9									92.6	83.7
10										93.3

i	Sample size *n*									
	11	12	13	14	15	16	17	18	19	20
1	6.11	5.61	5.19	4.83	4.52	4.24	4.00	3.78	3.58	3.41
2	14.9	13.7	12.7	11.8	11.0	10.3	9.75	9.22	8.74	8.31
3	23.7	21.7	20.1	18.7	17.5	16.4	15.5	14.6	13.9	13.2
4	32.4	29.8	27.6	25.7	24.0	22.5	21.2	20.1	19.0	18.1
5	41.2	37.9	35.1	32.6	30.5	28.6	26.9	25.5	24.2	23.0
6	50.0	46.0	42.5	39.6	37.0	34.7	32.7	30.9	29.4	27.9
7	58.8	54.0	50.0	46.5	43.5	40.8	38.5	36.4	34.5	32.8
8	67.6	62.1	57.5	53.5	50.0	46.9	44.2	41.8	39.7	37.7
9	76.3	70.2	65.0	60.4	56.5	53.1	50.0	47.3	44.8	42.6
10	85.1	78.3	72.4	67.4	63.0	59.2	55.8	52.7	50.0	47.5
11	93.9	86.3	80.0	74.3	69.5	65.3	61.5	58.2	55.2	52.5
12		94.4	87.3	81.3	76.0	71.4	67.3	63.6	60.3	57.4
13			94.8	88.2	82.5	77.5	73.0	69.0	65.5	62.3
14				95.2	89.0	83.6	78.8	74.5	70.6	67.2
15					95.5	89.7	84.5	79.9	75.8	72.1
16						95.8	90.3	85.4	81.0	77.0
17							96.0	90.8	86.1	81.9
18								96.2	91.3	86.8
19									96.4	91.7
20										96.6

i	Sample size *n*									
	21	22	23	24	25	30	35	40	45	50
1	3.30	3.15	3.01	2.88	2.77	2.31	1.98	1.73	1.54	1.38
2	7.97	7.61	7.28	6.98	6.70	5.59	4.80	4.20	3.74	3.37
3	12.6	12.1	11.6	11.1	10.6	8.88	7.63	6.68	5.94	5.35
4	17.3	16.5	15.8	15.2	14.6	12.2	10.4	9.15	8.14	7.33
5	22.0	21.0	20.1	19.3	18.5	15.5	13.3	11.6	10.4	9.32
6	26.7	25.5	24.4	23.4	22.5	18.8	16.1	14.1	12.6	11.3
7	31.3	29.9	28.6	27.5	26.4	22.0	18.9	16.6	14.8	13.3
8	36.0	34.4	32.9	31.6	30.3	25.3	21.8	19.1	17.0	15.3
9	40.1	38.8	37.2	35.7	34.3	28.6	24.6	21.5	19.2	17.3
10	45.3	43.3	41.5	39.8	38.2	31.9	27.4	24.0	21.4	19.2
11	50.0	47.8	45.7	43.9	42.1	35.2	30.2	26.5	23.6	21.2
12	54.7	52.2	50.0	48.0	46.1	38.5	33.1	29.0	25.8	23.2
13	59.3	56.7	54.3	52.0	50.0	41.8	35.9	31.4	28.0	25.2
14	64.0	61.2	58.5	56.1	53.9	45.1	38.7	33.9	30.2	27.2

15	68.7	65.6	62.8	60.2	57.9	48.4	41.5	36.4	32.4	29.2
16	73.3	70.1	67.1	64.3	61.8	51.6	44.4	38.9	34.6	31.1
17	78.0	74.5	71.4	68.4	65.7	54.9	47.2	41.3	36.8	33.1
18	82.7	79.0	75.6	72.5	69.7	58.2	50.0	43.8	39.0	35.1
19	87.4	83.5	79.9	76.6	73.6	61.5	52.8	46.3	41.2	37.1
20	92.0	87.9	84.2	80.7	77.5	64.8	55.6	48.8	43.4	39.1
21	96.7	92.4	88.4	84.8	81.5	68.1	58.5	51.2	45.6	41.1
22		96.8	92.7	88.9	85.4	71.4	61.3	53.7	47.8	43.1
23			97.0	93.0	89.4	74.7	64.1	56.2	50.0	45.0
24				97.1	93.3	78.0	66.9	58.7	52.2	47.0
25					97.2	81.2	69.8	61.1	54.4	49.0
26						84.5	72.6	63.6	56.6	51.0
27						87.8	75.4	66.1	58.8	53.0
28						91.1	78.2	68.6	61.0	55.0
29						94.4	81.1	71.0	63.2	57.0
30						97.7	83.9	73.5	65.4	58.9
31							86.7	76.0	67.6	60.9
32							89.5	78.5	69.8	62.9
33							92.4	80.9	72.0	64.9
34							95.2	83.4	74.2	66.9
35							98.0	85.9	76.4	68.8
36								88.4	78.6	70.8
37								90.8	80.8	72.8
38								93.3	83.0	74.8
39								95.8	85.2	76.8
40								98.3	87.4	78.8
41									89.6	80.8
42									91.9	82.7
43									94.1	84.7
44									96.3	86.7
45									98.5	88.7
46										90.7
47										92.7
48										94.6
49										96.6
50										98.6

For large n, u_i can be calculated from

$$u_i \approx (i-0.3)/(n+0.4)$$

18.2 Point Estimation

Definitions and basic results

Let $T=T(X_1, X_2, \ldots, X_n)$ be an estimator of the parameter $q(\Theta)$.

T is *unbiased* if

$$E[T]=q(\Theta).$$

T is *consistent* if

$$\lim_{n\to\infty} P(|T(X_1, X_2, \ldots, X_n)-q(\Theta)|\geqslant\varepsilon)=0 \text{ for every } \varepsilon>0$$

or

$$\operatorname*{limp}_{n\to\infty} T(X_1, X_2, \ldots, X_n)=q(\Theta)$$

T is *asymptotically efficient* if T is asymptotically normal with minimum variance.

T is *sufficient for* Θ if the conditional distribution of $(X_1, X_2, \ldots, X_n)$ given T does not involve Θ.

T is *uniformly minimum variance unbiased* (T is *UMVU*) if T is unbiased and has minimum variance among all unbiased estimates.

T is *maximum likelihood estimate* of Θ if T is that value of Θ which maximizes the probability function (discrete case) or the probability density (continuous case) $p(\boldsymbol{x}, \Theta)$ of $(X_1, X_2, \ldots, X_n)$

$b(\Theta, T)=E[T-q(\Theta)]=$bias of T

$R(\Theta, T)=E[(T-q(\Theta))^2]=$mean squared error of T

$R(\Theta, T)=\mathrm{Var}[T]+b^2(\Theta, T)$

The *Fisher information number* $I(\Theta)$ is defined as

$$I(\Theta)=E\left[\left(\frac{\partial}{\partial\Theta}\ln p(X_1, X_2, \ldots, X_n, \Theta)\right)^2\right]$$

where $p(x_1, x_2, \ldots, x_n, \Theta)$ is the probability function or density function of $(X_1, X_2, \ldots, X_n)$.

The information inequality

$$\mathrm{Var}[T]\geqslant\Psi'(\Theta)^2/I(\Theta),$$

where $\Psi(\Theta)=E[T]$.

The Cramér-Rao inequality

$$\mathrm{Var}[T]\geqslant 1/I(\Theta),$$

if T is unbiased.

UMVU* estimation with sufficient statistics *T

Let T be sufficient for Θ. Then, according to Lehmann-Scheffé, a *UMVU*-estimator of $q(\Theta)$ can be constructed in two ways.

1) Find $h(T)$ such $E[h(T)]=q(\Theta)$

2) Let S be unbiased and find $E[S|T]$.

Specific *UMVU* estimates

$X_1, X_2, \ldots, X_n$ are independent with the same distribution.

$$S=\sum_{i=1}^{n} X_i \qquad \bar{X}=S/n \qquad s^2=\frac{1}{n-1}\sum_{i=1}^{n}(X_i-\bar{X})^2$$

Distribution	Estimated parameter	*UMVU* Estimate T	Var[T]
Uniform $U(0, \Theta)$	Θ	$\frac{n+1}{n} X_{\max}$	$\frac{\Theta^2}{n(n+2)}$
Exponential $E(\lambda)$	λ	$\frac{n-1}{S}$	$\frac{\lambda^2}{n-2}$
Exponential $E[\lambda]$	$\frac{1}{\lambda}$	$\bar{X}$	$\frac{1}{n\lambda^2}$
Normal $N(\mu, \sigma)$	μ	$\bar{X}$	$\frac{\sigma^2}{n}$
Normal $N(\mu, \sigma)$	σ^2	s^2	$\frac{2\sigma^4}{n-1}$
Normal $N(\mu, \sigma)$	σ	$c_n s$ (For c_n see below.)	$\sigma^2(c_n{}^2-1)$
Poisson $P(\lambda)$	λ	$\bar{X}$	$\frac{\lambda^2}{n}$
$f(x)=p^x(1-p)^{1-x}$, $x=0, 1$	p	$\bar{X}$	$\frac{p(1-p)}{n}$

Estimation of moments

Unbiased estimation of μ, μ_2, μ_3 och μ_4

$$\mu_k=E[(X-\mu)^k] \qquad \mu_2=\sigma^2$$

$$m_k=\frac{1}{n}\sum_{i=1}^{n}(X_i-\bar{X})^k \qquad m_2=\frac{n-1}{n}s^2$$

$$E[\bar{X}]=\mu$$

$$E\left[\frac{n}{n-1}m_2\right]=\mu_2 \qquad E[s^2]=\sigma^2$$

$$E\left[\frac{n^2}{(n-1)(n-2)}m_3\right]=\mu_3$$

$$E\left[\frac{n(n^2-2n+3)}{(n-1)(n-2)(n-3)}m_4-\frac{3n(2n-3)}{(n-1)(n-2)(n-3)}m_2{}^2\right]=\mu_4$$

Estimation of skewness γ_1 and curtosis γ_2

$$\gamma_1=\mu_3/\sigma^3 \qquad g_1=m_3/m_2^{3/2}$$

$$\gamma_2=(\mu_4/\sigma^4)-3 \qquad g_2=(m_4/m_2^2)-3$$

$$\lim_{n\to\infty} E[g_1]=\gamma_1 \qquad \lim_{n\to\infty} E[g_2]=\gamma_2$$

For a normal $N(\mu,\ \sigma)$-distribution

$$\gamma_1=\gamma_2=0 \qquad E[g_1]=0 \qquad E[g_2]=-\frac{6}{n+1}$$

Gurland-Tripathi's correction factor for *s*

Let s be the standard deviation of a sample with n observations from the $N(\mu,\ \sigma)$-distribution. Then $c_n s$ with c_n from the table below is the *UMVU* estimate of σ.

Sample size n	c_n	Sample size n	c_n
2	1.2533	12	1.0230
3	1.1284	13	1.0210
4	1.0854	14	1.0194
5	1.0639	15	1.0180
6	1.0509	16	1.0168
7	1.0424	17	1.0157
8	1.0362	18	1.0148
9	1.0317	19	1.0140
10	1.0281	20	1.0132
11	1.0253	21	1.0126

18.3 Confidence Intervals

Definitions. Pivot variables

A 100 α% *confidence interval* for an unknown parameter Θ is an interval which has been calculated such that the probability that it contains Θ is α. Very often the value $\alpha=0.95$ is used.

Let $T=T(X_1, X_2, \ldots, X_n)$ be a statistic. A suitable method to find a confidence interval for a parameter Θ is to use a *pivot variable* $g(T, \Theta)$. Such a variable has a distribution which is *independent of the unknown parameter* Θ. Then x_1 and x_2 can be found such that

$$P(x_1<g(T, \Theta)<x_2)=\alpha$$

If

$$x_1<g(T, \Theta)<x_2 \Leftrightarrow h_1(x_1, x_2, T)<\Theta<h_2(x_1, x_2, T),$$

then

$$h_1<\Theta<h_2$$

is a 100 α% confidence interval for Θ.

If

$$P(g(T, \Theta)\leqslant x_1)=P(g(T, \Theta)\geqslant x_2)=(1-\alpha)/2$$

the confidence interval is *symmetric*.

Confidence intervals $\Theta<h$ or $\Theta>h$ are called *one-sided*.

Example. Let $X_1, X_2, \ldots, X_n$ be independent $E(\lambda)$ variables. Then $2\lambda S$ with $S=\sum_{i=1}^{n} X_i$ can be used as a pivot variable, because $2\lambda S$ has a $\chi^2(r)$-distribution with $r=2n$. We can use $2\lambda S$ to find a confidence interval for λ or $1/\lambda$ or other functions of λ. From a table of the χ^2-distribution we can find x_1 and x_2 such that

$$P(x_1<2\lambda S<x_2)=\alpha.$$

This gives the following 100 α% confidence interval for $1/\lambda$

$$\frac{2S}{x_2}<\frac{1}{\lambda}<\frac{2S}{x_1}$$

Specific confidence intervals

$X_1, X_2, \ldots, X_n$ are independent with the same distribution.

$$S=\sum_{i=1}^{n} X_i, \qquad \bar{X}=S/n, \qquad s^2=\frac{1}{n-1}\sum_{i=1}^{n}(X_i-\bar{X})^2$$

Distribution	Parameter	Pivot variable	Distribution of pivot variable	Two-sided confidence interval
Exponential $E(\lambda)$	λ	$2\lambda S$	$\chi^2(2n)$	$x_1/(2S)<\lambda<x_2/(2S)$
Exponential $E(\lambda)$	$1/\lambda$	$2\lambda S$	$\chi^2(2n)$	$2S/x_2<1/\lambda<2S/x_1$ (see also sec. 18.4)
Exponential $E(\lambda)$	λ	λS	$\Gamma(n, 1)$	$x_1/S<\lambda<x_2/S$
Exponential $E(\lambda)$	$1/\lambda$	λS	$\Gamma(n, 1)$	$S/x_2<1/\lambda<S/x_1$ (see also sec. 18.4)
Normal $N(\mu, \sigma)$ σ known	μ	$\frac{\bar{X}-\mu}{\sigma}\sqrt{n}$	$N(0, 1)$	$\mu=\bar{X}\pm x\sigma/\sqrt{n}$
Normal $N(\mu, \sigma)$	μ	$\frac{\bar{X}-\mu}{s}\sqrt{n}$	$t(n-1)$	$\mu=\bar{X}\pm xs/\sqrt{n}$ (see also sec. 18.4)
Normal $N(\mu, \sigma)$	σ^2	$(n-1)s^2/\sigma^2$	$\chi^2(n-1)$	$(n-1)s^2/x_2<\sigma^2<(n-1)s^2/x_1$ (see also sec. 18.4)
$f(x)=p^x(1-p)^{1-x}$ $x=0, 1$	p	$\frac{\bar{X}-p}{\sqrt{\bar{X}(1-\bar{X})/n}}$	approx. $N(0, 1)$	$p=\bar{X}\pm x\sqrt{\bar{X}(1-\bar{X})/n}$ (see also sec. 18.4)
X Poisson (ct)	c	$\frac{X-ct}{\sqrt{X}}$	approx. $N(0, 1)$	$c=\frac{X}{t}\pm x\frac{\sqrt{X}}{t}$

Sampling finite populations

Notation

Population mean $\bar{X}=\frac{1}{N}\sum_{i=1}^{N} x_i$

Population variance $S^2=\frac{1}{N-1}\sum_{i=1}^{N}(X_i-\bar{X})^2$

Population proportion P

Sample mean $\bar{x}=\frac{1}{n}\sum_{i=1}^{n} x_i$

Sample variance $s^2=\frac{1}{n-1}\sum_{i=1}^{n}(x_i-\bar{x})^2$

Sample proportion p

Simple random sampling

$$E[\bar{x}]=\bar{X} \qquad \text{Var}[\bar{x}]=\frac{1-f}{n}S^2 \qquad f=\frac{n}{N}$$

$$E[p]=P \qquad \text{Var}[p]=\frac{P(1-P)}{n}\cdot\frac{N-n}{N-1}\approx\frac{P(1-P)}{n}(1-f)$$

Confidence interval for P with confidence level approximately equal to α

$$P = p \pm u_\alpha \sqrt{\frac{p(1-p)}{n}(1-f)}$$

where u_α is calcalated from the $N(0, 1)$ distribution e.g. $u_{0.95}=1.96$.

Stratified sampling

$\bar{X}$ estimated by $\hat{\bar{X}} = \sum_{i=1}^{r} (N_i/N)\, \bar{x}_i$

$$\mathrm{Var}[\hat{\bar{X}}] = \sum_{i=1}^{r} \left(\frac{N_i}{N}\right)^2 \frac{1-f_i}{n_i} S_i^2$$

Proportional allocation $n_i = (N_i/N)n$.

Optimum allocation with c_i=cost per sampled unit from stratum i

$$n_i = \frac{N_i S_i/\sqrt{c_i}}{\sum_{i=1}^{r} N_i S_i/\sqrt{c_i}} \cdot n$$

P estimated by $\hat{P} = \sum_{i=1}^{r} (N_i/N)\, p_i$

$$\mathrm{Var}[\hat{P}] = \sum_{i=1}^{r} \frac{N_i^2}{N^2} \cdot \frac{P_i(1-P_i)}{n_i} \cdot \frac{N_i - n_i}{N_i - 1} \approx \frac{1}{N^2} \sum_{i=1}^{r} N_i^2 \frac{P_i(1-P_i)}{n_i}(1-f_i)$$

Confidence interval for P with confidence level approximately equal to α

$$P = \hat{P} \pm u_\alpha \frac{1}{N} \sqrt{\sum_{i=1}^{r} N_i^2 \frac{\hat{p}_i(1-\hat{p}_i)}{n_i}}$$

Optimum allocation with c_i=cost per sampled unit from stratum i

$$n_i = \frac{N_i \sqrt{P_i(1-P_i)/c_i}}{\sum_{i=1}^{r} N_i \sqrt{P_i(1-P_i)/c_i}} \cdot n$$

Confidence interval for unknown probability

Chart giving 95% confidence interval for unknown probability p from observed relative frequency f in n trials.

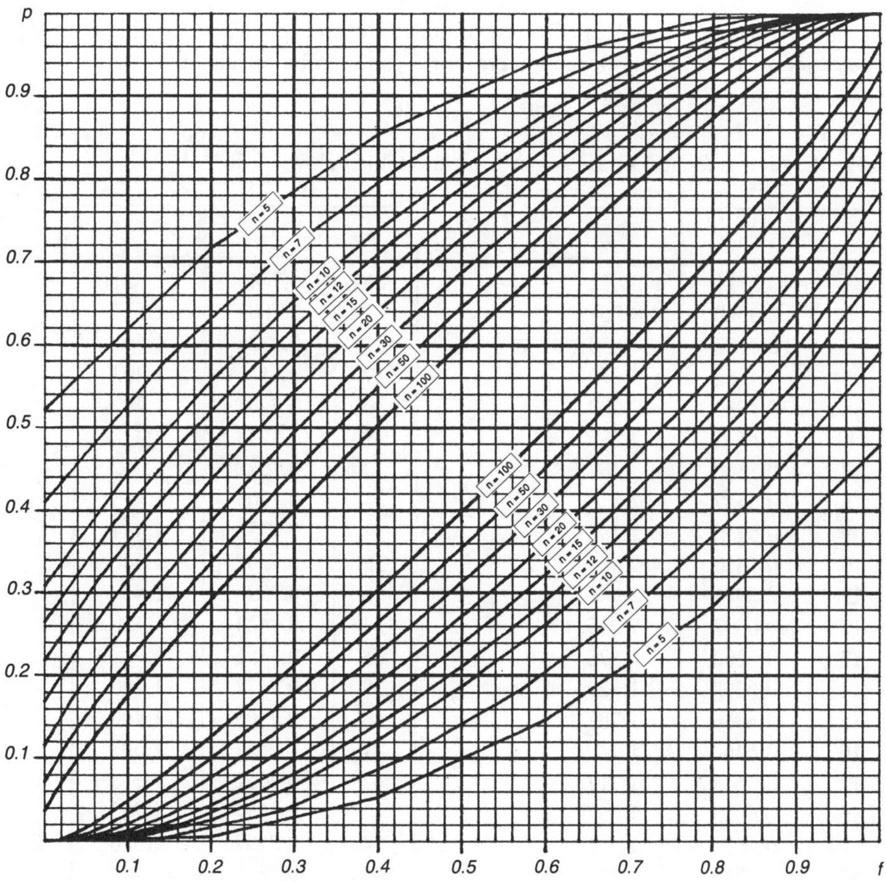

For large n, the confidence interval can be found from the pivot variable

$$(f-p)/\sqrt{\frac{f(1-f)}{n}},$$

which is approximately $N(0, 1)$ e.g.

$$p=f\pm 1.96\sqrt{\frac{f(1-f)}{n}}$$

is a confidence interval for p with approximate confidence level 0.95.

18.4 Tables for Confidence Intervals

Confidence interval for the expectation of an exponential distribution

Let $\overline{x}$ be the mean for a sample of n observations from an exponential distribution with expectation μ. A 95% confidence interval for μ can then be calculated by

$$k_1 \cdot \overline{x} \leqslant \mu \leqslant k_2 \cdot \overline{x} \qquad (0.95)$$

where the factors k_1 and k_2 are taken from the table below.

n	k_1	k_2
1	0.2711	39.22
2	0.3590	8.264
3	0.4153	4.850
4	0.4562	3.670
5	0.4882	3.080
6	0.5142	2.725
7	0.5360	2.487
8	0.5547	2.316
9	0.5710	2.187
10	0.5853	2.085
11	0.5981	2.003
12	0.6097	1.935
13	0.6202	1.878
14	0.6298	1.829
15	0.6386	1.787
16	0.6467	1.749
17	0.6543	1.717
18	0.6613	1.687
19	0.6679	1.661
20	0.6741	1.637
21	0.6799	1.615
22	0.6853	1.596
23	0.6905	1.578
24	0.6954	1.561
25	0.7001	1.545

n	k_1	k_2
26	0.7045	1.531
27	0.7087	1.517
28	0.7128	1.505
29	0.7166	1.493
30	0.7203	1.482
31	0.7238	1.472
32	0.7272	1.462
33	0.7305	1.453
34	0.7336	1.444
35	0.7367	1.436
36	0.7396	1.428
37	0.7424	1.420
38	0.7451	1.413
39	0.7477	1.406
40	0.7503	1.400
41	0.7527	1.393
42	0.7551	1.388
43	0.7574	1.382
44	0.7597	1.376
45	0.7618	1.371
46	0.7639	1.366
47	0.7660	1.361
48	0.7680	1.356
49	0.7699	1.352
50	0.7718	1.347

Confidence interval for the expectation of a normal distribution

Let $\bar{x}$ and s be the mean and the standard deviation for a sample of n observations from a normal distribution with expectation μ. A 95% or 99% confidence interval for μ can then be calculated by

$$\mu=\bar{x}\pm k_1\cdot s\ (0.95) \text{ and } \mu=\bar{x}\pm k_2\cdot s\ (0.99)$$

where the factors k_1 and k_2 are taken from the table below.

n	k_1	k_2
2	8.985	45.01
3	2.484	5.730
4	1.591	2.920
5	1.242	2.059
6	1.049	1.646
7	0.9248	1.401
8	0.8360	1.237
9	0.7687	1.118
10	0.7154	1.028
11	0.6718	0.9556
12	0.6354	0.8966
13	0.6043	0.8472
14	0.5774	0.8051
15	0.5538	0.7686
16	0.5329	0.7367
17	0.5142	0.7084
18	0.4973	0.6831
19	0.4820	0.6604
20	0.4680	0.6397
21	0.4552	0.6209
22	0.4434	0.6037
23	0.4324	0.5878
24	0.4223	0.5730
25	0.4128	0.5594
26	0.4039	0.5467
27	0.3956	0.5348
28	0.3878	0.5236

n	k_1	k_2
29	0.3804	0.5131
30	0.3734	0.5032
31	0.3668	0.4939
32	0.3605	0.4851
33	0.3546	0.4767
34	0.3489	0.4688
35	0.3435	0.4612
36	0.3384	0.4540
37	0.3334	0.4471
38	0.3287	0.4405
39	0.3242	0.4342
40	0.3198	0.4282
41	0.3156	0.4224
42	0.3116	0.4168
43	0.3078	0.4115
44	0.3040	0.4063
45	0.3004	0.4013
46	0.2970	0.3966
47	0.2936	0.3919
48	0.2904	0.3875
49	0.2872	0.3832
50	0.2842	0.3790
60	0.2583	0.3436
70	0.2384	0.3166
80	0.2225	0.2951
90	0.2094	0.2775
100	0.1984	0.2626

For large n, $k_1 \approx 1.9600/\sqrt{n}$ and $k_2 \approx 2.5758/\sqrt{n}$.

One-sided confidence interval for the standard deviation of a normal distribution

Let s be the sample standard deviation for a sample of n observations from a normal distribution with standard deviation σ. A one-sided confidence interval for σ with coefficients 0.95 and 0.99 is then given by

$$\sigma \leqslant k_1 \cdot s \quad (0.95) \quad \text{and} \quad \sigma \leqslant k_2 \cdot s \quad (0.99),$$

where the factors k_1 and k_2 are taken from the table below.

n	k_1	k_2
2	15.81	–
3	4.407	10.00
4	2.919	5.108
5	2.372	3.670
6	2.090	3.032
7	1.916	2.623
8	1.797	2.377
9	1.711	2.205
10	1.645	2.076
11	1.593	1.977
12	1.551	1.898
13	1.515	1.833
14	1.485	1.779
15	1.460	1.733
16	1.437	1.694
17	1.418	1.659
18	1.400	1.629
19	1.385	1.602
20	1.370	1.578
21	1.358	1.556
22	1.346	1.536
23	1.335	1.518
24	1.325	1.502
25	1.316	1.487
26	1.308	1.473
27	1.300	1.460
28	1.293	1.448

n	k_1	k_2
29	1.286	1.437
30	1.280	1.426
31	1.274	1.416
32	1.268	1.407
33	1.263	1.398
34	1.258	1.390
35	1.253	1.382
36	1.248	1.375
37	1.244	1.368
38	1.240	1.362
39	1.236	1.355
40	1.232	1.349
41	1.228	1.343
42	1.225	1.338
43	1.222	1.333
44	1.218	1.328
45	1.215	1.323
46	1.212	1.318
47	1.210	1.314
48	1.207	1.309
49	1.204	1.305
50	1.202	1.301
60	1.180	1.268
70	1.165	1.243
80	1.152	1.224
90	1.142	1.209
100	1.134	1.196

Two-sided confidence interval for the standard deviation of a normal distribution

Let s be the sample standard deviation for a sample of n observations from a normal distribution with standard deviation σ. A 95% two-sided confidence interval for σ is then given by

$$k_1 \cdot s \leqslant \sigma \leqslant k_2 \cdot s \quad (0.95),$$

where the factors k_1 and k_2 are taken from the table below.

n	k_1	k_2
2	0.4461	31.62
3	0.5206	6.262
4	0.5665	3.727
5	0.5991	2.875
6	0.6242	2.453
7	0.6444	2.202
8	0.6612	2.035
9	0.6754	1.916
10	0.6878	1.826
11	0.6987	1.755
12	0.7084	1.698
13	0.7171	1.651
14	0.7249	1.611
15	0.7321	1.577
16	0.7387	1.548
17	0.7448	1.522
18	0.7504	1.499
19	0.7556	1.479
20	0.7605	1.461
21	0.7651	1.444
22	0.7694	1.429
23	0.7734	1.415
24	0.7772	1.403
25	0.7808	1.391
26	0.7843	1.380
27	0.7875	1.370
28	0.7906	1.361

n	k_1	k_2
29	0.7936	1.352
30	0.7964	1.344
31	0.7991	1.337
32	0.8017	1.329
33	0.8042	1.323
34	0.8066	1.316
35	0.8089	1.310
36	0.8111	1.304
37	0.8132	1.299
38	0.8153	1.294
39	0.8172	1.289
40	0.8192	1.284
41	0.8210	1.280
42	0.8228	1.275
43	0.8245	1.271
44	0.8262	1.267
45	0.8279	1.263
46	0.8294	1.260
47	0.8310	1.256
48	0.8325	1.253
49	0.8339	1.249
50	0.8353	1.246
60	0.8476	1.220
70	0.8574	1.200
80	0.8655	1.184
90	0.8722	1.172
100	0.8780	1.162

Confidence interval for the difference of expectations of normal distributions

Let $\bar{x}_1$ and $\bar{x}_2$ be the means in samples of n_1 and n_2 observations from normal distributions with the same variance and expectations μ_1 and μ_2. Let

$$s^2=((n_1-1)s_1{}^2+(n_2-1)s_2{}^2)/(n_1+n_2-2)$$

be the pooled variance.

A 95% confidence interval for $\mu_1-\mu_2$ is then given by $\mu_1-\mu_2=\bar{x}_1-\bar{x}_2\pm ks$ (0.95), where the factor k is taken from the table below.

n_2 \ n_1	2	3	4	5	6	7	8	9	10
2	4.3027	2.905	2.404	2.151	1.998	1.896	1.823	1.768	1.726
3		2.267	1.963	1.787	1.672	1.591	1.532	1.485	1.449
4			1.730	1.586	1.489	1.418	1.364	1.323	1.289
5				1.458	1.370	1.305	1.255	1.215	1.183
6					1.286	1.225	1.177	1.139	1.108
7						1.165	1.118	1.081	1.050
8							1.072	1.036	1.006
9								0.9993	0.9694
10									0.9396

n_2 \ n_1	11	12	13	14	15	16	17	18	19	20
2	1.692	1.664	1.641	1.621	1.605	1.590	1.577	1.566	1.556	1.547
3	1.419	1.395	1.374	1.356	1.341	1.327	1.316	1.305	1.296	1.288
4	1.261	1.238	1.219	1.202	1.187	1.174	1.163	1.153	1.144	1.136
5	1.157	1.135	1.116	1.099	1.085	1.072	1.061	1.051	1.042	1.034
6	1.082	1.060	1.041	1.025	1.011	0.9986	0.9875	0.9776	0.9688	0.9607
7	1.025	1.003	0.9849	0.9689	0.9548	0.9424	0.9314	0.9215	0.9125	0.9044
8	0.9803	0.9589	0.9405	0.9245	0.9104	0.8980	0.8869	0.8770	0.8680	0.8599
9	0.9443	0.9229	0.9046	0.8885	0.8744	0.8620	0.8508	0.8408	0.8318	0.8236
10	0.9145	0.8932	0.8747	0.8587	0.8445	0.8320	0.8208	0.8107	0.8016	0.7933
11	0.8895	0.8681	0.8496	0.8335	0.8193	0.8067	0.7954	0.7852	0.7761	0.7677
12		0.8467	0.8281	0.8119	0.7976	0.7850	0.7736	0.7634	0.7541	0.7457
13			0.8095	0.7932	0.7789	0.7661	0.7547	0.7444	0.7351	0.7266
14				0.7769	0.7625	0.7496	0.7381	0.7278	0.7184	0.7098
15					0.7480	0.7350	0.7235	0.7130	0.7035	0.6949
16						0.7221	0.7104	0.6999	0.6903	0.6816
17							0.6987	0.6881	0.6784	0.6697
18								0.6774	0.6677	0.6589
19									0.6580	0.6491
20										0.6402

Tolerance limits for the normal distribution

The following table gives factors k_1 and k_2 such that the following kind of statements can be made.

At least the proportion P is less than $\bar{x}+k_1s$ with confidence 0.95.

At least the proportion P is greater than $\bar{x}-k_1s$ with confidence 0.95.

At least the proportion P is between $\bar{x}-k_2s$ and $\bar{x}+k_2s$ with confidence 0.95.

n=sample size $\bar{x}$=sample mean s=sample standard deviation

	P=0.90		P=0.95		P=0.99	
n	k_1	k_2	k_1	k_2	k_1	k_2
6	3.006	3.712	3.707	4.414	5.062	5.775
7	2.755	3.369	3.399	4.007	4.641	5.248
8	2.582	3.136	3.188	3.732	4.353	4.891
9	2.454	2.967	3.031	3.532	4.143	4.631
10	2.355	2.839	2.911	3.379	3.981	4.433
11	2.275	2.737	2.815	3.259	3.852	4.277
12	2.210	2.655	2.736	3.162	3.747	4.150
13	2.155	2.587	2.670	3.081	3.659	4.044
14	2.108	2.529	2.614	3.012	3.585	3.955
15	2.068	2.480	2.566	2.954	3.520	3.878
16	2.032	2.437	2.523	2.903	3.463	3.812
17	2.001	2.400	2.486	2.858	3.415	3.754
18	1.974	2.366	2.453	2.819	3.370	3.702
19	1.949	2.337	2.423	2.784	3.331	3.656
20	1.926	2.310	2.396	2.752	3.295	3.615
21	1.905	2.286	2.371	2.723	3.262	3.577
22	1.887	2.264	2.350	2.697	3.233	3.543
23	1.869	2.244	2.329	2.673	3.206	3.512
24	1.853	2.225	2.309	2.651	3.181	3.483
25	1.838	2.208	2.292	2.631	3.158	3.457
30	1.778	2.140	2.220	2.549	3.064	3.350
35	1.732	2.090	2.166	2.490	2.994	3.272
40	1.697	2.052	2.126	2.445	2.941	3.213
45	1.669	2.021	2.092	2.408	2.897	3.165
50	1.646	1.969	2.065	2.379	2.863	3.126

18.5 Tests of Significance

Basic definitions

Let $(X_1, X_2, \ldots, X_n)$ have a distribution which depends upon a parameter Θ. A *hypothesis* H specifies that $\Theta \in \Theta_0$,

$$H: \Theta \in \Theta_0.$$

The *alternative hypothesis* K specifies that $\Theta \in \Theta_1$,

$$K: \Theta \in \Theta_1.$$

A hypothesis H or K is *simple* if it contains only one point, otherwise it is *composite.*

In *significance testing* a *test statistic* $T=T(X_1, X_2, \ldots, X_n)$ is used to *accept* or *reject* H. The hypothesis H is rejected if $T \in C$, where C is called the *critical region.*

A *type 1 error* is to reject H when it should be accepted and a *type* 2 *error* is to accept H when it should be rejected.

The *power* of a test against the alternative Θ is the probability of rejecting H when Θ is true. The (significance) *level* α is the maximum probability to rejecting H when H is true. A test is *uniformly most powerful* (UMP) if it maximizes the power for all alternatives.

Specific tests

$X_1, X_2, \ldots, X_n$ are independent with the same distribution

$$S=\sum_{i=1}^{n} X_i,\ \bar{X}=S/n, \qquad s^2=\frac{1}{n-1}\sum_{i=1}^{n}(X_i-\bar{X})^2$$

α=(significance) level $\qquad x_\alpha$=α-fractile of T under H_0, $P(T\leq x_\alpha)=\alpha$

Distribution	Parameter	Hypothesis H	Hypothesis K	Test statistic T	Distribution of T under H	Critical region
$E(\lambda)$	λ	$\lambda=\lambda_0$	$\lambda>\lambda_0$	$2\lambda_0 S$	$\chi^2(2n)$	$T>x_{1-\alpha}$
$E(\lambda)$	λ	$\lambda=\lambda_0$	$\lambda<\lambda_0$	$2\lambda_0 S$	$\chi^2(2n)$	$T<x_\alpha$
$E(\lambda)$	λ	$\lambda=\lambda_0$	$\lambda\neq\lambda_0$	$2\lambda_0 S$	$\chi^2(2n)$	$T<x_{\alpha/2},\ T>x_{1-\alpha/2}$
$N(\mu, \sigma)$ σ known	μ	$\mu=\mu_0$	$\mu<\mu_0$	$\frac{\bar{X}-\mu_0}{\sigma}\sqrt{n}$	$N(0, 1)$	$T<x_\alpha$
$N(\mu, \sigma)$ σ known	μ	$\mu=\mu_0$	$\mu>\mu_0$	$\frac{\bar{X}-\mu_0}{\sigma}\sqrt{n}$	$N(0, 1)$	$T>x_{1-\alpha}$
$N(\mu, \sigma)$ σ known	μ	$\mu=\mu_0$	$\mu\neq\mu_0$	$\frac{\bar{X}-\mu_0}{\sigma}\sqrt{n}$	$N(0, 1)$	$\lvert T\rvert>x_{1-\alpha/2}$
$N(\mu, \sigma)$	μ	$\mu=\mu_0$	$\mu<\mu_0$	$\frac{\bar{X}-\mu_0}{s}\sqrt{n}$	$t(n-1)$	$T<x_\alpha$
$N(\mu, \sigma)$	μ	$\mu=\mu_0$	$\mu>\mu_0$	$\frac{\bar{X}-\mu_0}{s}\sqrt{n}$	$t(n-1)$	$T>x_{1-\alpha}$
$N(\mu, \sigma)$	μ	$\mu=\mu_0$	$\mu\neq\mu_0$	$\frac{\bar{X}-\mu_0}{s}\sqrt{n}$	$t(n-1)$	$\lvert T\rvert>x_{1-\alpha/2}$
$N(\mu, \sigma)$	σ^2	$\sigma^2=\sigma_0^2$	$\sigma^2<\sigma_0^2$	$(n-1)s^2/\sigma_0^2$	$\chi^2(n-1)$	$T<x_\alpha$
$N(\mu, \sigma)$	σ^2	$\sigma^2=\sigma_0^2$	$\sigma^2>\sigma_0^2$	$(n-1)s^2/\sigma_0^2$	$\chi^2(n-1)$	$T>x_{1-\alpha}$
$N(\mu, \sigma)$	σ^2	$\sigma^2=\sigma_0^2$	$\sigma^2\neq\sigma_0^2$	$(n-1)s^2/\sigma_0^2$	$\chi^2(n-1)$	$T<x_{\alpha/2},\ T>x_{1-\alpha/2}$
$p^x(1-p)^{1-x}$ $x=0, 1$	p	$p=p_0$	$p<p_0$	$\frac{S-np_0}{\sqrt{np_0(1-p_0)}}$	approx. $N(0, 1)$ n large	$T<x_\alpha$
$p^x(1-p)^{1-x}$ $x=0, 1$	p	$p=p_0$	$p>p_0$	$\frac{S-np_0}{\sqrt{np_0(1-p_0)}}$	approx. $N(0, 1)$ n large	$T>x_{1-\alpha}$

χ^2-tests and contingency tables

Test of specified distribution

The hypothesis H that the outcomes $u_1, u_2, \ldots, u_k$ have probabilities $\Theta_1, \Theta_2, \ldots, \Theta_k$ can be tested using the test variable

$$\sum_{i=1}^{k}\frac{(n_i-n\Theta_i)^2}{n\Theta_i}, \qquad n=\sum_{i=1}^{k} n_i.$$

Here n_i is the observed frequency of the outcome u_i, $n=\Sigma n_i$. Under H, the test variable is approximately χ^2_{k-1}.

Test of family of distributions

The hypothesis H that outcomes u_i for $i=1, 2, \ldots, k$ have probabilities $\Theta_i(\gamma_1, \gamma_2, \ldots, \gamma_r)$, $r<k-1$, can be tested using the test variable

$$\sum_{i=1}^{k} \frac{(n_i - n\Theta_i(\hat{\gamma}_1, \hat{\gamma}_2, \ldots, \hat{\gamma}_r))^2}{n\Theta_i(\hat{\gamma}_1, \hat{\gamma}_2, \ldots, \hat{\gamma}_r)}.$$

Here the estimates $\hat{\gamma}_1, \hat{\gamma}_2, \ldots, \hat{\gamma}_r$ are obtained from

$$\sum_{i=1}^{k} \frac{n_i}{\Theta_i(\gamma_1, \gamma_2, \ldots, \gamma_r)} \frac{\partial}{\partial \gamma_j} \Theta_i(\gamma_1, \gamma_2, \ldots, \gamma_r)=0, \; j=1,\ldots, r.$$

Under H, the test variable is approximately χ^2_{k-r-1}.

Test of independence in $p \times b$ contingency table

$C_1 \backslash C_2$	1	2	...	b	Σ
1	n_{11}	n_{12}	...	n_{1b}	r_1
2	n_{21}	n_{22}	...	n_{2b}	r_2
⋮	⋮	⋮	⋮	⋮	⋮
p	n_{p1}	n_{p2}	...	n_{pb}	r_p
Σ	c_1	c_2	...	c_b	n

$$r_i = \sum_{j=1}^{b} n_{ij}$$

$$c_j = \sum_{i=1}^{p} n_{ij}$$

$$n = \sum_{i=1}^{p} r_i = \sum_{j=1}^{b} c_j$$

Test of the hypothesis that two characteristics C_1 and C_2 with p and b possibilities are independent can be performed with the test variable

$$n \sum_{i=1}^{p} \sum_{j=1}^{b} \frac{\left(n_{ij} - \frac{r_i c_j}{n}\right)^2}{r_i c_j}.$$

Under the hypothesis of independence the test variable is approximately $\chi^2_{(p-1)(b-1)}$.

For a 2×2 table the test variable can be written

$$\frac{(n_{11}n_{22} - n_{12}n_{21})^2 n}{c_1 c_2 r_1 r_2}.$$

In this case the test variable is approximately χ_1^2. The expected frequencies for the four cells should be at least 5 for the approximation to be valid.

Test of homogeneity in $p \times b$ confingency table

The χ^2-test in the previous section can be used to test a homogeneity hypothesis for a $p \times b$ contingency table. This hypothesis states that each row represents a sample of the same random variable.

Sequential testing

Let $X_1, X_2, \ldots$ be independent equally distributed random variables with density or probability function f. A *sequential probability ratio test* (*SPRT-test*) to choose between H: $f=f_0$ and K: $f=f_1$ is defined as follows. Put $Y_n = \prod_{i=1}^{n} (f_1(X_i)/f_0(X_i))$ and observe X_1, $X_2 \ldots$ after each other. At each stage compute Y_n and continue sampling when $B<Y_n<A$. If $Y_n \leq B$ accept H and if $Y_n \geq A$ accept K where A and B, $0<B<1<A$, are suitable constants.

Let N be the number of observations, α the probability of an error of the first kind (to reject H when H is true) and γ the probability of an error of the second kind (to accept H when K is true). Then

$$P(N<\infty|H)=P(N<\infty|K)=1$$

$$\frac{1-\gamma}{\alpha} \geq A \qquad \frac{\gamma}{1-\alpha} \leq B$$

With $A=(1-\gamma_1)/\alpha_1$ and $B=\gamma_1/(1-\alpha_1)$ the following inequalities hold

$$\alpha \leq \alpha_1/(1-\gamma_1) \qquad \gamma \leq \gamma_1/(1-\alpha_1) \qquad \alpha+\gamma \leq \alpha_1+\gamma_1$$

In practice $A=(1-\gamma)/\alpha$ and $B=\gamma/(1-\alpha)$ are used.

Example. For $f(x)=p^x(1-p)^{1-x}$, $x=0, 1$, H: $p=p_0$, K: $p=p_1$, $p_0<p_1$ the *SPRT* test can be performed as follows.

Put $Y_n = \sum_{i=1}^{n} X_i$ and

$$b_n = \frac{\log \dfrac{\gamma}{1-\alpha} + n \log \dfrac{1-p_0}{1-p_1}}{\log \dfrac{p_1(1-p_0)}{p_0(1-p_1)}} \qquad a_n = \frac{\log \dfrac{1-\gamma}{\alpha} + n \log \dfrac{1-p_0}{1-p_1}}{\log \dfrac{p_1(1-p_0)}{p_0(1-p_1)}}$$

Continue to sample when

$$b_n < Y_n < a_n$$

Accept H when $Y_n \leq b_n$ and reject H when $Y_n \geq a_n$.

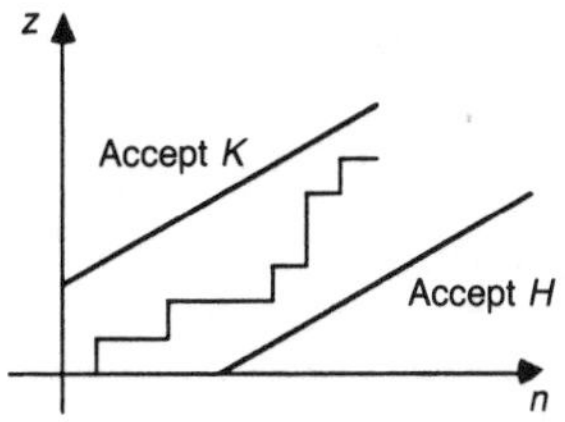

Use of Studentized range for test of normality

Let s be the standard deviation and R the range for a sample of n observations. Then $q_n = R/s$ is the *studentized range* of the sample. The following table gives $F(x) = P(q_n \leq x)$ for a sample from a normal distribution.

The hypothesis that the sample comes from a normal distribution can be tested with q_n as test statistic. The hypothesis is rejected for small or large q_n. E.g. for $n=15$ the hypothesis of normality is rejected on level 0.05 if $q_n < 2.88$ or $q_n > 4.29$.

Size of sample	$F(x)$							
n	0.005	0.01	0.025	0.05	0.95	0.975	0.99	0.995
6	2.11	2.15	2.22	2.28	3.012	3.056	3.095	3.115
7	2.22	2.26	2.33	2.40	3.222	3.282	3.338	3.369
8	2.31	2.35	2.43	2.50	3.399	3.471	3.543	3.585
9	2.39	2.44	2.51	2.59	3.552	3.634	3.720	3.772
10	2.46	2.51	2.59	2.67	3.685	3.777	3.875	3.935
11	2.53	2.58	2.66	2.74	3.80	3.903	4.012	4.079
12	2.59	2.64	2.72	2.80	3.91	4.02	4.134	4.208
13	2.64	2.70	2.78	2.86	4.00	4.12	4.244	4.325
14	2.70	2.75	2.83	2.92	4.09	4.21	4.34	4.431
15	2.74	2.80	2.88	2.97	4.17	4.29	4.44	4.53
16	2.79	2.84	2.93	3.01	4.24	4.37	4.52	4.62
17	2.83	2.88	2.97	3.06	4.31	4.44	4.60	4.70
18	2.87	2.92	3.01	3.10	4.37	4.51	4.67	4.78
19	2.90	2.96	3.05	3.14	4.43	4.57	4.74	4.85
20	2.94	2.99	3.09	3.18	4.49	4.63	4.80	4.91
25	3.09	3.15	3.24	3.34	4.71	4.87	5.06	5.19
30	3.21	3.27	3.37	3.47	4.89	5.06	5.26	5.40
35	3.32	3.38	3.48	3.58	5.04	5.21	5.42	5.57
40	3.41	3.47	3.57	3.67	5.16	5.34	5.56	5.71
45	3.49	3.55	3.66	3.75	5.26	5.45	5.67	5.83
50	3.56	3.62	3.73	3.83	5.35	5.54	5.77	5.93
55	3.62	3.69	3.80	3.90	5.43	5.63	5.86	6.02
60	3.68	3.75	3.86	3.96	5.51	5.70	5.94	6.10
65	3.74	3.80	3.91	4.01	5.57	5.77	6.01	6.17
70	3.79	3.85	3.96	4.06	5.63	5.83	6.07	6.24

18.5

Bartlett's test for equal variance

Let $s_1^2, s_2^2, \ldots, s_k^2$ be the variances for k independent samples with $n_1, n_2, \ldots, n_k$ observations from normal distributions $N(\mu_i, \sigma_i)$, $i=1, 2, \ldots, k$. Let $r_i=n_i-1$ and $f_i=r_i/\sum_{i=1}^{k} r_i$. The hypothesis $\sigma_1^2=\sigma_2^2=\ldots=\sigma_k^2$ can be tested with the test statistic

$$b=\prod_{i=1}^{k} (s_i^2)^{f_i} / \sum_{i=1}^{k} f_i s_i^2.$$

For $n_1=n_2=\ldots=n_k=n$ the hypothesis is rejected on level 0.05 if $b<b_k(n)$, where $b_k(n)$ is given in the table below. For unequal n_i the hypothesis is rejected on level 0.05 for $b<b^*$, where

$$b^*=\sum_{i=1}^{k} n_i b_k(n_i)/n \quad \text{(the Dyer-Keating approximation)}$$

and $n=\sum_{i=1}^{k} n_i$.

n \ k	2	3	4	5	6	7	8	9	10
3	0.3123	0.3058	0.3173	0.3299	-	-	-	-	-
4	0.4780	0.4699	0.4803	0.4921	0.5028	0.5122	0.5204	0.5277	0.5341
5	0.5845	0.5762	0.5850	0.5952	0.6045	0.6126	0.6197	0.6260	0.6315
6	0.6563	0.6483	0.6559	0.6646	0.6727	0.6798	0.6860	0.6914	0.6961
7	0.7075	0.7000	0.7065	0.7142	0.7213	0.7275	0.7329	0.7376	0.7418
8	0.7456	0.7387	0.7444	0.7512	0.7574	0.7629	0.7677	0.7719	0.7757
9	0.7751	0.7686	0.7737	0.7798	0.7854	0.7903	0.7946	0.7984	0.8017
10	0.7984	0.7924	0.7970	0.8025	0.8076	0.8121	0.8160	0.8194	0.8224
11	0.8175	0.8118	0.8160	0.8210	0.8257	0.8298	0.8333	0.8365	0.8392
12	0.8332	0.8280	0.8317	0.8364	0.8407	0.8444	0.8477	0.8506	0.8531
13	0.8465	0.8415	0.8450	0.8493	0.8533	0.8568	0.8598	0.8625	0.8648
14	0.8578	0.8532	0.8564	0.8604	0.8641	0.8673	0.8701	0.8726	0.8748
15	0.8676	0.8632	0.8662	0.8699	0.8734	0.8764	0.8790	0.8814	0.8834
16	0.8761	0.8719	0.8747	0.8782	0.8815	0.8843	0.8868	0.8890	0.8909
17	0.8836	0.8796	0.8823	0.8856	0.8886	0.8913	0.8936	0.8957	0.8975
18	0.8902	0.8865	0.8890	0.8921	0.8949	0.8975	0.8997	0.9016	0.9033
19	0.8961	0.8926	0.8949	0.8979	0.9006	0.9030	0.9051	0.9069	0.9086
20	0.9015	0.8980	0.9003	0.9031	0.9057	0.9080	0.9100	0.9117	0.9132
21	0.9063	0.9030	0.9051	0.9078	0.9103	0.9124	0.9143	0.9160	0.9175
22	0.9106	0.9075	0.9095	0.9120	0.9144	0.9165	0.9183	0.9199	0.9213
23	0.9146	0.9116	0.9135	0.9159	0.9182	0.9202	0.9219	0.9235	0.9248
24	0.9182	0.9153	0.9172	0.9195	0.9217	0.9236	0.9253	0.9267	0.9280
25	0.9216	0.9187	0.9205	0.9228	0.9249	0.9267	0.9283	0.9297	0.9309
26	0.9246	0.9219	0.9236	0.9258	0.9278	0.9296	0.9311	0.9325	0.9336
27	0.9275	0.9249	0.9265	0.9286	0.9305	0.9322	0.9337	0.9350	0.9361
28	0.9301	0.9276	0.9292	0.9312	0.9330	0.9347	0.9361	0.9374	0.9385
29	0.9326	0.9301	0.9316	0.9336	0.9354	0.9370	0.9383	0.9396	0.9406
30	0.9348	0.9325	0.9340	0.9358	0.9376	0.9391	0.9404	0.9416	0.9426
40	0.9513	0.9495	0.9506	0.9520	0.9533	0.9545	0.9555	0.9564	0.9572
50	0.9612	0.9597	0.9606	0.9617	0.9628	0.9637	0.9645	0.9652	0.9658
60	0.9677	0.9665	0.9672	0.9681	0.9690	0.9698	0.9705	0.9710	0.9716
80	0.9758	0.9749	0.9754	0.9761	0.9768	0.9774	0.9779	0.9783	0.9787
100	0.9807	0.9799	0.9804	0.9809	0.9815	0.9819	0.9823	0.9827	0.9830

18.6 Linear Models

The two-sample case

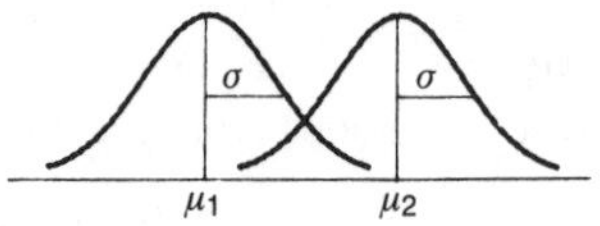

Let $X_{11}, X_{12}, \ldots, X_{1n_1}$ be independent and $N(\mu_1, \sigma)$ and let $X_{21}, X_{22}, \ldots, X_{2n_2}$ be independent and $N(\mu_2, \sigma)$

$$\bar{X}_1 = \sum_{i=1}^{n_1} X_{1i}/n_1 \qquad \bar{X}_2 = \sum_{i=1}^{n_2} X_{2i}/n_2$$

$$s_1^2 = \frac{1}{n_1-1}\sum_{i=1}^{n_1}(X_{1i}-\bar{X}_1)^2 \qquad s_2^2 = \frac{1}{n_2-1}\sum_{i=1}^{n_2}(X_{2i}-\bar{X}_2)^2$$

$$s^2 = \frac{(n_1-1)s_1^2+(n_2-1)s_2^2}{n_1+n_2-2} \qquad \text{(pooled variance)}$$

$$T = \frac{(\bar{X}_1-\bar{X}_2)-(\mu_1-\mu_2)}{s}\sqrt{\frac{n_1n_2}{n_1+n_2}}$$

The statistic T is $t(n_1+n_2-2)$ and can be used as a pivot variable to derive a confidence interval for $\mu_1-\mu_2$ e.g.

$$\mu_1-\mu_2 = \bar{X}_1-\bar{X}_2 \pm xs\sqrt{\frac{n_1+n_2}{n_1n_2}}$$

(see also table in sec. 18.4).

The statistic T with $\mu_1=\mu_2$ can be used to test the hypothesis $H\colon \mu_1=\mu_2$. For example let the hypotheses be

$$H\colon \mu_1=\mu_2 \qquad K\colon \mu_1<\mu_2$$

Then

$$T = \frac{\bar{X}_1-\bar{X}_2}{s}\sqrt{\frac{n_1n_2}{n_1+n_2}}$$

is $t(n_1+n_2-2)$ under H and H is rejected for $T<x_\alpha$.

If $X_{11}, X_{12}, \ldots, X_{1n_1}$ are $N(\mu_1, \sigma_1)$ and $X_{21}, X_{22}, \ldots, X_{2n_2}$ are $N(\mu_2, \sigma_2)$ then

$$T_1 = \frac{s_1^2/\sigma_1^2}{s_2^2/\sigma_2^2}$$

is $F(n_1-1, n_2-1)$ and can be used as a pivot variable to derive a confidence interval for σ_1^2/σ_2^2 e.g.

$$\frac{1}{x_2}\cdot\frac{s_1^2}{s_2^2} < \frac{\sigma_1^2}{\sigma_2^2} < \frac{1}{x_1}\cdot\frac{s_1^2}{s_2^2}$$

The statistic T_1 with $\sigma_1=\sigma_2$ can be used to test the hypothesis $H\colon \sigma_1=\sigma_2$. For example let the hypotheses be

$$H\colon \sigma_1=\sigma_2 \qquad K\colon \sigma_1>\sigma_2$$

Then $T_1=s_1^2/s_2^2$ is $F(n_1-1, n_2-1)$ under H and H is rejected for $T_1>x_{1-\alpha}$.

Simple linear regression

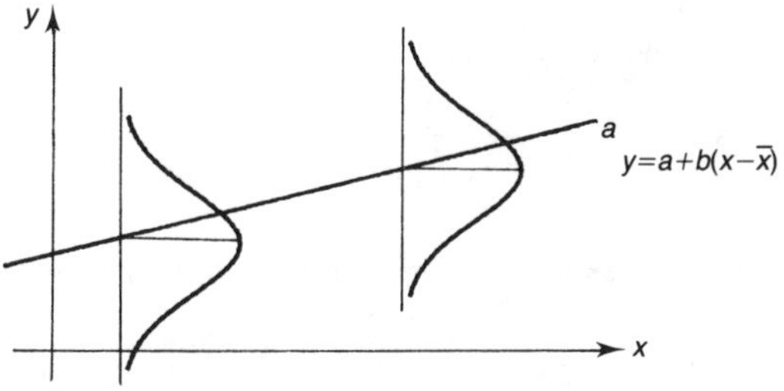

Mathematical model

$Y_1, Y_2, \ldots, Y_n$ for given $x_1, x_2, \ldots, x_n$ are independent and Y_i is $N(a+b(x_i-\overline{x}), \sigma)$, where $\overline{x}=\sum_{i=1}^{n} x_i/n$.

***UMVU* estimates**

$$\hat{a}=\sum_{i=1}^{n} Y_i/n=\overline{Y}$$

$$\hat{b}=\frac{\sum_{i=1}^{n}(x_i-\overline{x})(Y_i-\overline{Y})}{\sum_{i=1}^{n}(x_i-\overline{x})^2}=\frac{\sum_{i=1}^{n}(x_i-\overline{x})Y_i}{\sum_{i=1}^{n}(x_i-\overline{x})^2}=\frac{\sum_{i=1}^{n}x_iY_i-(\sum_{i=1}^{n}x_i)(\sum_{i=1}^{n}Y_i)/n}{\sum_{i=1}^{n}x_i^2-(\sum_{i=1}^{n}x_i)^2/n}$$

$$\hat{\sigma^2}=s^2=\frac{1}{n-2}\sum_{i=1}^{n}(Y_i-\hat{a}-\hat{b}(x_i-\overline{x}))^2=$$
$$=\frac{1}{n-2}\left(\sum_{i=1}^{n}(Y_i-\overline{Y})^2-\hat{b}^2\sum_{i=1}^{n}(x_i-\overline{x})^2\right)$$

$$\widehat{E[Y_i]}=\hat{a}+\hat{b}(x_i-\overline{x})=\hat{\mu}_i$$

Confidence intervals

Parameter	Pivot variable	Distribution of pivot variable	Two-sided confidence interval
a	$\frac{\hat{a}-a}{s}\sqrt{n}$	$t(n-2)$	$a=\hat{a}\pm xs/\sqrt{n}$
b	$\frac{\hat{b}-b}{s}\sqrt{\sum_{i=1}^{n}(x_i-\overline{x})^2}$	$t(n-2)$	$b=\hat{b}\pm xs/\sqrt{\sum_{i=1}^{n}(x_i-\overline{x})^2}$
σ^2	$\frac{(n-2)s^2}{\sigma^2}$	$\chi^2(n-2)$	$\frac{(n-2)s^2}{x_2}<\sigma^2<\frac{(n-2)s^2}{x_1}$
$E[Y_i]=\mu_i$	$\frac{\hat{\mu}_i-\mu_i}{s\sqrt{\frac{1}{n}+\frac{(x_i-\overline{x})^2}{\sum_{i=1}^{n}(x_i-\overline{x})^2}}}$	$t(n-2)$	$\mu_i=\hat{\mu}_i\pm xs\sqrt{\frac{1}{n}+\frac{(x_i-\overline{x})^2}{\sum_{i=1}^{n}(x_i-\overline{x})^2}}$

Prediction interval for Y for given x

$$\frac{Y-\hat{a}-\hat{b}(x-\bar{x})}{s\sqrt{1+\frac{1}{n}+\frac{(x-\bar{x})^2}{\sum_{i=1}^{n}(x_i-\bar{x})^2}}}$$

is $t(n-2)$, which gives the following prediction interval for Y

$$Y=\hat{a}+\hat{b}(x-\bar{x})\pm xs\sqrt{1+\frac{1}{n}+\frac{(x-\bar{x})^2}{\sum_{i=1}^{n}(x_i-\bar{x})^2}}$$

Testing the hypothesis H: $b=0$

The variable $T=\hat{b}\sqrt{\sum_{i=1}^{n}(x_i-\bar{x})^2}/s$ is $t(n-2)$ under the hypothesis H: $b=0$. If the hypotheses are

$$H: b=0 \qquad K: b\neq 0$$

H is rejected when $|T|>x_{1-\alpha/2}$.

Analysis of variance (*ANOVA*)

Completely randomized, one factor design

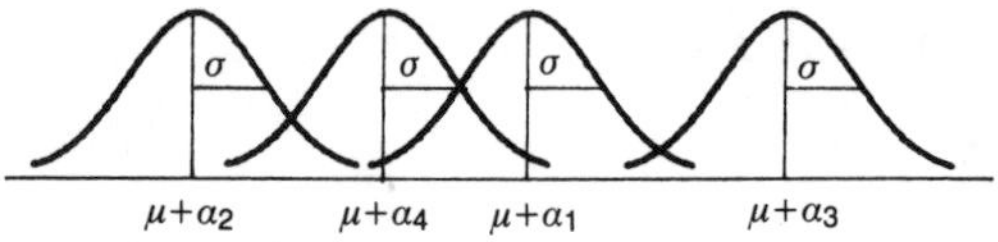

Mathematical model

Y_{ij} for $i=1, 2, \ldots, p$ and $j=1, 2, \ldots, n_i$ are independent and Y_{ij} is $N(\mu+\alpha_i, \sigma)$ with $\sum_{i=1}^{p}\alpha_i=0$

UMVU-estimates

$$\hat{a}_i=Y_{i\cdot}-\hat{\mu},\ \hat{\mu}=\sum_{i=1}^{p}Y_{i\cdot}/p \text{ with } Y_{i\cdot}=\sum_{j=1}^{n_i}Y_{ij}/n_i$$

$$\hat{\sigma}^2=s^2=\frac{1}{n-p}\sum_{i=1}^{p}\sum_{j=1}^{n_i}(Y_{ij}-Y_{i\cdot})^2 \quad n=\sum_{i=1}^{p}n_i$$

Confidence intervals

The pivot variable $(Y_{i\cdot}-\mu-\alpha_i)\sqrt{n_i}/s$ is $t(n-p)$, which gives the confidence interval

$$\mu+\alpha_i=Y_{i\cdot}\pm xs/\sqrt{n_i}$$

Simultaneous confidence intervals for all possible *contrasts* $\sum_{i=1}^{p} a_i\mu_i$ with $\mu_i=\mu+\alpha_i$ and $\sum_{i=1}^{p} a_i=0$ with simultaneous confidence level α are given by

$$\sum_{i=1}^{p} a_i\mu_i=\sum_{i=1}^{p} a_iY_{i\cdot}\pm As\sqrt{\sum_{i=1}^{p} a_i^2/n_i},$$

where $A=\sqrt{(p-1)x_\alpha}$ and x_α is the α-fractile of the $F(p-1,\ n-p)$ distribution.

***ANOVA* table**

Test of the hypothesis H: $\alpha_1=\alpha_2=\ldots=\alpha_p=0$ can be organized in an *ANOVA* table

Source	*Df*	Sum of squares	Mean squares	Expectation	*F*
Treatment	$p-1$	$SS_T=\sum_{i=1}^{p} n_i(Y_{i\cdot}-Y_{\cdot\cdot})^2$	$MS_T=SS_T/(p-1)$	$\sigma^2+\frac{1}{p-1}\sum_{i=1}^{p} n_i\alpha_i^2$	$\frac{MS_T}{MS_E}$
Error	$n-p$	$SS_E=\sum_{i=1}^{p}\sum_{j=1}^{n_i}(Y_{ij}-Y_{i\cdot})^2$	$MS_E=SS_E/(n-p)$	σ^2	
Total	$n-1$	$SS_T=\sum_{i=1}^{p}\sum_{j=1}^{n_i}(Y_{ij}-Y_{\cdot\cdot})^2$			

$$Y_{\cdot\cdot}=\sum_{i=1}^{p}\sum_{i=1}^{n_i} Y_{ij}/n,\quad n=\sum_{i=1}^{p} n_i$$

Completely randomized two factor design with equal number of observations per cell

Mathematical model

Y_{ijk} for $i=1, \ldots, p$, $j=1, \ldots, b$, $k=1, \ldots, c$ are independent and Y_{ijk} is $N(\mu+\alpha_i+\lambda_j,\ \sigma)$ with $\Sigma\,\alpha_i=\Sigma\,\lambda_j=0$. Instead of this *additive model*, a more general model with *interactions* Θ_{ij} can be used. Then Y_{ijk} is

$$N(\mu+\alpha_i+\lambda_j+\Theta_{ij},\ \sigma) \text{ where } \sum_{i=1}^{p}\Theta_{ij}=\sum_{j=1}^{b}\Theta_{ij}=0.$$

***UMVU*-estimates**

$\hat{\alpha}_i=Y_{i\cdot\cdot}-Y_{\cdot\cdot\cdot}$ $\qquad$ $\hat{\lambda}_j=Y_{\cdot j\cdot}-Y_{\cdot\cdot\cdot}$ $\qquad$ $\hat{\mu}=Y_{\cdot\cdot\cdot}$

$Y_{\cdot\cdot\cdot}=\sum_{i=1}^{p}\sum_{j=1}^{b}\sum_{k=1}^{c} Y_{ijk}/pbc$ $\qquad$ $Y_{i\cdot\cdot}=\sum_{j=1}^{b}\sum_{k=1}^{c} Y_{ijk}/bc$ $\qquad$ $Y_{\cdot j\cdot}=\sum_{i=1}^{p}\sum_{k=1}^{c} Y_{ijk}/pc$

***ANOVA* table**

Tests of the hypotheses H_1: $\alpha_1=\alpha_2=\ldots=\alpha_p=0$ and H_2: $\lambda_1=\lambda_2=\ldots=\lambda_b=0$ can be organized in an *ANOVA table.*

Source	Df	Sum of squares	Mean square	Expectation	F
First factor	$p-1$	$SS_1=cb\sum_{i=1}^{p}(Y_{i\cdot\cdot}-Y_{\cdot\cdot\cdot})^2$	$MS_1=\dfrac{SS_1}{p-1}$	$\sigma^2+\dfrac{cb\sum_{i=1}^{p}\alpha_i^2}{p-1}$	$\dfrac{MS_1}{MS_E}$
Second factor	$b-1$	$SS_2=cp\sum_{j=1}^{b}(Y_{\cdot j\cdot}-Y_{\cdot\cdot\cdot})^2$	$MS_2=\dfrac{SS_2}{b-1}$	$\sigma^2+\dfrac{cp\sum_{j=1}^{b}\lambda_j^2}{b-1}$	$\dfrac{MS_2}{MS_E}$
Error	$n-p-b+1$	$SSE=\sum_{i=1}^{p}\sum_{j=1}^{b}\sum_{k=1}^{c}(Y_{ijk}-Y_{i\cdot\cdot}-Y_{\cdot j\cdot}+Y_{\cdot\cdot\cdot})^2$	$MS_E=\dfrac{SS_E}{n-p-b+1}$	σ^2	
Total	$n-1$	$SS_T=\sum_{i=1}^{p}\sum_{j=1}^{b}\sum_{k=1}^{c}(Y_{ijk}-Y_{\cdot\cdot\cdot})^2$			

For $c>1$ the hypothesis H: $\Theta_{ij}=0$, $i=1, \ldots, p$, $j=1, \ldots, b$ can be tested using the testvariable

$$\frac{(n-pb)\,c\sum_{i=1}^{p}\sum_{j=1}^{b}(Y_{ij\cdot}-Y_{i\cdot\cdot}-Y_{\cdot j\cdot}+Y_{\cdot\cdot\cdot})^2}{(b-1)(p-1)\sum_{i=1}^{p}\sum_{j=1}^{b}\sum_{k=1}^{c}(Y_{ijk}-Y_{ij\cdot})^2},$$

which, under the hypotheses, is $F(n-pb, (b-1)(p-1))$.

The general linear model

$$\begin{bmatrix} Y_1 \\ Y_2 \\ \vdots \\ Y_n \end{bmatrix} = \begin{bmatrix} c_{11} & c_{12} & \cdots & c_{1p} \\ c_{21} & c_{22} & \cdots & c_{2p} \\ \cdots & \cdots & \cdots & \cdots \\ c_{n1} & c_{n1} & \cdots & c_{np} \end{bmatrix} \begin{bmatrix} \beta_1 \\ \beta_2 \\ \vdots \\ \beta_p \end{bmatrix} + \begin{bmatrix} \varepsilon_1 \\ \varepsilon_2 \\ \vdots \\ \varepsilon_n \end{bmatrix}$$

or

$$\boldsymbol{y}=C\boldsymbol{\beta}+\boldsymbol{\varepsilon}$$

where C is a known design matrix and $\varepsilon_1, \varepsilon_2, \ldots, \varepsilon_n$ are independent and $N(0, \sigma)$.

If Rank$(C)=p$, *UMVU* estimes of $\beta_1, \beta_2, \ldots, \beta_p$ are given by

$$\hat{\boldsymbol{\beta}}=(C^tC)^{-1}C^t\boldsymbol{y}$$

18.7 Distribution-free Methods

The Kolmogorov-Smirnov statistics D_n

$$D_n = \max_x |S_n(x) - F_0(x)|$$

S_n= empirical distribution function

F_0= distribution function under null hypothesis

F= distribution function of D_n, $F(x)=P(D_n \leq x)$

n	$F(x)=0.80$	0.85	0.90	0.95	0.99
1	0.900	0.925	0.950	0.975	0.995
2	0.684	0.726	0.776	0.842	0.929
3	0.565	0.597	0.636	0.708	0.829
4	0.493	0.525	0.565	0.624	0.734
5	0.447	0.474	0.510	0.563	0.669
6	0.410	0.436	0.468	0.520	0.617
7	0.381	0.405	0.436	0.483	0.576
8	0.358	0.381	0.410	0.454	0.542
9	0.339	0.360	0.387	0.430	0.513
10	0.323	0.342	0.369	0.409	0.489
11	0.308	0.326	0.352	0.391	0.468
12	0.296	0.313	0.338	0.375	0.450
13	0.285	0.302	0.325	0.361	0.432
14	0.275	0.292	0.314	0.349	0.418
15	0.266	0.283	0.304	0.338	0.404
16	0.258	0.274	0.295	0.327	0.392
17	0.250	0.266	0.286	0.318	0.381
18	0.244	0.259	0.279	0.309	0.371
19	0.237	0.252	0.271	0.301	0.361
20	0.232	0.246	0.265	0.294	0.352
25	0.208	0.22	0.238	0.264	0.317
30	0.190	0.20	0.218	0.242	0.290
35	0.177	0.19	0.202	0.224	0.269
Formula for large n	$\frac{1.07}{\sqrt{n}}$	$\frac{1.14}{\sqrt{n}}$	$\frac{1.22}{\sqrt{n}}$	$\frac{1.36}{\sqrt{n}}$	$\frac{1.63}{\sqrt{n}}$

The Wilcoxon statistic U (or W)

$n=$ number of X-observations
$m=$ number of Y-observations
$U=$ number of pairs (X_i, Y_j) with $X_i<Y_j$
$W=$ rank sum of the Y-observations

$$U=W-\frac{m(m+1)}{2}$$

$$E[U]=\frac{nm}{2} \qquad \mathrm{Var}[U]=\frac{nm(n+m+1)}{12}$$

The table gives

$$P(U\leqslant x)=P(U\geqslant nm-x)$$

$$r_1=\min(n, m) \qquad r_2=\max(n, m)$$

r_1	x	$r_2=3$	$r_2=4$	$r_2=5$	$r_2=6$	$r_2=7$	$r_2=8$
3	0	0.0500	0.0286	0.0179	0.0119	0.0083	0.0061
	1	0.1000	0.0571	0.0357	0.0238	0.0167	0.0121
	2	0.2000	0.1143	0.0714	0.0476	0.0333	0.0242
	3	0.3500	0.2000	0.1250	0.0833	0.0583	0.0424
	4	0.5000	0.3143	0.1964	0.1310	0.0917	0.0667
	5	0.6500	0.4286	0.2857	0.1905	0.1333	0.0970
	6	0.8000	0.5714	0.3929	0.2738	0.1917	0.1394
	7	0.9000	0.6857	0.5000	0.3571	0.2583	0.1879
	8	0.9500	0.8000	0.6071	0.4524	0.3333	0.2485
	9	1.0000	0.8857	0.7143	0.5476	0.4167	0.3152
	10		0.9429	0.8036	0.6429	0.5000	0.3879
	11		0.9714	0.8750	0.7262	0.5833	0.4606
4	0		0.0143	0.0079	0.0048	0.0030	0.0020
	1		0.0286	0.0159	0.0095	0.0061	0.0040
	2		0.0571	0.0317	0.0190	0.0121	0.0081
	3		0.1000	0.0556	0.0333	0.0212	0.0141
	4		0.1714	0.0952	0.0571	0.0364	0.0242
	5		0.2429	0.1429	0.0857	0.0545	0.0364
	6		0.3429	0.2063	0.1286	0.0818	0.0545
	7		0.4429	0.2778	0.1762	0.1152	0.0768
	8		0.5571	0.3651	0.2381	0.1576	0.1071
	9		0.6571	0.4524	0.3048	0.2061	0.1414
	10		0.7571	0.5476	0.3810	0.2636	0.1838
	11		0.8286	0.6349	0.4571	0.3242	0.2303
	12		0.9000	0.7222	0.5429	0.3939	0.2848
	13		0.9429	0.7937	0.6190	0.4636	0.3414
	14		0.9714	0.8571	0.6952	0.5364	0.4040
	15		0.9857	0.9048	0.7619	0.6061	0.4667

r_1	x	$r_2=5$	$r_2=6$	$r_2=7$	$r_2=8$
5	0	0.0040	0.0022	0.0013	0.0008
	1	0.0079	0.0043	0.0025	0.0016
	2	0.0159	0.0087	0.0051	0.0031
	3	0.0278	0.0152	0.0088	0.0054
	4	0.0476	0.0260	0.0152	0.0093
	5	0.0754	0.0411	0.0240	0.0148
	6	0.1111	0.0628	0.0366	0.0225
	7	0.1548	0.0887	0.0530	0.0326
	8	0.2103	0.1234	0.0745	0.0466
	9	0.2738	0.1645	0.1010	0.0637
	10	0.3452	0.2143	0.1338	0.0855
	11	0.4206	0.2684	0.1717	0.1111
	12	0.5000	0.3312	0.2159	0.1422
	13	0.5794	0.3961	0.2652	0.1772
	14	0.6548	0.4654	0.3194	0.2176
	15	0.7262	0.5346	0.3775	0.2618
	16	0.7897	0.6039	0.4381	0.3108
	17	0.8452	0.6688	0.5000	0.3621
	18	0.8889	0.7316	0.5619	0.4165
	19	0.9246	0.7857	0.6225	0.4716
6	0		0.0011	0.0006	0.0003
	1		0.0022	0.0012	0.0007
	2		0.0043	0.0023	0.0013
	3		0.0076	0.0041	0.0023
	4		0.0130	0.0070	0.0040
	5		0.0206	0.0111	0.0063
	6		0.0325	0.0175	0.0100
	7		0.0465	0.0256	0.0147
	8		0.0660	0.0367	0.0213
	9		0.0898	0.0507	0.0296
	10		0.1201	0.0688	0.0406
	11		0.1548	0.0903	0.0539
	12		0.1970	0.1171	0.0709
	13		0.2424	0.1474	0.0906
	14		0.2944	0.1830	0.1142
	15		0.3496	0.2226	0.1412
	16		0.4091	0.2669	0.1725
	17		0.4686	0.3141	0.2068
	18		0.5314	0.3654	0.2454
	19		0.5909	0.4178	0.2864
	20		0.6504	0.4726	0.3310
	21		0.7056	0.5274	0.3773
	22		0.7576	0.5822	0.4259
	23		0.8030	0.6346	0.4749

r_1	x	$r_2=7$	$r_2=8$
7	0	0.0003	0.0002
	1	0.0006	0.0003
	2	0.0012	0.0006
	3	0.0020	0.0011
	4	0.0035	0.0019
	5	0.0055	0.0030
	6	0.0087	0.0047
	7	0.0131	0.0070
	8	0.0189	0.0103
	9	0.0265	0.0145
	10	0.0364	0.0200
	11	0.0487	0.0270
	12	0.0641	0.0361
	13	0.0825	0.0469
	14	0.1043	0.0603
	15	0.1297	0.0760
	16	0.1588	0.0946
	17	0.1914	0.1159
	18	0.2279	0.1405
	19	0.2675	0.1678
	20	0.3100	0.1984
	21	0.3552	0.2317
	22	0.4024	0.2679
	23	0.4508	0.3063
	24	0.5000	0.3472
	25	0.5492	0.3894
	26	0.5976	0.4333
	27	0.6448	0.4775

r_1	x	$r_2=8$
8	0	0.0001
	1	0.0002
	2	0.0003
	3	0.0005
	4	0.0009
	5	0.0015
	6	0.0023
	7	0.0035
	8	0.0052
	9	0.0074
	10	0.0103
	11	0.0141
	12	0.0190
	13	0.0249
	14	0.0325
	15	0.0415
	16	0.0524
	17	0.0652
	18	0.0803
	19	0.0974
	20	0.1172
	21	0.1393
	22	0.1641
	23	0.1911
	24	0.2209
	25	0.2527
	26	0.2869
	27	0.3227
	28	0.3605
	29	0.3992
	30	0.4392
	31	0.4796

The Wilcoxon sign rank statistic

$n=$ number of differences $X_i - Y_i = Z_i$

$R_i = \pm$rank of $|Z_i|$ sign of R_i=sign of Z_i

$$W = \frac{1}{2}\sum_{i=1}^{n} R_i + \frac{n(n+1)}{4}$$

The table gives

$$P(W \leqslant x) = P\left(W \geqslant \frac{n(n+1)}{2} - x\right)$$

x	$n=1$	$n=2$	$n=3$	$n=4$	$n=5$	$n=6$	$n=7$
0	0.5000	0.2500	0.1250	0.0625	0.0313	0.0156	0.0078
1		0.5000	0.2500	0.1250	0.0625	0.0313	0.0156
2			0.3750	0.1875	0.0938	0.0469	0.0234
3				0.3125	0.1563	0.0781	0.0391
4				0.4375	0.2188	0.1094	0.0547
5					0.3125	0.1563	0.0781
6					0.4063	0.2188	0.1094
7					0.5000	0.2813	0.1484
8						0.3438	0.1875
9						0.4219	0.2344
10						0.5000	0.2891
11							0.3438
12							0.4063
13							0.4688

x	$n=8$	$n=9$	$n=10$	$n=11$	$n=12$	$n=13$	$n=14$
0	0.0039	0.0020	0.0010	0.0005	0.0002	0.0001	0.0001
1	0.0078	0.0039	0.0020	0.0010	0.0005	0.0002	0.0001
2	0.0117	0.0059	0.0029	0.0015	0.0007	0.0004	0.0002
3	0.0195	0.0098	0.0049	0.0024	0.0012	0.0006	0.0003
4	0.0273	0.0137	0.0068	0.0034	0.0017	0.0009	0.0004
5	0.0391	0.0195	0.0098	0.0049	0.0024	0.0012	0.0006
6	0.0547	0.0273	0.0137	0.0068	0.0034	0.0017	0.0009
7	0.0742	0.0371	0.0186	0.0093	0.0046	0.0023	0.0012
8	0.0977	0.0488	0.0244	0.0122	0.0061	0.0031	0.0015
9	0.1250	0.0645	0.0322	0.0161	0.0081	0.0040	0.0020
10	0.1563	0.0820	0.0420	0.0210	0.0105	0.0052	0.0026
11	0.1914	0.1016	0.0527	0.0269	0.0134	0.0067	0.0034
12	0.2305	0.1250	0.0654	0.0337	0.0171	0.0085	0.0043
13	0.2734	0.1504	0.0801	0.0415	0.0212	0.0107	0.0054

14	0.3203	0.1797	0.0967	0.0508	0.0261	0.0133	0.0067
15	0.3711	0.2129	0.1162	0.0615	0.0320	0.0164	0.0083
16	0.4219	0.2480	0.1377	0.0737	0.0386	0.0199	0.0101
17	0.4727	0.2852	0.1611	0.0874	0.0461	0.0239	0.0123
18		0.3262	0.1875	0.1030	0.0549	0.0287	0.0148
19		0.3672	0.2158	0.1201	0.0647	0.0341	0.0176
20		0.4102	0.2461	0.1392	0.0757	0.0402	0.0209
21		0.4551	0.2783	0.1602	0.0881	0.0471	0.0247
22		0.5000	0.3125	0.1826	0.1018	0.0549	0.0290
23			0.3477	0.2065	0.1167	0.0636	0.0338
24			0.3848	0.2324	0.1331	0.0732	0.0392
25			0.4229	0.2598	0.1506	0.0839	0.0453
26			0.4609	0.2886	0.1697	0.0955	0.0520
27			0.5000	0.3188	0.1902	0.1082	0.0594
28				0.3501	0.2119	0.1219	0.0676
29				0.3823	0.2349	0.1367	0.0765
30				0.4155	0.2593	0.1527	0.0863
31				0.4492	0.2847	0.1698	0.0969
32				0.4829	0.3110	0.1879	0.1083
33					0.3386	0.2072	0.1206
34					0.3667	0.2274	0.1338
35					0.3955	0.2487	0.1479
36					0.4250	0.2709	0.1629
37					0.4548	0.2939	0.1788
38					0.4849	0.3177	0.1955
39						0.3424	0.2131
40						0.3677	0.2316
41						0.3934	0.2508
42						0.4197	0.2708
43						0.4463	0.2915
44						0.4730	0.3129
45						0.5000	0.3349
46							0.3574
47							0.3804
48							0.4039
49							0.4276
50							0.4516
51							0.4758
52							0.5000

Confidence interval for the median

Let $x_{(1)} \leqslant x_{(2)} \leqslant \ldots \leqslant x_{(n)}$ be an ordered sample from a continuous distribution with median m. Then

$$x_{(k)} \leqslant m \leqslant x_{(n-k+1)}$$

is a confidence interval for m. The following table gives values of k such that the level of confidence is close to 0.95. The exact confidence level is also given in the table.

n	k	Confidence interval	Level of confidence
5	1	$x_{(1)} \leqslant m \leqslant x_{(5)}$	0.938
6	1	$x_{(1)} \leqslant m \leqslant x_{(6)}$	0.969
7	1	$x_{(1)} \leqslant m \leqslant x_{(7)}$	0.984
8	2	$x_{(2)} \leqslant m \leqslant x_{(7)}$	0.930
9	2	$x_{(2)} \leqslant m \leqslant x_{(8)}$	0.961
10	2	$x_{(2)} \leqslant m \leqslant x_{(9)}$	0.979
11	3	$x_{(3)} \leqslant m \leqslant x_{(9)}$	0.935
12	3	$x_{(3)} \leqslant m \leqslant x_{(10)}$	0.961
13	3	$x_{(3)} \leqslant m \leqslant x_{(11)}$	0.978
14	4	$x_{(4)} \leqslant m \leqslant x_{(11)}$	0.943
15	4	$x_{(4)} \leqslant m \leqslant x_{(12)}$	0.965
16	4	$x_{(4)} \leqslant m \leqslant x_{(13)}$	0.979
17	5	$x_{(5)} \leqslant m \leqslant x_{(13)}$	0.951
18	5	$x_{(5)} \leqslant m \leqslant x_{(14)}$	0.969
19	6	$x_{(6)} \leqslant m \leqslant x_{(14)}$	0.936
20	6	$x_{(6)} \leqslant m \leqslant x_{(15)}$	0.959
21	6	$x_{(6)} \leqslant m \leqslant x_{(16)}$	0.973
22	7	$x_{(7)} \leqslant m \leqslant x_{(16)}$	0.948
23	7	$x_{(7)} \leqslant m \leqslant x_{(17)}$	0.965
24	8	$x_{(8)} \leqslant m \leqslant x_{(17)}$	0.936
25	8	$x_{(8)} \leqslant m \leqslant x_{(18)}$	0.957
26	8	$x_{(8)} \leqslant m \leqslant x_{(19)}$	0.971
27	9	$x_{(9)} \leqslant m \leqslant x_{(19)}$	0.948
28	9	$x_{(9)} \leqslant m \leqslant x_{(20)}$	0.964
29	10	$x_{(10)} \leqslant m \leqslant x_{(20)}$	0.939
30	10	$x_{(10)} \leqslant m \leqslant x_{(21)}$	0.957
31	10	$x_{(10)} \leqslant m \leqslant x_{(22)}$	0.971
32	11	$x_{(11)} \leqslant m \leqslant x_{(22)}$	0.950
33	11	$x_{(11)} \leqslant m \leqslant x_{(23)}$	0.965
34	12	$x_{(12)} \leqslant m \leqslant x_{(23)}$	0.942
35	12	$x_{(12)} \leqslant m \leqslant x_{(24)}$	0.959
36	13	$x_{(13)} \leqslant m \leqslant x_{(24)}$	0.935
37	13	$x_{(13)} \leqslant m \leqslant x_{(25)}$	0.953
38	13	$x_{(13)} \leqslant m \leqslant x_{(26)}$	0.966
39	14	$x_{(14)} \leqslant m \leqslant x_{(26)}$	0.947
40	14	$x_{(14)} \leqslant m \leqslant x_{(27)}$	0.962

18.8 Statistical Quality Control

Factors for computing control chart lines

The table gives factors for computing central line (C_L) and control lines (C_u and C_l) for control charts according to the 3σ-rule. Sample size n.

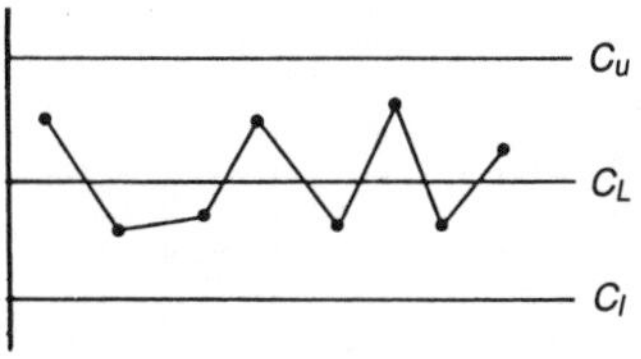

$\bar{x}$-chart

1. $C_L=\mu_0$
   $C_l=\mu_0-A\sigma_0$
   $C_u=\mu_0+A\sigma_0$

2. $C_L=\bar{\bar{x}}$
   $C_l=\bar{\bar{x}}-A_2\bar{R}$
   $C_u=\bar{\bar{x}}+A_2\bar{R}$

R-chart

1. $C_L=d_2\sigma_0$
   $C_l=d_2\sigma_0-3d_3\sigma_0=D_1\sigma_0$
   $C_u=d_2\sigma_0+3d_3\sigma_0=D_2\sigma_0$

2. $C_L=\bar{R}$
   $C_l=\bar{R}-3d_3\bar{R}/d_2=D_3\bar{R}$
   $C_u=\bar{R}+3d_3\bar{R}/d_2=D_4\bar{R}$

	$\bar{x}$-chart		R-chart					
n	A	A_2	d_2	d_3	D_1	D_2	D_3	D_4
2	2.121	1.880	1.128	0.853	0	3.686	0	3.267
3	1.732	1.023	1.693	0.888	0	4.358	0	2.575
4	1.500	0.729	2.059	0.880	0	4.698	0	2.282
5	1.342	0.577	2.326	0.864	0	4.918	0	2.115
6	1.225	0.483	2.534	0.848	0	5.078	0	2.004
7	1.134	0.419	2.704	0.833	0.205	4.203	0.076	1.924
8	1.061	0.373	2.847	0.820	0.387	5.307	0.136	1.864
9	1.000	0.337	2.970	0.808	0.546	5.394	0.184	1.816
10	0.949	0.308	3.078	0.797	0.687	5.469	0.223	1.777
11	0.905	0.285	3.173	0.787	0.812	5.534	0.256	1.744
12	0.866	0.266	3.258	0.778	0.924	5.592	0.284	1.716
13	0.832	0.249	3.336	0.770	1.026	5.646	0.308	1.692
14	0.802	0.235	3.407	0.762	1.121	5.693	0.329	1.671
15	0.775	0.223	3.472	0.755	1.207	5.737	0.384	1.652
16	0.750	0.212	3.532	0.749	1.285	5.779	0.364	1.636
17	0.728	0.203	3.588	0.743	1.359	5.817	0.379	1.621
18	0.707	0.194	3.640	0.738	1.426	5.854	0.392	1.608
19	0.688	0.187	3.689	0.733	1.490	5.888	0.404	1.596
20	0.671	0.180	3.735	0.729	1.548	5.922	0.414	1.586

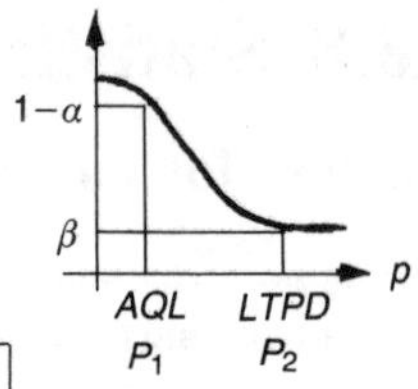

Table for construction of single acceptance sampling control plans

	Values of p_2/p_1 for:					Values of p_2/p_1 for:			
c	α=0.05 β=0.10	α=0.05 β=0.05	α=0.05 β=0.01	np_1	c	α=0.01 β=0.10	α=0.01 β=0.05	α=0.01 β=0.01	np_1
0	44.890	58.404	89.781	0.052	0	229.105	298.073	458.210	0.010
1	10.946	13.349	18.681	0.355	1	26.184	31.933	44.686	0.149
2	6.509	7.699	10.280	0.818	2	12.206	14.439	19.278	0.436
3	4.890	5.675	7.352	1.366	3	8.115	9.418	12.202	0.823
4	4.057	4.646	5.890	1.970	4	6.249	7.156	9.072	1.279
5	3.549	4.023	5.017	2.613	5	5.195	5.889	7.343	1.785
6	3.206	3.604	4.435	3.286	6	4.520	5.082	6.253	2.330
7	2.957	3.303	4.019	3.981	7	4.050	4.524	5.506	2.906
8	2.768	3.074	3.707	4.695	8	3.705	4.115	4.962	4.130
9	2.618	2.895	3.462	5.426	9	3.440	3.803	4.548	4.130
10	2.497	2.750	3.265	6.169	10	3.229	3.555	4.222	4.771
11	2.397	2.630	3.104	6.924	11	3.058	3.354	3.959	5.428
12	2.312	2.528	2.968	7.690	12	2.915	3.188	3.742	6.099
13	2.240	2.442	2.852	8.464	13	2.795	3.047	3.559	6.782
14	2.177	2.367	2.752	9.246	14	2.692	2.927	3.403	7.477
15	2.122	2.302	2.665	10.035	15	2.603	2.823	3.269	8.181
16	2.073	2.244	2.588	10.831	16	2.524	2.732	3.151	8.895
17	2.029	2.192	2.520	11.633	17	2.455	2.652	3.048	9.616
18	1.990	2.145	2.458	12.442	18	2.393	2.580	2.956	10.346
19	1.954	2.103	2.403	13.254	19	2.337	2.516	2.874	11.082
20	1.922	2.065	2.352	14.072	20	2.287	2.458	2.799	11.825
21	1.892	2.030	2.307	14.894	21	2.241	2.405	2.733	12.574
22	1.865	1.999	2.265	15.719	22	2.200	2.357	2.671	13.329
23	1.840	1.969	2.226	16.548	23	2.162	2.313	2.615	14.088
24	1.817	1.942	2.191	17.382	24	2.126	2.272	2.564	14.853
25	1.795	1.917	2.158	18.218	25	2.094	2.235	2.516	15.623
26	1.775	1.893	2.127	19.058	26	2.064	2.200	2.472	16.397
27	1.757	1.871	2.098	19.900	27	2.035	2.168	2.431	17.175
28	1.739	1.850	2.071	20.746	28	2.009	2.138	2.393	17.957
29	1.723	1.831	2.046	21.594	29	1.985	2.110	2.358	18.742
30	1.707	1.813	2.023	22.444	30	1.962	2.083	2.324	19.532
31	1.692	1.796	2.001	23.298	31	1.940	2.059	2.293	20.324
32	1.679	1.780	1.980	24.152	32	1.920	2.035	2.264	21.120
33	1.665	1.764	1.960	25.010	33	1.900	2.013	2.236	21.919
34	1.653	1.750	1.941	25.870	34	1.882	1.992	2.210	22.721
35	1.641	1.736	1.923	26.731	35	1.865	1.973	2.185	23.525
36	1.630	1.723	1.906	27.594	36	1.848	1.954	2.162	24.333
37	1.619	1.710	1.890	28.460	37	1.833	1.936	2.139	25.143
38	1.609	1.698	1.875	29.327	38	1.818	1.920	2.118	25.955
39	1.599	1.687	1.860	30.196	39	1.804	1.903	2.098	26.770

Numerical example

We want to design a single sampling plan with

$$p_1=0.013,\ \alpha=0.05,\ p_2=0.054,\ \beta=0.10$$

In this case $p_2/p_1=4.15$. Choose in the table for $\alpha=0.05$, $\beta=0.10$, the value of p_2/p_1 which is closest to 4.15, that is 4.057. This gives $c=4$ and $np_1=1.970$. Thus $n=1.970/0.013=151.2$. The plan is then to sample 151 units and reject if more than 4 units are defect.

18.9 Factorial Experiments

The complete 2^3 factorial design

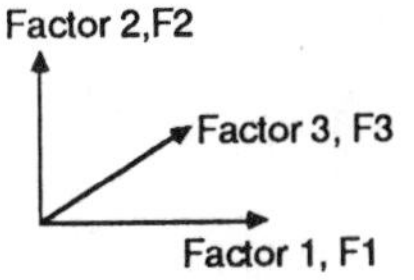

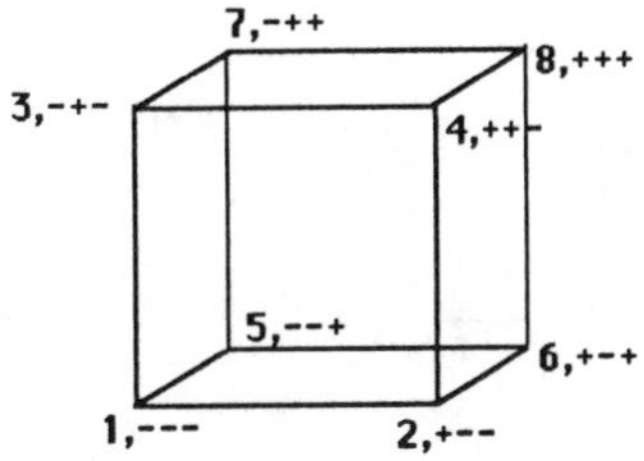

Standard order of runs

Run	$F1$	$F2$	$F3$	Observed effect
1	−	−	−	Y_1
2	+	−	−	Y_2
3	−	+	−	Y_3
4	+	+	−	Y_4
5	−	−	+	Y_5
6	+	−	+	Y_6
7	−	+	+	Y_7
8	+	+	+	Y_8

Estimation of effects

$F1$	$F2$	$F3$	$F1\times F2$	$F2\times F3$	$F1\times F3$	$F1\times F2\times F3$	Y
−	−	−	+	+	+	−	Y_1
+	−	−	−	+	−	+	Y_2
−	+	−	−	−	+	+	Y_3
+	+	−	+	−	−	−	Y_4
−	−	+	+	−	−	+	Y_5
+	−	+	−	−	+	−	Y_6
−	+	+	−	+	−	−	Y_7
+	+	+	+	+	+	+	Y_8

Effect	Estimator
$F1$	$(-Y_1+Y_2-Y_3+Y_4-Y_5+Y_6-Y_7+Y_8)/4$
$F2$	$(-Y_1-Y_2+Y_3+Y_4-Y_5-Y_6+Y_7+Y_8)/4$
$F3$	$(-Y_1-Y_2-Y_3-Y_4+Y_5+Y_6+Y_7+Y_8)/4$
$F1\times F2$ interaction	$(+Y_1-Y_2-Y_3+Y_4+Y_5-Y_6-Y_7+Y_8)/4$
$F2\times F3$ interaction	$(+Y_1+Y_2-Y_3-Y_4-Y_5-Y_6+Y_7+Y_8)/4$
$F1\times F3$ interaction	$(+Y_1-Y_2+Y_3-Y_4-Y_5+Y_6-Y_7+Y_8)/4$
$F1\times F2\times F3$ interaction	$(-Y_1+Y_2+Y_3-Y_4+Y_5-Y_6-Y_7+Y_8)/4$

A 2^3 factorial design with blocking

In the following design block levels *A* and *B* and the three factor interaction effect $F1 \times F2 \times F3$ are confounded.

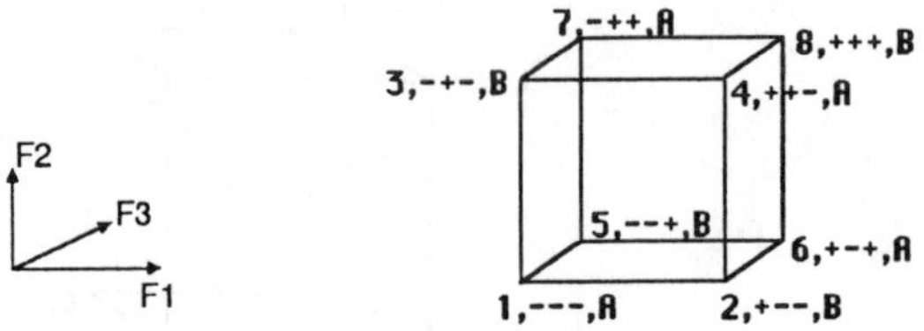

*F*1	*F*2	*F*3	*F*1×*F*2	*F*2×*F*3	*F*1×*F*3	*F*1×*F*2×*F*3	Block
−	−	−	+	+	+	−	*A*
+	−	−	−	+	−	+	*B*
−	+	−	−	−	+	+	*B*
+	+	−	+	−	−	−	*A*
−	−	+	+	−	−	+	*B*
+	−	+	−	−	+	−	*A*
−	+	+	−	+	−	−	*A*
+	+	+	+	+	+	+	*B*

Half fractional factorial designs

A $(2_{III})^{3-1}$ factorial design has $2^{-1} \cdot 2^3 = 4$ runs and the resolution is III, which means that main effects are not confounded with each other but only with two and three factor interactions. The following diagrams describe two such designs. The second is the *complementary* of the first design.

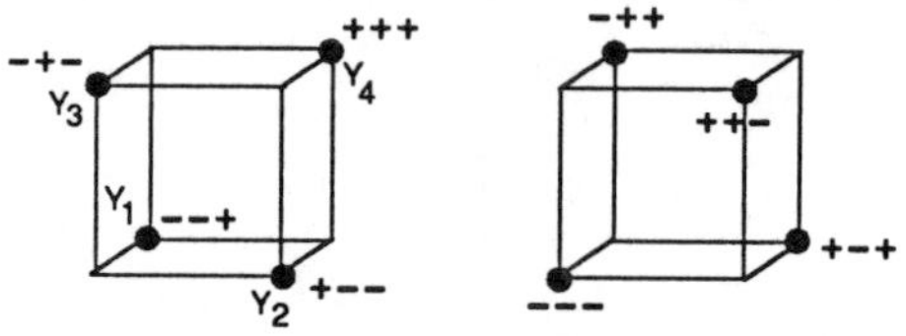

Design table for the first case.

*F*1	*F*2	*F*3	Observation
−	−	+	Y_1
+	−	−	Y_2
−	+	−	Y_3
+	+	+	Y_4

The confounding pattern is found from the *generator* 3=12 by multiplication first by 3 and then by 1 and 2.

I = 123   1=23   2=13   3=12

$(-Y_1 + Y_2 - Y_3 + Y_4)/2$ estimates the effect of $F1 + F2 \times F3$

$(-Y_1 - Y_2 + Y_3 + Y_4)/2$ estimates the effect of $F2 + F1 \times F3$

$(+Y_1 - Y_2 - Y_3 + Y_4)/2$ estimates the effect of $F3 + F1 \times F2$

The following table describes the best half fractional designs for 4 and 5 variables.

$(2_{IV})^{4-1}$

*F*1	*F*2	*F*3	*F*4
−	−	−	−
+	−	−	+
−	+	−	+
+	+	−	−
−	−	+	+
+	−	+	−
−	+	+	−
+	+	+	+

4=123

I=1234

$(2_{V})^{5-1}$

*F*1	*F*2	*F*3	*F*4	*F*5
−	−	−	−	+
+	−	−	−	−
−	+	−	−	−
+	+	−	−	+
−	−	+	−	−
+	−	+	−	+
−	+	+	−	+
+	+	+	−	−
−	−	−	+	−
+	−	−	+	+
−	+	−	+	+
+	+	−	+	−
−	−	+	+	+
+	−	+	+	−
−	+	+	+	−
+	+	+	+	+

5=1234

I=12345

$(2_R)^{k-n}$ fractional factorial designs

$(2_R)^{k-n}$ fractional factorial designs with *N* runs					
Design	*N*	Generators	Design	*N*	Generators
$(2_{III})^{3-1}$	4	3=12	$(2_{IV})^{6-1}$	32	6=12345
$(2_{IV})^{4-1}$	8	4=123	$(2_{III})^{7-4}$	8	4=12, 5=13, 6=23,7=123
$(2_{III})^{5-2}$	8	4=12 , 5=13	$(2_{IV})^{7-3}$	16	5=123, 6=234, 7=134
$(2_{V})^{5-1}$	16	5=1234	$(2_{IV})^{7-2}$	32	6=1234, 7=1245
$(2_{III})^{6-3}$	8	4=12 , 5=13 , 6=23	$(2_{VII})^{7-1}$	64	7=123456
$(2_{IV})^{6-2}$	16	5=123 , 6=234			

The resolution is *R* if no *p*-factor effect (*p*-factor interaction) is confounded with other effects with less than *R*-*p* factors.

Example. A $(2_{III})^{6-3}$ design has the generators 4=12, 5=13, 6=23. Multiplication gives

I = 124 = 135 = 236 = 2345 = 1346 = 1256 = 456

Multiplication by 1, 2,… gives the confounding pattern

1 = 24 = 35 = 1236 = 12345 = 346 = 256 = 1456
2 = 14 = 1235 = 36 = 345 = 12346 = 156 = 2456
and so on.

18.10 Statistical glossary

Admissibility, a statistical method is admissible if there does not exist in the considered class of methods another method which performs uniformly at least as well as the considered method.

AEDL, Average Extra Defectives Limit, for a continuous sampling plan CSP the AEDL measure is the expected extra number of defectives passed above the AOQL.

AIC, Akaike Information Criterion is definied as follows.

$$\text{AIC} = -2\times(\text{maximum loglikelihood of the model}) + \\ +2\times(\text{number of free parameters of the model})$$

AIC can be used as a basis of comparison and selection among several models. A model that minimizes the AIC can be considered to be the most appropriate model.

AOQ, Average Outgoing Quality, the expected quality of the outgoing product following the use of an acceptance sampling plan.

AOQL, Average Outgoing Quality Limit, maximum of AOQ over all possible levels of incoming quality.

AQL, Acceptable Quality Level, the maximum percent defective that can be considerd satisfactory for a production process.

ARL, Average Run Length, the expected length of time a production process will run before a control procedure will indicate a shift in the process level.

ASN, Average Sample Number, the expected number of observations until decision in a multistage statistical method.

ATI, Average Total Inspection.

Block, in experimental design the material is divided in a number of blocks to isolate sources of heterogeneity.

BLUE, Best Linear Unbiased Estimator.

Bootstrapping, computer intensive method based on resampling of an observed sample.

Brownian motion, a stochastic process $X(t)$ is a Brownian motion with drift parameter μ and variance parameter σ^2 if 1) $X(t)-X(0)$ is $N(\mu t, \sigma\sqrt{t})$ and 2) $X(t_1)-X(t_0)$, $X(t_2)-X(t_1)$, ..., $X(t_n)-X(t_{n-1})$ with $0 \leq t_0 \leq t_1 \leq \ldots \leq t_n$ are independent.

Burn in, a selection method used in reliability testing to improve the reliability of the items.

Censored data, a sample is censored when not all the observations in the sample are known. In type 1 censoring (to the right) observations above a certain level are not recorded. In type 2 censoring the sampling is stopped when the rth-ordered observation has been obtained. There exist special statistical methods to deal with censored data.

Chernoff faces, a graphical presentation of multivariate observations by a cartoon of a face. Up to 18 dimensions can be recorded.

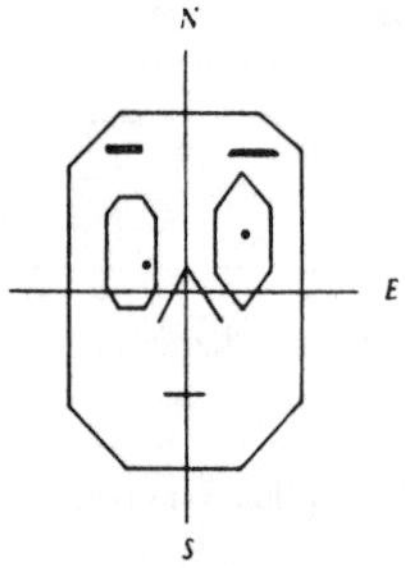

Variable	*Feature*
1	*NW* angle of face
2	*NE* angle of face
3	Left eyebrow
4	Right eyebrow
5	Socket of left eye
6	Pupil of left eye
7	Socket of right eye
8	Pupil of right eye
9	Shape of nose
10	Shape of mouth
11	*SW* angle of face
12	*SE* angle of face

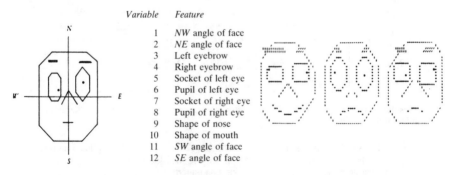

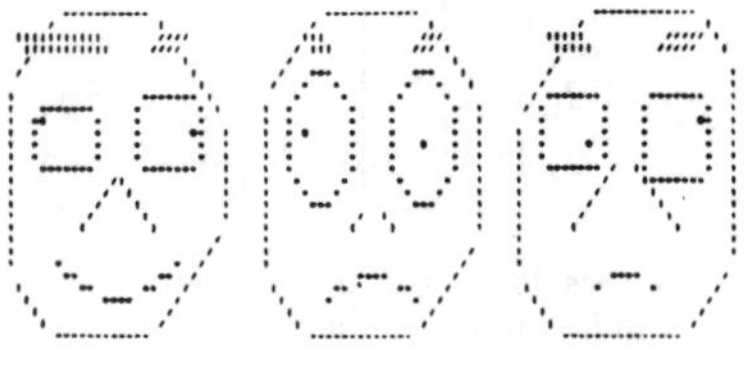

Crossvalidation, the method to divide a sample at random in two parts and to use one part for model fitting and the other part for model validation.

Confounding, in experimental design when certain comparisons can be made only for treatments in combinations and not for separate treatments, those treatment effects are said to be confounded.

Counter models, statistical models for counters used to register impulses such that registration of an impulse causes a certain dead time. In a type 1 counter process impulses arrived during a dead time do not cause any dead time but in a type 2 counter process each arriving impulse causes a dead time.

CSP, Continuous Sampling Plan, a type of sampling inspection when the production process is continuous.

DGP, Data Generating Process.

EDA, Exploratory Data Analysis.

EDF, Empirical Distribution Function, the EDF $\hat{F}$ of a sample $(X_1, X_2, \ldots, X_n)$ is defined as $\hat{F}(x) = (\text{Number of } X_i \leq x)/n$.

Efficiency, let $\hat{\theta}_1$ and $\hat{\theta}_2$ be consistent estimators of a parameter θ. Then the relative efficiency of $\hat{\theta}_1$ with respect to $\hat{\theta}_2$ is defined as $\text{Var}[\hat{\theta}_2]/\text{Var}[\hat{\theta}_1]$.

EVOP, Evolutionary Operation, a technique for the optimisation of a production process. The method is based on routines for systematic small changes of the process variables.

Extreme value distribution, distribution of the largest (smallest) observation in a sample. There exists three different types of asymptotic extreme value distributions.

FTA, Fault Tree Analysis, a graphical technique used for qualitative and quantitative reliability analysis of technical systems.

Graeco-Latin square, in a Graeco-Latin square with e.g. symbols A, B, C, α, β and γ each Latin letter occurs exactly once in each row and each column, each Greek letter occurs exactly once in each row and each column and a Greek letter occurs exactly once with each Latin letter.

$A\alpha$	$B\beta$	$C\gamma$
$B\gamma$	$C\alpha$	$A\beta$
$C\beta$	$A\gamma$	$B\alpha$

$A\alpha$	$B\beta$	$C\gamma$	$D\delta$
$B\delta$	$A\gamma$	$D\beta$	$C\alpha$
$C\beta$	$D\alpha$	$A\delta$	$B\gamma$
$D\gamma$	$C\delta$	$A\alpha$	$B\beta$

Graeco-Latin squares are used in experimental design to eliminate the effects of three blocking variables.

Homoscedasticity, the property of random variables to have equal variance. The contrary property is called hetereoscedasticity.

IID, Independently and Identically Distributed.

IQL, Indifference Quality Level, for a given sampling inspection plan IQL is that quality of an incoming batch for which the acceptance probability is 0.5.

Isotonic regression function, a function f is isotonic on a set A if $x<y \Rightarrow f(x) \leq f(y)$ for x, y in A. The isotonic regression function g with weights w on A with respection to f is the function g which minimizes $\sum_x w(x)(g(x)-f(x))^2$ in the class of isotonic functions on A.

Jackknife methods, certain methods to reduce bias in estimators. For example, let the estimator T_n for a sample of n observations be biassed to order $1/n$ and let $(T_{n-1}^{(i)})$ be the estimator when observation no i is omitted. Then the jackknife J calculated as the mean of $nT_n-(n-1)(T_{n-1}^{(i)})$ is biassed to at most the order $1/n^2$.

Kalman filter, a recursive unbiased least-squares estimator (predictor) of a Gaussian random signal, widely used e.g. in aerospace applications.

Kaplan-Meier estimator, a non-parametric estimator of the survival function based on censored data.

Latin square, a latin square of n symbols is an $n \times n$ arrangement such that each row and each column has each symbol exactly once. Latin squares are used in experimental design to eliminate the effects of two blocking variables.

$$\begin{array}{ccc} A & B & C \\ B & C & A \\ C & A & B \end{array} \qquad \begin{array}{cccc} A & B & C & D \\ B & A & D & C \\ C & D & A & B \\ D & C & B & A \end{array} \qquad \begin{array}{ccccc} A & B & C & D & E \\ B & C & D & E & A \\ C & D & E & A & B \\ D & E & A & B & C \\ E & A & B & C & D \end{array} \qquad \begin{array}{cccccc} A & B & C & D & E & F \\ B & A & F & E & C & D \\ C & F & B & A & D & E \\ D & C & E & B & F & A \\ E & D & A & F & B & C \\ F & E & D & C & A & B \end{array}$$

Permutations of rows or columns give another Latin square.

LTPD, Lot Tolerance Percent Defective, quality level at which acceptance probability is equal to the consumer's risk.

Martingale, a sequence of random variables $X_1, X_2, \ldots$ is a martingale if

$$E[X_{n+1}|X_1, X_2, \ldots, X_n]=X_n, \; n \geq 1.$$

Nominal data, statistical data which are numbers of observed classes in a classification.

OLSPCS, On Line Statistical Process Control System.

Ordinal data, statistical data that has a natural ordering of the possible values but for which the distances between the values is not defined.

Outlier, observation in a sample which is so separated from the other observations so it can be believed to be from a separate distribution. There exists special statistical outlier tests.

Pattern recognition, identification of patterns by automatic means.

PERT, Program Evaluation and Review Technique, a graphical method used for planning and controlling major projects.

Rank correlation, rank correlation measures the degree of correspondence between two rankings (orderings). The most used rank correlation coefficients are Kendalls τ and Spearmans ϱ.

Renewal process, a stochastic point process in which times between events (renewals) are IID.

Renewal theorem, in its simplest form it states that if $X(t)$ is the number of renewals in $(0, t)$ and μ is the expected time between renewals then $E[X(t)]/t \rightarrow 1/\mu$ $(t \rightarrow \infty)$.

Robustness, a statistical method is robust if it is not sensitive tŏ departures from those assumptions on which it was derived.

Semi-Markov process, a stochastic process that changes states according to a Markov chain, but in which the lengths of time spent in each state are random variables.

Sheppard correction, corrections applied to moments calculated from grouped observations. For the sample variance s^2 the Sheppard corrected value is $s^2-h^2/12$, where h is the length of the grouping interval.

SPRT, Sequential Probability Ratio Test.

Statistic, function of random variables $X_1, X_2, \ldots, X_n$.

Stress-strength model, statistical model of reliability based on stress and strength of material.

Subjective probability, probability interpreted as measuring degree of belief.

Taguchi method, a systematic approch advocated by Taguchi for use of statistical methods at the product and process design stage to improve product quality.

TTT-statistic, if $0=X_0 \leq X_1 \leq \ldots \leq X_n$ is an ordered sample the TTT-statistic of the sample is defined as $\sum_{i=1}^{n} (n-i+1)(X(i)-X(i-1))$. The statistic has an obvious interpretation in life testing.

TTT-transform, TTT-plot, useful tools in reliability for model identification, replacement theory and burn-in problems.

White noise, a stochastic process is a white noise process if it has a constant spectral density function.

19 Miscellaneous

Greek alphabet

α	A	alpha	ι	I	iota	ϱ	P	rho
β	B	beta	$\varkappa$	K	kappa	σ	Σ	sigma
γ	$\varGamma$	gamma	λ	$\varLambda$	lambda	τ	T	tau
δ	$\varDelta$	delta	μ	M	mu	υ	Y	upsilon
ε	E	epsilon	ν	N	nu	ϕ	$\varPhi$	phi
ζ	Z	zeta	ξ	$\varXi$	xi	χ	X	chi
η	H	eta	o	O	omicron	ψ	$\varPsi$	psi
θ	$\varTheta$	theta	π	$\varPi$	pi	ω	$\varOmega$	omega

Mathematical constants

Approximations to 25 decimal places and hexadecimal places

$\sqrt{2}=$	1.41421 35623 73095 04880 16887=1.6A09E	667F3	BCC90	8B2FB	1366F
$\sqrt{3}=$	1.73205 08075 68877 29352 74463=1.BB67A	E8584	CAA73	B2574	2D708
$\sqrt{5}=$	2.23606 79774 99789 69640 91737=2.3C6EF	372FE	94F82	BE739	80C0C
$\sqrt{10}=$	3.16227 76601 68379 33199 88935=3.298B0	75B4B	6A524	09457	9061A
ln2=	0.69314 71805 59945 30941 72321=0.B1721	7F7D1	CF79A	BC9E3	B3980
ln3=	1.09861 22886 68109 69139 52452=1.193EA	7AAD0	30A97	6A419	8D550
ln10=	2.30258 50929 94045 68401 79915=2.4D763	776AA	A2B05	BA95B	58AE1
1/ln2=	1.44269 50408 88963 40735 99247=1.71547	652B8	2FE17	77D0F	FDA0D
1/ln10=	0.43429 44819 03251 82765 11289=0.6F2DE	C549B	9438C	A9AAD	D557D
$\pi=$	3.14159 26535 89793 23846 26434=3.243F6	A8885	A308D	31319	8A2E0
$\pi/180=$	0.01745 32925 19943 29576 92369=0.0477D	1A894	A74E4	57076	2FB37
$180/\pi=$	57.29577 95130 82320 87679 81548=39.4BB83	4C783	EF70C	2A5D4	DFD03
$1/\pi=$	0.31830 98861 83790 67153 77675=0.517CC	1B727	220A9	4FE13	ABE90
$\pi^2=$	9.86960 44010 89358 61883 44910=9.DE9E6	4DF22	EF2D2	56E26	CD981
$\sqrt{\pi}=$	1.77245 38509 05516 02729 81675=1.C5BF8	91B4E	F6AA7	9C3B0	520D6
$e=$	2.71828 18284 59045 23536 02875=2.B7E15	1628A	ED2A6	ABF71	5880A
$1/e=$	0.36787 94411 71442 32159 55238=0.5E2D5	8D8B3	BCDF1	ABADE	C7829
$e^2=$	7.38905 60989 30650 22723 04275=7.63992	E3537	6B730	CE8EE	881AE
$\gamma=$	0.57721 56649 01532 86060 65121=0.93C46	7E37D	B0C7A	4D1BE	3F810
$\phi=$	1.61803 39887 49894 84820 45868=1.9E377	9B97F	4A7C1	5F39C	C0606
sin1=	0.84147 09848 07896 50665 25023=0.D76AA	47848	67702	0C6E9	E909C
cos1=	0.54030 23058 68139 71740 09366=0.8A514	07DA8	345C9	1C246	6D977

0.1=0.19999	99999	99999	99999	9999A
0.01=0.028F5	C28F5	C28F5	C28F5	C28F6
0.001=0.00418	9374B	C6A7E	F9DB2	2D0E5
10^{-4}=0.00068	DB8BA	C710C	B295E	9E1B1
10^{-5}=0.0000A	7C5AC	471B4	78423	0FCF8
10^{-6}=0.00001	0C6F7	A0B5E	D8D36	B4C7F

cot 355=33173. 70877 45785 70590 14882 75772

tan 573204=−3 402633. 79542 44036 22259 83658 79644

Famous numbers

The number π

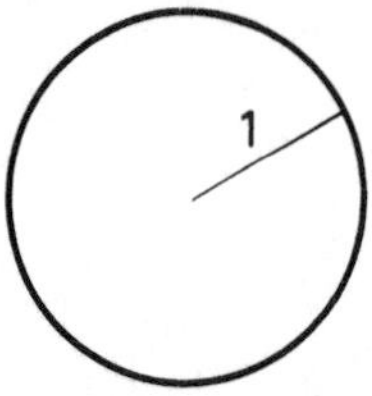

Circumference $=2\pi$ Area $=\pi$

$\pi \approx 3.14159\ 26535\ 89793\ 23846\ 26433\ 83279\ 50288$

(This approximation was calculated by Ludolf van Ceulen (1540–1610))

The first 1000 decimal places of π:

1415926535	8979323846	2643383279	5028841971	6939937510	5820974944	5923078164	0628620899	8628034825	3421170679
8214808651	3282306647	0938446095	5058223172	5359408128	4811174502	8410270193	8521105559	6446229489	5493038196
4428810975	6659334461	2847564823	3786783165	2712019091	4564856692	3460348610	4543266482	1339360726	0249141273
7245870066	0631558817	4881520920	9628292540	9171536436	7892590360	0113305305	4882046652	1384146951	9415116094
3305727036	5759591953	0921861173	8193261179	3105118548	0744623799	6274956735	1885752724	8912279381	8301194912
9833673362	4406566430	8602139494	6395224737	1907021798	6094370277	0539217176	2931767523	8467481846	7669405132
0005681271	4526356082	7785771342	7577896091	7363717872	1468440901	2249534301	4654958537	1050792279	6892589235
4201995611	2129021960	8640344181	5981362977	4771309960	5187072113	4999999837	2978049951	0597317328	1609631859
5024459455	3469083026	4252230825	3344685035	2619311881	7101000313	7838752886	5875332083	8142061717	7669147303
5982534904	2875546873	1159562863	8823537875	9375195778	1857780532	1712268066	1300192787	6611195909	2164201989

Decimal approximations with more than 29 millions decimal places have been calculated with computers.

π(in octal) $=3.11037\ 55242\ 10264$

Rational approximations to π:

$\frac{22}{7} \approx 3.142\ 857\ 143$ $\frac{355}{113} \approx 3.141\ 592\ 920$

The number e

The number e is the basis of natural logarithms.

$$e = \lim_{n \to \infty} \left(1+\frac{1}{n}\right)^n \qquad e = 1+\frac{1}{1!}+\frac{1}{2!}+\frac{1}{3!}+\frac{1}{4!}+\ldots$$

$e \approx 2.71828\ 18284\ 59045\ 23536\ 02874\ 71352\ 66249\ 77572$

Rational approximations to e:

$\frac{193}{71} \approx 2.7183\ 098$ $\frac{1264}{465} = 2.718\ 279\ 570$

Euler's constant γ or C

$$\gamma = \lim_{n \to \infty} \left(1+\frac{1}{2}+\frac{1}{3}+\ldots+\frac{1}{n}-\ln n\right)$$

$\gamma \approx 0.57721\ 56649\ 01532\ 86060\ 65120\ 90082\ 40243\ 10421\ 59335\ 93992$
$35988\ 05767\ 23488\ 48677\ 26777\ 66467\ 09369\ 47063\ 29174\ 67495.$

The golden section ϕ

Point P divides AB such that

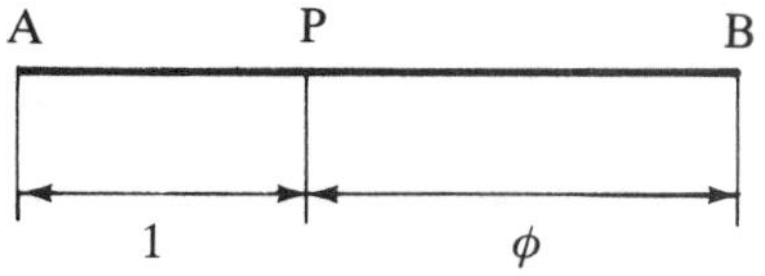

$$PB/AP=\phi=\frac{1+\sqrt{5}}{2}.$$

Then $PB/AP=AB/PB$.

$$\phi\approx 1.61803\ 39887\ 49894\ 84820\ 45868\ 34365$$

Physical constants

Name	Symbol	Value
Atomic mass unit	u	$1.66057 \cdot 10^{-27}$ kg
Mass of neutron	m_n	$1.6750 \cdot 10^{-27}$ kg
Mass of proton	m_p	$1.6726 \cdot 10^{-27}$ kg
Mass of electron	m_e	$9.1095 \cdot 10^{-31}$ kg
Electron charge	e	$1.6022 \cdot 10^{-19}$ As
Bohr radius	a_0	$5.2918 \cdot 10^{-11}$ m
Planck constant	h	$6.6262 \cdot 10^{-34}$ Js
Gravitational constant	G	$6.6720 \cdot 10^{-11}$ $Nm^2\ kg^{-2}$
Gravitational acceleration	g	9.80665 m/s^2
Speed of light in vacuum	c_0	$2.997925 \cdot 10^8$ m/s
Boltzmann constant	k	$1.3807 \cdot 10^{-23}$ JK^{-1}
Rydberg constant	R_H	$1.09678 \cdot 10^7$ m^{-1}
Stefan-Boltzmanns constant	σ	$5.670 \cdot 10^{-8}$ $W m^{-2} K^{-4}$
Constant in Wien's displacement law	b_λ	$2.898 \cdot 10^{-3}$ m · K
Avogadro's constant	N_A	$6.022 \cdot 10^{23}$ mol^{-1}
Universal gas constant	R	8.314 J $mol^{-1} K^{-1}$
Faraday constant	F	$9.6485 \cdot 10^4$ As mol^{-1}
Permittivity of vacuum	ε_0	$8.854 \cdot 10^{-12}$ As $V^{-1} m^{-1}$
Permeability of vacuum	μ_0	$1.25664 \cdot 10^{-6}$ Vs $A^{-1} m^{-1}$

Roman numeral system

I=1 V=5 X=10 L=50 C=100 D=500 M=1000
1=I 2=II 3=III 4=IV 5=V 6=VI 7=VII 8=VIII 9=IX 10=X

1989=MDCCCCLXXXVIIII=MCMLXXXIX

Prefix

Power of 10	Prefix	Notation
1 000 000 000 000 000 000 = 10^{18}	exa	E
1 000 000 000 000 000 = 10^{15}	peta	P
1 000 000 000 000 = 10^{12}	tera	T
1 000 000 000 = 10^{9}	giga	G
1 000 000 = 10^{6}	mega	M
1 000 = 10^{3}	kilo	k
100 = 10^{2}	hecto	h
10 = 10^{1}	deca	da, D
0.1 = 10^{-1}	deci	d
0.01 = 10^{-2}	centi	c
0.001 = 10^{-3}	milli	m
0.000 001 = 10^{-6}	micro	μ
0.000 000 001 = 10^{-9}	nano	n
0.000 000 000 001 = 10^{-12}	pico	p
0.000 000 000 000 001 = 10^{-15}	femto	f
0.000 000 000 000 000 001 = 10^{-18}	atto	a

Conversion factors. US and metric system (SI)

US	SI
1 inch	2.540 cm
1 foot	30.48 cm
1 yard	0.9144 m
1 mile	1.609 km
1 $inch^2$	6.452 cm^2
1 $foot^2$	929.030 cm^2
1 $yard^2$	0.836 m^2
1 $mile^2$	2.5899 km^2
1 acre	4046.9 m^2
1 $inch^3$	16.387 cm^3
1 $foot^3$	28317 cm^3
1 $yard^3$	0.765 m^3

SI	US
1 mm	0.03937 inches
1 cm	0.3937 inches
1 m	3.281 feet
1 m	1.094 yards
1 km	0.6214 miles
1 cm^2	0.155 $inch^2$
1 m^2	1.196 $yard^2$
1 m^2	10.764 $foot^2$
1 km^2	0.386 $mile^2$
1 cm^3	0.061 $inch^3$
1 m^3	1.308 $yard^3$

1 nautic mile=1.852 km=6080.20 feet

1 gallon=3.785 liters 1 liter=61.024 $inch^3$

1 hectare=2.471 acres

1 ounce=28.34952 grams

1 pound=453.59237 grams

1 radian=$\frac{180}{\pi}$ ($\approx$53.70) degrees 1 degree=$\frac{\pi}{180}$ ($\approx$0.01745) radians

F=Fahrenheit temperature C=Celcius temperature K=Kelvin temperature

$F=\frac{9}{5}C+32$ $C=K-273.15$

Abbreviations in computer science

A/D	Analog-to-Digital
ADP	Automatic Data Processing
AED	Automatic Engineering Design
AI	Artificial Intelligence
AL	Assembly Language
ALU	Arithmetic-Logic Unit
ASCII	American Standard Code for Information Interchange
ASL	Available Space List
BCD	Binary-Coded Decimal
BDP	Business Data Processing
bps	Bits per second
BSAM	Basic Sequential Access Method
CAD	Computer-Aided Design
CAI	Computer-Aided Instruction
CAM	Computer-Aided Manufacturing
CG	Computer Graphics
CIM	Computer-Integrated Manufacturing
CMS	Conversational Monitor System
CPU	Central Processing Unit
CTSS	Compatible Time-Sharing System
D/A	Digital-to-Analog
DAM	Direct Access Method
DMA	Direct Memory Access
DML	Data Manipulation Language
DOS	Disk Operating System
DRAM	Dynamic Random Access Memory
FF	Flip-Flop
FIFO	First-In-First-Out
FPS	Floating Point System
GIGO	Garbage-In-Garbage-Out
HLL	High-Level Language
IC	Integrated Circuit
I/O	Input/output
JCL	Job Control Language
LIFO	Last-In-First-Out
LP	Linear Programming
LSI	Large-Scale Integration
Megaflop	Million floating-point operations per second
MIPS	Millions instructions processed per second
MSB	Most Significant Bit (Byte)
PROM	Programmable Read-Only Memory
RAM	Random Access Memory
ROM	Read-Only Memory
SAM	Sequential Access Method
SJF	Shortest Job First
SNA	System Network Architecture
SP	Structured Programming
VDU	Video Display Unit
VM	Virtual Memory
WFF	Well-formed Formula

History

Famous mathematicians

Niels Henrik Abel (1802–1829)

Norwegian. Proved the impossibility of solving the general quintic equation in radicals. Died in poverty before news of a university post in Berlin reached him.

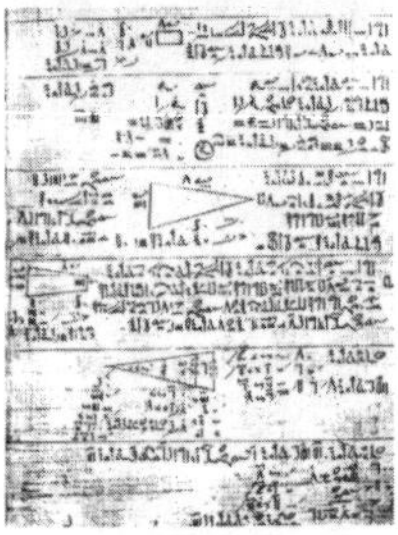

Ahmes (about 1700 B C)

The first mathematician the name of whom is known to us. He wrote Papyrus Rhind, which contains 84 mathematical problems. It is our most important source for knowledge about Egyptian mathematics.

Archimedes (287–212 B C)

The greatest mathematican in antiquity with important contributions to geometry and statics. He calculated the area of a parabolic segment and was thus a pioneer for the calculus. He was advizer to king Hiero of Syracuse and was killed by the Romans when they sacked the city.

Charles Babbage (1792–1871)

English mathematician and inventor. Worked on an analytical engine to be programmed to perform a sequence of arithmetical operations. Pioneer of the computer.

Jacques Bernoulli (1654–1705)

The Bernoulli family produced a large number of talented mathematicians. He contributed to the calculus, the creation of Newton and Leibnitz. Is also one of the founders of probability theory. Made the first proof of the law of large numbers.

Johan von Bolyai (1802–1860)

Hungarian mathematician who besides Lobachevski discovered non-Euclidian geometries.

Georg Cantor (1845–1918)

German mathematician. Founder of set theory, which is of great importance for the foundation of many mathematical theories.

Augustin Louis Cauchy (1789–1857)

French mathematician who made important contributions to group theory and the theory of functions. The first to give a stringent definition of the limit concept.

René Descartes (1596–1650)

French philosopher and mathematician. Creator of analytic geometry. Has been called 'the father of modern philosophy'.

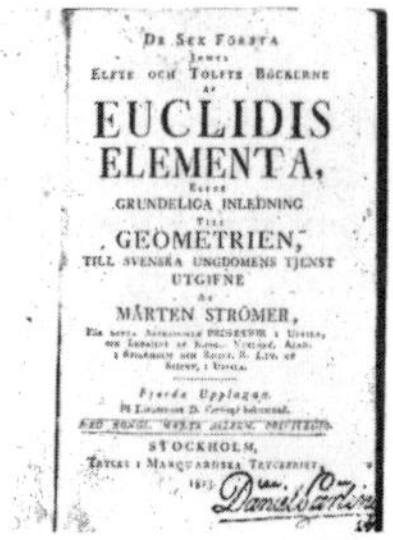
De Sex Första
Jemte
Elfte och Tolfte Böckerne
af
EUCLIDIS
ELEMENTA,
Eller
GRUNDELIGA INLEDNING
Till
GEOMETRIEN,
TILL SVENSKA UNGDOMENS TJENST
UTGIFNE
af
MÅRTEN STRÖMER,
Fjerde Upplagan.
STOCKHOLM,

Euclid of Alexandria

Euclid lived about 300 B C. He collected the mathematical knowledge of his time, especially in geometry, in his work *Elements,* the most well known mathematical book through all the times.

Leonhard Euler (1707–1783)

Swiss mathematician, professor in mathematics in Berlin and St Petersburg. Euler is one of the most productive mathematician of all times. Active even in his last years when he was blind.

Pierre de Fermat (1601–1655)

Fermat was Counsellor in the Parliamant in Toulouse and is considered to be the greatest of all amateur mathematicians. He made important discoveries in analysis and number theory. A famous correspondence with Blaise Pascal is the origin of probability theory.

Joseph Fourier (1768–1830)

French mathematician and administrator. He accompanied Napoleon to Egypt as scientific advisor. His book *Theorie analytique de la chaleur* has been described as "a great mathematical poem" and is the origin of Fourier series and Fourier transforms.

Evariste Galois (1811–1832)

French mathematician with a short and troubled life. Made important contributions to the theory of equations. Killed in a duel.

Karl Friedrich Gauss (1777–1855)

German, considered to be one of the greatest mathematicians ever. Has been called "the king of mathematicians". Contributed to many different areas of pure and applied mathematics. His most important work has the title "Disquistiones Aritmeticae".

David Hilbert (1862–1943)

German mathematician, professor in Göttingen. He is considered to be the last mathematican with a working knowledge in all areas of mathematics.

Felix Klein (1849–1925)

German geometer and algebraist. In his inaugural address in Erlangen he formulated his Erlangen program, which greatly influenced geometry and its teaching. He was also very interested in the didactics of mathematics.

Sonia Kovalevsky (1853–1891)

Daughter of a Russian artillery officer. Studied with Weierstrass in Germany. Professor of mathematics in Stockholm from 1884. Worked among other things with partial differential equations.

Joseph Louis Lagrange (1736–1813)

French mathematician, who worked in Turin, Berlin and Paris, where he was professor at École Polytechnique. Contributed to analysis, number theory, probability theory and theoretical mechanics.

Pierre Simon Laplace (1749–1827)

French marquis and mathematician, he has been called "The Newton of France". Once had Napoleon as a pupil. Napoleon later made him minister of the interior for a short time. Contributed to analysis, probability theory and celestial mechanics.

Gottfried Wilhelm Leibniz (1646–1716)

German mathematician and expert on law. Has been described as the last one to master all the knowledge of his time. Discovered besides Newton the differential and integral calculus. Is also founder of mathematical logic.

Nikolaij Lobachevski (1793–1856)

Russian mathematician working in Kazan. One of the discoverers of non-Euclidean geometries. Although blind, year before his death he dictated a large exposition of his non-Euclidian geometry.

Andrei Andrejevitch Markov (1856–1922)

Russian mathematician, working in St Petersburg. Founder of the theories of Markov chains and Markov processes, which are of great importance in probability theory.

Gösta Mittag-Leffler (1846–1927)

Swedish mathematician, one of Weierstrass' most important students. Contributed to analysis. Founder of the journal Acta Mathematica.

John Napier (1550–1617)

British mathematician, physicist and engineer, the founder of modern logarithms.

John von Neumann (1903–1957)

Born in Hungary but active as a mathematician in USA. Has been called "the Archimedes of our time" due to his great contributions to various areas of both pure and applied mathematics. Founder of the theory of games. Behind the construction of the first computers.

Isaac Newton (1642–1727)

English mathematician and physicist. Newton is one of the dominating figures in the history of science. He and Leinbitz discovered differential and integral calculus. Discovered the general gravitation law.

Blaise Pascal (1623–1662)

French mathematician and philosopher. Although only 16 he wrote a paper on new results for conics. Together with Fermat founder of probability theory. Constructed the first calculating machine.

Henri Poincaré (1854–1912)

French mathematician with important contributions to analysis, topology, probability and mathematical physics. Wrote in all about 500 mathematical papers.

Pythagoras (about 580–500 B C)

Greek philosopher and mathematician. Founded a brotherhood in Croton where he taught mathematics, astronomy, music, and philosophy. His name is linked with the theorem of Pythagoras about the squares on the sides of a right angled triangle.

Srinivasan Ramanujan (1887–1920)

Mathemtician from India, who taught himself with the aid of one textbook. Obtained astonishing mathematical results, which he wrote down in his diary. Came later to Cambridge in England, where he worked with a professional mathematician, G. H. Hardy.

Bernhard Riemann (1826–1866)

German mathematician who made important contributions to the theory of functions. With his lecture "Über die Hypothesen, welche der Geometrie zugrunde liegen" he prepared for Einsteins relativity theory.

Karl Weierstrass (1815–1897)

German mathematician who with Riemann founded the modern theory of functions. He was the first to give an example of a non-differentiable continuous function.

Norbert Wiener (1894–1964)

American mathematician. Was a mathematical infant prodigy. Although only 14 he took his first academic degree. Founder of the cybernetics.

19

Glossary of functions

$\lvert x \rvert$	sec. 2.1
$\sqrt[n]{x}$	2.1
x^a	5.3
a^x, $e^x = \exp(x)$	5.3
${}^a\log x$	5.3
$\ln x$	5.3
$\sinh x$	5.3
$\cosh x$	5.3
$\tanh x$	5.3
$\coth x$	5.3
$\operatorname{arsinh} x$	5.3
$\operatorname{arcosh} x$	5.3
$\operatorname{artanh} x$	5.3
$\operatorname{arcoth} x$	5.3
$\sin x$	5.4
$\cos x$	5.4
$\tan x$	5.4
$\cot x$	5.4
$\sec x$	5.4
$\csc x$	5.4
$\arcsin x$	5.4
$\arccos x$	5.4
$\arctan x$	5.4
$\operatorname{arccot} x$	5.4
$F(a, b, c, x)$	9.2
$F(b, c, x)$	9.2
$P_n(x)$	12.2
$P_n^m(x)$	12.2
$T_n(x)$	12.2
$U_n(x)$	12.2
$H_n(x)$	12.2
$h_n(x)$	12.2
$L_n(x)$	sec. 12.2
$L_n^{(\alpha)}(x)$	12.2
$l_n(x)$	12.2
$P_n^{(\alpha,\beta)}(x)$	12.2
$B_n(x)$	12.3
$E_n(x)$	12.3, 12.5
$J_p(x)$	12.4
$Y_p(x)$	12.4
$H_p^{(1)}(x)$	12.4
$H_p^{(2)}(x)$	12.4
$I_n(x)$	12.4
$K_n(x)$	12.4
$\operatorname{ber}(x)$	12.4
$\operatorname{bei}(x)$	12.4
$\operatorname{ker}(x)$	12.4
$\operatorname{kei}(x)$	12.4
$\Gamma(x)$	12.5
$B(p, q)$	12.5
$F(k, \varphi)$	12.5
$E(k, \varphi)$	12.5
$\pi(k, n, \varphi)$	12.5
$K(k)$	12.5
$E(k)$	12.5
$\operatorname{Ei}(x)$	12.5
$\operatorname{li}(x)$	12.5
$\operatorname{erf}(x)$	12.5
$\operatorname{Si}(x)$	12.5
$\operatorname{Ci}(x)$	12.5
$C(x)$	sec. 12.5
$S(x)$	12.5
$\theta(t)$	12.6
$\operatorname{sgn}(t)$	12.6
$\delta(t)$	12.6
$\varphi(x)$	17.2
$\Phi(x)$	17.8

Glossary of symbols

Symbol	Meaning	Section
$\wedge$	and	1.1
$\vee$	or	1.1
$\bar{\vee}$	exclusive or	1.1
$\neg$, $\sim$	negation	1.1
$\Rightarrow$	implies	1.1
$\Leftrightarrow$	equivalent	1.1
$\exists$	there exists	1.1
$\forall$	for all	1.1
$\uparrow$	NAND	1.1
$\downarrow$	NOR	1.1
$\because$	thus	
$\in$	belongs to	1.2
$\subset$	subset	1.2
$\supset$	superset	1.2
$\complement$	complement	1.2
$\cap$	intersection	1.2
$\cup$	union	1.2
$\setminus$	difference	1.2
$\triangle$	symmetric difference	1.2
	difference	6.1, 6.3
	Laplacian	11.2
$\times$	product set	1.2
$\varnothing$	empty set	1.2
D_f, R_f	domain, range of f	1.3
$\binom{n}{k}$	binomial coefficient	2.1
$\boldsymbol{N}, \boldsymbol{Z}, \boldsymbol{Q}, \boldsymbol{R}, \boldsymbol{C}$	number sets	2.2
$\boldsymbol{R}^n$	Euclidean space	10.1
$\boldsymbol{Z}_n$	set of congruence classes modulo n	1.4
$\boldsymbol{Z}_2{}^n$	set of binary n-tuples	1.6
$F[x]$	polynomial ring over field F	1.4
∂S, $\bar{S}$	boundary, closure of S	10.1
$\boldsymbol{u}\cdot\boldsymbol{v}$, $\boldsymbol{u}^t\boldsymbol{v}$, $(\boldsymbol{u}, \boldsymbol{v})$, (u/v), $\mathbf{u}^*\mathbf{v}$	scalar product	3.4, 4.1, 4.7, 12.7, 4.10
$\boldsymbol{u}\times\boldsymbol{v}$	vector product	3.4
$\lvert\boldsymbol{u}\rvert$	length (norm)	3.4, 4.7
$\lVert u\rVert$, $\lVert u\rVert_{m,p}$	norm	12.7, 12.8
$A=(a_{ij})$, $[a_{ij}]$	matrix	4.1
A^t	transpose of matrix	4.1
A^*	adjoint of matrix	4.10
$\lVert A\rVert$	norm of matrix	16.2
$y'=\dfrac{dy}{dx}=Dy$, $\dot{y}=\dfrac{dy}{dt}$	derivative	6.3

$f'_x = f_x = \frac{\partial f}{\partial x}$	partial derivative	10.4
f'_e	directional derivative	10.4
$\frac{\partial}{\partial n}$	normal derivative	11.4
$\frac{\partial(y_1, \dots, y_n)}{\partial(x_1, \dots, x_n)}$	Jacobian determinant	10.6
∇	gradient	11.2
$E, \Delta, \nabla, \delta, \mu$	difference operators	16.3
L^p	Lebesgues spaces	12.8
C^m	spaces of differentiable functions	12.8
$W^{m,p}, H^m$	Sobolev spaces	12.8
$O(\ ), o(\ )$	ordo	8.5
$[\ ,\ ], (\ ,\ )$	closed, open interval	6.1
$\sim$	row equivalence asymptotic equivalence Fourier series expansion	4.1 8.5 12.1
$*$	convolution	12.8
$=$	equality	
$\neq$	inequality	
$< (\leq)$	less than (or equal)	
$> (\geq)$	greater than (or equal)	
$\equiv$	identity	
$\approx$	approximately equal	
$\cong$	congruent to	
$//$	parallel to	
$\perp$	perpendicular to	
∞	infinity	
$\delta_{kn} = \begin{cases} 1, & k=n \\ 0, & k \neq n \end{cases}$	Kronecker delta	

Index